Kirk-Othmer

ENCYCLOPEDIA OF CHEMICAL TECHNOLOGY

Second Edition

VOLUME 22

Water (Desalination)

to

Zone Refining

Interscience Publishers
a division of John Wiley & Sons, Inc.
New York · London · Sydney · Toronto

Kirk-Othmer

ENCYCLOPEDIA

OF CHEMICAL

TECHNOLOGY

Second completely revised edition

VOLUME 22

Water (Desalination)
to
Zone Refining

CONTENTS

EDITORIAL STAFF FOR VOLUME 22

Associate Editor: E V A P A R O L L A D U K E S

Leslie Holzer   Anna Klingsberg   Carolyn C. Wronker

CONTRIBUTORS TO VOLUME 22

M. A. Amerine, *University of California, Davis, California,* Wine

A. J. Baker, *U.S. Department of Agriculture,* Wood

W. B. Blumenthal, *National Lead Company,* Zirconium and zirconium compounds

C. W. Bunn, *The Royal Institution, London,* X-Ray analysis

Frank F. Cesark, *American Cyanamid Company,* Xanthene dyes

H. E. Cier, *Esso Research and Engineering Company,* Xylenes and ethylbenzene

William J. Connick, Jr., *U.S. Department of Agriculture,* Waterproofing and water repellency

G. I. de Becze, *St. Thomas Institute,* Yeasts

H. W. Eickner, *U.S. Department of Agriculture,* Wood

W. E. Eslyn, *U.S. Department of Agriculture,* Wood

J. P. Faust, *Chemical Division, Olin Corporation,* Water (Treatment of swimming pools)

Edward A. Fenton, *American Welding Society,* Welding

A. H. Gower, *Chemical Division, Olin Corporation,* Water (Treatment of swimming pools)

G. J. Hajny, *U.S. Department of Agriculture,* Wood

R. A. Hann, *U.S. Department of Agriculture,* Wood

G. H. Harris, *The Dow Chemical Company,* Xanthates

D. Robert Hay, *Drexel Institute of Technology,* Zone refining

Donald B. Kendall, *Toledo Scale Company, Division of Reliance Electric Company,* Weighing and proportioning

R. C. Koeppen, *U.S. Department of Agriculture,* Wood

Bernard Kopelman, *Sylvania Electric Products Inc.,* Wolfram and wolfram alloys

Alan Lawley, *Drexel Institute of Technology,* Zone refining

M. Lipson, *Division of Textile Industry, CSIRO, Geelong, Victoria, Australia,* Wool

M. MacInnis, *Sylvania Electric Products Inc.,* Wolfram compounds

M. M. MacMasters, *Kansas State University,* Wheat and other cereal grains

E. S. McLoud, *Consultant,* Waxes

M. A. Millet, *U.S. Department of Agriculture,* Wood

W. E. Moore, *U.S. Department of Agriculture,* Wood

Donald F. Othmer, *Polytechnic Institute of Brooklyn,* Water (Water supply and desalination)

W. Parrish, *International Business Machines Corporation,* X-Ray analysis (Fluorescence spectrography)

James R. Pfafflin, *University of Windsor, Ontario,* Water (Sewage; Water reuse)

J. R. Plimmer, *U.S. Department of Agriculture,* Weed killers

Y. Pomeranz, *U.S. Department of Agriculture,* Wheat and other cereal grains

J. David Reid, *U.S. Department of Agriculture,* Waterproofing and water repellency

J. K. Rice, *Cyrus Wm. Rice Division, NUS Corporation,* Water (Industrial water treatment)

John D. Roach, *National Lead Company,* Zirconium (Zirconium hydrates)

A. W. Schlechten, *Colorado School of Mines,* Zinc and zinc alloys

D. E. Simon II, *Cyrus Wm. Rice Division, NUS Corporation,* Water (Industrial water treatment)

J. E. Singley, *University of Florida,* Water (Municipal water treatment)

J. S. Smith, *Sylvania Electric Products Inc.,* Wolfram and wolfram alloys

Harold Tarkow, *U.S. Department of Agriculture,* Wood

A. Paul Thompson, *The Eagle-Picher Company,* Zinc and zinc alloys; Zinc compounds

ABBREVIATIONS AND SYMBOLS

A	ampere(s)	APHA	American Public Health Association
A	anion (eg, HA)		
Å	Angstrom unit(s)	API	American Petroleum Institute
AATCC	American Association of Textile Chemists and Colorists		
		app	apparatus
		approx	approximate(ly)
abs	absolute	aq	aqueous
ac	alternating current	Ar	aryl
ac-	alicyclic (eg, ac-derivatives of tetrahydronaphthalene)	as-	asymmetric(al) (eg, as-trichlorobenzene)
		ASA	American Standards Association. Later (1966) called USASI
accel(d)	accelerated(d)		
acceln	acceleration		
ACS	American Chemical Society	ASHRAE	American Society of Heating, Refrigerating and Air-Conditioning Engineers
addn	addition		
AEC	Atomic Energy Commission		
AGA	American Gas Association		
Ah	ampere-hour(s)	ASM	American Society for Metals
AIChE	American Institute of Chemical Engineers		
		ASME	American Society of Mechanical Engineers
AIME	American Institute of Mining and Metallurgical Engineers		
		ASTM	American Society for Testing and Materials
AIP	American Institute of Physics	atm	atmosphere(s), atmospheric
AISI	American Iron and Steel Institute	at. no.	atomic number
		at. wt	atomic weight
alc	alcohol(ic)	av	average
alk	alkaline (not alkali)	b	barn(s)
Alk	alkyl	b (as in b_{11})	boiling (at 11 mm Hg)
AMA	American Medical Association		
		bbl	barrel(s)
A-min	ampere-minute(s)	bcc	body-centered cubic
amt	amount (noun)	Bé	Baumé
anhyd	anhydrous	Bhn	Brinell hardness number
AOAC	Association of Official Analytical (formerly Agricultural) Chemists	bp	boiling point
		BP	*British Pharmacopoeia* (General Medical Council in London)
AOCS	American Oil Chemists' Society		
		Btu	British thermal unit(s)

bu	bushel(s)	crystd	crystallized
C	Celsius (centigrade); coulomb(s)	crystn	crystallization
		cSt	centistokes
C-	denoting attachment to carbon (eg, *C*-acetyl-indoline)	cu	cubic
		d	density (conveniently, specific gravity)
ca	circa, approximately	*d*	differential operator
CA	Chemical Abstracts	*d-*	*dextro-*, dextrorotatory
cal	calorie(s)	D	Debye unit(s)
calcd	calculated	D-	denoting configurational relationship (as to *dextro*-glyceraldehyde)
cfm, ft³/min	cubic foot (feet) per minute		
cg	centigram(s)	db	dry-bulb
cgs	centimeter-gram-second	dB	decibel(s)
Ci	curie(s)	dc	direct current
CI	Colour Index (number); the CI numbers given in *ECT*, 2nd ed., are from the new *Colour Index* (1956) and Suppl. (1963), *Soc. Dyers Colourists*, Bradford, England, and *AATCC*, U.S.A.	dec, decomp	decompose(s)
		decompd	decomposed
		decompn	decomposition
		den	denier(s)
		den/fil	denier(s) per filament
		deriv	derivative
		detd	determined
		detn	determination
CIE	Commission Internationale de l'Eclairage (see also ICI)	diam	diameter
		dielec	dielectric (adj.)
		dil	dilute
cif	cost, insurance, freight	DIN	Deutsche Industrienormen
cl	carload lots	distd	distilled
cm	centimeter(s)	distn	distillation
coeff	coefficient	dl	deciliter(s)
compd, cpd	compound (noun)	*dl-*, DL	racemic
		dm	decimeter(s)
compn	composition	DOT	Department of Transportation
concd	concentrated		
concn	concentration	dp	dewpoint
cond	conductivity	dyn	dyne(s)
const	constant	*e*	electron; base of natural logarithms
cont	continued		
cor	corrected	ed.	edited, edition, editor
cp	chemically pure	elec	electric(al)
cP	centipoise(s)	emf	electromotive force
cpd, compd	compound (noun)	emu	electromagnetic unit(s)
		eng	engineering
cps	cycles per second	equil	equilibrium(s)
crit	critical	equiv	equivalent
cryst	crystalline	esp	especially

esr, ESR	electron spin resonance	hyg	hygroscopic
est(d)	estimate(d)	Hz	hertz(es)
estn	estimation	i, insol	insoluble
esu	electrostatic unit(s)	^{i} (eg, Pr^i)	iso (eg, isopropyl)
eu	entropy unit(s)	*i-*	inactive (eg, *i*-methionine)
eV	electron volt(s)	IACS	International Annealed
expt(l)	experiment(al)		Copper Standard
ext(d)	extract(ed)	ibp	initial boiling point
extn	extraction	ICC	Interstate Commerce
F	Fahrenheit; farad(s)		Commission
F	faraday constant	ICI	International Commission
FAO	Food and Agriculture		on Illumination (see also
	Organization of the		CIE); Imperial Chemical
	United Nations		Industries, Ltd.
fcc	face-centered cubic	ICT	International Critical
Fed, fedl	federal (eg, Fed Spec)		Tables
fl oz	fluid ounce(s)	ID	inner diameter
fob	free on board	IEEE	Institute of Electrical and
fp	freezing point		Electronics Engineers
frz	freezing	in.	inch(es)
ft	foot (feet)	insol, i	insoluble
ft-lb	foot-pound(s)	IPT	Institute of Petroleum
ft^3/min,			Technologists
cfm	cubic foot (feet) per minute	ir	infrared
g	gram(s)	ISO	International Organization
g	gravitational acceleration		for Standardization
G	gauss(es)	IU	International Unit(s)
G	Gibbs free energy	IUPAC	International Union of
gal	gallon(s)		Pure and Applied
gal/min,			Chemistry
gpm	gallon(s) per minute	J	joule(s)
g/den	gram(s) per denier	K	Kelvin
gem-	geminal (attached to the	*K*	dissociation constant
	same atom)	kbar	kilobar(s)
g-mol	gram-molecular (as in	kc	kilocycle(s)
	g-mol wt)	kcal	kilogram-calorie(s)
g-mole	gram-mole(s)	keV	kilo electron volt(s)
G-Oe	gauss-oersted(s)	kg	kilogram(s)
gpm,		kG	kilogauss(es)
gal/*min*	gallon(s) per minute	kgf	kilogram force(s)
gr	grain(s)	kJ	kilojoule(s)
h, hr	hour(s)	kp	kilopond(s) (equals kilo-
hl	hectoliter(s)		gram force(s)
hmw	high-molecular-weight(adj.)	kV	kilovolt(s)
hp	horsepower(s)	kVa	kilovolt-ampere(s)
hr, h	hour(s)	kW	kilowatt(s)
hyd	hydrated, hydrous	kWh	kilowatt-hour(s)

l	liter(s)	mm	millimeter(s)
l-	*levo-*, levorotatory	mM	millimole(s)
L-	denoting configurational relationship (as to *levo*-glyceraldehyde)	mM	millimolar
		mo(s)	month(s)
		mol	molecule, molecular
lb	pound(s)	mol wt	molecular weight
LC$_{50}$	concentration lethal to 50% of the animals tested	mp	melting point
		mph	miles per hour
		MR	molar refraction
lcl	less than carload lots	mV	millivolt(s)
LD$_{50}$	dose lethal to 50% of the animals tested	mμ	millimicron(s) (10^{-9} m)
		^{n} (eg, Bu^n),	
liq	liquid	*n-*	normal (eg, normal butyl)
lm	lumen		
lmw	low-molecular-weight (adj.)	n (as, n_D^{20})	index of refraction (for 20°C and sodium light)
ln	logarithm (natural)		
log	logarithm (common)	*n-*, ^{n}	normal (eg, *n*-butyl, Bu^n)
m	meter(s)	N	normal (as applied to concentration)
m	molal		
m-	meta (eg, *m*-xylene)		
M	metal	N-	denoting attachment to nitrogen (eg, *N*-methylaniline)
M	molar (as applied to concentration; not molal)		
		NASA	National Aeronautics and Space Administration
mA	milliampere(s)		
mAh	milliampere-hour(s)	ND	*New Drugs* (NND changed to ND in 1965)
manuf	manufacture		
manufd, mfd	manufactured	NF	*National Formulary* (American Pharmaceutical Association)
manufg, mfg	manufacturing		
		nm	nuclear magneton; nanometer(s) (10^{-9} m)
max	maximum		
Mc	megacycle(s)		
MCA	Manufacturing Chemists' Association	nmr, NMR	nuclear magnetic resonance
		NND	*New and Nonofficial Drugs* (AMA) (1958–1965). Later called ND
mcal	millicalorie(s)		
mech	mechanical		
meq	milliequivalent(s)	NNR	*New and Nonofficial Remedies* (1907–1958). Later called NND
MeV	million electron volt(s)		
mfd, manufd	manufactured	no.	number
mfg, manufg	manufacturing	NOIBN	not otherwise indexed by name (DOT specification for shipping containers)
mg	milligram(s)		
min	minimum; minute(s)	*o-*	ortho (eg, *o*-xylene)
misc	miscellaneous	*O-*	denoting attachment to oxygen (eg, *O*-acetyl-hydroxylamine)
mixt	mixture		
ml	milliliter(s)		
MLD	minimum lethal dose	Ω	ohm(s)

Ω-cm	ohm-centimeter(s)	rad	radian
OD	outer diameter	Rep	roentgen(s) equivalent
Oe	oersted(s)		physical
o/w	oil-in-water (eg, o/w	resp	respectively
	emulsion)	rh	relative humidity
owf	on weight of fiber	Rhe	unit of fluidity (1/P)
oz	ounce(s)	RI	Ring Index (number);
p-	para (eg, *p*-xylene)		from *The Ring Index*,
P	poise(s)		Reinhold Publishing
pdr	powder		Corp., N.Y., 1940.
PhI	*Pharmacopoeia Internation-*		See also RRI
	alis, 2 vols. and Suppl.,	rms	root mean square
	World Health Organiza-	rpm	revolutions per minute
	tion, Geneva, 1951, 1955,	rps	revolutions per second
	and 1959	RRI	Revised Ring Index (num-
phr	parts per hundred of rubber		ber); from *The Ring*
	or resin		*Index*, 2nd ed., American
pos	positive (adj.)		Chemical Society, Wash-
powd	powdered		ington, D.C., 1960
ppb	parts per billion	RT	room temperature
	(parts per 10^9)	s, sol	soluble
ppm	parts per million	^{s} (eg, Bu^s),	
ppt(d)	precipitate(d)	*sec-*	secondary (eg, *sec*-butyl)
pptn	precipitation	*s-*, *sym-*	symmetrical (eg, *s*-di-
Pr. (no.)	Foreign prototype (num-		chloroethylene)
	ber); dyestuff designa-	*S-*	denoting attachment to
	tion used in *AATCC*		sulfur (eg, *S*-methyl-
	Year Books for dyes not		cysteine)
	listed in the old *Colour*	SAE	Society of Automotive
	Index (1924 ed.; 1928		Engineers
	Suppl.); obsolete since	satd	saturated
	new *Colour Index* was	satn	saturation
	published (1956 ed.;	scf, SCF	standard cubic foot (feet)
	1963 Suppl.)		(760 mm Hg, 63°F)
prepd	prepared	scfm	standard cubic feet per
prepn	preparation		minute
psi	pound(s) per square inch	Sch	Schultz number (designa-
psia	pound(s) per square inch		tion for dyes from *Farb-*
(psig)	absolute (gage)		*stofftabellen*, 4 vols.,
pt	point		Akademie Verlag,
pts	parts		Leipzig, 1931–1939)
qual	qualitative	sec	second(s)
quant	quantitative	*sec-*, ^{s}	secondary (eg, *sec*-butyl;
qv	which see (quod vide)		Bu^s)
R	Rankine; roentgen;	SFs	Saybolt Furol second(s)
	univalent hydrocarbon	sl s, sl sol	slightly soluble
	radical (or hydrogen)	sol, s	soluble
		soln	solution

soly	solubility		(ASA changed to USASI in 1966)
sp	specific		
sp, spp	species (sing. and pl.)	USP	(*The*) *United States Pharmacopeia* (Mack Publishing Co., Easton, Pa.)
Spec	specification		
sp gr	specific gravity		
SPI	Society of the Plastics Industry	uv	ultraviolet
sq	square	V	volt(s)
St	stokes	*v-*, *vic-*	vicinal (attached to adjacent atoms)
STP	standard temperature and pressure (760 mm Hg, 0°C)	var	variety
		vic-, *v-*	vicinal (attached to adjacent atoms)
subl	sublime(s), subliming	vol	volume(s) (not volatile)
SUs	Saybolt Universal second(s)	v s, v sol	very soluble
sym, *s-*	symmetrical (eg, *sym*-dichloroethylene)	vs	versus
		v/v	volume per volume
t (eg, Bu*t*), *t-*, *tert-*	tertiary (eg, tertiary butyl)	W	watt(s)
		Wh	watt-hour(s)
t-, *tert-*, *t*	tertiary (eg, *t*-butyl)	w/o	water-in-oil (eg, w/o emulsion)
TAPPI	Technical Association of the Pulp and Paper Industry	wt	weight
		w/v	weight per volume
tech	technical	w/w	weight per weight
temp	temperature	xu (ca 10^{-11} cm)	x unit(s)
tert-, *t-*, *t*	tertiary (eg, *tert*-butyl)		
theoret	theoretical		
Twad	Twaddell	yd	yard(s)
USASI	United States of America Standards Institute	yr	year(s)

Quantities

Some standard abbreviations (prefixes) for very small and very large quantities are as follows:

deci (10^{-1})	d	deka (10^1)	dk
centi (10^{-2})	c	hecto (10^2)	h
milli (10^{-3})	m	kilo (10^3)	k
micro (10^{-6})	μ	mega (10^6)	M
nano (10^{-9})	n	giga (10^9)	G (or B)
pico (10^{-12})	p	tera (10^{12})	T
femto (10^{-15})	f		
atto (10^{-18})	a		

W continued

WATER

WATER SUPPLY AND DESALINATION

Of the surface of the earth, 71% is the area of the oceans—140 million square miles; their average depth is $2\frac{1}{8}$ miles, and their volume is 330 million cubic miles.

About 60% of the land area of the earth is arid or semi-arid, not generally considered habitable. Mountainous areas and the polar areas covered by ice must also be subtracted from that available for man's living and agriculture, leaving only a small fraction. Preferred land areas support the expanding world population more generously, so increasing amounts of better lands are used for living—urban areas, industries, roads, airfields, etc. Unit areas for living and industry usually require much more

1

water than farm lands. Thus, agricultural areas with adequate water become smaller as population and industry increase, and the demands for water multiply.

"Fresh" water means potable water, with not more than 500 parts per million or 0.05% of dissolved solids. Water with over 1000 ppm is usually considered unfit for human consumption. However, in some parts of the world, people are forced to survive with much higher concentrations of salts, as some land animals must also do.

Rain is the source of fresh water, and its precipitation of 340×10^{12} gallons per day (gpd) over the earth's surface averages about 50 inches per year. Extremes are the practical zero of North Chile's desert bordering the Pacific Coast, to over 1000 inches in some tropical forests and some high slopes where the high, cold mountains condense floods from the clouds.

Even rain is not pure water. Reports from the U.S. Geological Survey show it contains from 2.3 to 4.6 ppm of solids, or a yearly precipitation of from 7 to 14 tons per square mile, while in some islands of the Caribbean 5 ppm sodium, 7 ppm chloride, and 11 ppm of bicarbonate were found, to indicate respectively 18, 25, and 35 tons deposited per square mile.

The oceans hold about 97% of the earth's water. More than 2% is locked up as ice in the polar caps; and over 75% of the fresh water of the world is ice. Of the 1% of liquid fresh water, some is ground water at depths of over 1000 feet, and impractical to obtain; and only the very small difference, possibly 0.6%, of the total water of this planet, is ever available to man as it cycles from sea to atmosphere to land to sea. Only recently has man been able to regulate that cycle to his advantage, and then only infinitesimally in some few isolated places.

Wells give ground water, stored from previous rains. But in recent years, wells have had to be made deeper and deeper to reach water, which shows that ground water is being used faster than it is being replenished. Water lying in deep strata for millions of years, like other minerals, is being "mined," never to be replaced. In Libya, recent drilling by an oil company found a "lake" 300 feet below the dry sands, hundreds of square miles in area, and 2500 ft deep (1). It has been estimated that this lake will supply irrigation of 200,000 acres for 300 years. This pumping of water to the surface is as final an act as the pumping up of Libya's petroleum, which probably dates from the same lush geological era. Once pumped up, neither can ever be replenished within another millenium!

Transport of Fresh Water

Containers have carried fresh water, usually for longer distances than would be practical for conduits. Animals have carried goatskin bags, jugs, wooden barrels, or disused five-gallon oil cans. Trucks and railways have carried tanks. The bridge-tunnel over and under the Chesapeake Bay near Washington and Baltimore has a 100,000 gallon tank which is filled by a tank-truck, hauling for 12 miles. Ships have carried water in ballast tanks, halfway around the world; and tankers otherwise returning light from oil deliveries may make the return voyage with fresh water, which may be as precious to the oil-rich, water-poor country as the oil is to its market. Recently, tremendous plastic and textile "sausages," towed by relatively small tugs, have been used for both oil and water transport.

The ancients successfully developed dams, canals, and aqueducts to bring flowing fresh water considerable distances to growing cities, and to irrigate agricultural lands.

Records of ancient Rome indicate that 14 aqueducts, each averaging over 90 miles long, carried a total of 300 million gallons of water per day from the surrounding highlands by gravity. The aqueducts of Istanbul are even more dramatic. The capabilities of engineers developed with increasing needs for water, and conduits became longer. The Romans depended on gravity flow in open channels and had no concept of friction losses in pipes, even though Pliny lists standard lead pipe in circumferences up to about 100 inches. (This represented the widths of strips of lead formed around a mandrel and "burned" or welded in a single seam.)

New York City has considered bringing water from the Great Lakes, and Tokyo has planned pipelines hundreds of miles long, back to the high mountains. A system costing billions of dollars will carry very large amounts of water to Southern California. When all costs are considered they will total about $0.50/1000 gallons for 750 miles, over mountains 4000 feet high. This may be the largest and most costly project man has ever built to carry potable water to the seacoast; but it adds not one drop of water to California's supply (2).

Fresh water can be *produced* from the sea water more cheaply when all costs of the tremendous investment of dams, reservoirs, conduits, and pumps are considered—and this without robbing from some other location. Desalination of sea water will cost Southern California much less than the transport of water even if there was more in the mountains of Northern California. It has been shown that, under many conditions, if, say, 2 to 3 million gpd of water must be dammed and piped 100 miles or more for a seacoast town (or 10 million gpd 150 for 200 miles) the same amount of fresh water may be manufactured from the sea more cheaply (3). Furthermore, desalination is likely to become less costly with improved technology, whereas transporting water is not likely to improve substantially, in cost, and in any case it *never adds fresh water*.

The Water Problem

Many cities of the world do not separately tax water which is distributed, and even in those places where water is in shortest supply, a bare minimal ratio may be free to everyone. Wasted water and unmetered water add to the overall water demand now as in ancient Rome, where water was delivered free to the fountains. Connections from the fountains were made by privately connected and owned lead pipe lines; and historians say one-third of these were unrecorded, illegal, and hence untaxed.

Three times as much water as in 1900 is used per person in the continental U.S., and the country as a whole uses ten times as much. Individual usage in some southern cities is 10 or 20 times as much, with swimming pools, lawns, air conditioners, and other consuming demands. The great population increase multiplies the total withdrawal, particularly in cities where it may be as much as 250 gallons per day per person.

Water consumption expands with and makes possible increasing standards of living. Thus, in Kuwait, both have multiplied many fold in recent years due to the pumping of crude oil from the ground for refining and the pumping of water from the sea for desalinating.

An analysis of water use in the United States and an estimate for the next 50 years (4) predicts withdrawals will increase by 400%, and consumptive uses over 100%. Most of this increase must be met by more reuse, ie, in manufacturing, from 2.3 times reused in 1965, to 6.3 times reused in the year 2020.

Twice as much water is pumped from the ground as soaks into it in some areas of the U.S. Although there is estimated to be 47.5 billion acre-feet of ground water within one-half mile of the surface, the water table has dropped in some places 5 to 10 feet for each year of the present generation, which is thus exhausting a historical treasure. Often this withdrawal causes sinking of the ground level, as near Houston, Texas, and Mexico City. Las Vegas is growing rapidly upon a base or ground level which, in recent years, has sunk 3 feet due to a greatly increased mining of prehistoric water to supply many acres of swimming pools, many thousands of "tons of refrigeration" (see Vol. 17, p. 295) for air-conditioning, and other water uses, from wells in a desert where only 3 to 4 inches of rain falls per year. Long Island's magnificent underground supply of water is largely unavailable because pumping it extensively causes infiltration of sea water.

Water is far from evenly divided in the United States, with major shortages in some very populous areas; always California; and, for the first half of the 60s, the northeast. For the United States as a whole, the demand for water will become equal to the total supply before 1985. Some areas will be desperately short much sooner. Water shortages are acute in years of low rainfall, as in 1957 when over 1000 communities in 47 of the 48 contiguous states restricted the use of water. In oil-rich Texas, people lined up to buy water—at twice the price of gasoline!

Two out of five U.S. cities have inadequate water supplies, even ignoring the fact that city dwellers take much more than their share. A quarter of the U.S. population faces serious water shortages. Yet, half of the states of the Union with two-thirds of the industry and over half of the population—like many of the people of many of the developing countries of the world—have direct access to as much as they can draw of the one-third of a billion cubic miles of sea water. Its solids content, mainly salt, varies from 25,000 ppm (2.5%) in the Baltic Sea to over 40,000 ppm (4%) in some of the more confined gulfs of the Indian Ocean; while the waters of the wide oceans are almost constant at 35,000 ppm (3.5%). Also, in many places inland, there are large quantities of water too brackish to drink.

Some of the most attractive areas of the world, particularly islands and beaches, are almost devoid of fresh water. This living space—and space for resort hotels—is lost. The biggest and one of the fastest growing of the industries of the world is tourism. It is a particularly attractive industry to developing countries; and in some of these it may almost be the only nonagricultural industry. Tourism may account for a substantial use of the available fresh water, and it often demands even more. Tourism may be stifled or entirely prevented in otherwise attractive places if there is insufficient fresh water. But the cost of fresh water, if available, is small compared to the revenue of a hotel. Even if 250 gpd of fresh water were used for a room with two people, the cost per person is $0.25 at a fresh water cost of $2.00 per 1000 gallons. Some fine hotels, as in Bermuda, reduce to a fraction the fresh water which is required and produced in the hotel by distributing sea water to baths and toilets. Thus, if only one-fifth of the usage is fresh water, the cost becomes insignificant.

Bermuda has one of the largest percentages of its income of any country from tourists—150,000 per year. By law, each house must have for each occupant a roof area of 120 square feet to catch the average annual rainfall of 58 inches and drain to a basement reservoir. Hill slopes have vegetation cleared above the natural limestone and coral which is plastered to give drains and catchments for heavy seasonal rains. But even with all this, some has to be imported at about $15 per 1000 gallons.

Jedda, on the East Coast of the Red Sea, and the principal port of Saudi Arabia, is the gateway and stopover annually for 200,000 Moslem pilgrims to Mecca. Fresh water, formerly sold from donkey carts for $5.00/1000 gallons, is now supplied by a major desalination plant using the designs of an American firm, Burns & Roe.

Because of lack of water in many of the developing nations, profitable material resources cannot be exploited to improve a poor economy (5). Thus, proven mines on Egypt's Red Sea coast cannot be operated. Others on the same coast are operable only with desalinated water for processing phosphate ores and for mining lead and zinc; and the workers are rationed 4 to 10 gallons per day for their uses. Fishing industries on South America's arid Pacific Coast cannot be expanded for lack of water to process the haul. These represent major losses in the world's supplies of minerals and foods; and there are many other examples. Not all fresh water shortages are in the torrid tropics; a substantial iron ore deposit in a small waterless island off the coast of Iceland needs water for workers. Needed is a desalination plant or some means of utilizing as liquid water the heavy fogs which prevail.

Wells withdraw the water for reuse from treated sewage which has been pumped back into the ground by many communities in the industrial nations of the world. In dry 1957, at least one town near the middle of the United States survived through complete reuse of the water discharged in its sewage. Cities on the lower reaches of a major river, with many cities above, use water which has been through sewers upstream many times. On the lower Mississippi, a water inventory indicates such reuse averages 14 times, with dependence on biochemical oxidation of the wastes during the miles of flow between cities. The Rhine River, in passing through several countries, all of which drink from and dump into its waters, causes international problems of pollution.

Many so-called "hard" contaminants are present in sewage, and these increase almost proportionally with reuse of the water. Such materials do not ferment or oxidize under ordinary sewage treatments. See Surfactants; see also Eutrophication in Supplement Volume.

The U.S. Government spent approximately $150 million in fiscal 1969 on water resources research relating to: artificial rain-making, soil conservation, waste treatment, desalting, public health, and planning research. This continuance of large annual expenditures, compared to those of the rest of the world, is against a need of over $100 billion estimated simply for water facilities for our cities within 10 to 15 years (6). This may also be compared to a budget of over 25 times as much in recent single years for research in space. However, all of the reports and results of research on water are available from the Office of Saline Water to everyone of every country.

The United States is not the world's driest country by far, but there is a tremendous challenge to its chemical engineers both in and out of government to provide systems to separate water from one-thirtieth as much salt in sea water, or one-thousandth as much salt in brackish waters. Within 20 years, over $100 billion will be spent worldwide on plants for desalinating water, according to the estimate of the Director of the Office of Saline Water. At an average rate of $5 billion a year, this will be a much greater budget than for all equipment for the chemical industries. Many more billions will go for plants for recovering water from sewage for almost immediate reuse or discharge to streams.

The greatest, although possibly not the most dramatic, engineering achievement of this next generation, may well be the development of processes and the building of plants for producing potable water, the prime necessity of life. Ours is a *world* of water—which might better be called not *Earth*, but *Water*. This is emphasized as the ack of water, hence life, in other parts of the universe becomes more apparent.

While water is the most abundant material, its cheap production, with less than 500 ppm impurities, will be more important to the world than atomic energy, as we know it today. Its generous supply is certainly needed much more by the United States than putting men on the moon! The late President Kennedy said that the nation developing a successful process "would get a great deal more lasting benefit than those countries that may be even first in space."

Saline Water for Municipal Distribution

Most of the water piped to cities and industry is used for little else than to carry off extremely small amounts of waste materials or waste heat. For most of this service, sea water may be used. If chlorination is required, it may be accomplished by direct electrolysis of the dissolved salt. However, against the obvious advantage of economy, several disadvantages must be considered in such uses of sea water: different detergents must be used with saline water; sewage treatment plants must be modified to use sea water; the usual metal pipes, pumps, condensers, coolers, meters, and other equipment are corroded by sea water; dual water systems must be built and maintained.

The high cost of a second water distribution system may be prohibitive for present cities; but population in cities has been and is increasing very fast and much new urban construction is necessary.

Chemical industry has answered all but the last of these objections. Pipes, valves, fittings, and almost all other items of small equipment are now available in plastic, cement, or ceramic which are proof against corrosion by salt water and less expensive than the metals now used. Synthetic detergents are available for use with sea water, although a final rinse with fresh water may be desired. Salt water sewage can be treated successfully.

Dual water systems, using fresh water and sea water, are already in use on shipboard, and in many island resort hotels. This trend will grow. Only a very small amount of potable water is actually taken internally; and it is quite uneconomic to desalinate all municipally piped water, although all distributed water must be clear and free of harmful bacteria.

Some inland municipalities now distribute water so brackish as to be unpleasant to the taste, even if distributed as potable with over 1000 ppm salts. Each home may produce or have delivered the very small requirement of fresh water for drinking and cooking. Small membrane desalinators may be rented and periodically serviced by a service company in some communities. Desalting units of other types are available for purchase as a major appliance—in the class of domestic clothes-washing machines, refrigerators, air conditioners, etc. These produce the relatively few gallons of potable water required, as compared to the much larger amount presently supplied. The cost per gallon of potable water produced in the home will be several times as great as the cost in a central desalinating plant; but the amount of desalinated water needed will be only a small fraction of the total supply.

Home desalinators are probably possible only for countries which are relatively industrialized, with a central service organization preferably operating on a rental-service contract basis, as is standard practice in many communities for water softeners. Alternatively, small amounts of potable water may be delivered by truck to distribution centers or to tanks on house roofs. This is the system in Kuwait which has 22 "filling" stations, from which hundreds of tank trucks buy water at $1.00/1000 gallons for distribution at $2.40 to $3.00/1000 gallons.

Water in Industry

Fresh water for industry often may be replaced by saline or brackish water, usually after sedimentation, filtration, and chlorination (electrically or chemically) or other treatment. Such treatment is not necessary for the largest user, the electric power industry, which, by 1985, in the United States, will be passing through its heat exchangers about one-quarter of the total supply of surface water and by 1995 about one-half (7). Single stations of 1000 MW may heat as much as 3 billion gallons per day as much as 10–15°C.

Cooling towers circulate the warmed water downwardly against a rising stream of air which removes heat by evaporating a part of the water to cool the balance. In America, cooling towers are not so common because of the larger bodies of water available compared with Europe. American towers are usually rectangular, with many gridworks of redwood for the water to descend over in films. In Europe they are more often the shape of a venturi or a corseted waist.

Although 100,000 gallons of water is used to make a ton of steel, a half-ton of gasoline, or only 800 pounds of acetate fiber little if any is required chemically in any of these processes. Recycling will reduce most of industry's requirements by a factor of 10 to 50. Much of this water, particularly that for cooling, and often that for washing, may be saline. Some petroleum refiners have used salt water to remove heat (water's principal role in gasoline production) and some, by evaporation in cooling towers, have actually produced table salt!

The pulp and paper industry has tried, for many years, to use salt water for some of the 100,000 gallons required to make a ton of paper. Here, however, salt is disadvantageous to the chemical processes, either in pulping the lignocellulose or in the recovery of values from the black liquor after pulping. Also, corrosion is most undesirable in the expensive paper-making machines, with the tremendous throughput of "white water." Thus, only limited success has been achieved.

The great fluid reservoir of heat, the alternative to water, fresh or salt, is the atmosphere. Increasingly, industry is using air coolers for dissipation of the large quantities of heat from power plants, petroleum and other distillation, and other process use. In 1969, $50 million of air coolers were installed, and this amount is expected to increase by 20% each year in the near future to reduce correspondingly the pumping of cooling water (8).

The warm, recycling, condenser water is cooled in a conventional cooling tower as it is partially evaporated to humidify air passing in a counter current. The warm water may also be passed through tubes with air blown over them as a coolant. The air itself is cooled and humidified by spraying fresh water into it as it passes over the tubes.

Alternatively, dry condensers have a forced draft over vapor filled tubes to condense the vapors, particularly of high-boiling petroleum fractions. The heat transfer coefficient from tube to water is an order of magnitude higher than from tube to air, thus usually some form of extended surface on the air side is used to increase the effective area of heat dissipation. Ground level and elevated condensers have long been used; and now air-cooled, overhead condensers are available in standard design. They are installed on top of distillation towers for either reflux or total condensation, with forced-draft blowers adding to the natural breeze of their exposed location (9). Petroleum and petrochemical installations have thus been made possible on arid islands.

Air coolers almost invariably add considerably to plant cost, but not usually to operating cost, based on direct, once-through use of water which requires no treatment. If the alternative to air coolers is the use of water which must have substantial treatment or pumping costs, air coolers cost less to operate.

Water for Agriculture

A half-gallon of water in some form may be the daily requirement of the average human, depending on many personal and external conditions. However, at least several hundred gallons per day are required for growing the vegetables, fruits, and grain making up the absolute minimal daily food ration for a vegetarian; and some thousands of gallons are required to produce a pound of beef, based on the cereals required by the animal.

A loaf of bread contains little, if any, of the more than a ton of water necessary for the growth of the wheat therein. The water content of vegetables comes from 1000 to 2000 times as much water which must be supplied in their growth. Some of these major losses in agriculture may be reduced by agronomists and plant physiologists. Studies are being made to determine if less water can be supplied to the roots, with better absorption there and less losses by transpiration through the leaves. Certainly plant structures of an entirely different type will be necessary to prevent, for example, the loss by transpiration in a field of corn on an August day, which may be equivalent to a half-inch of rainfall. However, the recently developed high-yield grains use much less water than the conventional varieties, and thus are a great aid to the developing nations.

Is it possible to breed plants which have more efficient systems for utilization of water, or can man help existing crop plants by spraying on them impervious coatings? Extremely small amounts of long-chain, fatty alcohols reduce evaporation losses from quiet lakes or reservoirs to less than 5% of the normal surface evaporation (See Vol. 21, p. 664). Is similar treatment applicable to plants? Or is it possible to develop a plant with a built-in ability to produce a chemical for this purpose, as is the waxy, almost water-proof skin of the cactus which covers the succulent tissues within? Research to answer these questions is in progress; and its success may be as vital as the development of better desalination processes.

"Waterproofing" sandy soils to prevent drain-through has been successful in increasing crops as much as 400% with the same rainfall. A special plow lifts the soil to allow melted asphalt to be layered in overlapping impermeable strips 32 inches wide and 2 feet below the surface (10). Waterproofing the surface in Israel by compacting with chemicals increases runoff to basins or fields on slopes below. Barren slopes in many places have been coated with asphalt or concrete—as thin as $\frac{1}{8}$ in.—to catch rain which is conducted to catch basins for irrigation or other uses.

The one-seventh of the world's crop lands which are irrigated produce one-quarter of the world's crops. Irrigation loses much water: directly, by seepage, by evaporation from the open channels and the soil, and otherways. Only a small fraction of the water withdrawn from the irrigation ditch or pipe is absorbed by the plants. Plastic films, as ground covers through which the plants protrude, prevent some losses at great expense for film and labor. Cheaper systems are necessary to assure better water utilization by plants. Other possible goals would be food plants whose membranes separate fresh water from brackish water, to give a nonsalty crop. Progress has been made in both these directions; also some plants have been developed which accumulate salt from the ground.

Vegetables and fruits such as melons are profitably grown in four or five crops per year, made possible by hydroponics, the growth of plants without soil in large, shallow concrete tanks containing gravel and water with added nutrients. One major installation is in the Caribbean island of Aruba, using man-made fresh water from the sea. Much water is still necessary per unit weight of crop; but the largest losses of ordinary irrigation are prevented, as indeed they must be because of the comparatively high cost of the water. Such concentrated agriculture is very expensive in preparation of "land" area, but economical of water and labor requirements. Production is high in the tropics; and hydroponics offers a major opportunity to many developing countries.

One other use of desalinated water for agriculture was also reported in the Channel Island of Guernsey, where a borderline production of about 5% of the total water used, eliminates the danger of a crop failure (5).

Economics of Water

Economically, water is without counterpart among the world's resources of materials. It has by far the widest range of values and prices of any material in commerce, from zero to almost infinity, depending on the location, its availability, and man's need for it. Usually it sells for less than 1¢ per 1000 gallons for irrigation; 2 to 5¢ for industry; 20 to 60¢ for domestic use; and any necessary price for drinking! Shipped to dry areas in tank cars, it may cost $7 to $15 per 1000 gallons; and when distributed by peddlers, the mark-up over that high price is considerable. Potable water presently is delivered by ship to many islands of the world, also to ports on the Red Sea and elsewhere. As late as the 1930s, water went by steamship from New York to South America, and in the last century by sailing ship from England to Chile. Now it also goes in "dracones," huge sausages having a skin of plastic and fabric and towed by a tug.

Agriculture as practiced always uses many times the minimum amounts of fresh water and will always be its largest user. Fresh water streams and rivers are used for waste disposal; there are also many non-consumptive but necessary uses of water for cooling, washing, and other industrial uses. When all of these are considered, the daily requirement varies between hundreds and thousands of gallons per day per person. The larger volumes are those under the highly developed and more complex patterns of community life in countries with the higher standards of living.

Only the most specialized forms of agriculture can afford desalinated sea water under any probable costs. Conventional irrigation cannot use water profitably if it costs more than 2 to 15¢ per 1000 gallons. The lower value is about the maximum for most agriculture in the U.S., and it is too high for the economics of most developing countries. The only case of commercial crops now being grown with irrigation water from desalination is in the Island of Guernsey in the English Channel. Here the crops have a high value. Also there is a very nearly adequate rainfall, with only a small fraction of the total water being desalinated. These unusual conditions are not expected to be duplicated elsewhere. One analysis voices optimism for the near future, with lower water costs opening to agriculture in present desert lands (11).

Cooling water for once-through use in power generation may be in the same price range as water for irrigation. Raw water for other industrial purposes usually cannot cost more than 15 to 25¢ per 1000 gallons, although some few industries may afford very much higher prices.

There is no upper limit of value for water to quench thirst; but generally fresh water for municipal use should not cost, including distribution, more than $0.50 to $1.00 per 1000 gallons. It is sold for less in most major cities of the world. Desalination processes will supply water in the quantities required by large sea coast cities possibly at a cost as low as $0.35 to $0.40 by 1975; and some present processes will reach these costs now with plants of 5 to 10 million gpd under special conditions. Thousands of pages have been written analyzing these economics under many possible conditions.

The capital costs of production plants, or "water factories," also are going down, particularly for those for large capacities. Plants using evaporation processes may cost $500 to $1000 per 1000 gallons daily capacity. In very large units, these costs will be less, using currently available processes, and hopefully much less using processes of the improved technology which is foreseen within 10 years. Major reduction is also expected in plant costs through the use of fewer and cheaper materials for construction.

In smaller plants, the principal cost of the product water is the capital charges of the plant; in larger plants, this will be less than 50% of the total. Energy, either electrical or thermal, is the principal other cost in larger plants where labor costs are small because of large throughput per employee.

Hence, cheap energy will make cheap water. Geothermal energy may be available in a very few places. Cheap nuclear energy has been promised, but not yet delivered; and the cost of nuclear equipment is, as yet, too high.

Natural gas, almost valueless in Kuwait, supplies energy to evaporate 30 million gallons a day for the use of its quarter million citizens, and an equal number of foreign residents.

Simultaneous production of electric power and fresh water from a thermal-electric power station is often considered, because both power needs and water needs increase simultaneously with population growth and increasing standards of living. If the price of either power or water is regarded as fixed, the other is determinable. A price for either which is too low gives a high price for the other; but with large amounts of electric power simultaneously produced to sell at $0.006/kWh, sea water may now be desalinated in large plants for $0.50/1000 gal. This cost will be reduced considerably within the next few years by improvements in known processes, particularly when atomic energy demonstrates that it can compete economically with fossil fuels.

Major Desalination Development Programs

THE OFFICE OF SALINE WATER (OSW)

The Department of the Interior of the United States, long active in many areas of water resources and management, was authorized by Congress in 1952 to organize the OSW with a $2 million budget for 5 years to develop economical processes for desalination. This budget was repeatedly expanded to a total of almost $275 million authorized through fiscal 1972 for research and development efforts with demonstration plants to prove the technical and economic feasibility of desalination processes. One fundamental principle guiding these efforts and enunciated by one President of the United

States is that all scientific and engineering knowledge and skills developed in the wide reaches of this program are to be immediately and freely available to every nation and every person in the world as a generous contribution to the betterment of the lives of all.

Basically, OSW depends on contracting out its scientific and engineering work to outside contractors, even as regards the operation of its own facilities. The program of the OSW is divided into Research, Engineering and Development; Project Management and Plant Engineering; and Desalting Feasibility and Economic Studies. Research in desalination at OSW embraces all physical properties of water and related aqueous solutions; data of all types required to improve existing processes; study and evaluation of feasibility of new processes selected as warranting preliminary plant design; and thorough investigation of possibilities suggested for novel and different methods. Unsolicited proposals are even encouraged.

Research, Engineering and Development carries the results of research activities through the processes satisfactory for large scale use under those particular conditions for which they may be suitable.

Project Management and Plant Engineering administers the coordination of steps necessary in the development of a process. Besides the facilities of the many hundreds of outside contractors, largely academic or engineering or research companies or divisions of operating corporations, OSW operates five Research and Test Stations, one on the Atlantic Ocean, one on the Pacific Ocean, one on the Gulf of Mexico, and two at sources of inland brackish water. Besides small scale research and test apparatus, these test stations operate pilot plants and intermediate size test beds of relatively large sizes, up to 2.6 million of gallons of water per day. Prototype plants then may be built in cooperation with some municipality on a cost sharing basis to demonstrate the technical and economic feasibility of a new process. A prototype plant completes the OSW developmental sequence.

Desalting Feasibility and Economic Studies participates in the review and planning of programs and policies and arranges liaisons in desalination programs with other U.S. government agencies, with several States, and agencies of many other governments, including Russia, Japan, Italy, Israel, Spain, Saudi Arabia, and Greece.

The OSW sponsored the First International Symposium in October 1965 for 2500 persons from 55 nations. This reported in hundreds of papers the state of desalination. The OSW has published over 500 reports of its projects and very comprehensive annual reports of its many activities.

Besides many private and corporate programs on processes and materials and academically sponsored researches, various nations other than the United States are conducting substantial research and development programs, including Great Britain, Russia, Israel and Japan.

THE UNITED NATIONS

The Resources and Transport Division of the United Nations has sponsored numerous desalination projects with various of the developing nations. Some of these have been financed by the U.N. Special Fund. The Resources and Transport Division sponsored the Interregional Seminar on the Economic Application of Water Desalination in September, 1965, in which 34 countries participated, and the related economic study *Water Desalination: Proposals for a Costing Procedure and Related Technical and Economic Considerations;* also sponsored was a detailed survey of the water needs of 43 developing nations. The 325-page report examined the current and future needs in those water-short areas in relation to their economics and the technical and economic prospects for desalinating sea or brackish water at attractive prices under the particular conditions prevailing (5).

EUROPEAN FEDERATION OF CHEMICAL ENGINEERING

The European Federation of Chemical Engineering has a Working Party on Fresh Water from the Sea consisting of representatives from 13 countries under the Chairmanship of Prof. A. A. Delyannis of Greece. It has held two Symposia in Athens (1962 and 1967), and is holding the third International Symposium on Fresh Water from the Sea at Dubrovnik, Yugoslavia, in 1970. The papers of the two Symposia in Athens were published (Refs. 12–13).

Manufactured Fresh Water

The possibility of production of fresh water from sea water or brackish water by separation of the salts, ie, desalination, opens a new dimension in the supply of fresh water; those areas bordering the sea would be supplied with an available raw material without limit or cost of transportation to the water facility. The successful realization of desalination by the chemical engineer, as opposed to the search for and transport of existing fresh water, gives new hope for adequate water in many, but not all, cases.

Progress has been indicated: on January 1st, 1969, worldwide, engineering was in progress for plants for more than 1250 million gallons daily production (14). These will add to the approximately 250 million gpd productive capacity estimated on that date.

Fresh water is now one of the materials produced by the methods and equipment of the chemical engineer when and as it is needed. Water has major differences from other materials: the very low value of the commodity and the very large amount of the product which is needed. Thus, it might be expected that plant costs would be major, compared to unit sales value of production; and, in fact, the capital charges for the production equipment do make up a very large percentage, compared with other process industries.

Major problems attend this new solution of the fresh water problem. These problems are enlargements of those familiar to the chemical engineer in processing other cheap or valueless raw materials into valuable products by treatment with chemicals, or thermal or electrical energy, when the principal program is a separation rather than a chemical change. They are quite different from the previous major problem of water supply: the happenstance of finding a river or lake within a reasonable distance, or of making a fortunate geological strike.

Desalination thus becomes one of the major needs of our increasing population; in many places it is even more urgent than the production of food, which is so often limited by water shortages. These shortages exist in many of the least developed and least prosperous countries of the world, and completely prevent their development and prosperity.

Chemists and engineers have had their ingenuity and imagination stimulated by the problem of production of this most common and most useful, and yet most unusual, material—water—by separation from one of the next most common materials, salt. The chemist knows that if it were possible to cancel for an instant ionization in water, as may be the case with the newly discovered "polywater" (see under Water, Properties), salt would be quite insoluble and would precipitate out of solution. And the engineer knows that if he were able to operate one of the reversible processes which he likes to consider, he could separate the water for less than one-fiftieth the thermal cost of its evaporation.

Without consideration of these idealized chemical and thermodynamic concepts, chemists and engineers have proposed hundreds of processes based on the various properties of water and its saline solutions. Subsequent industrialization of the best of these has reduced the cost of manufactured water to a small fraction of that of 15 years ago. However, each year the differential percentage is smaller; and it may, in 1970, be about 10% lower per year. Complete books and hundreds of articles and patents each year describe these processes on which many millions of research dollars are being spent annually. The most active agency in this search and engineering development is the Office of Saline Water of the U.S. Department of the Interior. Its director, however, has said "It may be true that better processes will be available in the years ahead; but technology will have to run hard and fast just to stay even with escalating construction costs" (15). Construction costs have been increasing 4–6% each year.

Fresh water may be obtained from salt water by many processes, falling into two classes: (1) fresh water is separated away from sea water, which becomes more concentrated (*water from salt*) and (2) salt or more concentrated brine is separated away from sea water, which becomes less concentrated until an acceptable level is reached (*salt from water*). Usually, the first type of process, water from salt, is the easier.

Unfortunately, the mechanism of the separation always requires very much more energy than that of breaking the molecules of water loose from the solution: evaporation has the latent heat of evaporation to make steam simultaneously with the separation, while freezing has the latent heat of freezing in making ice. The theoretical thermodynamic energy requirement of separating water from brine, because of the heat of solution of the salt, is quite small. Starting from sea water having 35,000 ppm salts, the minimum energy requirement would be 41 kWh per 1000 gallons if dry salt were produced. This minimum energy requirement is greatly reduced if the evaporation is not carried so far: starting with 2000 gallons of sea water, to produce 1000 gallons of fresh water and 1000 gallons of brine of twice the original concentration would require only 5.80 kWh, but twice the volume of sea water would have to be handled than if salt were produced (16). The energy requirement could be reduced to a theoretical minimum of 1.98 kWh per 1000 gallons by using an indefinitely large volume of sea water, whose concentration would be changed only very slightly. However the pumping, deaerating, and possible chemical treatment necessary to accomplish the separation correspondingly become infinitely expensive in handling an infinite amount of the raw sea water. These costs go down and the thermodynamic separating costs go up as the amount of sea water processed decreases and the percent yield of product fresh water increases. Also, the much larger added energy costs of the necessary inefficiencies, eg, the change of phase of water, are usually not altered substantially by the strength of the solution; so the total energy required does not vary as much proportionally as does the much smaller heat of solution.

Balancing these factors gives the most economic concentration ratio for concentrating sea water; often it may be about 2 to 3 volumes of feed to 1 of product water, or about 3 to 8 of feed to 1 of product when brackish water is evaporated. Usually brackish water is available only in limited quantities; also the disposal of the concentrated brine may be a major problem. Thus, the product should have a much higher %, with the use of the brackish water feed, and have a lower amount of concentrate, than in the production of fresh water from the unlimited sea.

On average, there is about 35,000 ppm of dissolved solids, in seawater; more in the hot, desiccated Red Sea, less in the cooler, river-fed Black Sea. If these solids were to be separated dry, and added to the land masses from whence they came, they would be measured in millions of cubic miles! The top four metals—sodium, magnesium, calcium, and potassium,—and their combining ions—chlorides, sulfates, bicarbonates, and bromides—are present in amounts beyond comprehension. Salt is largely produced as the residue from solar evaporation of sea water in the United States and other countries. By adding phosphorus compounds and ammonia to brines from desalination plants, magnesium, a scale-former, may be precipitated; and mixed salts for fertilizer may be prepared of slightly greater value than that of the chemicals used.

Bromine and iodine have been economically removed from sea water for many years, usually while recovering salt. Currently, 150,000 tons/year of magnesium metal is produced from the sea. With the rising use of light metals in the automobile and other industries, this amount is increasing. New processes in Germany cheapen the cost of pure magnesia for use as such, or for making metal. Much other mineral value will be extracted from sea water.

Sea water "comes from springs which leach minerals from their deposits and carry these dissolved solids to the oceans from whence they are never lifted by the clouds which produce the rivers. So the sea now would be more salt in our times than it has ever been any time previously." Leonardo da Vinci, the artist-engineer, said this five centuries ago; and this increasing salinity has been a basis of estimating the age of the earth. At the present time, sea water contains an even higher concentration of magnesium, copper, gold, uranium, and almost every other metal than in Leonardo's day. For most of these, the amounts are too small to be recovered economically, as yet. As mines on land are exhausted, and the skill of the chemical engineer is sharpened in separating processes, the future may find concentrating plants for metals as by-products of fresh water production. There are 10 tons of uranium and 70 pounds of gold for each cubic mile of sea water. These amounts of very valuable metals, multiplied by the practically infinite volume of the ocean, represent tremendous values, all safely locked away, so far, by their dilution. (See Ocean raw materials.)

The saline water industry is already quite substantial. By 1980, it will be much more substantial, since population expands, and man's needs and desires expand, particularly in those areas which are not now as industrially sophisticated or as lavish in their use of materials as are the highly industrialized countries. Thus, within the next few years, fresh water production will pass the 1 billion gpd mark, and, by 1985, will be 15–20 billion. Usually twice as much sea water is processed as the fresh water produced, so that the solids concentration of the effluent brine is doubled. Chemical recovery will be cheapened not so much by this increase in concentration as by the pumping, purification, often heating, and simply the presence of the sea water in a processing plant.

Materials for Desalination Plants. This billion (10^9) gallons of fresh water which will soon be produced daily in the world at about an average of $1 per 1000 gallons (some much lower) will thus require the daily expenditure of about $1 million. This will be spent principally for two items: energy, and capital charges, ie, interest, depreciation, amortization and replacement due to corrosion and wear of the equipment and the materials therein. Relatively smaller items of cost will be for operating labor and for expendable materials, including chemicals used in the processing, ion-exchange resins, membranes, etc.

The first object in the design of processes is the minimization of costs, thus of the amounts of thermal or mechanical energy, and of the amount of equipment, and particularly of the materials required in its construction. Quite often, in optimization studies of several processes, as energy costs go down, equipment costs go up, and vice versa. As always, the most economic balance must be struck. Often under current conditions, the water costs least overall when the energy cost and the cost due to the capital costs of the equipment are about equal.

Of the $5 billion per year soon to be spent for plants, materials for the equipment might be regarded as at least 50%, with costs of engineering, fabrication, transportation, installation, etc accounting for the rest. Of materials used in equipment, by far the largest amount will be for metals, particularly those metals which are least corroded by sea water, eg, copper, nickel, and their alloys. These are not the most abundant in the world's resources or the least expensive. Although titanium has been successfully used, it cannot be specified generally at its present high cost. Fortunately, newer metallurgical processes promise very much lower costs for titanium from ores which are abundant throughout the world. This may well be the metal for the future desalination plants. The most economical fabrication of metals into sheets, tubes and other shapes, along with the optimization of design of processes and equipment, will be a great test of the skill of the metallurgist, the chemical engineer, and the mechanical engineer.

Many other materials besides metals are used in equipment. Concrete is used not only for foundations, but also, in the larger and more sophisticated present designs of equipment, in the major vessels and shells of processing units, tanks, pipelines, etc. The high mechanical strength necessary is obtainable with prestressed concrete. More recently, it has been made almost impervious to sea water at the temperatures of evaporators, 290°F, by impregnation with acrylic or other resins, then irradiated for curing. The use of concrete and plastics will reduce the vast weight of metals required which indeed, as a major class of man's resources, become more precious year by year, as does water itself. After metals and cement, the next major class of materials will be plastics and high polymers in many forms: pipes, sheets or membranes, ion exchange resins, gaskets, coatings, and linings, etc.

Other materials are charged and used as part of the operation, including various liquids for solvents, heat transfer agents, refrigerants and similar thermodynamic agents. These all are used in cycles, and in some cases they may have only such a small loss per cycle that their original cost may almost be regarded as an investment, part of the plant cost, and their makeup as depreciation. Chemicals which are strictly expendable are those used for treating the raw sea water, such as sulfuric acid and lime, algicides, phosphates, and other scale-prevention materials; smaller amounts of other chemicals for treating product water are used, such as chlorine or activated carbon, higher alcohols to stop evaporation in storage basins; and minor amounts of still others are used.

These materials will go into major plants weighing many thousands of tons and processing millions of gallons of water per day, with individual units having tens of thousands of miles of tubes, or many acres of sheet plastics. However, in the modern "do-it-yourself" world, many desalination plants will come into the home, the farm, the small apartment house or resort hotel, as major appliances, and will be priced in the range of automobiles, without their complexity, but hopefully with their dependability and low cost, thus allowing popular use. Desalination units of both a very large size and also a very small size will become familiar within the last quarter of this century. Household units will become as common in this period as automobiles and automatic home laundry machines have become in the first three quarters of this century.

Evaporation Processes for Desalination

Solar Evaporation. The solar constant, that is, the radiant heat from the sun which would fall on an unobstructed surface normal to the sun's rays at the earth's distance from the sun and without losses due to the atmosphere, is about 10,000 Btu per day per square foot. Most of the earth's surface, of course, is not normal to the sun's rays, and an average value over the whole earth would be divided by 4 to give about 2500 Btu per square foot in 24 hours, or somewhat over 100 Btu per square foot per average hour, day and night. Very large variations also occur with latitude and with the seasons.

Losses due to the atmosphere, clouds, etc must be considered. Average figures have little interest. But from measurements made in many stations, charts are available giving contour lines of different constant values of solar energy actually received during each month of the year.

The evaporation-condensation process of nature's cycle for the production of sweet water, solar evaporation, has long been at least partially controlled for water production. The equipment is very simple, but does require a condenser above the evaporating sea water. Many insignificant units have been used for centuries; but the first large solar unit was built in a mountain mining camp in Chile eighty years ago. It had about 51,000 square feet of surface, averaging a production of about 6000 gpd, used mainly for animals. Solar radiation passed through inclined glass plates, and the condensate formed on and then ran off the under surface. The available feed water had about four times the salinity of sea water.

Solar evaporators, as now designed in many countries, absorb the sun's radiant heat on blackened bottoms of shallow, narrow pans or troughs, in which sea water evaporates without boiling. Vapors are carried away from the water surface by air currents, to contact and condense on the lower surfaces of transparent plastic or glass sheets immediately above, which transmit the sun's radiant energy to the pan. Convective cooling by the atmosphere of the upper surface of the transparent glass or plastic condenser sheets removes heat and condenses the vapors. The transparent sheets are at an angle to allow the condensate on the sloping underneath surfaces to run down to gutters alongside the troughs. Assuming average conditions and 50 percent thermal efficiency, the production may approximate 0.5 to 0.8 gallons of water per square foot per day throughout the year in northern Africa, northern India, and southern United States. Dust on the transparent condensing surfaces absorbs solar heat and reduces both evaporation and condensing capacity. Proper design allows any rain which does fall to be collected.

Many variations have been studied, including rotating and angular surfaces of the evaporating sea water to absorb more of the sun's radiation, arrangements to use the evaporation in a multiple-effect arrangement, and condenser tubes carrying and preheating the sea water feed.

The most successful work is probably that in Greece, under the direction of Drs. A. and E. Delyannis, Professors of Chemical Engineering at the Technical University of Athens. Several designs of evaporators were used in the units on as many islands, culminating in the largest on Patmos, with an evaporating area of about 93,000 ft² in a plot of 130,000 ft². Substantially, this Delyannis design, after several years of success at Patmos and other islands, is now being used in a plant of twice the size in Pakistan (17,18). Figure 1 is a photograph of the Patmos installation. It consists of 71 double bays, each 10 ft × 131 ft (see Figure 2). The covers are of window glass, slightly less than ⅛ in. thick, the larger one with at 12° southward facing slope, sealed with plastic putty into frames of extruded aluminum alloy. The distillate troughs of the same material have a slope of 1 to 1000. The liner of the basins is ⅟₃₂ in. black butyl rubber sheet, formed to walls and curbs of concrete. Piping, valves, and fittings are mostly of polyvinyl chloride, some of polypropylene. Rain collection is about 24 in. per year.

Fig. 1. Delyannis type solar evaporator, Patmos Island, Greece. Largest of 7 built on Greek Islands. 93,000 ft² evaporator, 71 units, each 10 × 131 ft. Window glass covers approximately ¹/₈ in. thick, extruded aluminum frames, basins of ¹/₃₂ in. black butyl rubber sheeting, walls and curbing of concrete. Courtesy Drs. A. and E. Delyannis.

The basins are filled with sea water and discharged of concentrate periodically—every two days in summer, every week in winter. Earlier work on a smaller Delyannis plant had predicted 0.074 gal/(ft²)(day) or 27 gal/(ft²)(yr). In addition, the installation collects the rain that falls on it, averaging 24 in./yr or 15 gal/(ft²)(yr). The total yield is therefore 42 gal/(ft²)(yr) or 3,900,000 gal/yr.

This production rate of water, 1000 gpd for each 13,500 ft², is excellent considering the high latitude of Patmos, 27°20′, the same as that of the Richmond, Virginia, or of San Francisco Bay, where large solar evaporating areas produce many thousands of tons of salt from seawater. Usually solar evaporation programs have been in lower latitudes than Patmos. South Africa, Australia, India, and Israel have higher solar radiations.

Daily production for each 1000 gallons may require 5000 to 15,000 or more square feet of evaporating surface, and at least as much condensing surface. This requires a large plant site, also much material: clear plastic sheets or glass for the cover-condenser; concrete, asphalt, or black plastic for the long, narrow evaporator pans; plastic gutters for the condensate-water produced; also much piping—plastic rather than metal—for sea water distribution and fresh water collection. Total costs of materials and plants are very much higher than for other types of water desalination systems,

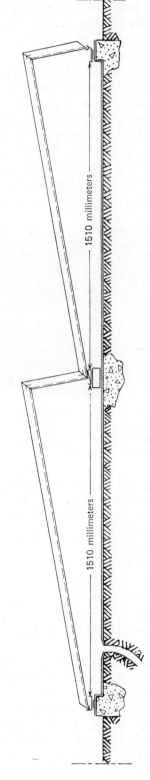

Fig. 2. Delyannis type solar still, cross-section of 1 double bay. Dimensions are in millimeters, large cover of glass faces south with 12° angle. Courtesy Drs. A. and E. Delyannis.

except in units for very small capacity. The large, covered, condensing areas are vulnerable to hail, wind, and dust storms. However, the thermal energy is free; and mechanical energy cost is low—largely for pumping. Operation of the very simple components requires only unskilled labor (19).

Solar evaporation has many advantages for small units in nonindustrialized regions, with daily capacities from a few gallons for a family unit, to a few thousand gallons for a community.

Solar Evaporation of the Ocean. To absorb solar heat for evaporation it has been proposed to use the vast volume and surfaces of tropic seas. Most of the solar heat received by the earth is stored there between latitudes 30° North and 30° South. The average temperature of the water at the surface rises (as also in the man-made solar evaporator) to that value where the vapor pressure of the sea water causes sufficient water to evaporate so as to maintain a dynamic heat balance of solar energy in, to latent heat of evaporation out. This water vapor is condensed by the cold upper atmosphere and returns ultimately in the fresh water cycle to the sea.

The heat of the sun is absorbed in the surface water which expands to a lower density and stays above the colder water below. Thus, over the tropic seas especially, there is a dramatic thermal syphon effect. Cold water comes from the polar areas, often flowing near the ocean floor. As it is heated, it rises. The water so warmed flows in vast currents back toward the poles, where it is chilled or frozen. Either as cold water at the bottom, or as a melting iceberg on the surface, it starts its return trip toward the equator, to repeat the cycle.

In the ceaseless ocean currents of many cubic miles daily of warm water flowing in one direction, and many other cubic miles of cold water flowing in another direction, often a current of warm water may be very close to a current of cold water. Water temperature may be as much as 35° or 45°F lower at a depth of 2000 to 3000 ft than temperatures at the surface, in currents at cross-flow, or even counter to each other.

To heat one cubic mile per day of the colder water to a temperature 45°F higher, would take an amount of thermal energy 25 to 30 times that of all of the electrical energy produced in the United States. The reverse is even more staggering; and it has been estimated that the heat in the warm water flowing in the Gulf Stream (540 cubic miles per day off the coast of Florida) could generate over 75 times the entire electric power used in all of the United States.

If one cubic mile of this warm water was cooled 20°F below the surface temperature by partial evaporation, and the vapors formed (with or without power production) were condensed by cold water from the deep, there would be evaporated about 20 billion gallons of water (20).

On a more reasonable scale, the processing and equipment have been engineered for a plant of 2 million gallons of water produced per day by such an evaporation. The estimated plant costs are high compared with other evaporating plants, but the water costs, without heat energy to be purchased, might be reasonable for those places where there is substantial difference in temperatures at the surface and far below, at a reasonable distance from land. Probably the only feasible system now known for such an evaporation is controlled flash evaporation, to be described later.

Evaporators with Surfaces for Boiling and for Condensing. Sailors have used simple evaporating apparatus to make drinking water for almost 400 years; and steamships have traditionally used evaporators, often multiple-effect, for the make-up water for the boilers, and for potable use.

Heat is transferred from prime steam coming from a boiler, or more often, from the exhaust or an intermediary stage of a turbine. It is condensed on one side of the first or heated metal surface, the evaporating surface, to boil the sea water contained in a suitable vessel on the other side of the tubular surface. Obviously the required area of heat transfer surface (hence the amount of metal required for a given duty) will be reduced by increasing the coefficient of heat transfer through the surface. For standard tubes, the heat transfer coefficient, reported in Btu/(hr)(ft²)(°F), is usually considerably below 1000, but values up to 2500 and more have been reached by Westinghouse, General Electric and others with tubes having special surfaces, expanded or corrugated, and with internal twisting baffles, and values up to 10,000 have been reached by Hickman (21) in rotating spheroidal surfaces. If these high heat transfer rates can be obtained, it will be possible to reduce the heat transfer area, and hence the material cost, provided no other considerations intervene.

The vapors from the evaporation pass to a second metal surface and are condensed to give fresh water on the tubular condensing surface. This condensing surface is cooled, for example, by circulating cooling water on the other side.

If the evaporation is violent, a small amount of mist from the boiling sea water will be carried over; and this entrainment may give 10 to 100 ppm or more of salt in the fresh water, depending on the design of the evaporator and the violence of ebullition. This entrainment may be reduced to 1 to 5 ppm (or even less) by proper design of the evaporation surface in relation to the evaporator vessel or body, or by passing the vapors through a demister or entrainment separator, which changes the direction of vapor flow in passing through a pad of metallic mesh, glass fibers, or other materials. The droplets of mist, due to their greater mass and hence greater inertia compared to that of the vapors, strike the metal or other parts of the demister instead of passing in the vapor streams around them; and are coalesced into films which drain back to the evaporating liquid.

In a once-through or single-effect evaporator the thermal energy consumption is high since all heat transferred through the evaporating surface is lost to the condensing surface. But, using the multiple effect principle, several units or effects are operated at successively lower pressures; the vapors from the evaporation at the higher pressure passing to condense and give up heat to the heating surface of the next effect, where the heat of condensation evaporates about the same amount of water in each successive effect from the liquid passing through the several effects in series. Allowing for heating of the liquid, losses, etc, a triple-effect may give a gain ratio of about 2 pounds evaporation per pound of steam used, and a quadruple-effect of about 3. See Evaporation.

Multiple effects have been used for many years in the salt, sugar, and other process industries. From 4 to 6 effects is the usual maximum number, balancing the lower steam costs against the increased capital costs with a larger number of effects, to give a minimum total cost per unit of evaporation.

This number of effects has been increased in producing fresh water from sea water to 12 effects in one demonstration unit of the OSW at Freeport, Texas, which gives a net evaporation or gain ratio of about 10 pounds fresh water produced per pound of boiler steam used. Figure 3 shows that a great mass of metal is required, since its amount increases per unit product somewhat faster than in direct proportion to the gain ratio. Particularly, this means a large number of tubular surfaces for evaporating and condensing. Also many heat-exchange surfaces are required for recovery of heat from the streams of salt water and fresh water to and from the many effects. In this

12 effect, 1 million gpd unit, more than thirty such separate bundles of tubes are used. This seems to be the maximum size and maximum number of effects of a multiple-effect which is practical; and other evaporator types are now regarded as more economical, considering both cost of plant and cost of water produced.

In the *vapor compression evaporator*, the vapors from boiling sea water inside the tubes, usually of a single effect or body which is, say, at atmospheric pressure, are compressed mechanically to a sufficiently high pressure to give a steam temperature high enough to allow condensation on the outside of the tubes. The single tubular surface is used now for both evaporating and condensing. Boiling continues as long as does the

Fig. 3. Multieffect long tube vertical (LTV) Evaporation: 12 Effects, 1 million gpd Demonstration Plant of OSW at Freeport, Texas. Some 20 heat-exchangers recovering heat in the fluid streams of the 12 effects of the evaporator. Courtesy of Office of Saline Water.

compression of the vapors formed. Mechanical power from an internal combustion engine may drive the compressor, and the waste heat from the engine block and the exhaust gas is used to preheat the sea water. This system is also used where electric power is available, or where power is generated by boiler steam which, after going through the turbine, is utilized ultimately for another purpose.

The largest vapor compression evaporator is a demonstration plant for 1 million gpd of the OSW in Roswell, New Mexico, evaporating alkaline brackish water having a mixed salt content of about 15,000 ppm. The compressor operates across two effects, to increase the amount of evaporation; and the energy required is about 60 kWh/1000 gallons fresh water produced. Figure 4 is a schematic flow diagram showing the evaporator with the water treatment facilities.

Power cost goes up rapidly with the increase of temperature drop across the tubes, which is required to cause heat transfer, since pressure difference, and thus compression

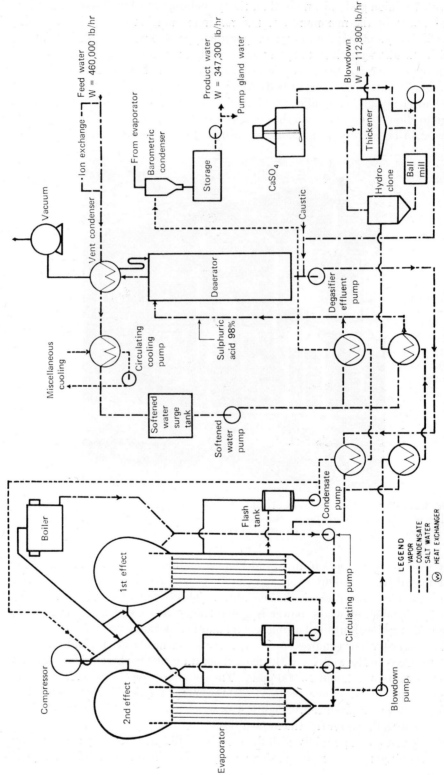

Fig. 4. Forced circulation vapor compression plant. Flow sheet of 1 million gpd OSW Demonstration Plant at Roswell, New Mexico. Flowsheet of evaporator plant using about 60KW electric power per 1000 gal fresh water with water treatment facilities. Courtesy of Office of Saline Water.

ratio on the compressor, increase even faster. To secure the necessary heat transfer without excessively large surface area, while minimizing ΔT, the heat transfer coefficient U must be as large as possible. U is increased by circulation of salt water through the evaporator tubes at a high velocity. This takes power for pumping. An economic balance shows that considerable energy may be so used, as it reduces the amount of heat transfer surface required, and thus the capital cost. (The thermal equivalent of this electric power is largely absorbed in the sea water.) The minimum cost occurs when the decrease in capital costs as velocity of water in the tubes increases per unit of product water just equals the increase in the total of the two power costs, for vapor compression and for liquid recirculation through the heating tubes.

As much as one-fifth or more of the total energy may be used for the recirculating pump; the balance compresses the vapors from the boiling sea water to the higher pressure on the other side of the tubes. From 200 to 800 times as much water may be circulated economically as is being evaporated in any one pass through the tubes, as compared to a film evaporator—also giving a high rate of heat transfer—where the liquid goes through the tubes only once in each effect (as, for example, in the LTV design of the 12 effect unit at Freeport, Texas).

For sea water desalination, Andersen has developed a vapor compression system in Denmark for which the installed plant costs about $1 per gallon of daily capacity in the 250,000 gpd size range. The fluid flow circuits are improved to minimize friction and head losses. About 50 kWh of electric power is required per 1000 gallons of product water; and the total water cost, including amortization, is estimated at about 60¢ per 1000 gallons. The Andersen patents in various countries provide an excellent example of a calculation for optimizing the several variables for the design of such a plant to minimize the total cost (22).

An additional advantage of the forced-circulation unit which makes it particularly useful with brackish water, often having large amounts of salts which form scale, is the high velocity of the liquid which minimizes scale formation. Furthermore, no cooling (condensing) water is required, another major advantage in arid areas where brackish water must be desalinated and where even brackish water may be available only in very small amounts. The brackish water may be concentrated three to eight-fold, rather than the usual two-fold, as with sea water, in order to minimize the amount of feed required and the amount of concentrate to be eliminated from the system. The disposal of this concentrate may represent a major problem, since there are usually no flowing streams of any consequence. If passed to a pond, it seeps to the water table below and pollutes wells. At Roswell, N.M. a lined 90 acre pond allows no seepage of the concentrate from the evaporator, but provides surface evaporation for the water from this brine.

Production of fresh water by vapor compression should require no prime thermal energy as such; since the evaporation is done entirely by the heat in the vapors boiled from the salt water and compressed to a temperature and pressure which will allow them to condense. The discharge of hot fresh water and of hot brine, together are equal to the salt water feed, and are used to preheat the saline water fed to the system. Heat losses and inefficiencies are made up by additional thermocompression.

More recently, newer design techniques have been developed which will give lower plant and operating costs than those available heretofore, by combination with vapor-reheat as mentioned below. The production of 1000 gallons of fresh water may require only about 25 kWh of power (23).

A limitation on the size of the vapor compression evaporator is the possible size of the compressor, a major cost item. Possibly the one at the Roswell unit is as large as practical. By operating at a higher pressure, the specific volume of the steam is greatly reduced; and the compression ratio for the same temperature difference across the heating tubes is also reduced, since steam pressure increases faster with temperature at higher temperatures. However, there are other problems: heavier equipment at higher pressures, increased tendency of scaling, increased heat exchanging of liquid streams from and to the ambient temperature, and increased heat losses. All of these factors must be considered and their relative effects balanced in selecting a design pressure.

Conventional Multiflash Evaporation. Flash evaporation is the vaporization, often violent, of a part of the water in a stream of hot sea water when it passes into a chamber at a lower pressure; vapors are formed adiabatically. The sensible heat given up by the cooling liquid as it comes to equilibrium with the lower vapor pressure is equated to the latent heat of the vapors formed. A high ratio of flash temperatures and pressures, as in the discharge of condensate from a steam trap to atmospheric pressure, causes an approach to equilibrium with almost explosive violence. The vapors formed are withdrawn and condense in preheating the raw sea water passing through a condensing system of tubes.

A large number of stages of such flash evaporations at successively lower pressures give what is called a multistage flash or MSF evaporation. The condensing of the vapors to heat the feed operates countercurrently to the MSF to give a cooling evaporation and heating-condensation heat interchange. There is no evaporation heat transfer surface; but there is a condensation heat transfer surface.

After passing from the last of the series of condenser-preheaters, the sea water is finally heated to its highest temperature in a prime heater by outside heat, before the first or highest pressure of the MSF. 97% of the present world capacity for desalination of sea water now uses this MSF system; and unless there is some unexpected breakthrough in technology, improvements of MSF will be the system of the future for making fresh water from the sea.

The process industries have used MSF for at least 50 years; the first patents were before 1900. Salt has been produced from heated brine which flash evaporates in steps to give vapors during its cooling. Only three stages were first used in the United States, in a counter-current cascade or ladder, in concentrating black liquor in the wood-pulp industry. See Pulp. Many more have been used in evaporating salt brines in France and in South America (24).

MSF was developed to allow the evaporation of liquids which normally form scale on tubular evaporation heat transfer surfaces. Sea water has bad scale-forming properties, as do almost all brackish waters. The greatest limitation to the use of MSF is still the presence of salts, such as calcium sulfate, which, because of lower solubility hot than cold, precipitate as scale in the tubes in the high temperature part of the system. Many chemical systems, often corrosive to tubes, and many mechanical systems, often abrasive to tubes, have been used in attempts to control scale, so that higher temperatures may be reached with consequent increase in thermal efficiency.

A mechanical system for scale removal has been described which may eliminate scale, to remove the necessity of chemical treatment of the feed sea water, while allowing a much lower ratio of blowdown of brine to sea water feed, and consequent saving of heat so lost (25).

The MSF operation cannot utilize all of the latent heat of the vapors which would be formed in cooling the hot sea water down to the temperature of the sea water feed. Desirably, this heat would be used in preheating the liquid feed under balanced conditions of heat flows and fluid flows. This is not possible, although substantial improvements in control of these fluid and heat flows are available to achieve a corresponding increase in thermal economy over the usual practice. In most plants, however, an additional amount of cold sea water is fed through the condensing tubes of the several stages at the low temperature end, to condense all of the vapors of the cooling brine in the last several flashers. This does produce some fresh water even though this heat of condensation is rejected to waste by discard of this additional stream of sea water.

The ladder of conventional MSF evaporation is diagrammed in Figure 5. This does not show the necessary pretreatment of the feed by addition of chemicals to eliminate scale formation and by vacuum deaeration to remove oxygen and carbon dioxide, which cause corrosion and which greatly reduce heat transfer in condensation of the steam (26). These are two essential operations in any but solar evaporation of sea water.

Most MSF evaporators built to date have a horizontal arrangement of the stages rather than the vertical arrangement diagrammed in Figure 5 and plants thus require a very much greater area than would be required with this vertical arrangement. More advanced present designs improve evaporation efficiencies by vertical arrangements, one of which will be discussed later; and it may be expected that vertical designs will be utilized increasingly in the future.

Sea water is usually heated to its highest temperature by boiler steam in the prime, or brine heater. The hot sea water passes downward on the left or cooling-evaporation side of a series of stages, each of which is at a lower saturation pressure, and hence temperature. The sea water flash-evaporates in each stage as it is cooled progressively, while being concentrated stage-by-stage until it discharges from the bottom stage. Some brine is discharged to waste; but a considerable fraction may be recycled through the right side to join the sea water feed. This recycle brine is at a higher temperature than the raw sea water; and it may be added at the stage where the sea water has already been preheated to its temperature.

From the thermodynamic standpoint, the objective in the flash chamber of a stage is to obtain adiabatically an equilibrium between the cooler sea water discharged from the stage and the vapor formed, the latent heat of which should equal the change in sensible heat of the sea water in cooling. All of the vapor leaving should be in equilibrium with all of the sea water leaving; and there should be no mist of droplets of sea water in the vapor stream caused by an ebullition which is too violent.

Mechanically, there are major obstacles to obtaining this equilibrium. The volume of sea water flowing, mostly recycle, is tremendous, usually at least ten times the rated fresh water capacity; for example, a standard plant with a capacity of 2.5 million gpd must handle a torrent of 25 million gpd or 17,500 gallons per minute. For any possible water velocity and any practical width of the stream, in the usual horizontal arrangement of stages end-to-end, this torrent must be at least one to several feet deep, but can only have a residence time in each flash chamber of some seconds. But the driving force toward equilibrium is usually only 4° or 5°F and sometimes is as low as 2°F in a 40 or more stage evaporator. This may be equivalent to a water pressure drop of less than a few inches of water in the low pressure stages and would be equalled by a hydrostatic head of that depth, although much more in the high pressure stages.

Thus bubbling or ebullition could not take place in any water which did not come at least that close to the surface; and this requires considerable agitation and turbulence of the whole depth of the water stream in order to approach equilibrium, which will not quite be reached in most practical cases. The maximum temperature drop of the water therefore can never quite be obtained in the stage; hence the maximum amount of vapor cannot be produced.

Of equal importance is the fact that the vapors issuing violently from the turbulent surface carry droplets of entrained sea water. To remove these droplets, which would carry sea water into the product, demisters are used. To force the vapors through these, there is an additional pressure loss. This means there is a loss of the effective temperature drop of the steam for its heating function in heating sea water to the high overall temperature range from the top to the bottom stage. However, as the temperature drop or flash temperature increases much over 5°F, the turbulence of the boiling becomes so violent that entrainment increases. Thus, a large number of flash stages, 40 or more, are usually used to reduce the individual flash temperature and violence of ebullition.

Better designs of flash evaporation chambers are being sought to minimize these and some related losses and inefficiencies which are compounded by the very large number of stages.

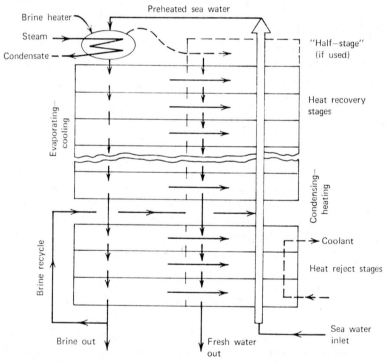

Fig. 5. Conventional multi-stage flash evaporation. After heating to highest temperature in brine heater at top, the sea water passes while flash evaporating and cooling at successively lower pressures: vertical arrows down on left side. The vapors leaving each flash evaporation: horizontal arrows, pass to preheat the sea water in the condensing-heating tubes: wide vertical arrow up on right side. The fresh water condensate also passes from higher to lower stages for evaporation and cooling, to discharge at bottom. Vapors may be also withdrawn from prime heater to be condensed in the halfstage (lines upper right) to increase the production of fresh water.

There are several measures of these inefficiencies of the MSF unit: (*1*) the lesser amount of vaporization, and condensation, from the flash evaporation (*2*) the lesser drop of temperature from sea water into and out of the stage; and better, (*3*) the great difference of temperature between sea water out of the stage and vapors out of the stage, which may be a substantial fraction of 1°F up to 2 or 3°F; this is called the transport loss per stage. This last is without reference to the small elevation of boiling point of the sea water due to the salt content which is not greatly different for each stage throughout the ladder of stages. The heating in each stage of the sea water is therefore to a temperature lower in each stage by this amount. Thus, the boiling point elevation is effective only at the top stage, where the water going to the prime heater must be heated by boiler steam through an additional temperature range equal to the boiling point elevation.

This effect of the boiling point elevation is quite different in the MSF compared to that in the multiple-effect evaporator, where the boiling point elevation is very important in reducing the available temperature drop, and hence efficiency in each of the effects. This is one of the advantages of the MSF.

The feed sea water passes counter-currently to the MSF through high velocity tubular heaters in the right or condensing-heating half of each stage, where it is preheated by heat given up by the condensation of the flash vapors. Condensate so formed in each stage passes to the next lower stage and is cooled, also by flash evaporation. Its sensible heat in cooling is thus converted to latent heat and, by condensation on the tubes, is added as sensible heat to help preheat the sea water.

An additional cooling stream of sea water is used to condense the vapors formed in the lowest stages, often as in Figure 5, those below the point where recycle brine is added back to the condensing-heating section. Here, in the lower right side, the dotted line is a stream of coolant sea water which adds to the condensing capability of the cold sea water feed. This coolant is passed to waste with corresponding heat reject from the latent heat of the vapors in the lower stages.

Leaving the heating tubes of the condensing half of the top stage, the preheated sea water goes to the prime or brine heater, thence to the vaporizing-cooling side of the top stage. The succession is repeated, stage by stage; and evaporation always takes place out of contact with a heat transfer surface; and condensation always takes place on the heat transfer surface while preheating the feed.

Figure 6 is a flow sheet of a 2.5 million gpd plant designed by Burns & Roe, Inc. for the OSW as a standard or Universal Desalination Plant (27) to incorporate the best of current practice in conventional MSF evaporation. The first plant of this design was constructed in Jedda, Saudi Arabia, under an agreement between agencies of two governments, and will be on-stream in early 1970. The flow sheet shows the raw sea water supplied at 85°F going to the heat rejection stages where it is heated to 97.7°F, partly discharged to waste, and partly treated chemically against scale formation before passing to the evaporation system. First there is an addition of sulfuric acid to decompose carbonates and liberate the carbon dioxide, and then a deaeration. Next, there is a treatment with a defoaming agent and also caustic soda to neutralize any excess acid.

Table 1 shows flows, temperature, and other data at the points numbered in Figure 6.

This treated sea water joins the discharge from the heat reject stages and passes to the lowest temperature stage of the heat recovery system. After passing all of these, it goes to the brine heater and is heated with prime steam from the boilers. The

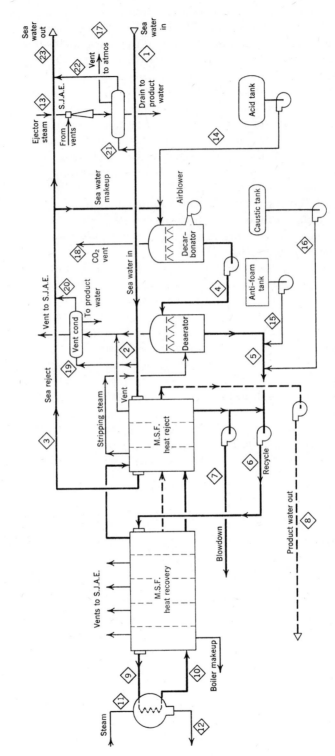

Fig. 6. 2.5 mgd Universal Desalination Plant.

Table 1. Material Balance for Fig. 6, 250°F Maximum Brine Temperature

Location	Flow, lb/hr × 10³	Flow, gpm	Temp, °F	Dissolved solids, ppm	Specific gravity
1 seawater supply	7880	15,250	85	35000	1.022
2 coolant seawater to evaporator	6293	12,300	85	35000	1.022
3 seawater reject	6293	12,400	97.73	35000	1.019
4 decarbonated make-up	2044.119	4,000	97.73	35000	1.02
5 deaerated make-up	2044.119	4,000	97.73	35000	1.02
6 recycle brine	6409	12,400	97.86	52500	1.031
7 blowdown	1178.719	2,520	97.86	60697	1.041
8 product water to storage	865.4	1,730	93.96	25 Max	1.0
9 brine heater inlet	6409	13,050	237.41	52500	0.982
10 brine heater outlet	6409	13,000	250	52500	0.980
11 brine heater steam	83.082		264.9		
12 brine heater condensate	83.082	177	262	2 Max	0.935
13 ejector steam	2000				
14 acid, H_2SO_4	0.240				1.84
15 antifoam	0–3ppm				
16 caustic	0–3ppm				
17 noncondensibles	0.225		210		
18 decarbonated vent CO_2	0.20		95		
19 seawater to vent condenser	1100	2,150	85	35000	1.022
20 seawater from vent condenser	1100	2,160	92	35000	1.021
21 seawater to steam jet air exhaust (SJAE)	500	980	85	35000	1.022
22 seawater from SJAE	500	985	95	35000	1.02
23 seawater discharge	7880	15,300	97	35000	1.02

hot brine returns to the MSF evaporation zones where it evaporates and cools, first in the heat recovery stages, then in the heat reject stages. Finally, after leaving the stage of lowest temperature, the brine stream is divided, part for recycle and part for blowdown. The distillate similarly flash-evaporates in cooling from the high temperature to the low temperature stages.

Figure 7 shows the first demonstration plant of the OSW. This MSF unit, rated at 1.0 million gpd, was first installed near San Diego, California and operated there from January 1962 to February 1964 under OSW contract. A dramatic need arose for

Fig. 7. OSW Demonstration MSF Evaporator, 1 million gpd. First installed and operated for some years near San Diego, California; then removed to supply emergency needs at Guantanamo Bay, Cuba, and operated there, all by Burns & Roe, Inc. Courtesy of Burns & Roe, Inc.

fresh water at the United States Navy Base in Guantánamo Bay, Cuba; and in less time than a new plant could be built, this was moved, installed, and is still operated by the contractor, Burns & Roe, Inc.

Many pounds of fresh water may be produced per pound of boiler or prime steam used; and units with 40 to 50 stages may have gain ratios as high as 8 to 11 pounds of water evaporated per pound of prime steam used, as also may units with a lower number of stages if a larger heat transfer surface is supplied on the condensing-heating side. By placing two or more MSF units in series as a multi-effect MSF evaporator (MEM-SF), the gain ratio may reach 20. The larger cost of the unit for these high gain ratios usually makes a plant more expensive in capital costs than can be balanced by the savings in energy costs.

The total cost for product water is largely made up of energy cost plus capital charges. When a unit of energy is cheap, water of lowest cost comes from a cheaper plant because the high cost of an expensive plant with a high gain ratio cannot be afforded. When a unit of energy is dear, a plant with a high gain ratio, usually more expensive, gives the lowest cost for water, still relatively high.

Evaporation in Brine Heater of MSF—Vapors to Half-Stage. Besides the conventional function of *heating* the sea water to its highest temperature, the brine heater may also evaporate some part of it; the vapors formed being condensed in a half-stage to give a substantial increase in capacity and lowering of heat costs (24).

The half-stage, when used, is one more of the series of the condensing-heating zones of the stages which preheat the sea water; it is located above the top stage, as shown by the broken lines in the upper right of Figure 5. Vapors withdrawn from the prime heater, now also an evaporator, are condensed here to preheat sea water even further.

The half-stage and brine heater together do exactly the same job as did the brine heater alone, ie, heating the brine from the temperature it leaves the condensing-heating zone of the top stage to the temperature at which it reenters the evaporation-cooling zone of this same top stage. Exactly the same amount of prime heat is supplied and used, but vapors formed in the brine heater-evaporator now pass to the half-stage to heat the brine leaving the top stage through some part of the temperature increment, necessarily done by prime heat. No less boiler steam is used, but there is an additional amount of condensate produced in the half-stage, which makes that much more fresh water. The use of the half-stage with vapors from the prime heater-evaporator would give no advantage if the object was simply to concentrate the feed.

The optimized design calculations of the Fluor Corporation for the OSW for the first 50 million gpd unit of the planned 150 million gpd plant in California may illustrate this use (28). The same amount of prime steam, 1,787,500 lb/hr required in the Fluor conventional operation is used also with evaporation in the prime heater supplying steam to a half-stage, as shown in the dashed lines on the upper right of Fig. 5. In both cases, prime steam heats the sea water leaving the condensing-heating half of the top stage at 235°F through a range of 150°F to its temperature of 250°F as it enters into the flashing zone of the top stage; but the heating is in two steps of which the half-stage is the first. However, the brine is evaporating from and at 250°F to supply vapors at this final temperature (less the boiling point elevation) to the half-stage for preheating the liquid some part of the 15°F temperature rise needed.

For example, the half-stage may be designed to heat the brine to 242.5°F at a terminal temperature difference of 5.1°F. Beside the same condensate in the brine heater as before, 1,787,500 lb/hr, there is additional condensate in the half-stage of one-half this amount, or 893,800 lb/hr. This increases the amount of product and decreases the unit steam cost by the ratio of this to the design product figure, or 5.15%. The gain ratio of the Fluor design is 9.72 and this goes up by 5.15% to 10.22 because of this additional fresh water produced, 893,800 lb/hr. This is 2.5 million gpd, for this first third of the plant, or 7.5 million gpd for the complete plant as planned, at no cost for heat.

The use of the half-stage lowers the effective temperature drop in the brine heater, but it increases U there, because heat is transferred to a boiling solution. However, the brine heater-evaporator will require a slightly larger heat transfer surface, and also a vaporizing space for evaporation. Also needed is the heating tube bundle of the half-stage. The total plant cost by adding the half-stage has been calculated to increase by 5%, with a simultaneous increased water production of 5.15%. Thus, the plant cost per unit of production is unchanged and the thermal advantage costs nothing.

Only a fraction of one percent of the liquid flow which is being heated is evaporated in the brine heater; and this does not cause any appreciable change in the scale-forming tendencies.

This improvement in steam economy decreases with an increasing number of stages, and there are 39 in the Fluor design. A small MSF evaporator with only 5 stages, designed for an industrial plant, indicated a heat saving of approximately 20%, and a considerable reduction in its cost using the half-stage and a prime evaporator-heater instead of the conventional prime heater alone.

Controlled Flash Evaporation (CFE). Flash evaporation and condensation of the vapors formed on the metal surfaces of the heating tubes are the two operations of the MSF, and, as many and larger MSF plants are being built, the inefficiences noted in each operation have become more evident and more expensive.

Major energy and capacity losses are caused by the high turbulence of the hot sea water in the flash chambers, where attainment of equilibrium may be almost explosive in its violence. There is a rapid increase of these losses with a larger temperature range of the flash evaporation.

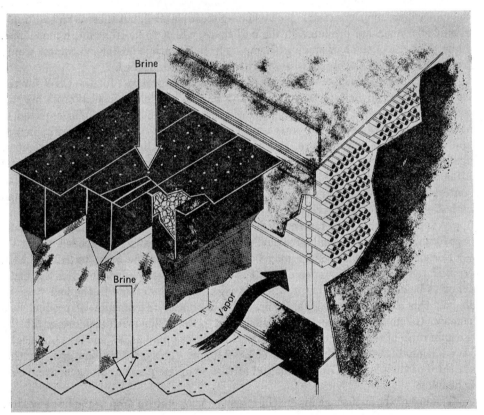

Fig. 8. Controlled flash evaporation (CFE), single stage, each with many short rectangular columns: Packed with Berl Saddles over which brine descends as films, while vapors disengage quietly. V-bottoms of wire mesh supporting packing allow exit of vapors formed and permit smooth flow of brine to drip-screen guiding it to next lower stage. Vapors pass to condenser tubes in rear and condense to preheat inlet sea water. Courtesy of Burns & Roe.

The desired evaporation would allow all of the heated sea water to pass in intimate contact with vapors under conditions which are controlled so as never to be far from equilibrium. This would require a continuous, but gradual, reduction in the pressure between two stages. Thin layers of sea water should be allowed to come to equilibrium in a co-current relation with the vapors. The term flash would then be a misnomer.

Evaporation and condensation are reciprocal phenomena at the phase interphase. Very large numbers of water molecules and large masses of water pass from either phase to the other, due to even an extremely small difference in the temperatures on either

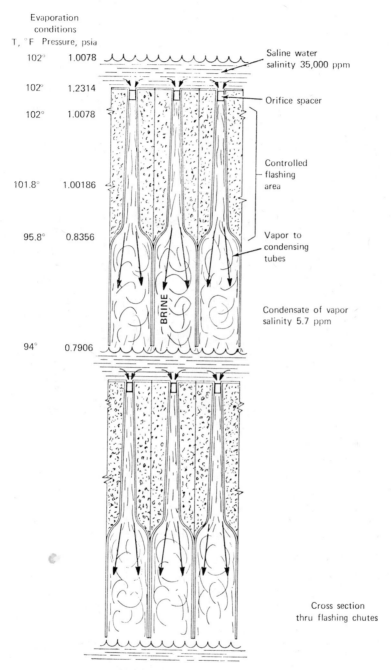

Fig. 9. Controlled Flash Evaporation (CFE). Cross section through Flashing Chutes. Sea water at top passes through orifices between orifice spacer blocks. As pressure lowers, water evaporates and cavitation forms a core stream of vapor, with films of liquid descending on each side wall of chute. Vapor separates free of sea water in lower expanded section, and passes frontwards and backwards to condenser tubes. The evaporating conditions are shown at left at respective distances of passage of liquid. Courtesy of Burns & Roe.

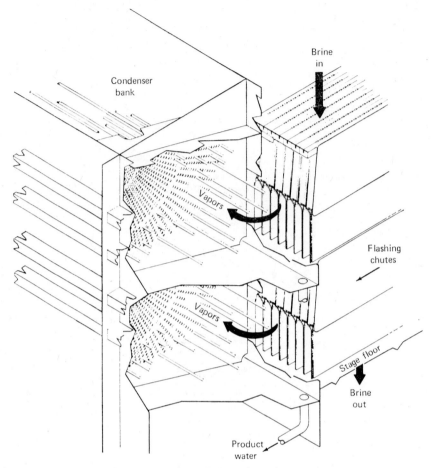

Fig. 10. CFE—Flashing Chutes and Condensers of Two Stages: Brine from plate above descends through rectangular orifices on right formed by spacers at top of 8 ft long chutes; vapors disengage and pass, first in line with chutes, and then between the skirts to condenser tubes on left, also to right, not shown. Product water condensate goes to water side of next lower stage. Courtesy of Burns & Roe.

side of the interface; and this ΔT will be measured in hundredths of a degree for any rate of heat flow which would ever be observed in practice (29). The losses caused in the usual turbulent flash chambers can be eliminated in a system devised and described as Controlled Flash Evaporation (CFE) in many U.S. and foreign patents. Recently, it has been developed in a substantial engineering research and design program supported by the OSW (30). Figure 8 shows a stage of one design of the CFE; the heated sea water is filmed in passing downward from a higher to a lower pressure zone. The hot brine flows through orifices in the plate of an upper stage and goes down into long rectangular column sections packed with Berl saddles or other suitable shapes, just as reflux liquid descends in films over the same packing in a distillation column. The bottom support for the 16 in. to 24 in. of column packing is a V-shaped trough from the bottom ridge of which liquid runs down to the next stage, while the vapors formed and passing cocurrently to the bottom of the packing are withdrawn laterally to the condenser.   Entrainment is so low that the condensate may be only 1 ppm, with a flash temperature range of as high as 15°–20°F.

Another CFE system shown in Figure 9 also maintains a substantial balance between the saturation pressure corresponding to the temperature of the sea water at

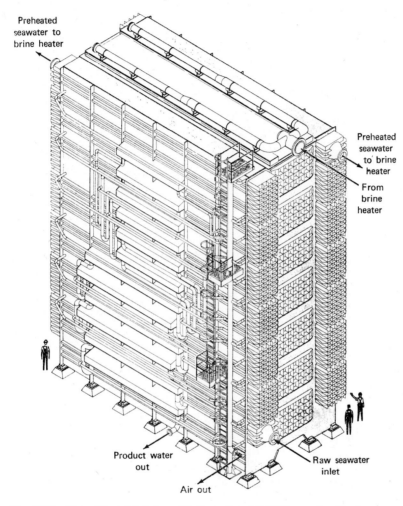

Preheated seawater to brine heater

Preheated seawater to brine heater

From brine heater

Product water out

Air out

Raw seawater inlet

Fig. 11.   CFE—Stage Plant Modulus with Capacity of 3 Million gpd.   Condensing-preheating tubes on each stage are approximately 48 ft long on either side of 6 sets of 8 ft long flashing chutes, each set about 8 ft wide; total width about 20 ft, total height about 58 ft.   Cold sea water enters bottom at both nearest corner and farthest corner and climbs back and forth through 13 superimposed tube bundles to top, then the two streams join to go to the brine heater, thence back to top distribution header to two points on each set of flashing chutes.   Courtesy of Burns & Roe, Inc.

every point, with the pressure of the flash vapors at that point as the sea water passes downward through narrow vertical channels or chutes, between a higher pressure stage and the next lower pressure stage.   These chutes, shown in cross section in Figure 9, may be formed by vertical plates at angles of a few degrees on either side of the vertical; thus, the distance between increases gradually.   After two or more feet, the width of the chute suddenly expands considerably at the lower edges of the plates.

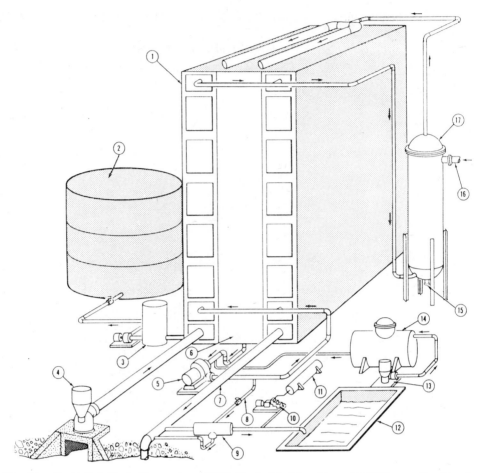

Fig. 12. Controlled Flash Evaporator (CFE) Single Modulus Plant (3 Million gpd). Showing arrangement and piping of accessory equipment. Courtesy of Burns & Roe, Inc.

1	First Stage (High Temp.)	7	Heat Rejection	13	Makeup Pump
2	Product Water Storage	8	Blow Down	14	Deaerator
3	Head Tank	9	Heat Exchanger	15	Condensate Out
4	Inlet Seawater Pump	10	Acid Pump	16	Steam
5	Recycle Brine Pump	11	Acid Storage	17	Brine Heater
6	Last Stage (Low Temp.)	12	Degassing Pond		

A large number of such chutes, in parallel, have their top narrow entrances or orifices for heated sea water at the bottom of the pool section of the upper stage. The heated sea water flows into the top of the chutes, through the orifice system, then downward due to the combined forces of gravity and the higher pressure of the stage above. Almost immediately, cavitation develops on the axial plane at the center of the chute. Vapors formed proceed downward as an expanding core or self-formed channel, between the two films of the sea water, one on each side of the chute, and about 1 mm thick. These thin films allow near-equilibrium of vapor and liquid to be reached quietly, more like a diffusion, since there is no breaking of the surface by bubbles or ebullition.

The expanding cross-section of the chute accommodates the increase in vapor volume, due to evaporation, and also to its expansion as the pressure decreases.

At the lower edges of the planes forming the chute, there is a considerable in-

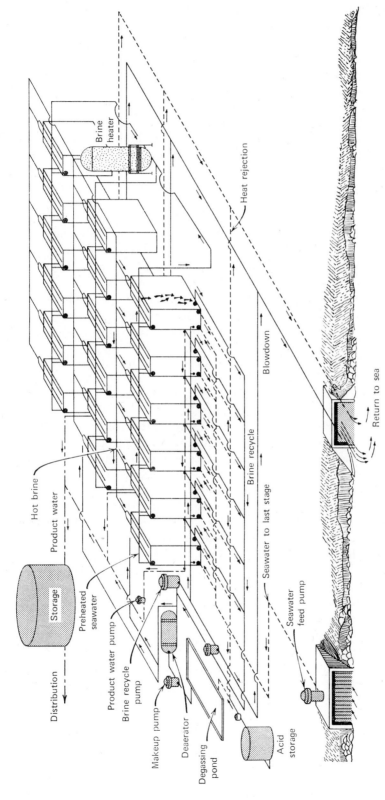

Fig. 13. Controlled Flash Evaporator (CFE) Plant with 18 Moduli in Three Banks of 6 each. Total capacity 50⁺ million gpd. Enlarged ground area to show piping. Courtesy of Burns & Roe.

crease in cross-section, so the velocity of the vapors decreases substantially as they go out laterally to the heating-condensation side of the stage. Simultaneously, the films of sea water run down the side walls of the greatly widened liquid passage to the pool of the next lower stage. From this, the sea water passes the orifices and enters the subsequent chutes. At every point, the sea water and vapors are in intimate contact, and at the same temperature. A detailed isometric broken view in Figure 10 shows the positioning of two sets of flashing chutes with the corresponding condenser tubes.

Proper design and operation of the CFE will allow a very high temperature range of flashing, up to 25°F or more, without losses of pressure and temperature due to turbulence or by-pass of the stage without flashing, and also without entrainment. The transport losses per stage are only about one-fifth that of the conventional MSF, or per degree of flash temperature, only about one-tenth. A true optimization may therefore be calculated between capital costs and operating costs, largely steam costs. It is not necessary to tolerate the high costs of the very large number of stages necessary in the conventional MSF which are required to insure a low flash temperature difference of not more than about 5°F. Thus, a lower number of stages and a lower capital investment are required for the CFE for the same thermal efficiency or gain ratio. Or, for the same capital cost, the gain ratio will be much improved.

Compared to conventional MSF with its relatively variable operation and efficiency, particularly with changing duty or throughput, the flow in the CFE from stage to stage is uniformly efficient over wide operating ranges. It is regulated by the height of the sea water in the pool of a stage above the orifices feeding the chutes. No liquids can pass from stage to stage without coming to equilibrium with the vapor leaving the stage, and this without the violent turbulence necessary in stages of conventional design. Also, it is impossible to have interstage flooding, a problem which has not been always satisfactorily solved in conventional plants.

CFE plants are designed with multiple stages arranged in a tower, up to 13 or more, one above the other, as shown in Figure 11. This module, approximately 20 ft by 48 ft by 58 ft high, contains the CFE stages and condensers for producing 3 million gpd. Figure 12 shows the unit with accessories; for large installations these standard 13 stage modules may be assembled on the plant site in both series and parallel. The area of the site required may be less than half as much as for the conventional MSF, which spreads over considerable area and requires a correspondingly large amount of piping. Desalination plants must usually be located near the demand of the dense population of urban areas, where land is expensive. Costs of plant sites have, in exceptional cases, ranged as high as 10 to 15% of the total calculated investment. The very substantial reduction in real estate requirements of CFE gives an important reduction in the total cost. Figure 13 is a schematic layout of 18 modules of 3 million gpd each, a plant for a nominal 50 million gpd of product.

One CFE low pressure stage with a 10°F flash range, which is part of a continuously operating commercial plant desalinating sea water, and which has no demisters, showed by single stage test a production of 15,000 gpd of water with only 1 to 2 ppm of salts. The operating conditions of this stage are as difficult as are ever encountered, since the pressure is less than 1 psia, the temperature is below 100°F, and over 350 cubic feet of vapor must be disengaged from the liquid surface for every pound of fresh water produced.

A major consideration in all evaporation of sea water is the removal of noncondensible gases, particularly air, but also carbon dioxide if there is an acid treatment for scale prevention of sea water which will be raised to the high temperature of an MSF evaporation. As a by-product of the CFE process, deaeration may be accomplished with a minimum of equipment compared to that normally required, and with practically no cost of operation.

Another development in the design of the CFE stage and the interrelation of the evaporation and condensation parts of the stage has resulted in a substantial increase in the coefficient of heat transfer of vapors in condensing on the preheating tubes which are a part of the stage. A saving in this important item of cost is thus possible.

Waste Heat from Condenser Water. CFE has also opened another field for desalination, and simultaneously has solved another major water problem of many large electric power stations, particularly those using nuclear fuels. A large part of the thermal energy must be rejected to cooling water in the condensers. The increased temperature of the water discharged back to the body from whence it came, usually 15 to 25°F, may be a nuisance to fish and other life, and causes thermal pollution.

For each pound of steam generated and passed through the turbines to the condenser, with a latent heat of about 1000 Btu, some 50 pounds of cooling water must be pumped to the condenser if the temperature rise there is 20°F. As already noted, the CFE has the ability to handle, in a relatively small and simple unit, large volumes of warm water, at a large temperature range of flashing, and with high efficiency. At least 10°F, or one-half of the temperature rise of the condenser water, can be recovered by CFE of the warm condenser water. Thus, if one-half of the heat added to the condenser water is utilized, one-half of the weight of the steam generated in the boilers may be produced as vapor, and hence fresh water, from the heat in the condenser water, with substantially no cost for thermal energy. Even if the condenser water has a much smaller temperature rise, only 10 to 12°F, it may be possible economically to use its heat in this way. Without demisters, the solids content will be from 1 to 2 ppm; very pure boiler feed water, 0.3 ppm or less, is made with demisters.

Meanwhile, the temperature of the heated cooling water discharged will be reduced, with less danger of thermal pollution, which may be important where the power plant is located on a small body of fresh water, a river, or a lake. CFE is particularly applicable in this service, also where the power plant is on tide water and uses saline water for cooling, with fresh water for boiler make-up at a premium. Similarly, it will save the use of expensive prime heat in producing fresh water on steamships.

The thermodynamic capability of a CFE unit working with the warm condenser waters is many times the requirement of make-up boiler water. Nevertheless, up to the present, no plants have been built with greater capacity than their own requirements so that fresh water could be produced for distribution outside. However, factory assembled CFE units, with all parts self-contained within a single shell, are available in standard designs, in sizes up to 100,000 gpd. The cross section of one is shown in Figure 14. Any desired larger sizes may be field-assembled.

If, and when, the vast amount of thermal energy which may be obtained from temperature differences between surface and deep currents in tropic seas is utilized in producing fresh water and/or electric power, as suggested above, CFE will be a preferred system for abstracting energy as vapors to give heat and/or power from the large volumes of warm surface water (20).

Multiflash Evaporation: Vapor Reheat Process (VR). Evaporators have heat transfer surfaces to boil the liquid and to condense the vapors. In a vapor-compression unit, and in a multiple-effect, these functions are combined by condensing vapors on one side of the heat transfer surface to boil the liquid on the other. The usual MSF evaporator eliminates the heat transfer surface for the evaporation, but not for the condensation. The vapor reheat (VR) process, a modified MSF, eliminates the metallic heat transfer surface in *both* the evaporation and the condensation. VR has, so far, been used on only a pilot scale.

Various descriptions of the VR form of MSF have indicated the reasons for its fundamental heat economies, lowered plant cost, and other advantages, (23,24,38).

The cooling-evaporation side of VR is shown in Figure 15 to be the same as for the conventional MSF, as is also the passage of the vapors so formed to the heating-condensating side. The CFE system could also be used on the left. The usual recycle of part of the blow-down brine is not shown.

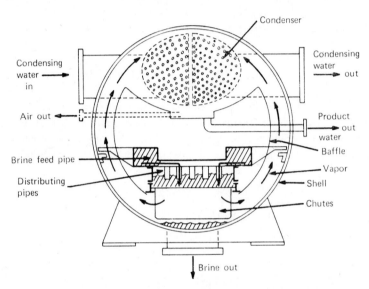

Fig. 14. Cross section of Small CFE Unit, Assembled in Tubular Steel Shell to Produce Make-Up Water for Boiler Feed. Sea water, warmed in turbine condenser, flows in cross hatched distribution system at right angles to section, and supplies flash chutes, parallel to section, with cooled brine, which flows out at bottom. Vapors go up to 2-pass condenser tubes cooled by same cooling water: entering rear left, leaving at rear right. Condensate fresh water discharge at right center. Courtesy of Burns & Roe.

In VR the vapors formed in each stage of the MSF are directly contacted with recycling colder fresh water in drops or other dispersed flow. The vapors condense directly on the surface of the fresh water to heat it as it flows counter-currently from stage to stage.

The open flow (sprays) of the cold fresh water stream on the heating-condensating side are indicated as arrow heads, pointing upward, just as the open flows for flash evaporation on the cooling-evaporating side are indicated by arrowheads pointing downward.

Only a very low temperature difference is necessary in the VR form of MSF for heat transfer in the condensation of vapors. These condense directly on the droplets of water sprays, as in the usual spray condenser. No intermediary heat transfer surfaces are necessary with accompanying stagnant water films on each side; a much closer approach temperature is always possible, measured values are in the 0.1°F range, and thermal equilibrium is reached very quickly. The condensation coefficient is extremely large across the very low resistance of the interface of vapor-liquid compared to that, plus the very much higher resistance of the usual condensing tube with stagnant liquid films on either side of the tube wall and the tube itself (29).

This low temperature drop is particularly important in the top stage, since lowering this temperature difference reduces considerably the amount of heat to be supplied to bring the brine up to the temperature of the top flashing stage. Hence, the prime heat required is lower, and a higher gain ratio results than with conventional MSF. The hot fresh water stream leaving the top stage is cooled in giving up its heat to preheat the cold sea water feed, and an amount of it which is equal to the condensate formed in all of the stages is removed as product. The sea water, following the preheating, passes through a brine heater, which may have standard heat transfer surfaces, eg, tubes. Preferably prime heat is added in a half-stage with direct condensation of prime steam on the hot cycling condensate after it leaves the top stage and before it is cooled in the preheater exchanger to heat the feed.

A conventional preheater to transfer the heat from the condensate stream to the sea water would require a large surface, as tubes. The liquid-liquid-liquid heat exchanger, sometimes called LEX, may be used instead to eliminate the use of any metallic surface. As shown in Figure 16 this operates as two counter-current, liquid-liquid extractors, with a liquid cycling through both, such as an oil which is immiscible in each stream. The hot fresh water condensate transfer its heat in the first "extractor" by bubbling-liquid counter-current contact to heat the oil, while the condensate is cooled. In the second extractor, the oil, now hot, is contacted with the cold sea water, which is heated, while the oil is cooled for recycle to the first extractor.

The intimate contact of the streams of insoluble liquids directly with each other gives very low terminal temperature differences in LEX. This unit has not yet been developed to large industrial sizes, but very high heat transfer rates have been reported in the range of 10,000 Btu/(hr)(ft³)(°F) in pilot plant operations. Here the rating is on the bulk volume with practically no internals. Substantially, only empty vessels of relatively inexpensive materials may be required. The high efficiency and low cost of such a heat exchanger indicate that prime heat added in a half-stage by direct contact with the condensate may then be transferred by the LEX to the cold sea water. Thus, the heat transfer tubes of the prime heater are also eliminated. Recent studies by Donald Q. Kern (31) for the OSW indicate very substantial economies are possible with the LEX compared with tubular heat exchangers for this and other surfaces.

VR thus operates without any metallic heat transfer surfaces. Single large, conventional MSF plants now being designed require many thousands of miles of tubes of expensive alloys of copper and nickel. The 150 million gpd plant of the Fluor design will require over 30,000 miles of tubes, many thousands of tons. The cost of the tubes, tube sheets, and related parts varies from 40 to 55% of the equipment costs of conventional large MSF plants (32). A very large saving in cost of plant will therefore result with the vapor reheat MSF evaporator. In fact, the world's needs for fresh water can hardly be supplied by conventional MSF evaporators using the amounts of copper and nickel for the alloy tubes which can be spared from the world's resources. The elimination of condenser tubes for the MSF also eliminates the possibility of corrosion and of scale formation in the banks of tubes at highest temperature. Calcium sulfate and other scale formers at elevated temperatures precipitate out in the sea water as it is heated in the liquid-liquid heat exchangers. The soft sludge may be readily filtered or

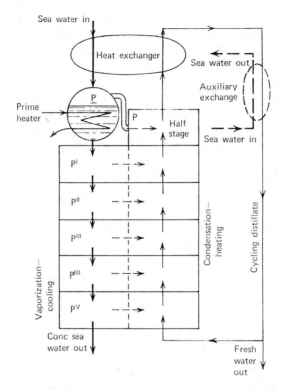

Fig. 15. Vapor Reheat MSF Evaporation, No Metallic Heat Transfer Surfaces. MSF evaporation-cooling of hot feed sea water on left as in Fig 5. Cooler fresh water on right, the condensing-heating side, passes in open flow, as by sprays, vertical arrows pointing upwardly. Vapors, horizontal arrows from MSF, are condensed by direct contact with these sprays. Heated fresh water is further heated in the half-stage before passing to the heat exchanger, where it preheats the sea water feed. Fresh water product is withdrawn on right, equivalent to condensation in the condensation-reheating stages. Heat may be rejected by auxiliary heat exchanger, with additional coolant.

settled out; or it may be allowed to cycle through the MSF and finally be blown down with the concentrated brine.

There is a separate fluid stream in each of the two halves of each stage of VR, instead of the single one of the conventional MSF evaporator; hence, there is greater flexibility of operation.

Here again, with VR, as a larger number of stages is used, the heat cost for the optimized plant decreases. However, it may be readily shown that the large number of stages (accompanied by the large increases in surface for condensation to increase the gain ratio) is not necessary with the VR. The advantage becomes less and less as the number of stages increases, and the optimum will be considerably less for the VR as also for the CFE in balancing the cost of plant, its corresponding capital charges, and the costs of operation, principally heat as steam. The total cost calculated for the product water is considerably less for the VR because of the markedly lower cost of the plant and the greater gain ratio.

The controlled flash evaporator gives a similar result by a major improvement on the evaporating-cooling side of the MSF ladder by improving the equipment, the operation, and the efficiency of obtaining heat transfer and equilibrium between the flashing liquid and the vapors formed. With VR the equipment, operation, and efficiency of obtaining heat transfer on the heating-condensating side of the MSF ladder are also improved.

Other Steps with VR. The same open condensation as in the other condensing zones with a half-stage gives the same higher gain ratio as noted with conventional MSF, when the brine heater is also used as an evaporator as shown in Figure 5. Alternatively, all prime steam may be added to the system in a direct-condensing half-stage to eliminate the heat transfer surface of the prime heater. Without metallic heat transfer surface for either the supply of prime heat or the condensation of MSF vapors, there is no possibility of scale formation; and the top temperature of the evaporation system may be increased substantially. The desired high top temperature may be optimized with VR based on considerations other than scale, such as increased cost of vessels with thicker walls to withstand higher pressures, and smaller volumes because of greater density of the steam, also processing factors related to heat balances. Thus, benefits may be secured which are not possible when there is a limitation of top temperature due to scale formation, as with usual MSF.

The arrangement of the VR indicated in Figure 15 with the stages fitted into a tower may be used. The stage of highest pressure may be at the top; also there are other arrangements; eg, the highest pressure stage may be at the bottom, with the flash expansion lifting the brine stage-by-stage to the top stage of lowest pressure.

It may be shown that, under some circumstances, a part of the evaporation may most economically be done, from a thermal standpoint, by a multi-effect unit, while another part of the evaporation is done by the VR unit. The two may be designed within the same tower shell to give what has been called a hybrid evaporator. Only a very large plant might find the added capital cost would be balanced by the heat savings. The heat savings depend, in part, on the fact that since no recycle of brine may be necessary, a once-through system of feed may be used economically.

Other improvements of MSF may be considered, and while many of these are applicable to all forms, conventional, CFE, and VR, they may be related to VR, which, with its two fluid streams, has a greater flexibility. One is the direct use of submerged combustion of a fluid fuel in the brine heater when boiler steam is not available. This eliminates heat transfer surfaces in the prime heater. Another is the use of two or more VR units in series in a multi-effect MSF arrangement. The flexibility given by the two streams instead of the single one, as in conventional MSF, allows the variation of flow patterns so as to improve the gain ratio considerably. However, while such large numbers of stages and so much equipment may be necessary in conventional MSF, the advantages and relative efficiencies of CFE and of VR will secure the same high gain ratios at very much less plant cost.

Vapor Reheat with Vapor Recompression. Vapors withdrawn from a lower stage of a VR evaporator may be recompressed to be used again as steam at the high pressure of the prime heater or the half-stage (24). Total electric power consumption may be about 25–30 kWh per 1000 gallons fresh water, using 15 stages of VR across a temperature range of 232°F to 214°F. By comparison, the OSW demonstration unit at Roswell, New Mexico, requires total input of almost 60 kWh. It operates on brackish water with a double-effect evaporator across this same recompression: from 15.12 psia to 23.65 psia in both cases, giving a ratio of 1.43. Optimization in other pressure ranges, possibly higher, may show even lower power costs in the combination of VR with vapor recompression.

The compressor requires power; a usual MSF or a VR modification thereof may use heat, such as steam. A boiler-power plant supplies both electric power and water as the two products produced by such a dual-purpose plant. Relative requirements of each may vary with time of day and season of the year. However, a turbo-generator may supply electric power and a very large VR evaporator may produce fresh water using low pressure steam heat from the turbine exhaust or from steam passed out at an intermediate pressure.

This evaporator may also be connected to a turbine or motor-driven vapor compressor. The relative amounts of mechanical energy (vapor compression) and of thermal energy (steam) supplied to the VR evaporator may be varied readily, depend-

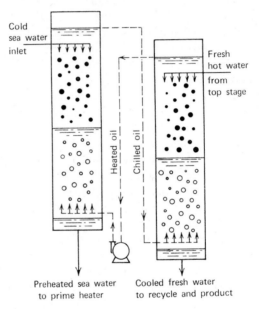

Fig. 16. Liquid-Liquid-Liquid Heat Exchanger. Hot fresh water condensate from the top stage enters top of right vessel, descends as droplets against a rising stream of droplets of colder oil, which is heated thereby. The fresh water, now cooled, is removed from the bottom of the right vessel, the heated oil enters as droplets the bottom of the left vessel, rises against the descending stream of the droplets of cold sea water, which is preheated thereby to go to the brine heater; and the chilled oil is recycled back to the right side.

ing on the variations in the demand load of electric power for sale. Thus an economic balance results to optimize the amount of water and of electrical energy produced over the hours of a day and the seasons of a year.

Absorption of Vapors. This method may be used instead of a mechanical compressor for the basis of a system of recompression, as in the usual absorption refrigeration machines (24). Low pressure water vapor is absorbed in a solvent of much lower vapor pressure; the charged solvent is heated at a higher pressure to give water vapor back at this higher pressure and to regenerate the solvent for reuse. In Figure 17, vapors are recompressed from the low pressure VR stage to a half-stage by an absorber-generator compression. The heat in the bottom stage vapors is thus elevated to the temperature and pressure of the half-stage, or of the prime heater, as would be done by a mechanical compressor.

Such a system uses as an absorbent a hydrophilic aqueous solution; inorganic solute examples are lithium bromide or caustic soda; organic solute examples are ethylene glycol, other hydroxy compounds, and some other highly soluble compounds. The principle is the same as that used in absorption refrigeration, wherein water is used as the refrigerant in a cycle between an absorber at a low pressure, and an evaporator or regenerator at a higher pressure.

In an absorber of any usual type, vapors from a bottom stage are absorbed into this concentrated aqueous solution. When diluted, it is pumped through a heat exchanger to the generator, which is simply an evaporator for concentrating it for re-use while supplying vapors as prime heat to the half-stage. This generator utilizes prime

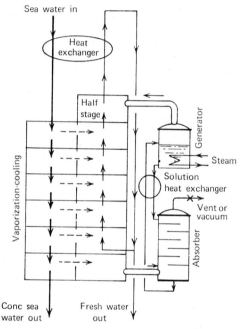

Fig. 17. Vapor Reheat Evaporation with Vapor Recompression by Absorption. Vapor Reheat operates above the lowest stage, the same as in Fig. 15; vapors formed in the lowest pressure stage pass to the absorber, and are absorbed in a strong hydrophilic liquid which, after dilution, cycles through the heat exchanger to the generator for its concentration. The generator or evaporator operates at higher pressures, the vapors leave, pass to the half-stage, to supply the prime heat necessary for the Vapor Reheat-MSF system.

thermal energy much more efficiently than if supplied directly to the half-stage. Thus, the gain ratio is 20 to 30% higher when prime steam goes to the regenerator-evaporator, concentrating the hydrophilic absorbent solution, than when the prime steam goes into the half-stage or the prime heater, due to the heat-pump effect.

The use of a multiple-stage absorber for absorbing-condensing, as in Figure 18, increases the gain ratio still more, particularly with a vaporization heat exchanger, ie, a small MSF unit to preheat the weak solution as it goes to the generator and to cool it when strong, as it goes back to the absorbing-condensing zones of the stages (24). The gain ratio is now 40–50% higher than that of the conventional MSF evaporator. However, more equipment is required, thus capital costs will be higher.

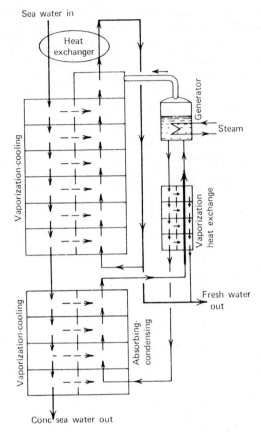

Fig. 18. Vapor Reheat Evaporator with Vapor Compression by Absorption: This is the operation of Fig. 17, wherein the absorber is a multistage vapor reheat type. The hydrophilic liquid absorbs vapor in vapor reheat stages below the regular evaporator. The hydrophilic liquid, after becoming dilute, is passed through a vaporization heat exchanger operating like a standard MSF evaporator, and is preheated in going to the generator. The generator concentrates the hydrophilic solution, passing vapors at its higher pressure as prime heat to a half-stage. The concentrated hydrophilic liquid leaving the generator is evaporated by MSF as it is cooled in the heat exchanger, and goes to the lowest stage of the Vapor Reheat absorber. Concentrated sea water is removed as blowdown as usual, and fresh water is removed from the circulating condensate stream of the main Vapor Reheat unit, and from the condensate in the vaporization heat exchanger.

General Comparison of Evaporation Processes. Multiple-effect evaporation has been developed and improved for generations in the process industries. It has been used almost exclusively in the United States when the volume of water to be evaporated demanded economy of heat utilization. The elevation of boiling point and other considerations limit the number of stages for desalinating sea water to about 12. This is two or three times the usual number for concentrating solutions in the process industries, where the higher concentrations of solute give higher elevations of boiling point, hence a greater loss of effective overall temperature drop.

The elevation of boiling point, ie, decrease of effective vapor pressure of a solution at a given temperature, also greatly increases the power cost in a vapor compression system, by increasing the necessary compression ratio. In multiple-flash evaporation systems, including controlled flash evaporation and vapor reheat, regardless of the number of stages, the boiling point elevation is only effective once, at the highest stage, in directly adding to the heat requirements.

In all evaporation systems, the disadvantages of boiling point elevation, plus those of pumping, deaerating, and possibly chemically treating sea water to prevent scale formation, must be balanced against the advantages of evaporation to a higher concentration. When these factors are considered, the brine discharge has usually been found to be optimum at about twice the concentration of the sea water feed. In handling brackish waters, a ratio of concentration of 3 to 6 may be used because (a) scaling problems, hence chemical treatment, may be expensive; (b) feed water may be scarce; and (c) disposal of concentrate may be difficult.

Controlled flash evaporation has major advantages in minimizing inefficiencies in the flash evaporation side of the stages, thus securing lower heat costs for the same number of stages and capital investment, also greater capacities for heat transfer of the condenser tubes, also elimination of substantially all of the costs for deaeration.

Vapor reheat has no scale problems since there are no metallic heat transfer surfaces. A higher operating temperature may thus be used with its attendant increase of thermal efficiency. Also, vapor reheat has a much lower loss in effective temperature drop on the condensing-heating side of the stages. This relatively close approach temperature is particularly important at the top stage, since this means a lower heat requirement. Thus, many more stages may be feasible, particularly since the expensive parts, the tubes, are not required. Optimization calculations, however, usually show fewer stages to be required for a given gain ratio. Fundamentally, VR has another process advantage in that the condensate in each stage passes to the next higher stage with the recycled fresh water, rather than to the next lower stage as in the multiflash system; and thus it is used more effectively in the condensing operation, to give a higher thermal efficiency.

For large installations, capital costs for the VR plant appear to be lower; heat and other operation costs may also be lower. A multiplicity of pumps may be required to pass the fresh water from stage to next higher stage. With very small capacity plants, capital costs increase per unit throughput for VR because of this number of pumps; and the forced circulation-vapor compression system with a high ratio of water circulated to water distilled may well give the lowest cost, simplest, and most readily automated unit.

For moderate and large capacity plants, apparently the CFE, combined with the VR, will show lower plant costs and lower water costs. In very large installations, ie,

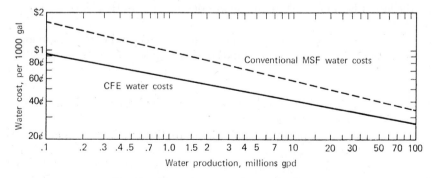

Fig. 19. Water costs per thousand gallons.

of tens or scores of millions of gpd, a combined heat and power production unit could utilize the electric power to evaporate water by modified vapor compression unit alongside of a multi-effect combined with a multi-stage flash, probably again a combined CFE and VR. The total costs would be for the single product, water, and would eliminate the arbitrary division of the costs between power and water.

Using the system developed by the OSW for calculating costs of desalination plants and desalinated water, the costs of water from many conventional MSF plants have been determined and published in various articles over a period of years (32,33).

While there has been a considerable variation in the value of the dollar and thus in the real values indicated during this inflationary period, there has also been a lowering of real costs from year to year due to improved technology, so that these factors tend to balance. The upper diagonal line in Figure 19, representing the cost of water in conventional MSF plants of different sizes, may be expected to represent an approximate value useful for estimating purposes. A lesser number of plants using multiple effects have been designed and water from them has been estimated similarly to cost about 15% more than that indicated by this upper line.

Also using the same OSW guide for costing, the same techniques were used to determine the costs of water under similar conditions as for the upper line for a like number of CFE plants ranging in capacities between 0.16 and 70 million gpd. These data points spread over about the same deviations as did the data for the upper curve, and are indicated by the lower line of the figure, which shows water costs about 60% as much by CFE in large plants, and about 50% in plants below about 1 million gpd.

Because of the large economies expected of VR, both in capital costs, due to elimination of heat transfer surfaces, and in thermal costs, it may be expected that costs by VR and its several modifications, particularly when combined with CFE, will be substantially below those indicated by the lower diagonal line of Figure 19.

Freezing

Aqueous solutions freeze to give fresh water ice. Separating the frozen ice to give fresh water, and remelting it with recovery of the "cold" in freezing other ice, has been used in many types of equipment, but no process has yet operated on a successful plant scale larger than about 25,000 to 50,000 gallons per day.

Freezing has several fundamental advantages over evaporation in separating water by change of phase:

(a) the latent heat of fusion is only about one-seventh of the latent heat of vaporization;

(b) sea water corrosion of mild steel or of prestressed and plastic-impregnated concrete, desirable construction materials, is much less at the freezing point than at the boiling point, and very much less than at the higher temperatures of flash evaporators;

(c) scale is not formed in freezing processes, particularly if the brine is not concentrated more than 2 or 3 times;

(d) freezing processes operate much closer to the ambient temperature than do evaporation processes. However, the cost of a Btu in refrigeration is much greater than in heating; and hence it must be safeguarded more effectively, thus insulation costs per square foot may not be greatly different;

(e) heat transfer surface has been eliminated through the use of direct contacting of two or more fluids at different temperatures in freezing processes to a much greater extent than in any of the evaporation processes except VR; hence, freezing processes usually have much less heat transfer surface.

On the other hand, freezing has several basic disadvantages:

(a) the time required for phase transition from liquid to solid is many orders of magnitude greater than that required for change from liquid to vapor;
(b) the separation of the last of the sea water phase is difficult, and it would be quite impossible to carry it out to the same extent as compared to a vapor phase;
(c) the crystals of ice are much more difficult to handle than the fluids in evaporation processes;
(d) the operation of a freezing process must be at a fixed temperature, since varying the pressure within practical limits does not allow flexible processing temperatures, as in evaporation; and thus, the freezing process cannot be multi-staged or multi-effected to reuse the latent heat of phase transition; although the heat energy given up in freezing may be reused once only by a suitable heat pump or vapor compressor to cause melting—just as in the vapor compression system of evaporation where the heat of condensation may be used once for supplying latent heat of evaporation.

Even so, it is probable that better thermodynamic techniques and better equipment design and operation of the crystallization process will be developed within these general limitations. Improved processes for freezing ice at the optimum rate to give larger crystals are available (34), and processes have been devised to handle the crystals formed as a fluid, first in a brine slurry from the freezer, then in a fresh water slurry to and through the condenser-melter (35).

Successful demonstration of the general advantages of freezing by overcoming the disadvantages might represent a breakthrough, and make practical a great lowering in the cost of desalinated water because of the much smaller amount of cheaper materials required in building freezing units as compared to evaporating units.

All workable freezing processes produce relatively small crystals, always much smaller than a cube one millimeter on the side. The desire is to produce the largest possible uniform crystals in a reasonable time. Larger crystals make washing easier because of more ready flow of wash water through the interstices; also, larger crystals minimize the surface area, which must be washed clean of the film of brine.

Water Vapor as Refrigerant. In the Zarchin-Colt Freezing Process, Figure 20, water vapor itself is the refrigerant, even though the specific volume at the freezing point of the brine is about 150 times that at the boiling point. An ingenious and inexpensive compressor has been developed, with flexible vanes on a rotor of large capacity, which is suitable only for the low pressure and compression ratio involved. The vapor pressure of water at the freezing point of the brine is only about 1/200th of atmospheric pressure (36).

A large, stirred freezer is fed with the sea water, prechilled in a plate heat exchanger almost to the freezing point of the brine, slightly below that of pure water. Further chilling is by lyophilization, inducing freezing by the removal of heat through the evaporation of vapors due to the suction of the compressor, through a mist separator.

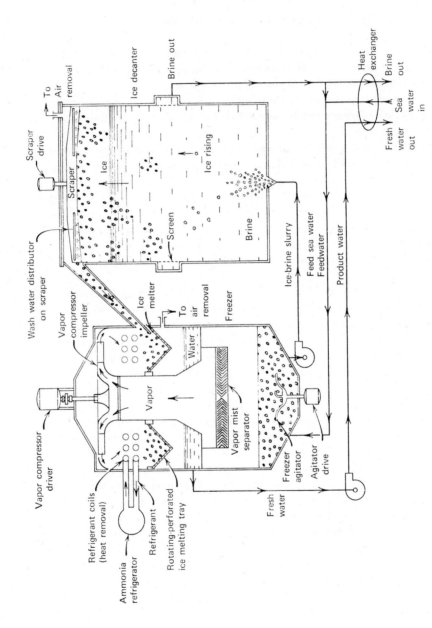

Fig. 20. The Zarchin-Colt freezing process.

Small crystals are formed when the evaporation of water cools the liquid to the freezing point. These grow in from 20 to 60 min to about 0.5 mm on a side. A slurry of these in the brine is pumped to an ice decanter where the crystals float over the liquid, and even above its surface. There they are washed by sprays of fresh water which works down to join the brine, and the top layer is continually trimmed by a rotating scraper. The crystals pass to a space supplied with the compressed vapors, which condense to melt the crystals suspended on a rotating perforated tray and give fresh water. The cold, fresh water and the cold brine are heat interchanged to prechill the sea water feed.

The total power cost has been estimated, for the improved operation, to be as low as 27.3 kWh/1000 gal in a combination of two ½ million gpd units. Accessories are a deaeration system, vacuum pump and blower, and also an ammonia refrigeration system for removing heat gains through the insulation and other heat not taken care of in the freezing-evaporation and condensing-melting cycle, such as heat of solution of salt in water, energy input by agitators, etc. These auxiliary energy demands account for approximately 30% of the input, pumps use another 10%, the agitators in the freezer use another 10%, while the compressors themselves use about half of the power in cooling for the lyophilization.

In this plant, designed for 1 million gal/day, the capital cost is estimated at about $1.26 million, and the product cost at 61¢/1000 gal, with electric power at 0.7¢/kWh.

In some processes, absorption of the water vapor, rather than compression, is used. This may be by a solution of lithium chloride as in conventional absorption refrigeration processes. Concentrated lithium chloride solution has a very low vapor pressure of water, thus it absorbs or withdraws water vapor from the brine to freeze ice crystals which are again handled in a brine slurry. The lithium chloride solution is diluted by the water absorbed as vapors; it is pumped through a heat exchanger to a regenerator or evaporator. Here the water is evaporated off, the concentrated solution which is obtained is passed back through the heat exchanger to be cooled, and thence to the absorber to repeat the cycle.

Direct Contact Refrigerants. In other freezing processes, liquid refrigerants other than water are used in direct contact with the brine. Hydrocarbons are used because of their cheapness, and because those of suitable boiling point allow operation at about the freezing point of the brine. Their make-up losses are noticeable, but this does not need to add greatly to the cost of the water.

A typical secondary, direct contact, refrigerant is n-butane. It is only slightly soluble in sea water. Simultaneous with the freezing of ice from sea water in a large, agitated vessel, there is a transfer of the latent heat of freezing of the ice to supply the latent heat of the evaporation of the butane caused by a lowering of the pressure by an evacuator-compressor. This direct contact refrigerating-freezing system was first suggested as a workable process in Norway (37) as a process of desalination of sea water. In Japan, this was the basis of one of several different processes utilizing refrigeration developed by government research laboratories together with industrial firms for producing concentrated salt brines from the sea to minimize the importation of salt.

A typical secondary refrigerant process (34,35) is shown in Figure 21. It uses a mechanical compressor which exhausts the butane vapors in the freezer at the saturation vapor pressure corresponding to the temperature of the ice freezing from the sea water. As always, a maximum production rate of ice is desired per unit volume. However, too high a cooling rate uniformly throughout a freezer produces many times too many seed crystals at the entrance of the feed, and very few crystals grow to the de-

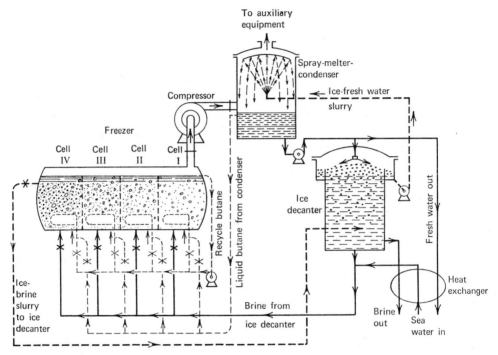

Fig. 21. Freezing by Direct Contact with Secondary Refrigerant. Prechilled sea water is refrigerated in the freezer by evaporation of butane. Ice crystals in a brine slurry are removed to the ice decanter, and brine is returned. Ice floats above liquid and is trimmed by rotating jet spray of fresh water from which some water is intercepted as wash for the crystals. Ice-fresh-water slurry passes through spray in melter-condenser supplied with compressed vapors of butane exhausted by compressor from freezer. Refrigeration effect of cold brine and cold fresh water is recovered in prechilling sea water, preferably by a liquid-liquid-liquid heat exchanger (LEX).

sired size, although there is a large weight of small ice crystals frozen. As small crystals grow, their surface area increases, and a greater rate of growth, and hence of refrigeration, per unit volume is possible. Also, as the freezing point of ice goes down with increased brine concentration, the temperature of the refrigerant must be reduced to give the temperature difference causing heat flow from the brine-ice mixture, and hence freezing.

The production of large crystals may be accomplished in a freezer divided into several compartments or cells, as Cell I, II, III and IV, from right to left, as in Figure 21. The walls dividing the cells are perforated to allow a slow movement of brine and crystals from right to left, with a minimum of back mixing, and they are cut off to allow a layer of butane on the surface to flow slowly over their tops from left to right.

If the refrigerant had the same boiling point in all cells, the temperature difference between it and the ice freezing would be much greater in the first cell, and much less in the last cell. However, by controlling the temperature difference, hence the rate of heat transfer, between the boiling butane and the ice freezing from the brine, the amount of refrigeration and ice formation may be controlled in each cell. In the first cell, the freezing point is higher, since the brine is more dilute. Concentration increases from cell to cell as the brine moves with crystals of increasing size toward the left. The last cell has the highest concentration and lowest freezing point. By elevating the boiling point of the butane at the same pressure in the first cell, the rate of refrigeration is reduced, and the desired number of seed crystals can be started. Progressively less elevation of boiling point in the cells, going from right to left, will give a rate of freezing controlled at the maximum possible without production of fresh seed crystals.

This increase of the refrigerant's boiling point may be accomplished by the addition of a higher boiling liquid, in the several cells: most of it in the first and varying down to zero in the last. A naphtha fraction, corresponding in vapor pressure to pure octane, may be used; and octane itself has a negligible vapor pressure compared to that of butane at these temperatures. The compartment on the right, where seed crystals form, has the largest amount of naphtha, hence the lowest rate of evaporation, cooling, and freezing. Here seed crystals are forming, and here is the minimum crystal surface area.

Crystal size and surface increases as the brine-ice mixture passes through successive compartments toward the left, with less and less naphtha in the butane and higher rates of evaporation and ice formation. At the left, there is no naphtha, and the maximum rate of butane evaporation gives the highest rate of cooling and of ice build-up on the largest area of crystal surface. This allows a maximum weight of ice of maximum size crystals to be produced in a minimum freezer volume with controlled rates of cooling during the steps of crystal growth. Thus, the controlled elevation of boiling point of the butane from cell to cell more than balances the depression of freezing point by salt concentration, so as to give increasing temperature differences in later cells.

In this freezer, agitation is accomplished entirely by the butane bubbling up through the sparger shown as a dashed line loop at the bottom of each cell and partially evaporating as it rises through the brine. This butane, as it is returned from the condenser to each cell, has added to it controlled amounts of butane off the supernatant layer in the freezer by the recycle pump. This recycle butane has naphtha accumulated in it because of its negligible volatility. Very little, if any, recycle butane goes to the cell farthest left, and more successively to each cell further right.

Thus, the amount of naphtha in each cell is controlled by the total amount added to the freezer in relation to the amount of pure butane added to each cell from the condenser, and the amount added by recycle, which contains naphtha. A temperature gradient to the coldest cell at the left is readily maintained by this inverse gradient of naphtha concentration. Similarly, from right to left, there is maintained a gradient of increasing: (a) rate of cooling, (b) concentration of salt in brine, (c) size of crystals, (d) mass of ice present per cubic foot; while desirably the *number* of crystals would be constant. Other controls which help to establish these optimum conditions are the volume of vapors drawn by the compressor suction, the feed rate of sea water, and the ratio of ice in brine discharge which, in turn, controls the rate of brine return from the ice decanter.

The latent heat given up in fusion supplies latent heat of vaporization of the butane, and little or none of the higher hydrocarbon vaporizes. Butane vapors formed are exhausted and compressed to a pressure corresponding to a temperature slightly above the melting point of pure ice. A slurry of ice crystals in brine is pumped from the left of the freezer to an ice separator, where a "plunger" or mass of ice crystals floats. The brine beneath is withdrawn, part for recycle to the freezer, and part to pass through the heat exchanger and back to the sea.

In this system, ice crystals are transferred from a slurry in brine to a slurry in fresh water. The mass of ice crystals rises above the liquid in the ice-decanter, and is "trimmed" by a jet stream of pure water at the freezing point, instead of a blade. A very small part of this pure water is caught at the surface to act as a wash, and this amount is controlled by the angle, velocity, and amount of the stream. It passes down through the ice crystals and washes their surfaces. The flat cone of water, formed by a rotating jet stream of high velocity water, cuts off the top crystals and mixes them into a slurry in a peripheral overflow trough. The slurry is pumped to the condenser-melter. Here, a nozzle with a wide orifice sprays the fresh water slurry into a closed vessel supplied with vapors of butane from the compressor discharge. The butane vapors condense, giving up their latent heat to melt the ice crystals carried in the spray. The condensed butane is decanted from the fresh water beneath the spray-condenser-melter and is returned to the freezer. Part of the decanted fresh water is withdrawn as product, the balance is recycled to the jet stream for trimming ice crystals from the top of the decanter and making the fresh water-ice crystal slurry. Any vapors not condensed pass out the top of the condenser to auxiliary equipment, which includes a tubular condenser refrigerated with ammonia. This removes the heat gains as inefficiencies in the balancing of the heats of freezing and of melting.

Liquid-liquid-liquid heat exchangers (LEX) are used in this and other freezing processes, as described under the Vapor Reheat MSF Process. The thermal liquid which receives and then transmits heat may be a petroleum fraction, such as a specially purified kerosene. Its solubility in water (0.01%) is much less than that of butane, thus in the direct contacting in the LEX, the kerosene extracts the butane, which may be periodically recovered by a simple distillation (34).

Pressure Freezing Process. As a corollary of its expansion on freezing, an increase in pressure lowers water's freezing point about $-0.01°C$ per atmosphere of increased pressure.

By contrast organic liquids such as a hydrocarbon mixture boiling at about 250°C and selected to freeze at a temperature somewhat lower than the freezing point of the brine solution will have a freezing or melting point which is higher at a higher pressure. Cheng and Cheng in Taiwan used this principle and devised a cyclic process (39,40).

Ice from prechilled sea water freezes at atmospheric pressure in a slurry of the liquid and frozen hydrocarbon with the hydrocarbon solid melting and later decanted as a liquid. The ice is separated from the sea water and washed as described for usual processes. The pure ice is remixed with the liquid hydrocarbon; and the pressure is now increased to about 2000 psi. At this pressure, the hydrocarbon liquid freezes small crystals, giving off heat; the temperature rises to between 3 and 4°C, and the ice is melted by absorbing heat, as part of the hydrocarbon liquid is frozen to crystals which make a hydrocarbon liquid slurry. The pure water is decanted as product, and the hydrocarbon liquid slurry is recycled to freeze more ice from raw, pre-chilled sea water.

Besides the usual advantages of freezing processes, several others have been noted, although no sizable pilot plant has been operated to prove them:

(1) the freezers (for ice and for the hydrocarbon) and the melters have small volume, because no vapor is present and the rate of crystal formation is high;

(2) large ice crystals are formed, and this reduces washing costs;

(3) no gas compressor, therefore lower energy requirement;

(4) deaeration or refrigerant gas stripping is not required;

(5) the hydrocarbon cut used is inexpensive and its flash point is satisfactorily high (this makes it safer than butane).

Against these may be noted that the chief disadvantage of the freezing process is doubled: there are two operations of crystal separation and washing.

Hydrate Freezing Processes. An important variation of the freezing processes depends on the formation of hydrates (actually clathrate compounds, with no chemical bonds) of gases used as refrigerants, eg, propane, refrigerant-12 (CF_2Cl_2), refrigerant-31 (CH_2FCl) and others. Such hydrate complexes, some of them containing as many as 17 molecules of water to one of gas, solidify or freeze, with a rejection of all other molecules but water as in normal ice making. They may be stable (ie, solid) at temperatures as high as 64°F. Hydrates may be either more or less dense than water or ice. Some of the advantages and disadvantages of freezing processes are present to different degrees as compared to the freezing of ice crystals, their separation from sea water and melting to give ice, but the process steps are similar (41).

The reaction of the hydrating agent with water to give solid hydrate is even slower than the freezing of ice, and the separation of the solid crystals and their washing are more difficult operations than with ice crystals. The advantage is the higher freezing and melting temperatures. As in freezing of ice, there is a lowering of the temperature of phase change with increasing salt concentration

The theory of the hydrate process has been very promising, but the problem of translating this into practice has been most difficult. Pilot plants of the OSW, based on two different variations of the hydrate process, have been quite unsuccessful in indicating any possibility of economical production of water by these methods.

Extraction

Salt is a purely ionic substance, and no solvent has been found for a conventional extraction of salt from water. However, the reverse extraction, water from brine, can be carried out with many organic liquids of greater or less miscibility. In evaporation, the water changes into another phase, the gaseous; in freezing it changes to the solid phase. It may be noted that in extraction (qv) the water is taken into *another* liquid phase.

As in other extractions, the solvent layer containing the water may be distilled, in a plate column if necessary, to separate pure water either overhead or as bottoms, depending on the relative vapor pressure and amount of the solvent. The brine would have to be stripped of solvent in a second column. Both columns would involve azeotropic distillation because of the necessary partial immiscibility of the two liquids. (See Azeotropic distillation.)

Alternatively, the difference in the solubilities for water at different temperatures of the solvent may be utilized. The water is extracted by the solvent, at the temperature of maximum solubility; then the temperature is changed to that of minimum solubility so that some water is separated out. Again, solvent recovery distillations are required for both layers and these are costly in equipment and energy. Thermodynamically, it may be shown that most extraction processes are fundamentally less efficient, thermally, than evaporation or freezing; but recent combined extraction-evaporation processes may have lower costs.

Some uncommon solvents which are expensive and relatively unstable, eg, substituted amines, have been used in ranges of temperature close to the ambient; common liquids as hydrocarbons may be used when advantage is taken of their differences of solubility for water, hot and cold, which are important only in very wide ranges of temperature, and hence pressure.

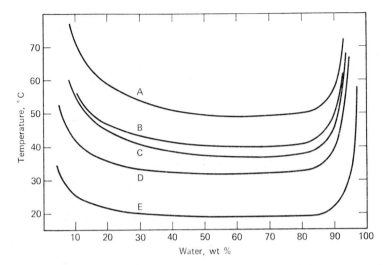

Fig. 22. Extraction Solvents for Water with Lower Mutual Solubilities Hot than Cold. Solubility diagrams of water of:

top curve	methyldiethylamine;
bottom curve	triethylamine
middle curve	1 part methyldiethylamine
	2 parts triethylamine;
next to top curve	1 part methyldiethylamine and
	3 parts triethylamine
next to bottom curve	1 to 1 mixture.

The very large change of solubility with lower temperature difference, indicated by the very flat bottoms of the curves, requires precise control of the operation of the system at the lower temperature. Courtesy of Office of Saline Water.

Solvents with Greater Dissolving Power for Water at Lower Temperatures. In many applications, the search for a solvent for extraction of a liquid from a solution must be carried out empirically. For water, however, the highly unusual properties give insight into possible suitable solvents (see OSW Annual Reports.)

A solvent containing in its molecule strong electronegative atoms will form hydrogen bonds with water molecules, and if the solvent also has hydrophobic characteristics, there is sufficient tendency to prevent complete miscibility, particularly with the salting out effect of the brine. With such specifications, a nitrogen-containing molecule is indicated with alkyl side chains on or near the nitrogen atom. There are many secondary and tertiary amines which comply with these specifications; and it has been found that their solubilities at different temperatures are suitable.

Using as the solvent a solution of two or more of these amines with 5 to 6 carbon atoms adds another dimension to the choice, as shown in Figure 22.

These amines are always completely miscible with water at a lower temperature, but at higher temperatures two liquid layers are formed, a water-rich layer and an amine-rich layer. Unfortunately, there are only small variations of solubility with temperature changes in solutions of relatively small amounts of either component (up to 10%) dissolved in a large amount of the other, while there is a very large change of solubility with temperature in the important range of over about 20% of either component dissolved in a corresponding amount of the other.

There is a heat of solution of water in the amine of about 50 Btu/lb of water. This is, however, small compared to the latent heats of fusion and evaporation of water in freezing and distilling processes. However, as in freezing, this cannot be recovered by a multiple-effect operation.

Operation of a pilot plant showed the wide change in capacity by a few degrees of temperature in the extractor, as would be expected from the flat bottoms of the solubility curves. An extract or solvent layer of 30–35% water and a raffinate or brine layer of 5% solvent is obtained, desirably at ambient temperature. The extract layer is then heated about 30°F to allow separation and decantation of a water phase with about 5% solvent, and a solvent phase for recycling to the extractor with about 10% water. A small amount of salt remains in the solvent; this must be washed out with some of the fresh water produced.

Especially, also, there is a considerable reduction in the solubility of water in the amine with increase of the concentration of the salt. Thus, the process is more advantageous for use with brackish waters than for use with sea water. Operation of a 2000 gpd pilot plant for some time studied brackish waters of 4000 to 10,000 ppm.

The high alkalinity of the solvents precipitates magnesium, which is often an important impurity in brackish waters. Of even more importance, the very small amount of solvent, a few parts per million, which remains tenaciously in the product after steam-stripping water, is a loss and a possible health hazard. Much work has been devoted to its complete elimination, but the processes suggested, ion exchange, fermentation, and others, add substantially to the cost of the product. Furthermore, the minimum solvent loss was determined as at least 11% per year, which would be a substantial cost item, considering the unit cost. The overall advantages of processes using these amines as solvents have not been demonstrated.

Another unusual solvent which preferentially dissolves water when cold is a mixture of specially selected hydrophilic and hydrophobic polymers (42). This solvent is expensive and has not been proven stable for long time use. It may produce fresh water cheaply if these objections can be overcome and the process design is improved.

Solvents with Greater Dissolving Power for Water at Higher Temperatures. Turning to more common, cheaper, and more stable compounds, petroleum hydrocarbons are the most available nonaqueous liquids or solvents, and their mutual solubility relations with water have been well studied. Thus, kerosene dissolves 8 mol percent of water at 350°F or 100 times as much as the 0.08 mol percent it dissolves at 90°F. Sea water and other dilute solutions of salt and related solids which are not themselves soluble in the hydrocarbons are almost without effect on the degree of solubility of the water. Since these solids do not dissolve in the hydrocarbon, the water dissolved is pure when the solvent is separated.

Thus, by contacting hot dilute sea water with kerosene, some of the water is dissolved and the more concentrated brine is separated by decantation for discard after heat recovery. On cooling, as shown by the solubility data, substantially pure water separates and can be decanted as product, containing an infinitesimal amount of kerosene. The kerosene layer, containing a very small amount of water, is heated for recycle.

The solubility of kerosene in the hot brine from the first contactor also is higher at this temperature than that in the pure water at the lower temperature, but this is always less than about 0.01%. This solubility may usually be neglected in the pure water if a carefully purified kerosene is used.

Several extraction processes have been suggested to utilize such inexpensive hydrocarbons as naphthas in the kerosene range, and various systems have been devised utilizing the simultaneous contacting or extraction taking place in a liquid-liquid-liquid heat exchanger (LEX).

The LEX is not immediately applicable, however, since the water which is dissolved in the hot kerosene leaving the one LEX which heats the sea water, would then separate when this hot stream of kerosene is cooled in the other LEX of the pair. Thus, while heat transfer would be excellent, the extraction would be negated.

Other systems of heat transfer have also been suggested, using an intermediate heat carrier such as solid particles in a fluidized bed or larger masses of solids with large surface area, which may be readily contacted with the fluid (43). Such solid interchangers of heat are well known, particularly for gas heat interchange; so far, no substantial progress has been made in their use in any water desalting process.

Hyperfiltration or Reverse Osmosis

The term hyper-filtration is more explanatory, but this method of separating water from brine is called reverse osmosis by most people concerned with it. Water molecules permeate and pass through some high polymer films or membranes under the influence of a very high pressure. A more concentrated brine is discharged from the membrane's up-stream side (see Osmosis). The pressure applied must be greater than the osmotic pressure of its brines to reverse the normal osmotic permeation. The osmotic pressure of sea water is about 25 atmospheres; for brackish water with 2000 ppm solids, it is about 1.4 atm; and the pressure for reverse osmosis must be greater (44).

There are indications that the permeation is not truly a reverse osmosis, and also that it is not an ultrafiltration of small molecules through holes of such size that larger molecules are physically screened out and retained. However, such a filtration concept has some substantiation. Hydration of ions may increase the effective size of the particles of solutes formed by the ionization to give a vastly larger size unit than single molecules of water. Also, high operating pressures do compact membranes and thus reduce flows of water, as if holes were partially closed due to the matrix being contracted by the operating pressure.

The transport of water through the membrane may be a hydrogen bonding phenomenon. It may be likened to an adsorption of water molecules on hydrogen bond sites of the polymeric membrane, with subsequent migration or diffusion of the water molecules from one such site to another in passing through the membrane under the influence of pressure.

Other suggested mechanisms are based on various other physical properties, including surface tension.

The very high pressures across the membranes (often 100 or more atm) and the development of suitable membranes and methods for their support, are major problems in the development of processes of industrial significance. An intriguing problem is the search for a polymeric material for membranes now usually 0.004 to 0.006 in. thick, with the essential property of allowing water molecules to permeate, while rejecting salt molecules or ions. Cellulose acetate is now the most used polymeric material; other polymers may be possible, and are being sought with large expenditures of research funds. The membrane has a very thin skin on the saline water side, about $\frac{1}{4}\%$ of its thickness (0.00001 in.). This skin is the effective rejector of all but water molecules, and once through it, water molecules readily pass the very porous structure of the balance of the membrane, which merely serves to support the skin.

Between about 10 and 100 ft² of efficient membrane surface (with suitable backing systems to give strength) allow the permeation of 1000 gpd of fresh water. The saline water should be free of particles which will foul the skin or otherwise disturb the transfer mechanism, and of bacteria which may, biologically, attack the membranes. Systems of agitation of the saline water and of wiping the membrane surfaces have improved the rate of permeation. Lives of the membranes have been measured in months; more recently, 1.5 years has been assumed as a possible life, and this is expected to increase to several years. However, membrane costs, both initial and replacement, are a major part of the costs of the product water.

Numerous ingenious mechanical devices have been designed for the combination of membranes and reinforcements which will allow the two aqueous streams, saline and fresh, to flow on respective sides of the membrane. Several are: (a) plates and frames, as in filter presses; (b) spiral, as rolls of alternate membranes for permeation and sheets of metal screen or mesh for passage of saline water and fresh water streams; (c) tubular, as a porous pipe, with a tubular membrane lining through the center of which saline water passes, while fresh water permeates the membrane and passes through the pores to an external collector; (d) hollow fibers, with saline water outside.

The last requires no reinforcement. Fine hollow, plastic fibers, bundled in very large numbers in a pressure tube, give a large surface for a unit volume. Nylon is stronger and more resistant to chemicals and bacteria than cellulose acetate, thus is preferred for these tubes which have an outer diameter less than a hair, in the range of 0.002 in., and a wall thickness of about 0.0005 in. (45). In this small tube, the strength is comparable to a tube with dimensions 1000 times as great, ie, a 2-inch tube with a 0.5

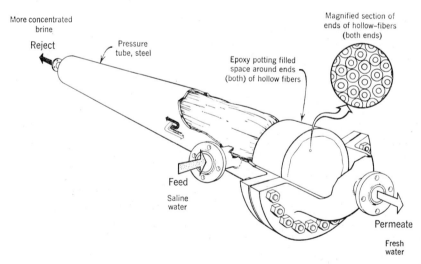

Fig. 23. Reverse Osmosis of Brackish Water Using as a Membrane Fine Nylon Hollow Fibers: Millions of fiber loops, like hairpins, have both ends encapsulated in a short epoxy cylinder enclosed in the flange on right. A magnified section of the ends of these thread-fine fibers is indicated. The saline water feed passes on the shell side from the feed at the right end to reject of more concentrated brine on the left. The permeate of fresh water discharges from the right nozzle. The steel casing is designed for the pressure necessary, usually 40–50 atmospheres. Courtesy of E. I. dupont de Nemours & Co., Inc.

in. wall. No reinforcement is necessary for these microscopic tubes. Some 25 million such tubes are assembled in a reverse or hairpin tube bundle in a shell made of a 14-inch pressure pipe. The single tube sheet is made by filling with an epoxy cement the spaces between the two ends of all of the tubes themselves, and the tubes and a circle with a flange (45).

The permeation of the fiber wall of this hair-thin tube is very much less than for other membrane surfaces, thus 6000 to 10,000 sq ft of membrane or fiber surface is required for the permeation of 1000 gpd of product. Saline water passes in the shell of the pressure pipe along the outsides of the fine tube walls acting as a membrane; some water permeates through the tube walls and flows out to a water box at one end, while a more concentrated brine discharges from the shell side. The advantage of this system is that there is no metal backing required to support the membrane, also that the tubular fibers may be drawn to give a vast surface area very simply and economically, and are then economically compacted in a relatively small pipe.

Systems of membranes in series, with suitable flows of brine, are usually necessary with brackish waters of more than 3000–4000 ppm solids. Sea water has also been desalinated in experimental units by two or more sequential permeations so that there is not too high a concentration difference across a membrane.

Large plants for brackish water are anticipated to have successful performance in the near future. Thus, Du Pont (see Figure 23), in placing in service its first major (100,000 gpd) plant using hollow nylon fibers in 16 pressure cylinders, announced its hope that larger plants would reduce the cost of desalinating brackish water to $0.50/ 1000 gal (46). In addition, a large number of companies are already offering relatively small plants, some for home use.

Ionic Processes

In ionic processes, the salt is separated as brine from the saline water, rather than the water being separated from brine as in processes previously discussed. This type of process is particularly useful for dilute saline or brackish waters, ie, with small amounts of dissolved salts. Since the mass of salt is replaced, either chemically or electrically, through removal of the ions separately, there is a direct and stoichiometric cost for the removal of each pound of salt, or each pound of sodium ion and of chloride ion, which is present in the solution. This is not the case with the physical processes of separation of water from a brine by evaporation, freezing, extraction, or hyperfiltration. The concentration of the salt in water to be desalinated is not so important for evaporation or freezing, wherein water is separated from the solution.

Ion Exchange. In the simplest ion exchange process, Na^+ is exchanged for H^+ in one pass through a bed of synthetic ion exchange resin. (See Ion exchange.) The dilute hydrochloric acid solution resulting passes through another bed of a different resin where the Cl^- is exchanged for OH^- giving deionized or fresh water. This continues until the respective beds are fully charged with Na^+ and Cl^- ions.

The two beds of resin are then separately reactivated by passing a solution of sulfuric acid through the first bed to replace the sodium ions with hydrogen ions; and the sodium sulfate formed runs off in solution to waste. Similarly, a solution of alkali as sodium hydroxide is passed to replace the chlorine ions in the second bed with hydroxyl ions and to form a solution of salt again, which is passed to waste. This process requires both sulfuric acid and caustic soda, and usually, at least 50 to 100% excess of each must be used. This will represent a very high and impossible chemical cost for large-scale desalination of sea water. Also there is a fouling of the resin beds, and re-placement of the resins periodically adds appreciably to the product cost. In special cases, particularly for desalinating brackish water, ion-exchange beds have been used, since only relatively smaller amounts of chemicals are required. Thus, water contain-ing 1750 ppm of salt is non-potable, but it would require only about 5 percent of the chemicals that sea water would require. On the other hand, the energy cost for pro-ducing fresh water by evaporation or freezing from the brackish water would not be so greatly different from that required for production from sea water.

This simple case with salt removal is typical of the removal of many other ionic substances, as in sea water. Also, other acidic materials besides sulfuric acid, and other basic materials besides caustic soda are used in various systems.

For emergency kits for aviators downed at sea, packages of ion-exchange resins have been made, to have only a single use. Sea water is passed through these small beds to make a small amount of drinking water. Provision for regeneration of the res-ins would be complicated, and the resins are discarded when charged with sodium and chlorine, respectively.

Various processes using less expensive regeneration systems are in relatively small-scale use with brackish waters. Sophisticated techniques allow the regeneration of resins by carbon dioxide, by lime, or by increasing their temperature to 80°C to bring about a marked change in the equilibrium conditions (47).

Fundamentally, for use in desalination, ion exchange resins may be considered as those working with: (a) strong acids, (b) weak acids, (c) strong bases, and (d) weak bases. Three or even four successive beds may be used, and each must be regenerated after it is spent in the particular exchange step in which it is used.

The Desal process (47) depends on the use of two Amberlite (Rohm and Haas) resins: IRA-68 (weak base anion) in a first and third bed, and IRC-84 (weak acid cation) in a second bed. In the first bed, the weak base resin (in a bicarbonate cycle) converts the anions, as Cl^- to bicarbonate, HCO_3^-, which is mildly alkaline. This solution then passes to the second bed, where the weak acid cation exchange is in the acidic or hydrogen form, to take out the cations, as Na^+, and give off CO_2 and water. The CO_2 dissolved in the water ionizes as the weak carbonic acid, and is recovered by passing through a third bed where Resin IRA-68 in the free base form is carbonated in recovering the CO_2 and passing off the deionized fresh water.

When the beds are exhausted, the first is regenerated with ammonia or lime, and the second with sulfuric acid. The third bed is ready for use for the first step, since the resin is in the bicarbonate form, and the cycle is reversed in the order of beds III, II, and I.

Plant costs for the Desal process are about $1/gpd capacity, and for waters with 1000 to 3000 ppm salts, the fresh water (50–100 ppm) may cost $0.25 to 0.50/1000 gal.

In the Sul-bi-Sul process, (47) a strong acid bed removes cations, Na^+, to give acids and is regenerated with sulfuric acid. The acidic effluent passes to the second bed having a resin which is a strong base type, now in the sulfate form. The acids in the effluent from the first bed convert the divalent sulfate to monovalent acid, or bisulfate, leaving an availability for removing an anion from the acid. All of the acid, H^+, ions are so removed due to the high acidity or low pH of the operating range of this resin. Most of the sulfate ions are converted to bisulfate ions. The anion regeneration may be done by reversing the equilibrium, sulfate-bisulfate. If there is sulfate in the brackish water, this may be used in a simple rinse operation.

One disadvantage is the low capacity of the strong base anion exchanger theoretically only half, because of the bisulfate ion formation. In practice, the capacity is much less than half.

It has also been suggested to operate a single bed having a mixture of two resins, one removing the sodium ion, and one removing the chloride ion at the ambient temperature. Some resins have different equilibrium conditions at temperatures of 175–190°F, and are regenerated by passing the brackish water available through at a temperature of 175°F. In the Sirotherm process, choice of resins gives possibilities of relatively long life, although rates of heat transfer and of ion exchange are low.

Other processes, including operation of continuous beds and of liquid ion exchangers are being tested on brackish waters and are available in commercial installations as large as 100,000 gpd.

Electrodialysis. Dialysis (qv) is the permeation of a membrane by molecules or ions. In electrodialysis (qv) this diffusion of ions is facilitated by the passage of an electric current. Electrodialysis utilizes two different types of specially developed plastic membranes, one that is much more permeable to anions, the other that is much more permeable to cations. The electric energy required is proportional to the concentration of salt in the saline water, so that this process, like ion exchange, is more suitable for brackish waters.

Polymeric materials for electrodialysis membranes have now been developed to be much more efficient materials than earlier ones, although their useful life is limited. Greater improvements may give a more rugged and more industrially dependable material with improved ionic separation capabilities and greater throughput. At best, their high replacement cost will be an important part of the cost of water by such processes.

A stack is formed of hundreds of such membranes, alternately spaced as to selectivity for positive and for negative ions. Surrounding frames make narrow compartments, the large surfaces of which are bounded on one side by one type of membrane and on the other side by the other type of membrane. Salt water is passed through all of the compartments in the same direction. Direct electric current is passed through the system, perpendicular to the flow of all the water of all of the compartments and of all the semipermeable membranes of the stack.

The cations (sodium) tend to migrate through the membranes permeable to positive ions in the direction of the current flow, and the anions (chloride) tend to migrate through the membranes permeable to negative ions in the reverse direction to the flow of the current. Since alternate compartments are right and left, one compartment will tend to lose sodium ions through the boundary membrane on one side, and also to lose chloride ions on the other side through its boundary membrane, while each of the intermediate chambers tend to acquire both ions from the respective adjacent compartments through the respective boundary membranes. Thus, streams through alternate compartments will have less ions, both positive and negative, and will be less saline, while streams through the intermediate compartments tend to become more concentrated in both ions and hence in salt. (See Vol. 7, pp. 849, 850.)

The process is governed by the distance of liquid flow in the compartments, the intensity of current flow, the permeability and other specifications of the two membranes used, their distance apart and some other considerations. Conductivity for electric current decreases with lower ionic concentration, and usually it is impractical to obtain water of less than 500 ppm of total dissolved solids.

Usually, the cost of the electrical power required to separate the ions from the product water will be lower than the cost of the chemicals used in ion exchange on the same feed water. But, the cost of electrical power is usually so high that, with the cost of equipment and/or replacing membranes, which are essentially fragile, only brackish water, can be desalinated economically by electrodialysis for large-scale use.

The success in South Africa with electrodialysis plants for individual farms, 1000 to 1500 gpd, and intermediate size units, up to 2500 gpd, when working with brackish water indicates that this type of unit may be dependable under such circumstances.

In both ion-exchange and electrodialysis processes, sediment and other impurities in the water to be treated can spoil the surfaces and greatly reduce the capacity. Careful pretreatment of the water to remove undesirable materials is usually necessary.

Most of the communal operating plants have been between 10,000 and 100,000 gpd, although the largest sustained operation has been at over 650,000 gpd at Buckeye, Arizona. Several other municipal plants are in the 250,000 gpd range (48).

Summary

Of the systems of desalination which are available, some, notably evaporation, are well developed and understood. All processes require energy, either thermal, mechanical, or electrical. With cheaper energy than is available from fossil fuels, costs will go down. Thus, cheap nuclear or geothermal energy will benefit this field, as it will benefit so many other of man's interests and needs. Fresh water costs at present are often made up of approximately 50% for the variable charges, largely for energy, and 50% for the fixed or capital charges of the hardware. The latter can be reduced to some extent by operation on a very large scale.

Although theoretically the energy costs of separating fresh water from sea water should be very small, large-scale evaporators produce a maximum of about 10 tons of water per ton of steam used. The usual multiple-effect evaporation seems to have come to its practical limit of minimum cost with this performance, while multiple-flash evaporation exhibits possibilities of considerable improvements.

A modified multistage flash evaporation process, vapor reheat, which has been demonstrated to be possible, but has not yet been operated on a large scale, will give 25 to 30 tons of fresh water per ton of steam. With controlled flash evaporation, this may be more, and, in addition, the plant costs will be substantially reduced, probably to the minimum under present technology, even for very large plants.

Recompression evaporators require 50 to 60 kWh/1000 gal fresh water produced, while the most economic which have been demonstrated, though not yet on a large scale, indicate the possibility of using only half this much energy, or about 25 to 30 kWh/1000 gal.

By freezing, the cost in kWh of demonstrated processes also seems to be about 25 to 30 kWh/1000 gal of fresh water, with the possibility of improved processes giving somewhat lower energy costs. Freezing plants appear to cost more for equipment per unit capacity of fresh water than do evaporation plants; they have some theoretical advantages, but these have not been fully demonstrated by large-scale operation.

Reverse osmosis, or hyperfiltration, is enjoying a large share of present research activities and expenditures on its membranes and mechanics. As yet, however, it has not indicated lower energy consumption than by the best of freezing and evaporation processes; it has expensive high pressure equipment and polymeric membranes which must be replaced often.   Membranes as capillary tubes may lower these costs considerably.

Salt separating processes, such as ion exchange and electrodialysis through membranes, depend on removal of the salt as positive and negative ions.   These have major advantages for brackish water where the concentration of salt is much less than in sea water, but are too expensive in chemicals and electric power for use with sea water.

In general, either very large units for large municipalities, or very small units for individual homes, appear to have advantages for future exploitation.   No basically new methods have been suggested which will offer advantages over the ones studied for years, although this does not preclude the possibility of future breakthroughs.

Besides enormous tonnages of metals and of plastics for plants, energy will be the important cost and supplies of fossil fuels will certainly not be adequate.   Indeed, fuel is often the predominating cost of fresh water.   Plants using nuclear fuels have not yet generated steam for power production as cheaply, when all costs are considered, as fossil fuels.   Until steam is made more cheaply by nuclear fuels, the discussion of nuclear fuels as a means of supplying the energy for water desalination is merely prophetic.   When steam costs are lowered by nuclear installations, their apparent disadvantages, eg, the large amount of materials required for shielding, and the remoteness of installations for safety, etc may be more than balanced.   However, nuclear energy or geothermal energy, or some other type of energy as yet untapped, must be made cheaply available if water is to continue to be plentiful, or at least sufficient for the numbers of people who are yet to be born, in this present century alone.

Bibliography

"Water, Demineralization" in *ECT* 1st ed, Suppl. 1, pp. 908–930 by E. R. Gilliland, Massachusetts Institute of Technology.

1. *Business Week* (Feb. 22, 1969).
2. J. W. Colton, *Trans. N.Y. Acad. Sci* [2] **23**, 625 (1961).
3. J. Koenig, *J. Am. Water Works Assoc.* **51**, 845, (1959).
4. *Environ. Sci. and Tech.* **5**, 115 (Feb. 1969).
5. *Water Desalination in Developing Countries*, U.N. Pub. 64 II. B. 5-1964.
6. S. F. Singer, *Environ. Sci. and Tech.* **3**, 197 (1969).
7. *Chem. Eng. News*, p. 50 (Feb. 24, 1969).
8. *Chemical Week*, p. 22 (July 5, 1969).
9. M. F. Dehne, *Chem. Eng. Progress* **65** (7), 51 (1969).
10. *Chem. Eng. News* (August 12, 1968).
11. G. Young, *Science*, p. 339 (Jan. 23, 1970).
12. *Dechema Monographien* Band 47, 781–834.
13. *Desalination*, issues during 1968–69.
14. *Chem. Eng. News*, p. 64 (Jan 20, 1969).
15. J. Hunter, *Chemical Week*, (March 1, 1969).
16. B. F. Dodge and A. M. Eshaya, *Advances in Chemistry*, American Chemical Society, Washington, D.C., 1960, p. 27.
17. A. Delyannis and E. Delyannis, *Chemic Ing. Technik* **41**, Heft 3, 90–96 (1969).
18. A. Delyannis and E. Delyannis, Private Communications.
19. *U.N. Solar Distillation as a Means of Meeting Small-Scale Water Demands*, U.N. Pub. 70 II. B. 1—1970.
20. D. F. Othmer, "Heat and Power from Seawater" in *Encyclopedia of Ocean Resources*, Reinhold Publishing Corp., N.Y., 1969.
21. K. C. D. Hickman, *Ind. Eng. Chem.* **49**, 786 (1957).
22. U.S. Pat. 2,619,453 (Nov. 25, 1952), R. Anderson.

23. U.S. Pat. 3,288,686 (Nov. 29, 1966), D. F. Othmer.

24. D. F. Othmer, "Evaporation and Desalination—Improved MSF Processes" in *Desalination* 6, 13–24 (1969).

25. U.S. Pat. 3,408,294 (Oct. 29, 1968), D. F. Othmer.

26. D. F. Othmer, *Ind. Eng. Chem.* **21**, 576 (1929).

27. Burns & Roe, *Universal Design—25 MGD Universal Desalting Plant*, OSW, Washington D.C., June, 1969.

28. *OSW Research & Development Report 233*, November 1966.

29. D. F. Othmer, *Condensation Coefficient of Heat Transfer*, Chemical and Process Engineering, London, June, 1968.

30. U.S. Pat. 3,214,348 (Oct. 26, 1965), J. Lichtenstein (to Saline Water Conversion Corp.). U.S. Pat. 3,214,349 (1965), E. C. Kehoe and E. C. Walker (to Saline Water Conversion Corp.). U.S. Pat. 3,214,350 (1965), Lichtenstein (to Saline Water Conversion Corp.). U.S. Pat. 3,214,-351 (1965), B. Bucalo and J. Lichtenstein (to Saline Water Conversion Corp.). U.S. Pat. 3,275,529 (1966), E. C. Kehoe and E. C. Walker (to Saline Water Conversion Corp.). U.S. Pat. 3,330,739 (1967) J. Lichtenstein and R. C. Roe (to Saline Water Conversion Corp.) U.S. Pat. 3,324,012 (1967), J. Lichtenstein and R. C. Roe (to Saline Water Conversion Corp.). U.S. Pat. 3,344,584 (1967), U.S. Pat. 3,385,770 E. C. Kehoe and E. C. Walker (to Saline Water Conversion Corp.) E. C. Kehoe, J. Lichtenstein, R. C. Roe, and E. C. Walker (to Saline Water Conversion Corp.). U.S. Pat. 3,438,202 (1969), R. C. Roe (to Saline Water Conversion Corp.). U.S. Pat. 3,454,471 (1969), E. C. Kehoe (to Saline Water Conversion Corp.). U.S. Pat. 3,471,178 (1969), R. C. Roe (to Saline Water Conversion Corp.).

31. D. Q. Kern and Associates, *OSW R. & D. Report 261*, Washington, D.C., 1967.

32. "MSF Desalting, State of the Art," *OSW R. & D. Progress Report 490*, 1969.

33. "A Study of Large Size Saline Water Conversion Plants," *OSW Progress Report 72*, 1963.

34. U.S. Pat. 3,250,081 (May 10, 1966), D. F. Othmer.

35. U.S. Pat. 3,377,814 (April 16, 1968), D. F. Othmer.

36. *Chem. Eng.* (July 6, 1964); *OSW R. & D. Progress Report 451*, 1969; *OSW R. & D. Progress Report 491*, 1969.

37. Norwegian Pat. 70,507 (June 3, 1946), Olav Jensen (to Norwegian Hydroelectric Co.).

38. U.S. Pat. 2,803,589 (1957), P. Thomas. U.S. Pat. 3,288,686 (1966), D. F. Othmer. U.S. Pat. 3,306,346 (1967), D. F. Othmer. U.S. Pat. 3,408,294 (1968), D. F. Othmer. U.S. Pat. 3,446,712 (1969), D. F. Othmer.

39. C. Y. Cheng and S. W. Cheng, *A. I. Ch. E. Journal* **13**, 41 (1967).

40. S. Y. Chin, L. T. Fan and G. R. Akins, *I. and E. C. Proc. Dev.* **8**, 347 (1969).

41. U.S. Pat. 2,904,511 (1959), (to Koppers Co.). U.S. Pat. 3,155,610, V. C. Williams (to Sweet Water Development Co.).

42. U.S. Pat. 3,386,913 (June 4, 1968), L. Lazare (to Puraq Co.).

43. P. Barton and M. R. Fenske, *Ind. Eng. Chem. Process Design Develop.* **9**, 18–26 (1970).

44. K. S. Spiegler *Principles of Desalination*, Academic Press, N.Y., 1966.

45. S. Walters, *Mechanical Engineering*, p. 104 (April 1968).

46. *Wall Street Journal*, August 20, 1969.

47. J. I. Bregman and J. M. Sackelford, *Environ. Sci. and Tech.* **3**, 336 (1969).

48. T. A. Kirkham, *Mechanical Engineering*, p. 47 (March 1968).

Donald F. Othmer
Polytechnic Institute of Brooklyn

INDUSTRIAL WATER TREATMENT

Water is drawn by industry from many different sources. It may be taken directly from a river, a lake, a well, or from a privately impounded supply, or it may be obtained from a neighboring municipality. Both the amount drawn by the industry and the degree of treatment accorded the water so withdrawn varies widely from industry to industry and from plant to plant. The quality of treatment may vary considerably within a given plant depending upon the particular uses to which the water is put. Table I shows the amounts of water withdrawn by various industries

Table 1. Industrial Plant and Thermal-Electric-Plant
Water Intake, Reuse, and Consumption, 1964[a]

	Water Intake, billion gal/yr (bgy)							
Industrial group	cooling and con- densing	boiler feed sanitary service, etc	pro- cess	total	water recycled,	gross water use, including recycling,	water con- sumed,	water dis- charged,
food and kindred products	392	104	264	760	520	1,280	72	688
textile mill products	24	17	106	147	163	310	13	134
lumber and wood products	71	24	56	151	66	217	28	123
paper and allied products	607	120	1,344	2,071	3,945	6,016	129	1,942
chemicals and allied products	3,120	202	564	3,886	3,688	7,574	227	3,659
petroleum and coal products	1,212	99	88	1,399	4,763	6,162	81	1,318
leather and leather products	1	1	14	16	2	18	1	15
primary metal industry	3,387	195	996	4,578	2,200	6,778	266	4,312
subtotal	8,814	762	3,432	13,008	15,347	28,355	817	12,191
other industries	571	197	271	1,039	1,207	2,246	71	968
total industry	9,385	959	3,703	14,047	16,554	30,601	888	13,159
thermal electric plants	34,849	b		34,849	5,815	40,664	68	34,781
total	44,234	959[c]	3,703	48,896	22,369	71,265	956	44,940

  [a] Source; census of manufacturers (Ref. 1).
  [b] Boiler-feed water use by thermal electric plants is estimated to be equivalent to sanitary service, in industrial plants etc.
  [c] Total boiler-feed water (excluding sanitary service in industrial plants).

for different uses, and Table 2 shows the quality of waters that have been used by different industries before being subjected to various degrees of treatment to meet their point-of-use requirements.

Pretreatment Systems

Suspended solids removal, particularly of the coarser materials from 5 μ up, such as sand and heavy silt, may be removed in sedimentation basins. Such basins usually serve a dual purpose, preliminary removal of suspended solids, and storage to balance variations in supply with the relatively constant demand of the plant processes. In these basins, detention time is measured in days, the amount depending upon the likelihood of interruption or reduction in the supply. A 30-day detention is not uncommon in some circumstances. Particles smaller than 1μ are generally not affected by the detention. The effect of continued aeration and sunlight in oxidizing organic peptizing substances may cause a certain amount of flocculation. Algal growth, particularly of the free floating type, occurs in warmer climates and may indeed contribute to the total turbidity emerging from the basin. Control of algal growth is usually accomplished by the addition of copper sulfate sprayed in aqueous solution on the water surface from a boat or spread by solution from solid material in burlap bags towed behind power boats that traverse the surface of the reservoir in a pattern. In the warmer climates, addition of copper sulfate every several months in the amount of one ppm, based on the top one foot of water, may be employed. In some exceptional circumstances in very arid regions with short water supply, evaporation control may be practiced by the addition of fatty alcohols which form a monolayer on the surface. (See Vol. 21, p. 664.)

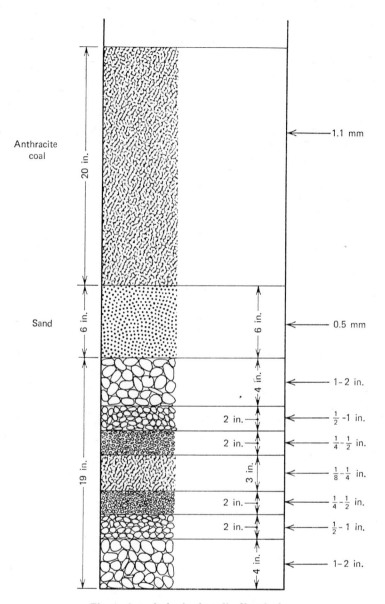

Fig. 1. A typical mixed media filter bed.

The degree of clarification applied to the water from the source of supply is governed both by the intended use of the water and by the level of turbidity that is to be removed. Two processes, singly or in combination, are generally employed for clarification. The first of these is filtration where turbidities are generally less than 50 ppm, and clarification can be accomplished simple by passing the water through a filter. Filters may be of three configurations, the most common being a single granular medium such as sand, typically of effective particle size of 0.4 mm at flow rates ranging from 1 to 3 gal/(min)(ft²). A small amount of a coagulant such as alum may be added ahead of the filter in amounts varying from 5 to 15 ppm. Such rapid

Table 2. Summary of Specific Quality Characteristics of Surface Waters That Have Been Used as Sources for Industrial Water Supplies[a]

| Characteristic | Boiler makeup water | | Cooling water | | | | Industrial process water | | | | | |
	industrial, 0–1,500 psig	utility, 700–5,000 psig	fresh once through	fresh makeup recycle	brackish[b] once through	brackish[b] makeup recycle	textile	lumber	pulp and paper	chemical	petroleum	primary metals
silica	150	150	50	150	25	25			50	50	50	
aluminum	3	3	3	3								
iron	80	80	14	80	1.0	1.0	0.3		2.6	5	15	
manganese	10	10	2.5	10	0.02	0.02	1.0			2		
copper							0.5					
calcium			500	500	1,200	1,200				200	220	
magnesium										100	85	
sodium and potassium											230	
ammonia												
bicarbonate	600	600	600	600	180	180				600	480	
sulfate	1,400	1,400	680	680	2,700	2,700				850	570	
chloride	19,000	19,000	600	500	22,000	22,000			200[c]	500	1,600	500
fluoride					5	5					1.2	
nitrate			30	30							8	
phosphate	50		4	4		5						

dissolved solids	35,000	35,000	35,000	1,000	35,000	35,000	150	1,080	2,500	3,500	1,500
suspended solids	15,000	15,000	5,000	15,000	250	250	1,000	d	10,000	5,000	3,000
hardness ($CaCO_3$)	5,000	5,000	850	850	7,000	7,000	120	475	1,000	900	1,000
alkalinity ($CaCO_3$)	500	500	500	500	150	150			500		200
acidity ($CaCO_3$)	1,000	1,000	0	200	0	0					75
pH	5.0–8.9	5.0–8.9	3.5–9.1	5.0–8.4	5.0–8.4	6.0–8.0	5–9	4.6–9.4	5.5–9.0	6.0–9.0	3–9
color units, APHA	1,200	1,200	1,200	1,200				360	500	25	
organics:											
methylene blue active substances	2[e]	10		1.3	1.3	1.3					
carbon tetrachloride extract	100	100	100[f]	100[f]	100	100				30	
chemical oxygen demand	100	500	100	100	200	200					
hydrogen sulfide					4	4					
temperature, °F	120	120	100	120	100	120		95[g]			100

[a] Unless otherwise indicated, units are mg/liter and values are max. No one water has all the max values shown.
[b] Water containing in excess of 1,000 mg/liter dissolved solids.
[c] May be ≤ 1,000 for mechanical pulping operations.
[d] No large particles ≤ 3 mm diameter.
[e] 1 mg/l for pressures up to 700 psig.
[f] No floating oil.
[g] Applies to bleached chemical pulp and paper only.

sand filters may either be operated by gravity, relying solely on a head of water over the filter medium to force the water through, or they may be completely enclosed in a cylindrical tank with pump pressure used to force the water through the medium. Pressure drops typically range from 1 to a maximum of 8 ft of water. The depth of bed employed is commonly from 2 to 3 ft. Finely divided anthracite coal of effective particle size of 0.6 mm may be used instead of sand. When anthracite is used, the addition of coagulation chemicals is usually required. The coagulation chemicals added serve the purpose of agglomerating the colloidally dispersed solids and aid in their adherence to the filter media and hence their removal. Clarification efficiencies measured as the ratio of suspended solids removed to suspended solids entering are commonly 0.90–0.99. The sand or anthracite coal in earlier designs was supported on a bed of graded gravel, but in current practice is more commonly supported directly on the filter bottom which is provided with strainers sufficiently fine to prevent sand or anthracite from passing through.

Previously, most filters, both pressure and gravity, were of a single medium, sand or anthracite. Currently, many new installations are being designed with a mixed media or graded density filter. In this case, filter media of different types such as sand and anthracite are employed together, with the anthracite, being the more coarse and lower density medium, appearing on the top of the filter; 20 in. of 1.0 mm anthracite coal may be placed on top of 6 in. of 0.4 mm sand. When the filter is backwashed to remove the suspended impurities that accumulate during the run, the less dense medium, the coal, is washed to the top by the upward flow of water and the heavier medium, the sand, even though finer, remains on the bottom. The suspended solids that have been removed, being much lighter than either medium and in flocculated form, are carried out by the upflowing water and washed to waste. In both types of filters, single medium and mixed media, the amount of backwash water required is approximately three percent of the total throughput of the filter during a run, the end of which is controlled by a given limit on the quality of the effluent. Figure 1 is a diagram of a typical mixed media filter bed. The advantages of the mixed media filter are that the coarser material on top causes removal of a large percentage of the suspended solids in the entering water and allows the solids removed to accumulate in the depth of the bed rather than forming a mat on the top. The remaining amount of suspended solids is removed on the finer media. The net result is an ability to handle an influent water with a much higher suspended solids content, and to process a greater quantity at higher flow rates without deterioration of effluent quality than is possible with a single medium filter. Where mixed media filters are employed, flow rates vary from 3 to 6 gal/(min)(ft^2). Total head loss is limited to approximately 8 ft as in a single medium filter. Increase in pressure drop beyond this point usually results in a penetration of suspended solids through the filter medium into the effluent.

Diatomaceous earth precoat type filters are also employed for low suspended solids supplies containing less than 50 ppm turbidity. In this case, diatomaceous earth, either without pretreatment or with a small amount of aluminum sulfate added, is caught on and supported by a fine mesh screen or porous stone base to form a layer that intercepts suspended solids in the influent. A precoat amount to establish this initial layer is usually added before each filter run commences. During operation a small amount of diatomaceous earth, 10–15 ppm, is added continuously depending upon the amount of suspended solids in the influent water to mix with those suspended solids as they are all intercepted on top of the original precoat layer. The continuing diluting feed of diatomaceous earth is called a body feed and prevents a rapid build up of pressure drop over the layer of material removed.

Diatomaceous earth filters are particularly advantageous for levels of influent turbidity less than 10 ppm since they are small in size and thus low in capital investment. Higher levels of turbidity cause excessive consumption of diatomaceous earth and require large installations since the length of time in service is shortened considerably and frequent backwashing is thus required. The quality of water produced is equal to that produced by single or mixed media granular filters under standard conditions.

For influent turbidities exceeding 50 ppm, clarification-flocculation is generally employed ahead of any filtration steps. In this process, chemicals such as alum or ferric sulfate or certain polymeric cationic organic chemicals are added to the raw water with rapid mixing and then passed into a chamber of 5–10 min detention time, wherein slow stirring allows formation of floc from the reaction of the coagulant and the turbidity. Following the flocculation chamber, water enters a sedimentation section which may be either horizontal flow or vertical flow. The so-called high rate clarification units employ vertical flow contacting the incoming water containing the flocculated turbidity with a bed of previously removed floc that is kept in agitation. This contact of fresh floc with previously collected floc is advantageous in certain circumstances as it achieves further agglomeration and produces more rapid settling. In horizontal flow clarification units, floc simply settles to the bottom as the flow moves slowly through the basin with the clarified effluent being removed via overflow launders at the far end or around the periphery in the case of circular basins. Horizontal flow clarification units generally employ rates based on approximately 90 min detention, whereas the vertical flow units employ rates based upon approximately 1 gal/(min)-(ft^2) of horizontal surface.

The flocculation process is complicated and often employs several different chemicals in sequence to achieve the desired results. T. A. Riddick (4) has contributed much to the technology of flocculation along with A. P. Black (5). A critical factor in the process is neutralization by the coagulant of the charge (zeta potential) on the colloidal particle which, in turn, allows the particle, with proper gentle motion of the fluid, to contact other similar particles and agglomerate into floc. Flocculation aids, which are water soluble long chain polymeric materials, may be added to adsorb on the particles after the particle charges have been neutralized by the coagulant. The flocculation aids form bridges between particles, and this speeds the growth of large floc particles which have high settling rates. "Activated silica" (prepared by treating a solution of sodium silicate with hypochlorous acid or with sodium bicarbonate) has been widely used as a flocculation aid. Recently the polyacrylamides have been used; they can be added to the water without any special equipment, such as that which is required to make the activated silica from the commonly supplied sodium silicate solutions.

It is well to point out that most colloidal matter appearing as turbidity in natural water carries a negative charge and, hence, requires positively charged or cationic coagulating agents. Sometimes cationic water soluble organic polymers such as aminated starches and high molecular weight (100,000) polyamines are employed in place of alum or ferric sulfate. In this case, the polymer because of its charge neutralizing as well as bridging capability acts as both the coagulant and the flocculation aid. The proper employment of flocculation aids with a coagulant can as much as double the throughput capacity of a clarification unit compared to use of the coagulant alone.

Control of the clarification process is generally achieved by measurement of the turbidity of the final product water. Turbidity measurements which are based on light scattering properties of the water (see Turbidimetry) are sometimes supplemented by microfiltration methods. These methods employ microporous filters of 0.45 μ pore size for a gravimetric determination of suspended matter. High quality clarified and settled water will contain between 1–3 ppm of suspended solids by the microfilter technique which corresponds roughly to a turbidity of 0.3–1.0 Jackson Turbidity Units. Where this quality of effluent is satisfactory for an industrial process, no further treatment is practiced. Most often, however, such clarification units are followed by filtration to further improve the quality and to compensate for times when the settling process is inefficient or upset. Single medium filters are generally employed although recent practice often uses mixed media filters to gain higher filtration rates and the consequent smaller and less expensive equipment. A high quality filtered water from the combined clarification-flocculation-filtration process will have a suspended solids by the microporous filter technique of 0.1–0.3 ppm and a turbidity of 0.03–0.1 Jackson Turbidity Units. There are times when a surface water supply contains iron and manganese in concentrations that are in excess of those tolerable in the final product water. The iron and manganese may have originated as acid mine drainage or may have come from natural soil minerals, stabilized by humic acids produced by soil bacteria or by live bark. These elements are difficult to remove by the normal clarification techniques of adsorption on or coprecipitation with alum floc. The color that is the result of such iron, manganese, and humic acid content most often must be removed by heavy oxidation prior to the addition of the coagulants. Breakpoint chlorination is sometimes effective in oxidizing the organic matter. Other times, stronger oxidants such as chlorine dioxide are required to achieve the desired decomposition of the stabilizing organic acids. Once this oxidation of the organic acids has been accomplished, iron, manganese, and much of the organic matter can be removed along with other turbidity by the clarification-flocculation processes previously described.

Where iron and manganese are present in a well water supply, the removal is more likely to be accomplished either by ion exchange, on acid- or salt-regenerated cation exchangers, or by the employment of permanganate filters. The latter process involves the addition of potassium permanganate to the water to oxidize the iron and manganese from their naturally occurring reduced states to hydrated ferric oxide and manganese dioxide which may be flocculated and removed. These oxidative reactions are catalyzed by previously oxidized precipitates collected on a sand or anthracite filter bed.

Where ion exchange processes are employed for iron and manganese removal, it is mandatory that oxygen be kept out of the system in order that the iron and manganese be maintained in their reduced states and thus exchangeable without precipitation in the ion exchange bed. Where a small amount of oxidation occurs in a sodium cycle ion exchange bed, it is necessary to add a reducing agent such as sodium hydrosulfite $Na_2S_2O_4$ (see Vol. 19, p. 419) to the sodium chloride brine to achieve reduction of the oxides and facilitate their removal by the regenerant. The oxidation filtration process as well as the sodium cycle ion exchange process can rarely remove iron and manganese to concentrations of less than 0.1 ppm (100 ppb), while the acid ion exchange process can routinely provide removal to this level or lower.

HARDNESS AND ALKALINITY REMOVAL

Many industrial water uses require that the hardness, the sum of the calcium and magnesium expressed as calcium carbonate, be less than one ppm, often less than 0.1 ppm. Where the supply water hardness is in excess of 100 ppm, partial reduction prior to ion exchange softening may be accomplished by precipitation of calcium as calcium carbonate and magnesium as magnesium hydroxide in either hot or cold reactors. This precipitation is accomplished by the addition of lime alone or in combination with soda ash, in accordance with the following equations.

$$Ca^{2+} + Na_2CO_3 \rightarrow CaCO_3 \downarrow + 2\ Na^+$$

$$Mg^{2+} + Ca(OH)_2 + Na_2CO_3 \rightarrow Mg(OH)_2 \downarrow + CaCO_3 \downarrow + 2\ Na^+$$

$$Ca(HCO_3)_2 + Ca(OH)_2 \rightarrow 2\ CaCO_3 \downarrow + 2\ H_2O$$

(The third of the above equations is often described as "add lime to remove lime.") It may be noted that lime will also precipitate the hydroxides of iron and aluminum.

In cold reactors, operating at ambient temperature, it is normal to use the add lime to remove lime reaction to remove calcium carbonate only. Where a hot process can be employed, lime and soda ash are usually used to remove both calcium and magnesium. Where these hot or cold process reactions can be employed, the raw water need not be preclarified but may be drawn directly from the source. In such a case, the turbidity is precipitated along with the calcium and magnesium. The cold reactors are essentially clarification-flocculation units with the addition of lime feeders. If an additional coagulant is required, these units employ either sodium aluminate or ferric sulfate instead of alum since the reactions take place at pH values up to 10. The addition of lime acts as a coagulant for the turbidity as well as a precipitant for the calcium bicarbonate. These reactors are generally followed by filters as previously described for turbidity removal. Where the effluent from the process can be hot, such as the makeup for the feedwater to steam generators, a hot process reactor is usually employed. At temperatures of 220–240°F, the precipitation reaction efficiencies are greatly increased and the level of hardness in the reactor effluent can be reduced to approximately 30–40 ppm with little difficulty. As previously mentioned, most requirements for hardness reduction specify reduction to less than 0.1 ppm. In such cases, the hot and cold reactors with their associated filters are followed by sodium cycle ion exchange equipment for final hardness reduction. The combination of the hot process with a sodium-chloride-regenerated ion exchange unit, commonly called a hot Z reactor, is frequently used since it achieves both hardness and alkalinity reduction. The ion exchange resins are sulfonated polystyrenes which perform satisfactorily up to temperatures of 250°F providing that, at such temperatures, the influent to the exchangers contains less than 0.03 ppm O_2. High oxygen concentration at high temperature causes rapid oxidative degradation of the resin and uneconomically short resin life. Detention time in the reactors is commonly in the order of 90 min, which allows for both completion of the reactions and sedimentation of the precipitates (see also Ion exchange).

Where hardness and alkalinities are less than 100 ppm, ion exchange of the previously clarified filtered water may be substituted for either one or both of the foregoing processes for both hardness and alkalinity reduction. Most commonly, only hardness reduction is practiced and this by the process of sodium cycle exchange in a cation exchange material. Since the ion exchange material tends to accumulate suspended matter, it is necessary that the influent water, if turbid, be clarified by flocculation, sedimentation, and filtration to a suspended solids level of less than one ppm and preferably less than 0.3 ppm. In the sodium cycle softening process, the ion exchange material is put in the sodium form by regeneration with sodium chloride brine. The influent water is passed through the ion exchange bed, generally at rates of 2.5–3.0 gal/(min)(ft³), where its hardness ions are replaced by the sodium ion in the exchange material. With proper design and operation, the resulting effluent can be reduced to hardness as low as 0.1 ppm. The exchange efficiencies are such that approximately 3.5 lb of salt are required per lb of hardness removed.

The sodium cycle softening process accomplishes no reduction in alkalinity. When this is important, the ion exchange process is modified by using an acid, usually sulfuric acid, instead of salt as the regenerant. In this instance, all the hardness, sodium and potassium ions in the influent are exchanged for hydrogen ion in passing through the ion exchange bed. The effluent is thus a very dilute acid solution, the acidity of which is equal to the dissolved solids (expressed in equivalent units). Where a significant portion, greater than 30%, of the anions in the influent water is bicarbonate, the effluent from the acid regenerated exchanger may be taken to a decarbonating tower where a counter current stream of air is used to strip the carbon dioxide from the water. Where deoxygenation is desired, a vacuum degasifier can be substituted for the decarbonator and both oxygen and carbon dioxide removed. The effluent from such a process is still weakly acid and generally requires the addition of an alkali or other alkaline water to produce a blend of near neutral pH. Where the bicarbonate is 40–60% of the total anions in the water, a split stream dealkalization unit may be employed. In this instance, roughly half of the water flow is passed through the acid regenerated ion exchange unit, while the other half is passed through a sodium chloride regenerated exchange unit, and then both flows are combined prior to the decarbonator. The result is that the sodium cycle softened water neutralizes the free mineral acidity in the hydrogen cycle water and, hence, no additional alkali is required. The effluent from these processes is a water containing 10–20 ppm of total alkalinity and less than 0.1 ppm of hardness. Where the total hardness is less than the total alkalinity, newly developed weakly acidic ion exchange resins may be employed, which will exchange for hydrogen in the resin only those cations equivalent to the total alkalinity, with the result that an effluent can be produced both low in alkalinity and of less than 0.1 ppm hardness by a single unit followed by a decarbonator.

GROSS DISSOLVED SOLIDS REMOVAL

Many industrial processes require some reduction in solids dissolved in raw water if these exceed 500 ppm. This reduction is particularly necessary where the supply contains several thousand ppm, such as is the case in many brackish wells, and even in some surface waters draining highly irrigated agricultural lands. Some bays and estuaries contaminated with sea water may also be sources of supply that fall in this category. Where a plant generates steam for process purposes, or particularly where a plant generates electric power by a steam turbine generator, multistage evaporators may be used for the removal of dissolved solids (see Distillation; Evaporation). Evaporators are particularly advantageous where the final product water must have a purity measured in the ppm range. Less severe requirements can be met by blending the relatively high purity distillate with a small flow of the raw water. The number of pounds of water evaporated per pound of steam originally supplied depends on the number of effects (see Evaporation; see also Water, desalination). A typical range for evaporators for industrial purposes is 5–10 lb of water per lb of steam. Depending upon the cost of steam as well as the temperature to which the raw water may be heated before introduction into the evaporator, costs from $0.50 to $2.00/1000 gal are common. The evaporation process has the advantage of being able, in most instances, to process the raw water without pretreatment. Sometimes, where heavy organic contamination, such as that from sewage, exists in the raw water, some organic matter removal may be necessary prior to feed to the evaporator. In addition, it may be necessary to feed antifoam agents to prevent excessive carryover of liquor into the distillate (see also Water, desalination).

In some cases where the water supply has high dissolved solids, but also high clarity (this may occur if the source is wells) electrodialysis (qv) may offer many advantages over distillation, particularly for flows less than 100,000 gpd and for plants where steam generation is not employed. Electrodialysis is most efficient when a relatively small percentage reduction in the raw water dissolved solids is required, for example, to reduce from 2000 to 500 ppm. Such reductions can be accomplished in 1 or 2 passes with costs ranging from $1.00 to $2.00/1000 gal. Where product purity less than 500 ppm, particularly less than 100 ppm, is required for the process, electrodialysis is not an economic choice. The maximum practical reduction in dissolved solids is approximately 90% of that in the supply. Electrodialysis has the disadvantage that since it depends upon very fine passages in the membranes it is necessary for the supply water to contain very low suspended solids, less than 1 ppm; low iron, less than 0.1 ppm; and low organic matter, less than 1 ppm, to avoid a rapidly increasing pressure drop. This drop, in turn, would lead to rupture of the membranes as well as to uneconomic pumping pressures. Nevertheless, electrodialysis is an excellent tool under the proper conditions, since it requires very little attention and is largely self-regulating.

Improvement in membranes for reverse osmosis (see Vol. 14, pp. 353–355) and particularly improvement in the means of mounting the membranes has led to increased application of this process to purification of water for industrial purposes. Reverse osmosis has many of the advantages of evaporation in that it can tolerate considerable quantities of suspended solids and other impurities in addition to dissolved impurities in the raw feedwater. Since it uses electric power for the pumps to supply the pressure, it is particularly applicable where steam generation is not practiced or the steam balance does not lead to an economic installation of evaporators. However, there is some transport of salt through the membrane and the product is not of the high purity obtainable by evaporation. Nevertheless, in many industrial situations, the purity requirements are measured in the hundred-ppm range and thus are readily met by the reverse osmosis process from supply waters of a wide range of qualities. The cost of processing water by reverse osmosis ranges from $0.25–$0.50/1000 gal including the cost of membrane depreciation. Since there is no phase change in the reverse osmosis process, waters containing high amounts of organic matter can be handled and not produce the foaming problems common to the evaporation process. Reverse osmosis thus appears to be able to assist in reuse of many process waters containing heavy organic contaminations where purification is impractical by other means.

An alternative to the foregoing processes is ion exchange using the newer weakly basic and weakly acidic resins together with recycling of inexpensive regenerants.

The chief example of this is the Desal process (6) which employs a macroporous weakly basic anion resin with a favorable chloride–bicarbonate selectivity coefficient in the primary exchanger, followed by a weakly acidic cation exchanger. The anion unit can be regenerated by a relatively inexpensive material such as ammonia and is then converted to the bicarbonate form by treatment with carbon dioxide or sodium bicarbonate. The cation unit is very efficiently regenerated by sulfuric acid at only 10% above stoichiometric amounts as compared with three times stoichiometric quantities for strongly acidic resins.

The raw water of high solids content enters the anion unit where mineral anions are converted to bicarbonate. The effluent from this unit then enters the cation unit where mineral cations are exchanged for hydrogen so that the effluent contains essentially only carbonic acid. Following this step, the bicarbonate can be removed by air stripping in a decarbonator, or by lime treatment, or, most commonly, by passage through a second ammonia-regenerated weakly basic anion exchanger for recovery of the bicarbonate. In the last case, the second anion unit is put into the bicarbonate form during the exhaustion cycle of the other two exchangers so that, at the end of this cycle, this second anion unit is in proper form to serve as the primary anion unit if the flow through the entire configuration is reversed.

The Desal Process can easily remove 90% or more of the dissolved solids in brackish or saline water. As with other ion exchange processes, this one requires a clarified influent. The effluent regeneration with inexpensive chemicals, the recovery of bicarbonate, and the possible recovery of ammonia regenerant in some configurations combine to allow this process to operate very economically in the range of $0.30–0.90/-1000 gal treated.

Another popular ion exchange process for gross dissolved solids removal is the proprietary SUL-biSUL (7) process. This process employs the conventional demineralization scheme of acid regenerated cation exchange followed by a strongly basic anion exchanger. The unique feature of this process is the operation of the anion unit. This operation depends upon the two stage ionization of sulfuric acid to monovalent bisulfate ion and to divalent sulfate ion. For operation, the anion resin is intentionally converted to its sulfate form. When it is exhausted by the acid effluent from the preceding cation unit, the sulfate ions, which occupy two exchange sites, are converted to bisulfate ions which occupy only one site. Thus, half the exchange sites on the resin are freed to exchange the anions in the influent water. When exhausted, the anion resin is regenerated by means of the cheapest possible alkaline material, namely a lime slurry or even the raw water if it contains sufficient alkalinity. This alkaline regenerant neutralizes the bisulfate ion and converts the resin back to the original sulfate form.

If the influent water contains 30% or more alkalinity, savings in acid regenerant consumption and waste disposal can be accomplished by adding a weakly acidic cation exchanger ahead of the strongly acidic unit. The SUL-biSUL process can easily reduce the total dissolved solids level of a brackish or saline water by 90% or more, at an estimated operating cost of from $0.40 to $1.25/1000 gal, depending upon the size of the plant, the quality of influent water, the regenerants used, and the difficulty for disposal of spent regenerant waste.

HIGH PURITY WATER PRODUCTION

Many industrial processes require water that contains less than 10 ppm dissolved solids, in some instances, less than 1 ppm, and in some extreme cases less than 0.1 ppm.

Where the purity requirements are in the one-to-ten-ppm range and steam is generated in the plant, evaporators may be the method of choice. The higher the dissolved solids in the raw water supply, the better the competitive position of evaporation. As pointed out above, the cost of the water will vary from $0.50–2.00/1000 gal depending upon the number of effects employed and the temperature to which the water can be heated without causing excessive deposition in the high temperature brine heater.

Evaporators have long been employed in seaboard power stations to produce the makeup for the steam generators. They have also been widely employed on ships for producing both makeup to the steam boilers and the supply of potable water. Since the evaporation process employs a phase change, a very wide range of substances can be tolerated in the feedwater and not transported into the vapor. This is particularly true of bacteria and viruses which may pass other processes not involving a phase change. Scaling remains the greatest single problem in the use of evaporators and necessitates special pretreatment to minimize it or periodic chemical and/or physical cleaning to remove it. Treatment of seawater with acid followed by degasification to remove carbon dioxide is commonly employed to reduce calcium carbonate scales. However, calcium sulfate scaling remains one of the major problems in seawater evaporators and one that is under considerable research and development at present. In many evaporation plants, it is possible to control scaling by a feed to the raw water of organic compounds that act as antinucleating agents, but the cost of this at present substantially increases the cost of the product water. In many industrial plants, to avoid scale formation the feedwater to the evaporators is passed through a sodium cycle cation exchanger to remove its hardness content. This softening allows a larger number of effects to be used. Ion exchange softening is practical in brackish waters but generally not practical in seawater. Evaporators are generally of three basic types: multistage flash, multistage long tube vertical, and vapor compression. The latter evaporation process does not require an outside source of heat since the heat input is obtained from power used to drive the vapor compressors employed. The vapor compression distillation process has the advantage of being able to be made in small compact selfcontained units which are able to operate with very little attendance and relatively low maintenance. On the other hand, the cost of water produced by the vapor compression process is the highest of the three evaporation processes employed.

Where the requirement for water purity is less than 1 ppm dissolved solids, demineralization by ion exchange is the commonly employed process. If the dissolved solids content of the raw water is greater than 500 ppm, some gross dissolved solids removal process, as described in the foregoing section, will be required before final demineralization by ion exchange can be accomplished economically. The ion exchange demineralization processes are of three basic types; individual beds of cation and anion exchange resin beads, beds of mixed resin beads, and beds of powdered mixed resins. The latter process is a proprietary one marketed under the name Powdex (8) and involves the use of very finely divided mixtures of cation and anion exchange resins in a completely regenerated form with the cation resin in the hydrogen form and the anion resin in the hydroxyl form.

The mixture is applied as a layer on suitable supporting filter elements. The layer is quite thin, of the order of 0.5–1 in. The exhausted resins are discarded rather than being regenerated. Hence, the process is limited to those situations where the influent water contains very little ionic contamination so that the resin consumption cost is economically justifiable. Such powdered ion exchange units are used for "polishing" (that is to say, further purification of) condensate and lower grade high purity water produced by conventional ion exchange processes. Because of the finely divided nature of the ion exchange material, high service flow rates and compact units are possible while still maintaining high effluent purities. The finely divided nature of the ion exchange material makes it an excellent filter for removing colloidal particles down to 0.1μ in addition to its ability to remove dissolved matter.

The more conventional ion exchange demineralization processes employ synthetic organic ion exchange resins in bead form, with bead sizes between 16 and 30 mesh. Where modest effluent purity levels of about 1 ppm can be tolerated, multibedded exchange systems are employed. These consist of an acid regenerated cation exchange resin in one vessel followed by anion exchange resin in the hydroxyl form in a separate vessel. Variations on this scheme involve separate beds of weakly acidic and strongly acidic cation exchange resin followed by separate beds of weakly basic and strongly basic anion exchange resins. Weakly acidic and weakly basic exchange resins have very high efficiencies approaching true stoichiometry between the regenerant and the ions removed. In most situations, the cation exchange resins are regenerated with sulfuric acid and the anion exchange resins with sodium hydroxide. Very complete data are available from the manufacturers of the resins giving all of the properties including expected ionic leakage from the beds for various regeneration levels of acid or alkali and the capacities of the exchange resins for their respective ions. The strongly acidic and strongly basic resins are able to remove all the cations and anions, both strong and weak acid, whereas the weakly acidic and weakly basic exchange resins are able to remove only cations equivalent to the weak acid content of the water and strongly acid anions. Carbonates and silicates, which both form weak acids, are not removed by the weakly basic resins. Where multiple beds employing both weak and strong resins are used, weakly acidic cation resin precedes the strongly acidic resin. The acid regenerant in such systems is passed first through the strongly acidic resin and then through the weakly acidic resin, gaining essentially 100% regeneration efficiency in the process. Similarly, the weakly basic anion resin precedes the strongly basic anion resin and the caustic soda at $120°F$ is used to regenerate first the strongly basic resin and then the weakly basic resin. When only strong acid and strong base resins are employed, regeneration efficiencies are typically three to four equivalents of acid per equivalent of cations removed and four to five equivalents of caustic per equivalent of anions removed. These efficiencies vary considerably depending upon the composition of the water, type of resin employed, and the purity of the effluent that is desired. The decision on whether or not to employ combination weak and strong resins is economic and involves the relative cost of the additional vessels and/or special regeneration systems versus the chemical savings gained by the higher efficiency.

When the product water must contain less than 0.1 ppm dissolved solids, beds of mixed ion exchange resin are employed (typically two parts of strongly basic anion resin to one part of strongly acid cation resin). The regenerated resins in acid and hydroxyl form are thoroughly mixed together by air or water action. For this reason, such a system is commonly called a mixed bed demineralizer. Following exhaustion of the bed, backwashing of the resins is employed to achieve hydraulic classification, the less dense anion resin becoming the top layer and the more dense cation resin remaining on the bottom. With known resin volumes, the interface separation point is known and a distributor grid is located at that point to allow separate passage of acid through the cation resin portion and caustic soda through the anion resin portion of the bed. Following regeneration, separate rinsing of each section is employed and then the beds are mixed in preparation for another cycle. Such mixed ion exchange beds are capable of producing an effluent having a specific conductance at 25°C of less than 0.1×10^{-6} ohm^{-1} cm^{-1} (greater than 10,000,000 ohm-cm specific resistance). Because the mixed ion exchange resins do not separate perfectly at the interface, there is a certain inherent inefficiency in the regeneration. Some of the anion resin is contacted by the acid used to regenerate the cation resin and thus is placed in the neutral salt state while some cation resin is contacted by the caustic soda and thus placed in the sodium form. In both cases, small amounts of resin in such forms, when mixed together in the bed, contribute to a loss of capacity and an increase in the leakage of cations and anions from the bed. Flow rates employed in such systems are commonly 2.5 gal/(min)(ft³) of resin bed volume. In high pressure steam power generation stations where ion exchange polishing of condensate is employed to achieve higher purity for feedwater to the steam generator, and where the contaminants in the condensate from the surface condenser are commonly less than 0.5 ppm, very high flow rates may be used. Flow rates up to 50 gal/(min)(ft²) or from 15 to 25 gal/-(min)(ft³) are not uncommon.

Since ion exchange beds, by their very nature, develop strong surface charges, they are able to act as excellent filter media. Ion exchange beds commonly remove 90% of the incoming colloidal matter down to extremely low levels measured in the ppb range. This same property requires that the water supplied to the ion exchange beds be very clear, containing, for best results, less than 0.5 ppm of suspended solids. Whatever suspended solids are removed by the filtering action of the beds must later be removed by backwashing. The more suspended solids that are removed by the bed, the poorer is the efficiency of the exchange process. This effect is true because the suspended solids tend to coat the resin beads and reduce the ion exchange rates. A heavy accumulation of suspended solids within the beds, in addition to coating the particles, causes poor hydraulic distribution of the water flow through the bed and a further decrease in bed efficiency.

In cases where calcium ion constitutes more than 40% of the total cations in the influent water and sulfuric acid is used for regenerating the cation exchange resin, it is necessary to employ a multistage regeneration technique. In such cases, the concentration of the first acid to contact the bed is 2% or less, the next portion of regenerant may be at 5%, and the final portion at 10%, or some other similar combination. It is most important when using sulfuric acid regenerant in such instances that the first concentration be quite low since the calcium removed from the bed tends to cause formation of calcium sulfate, which precipitates on the resin reducing both regeneration efficiency and service capacity. In small units (less than 100 gal/min), hydrochloric acid may be substituted for sulfuric acid for convenience even though the cost of the hydrochloric acid is greater. When using hydrochloric acid, no stepwise regeneration is necessary and the acid strength may be anywhere from 5 to 10% depending upon characteristics of the resin.

One of the more difficult problems associated with demineralization is fouling of the anion exchange resins by irreversible absorption of organic matter from the water. The organic acids responsible are generally described as humic acids and represent part or all of the color in surface waters. As previously mentioned, the humic acids also cause problems in the coagulation process by forming stable complexes with iron and manganese that may be present. The degree to which the anion resin can exchange the acid is dependent upon the porosity of the resin. Each regeneration of the resin removes most but possibly not all of the acid absorbed in the previous service cycle and so, after many cycles, the gradual accumulation of acid within the resin occupies exchange sites and decreases the capacity of the resin. Various treatments can be employed to remove the accumulated organic matter but none is entirely successful. In recent years, the greatest success in overcoming this problem has been achieved with the development of so-called macroporous anion exchange resins. These macroporous resins are able, because of large well defined pores within the beads, to handle high molecular weight anions such as the humic acids. Even with the increased ability to handle and remove organic acids from water, it is desirable to gain as complete removal of these acids as possible during the coagulation process. If possible, the organic acid content of the influent water should be reduced to less than one ppm before entering the demineralizer.

Where the best of the coagulation-filtration procedures are unable to remove sufficient organic matter to prevent undue fouling of anion exchange resins in demineralization processes, beds of activated carbon may be placed ahead of the demineralizer or, in some instances, between the acid regenerated cation unit and the anion unit. Such beds are able to remove the organic acids to concentrations in the ppb range, but the removal is irreversible and the carbon beds must be replaced, when exhausted, with beds of fresh material. The advent of the macroporous anion exchange resins has largely replaced the use of activated carbon for organic matter removal ahead of demineralizers. In some situations, the activated carbon filter is employed ahead of the demineralizer to dechlorinate the influent water completely. Such dechlorination is desirable because chlorine residuals in the influent water to a demineralizer result in oxidation in the cation exchange unit with consequent sloughing of degradation products into the cation effluent. These degradation products may foul the subsequent anion exchange resin or increase the organic matter content of the finished effluent or both. In larger installations where adequate controls can be employed, dechlorination is usually practiced by the addition of sodium sulfite to the influent water to the demineralizer. In smaller installations where such added controls are uneconomical, the activated carbon bed is most useful for this purpose.

Some industrial processes require that, in addition to complete removal of dissolved matter from the water, the water contain no bacteria or other microorganisms. This requirement is particularly severe in the electronics industry where the formation of slimy bacterial growths in the demineralized water lines can cause small amounts of such growths to be entrained by the water and deposited on the electronic components causing failures. One of the most practical methods employed for cold sterilization of the effluent from demineralizers involves the use of microporous membrane filters. Filters with pore sizes of 0.45 μ are commonly used for this purpose as they represent a good compromise between filtration rate and the ability to remove bacteria. Such microporous filters also have the advantage of removing any very finely divided ion exchange particles from the demineralizer beds and also removing any colloidal clay that may have passed through the entire pretreatment system. Thus, ultrapure water production in its most sophisticated version involves clarification, filtration, activated carbon adsorption, demineralization by ion exchange, and ultrafiltration with microporous filter media. The microporous filter membranes employed for this purpose do not filter in depth, but rather filter in a layer retained on the surface of the medium. When the pressure drop over the filter rises to the maximum tolerable value, the membrane must be discarded and replaced with a new one.

Post Treatment Systems

Sedimentation, coagulation, filtration, softening, and demineralization are all so-called pretreatment methods that are used to alter the quality of the water to that required by the process. These processes all involve the removal of something or a substitution of one ion for another. It is many times beneficial to supplement these pretreatment processes with post treatment which, in this broader sense, represents simply the addition of one or more items to the pretreated water in order to make it especially suited for a particular industrial process. Such post treatment may be as simple as the addition of an alkali or an acid to adjust the pH to a value required by the process itself or for control of corrosion or deposition in the piping. Where corrosion and deposition in the lines is the primary concern, the pH may be adjusted in accordance with either the Langelier (9) or the Ryznar Index (10). These indexes are empirical expressions of calcium carbonate solubility as related to total dissolved solids, temperature, alkalinity, calcium hardness, and pH. Use of these indexes is based upon the intentional deposition of calcium carbonate on the walls of the pipe to form a protective layer, while controlling the amount of this deposition closely enough that flow is not restricted.

Corrosion and deposits in distribution lines can also be controlled by the addition of polyphosphates such as sodium hexametaphosphate or sodium tripolyphosphate in less than stoichiometric amounts commonly referred to as threshold treatment. This addition of polyphosphate inhibits calcium carbonate deposition by sequestration, reduces tuberculation corrosion in the line, and also sequesters small amounts of iron oxide formed by the corrosion reactions, thus preventing the iron oxide from forming deposits in the equipment.

It is also frequently necessary to control the formation of biological slimes either in the distribution piping or in the equipment itself. Such slime formation is particularly troublesome in heat exchange equipment where it rapidly impedes heat transfer, necessitating frequent cleaning of the equipment. Such biological growths in pipes can enhance corrosion reactions in addition to restricting flow. Control is generally accomplished by chlorination, maintaining 0.3–0.5 ppm in the water, or alternatively "shocking" the water with a high chlorine dose of 1–2 ppm for several hours each day. Recently, high-molecular-weight polymers such as the polyacrylamides have become available which, when added to once-through distribution systems, cause flocculation of loose corrosion products as well as sediments and slimes and allow the normal velocity of the water in the line to carry the flocculated material away. Such mud removal agents have become economically attractive in once-through cooling water systems using turbid, untreated raw water.

Sometimes where corrosion control is required in once-through hot water systems, 10–30 ppm of sodium silicate is added to the water continuously to form a protective mixed oxide-silicate film on the steel pipe surfaces. This is also effective in preventing corrosion under certain conditions in copper distribution pipes.

Many industrial water utilizing systems such as steam generators, recirculated cooling systems, paper making systems, recirculating cooling sprays in steel plants, etc., require special corrosion and deposit control treatments that can be afforded only because such systems are essentially closed with only modest water loss to waste. See Corrosion inhibitors.

Bibliography

"Water (industrial)" in ECT 1st ed., Vol. 14, pp. 926–946, by E. P. Partridge, Hall Laboratories, Inc.

1. "Report of the Committee on Water Quality Criteria," Federal Water Pollution Control Administration, U.S. Department of the Interior, April 1, 1968, p. 188.
2. *Ibid.*, p. 189.
3. "Experiences with Multiple-Bed Filters," Kenneth E. Shull, *J. Am. Water Works Assoc.* **57,** (3) (1965).
4. T. M. Riddick, "Zeta Potential: New Tool for Water Treatment," *Chemical Engineering,* June 26, 1961, pp. 120–146.
   T. M. Riddick, "The Role of Zeta Potential in Coagulation Involving Hydrous Oxides," Paper #1 presented at 1st Water Conference, TAPPI, Netherland-Hilton Hotel, Cincinnati, Ohio, June 4, 1963, pp. 1–31.
5. A. P. Black, "Improvements in Instrumentation and Techniques for Microelectrophoresis," *J. Am. Water Works Assoc.* 485–491 (1965).
   A. P. Black, "Electrokinetic Behavior of Aluminum Species in Dilute Dispersed Kaolinite Systems," *J. Am. Water Works Assoc.,* 1173–1183 (1967).
   A. P. Black, "Electrophoretic Studies of Sludge Particles Produced in Lime-Soda Softening," *J. Am. Water Works Assoc.,* 737–747 (1961).
6. Desal Process Brochure IE-114-67, Rohm and Haas Co., Philadelphia, Pa.
7. Nalco Chemical Company, Ion Exchange Division, Chicago, Illinois "The Su-biSul Process," Bulletin Z-18, TD Index 680.93.
8. Graver Water Conditioning Company, Division of Union Tank Car Company, New York, "Powdex Process, Precoat Ion Exchange System," Technical Bull. 101.
9. W. F. Langelier, "The Analytical Control of Anti-Corrosion Water Treatment," *J. Am. Water Works Assoc.* **28** (10), (1936).
10. J. W. Ryznar, "A New Index for Determining Amount of Calcium Carbonate Scale Formed by a Water," *J. Am. Water Works Assoc.* **36** (4), (1944).

J. K. Rice and D. E. Simon II
Cyrus Wm. Rice Division,
NUS Corporation

MUNICIPAL WATER TREATMENT

One of man's necessities is pure water. He can exist for only a short time without it. His basic requirement is only about three quarts a day, but an urban civilization increases the actual amount utilized to a current average in the United States of about

Table 1. Municipal Water Prices in U.S. Cities (2)

City	¢/1000 gal
San Diego, Calif.	15.0
Des Moines, Iowa	49.3
Fort Worth, Texas	76.7
Little Rock, Ark.	53.3
Washington, D.C.	25.0
Baltimore, Md.	17.4
Philadelphia, Pa.	14.4
Milwaukee, Wisc.,	26.7

150 gal per capita per day. This, of course, includes much more than the water consumed by man. It considers not only such uses as household requirements and sanitary needs but also the urban industrial and municipal requirements. The total requirement including all industrial and agricultural uses in 1965 was about 1600 gal per capita per day or 2400 tons per year (1). This far exceeds the mass of all of man's other material requirements, such as food, housing, automobiles, etc, which total about 20 tons per capita per year. With the very large quantities of water required, it is obvious that the costs must be low. Table 1 shows costs to consumers in representative U.S. cities (2).

Water Quality

The principal sources of water are surface supplies such as streams, rivers, lakes, reservoirs, and groundwater supplies, ie, those obtained from deep wells. The general properties of the two sources are listed in Table 2.

Selection of a source of supply is made on the basis of the specific properties of those supplies available. The 1962 U.S. Public Health Service *Drinking Water Standards* (3) establishes criteria for the maximum levels of deleterious substances such as arsenic and lead. One set of levels defines waters as acceptable only in cases where no better

Table 2. General Properties of Natural Water Supplies

Property	Surface Supplies	Ground water Supplies
mineralization	low	high
dissolved oxygen	usually saturated	very low
H_2S	absent, unless polluted	may be present
pollution	common	uncommon
color	common	uncommon
turbidity	common	uncommon
Fe and Mn	uncommon	common
quantity	variable, unless impounded	constant

Table 3. Water Supply Selection Criteria (3)

Substance	Max desirable concn, mg/l	Max permissible concn mg/l
alkylbenzenesulfonate	0.5	
arsenic	0.01	0.05
barium		1.0
cadmium		0.01
chloride	250	
chromium, hexavalent		0.05
copper	1.0	
cyanide	0.01	0.2
fluoride	0.3	2.0
iron	0.3	
lead		0.05
manganese	0.05	
selenium		0.01
silver		0.05
zinc	5.0	

source is available and another establishes levels above which the water is considered unsafe. These criteria are summarized in Table 3.

There has been an increasing trend toward the use of ground-waters where they are available primarily because of the reduced pollution load and the assurance of a constant supply. The major problem is the low rate at which many such supplies are replenished. Historically, the choice favored surface supplies as they were more readily and obviously available. The current ratio of the use of surface to groundwater supplies is about 3:1 for the approximately 22,000 public water supplies in the United States.

Treatment Methods

When alternative sources of water are available to a city, the decision must be made based on such factors as quantity available, continuity of availability, treatment costs and capital costs, as well as assurance that the treatment process chosen will provide water that is biologically safe and esthetically acceptable. Other general criteria of a functionally ideal water are expressed in the *Quality Water Goals* of the American Water Works Association (4) as: "Ideally, water delivered to the consumer should be clear, colorless, tasteless, and odorless. It should contain no pathogenic organisms and be free from biological forms which may be harmful to human health or esthetically objectionable. It should not contain concentrations of chemicals which may be physiologically harmful, esthetically objectionable, or economically damaging. The water should not be corrosive or encrusting to, or leave deposits on, water conveying structures through which it passes or in which it may be retained, including pipes, tanks, water heaters, and plumbing fixtures. The water should be adequately protected by natural processes, or by treatment processes, which insure consistency in quality." The choice of treatment method is a function of the raw-water quality. The majority of surface waters must be treated for the removal of suspended solids. These solids include clay, bacteria, viruses, and organic debris resulting from partial breakdown of leaves and other plant materials. Some surface waters contain organic color which is usually removed for esthetic reasons. Waters obtained from impounded supplies may contain reduced iron and manganese which must be removed to prevent precipitation of the oxidized forms in the distribution system or at the point of ultimate use.

A complete sequence of treatment for a municipal plant would include (1) application of copper sulfate to a reservoir or lake for algae control; (2) addition of activated

Table 4. Treatment Processes Used by Municipal Plant in the 100 Largest U.S. Cities

Process	Number of cities	Percentage of total population[a]
chlorination	98	98.8
sedimentation and coagulation	68	63.8
filtration	76	56.8
softening	28	18.1
iron and/or manganese removal	10	5.6
no treatment	1	0.1
fluoridation[b]	34	35.0

[a] The total population as determined by the 1960 census was 60,000,000.

[b] At the end of 1969 the use of fluoridation had increased appreciably as several major cities had started fluoride addition, including New York City and Atlanta.

carbon and chlorine at the head of the plant for taste, odor, and bacteria control.   (3) addition of a coagulant in a rapid mix tank for turbidity removal; (4) passage through a flocculation tank to promote floc growth; then through (5) a horizontal or vertical sedimentation tank; (6) followed by pH adjustment; and thus (7) to the gravity filters.   The water from the filters, after the addition of chlorine, is usually stored in a clearwell until it is pumped to service or to an elevated or other storage tank.   Some, or all, of the above may be used in any given case for a surface water containing turbidity. If the water is also hard or is hard rather than turbid it may be softened by the lime, lime–soda or ion exchange process in addition to or instead of the coagulation-flocculation-sedimentation steps prior to filtration, chlorination, and distribution.   The types of treatment used by the 100 largest U.S. cities are summarized in Table 4 from data collected by the U.S. Geological Survey in 1962 and analyzed by Dufor and Becker (5).

DISINFECTION

As shown in Table 4, the majority of plants in the United States practice disinfection, most frequently chlorination.   Disinfection is the destruction of pathogenic organisms as contrasted to sterilization which is the elimination of all living organisms. It is not uncommon for small plants using uncontaminated deep wells to distribute the water untreated and undisinfected.   This practice can lead to serious consequences in the event of a cross connection or other type of contamination of the well or distribution system as occurred in 1965 in Riverside, California resulting in 18,000 cases of gastroenteritis, and where the contamination was found to have entered at a pump in the system.   The water was not disinfected because it was from deep wells and was considered bacteriologically safe.   The absence of a residual disinfectant led to distribution of a contaminated water.

Disinfection of a public water supply was first practiced in the United States by G. A. Johnson and J. L. Leal at Jersey City, New Jersey in 1908.   The agent used was a solution of calcium hypochlorite.   From that day on, the incidence of water-borne diseases such as typhoid fever, dysentery, and hepatitis has decreased in proportion to the extent of such disinfection, until today it is indeed rare that such diseases occur and then only when some adequate residual disinfectant is not carried in the system.

One of the major advantages of chlorination as a disinfection method over such disinfectants as ozone, radiation, gamma rays, etc, is the ability to carry a residual throughout the distribution system with the resultant capacity to inactivate any contamination that may be introduced after the initial disinfection.   When chlorine gas, the most common disinfectant in all but the smallest plants, is added to water the following reactions take place:

$$Cl_2 + H_2O \rightleftharpoons H^+ + Cl^- + HOCl \tag{1}$$

$$HOCl \rightleftharpoons H^+ + OCl^- \tag{2}$$

The equilibrium constant for equation (1) is 4.5×10^{-4} at $25°C$ and the K_a for equation (2) is 2.5×10^{-8} at $25°C$.   Figure 1 shows the distribution of chlorine species as a function of pH, based on the constants above.   Since the bactericidally active forms are Cl_2 and HOCl it can be seen that at pH values above 7.6 less than 50% of the total chlorine is available in an effective form.   Since many potable waters are treated and distributed at pH values greater than 7.6, it is frequently necessary to carry a high residual in order to provide adequate protection.   It is necessary to carry at least 0.3 mg/l of free available chlorine residual in the distribution system.   *Free available*

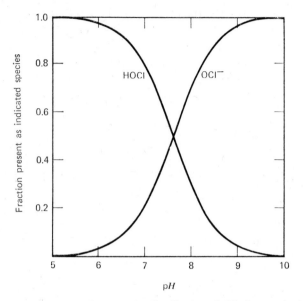

Fig. 1. Effect of pH on the distribution of chlorine species.

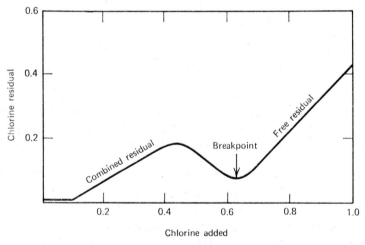

Fig. 2. Breakpoint chlorination curve.

chlorine residual is defined as that analyzed as HOCl plus OCl⁻. The *total residual chlorine* includes free and combined residuals, the combined being principally due to chlorine that has reacted with ammonia or amines to produce chloramines. The reactions may be illustrated by ammonia itself as:

$$HOCl + NH_3 \rightleftharpoons NH_2Cl + H_2O \qquad (3)$$
$$\text{monochloramine}$$

$$NH_2Cl + HOCl \rightleftharpoons NHCl_2 + H_2O \qquad (4)$$
$$\text{dichloramine}$$

$$NHCl_2 + HOCl \rightleftharpoons NCl_3 + H_2O \qquad (5)$$
$$\text{trichloramine,}$$
$$\text{(nitrogen trichloride)}$$

Chloramines, due to their reduced oxidation potential, are much less effective disinfectants than hypochlorous acid. They are frequently employed as disinfectants, however, in systems in which it is difficult or impossible to carry a free available chlorine residual because of taste and odor problems. In these cases ammonia or an ammonium salt is added to the plant effluent, after disinfection with Cl_2, to provide a combined residual in the system. The reduced effectiveness requires that higher residuals be carried, up to 3 mg/l.

Chlorine may also serve as an oxidant for H_2S and $Fe(II)$, as well as various organic impurities. These result in a chlorine demand by any given water, which must be satisfied before effective disinfection may be accomplished. Figure 2 shows the effect of adding chlorine to a sample containing reducing agents. The first horizontal portion of the curve represents the chlorine demand of the water, ie, reducing agents that react with chlorine. The next portion of the curve represents reaction with ammonia and its derivatives to yield a combined residual. This is followed by further oxidation to ineffective dichloroamine and nitrogen trichloride until the *breakpoint* is reached after which every additional part of chlorine added results in a part of free residual.

In some cases it is necessary to use high dosages of chlorine in order to oxidize organic materials that cause tastes and odors or high levels of bacteria. This is referred to as *superchlorination*. The large excess must be removed before distribution because of objections by consumers to the taste and odor of chlorine itself. Dechlorination is accomplished by adding a reducing agent such as sulfur dioxide, sodium sulfite, or sodium bisulfite. It may also be absorbed by passage through activated carbon.

Chlorine may be added to the raw water prior to any other treatment, *prechlorination*, or added at various points through the plant and possibly again after all other treatment, *postchlorination*. It is sometimes necessary to add chlorine at locations in the distribution system in order to maintain a residual throughout the system. Other disinfectants that have been used or proposed include bromine, iodine, ozone, and chlorine dioxide. Ozone has been used in Europe, principally because of the dislike for the taste and odor of chlorine residuals. It has never been a factor in the United States because of its very high oxidation potential which results in the inability to carry any residual. Ozone is produced by the corona discharge of high voltage electricity in air. Since the efficiency is generally less than 1%, the cost of ozone is high compared to chlorine.

Bromine has not been used to any appreciable extent but iodine has been studied extensively and used in many systems. It has the advantage of lower oxidation potential than chlorine, thus allowing a residual in the system. In addition, it is a good viricide and cysticide, and since it is solid, it is easier to handle than gaseous chlorine. Iodine is used extensively in the field by the armed forces of the United States.

Chlorine dioxide is a strong oxidizing agent and can be produced from sodium chlorite by reaction with chlorine.

$$2NaClO_2 + Cl_2 \rightleftharpoons 2NaCl + 2ClO_2 \qquad (6)$$

Chlorine dioxide has the advantage over chlorine in that it has a higher oxidation potential at high pH values, where Cl_2 is relatively ineffective. Chlorine dioxide has been used, particularly in Europe, for many years in applications where tastes and odors result from the use of chlorine and in the U.S. in the product from water treatment plants.

SEDIMENTATION AND FILTRATION

Since most surface waters contain varying amounts of suspended solids, including silt, clay, bacteria, viruses, etc, it is necessary to remove them prior to distribution to the domestic or industrial consumer. The principal processes are sedimentation and filtration, both of which have been utilized from the early days of recorded history for water clarification. Some of the earliest filters were unglazed pottery jars used by the Egyptians before the time of Christ. Sedimentation, alone, is rarely adequate for the clarification of turbid waters and is of little or no value for the removal of such very fine particles as clay, bacteria, etc. Table 5 shows the effect of particle size on the sedimentation rate of a solid having a specific gravity of 2.65 in water at 20°C.

Table 5. Sedimentation Rate as a Function of Particle Diameter (6)

Equivalent spherical radius	Approximate size	Sedimentation rate (time to settle 1 ft)
10 mm	gravel	0.3 sec
1 mm	coarse sand	3 sec
0.1 mm	fine sand	38 sec
0.01 mm	silt	33 min
10^{-3} mm (1μ, 1000 nm)	bacteria	55 hr
100 nm	colloid	230 hr
10 nm	colloid	6.3 yr
1 nm	colloid	63 yr

The design of many plants treating surface waters includes preceeding the plant with a reservoir in order to allow the larger particles to settle, as well as to buffer changes in water quality. Table 5 shows that the only particles that sediment within a reasonable period of time are silt, sand, and possibly some larger bacteria. The listed values are for an undisturbed system and would be much larger for actual cases where currents tend to upset the settling. This emphasizes the necessity for further treatment for potable waters. This treatment may involve only filtration through sand or multiple media filters or may require considerable pretreatment before filtration such as coagulation and flocculation.

The use of sand filtration as a method of clarifying water dates to the small sand beds designed for Paisley, Scotland in 1804 by John Gibbs and followed by the larger sand filters used by the Chelsea Water Company of London in 1828 to filter Thames River water. These were designed by James Simpson. The first use of sand filters in the United States was at Poughkeepsie, N.Y., as designed by James Kirkwood in 1872 after inspecting those in England. These were what we now refer to as slow sand filters, ie, the filter rates were low compared to current filter rates. Table 6 compares the properties of slow sand filters with rapid sand filters, and high rate filters. Slow sand filters require large surface areas because of the low rates that must be used. This led to the development of higher rate filters because of the very large areas required to filter water for the larger cities. There are slow sand filters still in use in the United States and in Europe.

The first patent relating to rapid sand filters was issued to Isiah Hyatt in 1884 (9) for the use of ferric chloride as a coagulant ahead of the filters in order to duplicate the slimy, gelatinous layer of bacteria, algae, and fungi that built up on the surface of slow sand filters and played an important part in the filtering action. This layer is called

Table 6. Typical Properties of Filter Types

Property	Type of filter		
	Slow sand	Rapid sand	High rate
design filtration rate, gal/(min)(ft²)	0.016–0.128	2	4–8
bed-size per million gal/day, ft²	43,600 ft²	350 ft²	120 ft²
type of media	ungraded	carefully graded sand with larger sizes at bottom	several types but usually carefully graded sand and carbon
bed depth	12 in. gravel beneath 30–42 in. of sand	18 in. gravel beneath 24–20 in. of sand	12 in. sand beneath 24 in. granular carbon
filter runs before cleaning	30 days	24–72 hr	24–72 hr
cleaning procedure	removal of surface layer for cleaning or washing surface in place	backwash by forcing water up through filter, may use surface jets or submerged jets to fluidize bed with water or air	sand as rapid sand
amount of water used for cleaning, % of water filtered	0.5	3–4	
Depth of filtration	surface only	first few inches	in depth
pretreatment	none	coagulation and sedimentation	coagulation and sedimentation

the *schmutzdecke* from the German Schmutz, "impurity, dirt, filth" and Decke, "layer." The first major municipal rapid sand filter was built at Little Falls, N.J. in 1909.

T. R. Camp (7) and W. R. Conley (8) designed and studied dual media filters for municipal and industrial application. An example of the use of high rate filters was at Hanford, Washington for treating a surface water for the Atomic Energy Commission facility. The advantage of higher filter rates was recognized at many municipal plants that were operating over capacity, the limiting factor commonly being the filter rate. Modification of the filters by changing the media delayed the necessity of adding new filters to the plant or possibly building a new plant. One of the first designs employed an inverted media loading in which very fine, dense, garnet sand was used at the bottom of the filter, covered by less dense, but coarser silica sand, and then by granulated carbon of even lower density and larger particle size. This type of construction provides the larger pore sizes for the initial contact with the water to be filtered and thereby allows deeper penetration of smaller particles into the bed and the consequent use of more of the filter medium for filtration. Backwashing returns the filter to the original particle size distribution because of the density differences.

The size of particles removed by such filters is less than the size of the passages themselves. The mechanism of removal has been shown to include adsorption of the impurities at the interface between the media and the water by specific chemical or Van der Waals attractions, or by electrostatic interaction when the media particles have surface charges opposite to those on the impurities to be removed. This contrast with the picture that was predominant for many years of a separation based upon simple straining.

Neither rapid sand nor mixed media filters will remove appreciable quantities of colloidal particles without adequate pretreatment. Even though it is widely believed that filters are an effective barrier against unsafe water, the effluent may be as colored, as turbid or as bacteriologically unsafe as the water applied. In contrast, slow sand filters require no pretreatment, as the slow passage through the bed allows the particles to contact and attach to the schmutzdecke.

The two steps in the removal of a particle from the liquid phase by the filter medium are (1) the transport step of the suspended particle to the surface of the medium and (2) interaction with the surface to form a bond strong enough to withstand the hydraulic stresses imposed on it by the passage of water over the surface. The transport step is influenced by such physical factors as concentration of the suspension, media particle size, media particle size distribution, temperature, flow rate, and flow time. These parameters have been considered in various empirical relationships that attempt to predict filter performance based on physical factors only (10,11). Most recent attention has been focused upon the second step, the interaction between the particles and the filter surface, postulating mechanisms based upon adsorption or specific chemical interactions (12).

The goal of filtration in the modern municipal treatment plant is a maximum of 0.1 turbidity units (these are arbitrary units called Jackson candle units from the standard instrument used, see Turbidity). The goals are expressed by the American Water Works Association in their *Quality Water Goals* adopted by the Board of Directors and published in December 1968 (13) as: "Today's consumer expects a sparkling, clear water. The goal of less than 0.1 unit of turbidity insures satisfaction in this respect. There is evidence that freedom from disease organisms is associated with freedom from turbidity, and that complete freedom from taste and odor require no less than such clarity. Improved technology in the modern treatment processes make this a completely practical goal."

Coagulation and Flocculation. The removal of colloidal particles such as turbidity, color, and bacteria requires agglomeration of these particles prior to filtration. Agglomeration may be carried out by chemical means by interaction of the colloidal particles with materials added as coagulants, or physico-chemically by neutralizing the charge on the particles or by interparticle bridging. The term "coagulation" is applied to the addition of any material that causes agglomeration of the colloids, and "flocculation" to the process of gentle agitation that builds floc particles large enough to settle rapidly. A physico-chemical mechanism was proposed by LaMer (14) for coagulation, based on charge neutralization by double layer compression allowing the particles to approach closely enough for short-range Van der Waals forces to cause agglomeration. Flocculation was seen as the growth of a three-dimensional structure by interparticle bridging.

The repulsive forces, that act to stabilize the suspension are; the charge on the particles, (like particles will have like charges and will therefore repel one another) and hydration. The attractive or destabilizing forces are; Brownian movement, Van der Waals forces, and gravity. As shown in Table 1, the effect of gravity is insignificant in the destabilization or sedimentation of colloidal particles. It is obvious that it is possible to destablize colloids by increasing the attractive forces or by decreasing the repulsive forces.

Among the various factors that result in charges on colloidal particles, are ionization of surface groups, such as acid groups on organic color particles and salt-like bonds on the surface of clay particles that result in ion exchange reactions with the solution. In general, the colloidal particles in natural waters are negatively charged. Clay, for example, has a net negative charge on the large surface with positive charges on the

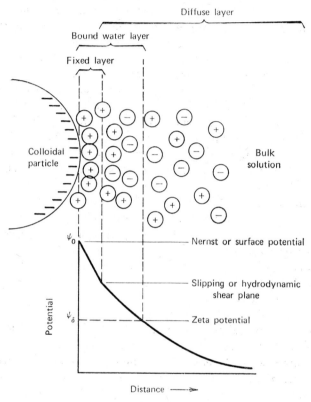

Fig. 3. Double layer model of colloidal particle.

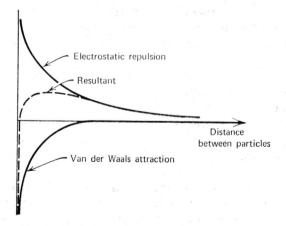

Fig. 4. Forces between two colloidal particles.

edges. Color particles are also negatively charged as are bacteria. This charge cannot be measured directly as the particles are surrounded by a sheath of counterions in a relatively fixed layer. This is shown in Figure 3. From this fixed layer the charge is gradually neutralized by the higher concentration of counterions in the diffuse double layer. The measurements that we can make of this charge are limited by the impossibility of getting the particle independent of the counterions. We are able to measure the rate at which the particle, and some of its entourage of counterions, moves under an impressed potential difference by various techniques such as microelectrophoresis but this only tells us the charge at the slipping plane or shear plane where the net force between the particle and the bulk solution is small enough so that the particle moves independently. The charge at this plane is referred to as the zeta potential and is shown in Figure 3 as ψ_ζ. The charge on the particle is ψ_0. The electrostatic replusion between two like-charged particles is shown in Figure 4 as a function of the distance between the particles. This force is shown in the top curve. The electrostatic repulsions are inversely proportional to the square of the distance between the particles and the Van der Walls attractive forces are inversely proportional to the fifth power of the distance. The resultant force is repulsive until the particles are relatively close together.

The other important stabilizing force is hydration, since these are hydrophilic colloids. The waters of hydration modify the exterior of the particle so that it approaches the properties of the bulk water. Owing to this the particles show no tendency to approach or to coalesce.

The destabilizing effect of the Brownian movement is that it brings the particles into contact simply because of their random wanderings through the solution. If two particles happen to collide or penetrate each others repulsive sphere they may be held together by the other destabilizing force, the Van der Waals forces. This force is proportional to the reciprocal of the distance between the particle to the 5th power. This means that the attractive forces generally act over a much shorter distance than do the repulsive forces, as shown in Figure 4, where the lower solid curve is the Van der Waals forces of attraction. The dotted curve in this figure is the resultant for a typical stabilized colloid.

It can be seen from these two factors, ie particle charge and Van der Waals forces, that we need to reduce the charge or compress the double layer in order to allow the particles to approach each other closely enough so that the Van der Waals forces can hold them together. There are two approaches to the accomplishment of this goal:

(1) reaction with the charged surface sites with an opposite charge on an insoluble material.

(2) Neutralization of the charge by opposite charges concentrated in the fixed layer or the immediate environment. It is difficult to distinguish between these two as they result in the same effect on the particle, insofar as this effect can be measured. Since the particles are negatively charged, they can be coagulated by cationic polyelectrolytes, organic, or inorganic. A particle that consists of an organic chain that is negatively charged, or an uncharged particle, may still be adsorbed, but this does not result in reduction of the particle charge.

Many metal ions react with water to produce hydrolysis products that are multiply charged inorganic polymers. These may react specifically with negative sites on the colloidal particles to form relatively strong chemical bonds, or they may be adsorbed at the interface. In either case the charge on the particle is reduced.

Figure 5 shows the effect of increasing the ionic concentration on the charge distribution. It can be seen, in comparing 5a and 5b, that increasing the total concentration of electrolyte reduces the effective distance through which the charge is manifested. This effect is even more pronounced when polyvalent cations are used as in 5c. Due to

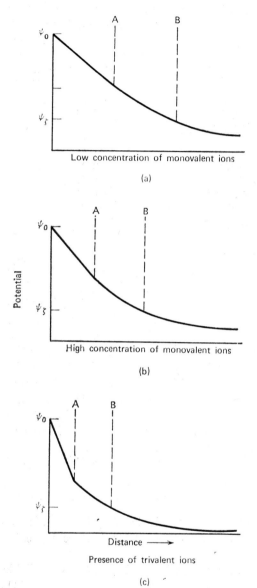

Fig. 5. Effect of ionic environment on the potential of a colloidal particle.

their increased charge density they can replace univalent ions in the fixed double layer and can be more closely packed in the diffuse double layer, thereby compressing the diffuse double layer and permitting the particles to approach each other more closely. In considering the overall destabilization reaction it is important to keep in mind that the Van der Waals attractive forces are essentially unaffected by the ionic environment, so that if the repulsive forces can be reduced sufficiently this will allow the resultant force to be attractive rather than repulsive for a given system.

There are two classes of coagulants used in municipal water treatment, inorganic metal salts, such as iron and aluminum, and organic polymers. The metal salts have been used since the early days of water treatment as the primary coagulants although they were first utilized as pretreatment for filtration. The metal salts hydrolyze when added to water according to the typical series of reactions shown in Table 7.

Table 7. Hydrolytic Reactions of Metal Ions

Reaction	pK	Reference
$Al^{3+} + H_2O \rightleftharpoons AlOH^{2+} + H^+$	5.03	15
$2Al^{3+} + 4H_2O \rightleftharpoons Al_2(OH)_4^+ + 2H^+$	6.27	15
$Al^{3+} + 3H_2O \rightleftharpoons Al(OH)_3 + 3H^+$	9.1	16
$Al^{3+} + 4H_2O \rightleftharpoons Al(OH)_4^- + 4H^+$	21.84	16
$Fe^{3+} + H_2O \rightleftharpoons FeOH^{2+} + H^+$	3.05	10
$Fe^{3+} + 2H_2O \rightleftharpoons Fe(OH)_2^+ + 2H^+$	6.31	10
$Fe^{3+} + 3H_2O \rightleftharpoons Fe(OH)_3 + 3H^+$	-0.5^a	
$Fe^{3+} + 4H_2O \rightleftharpoons Fe(OH)_4^- + 3H^+$	18.0	11

[a] Calculated from $pK_w = 14$, $pK_{sp} = 42.5$.

From these values it can be seen that the predominant equilibrium species for the hydrolysis of aluminum and iron (III) ions over the pH range of interest in water treatment, ie pH 5–9, are $Al(OH)_3$ and $Fe(OH)_3$ as shown in Figure 6 and 7 which express the concentration of the various species as a function of pH. These equilibrium conditions may obtain in the course of transit through the coagulation, flocculation, and sedimentation processes in treatment plants. It has been shown by Singley and Sullivan (17,18) that the reactive species may not be those that predominate under equilibrium conditions but less hydrolyzed species that are formed in the first few seconds or minutes of reaction with water. These species are positively charged and would effectively interact with the negative colloidal impurities.

The metal salts react with the alkalinity present in water, thereby reducing it, so that it may be necessary to add alkalinity in the form of lime or soda ash. One part of technical aluminum sulfate, $Al_2(SO_4)_3.14H_2O$ will reduce the alkalinity, as $CaCO_3$ by 0.55 parts and one part of technical ferric sulfate, $Fe_2(SO_4)_3.H_2O$ by 0.68 parts. The reaction may be shown as:

$$Fe^{3+} + 3HCO_3^- \rightleftharpoons Fe(OH)_3 + 3CO_2 \qquad (7)$$

The selection of a particular metal salt is based on such factors as local availability, convenience, economics, and effectiveness for the specific treatment problem. Those commercially available are listed below.

Aluminum sulfate, commercially $Al_2(SO_4)_3.14H_2O$, available as either a granular solid containing 17% Al_2O_3 or a solution containing 8.3% Al_2O_3. This is commonly referred to as alum or filter alum. The higher cost of transporting the solution may be more than compensated by the ease in handling and savings in labor and equipment required for feeding and dissolving the solid (19).

Ferric sulfate, commercially $Fe_2(SO_4)_3.H_2O$, containing a minimum of 20% iron (III). It is available only as a solid which must be dissolved immediately before use. The solution must be kept concentrated to avoid premature hydrolysis and precipitation of $Fe(OH)_3$. Such concentrated solutions have low pH values that prevent

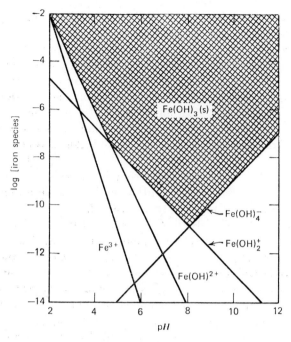

Fig. 6. Equilibrium solubility domain for $Fe(OH)_3$. The cross-hatching defines the area of stability of solid $Fe(OH)_3$.

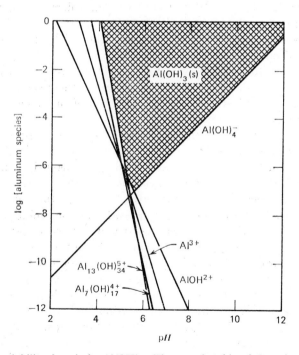

Fig. 7. Equilibrium solubility domain for $Al(OH)_3$. The cross-hatching defines the area of stability of solid $Al(OH)_3$.

hydrolysis but lead to very corrosive conditions. Containers must, therefore, be coated with or be constructed of corrosion proof materials.

Ferrous sulfate, commercially $FeSO_4.7H_2O$, referred to as "copperas." This may be the least expensive iron salt, particularly in localities where it may be available as a by-product of the manufacture of titanium dioxide or of the pickling of steel (see Vol. 12, p. 41). Its high solubility, as the hydroxide, about 4 mg/l as iron, precludes its use as a coagulant, necessitating oxidation to the $+3$ state. This can be accomplished readily at pH values above six by oxygen or chlorine. Each part of $FeSO_4.7H_2O$ stoichiometrically requires 0.03 parts of oxygen or 0.126 parts of chlorine for oxidation to the ferric state.

Sodium aluminate, $NaAlO_2$ normally contains an excess of sodium hydroxide or soda ash to maintain a sufficiently high pH to prevent precipitation of $Al(OH)_3$ in the solution prior to the addition to the water as a coagulant.

The organic coagulants in use today may be either natural or synthetic. Among the useful natural polymers are starches, gums, and gelatin. There are four classes of organic polymeric coagulants; cationic, anionic, ampholytic (these are frequently referred to as polyelectrolytes), and nonionic. The cationic class comprises quaternary ammonium compounds, such as polyvinylpyridinium butyl chloride; sodium polyethacrylate is an example of an anionic polyelectrolyte; poly(lysine–glutamic acid), having both NH_3^+ and COO^- groups, is an example of an ampholytic polyelectrolyte.

Nonionic polymers such as polyethylene oxide have no ionizable sites but do have sites that may be acidic or basic by distortion of the electronic distribution. Polyelectrolytes can act as charge neutralizing agents for oppositely charged species. However, only cationic polyelectrolytes are of significance because of the dominance of negatively charged natural colloids. It has been amply demonstrated that flocculation of colloids by polymers is predominantly due to adsorption of the polymer at the interface between the colloid and the solution followed by polymer–polymer cross-linking (20,21) to build a three-dimensional floc particle. The apparent anomaly of coagulation of a negative colloid by an anionic polyelectrolyte or a nonionic polymer may be explained on the basis of reduced free energy of the system as a result of hydrogen bonding or Van der Waals forces between the colloid and the polymer chain despite the obvious reduction in entropy.

The removal of turbidity by the coagulation–flocculation–sedimentation–filtration process is the principal chemical treatment, other than chlorination, used in the United States, since most surface waters contain colloidally suspended solids which cannot be removed by filtration alone. In a *conventional* or *horizontal plant* the coagulant is usually added in a rapid mix tank or is mixed by passage through a baffled chamber. The water, with its very fine microfloc, is passed into a flocculation basin where it is gently agitated to bring the microflocs into contact to build larger flocs that are large enough to settle during the transit of the water through the next unit, a settling basin. The transit time through the settling basin may vary from 4 to 12 hours, depending upon the design and the demand. The settled water then passes to the filters. Prior to filtration the pH may be adjusted and the water may be chlorinated. After filtration the water may be postchlorinated before distribution. In contrast, *up flow, solids-contact units* contain all of the coagulation–flocculation–sedimentation steps in a single tank. The residence time in such a unit varies from 2–4 hours. The greatly reduced time is due to the passage of the water up through a blanket of sludge thereby increasing the possibility of attachment of the microflocs to the heavier, previously precipitated floc. The coagulant is added to the raw water at the bottom of the vertical unit, flocculation may be accomplished in a section prior to passage up through the sludge. After overflowing the unit, it is treated the same as effluent from a horizontal unit. A typical up flow unit is designed for a rate of 1.5–2.5 gal/(min) (ft²). A unit about 10 ft deep and 32 ft in diameter would produce about 1.5 million gal per day, which corresponds roughly to the requirements for a city of 10,000 inhabitants.

The optimum pH range for the use of alum as the coagulant for turbidity removal is from 7–9. The dosage required varies, depending on the concentration of the colloid, from 5 to 50 mg/l. The optimum pH range for the use of ferric salts is from 6–10, and the concentrations are similar to those used for alum. The polymers are essentially insensitive to pH except that it may affect the structure and the charge on the polymer. The dosage range for the polymers may be as low as 0.05–0.1 mg/l.

Color may be removed effectively and economically using either alum or ferric sulfate. The optimum pH range is much lower than for turbidity removal. For alum it is from 5 to 6 and for ferric sulfate from 3 to 4. The reaction has been shown to be stoichiometric (22) and has been explained as a specific reaction of the coagulant with the color to form an insoluble compound. The dosage required may be as high as 100–150 mg/l.

Raw water colors may be as high as 450–500 units on the APHA color scale, (see Vol. 5, p. 809). The recommended maximum color in the finished water is 15 units, although most municipal treatment plants seldom exceed 5 units.

The same types of treatment units are utilized as in turbidity removal. One feature that must be remembered is that the pH must be increased prior to filtration so that the metal hydroxides are removed by the filters. At low pH values metal ions or their soluble complexes readily pass through the filters, and cause trouble by forming insoluble species in storage tanks and in the distribution system. For iron salts it is important that the pH be greater than 6 as the oxidation of iron (II) to iron (III) occurs rapidly above this pH in the presence of dissolved oxygen.

Softening

A water may be classified as hard it if contains more than 120 ppm of divalent ions, usually calcium and magnesium, expressed as $CaCO_3$. See Hardness under Water Analysis.

The disadvantage of hard water is not only the cost of the soap wasted and the inconvenience of lathering. The calcium and magnesium salts that cause hardness have negative solubility coefficients, and tend to precipitate when the temperature is increased. A deposit of scale forms on heat transfer surfaces, leading to localized overheating.

The American Water Works Association Quality Water Goals recommend a maximum total hardness of 80 ppm for municipal purposes. Municipal plants, though, distribute waters ranging from 70 to 120 ppm, the final quality being determined by considering such factors as public demand and economics. The hardness of waters provided in the 100 largest U.S. cities in 1960, treated or untreated, vary from 0 ppm to 738 ppm with a median value of 90 ppm. Only 13 had waters with a hardness exceeding 180 ppm. Twenty-eight of these cities soften their water supply.

There are two principal methods of softening water for municipal purposes; lime or lime–soda, and ion exchange. The choice between the two is dependent upon such factors as the raw water quality, the local cost of the softening chemicals, and disposal of waste streams.

The first softening plant in North America was in Winnipeg, Canada, in 1901, but England had over 50 plants by 1900. The first plant in the United States was built in Oberlin, Ohio in 1903 and was followed by other plants in Ohio and Florida. The early plants used the lime softening process with fill and draw units. Later, continuous treatment units were developed which greatly increased the amount of water that could be treated in a facility of given size. Today more than 1000 municipalities soften water; the majority of these are in the Midwest and in Florida.

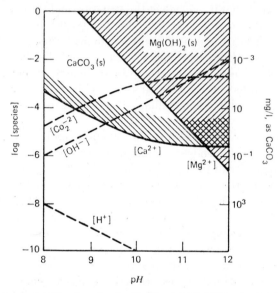

Fig. 8. Equilibrium solubility domains for $CaCO_3$ and $Mg(OH)_2$ at a total carbonic species concentration of $5 \times 10^{-3} M$. The shade areas above the Mg^{2+} and Ca^{2+} curves define the areas of stability of solid $Mg(OH)_2$ and $CaCO_3$, respectively.

The lime or lime–soda process is based upon precipitation of calcium as calcium carbonate and magnesium as magnesium hydroxide. The lower limits of softening by this process, then, are based on the solubility of these compounds. The solubilities are shown in Figure 8 as functions of pH. When lime, alone, is used only the carbonate hardness is reduced, (the carbonate hardness being that present as calcium or magnesium bicarbonate.) The additional use of soda ash can reduce the noncarbonate hardness by providing additional carbonate ion. This is shown by the reactions that cover the various steps in the process:

1. The reactions of any free CO_2 with the added lime.

$$CO_2 + Ca(OH)_2 \rightleftharpoons CaCO_3 + H_2O \tag{8}$$

This reaction provides no softening but takes place preferentially since CO_2 is the strongest acid present in the system.

2. The reaction of calcium carbonate hardness with lime, when the calcium may be represented as the bicarbonate at the pH values normally occurring.

$$Ca(HCO_3)_2 + Ca(OH)_2 \rightleftharpoons 2CaCO_3 + 2H_2O \tag{9}$$

3. The reaction of magnesium carbonate hardness with lime.

$$Mg(HCO_3)_2 + 2Ca(OH)_2 \rightleftharpoons 2CaCO_3 + Mg(OH)_2 + 2H_2O \tag{10}$$

4. The reaction of calcium noncarbonate hardness with soda ash. It is necessary to use soda ash to provide the CO_3^{2-} as the available CO_3^{2-} would be used up in the reactions represented by reactions (9) and (10). The noncarbonate hardness may be represented as sulfate, although any anion except carbonate or bicarbonate could be present.

$$CaSO_4 + Na_2CO_3 \rightleftharpoons CaCO_3 \rightarrow CaCO_3 + Na_2SO_4 \tag{11}$$

5. The reaction of magnesium noncarbonate hardness with lime and with soda ash. This is a two step reaction since reaction (12) produces a reasonably soluble calcium salt that must be reacted with CO_3^{2-} in order to precipitate the calcium.

$$MgSO_4 + Ca(OH)_2 \rightleftharpoons Mg(OH)_2 + CaSO_4 \tag{12}$$

$$CaSO_4 + Na_2CO_3 \rightleftharpoons CaCO_3 + Na_2SO_4 \tag{13}$$

It is to be noted that reactions (11) and (13) are the same but represent two different sources of the $CaSO_4$.

From these reactions it can be seen that the addition of lime always serves three purposes and may serve a fourth. It removes, in order, CO_2, calcium carbonate hardness, and magnesium carbonate hardness, reactions 8, 9, and 10, where magnesium noncarbonate hardness must be removed the lime converts it to calcium noncarbonate hardness, reaction 12. Soda ash, then, removes noncarbonate hardness according to reaction 11 or 13.

The quantities of lime and soda ash required can be calculated from the stoichiometry of the reactions. The effluent quality is a function of the solubilities of calcium carbonate and magnesium hydroxide and of the quantities of softening chemicals added. The acceptable level of total hardness can be decided upon; it normally ranges from 70 to 120 mg/1, expressed as $CaCO_3$. The sum of the solubilities of calcium carbonate and magnesium hydroxide is about 50–70 mg/1, depending upon the pH. The sum of the concentrations of the carbonic species, $HCO_3^- + CO_3^{2-}$, expressed as $CaCO_3$ or carbonate hardness, can be as low as 20 mg/1 but is usually in the range of 30–50 mg/l because of $CaCO_3$ supersaturation. It is desirable to reduce the carbonate hardness to the lowest practical value as this allows the maximum concentration of noncarbonate hardness to remain. This decreases the cost of treatment, as soda ash costs $50–60/ton compared to $15–20/ton for lime. Additionally, it requires 106/56 as much soda ash as lime to remove the same amount of hardness. A further factor in the economics of lime–soda softening is that the amount of magnesium to be left in is maximized, since it requires twice as much lime to remove magnesium as it does to remove calcium. Unfortunately, all of the hardness left in cannot be present as magnesium since magnesium hardness above 40 ppm causes scaling problems in hot water heaters at 140°F, the normal temperature setting. For this reason the magnesium concentration is usually reduced to this level.

Since the effluent from a softening unit is usually supersaturated with calcium carbonate at the high pH values normally resulting, it is necessary to reduce the pH to a value that allows the solution to be exactly saturated for the calcium ion and carbonate ion concentrations present. The equation controlling the relationship is;

$$CaCO_3(s) + H^+ \rightleftharpoons Ca^{2+} + HCO_3^- \tag{14}$$

The equilibrium expression is;

$$K = \frac{(Ca^{2+})\ (HCO_3^-)}{(H^+)} = K_{sp}/K_2 \qquad (15)$$

or

$$pH_s = pK_2 - pK_{sp} + p_{Ca} + p_{Alk} \qquad (16)$$

where pHs is the pH of saturation, pK_2 is for the second ionization step for carbonic acid, pK_{sp} is for the solubility of $CaCO_3$, p_{Ca} is the negative logarithm of the molar concentration of calcium ion and p_{Alk} is the negative logarithm of the bicarbonate ion concentration, which is measured analytically as the alkalinity, the total titratable bases present.

The reduced pH is usually attained by recarbonation, the addition of CO_2. The CO_2 is obtained from the combustion of fuel oil or from tanks of liquid CO_2.

Modifications of the basic process are undersoftening, split recarbonation, and split treatment. In *undersoftening* the pH is raised to between 8.5 and 8.7 to remove only calcium. No recarbonation is required. In *split recarbonation* the treatment is divided between two units in series. In the first, or primary, unit the required lime and soda ash are added, the water settled and recarbonated just to a pH of about 10.3, which is the minimum pH at which the carbonic species are present principally as the carbonate ion. The primary effluent then enters the second, or secondary unit, where it contacts recycled sludge from the secondary unit to precipitate almost pure calcium carbonate. The effluent is settled, recarbonated to the pH of saturation and filtered. The advantages over conventional treatment are reductions in lime, soda ash, and CO_2 requirements, very low alkalinities and reduced maintenance costs because of the stability of the effluent. The major disadvantages are the necessity for very careful pH control and the requirement for twice the normal plant capacity.

The principal use for *split treatment* is in cases where the magnesium content of the water is high. This process also requires two units in series, which doubles the size of the plant. The raw water stream is split so that the ratio of the fraction treated in the primary unit to the fraction treated in the secondary unit is such that when all of the magnesium is removed in the primary unit and none in the secondary, the mixed effluents will contain 10 ppm magnesium, as magnesium, or 41 ppm as $CaCO_3$. For example a raw water containing 40 ppm magnesium would treat 30/40 of the water in the primary unit and 10/40 in the secondary unit. Sufficient lime is added ahead of the primary unit to remove all of the magnesium, or at least to less than 1 ppm. This requires a pH of 11.1–11.3 with an excess of about 70 ppm of caustic alkalinity. The effluent from the primary unit is mixed with the bypassed raw water in the secondary unit with no further chemical addition. The excess lime in the primary effluent is used to remove the calcium hardness from the bypassed raw water. This process has the advantage of using the excess lime required for magnesium removal to soften raw water whereas in conventional treatment this excess must be wasted by recarbonation. Another advantage is the very low alkalinity of the finished water, which may be as low as 18–20 mg/l. This, of course, saves soda ash.

One of the major problems associated with lime or lime–soda softening is the disposal of the sludge produced. Depending upon the ratio of calcium to magnesium removed and upon the amount of soda ash used, the sludge produced is from 2.8 to 3.6 times the weight of the lime added. The principal methods of disposal have been lagooning, discharge into the nearest water course, or discharge to the sanitary sewer system. The later two are no longer acceptable methods of disposal because of the resultant pollution load.

A method of disposal for large plants is *recalcination*, ie regeneration of lime from the $CaCO_3$ by heating in a kiln to drive off CO_2. The lime can be reused in the plant, the CO_2 used for recarbonation and the excess lime sold. In cases where $Mg(OH)_2$ has been precipitated along with the $CaCO_3$, it is necessary to remove the magnesium prior to recalcination. This is accomplished by selectively dissolving the magnesium at a pH value of about 10.3. This is attained by the use of CO_2 from the kiln. At lower pH values $CaCO_3$ will dissolve and at higher values magnesium will remain in the sludge.

Ion Exchange

For waters containing high noncarbonate hardness or high magnesium, or both, the use of lime or lime–soda may be very expensive. In such cases ion exchange softening may be more economical, particularly if brines are locally available at little cost for regeneration of the resins.

Ion exchange softening involves exchanging sodium or hydrogen ions for the calcium or magnesium ions in the raw water. The exchange reactions are equilibria that favor the formation of polyvalent ion-resin complexes over monovalent ion-resin complexes. The raw water to be softened is passed through the resins that are initially saturated with sodium or hydrogen ions. The exchange is referred to as the exhaustion reaction and may be shown as $Ca^+ + 2NaR \rightleftarrows CaR_2 + 2Na^+$ (R represents the resin matrix). The regeneration step is simply the reverse of the exhaustion step with the concentration of the regenerant ion increased to 5–25% in order to reverse the reactions. For sodium ion regeneration sodium chloride is used and for hydrogen ion regeneration either sulfuric or hydrochloric acid. The theoretical requirement for 98% NaCl to regenerate the resin is 1.0 lb for each 400 g of hardness removed, the hardness expressed as $CaCO_3$. The actual requirements vary from 1.2 to 2.4 lb depending upon the specific resin used and the regeneration conditions. The water passing through the exchanger is reduced to 0–2 parts hardness. Since this is usually much lower than the final acceptable hardness, only a portion of the water is passed through the exchanger. Enough is by passed to give the desired final hardness.

The resins used are highly crosslinked organic polymers having acidic functional groups. The most common of the currently used resins are sulfonated copolymers of styrene and divinylbenzene.

Several modifications of the ion exchange process are possible in order to handle special cases. For example a water with high total solids can be treated partially by sodium cycle exchange followed by aeration to remove CO_2 and consequently part of the anions. It should be kept in mind that sodium cycle exchange increases the soluble solids content as two sodiums (atomic weight $2 \times 23 = 46$) are substituted for one calcium (40) or magnesium (24.3) Hydrogen cycle exchange replaces one calcium or one magnesium with two hydrogens and thus reduces the total solids. The disadvantage is the reduction in pH. The resultant water is very corrosive and must have the pH raised before distribution.

Another possible modification is the use of sea water as the regenerant. Even though it has only 2.7% NaCl and contains calcium and magnesium it sometimes can be purified by coagulation, filtration, and chlorination more cheaply than salt can be purchased. The lower concentration reduces the regeneration efficiency by 40–50%.

Two types of installations are in use today, fixed bed and continuous-regeneration units. The continuous units consist of a closed circuit containing two sections, one for softening and one for regeneration with the resin circulated countercurrent to the raw-water flow through the softening tank. The resin then is pulsed periodically into the regeneration tank where it is regenerated and rinsed, then pulsed back to the softening tank. The advantage is that there is continuous softening without the down time customary for fixed bed units.

TASTE AND ODOR CONTROL

Tastes and odors may be present in surface waters due to the action of biological organisms such as algae or may result from pollution by industry, domestic sewage, or agriculture. Ground waters may have taste and odor if they are polluted or if they contain gases such as H_2S or CH_4, which always contains associated impurities which have taste and odor. The removal of these may be accomplished by adsorption with activated carbon, oxidation with chlorine, potassium permanganate or ozone, or by aeration.

Organic materials are generally removed by addition of powdered activated carbon. The carbon may be added at any point in the plant although it is advantageous to have as much contact as possible. The adsorption reaction is relatively slow at room temperature since it is diffusion controlled. Oxidation with chlorine, $KMnO_4$ or ozone may destroy tastes and odors or it may intensify them depending upon the particular compounds involved. For example, chlorination of phenolic compounds leads to greatly increased tastes and odors. For this reason the system must be studied in the laboratory prior to plant runs.

Hydrogen sulfide and methane may be removed by aeration although the major reduction in hydrogen sulfide may result from oxidation by the dissolved oxygen introduced during the aeration. At low pH values the product is sulfate whereas at high pH values the product is free sulfur. Organic pollutants that may not manifest themselves as tastes or odors and yet may be undesirable, can be removed by activated carbon treatment. In cases of high concentrations or toxic compounds it is necessary to add the carbon at several places in the plant, in order to have a large excess of fresh carbon at all times.

IRON AND MANGANESE

Ground waters or water withdrawn from the depths of reservoirs may contain iron and manganese in the 2+, or reduced, state in which they are soluble. Either one in equilibrium with dissolved oxygen would exist in an oxidized state, which would be insoluble, $Fe(OH)_3$ for iron and MnO_2 for manganese. The disadvantage of allowing the reduced metal ions to remain in the finished water is that when it comes into contact with the atmosphere, the oxidized state precipitates upon domestic fixtures or clothes yielding a reddish-brown stain with iron and a dark brown to black stain with maganese.

Both ions can be removed by oxidation and subsequent filtration. Aeration is adequate for iron(II) oxidation at pH values above 6, but the oxidation of manganese (II) is much too slow, even at higher pH values, for effective removal. Potassium permanganate or chlorine dioxide are frequently employed for the oxidation of manganese but must be followed by coagulation prior to filtration because of the formation of colloidal MnO_2.

FLUORIDATION

The practice of adding fluoride ion to domestic water supplies has been increasing since the first experiments in the city of Grand Rapids, Michigan in 1945. It has been shown by Churchill (23) that dental decay was reduced significantly by maintaining a fluoride residual of about 1 mg/l in the public water supplies. The concentration recommended by the U.S. Public Health Service varies from 0.6 to 1.7 depending upon the annual average maximum daily air temperature (3). (The temperature affects the amount of water ingested).

The early programs were designed as demonstration studies and were followed closely by periodic dental examinations. When the Public Health Service formally endorsed fluoridation in 1950 many cities started the practice which has grown to the point that in 1969 there were over 2550 water systems serving fluoridated water to over 4500 communities and 84 million people in the U.S. There were over 7 million people in Canada and over 38 million people in 30 other countries with fluoridated supplies (24).

The sources of the fluoride ion are varied but the most common are sodium fluoride; sodium fluorosilicate, Na_2SiF_6; fluorosilicic acid H_2SiF_6; and ground fluorspar, CaF_2. The solids, NaF, Na_2SiF_6 and CaF_2 are fed with accurate dry feeders. Sodium fluoride can also be fed from a constant strength saturated solution containing about 4% NaF. Solution of H_2SiF_6 are also fed with liquid feeders. The accuracy of such feeders has been estimated to be about 99% (24). Careful control is possible because of the development of a specific ion electrode for F^- analysis.

The first state to require fluoridation by law was Connecticut in 1965. This has been followed by Michigan, Delaware, Illinois, South Dakota and Ohio. Several other states have such laws before their legislatures (1970) and the countries of Australia, Brazil, Chile, and Ireland have compulsory fluoridation. The practice has been approved and recommended by the American Dental Association, the American Medical Association, the U.S. Public Health Service, and by the American Water Works Association.

Bibliography

"Water (Municipal)" in *ECT* 1st ed., Vol. 14, pp. 946–962, by H. O. Halvorson, University of Illinois.

1. C. R. Murray, *J. Am. Water Works Assoc.* **61**, 567 (1969).
2. F. P. Linaweaver and J. B. Wolff, *Report, 1, Phase two, Residential Water Use*, Johns Hopkins University (1964).
3. *Public Health Ser. Pub. No. 956, Public Health Service Drinking Water Standards* (1962).
4. *J. Am. Water Works Assoc.* **60**, 1317 (1968).
5. C. N. Dufor and E. J. Becker, *J. Am. Water Works Assoc.* **56**, 23 (1964).
6. S. T. Powell, *Water Conditioning for Industry*, McGraw-Hill Book Co., New York (1954).
7. T. R. Camp, *J. Am. Water Works Assoc.* **53**, 1478 (1961).
8. W. R. Conley, *J. Am. Water Works Assoc.* **53**, 1473 (1961).
9. U.S. Pat. 293,740 (1884), I. S. Hyatt.
10. G. Biederman quoted by K. Schlyter, *Trans. Roy. Inst. Technol. Stockholm* 196 (1962).
11. H. Lengweiler, W. Buser, and W. Feitknecht, *Helv. Chim. Acta* **44**, 805 (1961).
12. C. R. O'Melia and W. Stumm, *J. Am. Water Works Assoc.* **59**, 1393 (1967).
13. *J. Am. Water Works Assoc.* **60**, 1317 (1968).
14. V. K. LeMar and T. W. Healy, *J. Phys. Chem.* **67**, 2417 (1963).
15. H. Kubota, Thesis, Univ. of Wisconsin (1956).
16. Z. G. Szabo, L. J. Czanyi, and M. Kaval, *Z. Anal. Chem.* **147**, 401 (1955).
17. J. E. Singley and A. P. Black, *J. Am. Water Works Assoc.* **59**, 1549 (1967).

18. J. H. Sullivan and J. E. Singley, *J. Am. Water Works Assoc.* **60,** 1280 (1968).
19. R. W. Ockershavsen, *J. Am. Water Works Assoc.* **57,** 309 (1965).
20. A. P. Black and M. R. Vilaret, *J. Am. Water Works Assoc.* **61,** 209 (1969).
21. F. B. Birkner and J. J. Morgan, *J. Am. Water Works Assoc.* **60,** 175 (1968).
22. A. P. Black, J. E. Singley, G. P. Whittle, and J. S. Maulding, *J. Am. Water Works Assoc.* **55,** 1347 (1963).
23. H. V. Churchill, *J. Am. Water Works Assoc.* **29,** 1399 (1931).
24. F. J. Maier, *J. Am. Water Works Assoc.* **62,** 3 (1970).

J. E. SINGLEY
University of Florida

SEWAGE

Sewage is the spent water supply of a community. Because of infiltration of ground water into loose sewer pipes, often the quantity of waste water treated is greater than the water initially consumed. Sewage is about 99.95% water, and about .05% waste material.

The greater the per capita use of water, the weaker (more dilute) will be the sewage. Industrial wastes will affect the strength of the sewage. Sewage flow, as can be expected, will be greater during the daylight hours. However, a larger city will exhibit less variation about a mean flow and many smaller communities in the late night hours will have a sewage that is almost all ground water.

The per capita production of sewage varies from less than 100 gal/day for a strictly residential community to around 300 gal/day for a highly industrialized area. Often the term "population equivalent" is used in evaluating industrial contributions to sewage. This is then applied to planning for hydraulic and BOD (biochemical oxygen demand) loadings.

Waste-water-treatment installations have high capital costs due to the need for large tanks, pumps, and grounds. If a community has combined storm and sanitary sewers it is often found necessary to bypass the flow during periods of heavy rains.

Strength of sewage is expressed in terms of the following parameters: turbidity, total solids, settlable solids, suspended solids, fixed solids, volatile solids, filterable solids, and BOD. See Water analysis.

Sewage Treatment

Why treat sewage?

Sewage-purification works were originally constructed for reasons based primarily on concepts of public health. However, to answer the above question on the grounds of prevention of disease is to give an incomplete answer. Protection of the oxygen resources of the receiving waters touches more on present thinking concerning the problem. Population is increasing and the demand for water is growing. The supply of fresh water that can be economically obtained is limited. (This is, of course, without recourse to obtaining water from the sea.) In inland areas river water may be used and reused twenty or more times before it reaches the sea. Lakes are essentially a closed system. With increased leisure time the question of water pollution has gone beyond the public-health aspects and the question of esthetics has entered the picture.

Private and Rural Disposal Systems. Disposal of human and kitchen wastes in areas not served by sewers and waste-water-treatment plants presents a special problem. Apart from entirely primitive privies, cesspools and septic tanks are possible.

Cesspools are simply pits into which waste is allowed to flow. The term leaching pit is sometimes used. Water seeps into the ground, leaving solid matter in the pit. Construction is of two types. A pit may be unlined, or it may be lined with sewer pipe laid on end. Almost nowhere in the United States are cesspools permitted by health authorities.

Septic tanks are widely used in smaller towns and outlying suburbs of larger cities. The tank is kept at all times entirely full of the waste, so that anerobic conditions prevail, in contrast to a cesspool; thus, it is a combination of sedimentation tank and anerobic digester. Sanitary and kitchen wastes flow into the tank and grease and light material rise to the top. Heavier particles settle to the bottom where anerobic stabilization occurs. Deflector plates are provided at inlet and outlet in order to minimize short circuiting. Effluent flows to a tile field where disposal is into the earth. The tile field is composed of perforated field tile fed by a manifold. The tile is underlain with granular material, usually gravel. Care must be taken that the earth does not become clogged by material carried over from the septic tank. Septic tanks are being replaced as more and more areas are served by municipal systems. Health authorities do not look with favor on septic tanks.

The capacity of a septic tank is a function of the number of persons or units to be served. In no case should the capacity be less than 1500 gal. Periodically it is necessary to employ a scavenger service for emptying the tank of accumulated solids. Solids thus collected may be discharged to a convenient collection system or directly to a waste-water-treatment plant.

With both cesspools and septic tanks, the effluent discharges in the immediate vicinity. If adequate controls are not exercised a closed system may result. An example of this took place in Suffolk County (Long Island), New York. Septic tanks were widely used, and there was strong local opposition to the considerable expense of installing sewers. Effluent from the septic tanks found its way into the ground water which is the supply for much of the county. Eventually, the problem was graphically pointed up by the appearance of foaming detergents in water issuing from the tap.

The Imhoff tank is similar in many respects to the septic tank. Its use is generally confined to small communities and isolated installations. Operation is a combination of sedimentation and anerobic digestion. The tank is composed of two chambers, one above the other. The surface configuration may be circular, square, or rectangular; the depth is 25–35 ft. Sewage flows through the upper chamber at a low velocity (ca 1 ft/sec.). Solids settle out and slide through a slot into the bottom chamber. Detention period is about 2 hr. The solids accumulating in the bottom, or digestion, chamber have an initial water content of 85–95%. After proper digestion of about 60 days the water content is reduced to about 50% and the volume is greatly reduced. Gases produced during digestion are vented to the atmosphere by gas vents located at the tank sides. Solids buoyed up by gas are prevented from escaping to the upper tank by deflector plates.

Community and City Disposal Systems

The above described expedients are totally insufficient for anything larger than a very small rural community.

Outline of Treatments. Figure 1 gives a very generalized picture of the arrangements necessary for treating sewage. The *primary* treatment consists of any operations, such as screening or sedimentation, that remove particles above colloidal size. It also

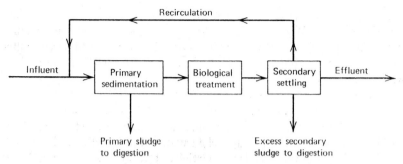

Fig. 1. Generalized diagram of sewage treatment steps.

usually removes some 30–60% of the BOD. Removal of colloidal or dissolved materials, and further reduction of the BOD, is accomplished by *secondary* treatment, which is biological, by encouraging the growth of microorganisms that utilize waste material in the sewage as food. One commonly used definition of *tertiary* treatment is any treatment in addition to secondary treatment. See Water reuse.

PRIMARY TREATMENT

Collection System. It is desirable, whenever possible, for the sewage to be collected by gravity flow. However, when the topography does not permit this, pumping is necessary. Flow velocities in the pipe must be maintained above a min of 2 ft/sec in order that solids do not settle out in the pipe. Low flow velocities, long detention times, and relatively high temperatures have caused treatment difficulties at the Hyperion plant in California and at many other places. It is necessary to have a means of by-passing the plant during periods of flow that exceed the hydraulic capacity of the plant.

Generally multiple units are provided for each stage of treatment. Thus, during periods of routine maintenance or repairs, it is not necessary to by-pass this treatment stage.

Protection must be given to pumps against large objects (wood, etc) in sewage. Coarse racks have clear openings greater than 2 in. Racks to be placed in advance of grit chambers and settling tanks ought to have clear openings 1–2 in. wide. These are cleaned by hand. A disadvantage is that loss of hydraulic head becomes excessive unless they are frequently cleaned. Disposal of rakings is usually by burial or incineration, digestion with sludge, or composting.

Mechanically cleaned racks allow smaller clear openings because head loss does not become so high. Mechanical cleaning can be either continuous or intermittent. Intermittent cleaning is cycled by float-operated switches controlled by a float in the influent channel.

Comminutors macerate floating material into sizes sufficiently small ($\frac{3}{16}$–$\frac{3}{8}$ in.) so that particles will not clog centrifugal pumps. Comminutors have almost completely replaced racks and screens with small clear openings.

Grit Chambers. These are sometimes placed ahead of sedimentation chambers, to remove grit without removing finer sediment. There are two reasons for the necessity of grit removal. The first is to provide against excessive wear in pumps. The second is the need for protection against loss of volumetric capacity of sludge digesters. It has been found that large digesters in plants serving low-lying sandy areas can lose as much as one third of their capacity in just a few years.

Grit chambers must operate in a fairly narrow velocity range. It has been found that at velocities above 1.25 ft/sec grit is scoured from the chamber bottom. Velocities lower than 0.75 ft/sec allow deposition of organic matter, called detritus, which is undesirable because the mixture is putrescible and unsuitable for such uses as fill.

Usually the grit chamber is designed to remove particles with a diameter of 0.02 cm and a specific gravity of 2.65. Grit chambers are frequently shaped to approximate a parabolic section because the velocity variation with varying flows follows a parabolic function. The amount of grit collected per million gal of flow varies from 1 to 12 ft^3. In smaller plants grit is periodically removed manually, while an automatic conveyor system is used in larger plants.

Sedimentation. In the sedimentation tanks (settling tanks) the smaller solids settle, and oil and grease, which are lighter than water, float and can be skimmed and taken to the sludge digester. Many types of settling and skimming tanks have been constructed. Depths vary from 7 to 15 ft, 10 ft being a commonly used value. Circular and rectangular tanks are used. The bottom is sloped at about 1% in rectangular tanks and 8% in circular tanks to facilitate removal of the settled sludge.

Design is usually on the basis of hydraulic loading, 1000 gal/(ft^2)(day) being a commonly used value for primary settling tanks, and 600–800 gal/(ft^2)(day) for humus tanks.

SECONDARY TREATMENT

There are two main processes utilized for biological secondary treatment of waste water. They are the *trickling filter,* and the *activated sludge* process.

Present-day biological treatment methods are a logical development from *sewage farms* (irrigation areas) to *intermittent sand filters* to *contact (fill-and-draw) beds.* Numerous modifications of the basic processes have evolved over the years, but the underlying principles remain unchanged. The basis is the formation of a suitable environment so that microorganisms may thrive under controlled conditions. The microorganisms come from the sewage itself. The suitable environment is one which is rich in food and maintained in an aerobic state. It has been truthfully said that a sewage treatment plant is a river in miniature.

Irrigation by sewage (*sewage farms*) provides water for irrigation and some treatment of the sewage. This means of sewage disposal conflicts with sound public-health practice, and cannot be used where the applied sewage will be brought into contact with food plants. It may be used for orchards, but one notable case involving olives has made spray irrigation by sewage a doubtful means of disposal. Odors cannot be avoided and treatment is not of a high degree. Removals decline markedly in cold weather. Irrigation by sewage is most logically applied in arid agricultural regions.

Intermittent sand filters are much like slow sand filters used in potable water treatment. The sewage is applied to the sandy area and allowed to flow slowly downward. Raw sewage may be applied at rates as high as 80,000 gal/(acre)(day), and 800,000 gal/(acre)(day) after biological treatment. The latter would be considered tertiary treatment.

Surface accumulations of solids must be periodically removed. This method is not recommended in areas underlain by limestone due to the danger of pollution of ground water. Biological films that form on the sand grains undergo continuous stabilization. It is necessary to rest the beds between dosings so that objectionable conditions do not develop.

Fill-and-draw beds operate as the name indicates. A tank filled with coarse, granular material is filled with sewage and allowed to stand full. It is then drained and allowed to rest. Air is drawn into the bed during emptying. Loadings of these beds are about 200,000 gal/(acre)(day). This process is little used today in urban sewage treatment, but it does find application in treatment of some industrial wastes.

Trickling Filters. The trickling filter is not a filter. It can be described briefly as a pile of stones (or other coarse material) over which the sewage flows. This process is probably the most widely used aerobic biological treatment method.

Sewage is distributed by slowly rotating arms equipped with nozzles and deflectors, and allowed to flow slowly over the filter stones. Air is drawn into the filter by temperature differential, thus keeping a supply of oxygen for the process.

The filter medium, until recently, was stone, diameters of 1–4 in. being the common sizes used. These stones permitted sufficiently loose packing to allow free flow of water and air and left sufficient openings to prevent clogging by biological slimes. There appears to be a trend toward the replacement of stone by plastic filter media. Filter depths range from 3 to 14 ft, 6 ft being a commonly used figure.

Sewage flows slowly downward over the filter medium and the effluent is collected in vitrified tile underdrains. The underdrains serve two purposes: (1) collection of filter effluent and (2) circulation of air into the filter. The underdrains discharge into a main collection channel which in turn discharges to a final settling tank.

The importance of the final settling tank, or *humus tank*, can be seen by an examination of what occurs in the trickling filter itself. A new filter is "broken in" by applying sewage as in normal operation. After a period of time the microbial mass will form on the filter medium and does its work of waste stabilization. Waste material will first be adsorbed, and then assimilated by the microorganisms.

Much of the organic waste material has been transferred to the microbial mass and will be utilized for cell synthesis and energy. However, there must be constant removal of microorganisms from the zooglea mass or the whole filter would become sluggish. It is necessary, therefore, to provide for removal of sloughed-off microorganisms. This is done in the secondary settling tanks. A rate of application lower than for primary settling tanks is required; 600 gal/(ft²)(day) is a common figure.

Trickling filters may be classified on the basis of (1) hydraulic loading per unit area, and (2) pounds of BOD/(day)(1000 ft³) filter volume.

Low-Rate Trickling Filters. This filter is the most commonly used in small treatment plants. The loadings are 2–4 million gal per acre per day (mgad) and 10–20 lb BOD/(1000 ft³)(day). With proper operation, BOD removals of 80–85% can be expected. Usually appreciable nitrate and nitrite concentrations are found in the effluent, indicating a high degree of organic stabilization.

A portion of the effluent is recirculated, as shown in Figure 1. This is done in order to (1) smooth out flow, (2) keep the food concentration more constant, (3) lower film thickness and control psychoda flies, and (4) reseed the applied sewage with acclimatized microorganisms.

In (3) reference is made to the psychoda, or filter fly. This very small fly breeds in thick trickling filter slimes. This fly does not bite, but can be extremely troublesome by getting into the ears and mouth of humans. Its radius of flight is small, but it can be carried great distances by the wind. Control of psychoda fly in its development phase is achieved by occasionally flooding the filter, or else by chlorination of the applied sewage.

Intermediate-Rate Trickling Filters. In the range of 4–10 mgad certain operational difficulties were frequently encountered, and this range was avoided for many years. It now appears that these difficulties stemmed mostly from the inability of the hydraulic load to keep the filter slimes from becoming too thick and thereby clogging the filter. This difficulty seems to be overcome by use of relatively large filter stones.

High-Rate Trickling Filters. When the hydraulic loading of sewage was increased to 10 mgad it was found that higher BOD removals per unit filter volume were achieved. However, because of the increased organic loading there was more organic material in the effluent. By the use of high recirculation ratios it was possible to maintain the high hydraulic loading and at the same time reduce the organic concentration in the effluent. Hydraulic loadings are 10–40 mgad and up to 90 lb BOD/(1000 ft³)(day), but removal efficiency is only 65–75%.

Super-Rate Trickling Filters. Experimental plants using plastic media have recently achieved very high removal efficiencies (97%) at hydraulic loadings of 100 mgad. These plants have been found to have much of the microbial mass in the recirculated effluent and are, in effect, a modification of activated sludge. Organic loadings are around 100 lb BOD/(1000 ft³)(day).

Trickling Filter Modifications. Over the years many modifications of the trickling filter have been proposed. These have dealt with improvement in media, improved air circulation, and loadings. In one modification, sewage was introduced at various levels of a very deep filter. This was an endeavor to distribute the load more uniformly over the entire filter depth. Hydraulic loading was up to 500 mgad.

Activated Sludge Process. Here, in contrast to the trickling-filter process, the activated sludge floc is suspended in the moving stream. This process originated in attempts to purify sewage by blowing air into it. It was observed that after prolonged aeration of sewage in a tank, flocs composed of voraciously feeding organisms developed. When aeration was stopped this floc settled. Addition of fresh sewage to the tank containing the sludge produced high purification in a reasonable time. Thus, the name given the floc was activated sludge. As was the case with the trickling filter, the activated sludge was first operated as a fill-and-draw system. Further research showed that continuous operation could be practiced.

The process involves: (*1*) the return of some of the activated sludge to the aeration tank influent and discharge of excess sludge to digestion, (*2*) aeration of the sludge-sewage mixture to attain purification, and (*3*) settling of the aeration tank effluent to remove floc from the plant effluent.

Step (*3*) is necessary for the same reason as in the case of the trickling filter—removal of organic material transferred from the sewage to the cell mass.

The floc is formed in sewage by aerobic growth of unicellular and filamentous bacteria. Protozoa and other organisms are found in the floc matrix. These organisms gain food and energy by feeding upon the sewage.

Aeration tanks are normally rectangular in cross section, 10–15 ft deep and 30 ft wide, and the surface length-to-width ratio should be greater than 5 to 1 in order to avoid short circuiting. Detention periods in the aeration tanks are commonly 4–8 hr for a conventional unit.

Because this is a strictly aerobic process, air requirements are high, and are difficult to satisfy because oxygen is only slightly soluble in water (10 mg/liter).

Two aeration systems are used, diffused air units and mechanical aeration units. Air diffusers are more commonly used in the United States, but some mechanical aeration units may be found in plants having capacities less than 1 million gal/day.

Both means of aeration perform three functions: (*1*) transfer oxygen to the sewage and maintain anaerobic conditions, (*2*) cause intimate mixing of sewage and floc, and (*3*) keep the floc in suspension.

Air is introduced from diffusers in such a way as to set up a spiral flow pattern, thus aiding mixing of floc and sewage. This also helps eliminate dead spaces. It was found that the oxygen demand decreases as the sewage flows through the aeration tank. Therefore, the number of diffuser units is increased at the head of the tank and decreased at the end of the tank. This process, known as tapered aeration, is now the common method in arrangement of diffusers.

Mechanical aeration has the same function as air diffusers but is accomplished by rotating paddles or brushes. Peripheral velocity in both methods is about 2 ft/sec.

Returned Sludge. The floc returned to the aeration tank has the same function as trickling-filter slime. However, the floc concentration can be varied as operational needs dictate. The amount of sludge returned varies from about 10 to 30%. The volume of suspended matter in the aeration tank will vary between 10 and 25% Usually the mixed-liquor suspended solids in the aeration tank vary from 600 mg/liter to 4000 mg/liter. This is on a dry-weight basis.

A very important parameter in routine control of the process is the ratio of the volume of MLSS to the dry weight of MLSS, called the sludge volume index (SVI). A well-operating plant has an SVI of 50–100, but operational difficulties occur when the SVI approaches 200.

The activated sludge process is microbiological in nature, and factors which promote or inhibit growth are of importance. The most important of these are pH, temperature, and ORP (oxidation-reduction potential). The pH will determine what organisms predominate in the system. Bacteria predominate from pH 6.5 to 9.0. Below pH 6.5 fungi become more important. If metabolic products are acidic, there must be adequate buffering capacity.

Step Aeration. New York City has plants scattered throughout the five boroughs treating a total of more than 1 billion gal of sewage every day. Two major modifications have resulted from operational experimention in New York. In conventional plants sewage is introduced at one end of the aeration tank and flows straight through. This gives a high initial microbial food supply and corresponding high oxygen requirements. In step aeration sewage is fed at various points along the aeration tank. This smooths out the food applied and results in lower oxygen demand. This modification has been successfully applied to increase the capacity of existing overloaded plants.

High Rate Activated Sludge. A low mixed-liquor suspended-solids (MLSS) concentration of 200–500 mg/l is maintained. This gives a high food-to-microbial-mass ratio, which keeps the floc in an active growth phase, but excess food will be discharged in the effluent. Although the BOD removal is not high, there are areas in which this is adequate. New York City applied this method successfully because of a weak sewage and low temperatures. Philadelphia and Los Angeles met with indifferent results because of stronger sewage or higher temperatures.

Biosorption. This process makes use of a phenomenon observed and ignored by many investigators. If activated sludge and raw sewage are mixed together in an aeration vessel there is noted a progressive reduction of BOD, then a rise, followed by another decrease. The first decrease had been attributed to experimental error. It was found that the dramatic reduction was due to adsorption of colloidal material onto the activated sludge floc. The plants at Austin, Texas and Bergen County, New

Jersey were converted from overloaded to underloaded by changing to the biosorption process.

SLUDGE DIGESTION

Putrescible material collected from the primary settling tanks and excess sludge from humus tanks must be disposed of cheaply and efficiently. This material is highly putrescible and a potential nuisance source. Because it is putrescible it can be stabilized by biological means, serving as food and energy sources for anaerobic microorganisms naturally found in the sludge. Raw sludge is about 95% water, but the water is not easily removed. As the sludge is broken down the water content is lessened, and the volume is markedly reduced. A rough rule is that sludge volume is reduced by half when water is lowered from 95 to 90%, and by two thirds when reduced from 95 to 85%. Fresh sludge has a gray color and can be pumped easily. Its odor is most disagreeable, being due principally to thiols. Digested sludge is black in color, granular, and has a slight tarry odor.

Sludge digestion is carried out in order to reduce the volume of sludge to be handled, and reduce the number of pathogens. Sludge is usually withdrawn at regular intervals from primary and secondary tanks and led by gravity to a sludge well. It is then pumped to the digester. Mixing is very important for efficient sludge digestion. Temperature is equally important. Destruction of sludge is carried on by microorganisms, and the kinetics of their life processes is temperature dependent. It has been found that a sludge temperature of about 95°F will give short detention times. Even shorter detention times for the same quantity of digested sludge can be achieved with temperatures of about 125–130°F, but this temperature range is not widely used for reasons of economics. Above 95°F an increase in detention time is noted, up to 110°F, and then again a decrease. The reason for this is the changing character of the predominant organisms.

Heating of sludge for efficient digestion is usually carried out in two ways. The older installations have hot-water coils in the periphery of the tank, and heat is transmitted to the digesting sludge. Mixing was believed to be adequately effected by turbulence due to gas generation. Mechanical mixers have been used. It was found, however, that mixing was not sufficient. In addition, heating of the entire tank contents was not achieved due to "baking" of sludge in the vicinity of the heating coils. A second method of sludge heating and mixing was developed, involving the use of external heat exchangers. Sludge is pumped from the digestion tank through a heat exchanger and returned to the tank. Two objectives are accomplished: (1) efficient mixing of sludge, thereby reducing the amount of inadequately digested sludge, and (2) more uniform temperature throughout the tank, thus reducing digestion time. The use of external heat exchangers has almost completely supplanted heating coils and internal mixers in new plant design.

Sludge gas generated during digestion is approximately 72% methane and 28% carbon dioxide. Hydrogen and H_2S are present in trace amounts. The gas thus generated has a calorific value of about 600 Btu/ft³. About 10 ft³ of gas is produced per cubic foot of raw sludge digested. Generally, the amount of sludge gas produced is sufficient to provide heat used to maintain the digesting sludge at the required temperature, heat the plant buildings, and provide hot water and incineration of digested sludge (when practiced), and fuel for generators.

Volatile acids, reported as acetic acid, are perhaps the most important parameter. The value should be below 1000 mg/liter in a healthy digestion process. A value of 6000 mg/liter or over indicates a malfunctioning process. pH values of 6.8 to 7.2 are optimum. Values less than 6.8 usually are due to excessive volatile acid production. In the past, liming of malfunctioning tank contents was practiced in an effort to adjust pH to about 7.0. However, the change in volatile acids production was due to a change in the dominating microorganisms. The lowered pH and high volatile acids concentrations were a symptom of a sick process, rather than the cause.

Sludge Disposal. Digested sludge is reasonably inert, but still has high water content. It can be dewatered further by drying beds, or by filtration. Alternatively, it can be disposed of by wet air oxidation (Zimpro process) or in the ocean.

Dewatering. Digested sludge can be further dewatered by drying on sand beds. These are of two types, open and covered. Open sand beds are exposed to the open atmosphere and drying is accomplished by both downward drainage and evaporation. Covered beds resemble a greenhouse. Temperatures are rather high as a result. In either case sludge is allowed to flow over beds of sand and then left to stand for the required time. The dried sludge is then scraped from the beds for ultimate disposal. Sludge may also be dewatered by vacuum filter drums. See Filtration.

Disposal of dried sludge from drying beds may be by incineration, or used as landfill or as fertilizer. Unfortunately, sewage sludge has not found wide use as fertilizer. This is due partly to esthetic reasons and partly to the comparatively low quality of the sludge as fertilizer. In some cities digested sludge is incinerated at elevated temperatures (1800–2000°F). This practice has fallen into disuse since it led to air pollution.

Wet Air Oxidation. The Zimmerman process (Zimpro) shows promise of solving some of the problems of sludge disposal. It is based on the principle that any organic compound can be oxidized in an aqueous solution if sufficient energy, in terms of heat and pressure, is supplied. Water in the operating temperature range of the process has a heat capacity eight times that of the flame gases typical of dry combustion. Thus, the flame temperature is greatly depressed.

The temperature must lie between 212°F and the critical temperature of water (705°F). Air supplied to the process becomes saturated with steam from contact with the water. The greater the quantity of air supplied, the more steam is carried from the reactor. There is, however, an upper limit to the amount of air input since all water must not be carried away. Pressure affects the amount of water vapor. The pressure must be increased in order to obtain a temperature increase if water loss is to be kept essentially constant. Air requirements for complete oxidation depend on the COD (chemical oxygen demand). A sludge with a high COD will require more air.

Capital costs for equipment are high and the large alloy-steel heat exchangers may represent more than half of the total cost. These are subject to very corrosive acids formed during the reaction, which with the high temperatures and pressures, reduce their service life. Recently developed are heat-recovery exchangers with no metallic heat-transfer surfaces, which by means of an intermediate phase, and other improvements, should reduce the cost of the equipment by about one half while allowing the recovery of a substantial amount of steam for power and process use. These and other improvements should expand the use of the wet combustion of sludges, which is the most promising way of handling sludges, particularly in the congestion of a large city.

Ocean Disposal. Disposal of sewage sludge by barging to sea has been practiced by a number of large seacoast cities for many years. The most notable example is New York City, and this city maintains a fleet of sludge barges. Disposal has been in a carefully chosen area about twelve miles from shore. Barging is under consideration by several other cities but the future of this practice is unclear. In order to avoid pollution of ocean water in the neighborhood of these cities it has been proposed to dump the sludge beyond the continental shelf.

CHLORINATION OF EFFLUENT

It is common practice to chlorinate effluent for bacterial control. Regulations vary from state to state, but most regulations require chlorination to a specified residue. Chlorine-content requirements usually vary from season to season, the most stringent rules governing the swimming season. A phenomenon not yet fully understood is that of aftergrowth; bacterial count is fairly low immediately after effluent discharge but then suddenly rises to a high figure.

In some plants chlorination of the influent is practiced for the purpose of odor control.

Chlorination of storm-water overflow is commonly practiced. In some cases storm-water overflow is subjected to simple sedimentation and/or screening, storage, and chlorination, and is discharged after cessation of the storm.

SPECIAL PROBLEMS

Watercraft Waste Disposal. Use of the waterways for pleasure boating increased enormously in the 1960s. With this increase came awareness that pollution from pleasure and commercial craft could cause serious problems. Regulatory agencies were faced with the problem of promulgating rational and enforceable standards. Requirements varied widely from state to state. Much confusion will, it appears, reign for many years to come.

Treatment devices for watercraft waste include macerator chlorination, recirculating toilets, incinerators, and holding tanks. Macerator chlorinators macerate the waste and dose the effluent with a chlorine compound, usually HTH (calcium hypochlorite, see Vol. 5, p. 16). Recirculating toilets recycle flush water in a closed system, with disposal of flush water and solids to a dockside facility. Such devices have been used on commercial aircraft for many years. Incinerators simply burn waste material to a stable residue. Safety and power requirements seem to limit use of this method. Holding tanks usually draw the flush water from overboard, and the waste material and flush water are retained aboard the vessel for discharge to a dockside facility. Little agreement has been reached by those holding opposing viewpoints. The type of treatment device to be used on tidal waters is the major point of disagreement.

Small Communities. In recent years manufacturers of waste-treatment equipment have endeavored to supply complete treatment plants for small communities or developments and isolated installations. Basically, these plants, called package plants, supply primary treatment and sometimes some biological treatment on a small scale without requiring extensive operating supervision. It is felt that such treatment is to be preferred to septic tanks or only primary treatment (Imhoff tanks, for example), but such installations are not the ultimate solution.

Surfactants. Soaps are usually long-chain molecules, and degradable during sewage treatment and sludge digestion. Around 1950 synthetic detergents (syndets) came into wide use. These substances had as their basic components alkylbenzene-sulfonates (ABS), and phosphates as builders. The early syndets were found to be quite resistant to biodegradation, and passed essentially unchanged through conventional waste-treatment facilities. As a result, excessive foaming was experienced in activated sludge tanks and at many points of high turbulence. It should be noted here that "suds" were not necessary for efficient cleaning in the home, but "suds" were used as a selling point. Thus, the problem could not be easily attacked by removal of the "suds" producer and substitution of a low-foam producer. Production of large amounts of foam created not only unsightly areas in treatment plants and receiving waters, but also a dangerous situation for workers in treatment plants.

In Europe sewer ordinances were enacted, forbidding discharge of nondegradable detergents (biologically hard) after a certain date. New compounds were substituted, and the problem was alleviated. In the United States less stringent action was taken and the approach was that of gentle persuasion. Much the same solution was achieved as that in Europe. See Surfactants.

Phosphates. With the solution of the surfactant problem, another problem has arisen. Household detergents now contain notable quantities of phosphates. These pass through the sewage-treatment plant unchanged, and are discharged into rivers and lakes. There, they may cause a pollution problem of an unusual kind, due, not to the discharge of a poison of any kind, but to the fact that the phosphates bring about an unusually prolific growth of algae. See Eutrophication, in Supplement Volume.

Federal Versus State Regulation

The modern approach to waste water treatment, that of protecting oxygen resources, and taking into account all aspects of the problem, from public health to esthetics, demands a systems approach, that is, considering this problem on an estuarine basis. ORSANCO (Ohio River Sanitation Commission) was formed in 1943, and is an early and excellent example of basin planning and control. A further step was taken in 1966, when the responsibility for water-pollution control was taken from the U.S. Public Health Service and placed with a new agency in the Department of the Interior, the Federal Water Quality Administration (FWQA). However, relations between the federal pollution control officials and state pollution control officials have not always been peaceful. Federal intervention is possible when pollution problems involving two or more states are, in the opinion of the Secretary of the Interior, not being acted upon with sufficient rapidity, or if the chief executive of a state invites federal participation. Such intervention is often not received with gratitude.

Recent Developments

UNOX is the trade name given a newly developed process designed to increase treatment efficiency in secondary (activated sludge) treatment facilities. The major difference from previous plants is the use of pure oxygen in place of atmospheric air. As in conventional activated sludge treatment, waste water and activated sludge travel concurrently in the aeration tanks. Tanks in the UNOX process are baffled for better mixing. Significant improvements in removals of BOD, COD, and suspended solids are reported for the new process.

The Z-M process is a new approach to waste treatment tested recently at the treatment-plant level. Instead of regarding pollutants on the basis of BOD, COD, etc, classification is made on the basis of waste-particle size and molecular weight as determined by chromatographic analysis. Passage of activated sludge effluent through activated carbon beds removes smaller sized material of lower molecular weight, but left essentially untouched material of higher molecular weight. By raising the pH to 11.5, it was determined that the heavier material (molecular weights of 1200 and above) was broken down to smaller size and weight, suitable for activated carbon removal. (See Water reuse.) In addition, lime used for pH adjustment can be easily regenerated by burning. Phosphate (see Eutrophication) is found to be concentrated in the resulting ash. It is necessary to control pH within quite narrow limits for optimum dosage results.

Bibliography

"Sewage," in *ECT* 1st ed., Vol. 12, pp. 191–207, by A. M. Buswell, Illinois State Water Survey Division.

General References

J. C. Stevens, ed., *The Hydrographic Data Book*, Lupold and Stevens, Inc., Portland, Ore., 1969.
Cleaning Our Environment, The Chemical Basis for Action, Am. Chem. Soc., Washington, D.C., 1969.
Ross E. McKinney, *Microbiology for Sanitary Engineers*, McGraw-Hill Book Company, New York, 1962.
G. M. Fair, J. C. Geyer, and D. A. Okun, *Water and Wastewater Engineering*, Vols. 1 and 2, John Wiley & Sons, Inc., New York, 1968.
Sewage Treatment Plant Design, ASCE Manual of Practice No. 36, Am. Soc. of Civil Engineers, New York, 1959.
J. W. Clarck and W. Viesman, *Water Supply and Pollution Control*, International Textbook Co., Inc., Scranton, Pa., 1966.
W. Hardenburgh and E. A. Rodie, *Water Supply and Waste Disposal*, International Textbook Co., Inc., Scranton, Pa., 1961.
T. Olson and F. J. Burgess, *Pollution and Marine Ecology*, Interscience Publishers, a division of John Wiley & Sons, Inc., New York, 1967.
L. Klein, *River Pollution*, Butterworths Scientific Publishers, Ltd., London, 1966, Chap. 3.
W. Eckenfelder and D. J. O'Connor, *Biological Waste Treatment*, Pergamon Press, Ltd., London, 1961.
N. Nemerow, *Theories and Practice of Industrial Waste Treatment*, Addison-Wesley, London, 1963.
Senate Subcommittee on Air and Water Pollution Control Hearing, Public Works Committee, 1st Session, 88th Congress, 1963.
Safety Standards for Small Craft, Committee A-8, Sewage Treatment, American Boat and Yacht Council, Inc., New York, 1968.
Chem. Eng. News, March 23, 1970, p. 15.
M. Zuckerman and A. H. Molof, "High-Quality Reuse Water by Chemical-Physical Wastewater Treatment," *J. Water Pollution Control Federation* **42** (3), 437 (1970).

JAMES R. PFAFFLIN
University of Windsor
Ontario, Canada

WATER REUSE

An essentially constant amount of water is available for utilization by all the living creatures on earth. Man is the only creature that has given consideration to the use of this resource. In earlier times the problem was to convey the water from one area to another. At present a serious complicating factor has entered into the calculations applying to water supply. This is the necessity to balance the total requirements against the total amount available. Increased population, higher individual consumption, and increased industrial demand have made it necessary to constantly seek new sources of supply. The concept of water reuse is receiving increasingly closer attention.

All water is reused. Along a river water may be used a dozen times from its deposition as precipitation until eventual mixing into the sea. At some future date the water will again follow the same river. Modern society requires great amounts of water and cannot wait patiently for the hydrologic cycle. Even if water were unlimited in quantity, economics requires that the supply be reasonably close. Otherwise, the cost of transportation becomes prohibitive.

Controlled water reuse is not a new practice. One of the most notable examples of controlled water reuse involves the Back River sewage works in Baltimore and the Sparrows Point plant of Bethlehem Steel. Well supplies for industrial purposes at the plant had shown increasingly high chloride concentrations due to intrusion of salt water. Difficulties were encountered in steel rolling and cooling processes. As a solution, Abel Wolman advised the use of secondary waste-water effluent from the Back River treatment plant for industrial process use downstream. This lessened the draft on the wells and reduced the intrusion of salt water. Initially, 120 million gallons per day of treated effluent was sold by the city to the company, and this figure has increased in twenty years. Little additional treatment was necessary in this case. Such an arrangement has proved advantageous to all concerned. Municipal waste water was looked upon as a product to be discarded. Unfortunately, it was necessary to spend money in order to make this discarded product acceptable on the basis of health and esthetics.

Water is seldom immediately available for reuse. The amount or degree of treatment necessary depends on the intended reuse. Often, in the economics of the process, there enters the factor of eliminating, or at least reducing, the amount of material that must be disposed of.

Regulations covering waste disposal are becoming ever more strict, and there is more stringent enforcement of these regulations. It is necessary, therefore, to apply sound engineering principles to the planning for waste disposal. Formerly, in designing of an industrial process, little attention was given to disposal of waste products. Not long ago, primary treatment of municipal waste water was acceptable. No longer does this apply. It is, therefore, becoming economically more attractive to practice waste water reuse.

Reuse of waste water can be practiced for industrial purposes and potable water. The degree of purity required for industrial water is in many cases greater than that required for potable water.

Public-health authorities have long accepted reused water as a source of potable water for water-poor areas. However, the widespread support of the medical profession will be necessary for acceptance of reused water for human consumption.

For some purposes water may be reused with a slight degree of treatment. However, it may be expected that for most reuse purposes conventional primary and secondary treatment (described above under Water, sewage) will not be sufficient. Further, or tertiary, treatment will be required.

Tertiary treatment of waste water involves additional treatment beyond the conventional trickling filter or activated sludge. Secondary, or biological, treatment plants may achieve a BOD and suspended-solids removal of 90% and reduce microbial contaminants to a presently acceptable level. Effluent contains BOD and COD which continue the natural cycle of decomposition. Some phosphorus and nitrogen compounds are not reduced during treatment and serve as nutrients for algal growth in the receiving water. Contaminating materials that cannot be removed by sedimentation and biological treatment are termed *refractory*. These range from simple inorganic salts to complex synthetic organic chemicals, such as pesticides, herbicides, insecticides, and surfactants. Recycling of waste water in an essentially closed system with no more than the conventional secondary treatment would soon produce an unacceptable concentration of many substances. Table 1 shows a comparison of a typical tap water and a secondary waste water effluent, with the increment of each component.

Table 1. Comparison of Tap Water and Waste-water Effluent (1)

Component	Tap water, mg/liter	Secondary waste-water, mg/liter	Increment, mg/liter
organic			
BOD	2	20	18
COD	12	100	88
alkylbenzenesulfonates	0.4	6.8	6.4
cations			
Na^+	50	125	75
K^+	3	13	10
NH_4^+	0.1	16	15+
Ca^{2+}	42	60	18
Mg^{2+}	16	19	3
Fe^{2+}			
anions			
Cl^-	66	143	77
NO_3^-	5	12	7
NO_2^-	0.15	1.5	1.3
HCO_3^-	198	296	98
CO_3^{2-}	0.1	1	1
SO_4^{2-}	56	84	28
SiO_3^{2-}	29	43	14
PO_4^{3-} (total)	8.1	25	17
PO_4^{3-} (other than orthophosate)	0.3	25	25
hardness[a]	158	235	77
alkalinity[a]	164	242	78
total dissolved solids	382	700	320
pH	8.0	7.4	−0.6

[a] As CaCo₃.

From Table 1 it can be seen that recycling of waste-water effluents would give an unacceptably high concentration of many substances after a short while. For reuse of municipal or industrial wastes it is necessary to achieve a higher removal of refractory contaminants than is now possible with conventional secondary treatment.

Tertiary Treatment Methods

In many cases the quality of a particular stream or other water source can be improved by merely requiring higher BOD or suspended solids removal. In other cases the aim might be to prepare effluent for ground water recharge, lower industrial use, or recreational use. Nutrient removal will, in some cases, be the major requirement. Table 2 gives a classification of waste-water-treatment processes.

Table 2. Classification of Waste-Water-Treatment Processes

Classification	Process	Substance removal
biological	conventional secondary treatment: trickling filter, activated sludge	suspended solids, BOD, bacteria, etc
	aerobic process modifications	conventional treatment plus nutrients
	anaerobic denitrification	nitrate nitrogen
	algae harvesting	nitrate nitrogen and phosphorus
chemical	ammonia stripping	ammonia nitrogen
	ion exchange	nutrients
	electrodialysis	salts
	chemical precipitation	suspended solids and phosphates
physical	activated carbon adsorption	organic compounds and suspended solids
	sedimentation	suspended solids and bacteria
	filtration	suspended solids and bacteria
	reverse osmosis	salts
	distillation	salts
	foam separation	detergents

Chlorination. This process is not truly a tertiary treatment process, but should be mentioned here. Normally, secondary waste-water effluents are chlorinated in order to achieve satisfactory bacterial quality. However, proper chlorine dosage may give effluents satisfactory for some reuse purposes.

Chemical Precipitation. Chemical precipitation alone should be classified as a primary treatment process, but when used as an addition to conventional secondary treatment it becomes an excellent tertiary treatment which finds wide acceptance for both municipal and industrial waste.

In this process an insoluble compound is formed, and the resulting precipitate allowed to settle from suspension. Charges on colloidal particles is at the same time neutralized. Removal of colloids is accomplished by neutralization, followed by agglomeration. The resulting floc sweeps other material, including bacteria, from suspension.

Removal of phosphate can be accomplished by addition of soluble salts of aluminum or iron. Insoluble aluminum phosphate or ferric phosphate is formed, but the removal mechanism may be more complex than this, involving adsorption of phosphate on a floc of metallic hydroxide.

Perhaps the cheapest and most easily controlled process for removal of phosphate is addition of lime. At a pH of 11 or above, the lime, in addition to its usual effects in removing magnesium and calcium, forms a precipitate consisting of hydroxylapatite, $Ca_5OH(PO_4)_3$.

Chemical precipitation has also been used for removal of heavy metals which form insoluble compounds in neutral or alkaline solutions. Fluorides and sulfides are also removed by this means.

The sludge resulting from chemical precipitation is often difficult to dispose of; it is seldom salable, and the best solution is probably to digest it with sludge from conventional treatment.

Filtration. Most tertiary treatment processes for removal of particulate matter involve filtration. Filtration is often a misnomer, since most removals are obtained by a process of adsorption on filter particles rather than true straining. If secondary effluents contain a high concentration of solids, binding will soon occur at the bed surface, giving high energy losses and passage of dirty water due to short circuiting. The best filtering materials are sand, mixed-media, and diatomaceous earth.

Rapid sand filters have a limited use in tertiary treatment. Suspended solids accumulate in the filter, reducing BOD at the same time. But it is necessary to backwash the filters in order to remove foreign material. Frequency of backwashing will depend on the surface application rate and concentration of contaminants in the applied water. Backwashing may be by water alone, at a rate of 24–30 gal/(min)(ft^2), or a combination of air and water, with water applied at a rate of 10–15 gal/(min)(ft^2) and air at about 4 ft^3/(min)(ft^2). Air serves to thoroughly agitate the filter. The medium is usually sand in the particle size range of 0.5–0.8 mm. The bed depth is usually about 24 in., and the rate of application of the influent water is 2–4 gal/(min)(ft^2). Most removal is near the top of the bed, and thus, since only a small part of the total void volume is utilized for particulate storage, clogging is relatively rapid.

One approach to increase the filter depth effectively is the use of a mixed-media filter. In this filter the coarsest material is at the top and the least coarse on the bottom. This is opposite to the stratification expected if all particles have the same density. In the mixed-media filter three or more materials of different specific gravities are intermixed. Coal, sand, and garnet are the materials most commonly used. Garnet has a specific gravity of 4; sand, 2.6; and coal, 1.6. Because of the differences in specific gravity, coarse coal occupies the upper layers, while the heavy and fine garnet occupies the lowermost portions. The resulting particle-size gradation decreases from about 1 mm at the top to 0.15 mm at the bottom. In this way, the entire filter bed is utilized for removal and storage of particulate matter. The fine garnet at the bottom forces the effluent to pass through a much finer barrier than is provided by sand, and the coarse upper layer reduces the possibility of clogging at the surface. Filter depths of 24–30 in. are common, and flow rates of 5–10 gal/(min)(ft^2) are normally used.

Diatomaceous earth filtration is also used. This is a form of mechanical separation that uses a layer of powdered filter aid, diatomaceous earth, built upon a relatively loose septum to screen out suspended solids. The filter will become clogged after a period of filtration, and pressure loss will become excessive. It is then necessary to backwash the filter. The minimum-size particle that can be removed by the filter is in the range of 0.5–1.0 μ.

Vacuum filters and pressure filters are both used. Turbidity is more easily removed by vacuum filters, with a removal efficiency of 85% a common figure. Flow rates are low, about 0.5–1.0 gal/(min)(ft²) and these filters are not feasible for treating effluents from large treatment plants.

Microstrainers. These have been used extensively in the past as a tertiary treatment method. Microstrainers are rotating drum screens with extremely fine stainless steel mesh, having 80,000 to 160,000 perforations per square foot. The flowing liquid enters the open end of the drum and passes through the mesh into the effluent chamber. The mesh traps solid impurities and rotates with the drum in order to come under a wash-water spray. The spray washes the trapped solids into hoppers for final disposal. The mesh is washed with filtered effluent discharged from jets fitted over the drum, and then passes under ultraviolet radiation for inhibition of microbiological growth. Usually it is necessary to wash the wire mesh with water containing an appreciable chlorine residue at 7- to 28-day intervals in order to control slime growth. The interval between washings depends on the quality of the effluent treated.

Removal efficiencies of microstrainers are on the order on 30–55% of the BOD applied, and 40–60% of the suspended solids applied.

Effluent Polishing. The term "polishing" is sometimes applied, in various contexts, to the preparation of a liquid of exceptional clarity. A polishing treatment is sometimes practiced on the effluent from "package" waste-treatment plants (see Water, sewage). The package plant effluent is collected in a sump and pumped through a tube-settling unit, followed by mixed-media filtration. Filter effluent is collected in a storage tank which is also used for backwashing the filter and serves the further purpose of chlorine contact chamber. The tube settler in the process serves as a supplemental solids-separation device which allows the filter to operate efficiently during severe operating upsets, such as a "shock" loading, either due to increased organic matter, or increased hydraulic flow. During such upsets the bulk of the suspended solids in the plant effluent will be removed in the tube settler and the filter will not be overloaded.

Operating data from such a system show that it is possible to produce an effluent with suspended solids and BOD of less than 50 mg/liter. The polishing unit continues to function well even when the suspended-solids content of the package-plant effluent exceeds 2000 mg/liter (0.2%).

This tube-settling, mixed-media combination can be used to remove the phosphate after it has been suitably treated for precipitation and coagulation.

Foam Separation. Conventional secondary treatment processes are unable to remove and decompose hard detergents such as ABS (alkylbenzenesulfonate). The foam-separation process was the first dependable means for removal of such substance. Foaming at points of high turbulence results when concentrations are appreciable, and, until foam separation, use of high-pressure water sprays was the only way of breaking the foam. This did not, however, remove the ABS, but only evidence of its presence.

The foam-separation process utilizes the ability of surface-active solutes, such as ABS, to collect at a gas–liquid interface, and promotes the reaction by artificially inducing foaming so as to form a large interfacial area. Solutions containing surface-active solutes, when foamed, transfer the ABS to the foam. The foam also concentrates suspended solids by a mechanism of flotation.

Flotation has been commonly used for industrial processes for clarification, but has been little used for waste treatment except in the reduction of ABS-bearing wastes. Studies have been made to determine suitability and effectiveness of foam separation as a treatment means on other constituents of waste water. These tests, made principally by AWTRP (Advanced Waste Treatment Research Program), indicate 80–90% ABS removals and 30% COD removals. Further, it was indicated that contaminant removals from municipal wastewaters could be improved by addition of selected cationic polyelectrolytes.

In the foam-separation process a sparger produces small bubbles of gas, usually air, that are dispersed through the waste container. Rising gases collect surface-active particles and suspended solids. Foam collects at the air–liquid interface. It is then taken out of the foamer and collapsed to yield a waste concentrate.

Figure 1 shows column foam separators and trough separators. In the column type, waste enters the column and flows downward. Gas spargers located at the bottom give countercurrent gas flow. Foam thus generated is carried upward from the gas–liquid interface by the airstream to a foam breaker and then to a foam collector. This type of separator is suitable for relatively small treatment installations.

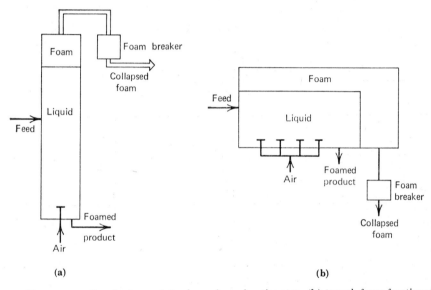

Fig. 1. Foam separation devices: (**a**) column foam fractionator, (**b**) trough foam fractionator.

For larger volumes of waste, such as from municipal or industrial treatment plants, the trough-type separator is more effective. The principles of operation are the same as in the column separator, but flow is through a horizontal trough.

The future of foam separation as a major treatment device is uncertain at this time. ABS is being replaced by LAS (linear alkylbenzenesulfonates) in detergents and foam-producing waste has been reduced accordingly. See Surfactants. However, foam-separation equipment has been applied to the removal of refractories and other pollutants such as organic hydrates and nitrogen compounds, with some success.

Activated Carbon Adsorption. Treatment of waste water by activated carbon shows promise of becoming one of the most important tertiary treatment processes. Activated carbon has long been used for removal of tastes, odors, and color, but it becomes inefficient for treatment of effluents containing high concentrations of organic matter due to decreased surface area.

Much applied research has recently been carried out on bed thickness, loading, flow rate, contact time, and also adsorption materials. A series of very comprehensive reports has been issued on these parameters by the Advanced Waste Treatment Research Program of the U.S. Public Health Service (4).

Activated carbon (see Vol. 4, p. 149) is available either as a fine powder or as granules having particle sizes from 8 to 40 mesh. The granular material is more easily and conveniently applied to waste waters. Usually, treatment of water by activated carbon is a continuous, rather than batch, process.

Water is passed through a column filled with activated carbon particles. The organic content of the water decreases along the path of flow. However, the carbon at the influent end becomes exhausted and requires regeneration.

Powdered activated carbon has a much more rapid adsorption rate than granular carbon because of the smaller particle size—about 10 μ. But the small-sized particle bed will have a much higher pressure drop for fixed bed operation, and there is need for further research and development in this area.

Activated-carbon adsorption can remove 80% of the COD, 70% of the BOD, and 90% of influent ABS. Operating difficulties may be encountered. Typical results are shown in Table 3.

Table 3. Results of Activated-Carbon Adsorption Treatment[a]

Parameter	Secondary effluent	Carbon column effluent	removal, %
suspended solids, mg/liter	10	1	90
COD, mg/liter	47	9.5	80
NO_3^- as N, mg/liter	6.7	3.7	45
turbidity, Jackson units	16.3	1.3	85
color, APHA	30	3	90
odor[b]	12	1	92

[a] Source: J. D. Parkhurst and F. D. Dryden, "Pomona Activated Carbon Plant," *J. Water Pollution Control Federation* **39**(10), 70 (1967).

[b] Number of dilutions till odor is no longer discernible.

Regeneration of the carbon is accomplished in multistage-hearth furnaces in a steam and air atmosphere at 170°F. Powdered carbon has been regenerated in a steam atmosphere at 750°F in electrical resistance regenerators. Both methods show a carbon loss of about 5% per cycle.

This method of contaminant removal is considered economically acceptable as a tertiary treatment method, according to the U.S. Public Health Service definition, which set a cost of 10¢ or less per 1000 gal of treated waste as the acceptable figure in 1969.

Ion Exchange. Ion exchange (qv) has been used for many years in softening hard waters. Calcium and magnesium ions are exchanged for a cation, usually sodium. Ion-exchange beds can remove 95% of the applied phosphate, 85% nitrate, 100% sulfate, and 45% COD. Cationic and anionic mixed beds have given good removal of color and organics, but organic materials tend to foul the bed. Cost and frequency of regeneration are now the major disadvantages of ion exchange as a tertiary treatment method.

Oxidation Ponds. Oxidation ponds, or sewage lagoons, have been used for many years in Europe for treatment of domestic and industrial waste. The practice has not become popular in the United States, due partly to public resistance to such facilities close to populated areas and partly to the very large area required. However, as a tertiary treatment device there is promise of greater utilization.

An oxidation pond is simply a large shallow pond into which waste is led and, after suitable detention, discharged to be reused or lost in a receiving water.

The capacity of an oxidation pond is, in effect, controlled by the rate of oxygen transfer into the pond. A common design figure is 20–30 lb of BOD per acre per day. Mixing is important; if the material were completely dispersed throughout the pond, oxygen demand would be uniform, but since this is not the case, higher demands exist in some parts than in others.

The process is aerobic and oxygen transfer is essentially through surface agitation. Surface agitation and flow patterns will depend greatly on wind currents, and this must be taken into account in design.

Algae were commonly thought to be the important organisms in waste stabilization, but this is not strictly true. They contribute to maintenance of aerobic conditions, but surface aeration is equally, if not more important.

Depth is an important consideration in design and operation. Four feet is considered a good depth, but much deeper ponds are reported to have operated successfully. A deposit is formed at the bottom, and provision must be made for its occasional removal. The pond must be shallow enough for sunlight, important for algae, to penetrate. At the same time it must be deep enough to keep down weed growth. Much of the waste stabilization is done by bacteria, and provision must be made for removal of microbial cells from the effluent. The pond functions as a sedimentation chamber, as well as a stabilization chamber. Current American practice is the use of a circular pond, with an influent feed at the center. There is some controversy concerning the use of a single pond as opposed to a series of ponds, in which the final pond functions as a settling tank.

Reverse Osmosis. Reverse osmosis seems to be the only membrane process applicable to preparation of used water for reuse. See Water, desalination.

Bibliography

1. *Studies of Waste Water Reclamation and Utilization, Publ. No. 9*, State Water Pollution Control Board, Sacramento, Calif., 1954.
2. *Preliminary Appraisal of Advanced Waste Treatment Processes, AWTR-2*, Advanced Waste Treatment Research Program, U.S. Public Health Service, Washington, D.C., 1926.
3. *Contaminant Removal from Sewage Plants by Foaming, ATWR-5*, Advanced Waste Treatment Research Program, U.S. Public Health Service, Washington, D.C., 1963.
4. *Feasibility of Granular Activated-Carbon Absorption for Waste Water Renovation, AWTR-10*, Advanced Water Treatment Research Program, U.S. Public Health Service, Washington, D.C., 1964.
5. *Evaluation of the Use of Activated Carbon and Chemical Regenerants in Treatment of Waste Water, AWTR-11*, Advanced Waste Treatment Research Program, U.S. Public Health Service, Washington, D.C., 1964.
6. *Evaluation of Various Absorbents and Coagulants for Waste Water Renovation, AWTR-12*, Advanced Waste Treatment Research Program, U.S. Public Health Service, Washington, D.C., 1966.
7. *Studies Relating to Market Projections for Advanced Water Treatment, AWTR-17*, Advanced Waste Treatment Research Program, U.S. Public Health Service, Washington, D.C., 1967.
8. W. C. Yee, "Selective Removal of Mixed Phosphates by Activated Alumina," *J. Am. Water Works Assoc.* **58**(2), 239 (1966).
9. L. K. Cecil, "Water Reuse and Disposal," *Chem. Eng.* May 5, 1969, p. 92.

10. L. K. Cecil, ed., "Water Reuse," *Am. Inst. Chem. Engrs. Symposium Series,* **78,** 63 (1967).
11. C. F. Garland and L. K. Cecil, "Industrial Reuse of Municipal Wastewater," *Proc. 7th Water and Waste Conf.*, Austin, Texas, 1967, pp. IV.1–10.
12. J. G. Brown, *Filtration and Separation* **2**(5), 378 (1965).

JAMES R. PFAFFLIN
University of Windsor
Ontario, Canada

TREATMENT OF SWIMMING POOLS

Water in swimming pools, except in flow-through pools and very small pools that are repeatedly filled and emptied, is used over and over. Continuous treatment of the pool water by mechanical filtration and chemical additions makes it possible to use the same water (except where total dissolved solids are very high) for a number of years.

The reuse of pool water makes it necessary to remove solid materials, such as grass, leaves, wind-blown dirt, etc, that enter the water. Solid particles are easily removed by recirculating the water through a filter.

There are many different filter designs using sand, diatomaceous earth, or a selected grade of anthracite coal (anthrafil) as the filter medium. In addition, cartridge-type filters may be used. These feature a disposable septum usually composed of nonwoven fabrics or treated paper. Vacuum cleaners, deriving their suction from the filter pump, are used to remove solids from the bottom of the pool. A long wand, similar to that on tank-type home vacuum cleaners is passed back and forth over the bottom of the pool. The dirt is removed by the filter.

Sanitizing chemicals must be added regularly to the pool water since bacteria are introduced by swimmers and dirt entering the water. The spores of algae are also carried into the water by wind and rain. Some chemicals serve as both bactericidal and algicidal agents, while others serve primarily as algicides.

For health reasons it is essential that a bactericidal condition be present in the pool water, and it is necessary to kill algae to avoid an algal bloom (green cloud) in the water and unsightly growth on walls and bottom of the pool. Algal growth in pool water will also rapidly clog the filters.

For sanitizers to be effective, it is necessary to control the pH of the pool water. Proper pH is also necessary for the comfort of the swimmer and to prevent corrosion of pool equipment. Thus pH-control chemicals are used to regulate the pH of pool water.

Some waters contain metallic ions that discolor the water. Chemicals to flocculate such elements or chelating agents to keep the metal ions in solution in an inactive form may be added to combat this condition.

Sanitizers

Chlorine has been, and continues to be, the accepted chemical for disinfecting swimming pool water (1). Chlorine gas, sodium hypochlorite solution, calcium hypochlorite, lithium hypochlorite, and chlorinated isocyanurates all supply available chlorine in the form of hypochlorous acid, HOCl, which is the chemical that acts as the bactericide and algicide in pool water. Each of these chlorine compounds will be discussed in some detail.

Other halogen compounds, namely those of bromine and iodine, are finding limited application as pool sanitizing chemicals. They will be discussed briefly, along with other products and processes for sanitizing pool water.

Chlorine Gas (see Alkali and chlorine industries; Chlorine). Chlorine gas (cylinder chlorine) is used as a biocidal agent in large pool installations and in public pools where it is required by law to have continuous feed of a sanitizing chemical. Some codes specify chlorine gas as the required or preferred sanitizer for large public pools.

Chlorine gas dissolves in water with reaction according to the following equation:

$$Cl_2 + H_2O \rightleftharpoons HOCl + H^+ + Cl^-$$

The hypochlorous acid then ionizes:

$$HOCl \rightleftharpoons H^+ + OCl^-$$

It can be seen that, as the pH increases, the amount of hypochlorous acid decreases.

In pools of 100,000–300,000 gal and larger, chlorine gas is almost universally used because of its low cost. For pools below 100,000 gal, it may be easier to use sodium hypochlorite solution, but the economics as compared to the use of chlorine gas will depend upon the availability of hypochlorite solution.

Chlorine gas requires special metering and feeding equipment plus experienced supervision and elaborate precautions against escape. Because of the hazards involved, this method of disinfection is impractical for small public pools, semiprivate pools, and the home pool owner. In addition to chlorine, it is necessary to add soda ash (sodium carbonate, Na_2CO_3) regularly to maintain the proper pH range.

Chlorine gas is generally packaged in steel cylinders of varying size, 100–150 lb being most common.

Sodium Hypochlorite Solution (see Vol. 5, p. 12). Sodium hypochlorite solution, usually 10–15% available chlorine (see Vol. 5, p. 9) by weight, is a widely used disinfecting agent for swimming pools. It is often referred to in the trade as "liquid bleach" and also, erroneously, as "liquid chlorine." Sodium hypochlorite in pool water reacts as follows:

$$NaOCl \xrightleftharpoons{H_2O} Na^+ + OCl^-$$

$$OCl^- + H^+ \rightleftharpoons HOCl$$

Made by a number of small bleach producers and chemical specialty manufacturers around the country, liquid bleach is used in place of chlorine gas and other chlorine compounds. Bleach plants are concentrated in the large metropolitan areas where they serve the industrial market and the laundry field. Transportation costs prohibit the shipment of liquid bleach for any appreciable distance.

Sodium hypochlorite is widely used in California and Florida, the two most highly populated in-ground pool areas. In these areas, it is a very popular sanitizer with the home pool owner and the pool-service companies. It is also widely used for the smaller public pools, and for semicommercial pools, such as those at motels and camps. In most cases, a hypochlorinator (chemical feeder) is required by law; sodium hypochlorite is an easy chemical to use. In some areas it is delivered to the user in bulk.

There has been a growing trend toward the use of sodium hypochlorite instead of chlorine gas in new, large public and semipublic pools because of the increasing awareness of the hazards in using chlorine gas, particularly in indoor pools. In general, pools of less than 100,000 gal use sodium hypochlorite. Over 100,000 gal the cost of treatment becomes a factor and chlorine gas is the choice if it is a question of economics and not safety.

A disadvantage of sodium hypochlorite is its limited storage stability. It is not stable at high temperatures, resulting in rapid loss of strength. Another disadvantage is the alkalinity of sodium hypochlorite which necessitates substantial additions of acid for pH control.

Calcium Hypochlorite (see Vol. 5, p. 16). This is one of the most widely used disinfectant chemicals for treating pool water and has been on the market since 1928. It contains 70% available chlorine, is a dry, stable product, and is easy to handle. It is a very inexpensive source of dry chlorine and is widely used in home pools, small commercial pools, and semiprivate pools.

Because of its economy and convenience, calcium hypochlorite is widely used for superchlorination and as a general sanitizing agent in the pool area. Even when chlorine gas or sodium hypochlorite solution is the primary sanitizer, calcium hypochlorite is added to supply the chlorine necessary for peak pool demands.

Calcium hypochlorite does not change the pH of the pool water as much as sodium hypochlorite or chlorine; thus, it is much easier to control the pH and much less acid or soda ash is needed. The product does add calcium ions to the pool water which may be undesirable in very hard water areas.

Lithium Hypochlorite (see Vol. 5, p. 15). Lithium hypochlorite, 35% available chlorine, was first introduced in 1963 (1). The product has all the advantages of calcium hypochlorite and is completely soluble. Lithium hypochlorite has a disadvantage in that it is a relatively expensive source of chlorine.

Chlorinated Isocyanurates (see Triazinetriol). Chlorinated isocyanurates were first introduced for pool use in 1958. They are dry, stable compounds with high available-chlorine contents, but they are an expensive source of chlorine. They have had to compete with the less expensive, inorganic hypochlorites, and with other types of halogen compounds. Their acceptance as pool sanitizers was initially delayed by questions as to their toxicity. However, work at Robert A. Taft Sanitary Engineering Center showed that chlorinated isocyanurate compounds and combinations of cyanuric acid and hypochlorite, would present no public health hazard when used in swimming pools under ordinary use conditions, provided the cyanurate concentration did not exceed 100 ppm. Their effectiveness also has been demonstrated (2), and the products are now finding wide acceptance.

The chlorinated isocyanurates used as swimming-pool disinfectants include: potassium dichloroisocyanurate (59% available chlorine); sodium dichloroisocyanurate (63–64% available chlorine); and trichloroisocyanuric acid (89–91% available chlorine).

Sodium dichloroisocyanurate is the preferred product for swimming pools, as it is more readily soluble than the potassium salt. Further, the sodium salt has approximately 7% more available chlorine per pound than the potassium salt, and the price per pound is usually similar. Tablets of trichloroisocyanuric acid are used with feeders or added to skimmers, since they are very slow dissolving and have a high chlorine content. Sodium carbonate has to be added to the pool to offset the added acidity.

When chlorinated isocyanurates hydrolyze they produce not only hypochlorous acid but also cyanuric acid or its salt. The cyanuric acid acts as a sun screen for the hypochlorous acid by a mechanism that is believed to involve absorption of ultraviolet radiation. This stabilizing effect (3) is the major advantage gained in using chlorinated isocyanurates, since a chlorine residue is maintained in the pool water for relatively long periods of time.

This stabilizing effect can be obtained either by initially adding extra dosages of chlorinated isocyanurates to the pool water, or by a more economical route, namely adding cyanuric acid as the stabilizer to the pool water and adjusting the pH. A minimum level of 25–30 ppm of cyanuric acid is added to the pool water. There are no positive benefits derived from cyanuric acid residue in excess of 60 ppm.

With cyanuric acid as the stabilizer other less expensive sources of chlorine, such as the hypochlorites, can be used. The amount of chlorine required for stabilized pool water is considerably less than that needed in a comparable nonstabilized pool.

Studies (2,4) have shown that as the concentration of cyanuric acid increases in a solution the activity of the chlorine decreases. Studies were made with concentrations from 0 to 100 ppm cyanuric acid. Excessive amounts of cyanuric acid in pool water can cause overstabilization, and a sharp decrease in the effectiveness of the chlorine. Thus, higher minimum standards for chlorine concentration are recommended for stabilized pools (5). The Los Angeles County Health Department *Guide for the Use of Chlorinated Cyanurates* (March 7, 1966), sets the following standards for chlorine concentrations, where cyanurates are used:

Cyanuric acid, ppm	*Chlorine residue,* ppm
25–60	1.0
61–100	1.5
over 100 not recommended	

These are accepted as equivalent to 0.4 ppm free residual chlorine in a nonstabilized pool.

The U.S.D.A. Agricultural Research Service Pesticides Regulation Division recommends that the chlorine residual should be 1.0 ppm and preferably 1.0–1.5 ppm if the pool water is stabilized.

Amidosulfuric Acid (see Sulfamic acid). Amidosulfuric acid had very limited use as a pool stabilizer during the 1950s and early 1960s. It did not find widespread acceptance because it consumed some of the available chlorine and sharply increased the acidity of the pool water.

Bromine (qv). Bromine and bromine compounds are used for the disinfection of swimming-pool waters (6). Although bromine has a lower oxidation potential than chlorine it is a highly effective bactericide; however, it has not become very popular. One reason for this is the higher cost of bromine as compared with chlorine. Another reason has been the difficulty in handling liquid bromine.

The development of "solid bromine" or "stick bromine," however, has helped to overcome the handling problem and more pools are now using it. The product contains 1-chloro-3-bromo-5,5-dimethylhydantoin (**1**). Feeders employing stick bromine

$$\underset{\textbf{(1)}}{\underset{\displaystyle O=\overset{\displaystyle}{C}-N-Br}{\overset{\displaystyle Cl}{\underset{\displaystyle}{\overset{\displaystyle H_3C}{\underset{\displaystyle H_3C}{>}}C-N-C=O}}}}$$

are installed in a by pass line of the pump circuit and the product slowly dissolves into the water. Buffers are incorporated into the sticks which help to keep the pH in control. Feeders have also been developed for liquid bromine in recent years. (See also Hydantoin.)

One of the advantages of bromine is the fact that the bromamines, as opposed to the chloramines, do not cause eye irritation.   A residue of 1–2 ppm is recommended for bromine and the pool can be maintained in a slightly wider pH range, 7 to 8, than when chlorine is used.

Iodine (qv).   A great deal of work has been done on the use of iodine as a disinfectant in swimming pools, particularly by A. P. Black and his associates (6,7).   Iodine is an effective bactericide but nevertheless has not become widely used for pools.   It is not as strong an oxidizing agent as chlorine, with the result that it does not fully oxidize organic matter.   Chlorine must be used for this.   It does not combine with ammonia or organic nitrogen to form iodamines and consequently causes no eye irritation. Pools treated with iodine may have a yellow tinge, or, if the control is not proper, a green or aquamarine color.   If the control gets out of hand, the pool may turn brown which usually means that the water must be changed.

Iodine is generally added to a pool in one of two ways.   The first is to prepare a saturated solution, by passing water over a bed of solid iodine, and then add it to the pool.   The second, and more common, especially for small installations, is to add a soluble iodide, such as NaI, to the pool and then add an oxidizing agent to form I_2 and HOI.   Both are effective bactericides.   Chlorine, hypochlorites, chloramines, or non-chlorine-containing oxidizing agents can also be used, hypochlorites being the most common.   Care must be taken not to oxidize the iodine to iodate (IO_3^-) which has no bactericidal power, and cannot be brought back down to OI^-, I_2, or I^-.

Although iodine is very expensive, Black has demonstrated (7) that it becomes converted to iodide, which can be used again, and the cycle then repeated many times, leading to good economy.   Studies at Olin Corporation, however, have shown that an appreciable amount of iodine can be lost simply by evaporation.

Loss of Available Halogens by Sunlight.   Two research studies have been made on the effect of sunlight on chlorine, bromine, and iodine.   The first showed the stabilizing effect of cyanurates for chlorine and also pointed out that bromine and iodine are not stabilized by cyanurate (8).   The second was a comparison of the rates of attack on solutions of chlorine, bromine, and iodine (9).   This demonstrated that hypochlorite and hypobromite are destroyed by ultraviolet radiation (sunshine) at the same rate, but that hypoiodite is attacked somewhat more slowly.   (The hypohalites are converted to halates and halides which have no oxidizing or germicidal action.)   The data are summarized in Table 1.

Table 1.   Loss of Available Halogen from Open Systems under the Xenon (UV) Arc, hr

Active species	Half-life	Time for 100% loss
$HClO$, ClO^-	0.6	2.5
$HBrO$, BrO^-	0.6	3.0
HIO, I_2	2.0	5.0

Other Sanitizing Agents.   There are a number of other sanitizing agents that can be used for purifying pool water but these account for an extremely small part of the market.

Ultraviolet radiation (see Photochemical technology) can purify water but it is difficult, expensive, and for all intents, impractical to use for swimming pools.   Residual protection is impossible to maintain.

Chlorine dioxide, ClO$_2$ (see Vol. 5, p. 35), is used in some parts of the world, particularly Europe, but finds practically no application in the United States. It is an effective germicidal agent and is reported to be superior to chlorine as a viricide. It also is effective for burning out organic matter, but it is more difficult to use and control than chlorine. The sodium chlorate, which is needed to generate chlorine dioxide, is dangerous, especially in small installations.

Silver and some of its compounds can be used in water treatment. Such products are bactericidal, but are relatively expensive and hard to use. They are sensitive to chemicals and impurities in the water.

Bleaching powder (see Vol. 5, p. 18) is used in some of the underdeveloped areas in the world for treating swiming pools, but finds little use elsewhere. The material can be put in a holding tank to let the lime settle out and then the clear supernate used, or it can be added in front of the pool filter which retains the lime.

Ozone is a bactericide that finds some use in water purification, but it is practically impossible to maintain effective residual levels in a swimming pool. Because of this and difficulties in generation and handling, ozone is not widely used.

Chlorinated hydantoins (10) have found limited use as sanitizing agents. One product marketed commercially contains 1,3-dichloro-5,5-dimethylhydantoin. (See Vol. 11, p. 154.)

The glycourils, numbering about fifty, are N-chloro compounds which have found some use in swimming pools. One example of the class is tetrahydroimidazo-[4,5-d]-imidazole-2,5-(1H,3H)-dione. The structure (2) is that of the tetrachloro derivative.

$$
\begin{array}{c}
\text{Cl} \qquad\qquad \text{Cl} \\
| \qquad\qquad\quad | \\
\text{N} \qquad \text{R} \qquad \text{N} \\
O{=}C \qquad\quad C \qquad\quad C{=}O \\
\text{CH}_2 \\
\text{Cl—N———C———N—Cl} \\
\text{R}
\end{array}
$$

(2)

Electrolytic cells, built in as an integral part of the pool installation, have been used to generate chlorine for swimming pools. This type of treatment has some utility, but has not found widespread use because of difficulties with the electrodes. It may necessitate a high concentration of sodium chloride in the pool water, which can be corrosive to the filter and piping. Some operating systems avoid this problem by using a separate stream of sodium chloride solution.

Superchlorination

In addition to the normal requirements for the various pool sanitizing agents, it is necessary at times to add supplementary amounts of chlorine to superchlorinate the pool. This shock treatment burns out any accumulation of organic material that builds up in the pool which would cloud the water and serve as a nutrient for algae and bacteria growth.

It is recommended that the pool be superchlorinated every other week when the average afternoon temperature is below 80°F and once every week when the temperature is above 80°F. This treatment calls for the addition of 10 ppm available chlorine. Superchlorination is essential for a stabilized pool containing combined chlorine.

Calcium hypochlorite, because of its convenience and economy, is the most widely used product for superchlorination. Sodium hypochlorite is used in some areas.

Algicides

Algicides (11–13) find use in swimming pools in special situations rather than as regular, routine treatments. Many times they are used in off-season care of the pool water. In a pool that has not been properly maintained algae can grow and present problems. Sides of the pool, bottom, and ladders become slippery; bad odors develop; the chlorine demand increases; the water becomes turbid; bacteria growth is fostered and hard-to-remove "stains" occur because of algal growth. If the proper level of sanitizer and pH are maintained, there is seldom any algal growth and little need for algicides. Situations may develop, however, particularly when the temperature is over 80°, where algicides are required. Care must be taken not to select an algicide that will react with other pool chemicals or possess too high a chlorine demand (14).

Kinds of Algicides. The chlorine, bromine, and iodine chemicals that are used as pool sanitizers are excellent algicides and should prevent most problems with algae.

Quaternary ammonium compounds (qv) have become popular in more recent years, and a number of "quats" are known. One of several groups is the alkyldimethylbenzylammonium chloride class.

Copper sulfate has been used for many years as an algicide, but is not generally effective in a swimming pool. At normal pool-water pH, some of the copper precipitates and can cause hard-to-remove stains. Copper chelates have now been developed and these are finding some utility.

Mercury derivatives are effective as algicides, but many are toxic to humans and animals, and care must be exercised in their application. One of the most popular is phenylmercuric acetate, but it is also a poison.

Derivatives of other heavy metals, such as silver, zinc, and lead, are algicides, but they find a rather limited use in swimming pools.

Algicides, especially the quaternaries, are liable to be removed by diatomaceous earth pool filters. This must be taken into account when trying to maintain a definite level in a pool for corrective purposes. Sometimes the filter is turned off to permit the quats to function.

When all corrective treatment has failed and the algae have become "old" (new or young algae are more readily destroyed), there is another procedure that is used to remove the algae. The pool can be drained and scrubbed down with a 5% hypochlorite solution or a strong copper sulfate solution, which is highly effective under these conditions. This requires a great deal of work and results in "down time" for the pool. The best procedure is to keep the algae from gaining a foothold in the first place.

pH Control

The control of pH in a pool is very important for a variety of reasons. The optimum pH range for pools is 7.2 to 7.6. In this range half of the available chlorine exists as HOCl and half exists as the OCl⁻ ion. At pH 7, the available chlorine is 75% in the form of HOCl, which is desirable from a germicidal standpoint, but at this low a pH the bathers will begin to notice irritation of the eyes and mucous membranes. Also, at low pH levels the metallic parts of the pool, pump, filter, and ladder, will corrode rapidly.

At high pH levels the available chlorine is mainly in the form of the OCl^- ion which has practically no bactericidal effectiveness. Since the OCl^- ion is attacked and destroyed by ultraviolet radiation, it is undesirable to have too much of the available chlorine in this form. At a pH of 8, only 23% of the available chlorine is present as HOCl, the bactericidal agent. High pH contributes to scale formation by causing the hardness in the water to precipitate. Scale formation in the pool is unsightly and has a bad effect on pool heaters because of reduced efficiency and even failure.

The pH is readily controlled by the use of available, inexpensive chemicals. Hydrochloric acid is frequently used to lower the pH of a pool. When a solid is preferred, sodium bisulfate ($NaHSO_4$) can be used. Sodium carbonate (Na_2CO_3) is normally used to increase the pH of pool water.

The role of pH is again emphasized in the Langlier or saturation index (15) which relates pH, temperature, dissolved solids, alkalinity, and hardness. By means of this index one can adjust these variables to maintain excellent control in a pool.

Hardness and Alkalinity

Hardness in water is a measure of the dissolved solids present, especially calcium and magnesium salts. In many areas the natural waters contain high concentrations of solids and this tends to build up in a pool by evaporation and the addition of chemicals. Excessive hardness can limit the choice of pool sanitizers, cause scale formation, reduce pool-heater efficiency, and lower filter effectiveness.

Alkalinity is a measure of the buffering capacity of the water, or the degree of resistance to change in pH of the water. In some parts of the world the alkalinity and pH of natural waters are high. In order to make these waters suitable for pool use some of the alkalinity (16) must be destroyed so that the pH can be lowered to the proper levels. This normally is done by the addition of acid, usually HCl. In the cases where the alkalinity is low pH control is difficult because the water has very little buffering capacity and the pH will change rapidly. Sodium bicarbonate is the agent of choice to increase alkalinity. Total alkalinity ideally should be in the range of 80 to 120 ppm, but many pools operate beyond these limits.

Feeders

Feeding devices (17) to disperse small amounts of chemicals into pools at continuously controlled rates are becoming more popular. The reason for feeders is to provide a proper, continuous level of protection at all times rather than have wide swings of sanitizer concentration. Feeders serve a real function in small pools as the owner can go away for a few days and leave the pool unattended.

There are devices which float at random around the pool and the chemicals, usually solids, are slowly leached out. Some of these devices use the sink-float principle whereby they sink to the bottom of the pool when full, and then float to the surface when empty.

There are pumps which meter chlorine gas or liquid bromine into pools at low rates. A number of pools use small proportioning pumps to deliver dissolved chemicals from prepared solutions. A word of caution on this type of feeder: sometimes these have been scaled down for small installations to the level where a bubble can block a line or a piece of dirt stop the flow.

Another method of feeding chemicals is to have a by pass stream from the pool leach out sanitizing agents from a bed or hopper of solid product. Still other devices use a small screw or auger-like arrangement to deliver solids. There are other means of adding chemicals to pools, but these are the principal types.

Test Kits

A test kit is a "must" for the satisfactory operation of a pool. They vary widely in testing capabilities—and cost—but all determine chlorine (18) and pH. These should be checked frequently, especially during periods of heavy bather loads.

Some kits also provide a means of determining hardness in the water. This is an important factor for pool control in many areas. Alkalinity is another quantity that can be determined with a test kit. This is a determining factor in proper control of a pool.

Some kits may provide a test for cyanuric acid which should be measured from time to time in stabilized pools (every few weeks if chlorocyanurates are being used as the source of available chlorine) to keep the level from becoming too high. Several kits also are available which utilize a photometric measurement to determine color quality and clarity of the water.

As a rule copper, nitrogen, iron, and manganese are not measured with a test kit. These are determined by sending a water sample to a local analytical laboratory.

The chemicals in test kits do deteriorate with age and refills are inexpensive. It is unwise to use the chemicals for more than one season.

Other Chemical Treatments

Occasionally a pool contains other impurities that need to be removed (19). One common foreign element to be reckoned with is iron. The presence of iron in pool water makes it unattractive and there are several means of controlling or removing it.

A common means of removing the iron is to floc the pool. This is done by adjusting the pH to 7.2–7.6, superchlorinating and then adding alum over the surface of the water (2 oz alum for each 1000 gal of water). A floc of metallic hydroxides will form, which is allowed to settle and then vacuumed from the pool. The pH is again adjusted to 7.2–7.6. Alternatively a number of chelating solutions are on the market based on ethylenediamine and similar compounds (see Complexing agents). These can complex the iron and other metal ions, particularly copper and manganese, and keep them in a soluble form.

Maganese is a problem in some areas and it can be removed by chlorination and precipitation at pH 8.5.

Copper in the water generally does not cause problems unless the pool gets out of balance. Sometimes it will precipitate as black deposit if the pH falls below 7.

Indoor Pools

Indoor pools must be maintained in a clean, sanitary condition just as outdoor pools, but there are some points of difference which must be noted (20).

Many times the pH is maintained between 8.0 and 8.5 even though chlorine becomes less effective as the pH increases. This relatively high pH is the optimum to control eye irritation, and unpleasant odors. As a result these pools require higher chlorine residuals (1.5–2.5 ppm). The increased chlorine level compensates for the lowered effectiveness.

These pools require shock treatment, just like outdoor pools, to remove organic and nitrogenous materials. In the absence of breezes to dispel odors the atmosphere can become distinctly unpleasant if the pools are not cleaned up and closely controlled. The time required for shocking, however, is significantly longer, 24 hr, because of the high pH which keeps down the HOCl concentration.

Scale is a problem in indoor pools because of the higher pH levels, but it can be controlled with polyphosphates. Scale can arise from hardness in the water, leaching of the pool walls, and the chemicals used. Phosphates provide food for algae, and this precludes their use for scale-control in outdoor pools (see Eutrophication, in Supplement Volume), but since algae do not grow in the absence of sunlight, phosphates can be used for indoor pools.

Bibliography

1. B. A. Kirk and W. A. Lindeke, "A New Pool Sanitizer Discussed by Maker," *Swimming Pool Weekly and Swimming Pool Age* **42**, 36–38 (1968).
2. J. R. Anderson, "The Effect of Cyanuric Acid on Chlorine's Killing Power," *Swimming Pool Data and Reference Annual* **32**, 86 (1965).
3. E. S. Roth, "How the Cyanurates Act as a Stabilizer, Sanitizing Instrument in Swimming Pools," *Swimming Pool Weekly and Swimming Pool Age* **43**, 37–40 (1969).
4. E. D. Robinton and E. W. Mood, "Cyanuric Acid's Effect on Swim Pool Sanitation," *Swimming Pool Weekly and Swimming Pool Age* **42**, 104–111 (1968).
5. L. S. Stuart and L. F. Ortenzio, "Swimming Pool Chlorine Stabilizers," *Soap Chem. Specialties* **40**, 79 (1964).
6. A. S. Behrman, *Water is Everybody's Business*, Doubleday & Co., Inc., New York, 1968, Chap. 7.
7. A. P. Black et al., *Swimming Pools and Natural Bathing Places*, an annotated bibliography, *Bull. No. 1586*, U.S. Public Health Service, U.S. Government Printing Office, Washington, D.C., 1966, pp. 30–31.
8. G. D. Nelson, "Swimming Pool Disinfection with Chlorinated Isocyanurates," *Special Report 6862*, Monsanto Co., March 1967.
9. J. J. Bishop, J. P. Faust, and S. I. Trotz, "Comparative Light Stability of Aqueous Chlorine, Bromine, and Iodine, *National Swimming Pool Institute Paper*, 12th Annual Convention, New Orleans, January 1969.
10. Israel, B. N. *Hydrolysis of Some Organo Halogenating Agents*, Ph.D. thesis, University of Wisconsin, 1962.
11. R. M. Stern, "The Basic Properties of Algae, How They Determine Chemical Treatment," *Swimming Pool Data and Reference Annual* **32**, 78 (1965).
12. J. M. Michael, "Algae Control," *Swimming Pool Weekly and Swimming Pool Age* **33**, 54 (1959).
13. S. Frances, "How Good is Your Present Algicide?" *Swimming Pool Weekly and Swimming Pool Age* **36**, 25 (1962).
14. G. P. Fitzgerald, "Compatibility of Swimming Pool Algicides and Bactericides," *Water Sewage Works* **115**, 65–71 (1968).
15. W. F. Langelier, "Chemical Equilibria in Water Treatment," *J. Am. Water Works Assoc.* **38**, 169 (1946).
16. F. J. McIntyre, "Alkalinity, pH, and Chlorination," *Swimming Pool Data and Reference Annual* **28**, 168 (1960).
17. A. Long, N. N. Coe, and T. L. Heying, "Erosion Feeders for Swimming Pools," *National Swimming Pool Institute Paper*, 12th Annual Convention, New Orleans, Jan. 1969.
18. J. D. Paton, "New Free Chlorine Test Method is Spreading Rapidly in Canada," *Swimming Pool Weekly and Swimming Pool Age* **43**, 37–38 (1969).

19. R. S. Ingols, "Metal Stains in Pools," *Swimming Pool Weekly and Swimming Pool Age* **42,** 41 (1968).
20. R. S. Ingols, "Indoor Swim Facilities Need Different Chemical Method," *Swimming Pool Weekly and Swimming Pool Age* **43,** 40 (1969).

General References

Swimming Pool Weekly and Swimming Pool Age, published three times each month except September.

Swimming Pool Data and Reference Annual, published each year in September.

Swimming Pool and Natural Bathing Places, an annotated bibliography, 1957–66, U.S. Department of Health, Education and Welfare, Public Health Service, Bureau of Disease Prevention and Environmental Control, National Center for Urban and Industrial Health, Washington, D.C.

L. C. Bell, "Don't Let Your pH See-Saw," *Swimming Pool Age* **36,** 28 (1962).

C. W. Chambers and N. A. Clarke, "Control of Bacteria," *Adv. Appl. Microbiol.* **8,** 105 (1966).

G. P. Fitzgerald and M. E. DerVartanian, "Factors Influencing the Effectiveness of Swimming Pool Bactericides," *Appl. Microbiol.* **15** (3), 504–509 (1967).

G. P. Fitzgerald and M. E. DerVartanian, "Ammonia Level Can Determine the Rate of Kill," *Swimming Pool Weekly and Swimming Pool Age* **42,** 92–103 (1968).

L. F. Ortenzio and L. S. Stuart, "A Standard Test for Efficiency of Germicides and Acceptability of Residual Disinfecting Activity in Swimming Pool Water," *J. Assoc. Offic. Agr. Chemists* **47,** 540–547 (1964).

E. D. Robinton and E. W. Mood, "An Evaluation of the Inhibitory Influence of Cyanuric Acid upon Swimming Pool Water Disinfection," *Paper, Public Health Association Annual Meeting, Chicago, Ill., October 1965.*

F. L. Strand, "Why Chlorinate Swimming Pool Water?" *Swimming Pool Weekly and Swimming Pool Age* **41,** 56 (1967).

F. L. Strand, *"Swimming Pool Operation Manual,"* National Swimming Pool Institute, Washington, D. C., 1969.

F. E. Swatek, H. Raj, and G. E. Kalbus, "Cyanuric Acid—A Laboratory Evaluation Using Swimming Pool Water," *Paper, National Swimming Pool Institute 10th Annual Convention, Las Vegas, Nevada, January 1967.*

G. C. White, "The Background of Today's Pool Water Chlorination Methods," *Swimming Pool Weekly and Swimming Pool Age,* Part 1: **43,** 28–30 (July 7, 1969); Part 2: **43,** 28–29 (July 14, 1969); Part 3: **43,** 34 (July 28, 1969); Part 4: **43,** 28–29 (Aug. 4, 1969).

Residential Pool Care Guide, National Swimming Pool Institute, Washington, D. C., 1969.

"Swimming Pool Chemicals," *Consumer Reports,* 367–371 (July 1969).

Albert Dietz and Robert Walsh, "Calcium Hypochlorite," *Swimming Pool Data and Reference Annual* **34,** 24–30 (1967).

Swimming Pools, Hoffman Publications, Inc., Fort Lauderdale, Fla., 1969.

J. P. FAUST and A. H. GOWER
Chemical Division, Olin Corporation

WATER GAS.   See Gas, manufactured.

WATERPROOFING AND WATER REPELLENCY

Among the basic needs of man, along with food, drink, clothing, and shelter, is the necessity of keeping dry. To this end, much research has been done and waterproofing treatments are applied to a great variety of products ranging from clothing to building structures and including such varied materials as textiles, paper, leather, cement, and masonry. Strictly speaking, the term waterproofing denotes a process that confers complete impermeability to the passage of water. It is also used, however, in a looser sense to include water-repellency (nonwetting) treatments which produce porous hydrophobic surfaces without appreciably changing the original texture and pore structure.

Textiles

A large amount of research has been conducted on treatments to make textiles water resistant. Early patent literature is cited in the books of Mierzinski (1) and Pearson (2). A review of chemical methods for imparting water repellency contains 332 references (3), while a bibliographical survey of the field from 1937 to 1947 (4) contains 250 references. Eighty-eight PB (Publication Board) reports, received from both domestic and foreign, civil and military agencies, are cited in the suggested list of references on water repellency and waterproofing of textiles issued in 1954 by the Office of Technical Services, U.S. Department of Commerce (5). Articles of practical interest have been written by several authors (6–8). A very thorough coverage of the subject of waterproofing and water repellency to 1963 is contained in the book edited by Moilliet (9). Also a large number of products are offered on the market for application to fabric. See also Coated fabrics; Textile technology.

Terminology. Some confusion prevails in the use of terms pertaining to waterproofing processes and resulting products. In the Technical Manual and Yearbook of the American Association of Textile Chemists and Colorists (AATCC) (10) and in the American Society for Testing and Materials (ASTM) standard D583-63 (11) are the definitions: *water resistance (fabric)*, a general term denoting the ability of a fabric to resist wetting and penetration of water; *water repellency (textile)*, the ability of a textile fiber, yarn, or fabric to resist wetting.

Although the term *waterproofing* is commonly applied in a broad sense to all types of treatments for rendering textiles resistant either to the passage of water through them or to wetting by contact, the tendency is to limit its application to the production of coated and impregnated fabrics that are highly impervious to water in both the liquid and vapor states. The term *water-repellent treatment* is then applied to those processes that produce a hydrophobic surface, characterized by high resistance to wetting. Such surfaces are usually permeable to water vapor and air, and they may or may not be highly resistant to the passage of liquid water under pressure.

Use of the term *waterproof* is generally discouraged because nothing is really waterproof in the truest sense of the word. However, for want of a better descriptive term to connote high resistance to penetration, and to distinguish between porous and nonporous surfaces, it is often applied in a purely relative sense to highly impervious fabrics produced by any means whatsoever. In this sense, it is applicable also to untreated fabrics such as heavy ducks and other swellable or dense fabrics which owe their resistance to tightness of the weave and swelling ability. Such fabrics possess little resistance to surface wetting unless they have been given a water-repellency treatment.

The *water-repellent* surface, in contrast to the *waterproof*, impervious, wall-like structure, is resistant to both the spreading of water on its surface and to wetting by raindrops. *Spot-resistant* and *stain-resistant* surfaces are similar but resistant to a lesser degree, for they are primarily intended to prevent staining by accidental spills of water-borne soiling agents.

The term *water resistance* is a convenient general term for covering all aspects of resistance to water, individually or collectively, and including resistance to penetration of water, wetting of the surface, and absorption. Under conditions other than those discussed here, it even covers resistance to damaging effects caused by water.

Other terms often encountered are *permeable* or *pervious-to-air and water-vapor*, and *porous*, to distinguish fabrics said to "breathe" from their relatively impenetrable counterparts described in opposite terms.

The terms *durable* and *nondurable* (also called *renewable* or *retreatable*) are applied to repellents and finishes to differentiate between those that are designed to withstand cleaning processes (durable) and those that must be reapplied (nondurable) after dry cleaning or laundering.

Impermeable Fabrics. The majority of the textiles made resistant, by man, to the passage of water are used to keep him or his property dry. When the textile is not to be worn, as a tent or a tarpaulin, coated impermeable fabrics are often satisfactory and generally more economical. In these cases, waterproofing is accomplished by filling the pores of the fabric with materials such as oils, varnishes, pigments, rubber, plastics, tar, and many other materials that interpose a physical barrier to prevent water from passing through.

The impregnating material is sometimes modified so that several benefits may be derived from one application. For example, chlorinated paraffin may be substituted for paraffin to contribute toward the flame resistance as well as the water resistance of a fabric. An example of such multipurpose treatment is used in the Army's fire, water, weather, and mildew resistant duck (12).

Dean (13) has described a simple waterproofing treatment for canvas, suitable for home application, which also gives some measure of protection against mildew and sunlight by incorporation of copper naphthenate and pigments with the waterproofing material. A typical formulation specifies 8.5 lb petroleum jelly, 1.5 lb beeswax, 5 lb dry earth pigments (such as ocher, sienna, or umber), 1.5 lb copper naphthenate (10% Cu), and 5 gal of mineral spirits. A pound of lampblack may be substituted for the earth pigments and 7.5 lb medium-hard asphalt together with 2.5 lb of petroleum jelly may be substituted for the petroleum jelly–beeswax mixture. The materials are dissolved or mixed in the open with great care to be sure no flames or sparks are near. The solvents may be warmed by placing the containers in a tub of hot water. The mixture is somewhat poisonous, and due precautions should be observed. The well-stirred mixture can be applied to canvas with a paint brush. Lighter fabrics may be soaked in the diluted solution, then wrung out and dried. The hands should be protected from contact with the liquid as far as possible.

The early oilskin slickers lost popularity because of their weight, stiffness, and general lack of comfort. Some of the coated fabrics used in the manufacture of rainwear have the advantage of extreme lightness and flexibility, but are almost as impermeable to air as they are to water. Thus, because of inadequate ventilation, the wearer may become almost as wet from perspiration as he would be from rain without the coat. Coated fabrics have lost much of their market to plastic films.

Another type of impermeable cloth, called self-sealing fabric, is exemplified in tightly woven tentage, water bags, and linen fire hose. For example, fire hose is woven from low-twist linen yarn which has low elongation. The fibers wet readily and then expand with the absorbed moisture to give a tightly packed structure with little stretchability which resists leakage under water pressure. Untreated cotton fibers tend to make a less favorable type of yarn structure for self-sealing. However, in England, Peirce et al. (14) produced a cotton fire hose, and further development gave a fine light-weight fabric of Oxford weave which was produced under the name of Ventile cloth. Some work on self-sealing fabrics at the Southern Regional Laboratory in the U.S. has been reviewed by Reid and Goldthwait (15) and includes: (a) the addition of hydroxyethylcellulose to the fibers as a permanent finish which swells when wet, thus preventing passage of water (16), (b) the use of fine, immature cottons to obtain better closing, and (c) the use of a special loom attachment to greatly increase the number of picks per inch (17). Although the self-sealing fabrics ordinarily require wetting to resist further passage of water, they are essentially impermeable to air or to large amounts of water after wetting. A certain amount of seepage occurs, of course, but this is often beneficial. In the case of fire hose, it protects the fabric from burning, and with water bags, evaporation keeps the contents comfortably cool. In modern practice some self-sealing fabrics are given an additional chemical treatment to make them water repellent.

Water-Impermeable—Water-Vapor-Permeable Coated Fabrics. An entirely different type of water-resistant fabric has been developed in work done for the U.S. Dept. of the Army (18) in that a coated fabric that is both permeable to water vapor and impermeable to water has been originated. This fabric is produced by incorporating a solid pigment into a synthetic rubber binder in such manner that a network of three-dimensional microscopic pores is produced. This material may be coated on one or both sides of a water-repellent cloth to give a coated fabric that has been found to withstand more than 277 cm of water pressure by a modified Suter test but that will also transmit at least 75% as much moisture vapor as the original fabric.

In explanation of the theory of this new fabric, an illustrative example is given: 100 g of a pigment of specific gravity of 2.70 was found to occupy 127 cm³ as compared to a theoretical volume of only 37 cm³. This meant that 90 cm³ of void space was available. If 50 cm³ of binder were used to hold these particles together, 40 cm³, or about 31.5% of the coating, would consist of voids available for transfer of moisture vapor. It was found that for the same pore area, moisture vapor transfer was not greatly affected by the size of the pores but hydrostatic head was greatly influenced at sizes smaller than 30 μ.

Another type of microporous finish is made by mixing pore-forming material with the coating polymer and then heating the mixture. The inserted material swells oa form small cells which are then leached out. Microscopic holes in the polymer remain which permit air, but not water, to pass (19).

A different approach to the problem of combining comfort and rain protection involves a change in the fabric porosity in response to changes in the environment (20). This change in porosity is derived from a water-swellable elastomer [based on poly-(ethyl acrylate)] dispersed in the fabric structure. The elastomer is deposited by a unique coating process as a discontinuous layer which is located below the surface and above the midplane of the fabric. When contacted with water, the elastomer rapidly swells and forms a continuous layer or barrier. Because the elastomer is placed below the surface, the fabric retains its texture and does not have the typical coated appearance.

Water-Repellent Fabrics. *Theory.* The discomfort entailed by the wearing of impermeable fabrics led to the development of fabrics that repel water but are permeable to both air and water vapor. These fabrics have been treated with a chemical so that the individual fibers repel the water and cause it to collect in drops on the surface and to run off instead of penetrating the yarns. The interstices between the fibers and yarns remain essentially unchanged, thus allowing the ventilation necessary for body comfort, but the surface tension of the water droplet prevents it from breaking up and going through the holes.

The theory of water repellency has been reviewed in considerable detail by Baxter and Cassie (21) and also by Rowen and Gagliardi (22). It was pointed out that when a drop of water touches a solid it may form an almost perfect sphere in the case where it does not wet the surface, or it may flatten and completely cover the surface if it wets it. The tendency of a solid to resist wetting is a function of the chemical nature of the solid surface, the presence of other molecules adsorbed on this surface, the roughness of the surface, and the porosity of the surface. When these factors are applied to a consideration of textiles, it is apparent that the hydrophilic surface characteristic of most textile fibers must be made more hydrophobic. Fabric and garment structure are also very important.

The presence of other molecules on the surface of the solid is often detrimental to the water-repellency of fabrics (23). Dirt, dust particles, greasy soil, residual soap after cleaning, and the like all affect the shape of the water drop. Also, impurities in the water itself lower the surface tension and enable the water to wet the fabric. The addition of wetting agents to water enables it to penetrate and wet almost any commercially treated water-repellent fabric (24). Both roughness and porosity of the surface affect wettability of the fibers. The principles involved have been applied to the postulation of an ideal fabric (21), and it is shown that a duck's feathers admirably satisfy the theoretical requirements.

There are several considerations in the engineering of a fabric that is to be made water repellent (7). The twist in the yarn and the twist in the ply should be enough so that no fuzzing or fraying is caused in weaving. The fabric should be tightly woven

Table 1. Comparison of Rainfall Resistance of a Single and Double Layer of Cotton Poplin Fabric

Condition	Resistance of cotton poplin to rainfall, sec	
	single layer	double layer
no water-repellent treatment	6	8
slight water-repellent treatment	15	44
more water-repellent treatment	40	700
better water-repellent treatment	51	8,500
best water-repellent treatment	65	10,000

with subsequent processing controlled to maintain tightness of weave. Garment design as well as fabric design is important. The effective resistance of two adjacent layers of water-repellent fabrics is many times that of either layer by itself. This is illustrated in Table 1 by Sookne (25) which shows time of water penetration under various conditions.

The greatly increased efficiency of two layers is attributed to the fact that a double layer affords greater energy absorption under impact; furthermore, a very low order of probability exists that two holes in the fabrics will coincide so as to form a continuous path for the raindrop.

It has also been reported (26) that the supporting surface markedly affects the water resistance of water-repellent cloth. Thus, it may be necessary for coat linings to be made water repellent. Finally, the sewing thread used in raincoats must also be made water repellent to prevent the garment from leaking at the seams in heavy rains.

General. The principal commercial processes are discussed under the following headings: metal salts and oxides; proteins and other nitrogenous compounds; esterification and etherification reactions of cellulose; silicones; and fluorochemicals.

Details of application are easily obtained from manufacturers' bulletins, but the exact composition of the various formulations offered is often a trade secret. After World War II, an extensive survey was made of the German chemical developments in improving the water resistance of textiles (27) and detailed formulas have been published on many of the German processes. German effort was aimed mainly at rayon treatment, and it is stated that the wax-aluminum salt nondurable treatments are conventional except for the substitution of zirconium oxychloride for aluminum salts. In the field of durable repellents, treatments were developed (Persistol VS and Persistol KF) which, although quite different from the long-chain pyridinium complexes commonly used in the United Kingdom and in the U.S., were claimed to be even more resistant to laundering.

A number of general precautions are necessary in the application of water-repellent agents to cloth (26,28). First, the cloth must be selected with considerable care both as to composition and construction. At one time cotton and wool garments were the only ones treated, but with advances in the use of synthetic fibers, nylon, polyester, viscose rayon, acetate, and others are also used for rainwear. The effect of construction and garment design has already been mentioned.

Secondly, the cloth must be carefully freed from natural impurities, residual finishing agents, and other chemicals such as starch size, alkalis, and soaps prior to treatment with the water-repellent agent.

Thirdly, the cloth must be thoroughly impregnated, particularly when durable repellents are used. The assistance of wetting agents is not feasible in many cases since residual amounts counteract the water-repellent effect.

Finally, the effect of other finishing agents applied with the water-repellent agent may lower the repellency. Water-repellent agents are sometimes used in conjunction with other finishing agents to give special effects. In particular, durable-press and water-repellency treatments are applied to produce raincoats which retain their good appearance. Wax dispersions such as Norane 255 (Warwick Div., Sun Chemical) and Cravenette N15 (Crown Metro Corp.) which are essentially free of metal salts, are frequently used on hydrophobic synthetics such as nylon or polyester taffeta.

Metal Salts and Oxides. One of the earliest methods of producing water-repellent cloth involved the precipitation of basic aluminum acetate of varied and indefinite composition within the fibers.   In one process (1,2,29) an aluminum acetate solution was prepared by the double decomposition of a 9% solution of aluminum sulfate with a 14% solution of lead acetate.   Cloth was then impregnated with the solution and rolled up wet without squeezing.   This was repeated four times and the cloth was dried by moderate heat in a festoon chamber.   In another process the cloth treated with aluminum acetate was made alkaline so that alumina was produced to make the cloth water-repellent.

The use of soap instead of alkali to precipitate the aluminum was an obvious step and the presence of an aluminum soap, mostly aluminum stearate, improved the product.   The aluminum soap may be applied from an emulsion as such or the components may be emulsified together; the reaction to form the soap takes place when the cloth is dried.   The aluminum soap may also be dissolved in Stoddard solvent or in benzene and applied from the solution, but this process involves the usual disadvantages of using flammable organic solvents.

Paraffin wax, emulsified with aluminum acetate or formate for use in one-bath treatments, was the predominant finish used in the mid-1950's.   The sale of these products has declined as more durable treatments have become available.   Among the commercial products reported to contain wax emulsions with aluminum salts are Impregnole SP and 337, (Sun Chemical), Aridex WP (DuPont), Cravenette FFS (Crown Metro Corp.), Paramul Repellent 115 (American Cyanamid), as well as several others.

The use of zirconium in place of aluminum assumed importance in Germany during World War II and has since been used to a considerable extent in the U.S.   The zirconium-wax treatment is suitable for a wide variety of fibers including the natural, spandex, polyester, acrylic, and polyamide types.   Application methods are identical with those just described for aluminum but the fabrics so treated, particularly wool and the hydrophobic synthetics such as nylon and polyester, will retain a considerable amount of water repellency after several launderings.   Zirconium oxychloride, $ZrOCl_2$, may be used and this must be buffered with sodium acetate to prevent tendering of the cloth due to evolution of hydrogen chloride when the oxychloride hydrolyzes. The superiority of the zirconium compounds over those of aluminum is ascribed to differences in solubility.   White (30) has pointed out that zirconium stearate is only slightly soluble in drycleaning solvents and that it is also highly resistant to caustic. Aluminum soaps dissolve in caustic, whereas the zirconium soap, if attacked, is converted to insoluble hydroxides or carbonates of zirconium.   The use of the rare earth soaps, as well as those of thorium, zirconium, and uranium, has been patented (30). Among the commercial products which are reported to contain emulsified waxes and zirconium salts are Impregnole FH, Norane 1-2, and Atcodri Z, as well as several others.

The use of tetraalkyl orthotitanates, where the alkyl may range from ethyl to octyl, has been suggested for making cloth permanently water-repellent but these compounds are not known to be used commercially (31).

One of the metal salts employed commercially is "stearatochromic chloride" (approximately hydroxostearatodichromium(III) chloride, $[Cr_2(OH)C_{17}H_{35}COO]Cl_4$) (32,33) (Quilon, Du Pont).   The compound is a Werner complex which attaches to the fiber surfaces by nonionic coordination linkages.   When the solution of cationic agent is applied to the textile and heated to about 120°C, the compound hydrolyzes and then dehydrates, forming a chelate ring which holds the organic portion of the molecule in an oriented position to the surface to give water repellency.

Proteins and Other Nitrogenous Compounds. Proteins have been used extensively to give water-repellent finishes. Solutions of proteins such as glue were insolubilized with aluminum compounds or formaldehyde, tannin, or dichromates. Marsh (29) reviews a number of patents on the use of proteins, and Harding (28) states that the "Mystolenes" manufactured in the United Kingdom are protein-aluminum complexes. Glue is used in many formulations (27).

The use of condensation nitrogen-base resins has led to durable water-repellent treatments principally for cotton- or rayon-containing fabrics. These resins are often based on the melamine molecule which has been modified with formaldehyde and contains various types of hydrophobic groups. Some finishes of this type are Permel Resin B (American Cyanamid), Norane 16 (Sun), and Phobotex FTC (Ciba). The usual sequence of operation for fabric treatment is impregnation, drying, and then heating at 150°C for 3 min. Morel (34) has described two treatments using urea-formaldehyde resins. A number of water repellents, including the silicones described below, are reported to be compatible with the nitrogenous resins used for creaseproofing. Some are reported to contain methylol stearamide. Those resin-base formulations which can be used as fluorochemical extenders (see below, Fluorochemicals) have maintained their sales, but their function is altered; they have become extenders rather than straight repellents.

Nitrogen-containing compounds which are assumed to etherify cellulose are considered in the following section.

Esterification and Etherification Reactions of Cellulose. The idea of combining a water-repellent group with cellulose by chemical reaction, for example, replacing the hydrogen of a hydroxyl group with a long-chain compound, is one that has intrigued many investigators of textile treatments. The special advantage of such a process would be the relatively stable character of the resultant compound, but the obvious difficulty has been carrying out such reactions in an economical manner for commercial exploitation.

The reactions of cellulose are mainly esterifications or etherifications. Schuyten et al. have reviewed a large number of such reactions claimed to give water repellency (3). See also Cellulose derivatives.

Examples of esterification include the formation of a partially substituted cellulose ester such as the stearate made by the use of stearic anhydride or stearoyl chloride. Substituted ketenes and long-chain isocyanates that might add directly to the cellulose have been suggested. Hamalainen et al. (35) found that *n*-octadecyl isocyanate imparted only mild water-repellent properties to cotton textiles and suggest that much of the water-repellent effect may be due to decomposition of the octadecyl isocyanate to yield a waxlike coating of dioctadecylurea rather than to a true reaction with the cellulose. Such processes do not appear promising for commercial use.

It is in the field of cellulose etherification that most progress has been made along the lines of developing durable water repellents. Cellulose ethers as a class are generally stable toward alkali and, therefore, finishes made by formation of ethers are usually durable toward laundering. The historical development of the water-repellent ethers has been reviewed by Marsh (29). Briefly, from 1931 to 1937 a number of materials were produced to give cotton a softer hand or improve its dyeing characteristics.

These compounds were designed to react with cellulose to form an ether through the agency of a chloromethyl group. Reynolds, Walker, and Woolvin, in 1937 (36), modified the method to produce durable water repellency using compounds such as 1-(octadecyloxymethyl)-pyridinium chloride to react with the cellulose. With a number of modifications (37) this process is the basis for the compounds known in England as Velan and in the U.S. as Zelan, and believed to be mainly stearamidomethylpyridinium chloride, although 1-(octadecyloxymethyl)pyridinium chloride is also used (28).

Stearamidomethylpyridinium chloride may be prepared by heating stearamide with paraformaldehyde and pyridine hydrochloride in solution in pyridine (37). Some of the reactions of this compound with alcohols and the formation of various stearamidomethyl ethers have been investigated by Weaver et al. (38).

$$[C_{17}H_{35}CONHCH_2NC_5H_5]^+Cl^- + ROH \rightarrow C_{17}H_{35}CONHCH_2OR + HCl + C_5H_5N$$

While the compounds formed are ethers, they do not possess the stability of the usual ethers because of the presence of the —CONH— group. For example, the amidomethyl ethers may be split by dilute aqueous solutions of strong acids.

This work has been further amplified (39) and applied to the reaction with cellulose where R in the above formula indicates a cellulosic residue attached to the hydroxyl group. Davis (40) had investigated the possible reaction of stearamidomethylpyridinium chloride with cotton and concluded that about 1–2% would react with cellulose to form the ether. He proposed that the mechanism involved hydrolysis to form methylolstearamide (N-(hydroxymethyl)stearamide), $C_{17}H_{35}CONHCH_2OH$, which then reacted with cellulose. In the later investigation (39), the workers agreed with the amount of reaction but found that methylolstearamide would not react with cellulose under the conditions of curing. They believe that the reaction as given above takes place to the extent of about one substituent for every 150 anhydroglucose units of the cellulose. They also point out that the conditions of curing during application of the compound to textiles favor the formation of distearamidomethane (N,N'methylenebissstearamide); $(C_{17}H_{35}CONH)_2CH_2$. This compound is deposited on the fibers and aids greatly in giving good water repellency. It is somewhat resistant to laundering, but the textiles made water repellent by distearamidomethane alone require ironing after laundering to recover their water repellency. Probably the waxlike material must be melted and redistributed on the fiber surface after the disruptive swelling effect of the laundering.

Commercial application of water repellents of the stearamidomethylpyridinium chloride type varies with the particular product, and a number of competitive products are on the market. A typical product has been described (39) as a paste composed largely of stearamidomethylpyridinium chloride. The paste is applied to the fabric in from 2–6% pickup, based on the weight of the fabric, the amount varying according to the properties of the individual fabric. It is used on cotton, linen, viscose rayon, cellulose acetate, nylon, acrylic fibers, silk, and wool. The paste is first dissolved in approx half of the water to be used at 40°C, using good mechanical stirring to disperse it. About 10% of the weight of the paste of anhydrous sodium acetate is separately dissolved in warm water and added. The whole is then made up to volume with more water. The sodium acetate is very important to the success of the treatment because it serves to buffer the solution and neutralize the hydrogen chloride evolved during the reaction and thus protects the cloth from tendering. The buffered solution is slightly acid and this acidity must be maintained during application. The opalescent solution is stable for only a few hours and should be applied to the cloth fairly soon at about 40°C.

The solution is impregnated into the cloth and excess solution squeezed out by passing the cloth through rubber squeeze rolls to leave the desired pickup. After the cloth is dried at 120–140°C, so as to remove the water but not to decompose the compound, it is cured at 140–200°C, the optimum temperature varying from about 177°C for cotton and linen to about 150°C for the synthetic fibers. Curing times vary inversely with the temperatures used, and conditions are usually established experimentally for a given fabric and equipment. During curing, part of the agent reacts with cellulose to give the cellulose ether and part decomposes to yield distearamidomethane.

Fumes of formaldehyde and pyridine are given off from the padding baths and on drying and curing the cloth. Therefore, good ventilation must be provided at all points. The fumes are slightly acidic, so acid-resistant materials must be used in the equipment. The cloth is also slightly acid at this point and contains residual decomposition products of the reaction. It must be neutralized, for example with small amounts of soda ash solution, and thoroughly washed. The cloth is then dried at the highest practicable temperature (135–175°C) to develop maximum repellency.

1-*n*-Octadecyl-3-ethyleneurea, developed as Persistol VS or Primenit in Germany, was claimed to be superior to Velan PF on rayon. Details of its preparation and use are given by Dahlen and Pingree (27). The repellent agent was prepared by treating *n*-octadecyl isocyanate with ethylenimine in emulsion:

$$\text{C}_{18}\text{H}_{37}\text{NCO} + \text{NH}\cdot\text{CH}_2\cdot\text{CH}_2 \rightarrow \text{C}_{18}\text{H}_{37}\text{NHCON}\cdot\text{CH}_2\cdot\text{CH}_2$$

This compound is claimed to react with cellulose (R_{cell}OH) on drying at about 80°C to give the corresponding ether:

$$\text{C}_{18}\text{H}_{37}\text{NHCON}\cdot\text{CH}_2\cdot\text{CH}_2 + \text{R}_{cell}\text{OH} \rightarrow \text{C}_{18}\text{H}_{37}\text{NHCONHCH}_2\text{CH}_2\text{OR}_{cell}$$

A small amount of a reagent capable of cross-linking the cellulose was also added. For this, compounds such as hexamethylenebis(*N'*-ethyleneurea), $\text{CH}_2\text{CH}_2\text{NCONH-}$ $(\text{CH}_2)_6\text{NHCONCH}_2\text{CH}_2$, were used.

Silicon Compounds. In 1942 a patent was granted to Patnode (41), and assigned to the General Electric Company, on the use of vaporized methyltrichlorosilane, CH_3SiCl_3, for treating various materials including paper, cotton cloth, and ceramics to render them water repellent. With this patent a new type of water-repellent agent was introduced. Until that time, most repellent agents had depended upon a long hydrocarbon chain for efficacy, whereas the silicon compounds needed only methyl groups attached to the silicon atom to prevent wetting. Most of the methyl silicones now on the market are made of relatively long chains of silicon and oxygen atoms to which are attached the methyl groups, which are hydrophobic and prevent wetting.

Silicon Compounds (organic). Methyltrichlorosilane did not prove practical for use on cotton textiles because the by-product, hydrogen chloride, tendered the textile. Barry (42) produced long-chain silicon acetates which gave water repellency without production of harmful strong acids. Eventually, silicone resins were found suitable for this purpose. The physical state of these compounds may be varied from low-viscosity fluids to rubberlike solids with suitable variation in the substituents and the degree of polymerization. The three largest producers are the Dow Corning Corporation, General Electric Company, and Linde Air Products Company. A review of the field in 1954 (21) states that the first large use of silicone resin water repellents was in 1950.

The silicones are produced by hydrolysis of alkylsilicone chlorides (alkylchlorosilanes) and polymerized to give poly(dimethylsiloxanes). A typical reaction may be expressed as follows:

$$3 (CH_3)_2SiCl_2 + 6 H_2O \rightarrow 3 (CH_3)_2Si(OH)_2 \rightarrow HOSi(CH_3)_2OSi(CH_3)_2OSi(CH_3)_2OH$$

The polymers are sometimes end-blocked with methyl groups. The application of these compounds alone to textiles is not practical because a temperature of 200–250°C, maintained for several hours, would be required to give durable water repellency. To overcome this difficulty, silicone manufacturers have altered the structure by introducing some reactive groups such as hydrogen atoms to give products of the type:

$$(CH_3)_3SiO[Si(CH_3)_2O] [SiHCH_3O] Si(CH_3)_3$$

The use of these poly(methylsiloxanes) by themselves gives good water repellency at cure temperatures of about 150°C, but a chief disadvantage is the change in handle; it stiffens the fabric. However, the combination of the fully methylated with the methyl hydrogen silicones results in products which are particularly suitable for textile application (23,43). Application of the silicones to cloth may be either from organic solvents or from aqueous emulsions, and examples of both are numerous in the patent literature. Some of the commercial products require a metallic salt catalyst and some do not. In general, curing at a fairly high temperature such as 155°C for 4–8 min is required in order to complete the polymerization of the silicone. The product is somewhat durable to laundering and drycleaning because of the insolubility of the cured film in both water and organic solvents. The use of a reactive polymer (coupling agent) that contains silanol groups for condensation with siloxanes and amidomethylpyridinium groups to react with cellulose to increase laundering durability has been reported (44). A silicone "alloy" made by the simultaneous polymerization of tetravinylsilane and methyl hydrogen siloxane and applied from an organic solvent gave good results (45).

Organo-titanium compounds have been shown to give a synergistic effect on water repellency when mixed with some silicone compounds (46). It is possible to prepare stable solutions of silicones and alkyltitanates in suitable organic solvents that can be applied to textiles and leather garments by a variety of methods, eg, padding, brushing, spraying, and so forth. The treated materials, such as tarpaulins, garments, or piece goods then only require air drying. After drying, the water repellency continues to develop and optimum results are obtained after about 48 hr. A chief advantage of the method is that it allows retreatment of goods, such as tents, under field conditions.

Most of the water-repellent silicone resins on the market are compatible with the creaseproofing resins of the urea- or melamine-formaldehyde type, and it is even claimed that enhanced water repellency is obtained with such resins (47). In an example cited, one part of an organopolysiloxane such as methyltris(trimethylsiloxy)-silane, $CH_3Si(OSi(CH_3)_3)_3$, is mixed with about one-fourth part of the hydrolyzate of methyldichlorosilane, CH_3SiHCl_2, and with a precondensate of a melamine resin. When applied to cloth, dried, and cured for two minutes at 150°C, the product is claimed to be water repellent. Crosslinked silicone films have been investigated as possible washwear, water-repellent finishes for cotton (48).

The silicones are claimed to give a number of desirable properties to the fabrics in addition to durable water repellency. They are especially noted for the durable, soft, smooth hand they impart. It is said (49,50) that on cellulose acetate fabrics the silicone completely covers the individual fibers and that the wrinkle recovery, abrasion resistance, tear and bursting strength, sewability, and appearance are all significantly improved by less than 2% add-on of silicone resin.

Fluorochemicals. Fluorochemical water and oil repellent finishes came into prominence in the early 1960s. The fluorochemicals used are organic compounds in which a high percentage of the hydrogen attached to carbon has been replaced by fluorine, as for example in perfluorobutyric acid, C_3F_7COCH.

The very low surface energy of many fluorochemicals, particularly those compounds with a perfluoroaliphatic group of at least four carbons, uniquely enables them to provide both water and oil repellency when applied to textiles (51). This combined repellency is the main reason for the remarkable consumer acceptance of these finishes. See Fluorine compounds, organic; see also Surface chemistry of fluorochemicals, under Fluorine, Vol. 9, p. 707.

Proper orientation of perfluoro groups is important as they should present a fluorinated surface or shield to water and oil molecules. One way of accomplishing this is to incorporate the fluorinated molecule into a polymer so that the perfluoro groups constitute the side chains as, for example, in the polyacrylate:

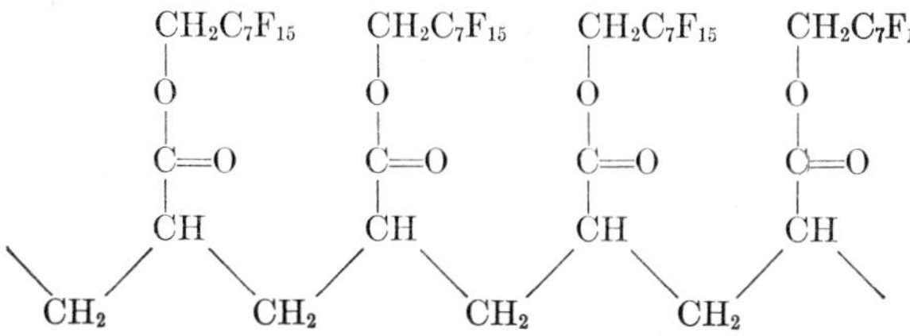

Successful commercial finishes such as Scotchgard 208 (3M Company) and Zepel B (Du Pont Company) are generally believed to be based on fluorinated polyacrylates. These finishes are supplied as fluorochemical resin emulsions and are applied to fabric by conventional pad-dry-cure techniques to deposit from about 0.4–1% fluorochemical solids on the fabric. Because fluorochemical finishes offer a combination of water and oil repellency that gives more complete stain resistance than water repellents alone, they are normally marketed as stain-repellent finishes. Formulations are also available in organic solvents for spray application to upholstery and drapery fabrics.

Fluorochemicals, being very costly, are normally not used alone but are combined with either wax, pyridinium salts, or resin-type water repellents. These additives act as extenders and pad-bath stabilizers and, in some cases, synergistically improve water repellency. The U.S. Army Quartermaster Corps developed a synergistic combination of fluorochemical resins and pyridinium water repellents and named this treatment Quarpel (52). They showed that outstanding water repellency and durability could be obtained on tightly constructed cotton fabrics by the application of a combination of 2% fluorochemical solids and 2% stearamidomethyl pyridinium chloride. When applied to 9 oz cotton sateen, the fabric withstood seven days continuous exposure to rain falling at a rate of 1 in./hr. After 15 launderings, results with this finish were claimed to be at least equal to the initial effects produced using the pyridinium water repellent alone. Combination finishes of this type have been investigated (53) with the objective of reducing costs but maintaining effectiveness.

Some other types of fluorochemical textile treatments that have been investigated include chromium coordination complexes of perfluorocarboxylic acids (54), polyethylenimines having perfluoroacyl side chains (55–58), and a fluoroalkyl siloxane polymer treatment for wool (59). Early commercial fluorochemical finishes that were highly water repellent retained ground-in oily stains even after repeated laundering. A subsequent trend in fluorochemical finishes has been the preparation of hybrid finishes which give less water repellency. These contain, in addition to perfluoroaliphatic groups that provide stain repellency in air, hydrophilic groups to allow better soil release during laundering (58,60). The hydrophilic groups permit the detergent solution to wet the soiled fabric more effectively.

Table 2. Methods for Evaluating Water-Resistant and Closely Related Properties of Textile Fabrics

| | Test reference | | | |
Test method	AATCC (10)	ASTM (11,63,64)	Federal test method (65)	Other
class 1				
spray	22-67	D583-63	5526	
rain	35-67	D583-63	5524	
impact penetration	42-67	D583-63	5522	
drop penetration		D583-63	5520	
Bundesmann water repellency				(66)
class 2				
hydrostatic				
low-range (Suter-type)	127-68	D583-63	5514, 5516	
high-range (Mullen)			5512	
class 3				
dynamic absorption	70B-67	D583-63	5500	
static absorption	21-67	D583-63	5502	
class 4				
contact angle				(67)
rolling drop				(24)
supplementary tests				
air permeability				
NBS-Frazier apparatus		D737-67	5450	
Gurley permeometer				Mfg. Instr.
densometer			5452	
water-vapor permeability		E96-66		

Test Methods. The ability of textile products to resist penetration and wetting (either spread of water over the surface or absorption) depends not only on the construction of the base fabric or yarn, the nature of the coating or repellent, and the method and efficiency of application, but also on "such external factors as the amount of water in contact with the material, the time of contact, the temperature of the cloth and the water, the humidity-regain condition of the cloth, the force and energy with which the rain-drop (or the laboratory equivalent) strikes the fabric, the angle of strike, and the amount of abrasion, weathering, cleaning and other general service conditions to which the material is subjected" (61). Therefore, in addition to testing a new material for its water-resistant properties, the same tests should be performed after it has been subjected to appropriate service conditions, either actual or simulated in the laboratory. Water-repellent fabrics intended for rainwear should be evaluated also for air and water-vapor permeability since both are important comfort factors and do not necessarily correlate with each other (18).

Innumerable methods for evaluating the water-resistant characteristics and utility of textile products have been described (26,62). They vary greatly in procedure, complexity, and interpretation. Methods for fibers and yarns usually employ rate of sinking or absorbency principles. Those for fabrics may be arranged in four classes as follows: Class 1—spray or sprinkling and drop tests to simulate the effect of rain in wetting and/or penetrating the fabric; Class 2—hydrostatic pressure tests to simulate the effects of the pressure of water standing in a fabric, such as a tent roof, or in a bag or hose; Class 3—immersion tests, with or without mechanical action, to check the effectiveness of the water repellent in reducing absorbency; and Class 4—surface tension methods, involving measurements of contact angle or some function of it, to evaluate the repellency or wettability of the surface.

Brief descriptions are given below of the present most commonly used tests for fabrics. For the most part, these have been fairly well standardized in the U.S. by the AATCC, the ASTM, and the branches of the Federal Government. Designations and references for such tests are given in Table 2.

Class 1. The *spray test* is one of the most widely used and simplest to perform. It distinguishes between untreated and treated fabrics, but does not detect small differences in water repellency. Results are independent of fabric construction, for it is a measure of resistance to wetting only and not of resistance to penetration. Essentially, 250 ml of water is sprayed on the test specimen which has been mounted on an embroidery hoop at a 45° angle in such a manner that the center of the specimen is 6 in. below the spray nozzle. The amount of wetting is determined by visual comparison with photographic standards.

The *A.A.T.C.C.-Slowinske raintester* simulates the effects of rainfalls of different intensities by control of the hydrostatic head acting on the spray. The amount of water penetrating the fabric within a specified time at any fixed head is found by determining the increase in weight of a blotter placed behind the fabric. From comparative data, a series of fabrics can be rated for resistance to penetration.

The *impact penetration tester* simulates the effect of a mild rainfall by 500 ml of water falling through a standard spray head from a height of 2 ft against the taut surface of a test fabric. The amount of penetration is determined by the change in weight of a blotter placed behind the specimen.

The *drop penetration tester* simulates the effect of heavy rain by successive drops falling from a 68 in. height and striking the fabric in the same spot. The amount of water passing through the fabric in a fixed time is determined. This tester, which was developed for the United States Quartermaster Corps, provides a relatively severe test but one that has been found especially useful for predicting order of protection in heavy rains. The energy of the drops is approximately 10,000–15,000 ergs, whereas the energy of drops in a cloudburst is about 3000 ergs.

The *Bundesmann water-repellency tester* is little used in the U.S., but it is used in a standard method in Great Britain (66) and Germany. It deserves mention for it appears to imitate service conditions for rainwear better than any of the other tests described. The rubbing and flexing action on the inside of a garment during wear is simulated by the action of wiper arms rubbing on the underside of the test specimen while it rotates under the standard shower designed to imitate a heavy downpour. The amount of water passing through in a fixed period of time is measured and the amount absorbed by the fabric is determined by weighing.

Class 2. *Hydrostatic pressure testers* determine either the head required to cause penetration or the amount of water passing through in a given time at a fixed head. This test is especially applicable to coated and to heavy, closely woven fabrics, such as ducks and canvas, which are to be used in contact with water. It is also very useful for detecting pinholes in coated fabrics and for studying the penetration behavior of untreated swelling-type fabrics. Results are largely dependent on the pore sizes in the cloth. On water-repellent rainwear fabrics, test results do not necessarily correlate with behavior in use—for even though such fabrics permit passage of water under a hydrostatic head, they are highly resistant to wetting and to water transmission by capillarity.

The *Mullen-type bursting-strength tester* has been adapted for testing very heavy fabrics, including fire hose, for resistance to leakage under high pressures. For these tests, water is substituted for the usual glycerin and the test specimen replaces the rubber diaphragm.

Class 3. The *dynamic absorption test* and the *static (immersion) absorption test* consist of tumbling (dynamic) or simply immersing (static) preweighed specimens in water for a fixed period of time and reweighing after the removal of the excess water. The percentage increase in weight indicates the effectiveness of the water-repellent finish.

Class 4. *Surface tension* measurements are helpful in evaluating new water repellents but have not been proven practical for general use. Test procedures are variously described in the literature but have not been standardized. Measurement of the contact angle is the most usual procedure (67), but later a surface-tension index or "rolling-drop" method was developed by Schuyten, Weaver, and Reid (24). A series of known surface tensions is used to determine the surface tension of the solution necessary to cause wetting of the sample. This method has special advantage over the spray test in that it permits greater differentiation among different repellents or among different methods of application to a standard cloth.

Supplementary Tests. *Air permeability* is generally measured in terms of cubic feet of air per square foot of fabric at a pressure drop of 0.5 in. of water, using either the Frazier air-permeability apparatus, developed at the National Bureau of Standards, or the Gurley Permeometer. Tightly woven, lightweight, nonresilient fabrics below the capacities of these instruments can be tested by the falling-cylinder method, using the Gurley Densometer. With this instrument, the time required for a known volume of air to be forced through the fabric under a continually falling pressure is measured.

Water-vapor permeability is a measure of the ability of the fabric to permit transfer of sensible and insensible perspiration from the body to the surrounding atmosphere. Either of two general principles can be employed for this test, namely: determination of (a) the rate at which moisture evaporates from a standard dish containing water and covered with the fabric or (b) the rate at which moisture is absorbed by a desiccant in the standard dish covered with the fabric (64).

Durability tests for resistance to laundering, drycleaning, weather, flexing, etc, are conducted in accordance with conventional test methods. However, it should be noted that commercial drycleaning is more severe than laboratory methods with the result that few finishes, if any, have satisfactory resistance to commercial drycleaning. See Detergency; Drycleaning; Laundering; Textile testing.

Paper

Waterproof Paper. Broadly speaking, the term waterproof paper is purely relative and includes any type of paper that has a high resistance to penetration of water in liquid form. In the paper industry, however, waterproof paper has a specialized meaning, as defined in ASTM D996-59T, *Definitions of Terms Relating to Shipping Containers:*

> "A flexible paper, laminated, coated, or infused to give water or water-vapor resistance, may be creped for added flexibility and/or reinforced with strands of sisal or other fiber or backed with fabric for added strength. Used as box, bag, crate, barrel, and case liners, and bale wrappers."

The term refers usually to kraft papers that have been laminated, coated, or impregnated with a thermoplastic material capable of resisting the penetration of water in liquid form. One of the oldest methods of producing waterproof paper is by laminating two or three plies of plain or extensible kraft paper with asphalt. However, polyethylene, microcrystalline wax, or blends of waxes are frequently used today, especially where any bleeding of the asphalt would damage the packaged article (such as furniture wrap) or where odor from the asphalt would be objectionable (such as food wrap). Polyethylene has the further advantage of permitting heat sealing and it is more suitable for low-temperature applications. Other methods of producing waterproof papers include coating single-ply kraft with polyethylene, asphaltic blends, or with other waterproof resins and by impregnating single-ply kraft with resins, waxes, or latex.

In addition to the uses mentioned in the definition above, waterproof papers serve as water and vapor barriers in roofing materials, temporary tarpaulins, covers for concrete during curing, linings for freight cars and trucks, covers and shrouds for goods and machinery in transit, and interior wraps, especially for export shipments. General requirements and test methods for papers of this type are covered in Federal Specification PPP-B-1055, *Barrier Material, Waterproofed, Flexible.* Trials under service conditions are recommended when selecting papers for specific end uses.

Waterproof paper should not be confused with moistureproof paper, designed to be resistant to water vapor; nor with wet-strength paper, which retains its strength completely or partially after immersion in water. The latter has no special resistance to water penetration in either liquid or vapor form unless it has been coated or laminated to other components that impart water resistance. Wet-strength paper owes this characteristic to treatment of the stock with resins such as urea-formaldehyde, melamine-formaldehyde, and polyamide.

Water-resistant papers. There are many kinds of papers that depend, in some degree, on water resistance for their utility; in fact, even writing and printing papers must have some degree of water resistance, imparted by rosin-sizing treatment, for, otherwise, they would absorb ink like a blotter. Many of the more resistant types are found in paper and paperboard products used for packaging materials and dispensing foods and drinks. Common examples are butcher's wrapping paper; household waxed paper; glassine; butter, milk, and ice cream cartons; drinking cups; soufflé cups; and straws. Other much-used water-resistant papers are agricultural papers, such as plant protectors, celery-bleaching paper, and mulch paper; adhesive tapes; artificial leather; and vulcanized fiber. The water-resistant properties are achieved by many kinds of treatments, such as waxing; coating with resins, asphalt, rubber latex, and plastics; laminating with plastic or metal foil; vulcanizing with zinc chloride; treating with silicones or fluorochemicals; and parchmentizing with sulfuric acid (68–71). Treatment with a combined solution of blocked polyisocyanates and polyvinylacetate polymers (72) and polymerization of hydrocarbons on the paper surface (73) are methods reported to be effective in making paper watertight. [See also Paper, Vol. 14, p. 494]. Research on paper is reviewed annually (69,71).

Leather

The deleterious effect of water on leather is well known. When leather is wet and then allowed to dry it usually becomes stiff and uncomfortable. Greases, alone and in combination with oils and waxes, have long been applied to leather to protect the fibers against damage by water and to improve water repellency. For effective waterproofing, however, large amounts of grease (more than 50% by weight) are required, resulting in a slab of grease held together by leather (74). Consequently, the pores of the leather are largely filled so that the permeability to water vapor is extremely small and the leather is uncomfortable to wear. Furthermore, because the fats and greases tend to work out of the leather during flexing or on repeated wetting and drying, the waterproofing effect is usually temporary so that frequent reapplication of the treatment is necessary.

Impregnation of leather with degraded natural rubber reduces the water absorption and transmission of leather about 50%, but does not seriously impair its water-vapor permeability (75,76). Sole leather can also be waterproofed by incorporating vulcanized linseed oil.

For both military and civilian wear, leather articles, such as shoes, gloves, and jackets, are often required to have high resistance to absorption and permeation by liquid water, but low resistance to the passage of water vapor. They are further required to have essentially the appearance, flexibility, and feel of the original leather. Under the auspices of the Office of the Quartermaster General, U.S. Army, a comprehensive investigation has been made of hydrophobic materials to replace the fats and greases used in tannery practice "so as to increase the water repellency of the leathers and at the same time retain or enhance their existing desirable characteristics" (77,78).

Silicones, fluorochemicals, and long-chain fatty acids and their chrome complexes are types of products developed for imparting water repellency to leather. They may be applied in solution form by impregnation in a drumming or dipping operation or by spraying, brushing, and swabbing, usually after the last wet operation given to the leather. Isopropyl alcohol-water mixtures, perchloroethylene, and mineral spirits are among the solvents that may be used with these products. Some of the products available for treating leathers are: alkenyl succinic acids [Bavon (79)]; silicones [Dow Corning 1109 (80)]; fluorochemicals [Scotchgard (81), Pentel (82)]; and stearato chromic chlorides [Quilon (83)]. These are used for water-repellent leather principally for footwear and garment leather applications, but their use is quite limited due in part to economics.

The efficiency of these water-repellent agents can be markedly improved, at least for shoe upper leather, by retannage of chrome leather with glutaraldehyde (84,85). These treatments do not appreciably affect the softness, feel, flexibility, and porosity of the leather, and the water-repellent leathers retain their water-vapor permeability though inhibiting penetration of liquid water.

Methods for testing the water resistance of leather and its products are included in the revised physical testing methods of the American Leather Chemists' Association (86), in Federal Specification KK-L-311a (87), in ASTM (Maeser) D2099-62T, in ASTM (Dow Corning) D2098-62T, and in ASTM D1913-63 (88). Resistance to water permeation is best tested under conditions simulating use. The Maeser dynamic water penetration machine, which imitates the effect of walking, has been used in the research program of the Quartermaster Corps to improve the water resistance of service shoes and combat boots (89,90). The Dow Corning Leather Tester ASTM imparts a more severe flexing action to the leather thereby accelerating the test, and has been used by many laboratories to evaluate water repellency of shoe upper leathers. Cheronis (77,78) describes a "rain-brick" performance test for leather gloves. Mann (91) developed a tap tester that was suggested as a laboratory method to evaluate repellency of soft leathers such as glove leathers. In the National Bureau of Standards test (87), the sample of leather is flexed during the test for resistance to penetration by water under atmospheric pressure. See also Leather.

Concrete and Masonry

In many uses it is highly important that concrete be watertight, particularly for basements or for outside surfaces exposed to the weather. Much research work has been carried out by the Portland Cement Association and information is disseminated in the form of bulletins (92a–f).

To eliminate and prevent unfair trade practices, the Federal Trade Commission has promulgated *Trade Practice Rules for the Masonry Industry*. These relate among other things to the use of such significantly descriptive phrases as waterproof, dampproof, weatherproof, watertight, and vaporproof.

The term waterproof applied to concrete and masonry implies complete imperviousness to water and to water vapor, whether the liquid and vapor are under pressure or not. To be waterproof, protection with a waterproof barrier or coating, such as a bituminous membrane (see above under paper,) is usually necessary because nearly all untreated concrete or masonry is more or less permeable to the passage of water and water vapor. However, the rate of capillary penetration of water under pressure through concrete is very small (93). Most difficulty occurs because of the leakage of water through openings larger than capillaries. Concrete or masonry which does not leak in this manner is considered water resistant. If it neither leaks nor is capable of transmitting water by capillarity, it is described as "water repellent." The latter, however, may be highly permeable to water vapor and may allow the transmission of water under pressure through its capillaries.

An annotated bibliography of 151 references on permeability and waterproofing of concrete has been published (94).

Standards for use in producing concrete have been established by the American Concrete Institute. When concrete mixtures are properly proportioned (92a–c) and placed by qualified workmen, the resulting concrete structure should be sufficiently watertight for most purposes without further treatment. Proper construction practice includes the installation of an effective drainage system at the building site, and the placing of adequate flashing under all vertical joints in sills, copings, and caps or other horizontal surfaces which permit the accumulation or passage of water (92d). Repair after construction is difficult; suggestions have been listed by the Portland Cement Association (92e).

Waterproofing below grade, or in water-holding structures may require the inclusion of waterstops placed within the concrete during construction. A waterstop is a length of flexible waterproof material placed at a joint in concrete to prevent passage of water (95). As a precaution against extreme groundwater conditions, basement floors are usually laid in two sections with a waterproofing membrane sandwiched between. This membrane can consist of two layers of roofing felt sealed together and to the concrete with hot bituminous material. It should cover the entire floor and extend some distance up the sides of the wall.

Concrete exposed to freezing weather may spall or chip off due to the freezing of absorbed water. Admixtures such as fatty acid salts or asphalt emulsions are sometimes incorporated into fresh concrete mixes to increase their water tightness. However, their long-term effectiveness has been questioned (96). Preformed, air-entrained concrete and mortar can be impregnated with a monomer such as methyl methacrylate which is then polymerized *in situ* either by radiation or by thermal catalytic techniques. The resulting concrete is reported to reduce significantly water permeability and water absorption (97).

Above-grade walls are usually required only to be resistant to the leakage of wind-driven rain, because the seepage of rain water by capillarity through such structures occurs very slowly and is usually of no practical importance. Water-repellent external surfaces are therefore preferable since they are highly water resistant but not waterproof. Waterproof coatings, such as oil-base paints, are vapor barriers and may be unsuitable because of possible condensation and trapping of moisture in the wall.

Proper selection of a coating for concrete masonry walls depends on exposure conditions, the permeability of the coating, and whether application will be above or below grade (92f). Above grade, moisture travels from the interior to the exterior of a wall. The interior surface should be an impermeable barrier. Because of imperfections in the coating or in construction, some moisture will enter the wall. Thus the exterior coating should be permeable enough to allow trapped moisture to escape but repellent enough to resist inward passage of water.

The opposite conditions exist for below grade masonry walls. Moisture travels from the exterior underground surface to the interior so that the interior coating should be somewhat permeable to prevent blistering and flaking. Interior coatings appreciably reduce the flow of water vapor through masonry walls but do not stop a flow of water under pressure that might arise from faulty construction.

Examples of permeable coatings are: fill coats, Portland cement latex paints, and silicone coatings. Bituminous coatings, oil-base and rubber-base paints form impermeable films.

For masonry walls that require waterproofing, a variety of protective coatings are available which not only seal the surface against moisture but also provide a decorative finish (98). Fill coats are used to fill voids and smooth out imperfections in masonry before the application of finish coats. Fill coats contain regular Portland cement as a binder combined with a silicious sand filler or an acrylic or polyvinyl acetate latex. Portland cement paints have a long history of success in water proofing masonry. They are available in powdered form in a variety of colors and are freshly mixed with water before application. The surface to be painted must be moist and the paint spray-dampened for up to 72 hr until the cement cures.

Latex paints are alkali resistant and have demonstrated excellent resistance to penetration by rain. They are water emulsions of materials such as styrene-butadiene, polyvinyl acetate, and acrylic resins and dry as soon as the water has evaporated. Oil-base paints are manufactured from natural oil resins or synthetic alkyd resins. Alkalinity of the masonry surface must be reduced through aging or by application of acidic pretreatments before painting with oil-base paints. Rubber-base paints are normally composed of chlorinated natural rubber blended with pigments and resins. They have good alkali resistance and are highly impermeable to water vapor. Because of their heavy consistency, application by roller is recommended. Epoxy coatings have been used for special purposes, but high cost coupled with application problems have prevented their general use on concrete masonry.

Silicone-based coatings do not alter the color or texture of masonry surfaces but provide good repellency to rain water while allowing passage of water vapor so that the structure can breathe. Bituminous coatings are low in cost and afford excellent resistance to penetration by water. They are produced from coal-tar or asphalt and are available either in solid form to be melted for hot application, or in liquid form—dissolved in solvent or emulsified in water. Use of bituminous coatings is favored where appearance is not important.

Waterproofing treatments have been applied to bridge decks to improve their resistance to freeze-thaw exposure and salt scaling. Silicones at first were thought to be effective, but this was later refuted (99). Linseed oil is fairly effective in reducing scaling and has a good service history (100). Good test results have been reported on the use of a penetrating epoxy (101).

Bibliography

"Waterproofing and Water Repellency" in *ECT* 1st ed., Vol. 14, pp. 862–980, by J. David Reid and Ruby K. Worner, U.S. Dept. of Agriculture.

1. S. Mierzinski, *The Waterproofing of Fabrics*, D. Van Nostrand Co., New York, 1922.
2. H. P. Pearson, *Waterproofing Textile Fabrics*, Chemical Catalog Co., New York, 1924, pp. 27–31.
3. H. A. Schuyten, J. D. Reid, J. W. Weaver, and J. G. Frick, Jr., *Textile Res. J.* **18**, 396–415, 490–503 (1948).
4. F. D. Horigan and C. R. Sage, "*A Literature Survey on Waterproofing and Water-Repellency of Textiles*," U.S. Quartermaster Corps. Tech. Libr. Bibliog. Ser., 2, 81 pp. (1947).
5. U.S. Dept. Comm. Office Tech. Serv., *Bibliography of Reports on Water Repellency and Waterproofing of Textiles*, CTR-151 (April 1954).
6. E. B. Higgins, *Textile Inst. Ind.* **4**, 255–257 (1966).
7. C. A. Davis, *Am. Dyestuff Reptr.* **56**, 555–558 (1967).
8. J. M. May, *Am. Dyestuff Reptr.* **58** (20), 15–19, 45 (1969).
9. J. L. Moilliet, ed., *Waterproofing and Water-Repellency*, Elsevier Publishing Co., Amsterdam, 1963.
10. *Water Resistance and Water Repellency of Fabrics 1968 Yearbook*, Vol. 44, American Association of Textile Chemists and Colorists, Durham, N.C., pp. B149–158.
11. *Textile Materials, Part 24*, ASTM Standards, pp. 126–145 (1968).
12. U.S. Defense Dept. Mil. Spec., Tarpaulin, Duck, Cotton; Fire, Water, Weather, Mildew Resistant Treated, MIL-T-82120 (Marine Corps) (May 16, 1966).
13. J. D. Dean, *Preserving Cotton Fabrics in Outdoor Use*, U.S. Department of Agriculture Circular, 790, 17 pp. (1948).
14. Brit. Pat. 549,974 (Dec. 16, 1942), F. T. Peirce and W. C. Gardiner, (to British Cotton Industry Research Assoc., and F. Reddaway and Co.). U.S. Pat. 2,350,696 (June 6, 1944), F. T. Peirce and W. C. Gardiner (to British Cotton Industry Research Assoc., and F. Reddaway and Co.).
15. J. D. Reid and C. F. Goldthwait, *Crops in Peace and War*, U.S. Dept. Agr. Yearbook, 1950–51, pp. 411–418.

16. C. F. Goldthwait and H. O. Smith, *Textile World* **95** (7), 105, 107, 196, 198 (1945).

17. M. Mayer, Jr., G. J. Kyame, and J. J. Brown, *Textile World* **102** (7), 114–115, 210 (1952).

18. G. E. Martin, H. S. Sell, and B. W. Habeck, *Rubber Age* **66,** 409–415 (1950); *Textile Res. J.* **20,** 123–32 (1950).

19. *Mod. Textiles Mag.* **44** (4), 28, 85 (1963).

20. J. R. Caldwell and C. C. Dannelley, *Am. Dyestuff Reptr.* **56,** 77–81 (1967).

21. S. Baxter and A. B. D. Cassie, *J. Textile Inst.* **36,** T67–90 (1945).

22. J. W. Rowen and D. Gagliardi, *J. Research Natl. Bur. Standards* **38,** 103–117 (1947).

23. R. L. Wayland, Jr., et al., *Amer. Dyestuff Reptr.* **52** (26), 17–25 (1963).

24. H. A. Schuyten, J. W. Weaver, and J. D. Reid, *Am. Dyestuff Reptr.* **38,** 364–366 (1949).

25. A. M. Sookne, "The Problem of Water-Repellent Fabrics," in *Symposium on the Functional Properties of Clothing Fabrics*, Textile Research Institute, New York, 1943, pp. 12–15.

26. G. A. Slowinske and A. G. Pope, *Am. Dyestuff Reptr.* **36,** 108–121 (1947).

27. M. A. Dahlen and R. A. Pingree, "German Chemical Developments in Improving Water Resistance of Textiles," U.S. Dept. Comm. Office Tech. Serv. P.B. Rept., PB-1576, p. 91 (1945).

28. T. R. Harding, *J. Textile Inst.* **42,** P691–P702 (1951).

29. J. T. Marsh, *An Introduction to Textile Finishing*, Chapman and Hall, London, 1966, pp. 458–494.

30. U.S. Pat. 1,536,254 (May 5, 1925), C. B. White (to F. S. Bennett, Inc.).

31. R. J. Speer and D. R. Carmody, *Ind. Eng. Chem.* **42,** 251–253 (1950).

32. R. K. Iler, *Ind. Eng. Chem.* **46,** 766–769 (1954).

33. U.S. Pat. 2,273,040 (Feb. 17, 1942), R. K. Iler (to DuPont).

34. A. Morel, *Ind. textile* **67,** 450–451 (1950).

35. C. Hamalainen, J. D. Reid, and W. N. Berard, *Am. Dyestuff Reptr.* **43,** 453–57 (1954).

36. Brit. Pat. 466,817 (June 7, 1937), R. J. W. Reynolds, E. E. Walker, and C. S. Woolvin (to Imperial Chemical Industries, Ltd.).

37. Brit. Pat. 475,170 (Nov. 15, 1937), A. W. Baldwin and E. E. Walker (to Imperial Chemical Industries, Ltd.).

38. J. W. Weaver, H. A. Schuyten, J. G. Frick, Jr., and J. D. Reid, *J. Org. Chem.* **16,** 1111–1116 (1951).

39. H. A. Schuyten, J. W. Weaver, J. G. Frick, Jr., and J. D. Reid, *Textile Res. J.* **22,** 424–432 (1952).

40. F. V. Davis, *J. Soc. Dyers Colourists* **63,** 260–263 (1947).

41. U.S. Pat. 2,306,222 (Dec. 22, 1942), W. I. Patnode (to General Electric Co.).

42. U.S. Pat. 2,405,988 (Aug. 20, 1946), A. J. Barry (to Dow Chemical Co.).

43. U.S. Pat. 2,588,365 (March 11, 1952), F. L. Dennett (to Dow Corning Corp.).

44. J. W. Gilkey, *Textile Res. J.* **33,** 129–137 (1963).

45. C. J. Connor, W. A. Reeves, and L. H. Chance, *Textile Res. J.* **30,** 171–178 (1960).

46. G. W. Madaras, *J. Soc. Dyers Colourists* **74,** 835–840 (1958).

47. U.S. Pat. 2,612,482 (Sept. 30, 1952), T. J. Rasmussen (to General Electric Co.).

48. J. B. Bullock and C. M. Welch, *Textile Res. J.* **35,** 459–471 (1965).

49. P. Duggan and F. Fortess, *Mod. Textiles Mag.* **34** (3), 31, 60, 62, 64, 66 (1953).

50. C. E. Gibson and F. L. Dennett, *Am. Dyestuff Reptr.* **42,** 275–278 (1953).

51. E. J. Grajeck and W. H. Petersen, *Textile Res. J.* **32,** 320–331 (1962).

52. C. G. DeMarco, A. J. McQuade, and S. J. Kennedy, *Mod. Textiles Mag.* **41** (2), 50–56 (1960).

53. H. B. Goldstein, *Textile Res. J.* **31,** 377–387 (1961).

54. F. J. Philips, L. Segal, and L. Loeb, *Textile Res. J.* **27,** 369–378 (1957).

55. A. G. Pittman and W. L. Wasley, *Am. Dyestuff Reptr.* **53** (10), 30–33 (1964).

56. J. P. Moreau, S. E. Ellzey, Jr., and G. L. Drake, Jr., *Am. Dyestuff Reptr.* **56** (4), 38–42 (1967).

57. W. J. Connick, Jr., and S. E. Ellzey, Jr., *Am. Dyestuff Reptr.* **57** (No. 3), 17–19 (1968).

58. J. P. Moreau and G. L. Drake, Jr., *Am. Dyestuff Reptr.* **58** (4), 21–26 (1969).

59. A. G. Pittman and W. L. Wasley, *Am. Dyestuff Reptr.* **56** (21), 23–24 (1967).

60. P. O. Sherman, S. Smith, and B. Johannessen, *Textile Res. J.* **39,** 449–459 (1969).

61. E. R. Kaswell, "Water Repellency and Water Resistance," in *Textile Fibers, Yarns and Fabrics*, Reinhold Publishing Corp., New York, 1953, pp. 236–255.

62. A. Klingelhöfer, H. Mendrzyk, and H. Sommer, *Wiss. Abhandl. Deut. Materialpruefungsanstalt* **1**, 1–41 (1940).

63. Textile Materials, Part 24, ASTM Standards, pp. 148–151 (1968).

64. General Test Methods, ASTM Standards, Part 30, pp. 198–207 (1968).

65. Federal Test Method Standard No. 191, "Textile Test Methods," General Services Administration, Washington, D.C., December 31, 1968.

66. "Textile Institute Tentative Specifications Nos. 7 and 8," *J. Textile Inst.*, **38**, S1–14 (1947).

67. H. Wakeham, W. B. Strickland, and E. L. Skau, *Am. Dyestuff Reptr.* **34**, 178–82 (1945).

68. R. H. Mosher, *Specialty Papers; Their Properties and Applications*, Chemical Pub. Co., New York, 1950, 520 pp.

69. *Paper Year Book*, Ojibway Press, Duluth, Minn.

70. J. N. Stephenson, ed., *Pulp and Paper Manufacture: Manufacture and Testing of Paper and Board*, McGraw-Hill Book Co., New York, 1953, Vol. III, 945 pp.

71. J. Weiner, *Pulp and Paper Manufacture; Bibliography and Patents.* Technical Association of the Pulp and Paper Industry, New York.

72. Brit. Pat. 940,771 (Nov. 6, 1963), (to W. R. Grace & Co.).

73. G. Rausing and S. Sunner, *Tappi* **45** (1), 203A–205A (1962).

74. C. E. Weir, J. Carter, S. B. Newman, and J. R. Kanagy, *J. Am. Leather Chemists' Assoc.* **43**, 69–95 (1948).

75. J. R. Kanagy and R. A. Vickers III, *J. Am. Leather Chemists' Assoc.* **45**, 211–242 (1950).

76. R. Oehler, T. J. Kilduff, and S. Dahl, *J. Am. Leather Chemists' Assoc.* **45**, 349–377 (1950).

77. N. D. Cheronis, "The Development of Water Resistant Leathers," *U.S. Quartermaster Research Development Rept., Footwear Leather Ser., No. 8*, 333 pp. (1954).

78. N. D. Cheronis, et al., *J. Am. Leather Chemists' Assoc.* **44**, 282–308 (1949).

79. Brit. Pat. 694,586 (July 22, 1953), G. H. von Fuchs; U.S. Pat. 2,693,996 (Nov. 9, 1954), G. H. von Fuchs.

80. *Silicone Notes, Preliminary Data, 8-211*, Dow Corning Corp., Midland, Mich., September 1954.

81. *3M Brand Leather Chemical FC-146*, Minnesota Mining & Manufacturing Company, St. Paul, Minn., May 1, 1960.

82. *Pentel*, Pennsalt Chemical Co., King of Prussia, Pa.

83. *Quilon Chrome Complex, Product Information Bulletin*, E. I. du Pont de Nemours & Company, Inc., Wilmington, Del.

84. F. P. Luvisi, W. J. Hopkins, J. Naghski, and E. M. Filachione, *J. Am. Leather Chemists' Assoc.* **61**, 585 (1966).

85. W. J. Hopkins, F. P. Luvisi, and E. M. Filachione, *J. Am. Leather Chemists' Assoc.* **62**, 162 (1967).

86. M. Maeser, et al., *J. Am. Leather Chemists' Assoc.* **48**, 218–226 (1953).

87. U.S. Federal Supply Service, Federal Spec. Leather; Methods of Sampling and Testing, KK-L-311a (Jan. 19, 1953).

88. ASTM Standards, Part 15, 665–668 (1966).

89. M. Maeser, *J. Am. Leather Chemists' Assoc.* **42**, 390–408 (1947).

90. W. T. Roddy, *J. Am. Leather Chemists' Assoc.* **43**, 419–422 (1948).

91. C. W. Mann, *J. Am. Leather Chemists' Assoc.* **51**, 634 (1956).

92. The following pamphlets are available from the Portland Cement Association, Skokie, Ill. Concrete Information: (a) Fundamental Facts about Concrete, ST-101; (b) Watertight Concrete, ST-33; (c) Building Watertight Concrete Masonry Walls, CP-13; (d) Recommended Practice for Building Watertight Basements with Concrete, CP-24-7; (e) Repairing Damp or Leaky Basements in Homes, CP-12; (f) Effect of Various Substances on Concrete and Protective Treatments, Where Required, ST-4-2.

93. P. T. Norton and D. H. Pletta, *Am. Concrete Inst.* **27**, 1093–1132 (1931).

94. *Annotated Bibliography on Permeability and Waterproofing of Concrete*, Canada Natl. Res. Council, Div. Bldg. Res. Bibliog. No. 13 (46 pp.) (Jan. 1958).

95. "Waterstops," Concrete Construction, p. 105 (April 1968).

96. American Concrete Institute Committee 212, *J. Am. Concrete Inst.* **60**, (11), 1481–1524 (Nov. 1963).

97. *Concrete-Polymer Materials, BNL 50134 (T-509)*, Brookhaven National Laboratory, 83 pp. (Dec. 1968).

98. "Decorative Waterproofing of Concrete Masonry Walls," *NCMA TEK, No. 10*, National Concrete Masonry Association, Arlington, Va. (June 1968).
99. P. Klieger and W. Perenchio, *Highway Research Record* **18**, 33–47 (1963).
100. W. E. Grieb and R. Appleton, *Public Roads*, **33** (1), 1–4 (April 1964).
101. R. Brink, W. E. Grieb, and D. O. Woolf, *Highway Research Record* **196**, 57–74 (1967).

J. David Reid and William J. Connick, Jr.
U.S. Dept. of Agriculture

WATTLE BARK.   See Leather, Vol. 12, p. 317.

WAXES

The word wax, originates from the Angle-Saxon word, weax, which was first applied to the natural material in the honeycomb of the bee (1).

The term wax is used very broadly and is applied to a wide variety of materials both as components and as finished preparations. The materials either have properties resembling certain well-known natural waxes or can be used to provide physical performances similar to those associated with some of the time-honored functional uses of wax, such as filling, sealing, polishing, candle making, etc. Waxy substances have been known and found useful by man for a long time. The properties and usefulness of beeswax probably predate written history. Originally the term wax would have denoted beeswax almost exclusively, but the following can be given as a definition: Historically, waxes are substances having characteristics somewhat like beeswax; chemically, waxes are esters of fatty acids and monohydric fatty alcohols; physically, waxes are water-repellent solids having useful degrees of plastic character; functionally, waxes now include many substances that can be used in place of natural waxes as ingredients in preparations, and the term "wax" also identifies the finished compositions providing the time-honored services associated with waxing treatments.

Waxes are widely distributed in nature with commercially important representatives in each of the following classifications: animal (including insects), vegetable, and mineral. Through chemical analysis it was found that beeswax and many of the other waxes had a similarity in the molecular structure of the principal compounds. Although it has been restrictive, the classical chemical definition of wax has been the esters of fatty acids and monohydric fatty alcohols. Many other chemical structures also are common in waxes and long-chain hydrocarbon compounds are particularly important in the mineral-wax classification. The organic structures that cause the water repellancy and many of the other fatty characteristics of the waxes are the long alkyl chains. These carbon chains are largely unbranched, but there may be one or more chains with from 20 to more than 70 carbons per molecule, without any repeating unit as in a polymer. The major classes of waxes are plant, animal (including insect), mineral (including petroleum), and synthetics. Waxes are used in many industries for a wide variety of applications such as paper coating, polishes, electrical insulation, carbon paper, textiles, leather, precision casting, pharmaceuticals, etc.

Properties.   The physical characteristics generally dictate the nature of the application and control man's appreciation of waxy substances. A cold-working, plastic nature practical for hand-filling, sealing, and polishing functions, has always been important and is associated in identifying waxlike character. Physically, therefore, waxes are plastic solids that have cold-flow yield values within the force range practical for manual working at room temperature. Like beeswax with its fatty chemical structure, waxes as a group have the hydrophobic or water-repellent character of long alkyl compounds.

Waxes together with oils, fats, gums, resins, and pitches are in the large grouping of fat-solvent soluble substances. Within the group the terms "waxy" and "waxlike" refer to the comparative hardness, brittleness, and lack of greasy feel of the waxes. The carbon chains of the alkyl structures in the waxes usually are longer than those in the oils and fats, and generally waxes are harder and higher melting than fats. It is true that both bayberry and Japan wax are glycerol esters, and thus chemically like fats, but these are exceptions; they are called waxes since they are relatively high melting solids with firm structures. Like many vegetable waxes they are both cuticular or exudates on the surface of plants and being relatively free from lower-molecular-weight oily components have the harder and drier feel common to waxes. On the other hand, sperm oil and jojoba bean oil are fatty esters of monohydric alcohols and thus would be classed as waxes by the chemical classification applied strictly, but they are fluid and for this reason are not generally regarded as waxes.

The solubility of the waxes varies widely in different fat solvents, and is dependent on the temperature. Generally, however, complete solution is obtained at temperatures approaching the melting point of the wax or the boiling point of the solvent. Other solubility properties within the fat-solvent-soluble group are valued highly and are associated with certain waxes. The ability to bind naphtha in a firm paste structure is valued and associated particularly with carnauba and certain other waxes. The ability to wet and disperse pigments in some cases, or to dissolve and carry concentrations of dyes is highly appreciated and associated with specific waxes. Emulsifiability with water for self-lustering floor polishes and aqueous creams for pharmaceutical products and furniture polishes also is a highly valued property of some waxes. Other properties that are associated with some waxes but are somewhat less characteristic are the high yield in plastic flow—important in high-gloss buffing wax and in high-pressure lubricants—a high dielectric constant for insulators and potting compounds, and combustibility with low ash content candles and solid fuel preparations.

Most of the natural substances in the fat-solvent-soluble group are fairly complex mixtures of compounds, and it is difficult to generalize as to structural chemical differences between resins, gums, waxes, etc. There are fairly diverse materials in most of the classifications. Many of the waxes, however, have fairly sharp melting and solidification temperature ranges, and often the major components are of similar structure, such as an aliphatic ester series varying over only about a two-fold range in molecular weight. In comparison to many polymeric gums, resins, and pitches, the waxes are less viscous at temperatures immediately above their melting points. Some macrocrystallinity may even be present and many typical waxes exhibit cold flow only after substantial stress has built up and the deforming forces have reached the yield value. Generally waxes have a more crystalline structure than gums and resins and may have even well-developed macrocrystals apparent in waxes like chinese insect wax and spermaceti. Waxes have little or none of the solid glassy appearance of the supercooled liquids, or the characteristic conchoidal fracture of the gums and resins.

Composition. The esters of long-chain fatty alcohols and acids have been considered the most typical chemical constituents of natural waxes as a group. Petroleum waxes and certain other mineral waxes, however, consist of hydrocarbons. Aliphatic or open chain structures with relatively little branching or side chains can be considered typical for both ester and hydrocarbon waxes.

Extensive studies in recent decades have provided almost complete characterization and quantitative analysis of a number of wax compositions. Examples of nearly all of the structural types likely to occur in the chemistry of aliphatic organic compounds have now been found in some natural wax. Monofunctional reactants and unbranched alkyl chains are characteristic, but polyfunctional compounds and some relatively short chain branching are present in certain waxes. The following compounds have been identified in the analysis of waxes, and some attempt to estimate frequency of occurrence and quantitative aspect has been considered in the listing order: (1) hydrocarbons (large tonnage of petroleum waxes, alkanes predominating with usually only minor amounts of unsaturates and branching); (2) alkyl esters (a short-chain acid or alcohol found in some cases, but roughly equal acid–alcohol chain lengths predominating); (3) primary alcohols; (4) acids; (5) ketones; (6) aldehydes; (7) secondary alcohols; (8) hydroxy acids; (9) lactones; (10) acetals; (11) diols (α-ω predominant); (12) dicarboxylic acids; (13) diketones; (14) polyesters.

A substantial resinous fraction can be separated from several of the natural waxes by solvent extraction. Such a deresinating would be a desirable refining step if objectional properties in the total natural product were due to its resinous character. The resinous fractions often represent a more polymeric and/or cyclic structure than the remaining wax. Along with the condensation and polymerization products possible from open-chain aldehydes, etc, some alicyclic compounds like cyclic terpenes and aromatic derivatives, such as those of hydroxybenzaldehyde and hydroxycinnamic acid, etc, are likely to be found in the resinous fractions.

Alicyclic and aromatic compounds indeed are found in several of the natural waxes, but usually they represent only a small portion of the total except when there is the substantial resin fraction. Carnauba wax, however, is approximately 30% esters of hydroxy- and methoxycinnamic acids. The esters have long alkyl chains in the alcohol group and are distinctly waxy and not resinous. The only other aromatic or ring structures in carnauba wax would be in a very minor resinous or gummy (alcohol soluble) fraction. The plant pigments and a fraction of 1% of sterols and triterpene diols would be in the small resin fraction.

Because of the exceptional properties and commercial value of the wax (1968 U.S. imports $4,490,000) from the leaves of the carnauba palm, its chemical composition has received particular attention for many years. Recent work at S. C. Johnson & Son, Inc., Racine, Wisconsin, as yet largely unpublished, now provides a fairly complete insight into the structural constituents and their approximate proportions in carnauba wax. Naturally, some variations exist due to the age of the leaf and site source of the tree, differences in trees possible within the species, and the harvesting process, but an approximate composition of carnauba wax is listed in Table 1.

Candelilla wax also is a plant wax of commercial importance (1968 U.S. imports $1,140,000) that has received considerable chemical attention. Approximately 50% of candelilla wax consists of paraffin-type hydrocarbons. Schuette and Baldinus in 1949 reported (2) that the candelilla hydrocarbons are essentially straight chain, and that about 90% is straight-chain hentriacontane, $C_{31}H_{64}$. Candelilla wax has a resin fraction soluble in cold alcohol that is about 20% of the plant's natural wax exudate.

Beeswax has also received considerable chemical attention and is important commercially (1968 U.S. imports $3,000,000, domestic production $3,960,000). Beeswax

Table 1. Approximate Composition of Carnauba Wax

Composition	Amount, %	Description
free acids	5.5	primarily C_{24}, C_{26}, C_{28} normal saturated monocarboxylic fatty acids
free alcohols	11	primarily C_{30}, C_{32}, C_{34} normal saturated monohydric primary fatty alcohols, C_{32} dominant (3)
hydrocarbons	1	primarily C_{27}, C_{29}, C_{31} normal saturated hydrocarbons
esters	82	
aliphatic, 40%		45.3% normal saturated monofunctional acids, primarily C_{24}, C_{26}, C_{28} with 54.7% normal saturated monofunctional primary alcohols, primarily C_{30}, C_{32}, C_{34}, with C_{32} dominant
ω-hydroxy[a], 13.2%		47.0% normal saturated monocarboxylic acids. 90% ω-hydroxy acids, primarily C_{22}, C_{24}, C_{26}, C_{28}, and 10% normal saturated monocarboxyl acids, primarily C_{24}, C_{26}, C_{28} with 53.0% normal saturated primary alcohols 90% monohydric alcohols, primarily C_{24}, C_{26}, C_{28}, C_{30}, C_{32}, C_{34}, and 10% α-, ω-dihydric alcohols, (glycols[b]) C_{24}, C_{26}, C_{28}, C_{30}, C_{32}, C_{34}.
cinnamic aliphatic diesters (4), 28.8%		23.0% hydroxycinnamic acid, 5.8% p-methoxycinnamic acid, in each case esterified with α-ω glycols as described above
miscellaneous	0.5	a triterpene (5), plant pigments, etc

[a] Average mol. wt. 860.   [b] The glycols were esterified with the monofunctional acid.

is a good example of the ester type of wax. Although as much as 13% of the wax is hydrocarbon, the principal constituents are the esters. The predominant hydrocarbon is hentriacontane as in candelilla. It is estimated that as much as 23% of the total wax is the one ester, myricyl palmitate, $C_{15}H_{31}COOC_{30}H_{61}$. Among the other esters present are myricyl alcohol compounds with acids, such as cerotic acid, $C_{27}H_{55}COOH$, estimated at 12% of the total, and other palmitates, such as that of lacceryl alcohol, $C_{32}H_{65}OH$, estimated at 2% of the total.

There is an impressive uniformity in the composition of beeswax from the different strains, races, and varieties of bees. However, an important group other than our hive or *Apis mellifica* bee is the ghedda group of Apis bees; their wax is reported to have the ceryl, C_{28}, alcohol grouping in place of the myricyl as the principal palmitic ester and also that a few % of fatty glycerides are present.

Testing. Physical properties, such as melting temperature and density, and chemical properties like free acid and ester content, vary over a range of values, but are useful in quality control and characterizing tests for different waxes. Tests fairly common throughout the industry include: hardness tests, melting point determinations, refractive index, specific gravity, moisture, suspended content, ash, unsaponifiables, hydrocarbons, resin content, and acid, saponification, iodine, and acetyl numbers. Some of these are listed in Table 2. Numerous other tests are designed to test the suitability of a wax in special performance applications, usually in combination with other ingredients of a formulation. Examples are the naphtha-paste-strength and water-emulsification tests in the polish industry, and the ink-oil-binding and pigment-dispersion-strength tests in carbon paper applications.

Processing and Refining. Certain physical operations are used in connection with the initial steps in harvesting waxes, such as drying, shredding, and beating

Table 2. Chemical and Physical Properties of Natural Waxes

Wax	Melting point	Acid number[a]	Saponification number[b]	Iodine number[c]	Acetyl number[d]	Specific gravity at 25°C	Hardness ASTM D-5
beeswax	62–70	17–21	85–95	8–12	15	.955–.975	15
bayberry-myrtle	48–50	4–30	205–215	1–4		.978–.980	7.5
candelilla	65–69	15–18	45–65	14–35		.969–.993	1.5
caranday	82–85	3–10	62–80			.990–.999	1
carnauba	83–86	3–8	72–85	8–12	55	.990–.999	1
castor bean wax	86	2	175–185	2–10	155–165	.980–.990	2
esparto grass wax	78	24	70			.988–.990	1.5
Japan wax	50–56	20	205–235	4–15	25–30	.975–.990	
montan crude wax	76–86	25	50–60	16		.990–.999	8
ouricury	83–85	10–20	85–95	7–10		1.05–1.06	1
retamo- ceri nimbi	76–78	45–50	88	5	16–17	.983–.987	2
shellac wax	74–78	12–16	100–125	1–2		.970–.980	2
spermaceti	40–50	1	120–135	2–4	2–3	.905–.960	16
sugar cane wax[e]	75–79	6–10	25–35			.985–.990	3
wool wax-lanolin	31–42	1–40	80–140	15–45		.924–.960	

[a] mg of KOH/g to neutralize.
[b] mg of KOH/g to saponify.
[c] mg of I_2/100 g absorbed.
[d] mg of KOH to neutralize acetic acid obtained from saponification of one g of acetylated wax.
[e] Deoiled and deresinated wax from fresh press cake.

certain plant materials, the screening of crude wax powders and the melting, filtering, and casting of the wax. Numerous other separation and refining processes are employed with some waxes, such as distillation fractionations, "sweating" (partial melting), adsorbent refining, and decolorization, and the use of solvents in deoiling and deresinating steps. The chemical processing steps commonly used with waxes include oxidation with air and peroxides, hydrogenation, hydrolysis, esterification, chlorination, etc.

Taken as a whole, it may be said that waxes, animal and vegetable, are indigenous to all climates; those which find technical application however, are mainly tropical and subtropical vegetable products (6).

Vegetable Waxes

Plant wax occurs in a cuticle layer on nearly all the external portions including leaves, stems, fruit, petals, and root structures. The amount of epicuticular wax varies greatly with species, and the concentrations in certain locations and on certain portions of the plants may be strikingly disparate. For example, the amount of wax may be tenfold higher on one side of a leaf, or the only concentrations of wax on a stem may be in the internodal zones in the lower portion of the plant.

The presence of wax undoubtedly serves protective functions in helping to preserve the water balance of the plant, minimize mechanical damage and sun scald, and inhibit fungal and insect attack. In some plant species, there appears to be some connection between the water stress of drought and the production of an increased total amount of wax. The amount of wax produced by the individual members of the same species can differ consistently and substantially even when growing under similar conditions.

Thus it is unlikely that a single environmental change, such as the precipitation rate, can be expected to produce either a completely uniform or a quantitatively similar effect on wax production across the wide variations in strains and species of plants. Although there may be variations in wax production with changes in the character of a growing season and the amount of rainfall, a plant that is representative of a heavy wax-producing strain can be expected to retain that characteristic and produce a notable amount of wax regardless of the amounts of rainfall. Observations of the innate or inbred wax-producing character under widely varying climates have been noted for the waxy members of the broadly distributed palms in the *genus Copernicia*, especially the carnauba palm, or *Copernicia prunifera*, in particular. Good yields of wax, for example, have been obtained from the leaves of carnauba palms growing in the heavy rainfall areas of Florida, Cuba, Rio de Janeiro in Brazil, and Georgetown in Guyana, as well as from the large native carnauba palmars (ie, natural palm stands) in the arid northeast area of Brazil. In the harvesting of carnauba wax, a higher average yield per leaf may be associated with drought, however, in the palmars that have good rains during the growing season and when a relative drought occurs well along in the development period of the new leaves.

Many vegetable waxes have received attention from time to time. Some remained of laboratory interest only, some have been produced commercially for limited periods only, and a few have maintained substantial levels in trade volume across the years. The following waxes in their approximate order of importance can be considered in the first ten commercially at present: carnauba, candelilla, hydrogenated castor oil, ouricury, Japan wax, bayberry wax, esparto grass wax, retamo, sugar canewax, and jojoba bean oil wax. Other vegetable waxes identified in the literature and in some cases of local or temporary commercial interest are: henequen agave (sisal hemp), raffia hemp (*Cannabis sativa*), palm wax (*Ceroxylon andicola*), caranday, cauassu, douglas fir bark, ucuhuba, flax, cotton, pisang, gondang, tea, apple, cranberry, rice bran oil wax, soya oil wax, and the wax from the flowers rose, mimosa, jasmine, lavender, and orange.

Carnauba wax is obtained from the leaves of the carnauba palm, *Copernicia prunifera* (*cerifera*) in the semiarid northeastern part of Brazil. While there has been some planting in plantations, selective studies on wax yields and even some hybridization work, nearly all of the tonnage of commerce comes from the palmars in the areas where the palms are indigenous. Some of the finest stands of carnauba are on the banks of streams and ponds, alluvial plains or fans in river basins, or on other flat lands where the water table is not too deep during most of the growing season. The trees of commerce are distributed over a wide area, but generally are centered around the Ceará-Piauí region where in some years the leaves are cut from most of an estimated total of nearly one hundred million trees. Many of the carnauba palmars are almost solely Copernicia prunifera and occur in fairly dense stands.

The wax is present even on the unopened young leaves in a fairly continuous flexible film over the areas of photosynthesis and on both sides of the leaf, but substantially thicker on the upper or "sun-side." After the leaf is cut and a major weight loss as moisture has occurred, the wax is no longer flexible and adherent, but has become brittle and will flake free of the surface. It then can be worked off as flakes or smaller particles of powdered wax from the entire fan-shaped leaf surface.

After harvesting the crude powder by drying the leaves, splitting open the webs, cutting the ribs, beating and screening by hand, or more commonly now, with carnauba

leaf-threshing machines, effort is made to reduce the amount of fiber and cellular fragments of the leaf in the total powder before the wax is melted. The potential grade
and color of the wax is controlled primarily by the stage of development and age of
the leaf after it emerges from the central stem tip. The youngest unopened leaves
produce the lightest yellow color and top grades of wax. Wax from the opened leaves
not only is somewhat darker, but additional strong green and brown plant pigments
readily dissolve from any residual leafy material still present when the powder from
the older leaves is melted. For this reason extra screening and any other steps practical to clean the powder, such as air and water flotations, are taken before the initial
melting and filtering.

An official grading system has been in effect for many years based on specifications
set by the Brazilian Government: number one and two (yellows) are similar and
bring top prices; number three (light fatty grey) is intermediate in value; number
four (fatty grey) and number five (chalky) are lower in price, but represent the bulk
of the trade volume.

Originally the dried leaves were shredded by drawing them across a trencher, a
series of toothlike blades that ripped open the leaf webs. At this step, flakes separated
and immediately fell to the floor from the areas of the leaf having a thicker layer of
wax. The leaves then would be beaten against wooden railings to dislodge as much as
possible of the remaining wax powder and dust. All of this work (trenchering, beating,
sweeping, and screening) was done by hand and usually in fairly closed, hot, and dusty
rooms. Now nearly all of the leaf dewaxing is performed by machines that have been
designed for the purpose. The powder from the machines has a large amount of leafy
material, however, and it is in a form that is very difficult to separate from the wax.
The color and grade of the wax therefore has not improved, but two men can process
between 10,000 and 20,000 leaves per hour.

Some water is added to the powder during the melting step to reduce scorching
and to increase heat transfer. Most of the lump wax of commerce has only about 1%
or less moisture content. This is true of the yellow grades from younger leaves
numbers one and two, and also grades three and four from the older opened leaves.
The chalky carnauba grade five however has a moisture content of perhaps 5% or even
higher. Across the years the demand for these grades has been maintained through
some fairly wide variations in availability and price. There is excellent uniformity
within the grades and all are sharp-melting (84°C) hard carnauba wax.

The chalky wax is prepared by using enough water during the melt to provide
some separation and cleaning by a floating action during the solidification of the wax
on a foam and water layer containing some sand, leafy matter, and dirt. A second
melting and filtration takes place after the lower layers have been removed. The
residual moisture in chalky lump wax tends to drop on prolonged storage in dry warehouses, and often an initial water content of near 10% may drop to 5% or under depending on conditions and length of time in storage. For some purposes the chalky
grade has been considered to have properties sufficiently desirable to more than compensate for the higher moisture content. There probably would be as much plant
pigment and resin present in chalky as in fatty gray number four grade, but the
chemical structures may not have been altered as much by thermal exposure in the
chalky and less polymerization may have occurred, particularly of the aromatic
esters involving cinnamic double bonds.

Adsorbent refining and some (as by peroxide) bleaching has been carried out in subsequent operations, but the only other treatment considered common for the crude wax would be the addition of small amounts of oxalic acid to the water in the initial melt for a relatively minor bleaching effect due to reduction. The final press cake residue can be solvent-extracted to recover more wax. Such a product is not equivalent to the quality of the regular grades and blends may be encountered that may or may not be properly designated.

The main uses for carnauba wax are in floor finishes and in carbon paper. Many formulations for furniture, shoe, automobile, and other polishes also use carnauba wax. Large amounts of carnauba wax were used in water-based floor waxes from the time of their introduction in the eary 1930s to the early 1960s. The relatively high proportion of hydroxy acid esters and free alcohols in carnauba wax probably were responsible for the excellence of the aqueous polishes. Because of the advances in polymer technology and its contributions to water-emulsion floor polishes, there has been some drop in the total demand for carnauba wax in recent years. It still is the most important plant wax, however, and except for the petroleum waxes it represents the greatest tonnage and dollar value of all the commercial waxes. It is widely used with other waxes to raise the melting point, gloss, hardness, etc, in the manufacture of paper, textiles, insulating materials, batteries, candles, matches, soap, salves, chalk, and crayons.

A close relative of the carnauba palm representing an even larger potential resource for a similar wax, should be mentioned; namely the *Copernicia alba* or *Caranday palm* of Paraguay. The carnauba and caranday palms are very similar and the waxes are very comparable. The wax layer of the caranday is somewhat more adherent to the leaf surface when the leaves are dried, so that there is less "free flaking." In the northeastern section of the Paraguayan Chaco there are enormous stands of the *C. alba*, and the total number of caranday palms may be even ten times that of the carnauba of Brazil. There has been almost no commercial development so far of the caranday wax.

Candelilla wax is obtained from the stem of *Pedilantus pavonis*, a rushlike leafless perennial weed growing in the semiarid regions of Northern Mexico. It is a hard, brittle wax like carnauba and has been important in commerce even before World War I. The candelilla wax of commerce actually is derived also from several other related Euphorbaiacceae plants including *Euphorbia antisyphlitica*. For most purposes the difference due to source apparently are small, and all the wax from very large areas of Chihuahua and Coahuila is sold as candelilla. The plants all produce numerous slender cylindrical green stems ranging from 2–4 ft in height. In harvesting the wax, the whole plant is pulled from the ground, submersed in water and heated until the wax separates and rises to the surface of the boiling water. Sulfuric acid is added to assist in separating the wax and to reduce foaming. The wax then is skimmed off and the excess water in it is boiled away in a separate tank. Any remaining dirt settles effectively and after solidification is scraped off the bottom of the cakes. The product is a hard, light brown wax with a melting point of about 70°C.

Solvent processes for the extraction of candelilla wax have been developed, but even with the higher wax yields that can be obtained, they apparently are not competitive at recent price levels, with the older water–sulfuric acid method. Candelilla wax has been the preferred vegetable wax in chewing gum formulations, and in general is used in applications similar to those of carnauba wax. It always has been considered to be inferior to carnauba for some applications, however, probably because of properties associated with its terpene resin content, 20%, and the large hydrocarbon content, 50%.

Hydrogenated castor oil is waxlike and need not be classified strictly as a synthetic product. It has a relatively high melting point, 86°C, for vegetable waxes and has solvent binding and solubility relationships that have helped establish its present position as a commercial wax. Chemically, it is fairly pure tri(12-hydroxy)stearin (glycerol tri(12-hydroxy)stearate) and several companies produce it by the catalytic hydrogenation of the oil from the castor bean. One of the best known trade names is Opalwax produced by E. I. DuPont de Nemours & Company. Castor waxes are used in polishes, textile finishes, carbon paper, cutting oils, lubricants, and waterproofing compounds.

Ouricury wax is obtained from the leaves of the *Cocos coronata* palm in the State of Baia, Brazil. There is a large potential resource of this wax from the number of native trees available, but unlike carnauba wax, the ouricury wax does not flake free of the leaf surface on drying, and must be scraped from the leaf by knife or glass edge, and then melted, and strained. The vitreous tightly adherent character results in the introduction of a substantial amount of plant pigment and resins from the leaf into the commercial wax. Ouricury is somewhat similar to carnauba, has almost the same melting point, 84°C, but there are no light-colored grades. Ouricury has a much higher hydrocarbon content, 17%, primarily as C_{31}, and more resinous material than carnauba. When there were periods of strong demand for carnauba and prices were high, larger tonnages of ouricury wax were produced.

Although ouricury has good carbon dispersing, oil binding properties, etc, appreciated in carbon paper formulation, most of the applications have been the same as those of carnauba wax, and the latter generally is preferred. There has been a decrease in both trade volume and price for carnauba in recent years, and this has all but wiped out the production of ouricury wax.

Japan wax is a fat from the berries of a sumaclike tree *Rhus succedanea* cultivated in Japan and China for its wax. The Japan wax occurs as a greenish coating on the fruit kernels, and several related varieties of sumac also have berries containing this waxy glyceride. It is primarily tripalmitin and in spite of over 5% glycerol ester of dibasic acids as high as C_{21} it has a relatively low melting point of only 51–55°C. Japan wax is used in textile finishing formulations, candles, leather dressings, etc.

Bayberry or myrtle wax is derived from the fruit of several species of Myrica, myrtle shrubs, with the *Myrica carolinensis* of East Coast of the United States and the myrtles of Colombia, South America, furnishing much of the bayberry wax of commerce. It is low in melting point, 42–48°C, but firmer, drier, and physically more waxlike than Japan wax. The myrtle waxes also are largely triglycerides, however, and chemically are fats with palmitic and myristic acids dominant. The ripe drupes, or berries, are submersed in boiling water, the wax rises to the surface, is skimmed off, and strained. The principal use of bayberry wax is in candles. In addition to the good burning properties, the characteristic fragrant odor is highly prized. Myrtle wax also can be used to make firm, white, high quality soaps.

Esparto grass wax is derived from North African grasses, *Stipa tenacissima* and *Lygeum spartum*, cultivated for use of the fibers in high-grade papers. The practice has been for the grass to be dewaxed by flailing before being made into paper primarily in the British Isles. The dust from the flailing is the crude wax powder. Esparto wax produced from the powder is a hard wax, melting point 78°C. It has a higher hydro-carbon content, primarily C_{31}, than carnauba wax, but essentially is an aliphatic ester wax that can find use as a substitute for carnauba wax in polish, carbon paper, textile applications, etc.

Retamo or ceri mimbi wax is derived from a semiarid shrub, *Bulnesia retamo*, growing in the foot hills of the mountains in central and northern Argentina. Retamo wax is like candelilla wax in several respects. It is a cuticular coating on the young branches or stem surfaces of the upper portion of the shrub. The branches are cut off and the wax recovered by floating on boiling water following the submersion in water and cooking a fresh charge of cuttings. The hydrocarbon content of the wax is about 18% and the resin or cold-acetone-soluble content is about 7%. The crude unbleached wax has a fairly high acid content with some variation, 45–50, and a high melting point, 78°C. Retamo wax has been produced and used to advantage in Argentina, particularly in the 1960s with volumes in 1969 at least as high as 50,000 lb/yr. There is a large resource and potential production for this hard vegetable wax in Argentina with the wide distribution and good stocking in some areas of the brushlike source along the slopes of the hills and mountains. For a large export volume, however, it would have to compete with both carnauba and candelilla. It now is used within Argentina commerce as the prime hard vegetable wax, and has broad applications including those in the various polish formulations.

Sugar cane wax occurs on the surface of the cane-stems of the giant grasses, *Saccharum officinarum*, etc, cultivated for sugar. The wax of commerce is obtained as a by-product in the production of raw sugar. Much of the cuticular wax is carried into the juice stream in the cane crushing and washing operation along with the sugar, other cellular material, and dirt. The suspended materials including the wax separate from the juice and concentrate in a mudlike layer in the lower portion of the large clarifiers. After the mud is filtered and washed for residual sugar, the wax is extracted from the filter cake along with oils and resins by the use of solvents such as petroleum naphthas. Because of differences in the varieties of cane, the season, the flocculating treatments, the storage conditions, and the age of the press cake, there are some wide variations in the character of different crude wax production. Some steps such as the use of fresh press cake to increase the uniformity of the crude wax are practical and both deoiling and deresinating can be carried out. The economics of the production of cane wax, however, is likely to be marginal when strong price competition is encoun-tered from other hard waxes.

The first commercial development of cane wax probably was at Natal, South Africa during the first World War. Several million pounds of the wax were produced, and records indicate that as much as twelve million pounds were exported from Natal in 1924. Cane wax has been produced since then in different places at different times. Examination of the products, however, nearly always indicated that some of the changes had taken place that are to be expected if there has been prolonged atmospheric exposure and drying out of the press cake before extraction. In Cuba in 1946 and 1947 continuous feed liquid–liquid extractors were installed to operate on fresh press cake during the crushing season (7). They were used at two mills for several years, and the crude wax, containing about 33% of a fluid fraction and 17% resin, was refined at Gramercy, Louisiana (8). A light colored hard wax and other products were produced through the 1950s. In the early 1960s there was a substantial drop in price of carnauba wax and for this reason and on account of political developments the sale of the products was gradually discontinued.

The cuticle wax on the cane stalks is largely a polymer of long chain waxy aldehydes, principally C_{28} (9). The trioxane type of polymer has been proposed and the presence of some aldol condensation products also would appear to be likely, particularly if the thermal exposures in the recovery and refining steps were sufficient to temporarily depolymerize the polymeric aldehydes in the initial cuticle wax from the sugar cane. The aldehyde structure lends itself to oxidation and other modification reactions. This could be of potential value, if carried out under controlled conditions, but also, as has been indicated, it can contribute to wide variations in the products of extraction if the press cake is held in storage for any extended period of time.

Jojoba wax is derived from the seeds of the *Simmonsia californica*, a shrub growing in the southwest in Arizona and California, and Sonora in Mexico. This natural oil is classed as a wax both because it is a fatty acid ester of a monohydric fatty alcohol and because on hydrogenation it becomes a solid waxy substance that melts as high as many waxes, 70°C. The oil is over 50% of the seeds, and is extracted from the seeds after they are thoroughly ground or crushed. Over 90% of the wax is believed to consist of esters, or monounsaturated monocarboxylic straight chain acids, C_{20} and C_{22} (mainly C_{20}), with monounsaturated monohydric straight chain alcohols, C_{20} and C_{22} (mainly C_{22}). Because of the clear cut composition, stability, and freedom from objectionable components, jojoba oil has received attention as a fatty carrier or medium for certain medical injections, for cosmetics, and as a fine lubricant for delicate mechanisms. Studies have been carried out indicating that systematic planting and harvesting might be practical.

Animal and Insect Waxes

Beeswax, the use of which has been traced back to the early Egyptians, is the most important commercially of the animal and insect waxes. The domesticated honey bee, *Apis mellifica*, furnishes much of the beeswax of commerce, but imported beeswax includes East Indian and African (ghedda bees) and South American (*Melipona*). Wax scales are secreted by eight wax glands on the underside of the abdomen of the worker bee as a digestive secretion formed in the stomach from the honey and flower pollen. The scale is removed from the abdomen and received by the mandible of a coworker, where it is chewed before being placed in the cell of a comb. Bees are believed to secrete 1 lb of wax for every 8 lb of honey they produce.

Crude beeswax varies in color and quality with the type of bee, its food, and the processing of the wax. Crude beeswax is usually rendered from the frames and scrapings by melting in boiling water that often contains a small amount of sulfuric acid. It is then strained and solidified. The wax is usually bleached, often by sunlight, yielding a white or light yellow product. Domestic beeswax does not bleach as readily as does much of the imported wax; consequently the domestic price is lower.

The main use of beeswax is in cosmetic preparations, because of its relative cheapness, plasticity, and ease of emulsification. Next largest use is as a major component of church candles (qv). Candles used for Mass and Benediction in Roman Catholic churches must contain at least 50% beeswax; many contain a higher percent. It is believed that beeswax was originally chosen because on burning it does not give off acrolein. Other uses include polishes, modeling, and pattern making.

Sperm oil and spermaceti wax. Sperm oil is obtained from the blubber and cavities in the head of the sperm whale, *Physeter macrocephalus*. The sperm whale is one of a family order largely confined to tropical waters, but stragglers reach both polar seas. Sperm oil is everywhere mixed with spermaceti, but the latter is largely obtained from the whale head cavity, yielding almost 500 gal of mixed sperm oil and spermaceti. The oil consists largely of esters of fatty acids and monohydric fatty alcohols, a large proportion of which are unsaturated.

There are distinct differences in the chemical composition of spermhead oil and sperm-blubber oil, in that the head oil has a higher content of lauric and myristic acids and the blubber oil is less saturated in its alcoholic and acidic components. It is therefore the more saturated hard oil which on cooling deposits spermaceti wax (cetyl palmitate). The oil can be readily hydrogenated and this product is capable of producing far higher yields of spermaceti wax of higher melting point and hardness than the usual commercial product. See also Vol. 1, p. 551.

The sperm oils have been graded by the refineries as winter sperm oil, congealing below 38°C, about 75% yield; spring sperm oil, congealing at 10–15.6°C, 9% yield; taut-pressed oil, melting at 32.2–35°C, 5% yield; and crude spermaceti, melting at 43.3–46.1°C, 11% yield. Sperm oil on saponification yields oleyl alcohol and saturated higher monohydric alcohols; hence sperm oil is classified as liquid wax. Sperm oil yields 39.2% wax alcohols; cachalot, 41–44%; and Arctic sperm, 38–39%.

The present primary use of sperm oil in the U.S. is as a specialized lubricant. The oil is treated with sulfur; this product is used as an additive for extreme-pressure lubricants where the mixture produces a high film strength. These lubricants are used both for industrial machinery and automotive uses. The sulfurized sperm oil is also used in preparing drawing lubricants. See Vol. 19, p. 499.

Spermaceti wax is obtained by allowing the head oil to cool and stand; the standard candle for measuring light intensity uses spermaceti because of the bright flame produced on burning. It is used for candles, cosmetics, and as a source for cetyl alcohol.

Chinese insect wax. This wax is deposited on the leaves and stems of certain types of ash and privet trees in China by a scale insect, *Coccus ceriferus*. The insect does not breed in the same district as the trees, and the eggs are transported yearly about 200 miles and suspended on the shoots of the trees. The branches of the tree, which are covered with a thick deposit from the insect, are placed in boiling water to release the wax. About 1500 insects are required to produce 1–2 g of wax. The wax is used in candles, furniture polishes, and coatings for paper and cloth.

Shellac wax. Shellac wax is obtained as a by-product in the preparation of spirit shellac. The shellac is deposited in a fashion similar to the Chinese wax by the lac insect. The shellac is dewaxed by a solvent treatment, and the wax used as a component of insulating material and in polishes. See *Shellac*.

Wool wax—lanolin. See also *Wool*. Lanolin is a refined waxlike product derived from the wool of sheep. It melts at about 40°C, is soluble in the common wax solvents, and consists primarily of waxy esters, over 70%, free monohydric alcohols, about 25%, with diols, lactones and free acids about 5%. Normal saturated monohydric alcohols from C_{16}–C_{26} and cholesterol and related sterols are present in substantial quantities in the wax esters and free alcohol portions. The acids present are primarily members of two branched chain saturated monocarboxylic fatty acid series, the "isoacid" series, C_{12}–C_{26}, with one methyl group on the $\omega-1$ carbon atom, and the "anteiso" series, probably with odd numbers of carbon atoms, $C_{15}-C_{27}$, with one methyl group on the $\omega-2$ carbon atom. There are also smaller amounts of normal acids, alpha hydroxy acids, and unsaturated acids.

The high proportion of free alcohols, particularly cholesterol, and the hydroxy acid esters in wool wax provide an unusual degree of hydrophilic character for an otherwise water-repelling fat or wax. This undoubtedly is a key property in the use of lanolin in the pharmaceutical and cosmetic applications. Lanolin can supply the needed softening to dry skin without interfering with the natural moisture transpiration. About 1,500,000 lb of wool wax is used per year, primarily for cosmetic and biological cream preparations.

Mineral Waxes

Peat wax is the fraction of peat that can be extracted with fat solvents such as petroleum naphthas. Although almost all peats have some wax content, only a few of the larger deposits have wax contents of as much as 10%. The crude peat waxes usually have a substantial portion of a resinlike fraction soluble in cold solvents. The wax portion is soluble hot and precipitates on cooling the solvent. In some crudes an asphalt or pitch component gradually separates from the other materials even in the hot solvent and can be removed. Naturally it is the peat waxes from the higher yield deposits, such as Irish peat and the peat from Chatham Island (New Zealand) that have been studied most extensively. There are differences in the crude waxes to be associated with the sources, but the wax fractions are similar and have many of the characteristics of montan wax. None of the peat waxes, however, has ever become firmly established in commerce.

Montan wax is a bituminous wax, occurring in brown coals or lignites (qv), from which it can readily be extracted. Wax-containing brown coals have been mined in Australia, New Zealand, Czechoslovakia, Russia, and the U.S. (California, Arkansas), in addition to the main source of supply in central Germany, where its extraction and processing is an old and established industry. The crude wax is complex chemically, but its constitution is similar to other natural waxes.

The coal is granulated, dried, and solvent-extracted to remove the wax. The crude wax is usually further processed to produce an acceptable market product. One type of refining is vacuum distillation using steam and producing the so-called "double-refined" montan wax. In a second and more important refining process, the wax may be deresinified by solvent processing, and then treated by chromic acid oxidation. The oxidation product may then be esterified with aliphatic glycols, or otherwise treated, to give a variety of highly useful waxes.

The brown coal from Germany has an average wax content of 10–15%, although some samples may run as high as 18%. After granulation and drying to a moisture content of about 10%, the wax is extracted from the coal with a solvent mixture of benzene and methanol. After solvent removal, the crude wax contains about 15–20% of resin, which is largely removed by a second extraction with a solvent such as ethyl alcohol. Alternatively, the wax may be vacuum-distilled at 50 mm pressure, giving a distillate which is a useful wax product. Yields are not very high in this process, however.

The crude montan wax is dark brown in color and melts at about 82°C. It is primarily an ester wax, nearly 60% esters, but the free acid content is substantial and is likely to exceed 15% of the total. The carbon chain length for both the acid and alcohol components apparently is about the same, about C_{26}. There probably are some hydroxy acid esters present and there may be some odd-numbered fatty acids in the C_{27}–C_{31} range either free or combined. In addition to a resin content of 10–12%, a ketone C_{28}–C_{30} content up to 10% is frequently reported.

The Bureau of Mines has investigated the possibilities of domestic production of montan wax. Using a solvent-extraction process, it found that yields from California and Arkansas lignite were comparable to those from German coal, but the resin content of the wax was higher.

The American Lignite Products Company produced domestic montan wax products at Ione, California for some time and now are continuing their operations as a part of Interpace Corporation. A series of high melting point ester waxes are produced by extraction and refining from the domestic lignite and are called Alpco Waxes.

Montan wax is used as an ingredient in polishes, carbon paper, electrical insulation, leather dressings, inks, and greases. It is also the major raw material for a series of synthetic *Hoechst waxes*. See below under Synthetic Waxes.

Ozocerite (ozokerite) is a hydrocarbon wax which is mined mostly in eastern Europe from veins of waxy shale that usually occur close to the surface of the earth. Some mining of the wax has been done in Utah and Texas. It closely resembles paraffin wax in appearance and composition, and it is therefore relatively easy to adulterate it with the cheaper paraffins. It is separated from earthy matter by melting in boiling water and then drawing off the wax layer. The crude wax may be purified by treatment with sulfuric acid and clay filtering. Ozocerite owes its wide use to its great compatibility with many substances and its affinity or absorptivity for solvents. In this respect it is superior to paraffin wax, which tends to crystallize from admixture and in so doing allows fluid components present in the mixture to "sweat out." Refined ozocerite is often called *ceresin*, although a mixture of paraffin wax and beeswax is also known as ceresin, as is a mixture of ozocerite and paraffin.

Ozocerite is harder and tougher than the crystalline paraffins from petroleum and the addition of 25% ozocerite to paraffin destroys the macrocrystalline structure. It is thought to be mixture containing hydrocarbons with ring structures embedded in a long polymethylene chain, short branched open chained structures, and the higher molecular weight C_{37}–C_{50} members of the normal hydrocarbon series instead of the C_{17}–C_{35} members of that series as in the crystalline paraffins.

Petroleum waxes are coproducts obtained in the manufacture of lubricating oils. The volumes of these waxes are enormous and the average production for the last ten years in the United States has been in excess of one and a half billion pounds. See Petroleum Waxes.

Synthetic Waxes

Many different molecular structures are represented in the materials having waxy character that have been synthesized or prepared by modification of intermediate natural raw materials.

The synthetic and modified ester type waxes are an important group valued both for their own individual properties and because they can be used to replace the hard high priced waxes. Some of the Hoechst waxes based on montan wax fall in the modified ester grouping. The Hoechst waxes were formerly called Gersthofen waxes and were the original I.G. waxes made at Oppau and Gersthofen by I.G. Farben A.G. before World War II. A series of acid and ester type waxes made from partial and from completely deresinated crude montan wax continue to be available and are produced at Gersthofen near Augsburg, Germany.   The present Hoechst Waxes S, L, E, and OP also are representative of the earlier products.

After solvent deresinating steps the montan wax is subjected to a series of exposures to hot chromic, sulfuric acid solutions.   This vigorous acid oxidation results not only in the bleaching of the dark color but also in the partial hydrolysis of the esters and the conversion of alcohols to monobasic and dibasic acids.

Hoechst wax S melting point 78–83°C, acid number 135–155, saponification number 155–175 represents only about 15% of the unreacted esters with about 68% monobasic montanic acids (C_{18}–C_{34} with the majority in the C_{28} category) and about 17% dibasic acids from the —OH containing esters.   Hoechst wax L melting point 80–85°C represents the product of a milder oxidation and contains a little more of the unreacted esters with acid number 120–140 and saponification number 140–160. Hoechst wax E is an ethylene glycol ester of Hoechst S with melting point 76–81°C, acid number 15–20 and saponification number 140–160.   Hoechst wax OP is an example of a mixture of the butylene glycol ester and the calcium salt product from Hoechst S with melting point 100–105°C, acid number 10–15 and saponification number 100–115. The oxidation and esterification processes relating to the Hoechst waxes are considered to be covered in the descriptions of the I.G. waxes by Albin H. Warth in *The Chemistry and Technology of Waxes* (10).

The Hoechst waxes are used in many different applications and are of particular interest in polishes and lubricants.   Two other ester waxes from montan that are of interest in selflustering polishes are Hoechst KSL and KPS.   Like Hoechst E they represent esterification of acid waxes with alcohols or glycols.   Also of interest in the polish industry are two Hoechst waxes KSE and KLE containing a nonionic emulsifier.

Some very useful waxy polyethylenes were developed in the 1960s, particularly for the selflustering formulations.   A number of easily emulsified polymer products were developed including some of the high density types of polyethylenes. In some cases the hydrocarbon homopolymers were partially oxidized and in other cases the improved emulsifiability and other properties of interest were obtained by introducing organic acids and other comonomers with polar groups in the ethylene polymerization process (11).

Some other polyolefins particularly polypropylenes (see Olefin polymers) are waxlike and so also are the chlorinated paraffins (see Vol. 5, p. 231).

Also during the 1960s some valuable waxy film formers were produced by the emulsion polymerization of ethylene and a number of relatively high molecular weight (15,000–20,000) aqueous dispersions of nonoxidized polyethylenes are available.   The products are very finely divided particles of a branched medium density ethylene polymer suspended in water usually with anionic or nonionic emulsifying agents.   The emulsion polymerization products can have excellent uniformity and the absence of oxygen in the polymer wax chain can help protect the film of the finished product from discoloration (12).

Other emulsion polymerization products such as the acrylate and styrene polymer dispersions have become important film forming ingredients in coatings and polishes but their properties usually are not like the waxes or the waxlike properties are only marginal.

The oxidized hydrocarbon wax products such as those from the Fischer-Tropsch paraffins and the microcrystalline petroleum waxes also fall in the ester-type group and are valued in polishes and other applications. The glycol esters, the hydrogenated oils, and particularly hydrogenated castor oil are part of the ester group and are valuable in paper and textile applications.

Other synthetics of entirely different structure such as the fatty amides, imides, amines, and nitriles can be waxlike and very useful particularly when special surface activity or coordinate solubilities are needed. The polyoxyethylenes, carbowaxes (Union Carbide, see Vol. 10, p. 655) are a unique and important group because of their water solubility and compatibility with fatty materials common to cosmetic. pharmaceutical, etc, formulations.

Because of the difficulties that would be encountered in trying to duplicate most of the natural waxes, it is likely that the important synthetic waxes will be those that are valued for their own individual characteristics.

Economic Aspects

The United States production of waxes has exceeded 1.6 billion pounds per year since 1965. Over 95% of that production however has been the relatively low-priced petroleum waxes so the total dollar value of the raw wax production was about $100 million per year. United States is by far the world's largest producer of petroleum waxes, and the exports for the last five years exceeded 440 million pounds per year (13). The United States also is by far the largest consumer of waxes, and it can be estimated that over 66% of the world's available tonnage of wax is used in the United States.

The wax imports for the 1964–1969 period have ranged from 24.5 to 26.7 million pounds per year, excluding wool wax, petroleum waxes, and mineral waxes (not specifically provided for in Dept. of Commerce classification) (13). The largest tonnage of a single wax above is the carnauba wax from Brazil. In 1969 slightly over 11.5 million pounds were imported at a cost of about 3.67 million dollars. From 1965–1970 the prices for carnauba wax have been at the lowest levels of several decades. The U.S. consumption of beeswax has exceeded 8 million pounds per year for over 5 years and about half has been from domestic production. In the last 4 years the price of beeswax has climbed from $0.44 to over $0.73 per pound (13). Other than hydrocarbon waxes, carnauba wax consumption had represented the highest dollar value in commercial waxes for many years. Now, however, beeswax has taken a definite lead in that rating.

Table 3. U.S. Imports of Wax (14)

	1968		1969	
	Thousands of pounds	Thousand of dollars	Thousands of pounds	Thousands of dollars
beeswax	4,130	3,000	4,440	3,250
spermaceti	260	60	170	40
animal wax NSPF[a]	460	310	520	370
candelilla	2,570	1,140	3,240	1,560
carnauba	14,190	4,490	11,530	3,670
Japan wax	320	110	160	70
ouricury	200	80	160	70
vegetable waxes NES[b]	56	17	410	120
montan wax	4,020	680	3,420	580

[a] NSPF (not specifically provided for).   [b] NES (not entered separately).

Montan wax imported from Germany is of continuing importance and slightly over 3.4 million pounds were received in the United States in 1969, and valued at $580,000. Candelilla wax from Mexico was next in volume at 3.24 million pounds for 1968, but was valued at $1,560,000 or 270% of the montan wax import value. Wool wax usually is not listed directly in the wax statistics, but it should be listed next with a U.S. consumption level at about 1.5 million pounds per year. Table 3 U.S. Imports of Wax.

Uses

The end use representing the largest volume application of waxes without question would be paper coating (see Paper). The petroleum waxes, of course, represent by far the largest proportion of that volume. Paper coating represented the end use in 1969 for 53% of the petroleum wax sales of one billion pounds in the United States. In the late 1950s paraffin for milk cartons represented 450 million pounds of wax. Plastic coated containers, however, took over most of that application between 1960 and 1964 (14). Although used in much smaller quantities, both carnauba wax and hydrogenated castor oil should be mentioned in the paper-coating-wax application.

The second largest volume application of waxes is in candles (qv) and molded novelties. Again the petroleum waxes would represent by far the largest proportion of that volume. In 1968, 136.5 million pounds of petroleum wax was used in the candle and novelty application. Beeswax also is important in candle production with candelilla and bayberry wax substantially smaller in volume.

The consumption of wax in the various applications for manufacture of electric equipment is the third largest classification. Paraffin wax itself, chlorinated paraffin, and chlorinated naphthalenes represent the great bulk of the electrical equipment waxes.

The application of wax in textile and leather sizing, waterproofing, etc may well be the fourth largest classification. Again paraffin wax would represent the bulk of the volume with candelilla, carnauba, and montan wax deserving mention in the leather applications.

The various polish applications; floor, auto, furniture, and shoe, use the greatest variety of waxes and the largest tonnage of the hard, high-priced waxes but probably would fall in the fifth largest rank for volume of wax consumed. Here too the volume proportion of paraffin wax is impressive, but the dollar investment in the other waxes definitely exceeds that for the paraffin.

Fruit and vegetable coatings probably would be sixth in volume application for wax with paraffin again being predominant. Carnauba wax is frequently used as a hardening agent and to increase the melting point.

Cosmetic and pharmaceutical production are estimated to be in the seventh place for volume of wax consumed if wool wax is included. Beeswax and lanolin are the two major waxes in the formulation of cosmetic and pharmaceutical creams.

Carbon paper, ribbons, and printing inks like the polishes can use a wide variety of waxes and probably rank next, or in eighth place, by tonnage consumed. Carnauba is likely to represent the major investment. Ouricury wax is an excellent hard wax in this group application, but cannot compete when the prices for carnauba are in a lower range.

Greases, lubricants, and mold releases represent perhaps the only remaining large group application. Hydrogenated oils, such as castor oil, are important in the formulations of a number of greases and the metal soaps are of course very important, and while not waxes are waxlike in many properties.

There are many small group applications such as the candelilla wax in chewing gum; paraffin and Japan wax in matches; carnauba, beeswax, and paraffin in dental wax, ski wax, and precision casting waxes; petroleum waxes in explosives; beeswax and spermaceti in crayons; and microwaxes in adhesives, sealing, and grafting waxes. These and other miscellaneous uses, however, are likely to represent less than 1% of the total commercial wax applications.

Bibliography

"Waxes" in *ECT* 1st ed., Vol. 15, pp. 1–17, C. J. Marsel, New York University.

1. H. J. Deuel, Jr., *The Lipids,* Wiley Interscience Publishers, New York, 1951.
2. H. A. Schuette and J. G. Baldinus, "Studies in Candelilla Wax II, Its Normal Paraffins," *J. Am. Oil Chemists, Soc.* **26,** 651–652 (1949).
3. D. T. Downing, Z. H. Kranz and K. E. Murray, "Studies in Waxes XX The Quantitative Analysis of Hydrolyzed Carnauba Wax by Gas Chromatography," *Australian J. Chem.* **14** (4), p. 622 (1961).
4. L. E. Vandenburg et al., "Aromatic Acids of Carnauba Wax," *J. Am. Oil Chemists, Soc.* **44,** (11), (1967).
5. C. S. Barnes et al., "Carnaubadiol, A Triterpene from Carnauba Wax," *Australian J. Chem.* **18,** 1411–1422 (1965).
6. T. P. Hilditch, *The Industrial Chemistry of the Fats and Waxes,* Bailliane, Tindall, and Cox, London, 1949.
7. E. S. McLoud, *Sugar Cane Waxes,* Official Proceedings C.S.M.A. pp. 162–166 (1950).
8. D. E. Whyte and B. Hengeveld, "Chemical Examination of Sugar Cane Oil," *J. Am. Oil Chemists, Soc.* **27** (2) (1949).
9. J. A. Lamberton and A. N. Redcliffe, "Polymeric Aldehydes in Sugar Cane Cuticle Wax," *Chem. Ind.* p. 1627 (1959).
10. A. H. Warth, *The Chemistry and Technology of Waxes,* Reinhold Publishing Co., New York, 1947, pp. 267–273.
11. P. G. McQuillan and S. J. Gregg, "Emulsifiable Polyethylene," *Detergents and Specialties,* Oct. 1969, pp. 62–68.
12. G. W. Douglas, "Polyethylene Emulsions in Metal Cross Linked Floor Waxes," *Detergents and Specialties,* Oct. 1969, pp. 52–56.
13. U.S. Department of Commerce, *Summary of Information on Waxes and Candles,* April 1968, revised.
14. American Petroleum Institute, Dept. of Statistics, *Annual Sales of Wax in the United States 1963–1968,* April 1969.

E. S. McLoud
Consultant

WEED KILLERS

One important factor contributing to significant increases in yields of major crops in the United States is the use of agricultural chemicals for pest control. It was stated in 1967 that, if losses due to all pests could be reduced by half, increases in total annual production of maize, wheat, rice, grain sorghums, soybeans, and potatoes would be about 25 million metric tons at current rates of production (1). As an example of the extent to which herbicides are used to control weeds, 53 million acres were treated with pre- and postemergence herbicides in the crop year of 1959 at a cost of $128 million.

The overall figures for pesticide production reflect the continuing growth of herbicide use. In 1967, U.S. sales of herbicides reached 287.6 million lb valued at $400 million, compared with 221.4 million lb valued at $257.4 million in 1966 (2). This increase from 95.2 million lb worth $92.1 million in 1962 is dramatic. The value of

Table 1. Estimated Use of Herbicides in the United States,[a] million lb (3a)

Herbicide	1964	1965	1966
sodium chlorate		24–28	
sodium tetraborate			18
sodium arsenite	1–2		
arsenic acid[b]		8[c]	
DNC ⎱ dinoseb ⎰	3.24		
PCP		1–2	
2,3,6-TBA	2.21		
dalapon	2.25		
CDAA	3.66		
propanil	3.85		
trifluralin	5.0		
diuron	1.12		
atrazine	10.9[d]		
MSMA	1.08	16[e]	

[a] Courtesy, Stanford Research Institute, Stanford, Calif.
[b] Cotton desiccant.
[c] This figure was estimated at 6–8 million lb in 1968.
[d] A significant nonagricultural market also exists.
[e] For weed control in cotton.

Table 2. U.S. Herbicide Production in 1968

Herbicide	Production, 1000 lb
2,4-D	
acid, esters, and salts	402,781
2,4,5-T	
acid, esters, and salts	94,116
pentachlorophenol	
herbicide, desiccant, and	
wood preservative	48,575
sodium chlorate	
herbicide and defoliant	30,000

SOURCE: *The Pesticide Review 1969*, Agricultural Stabilization and Conservation Service, U.S. Dept. of Agriculture, Washington, D.C.

Table 3. Major Herbicide Imports in 1968

Herbicide	Import, lb
chloroxuron	145,502
2,4-D	2,473,578
diquat dibromide	374,100
mecoprop	156,055
paraquat (salts)	1,663,685
pyrazon	144,472
2,4,5-T	315,556

SOURCE: *The Pesticide Review 1969*, Agricultural Stabilization and Conservation Service, U.S. Dept. of Agriculture, Washington, D.C.

Table 4. U.S. Exports of Herbicides, 1968

Herbicide	Value, $1000	Export, 1000 lb
2,4-D and 2,4,5-T[a]	1,234	3,391
other organic herbicides,		
technical	19,693	20,882
herbicide formulations	44,213	46,327
inorganic herbicides, technical	606	2,325

SOURCE: *The Pesticide Review 1969*, Agricultural Stabilization and Conservation Service, U.S. Dept. of Agriculture, Washington, D.C.

[a] Includes salts and esters based on technical acid content.

herbicide sales now exceeds that of insecticides and fungicides combined ($357.2 million in 1967) (3). Tables 1–4 illustrate the economic picture.

There are several reasons for the increase in sales of herbicides. The potential savings from agricultural and nonagricultural use which result from chemical control of weeds provide sufficient incentive. In addition, the limited supply and increasing cost of hand labor has created a demand for mechanized processes in agriculture.

In contrast, the sales of insecticides have leveled off in recent years. This was to be expected as the cost of control programs reaches a point when no further economic benefit is to be anticipated. New techniques of insect control are being studied and some are achieving success, for example insect sterilization.

The herbicide market developed later than the insecticide market. There is still a great need for herbicides demonstrating greater selectivity between desirable and undesirable plants. A lower effective dosage in lb per acre would result in lower application costs and possibly decreased residue problems. An important part in spurring the growth of herbicide sales is being played by herbicides which are nonpersistent and highly selective, such as those of the thiocarbamate group.

The remarkable increases in use are due to the discovery of the growth-regulating activity of the chlorinated phenoxyalkanoic acids. Kögl's discovery in 1934 that indoleacetic acids promote cell elongation in plants, led to the synthesis and evaluation of many structurally related organic compounds (4). These studies eventually revealed the extremely high activities of indoleacetic acids and halogenated aryloxyacetic acids.

Historically the use of weed killers may be traced to ancient times. A variety of organic and inorganic chemicals has been used since the beginning of the twentieth century. Sulfuric acid was first recommended as a weed killer in 1909 (5). Sodium chlorate, arsenical compounds, copper sulfate, and other inorganic compounds have also been known as weed killers since the early twentieth century. Dinitrocresol (3,5-dinitro-o-cresol) was used as an insecticide since 1928, but it was not until the 1930s that the selective herbicidal properties of this and related compounds were discovered (6). The discovery of the phenoxyalkanoic acids as weed killers, in the early 1940s, followed by their successful development and commercial application, provided the stimulus required by industry to undertake the search for new synthetic chemicals. Currently more than 127 of these are commercially available.

The stages of discovery and development of a herbicide may be long and expensive. Programs of synthesis, evaluation, field testing, analytical and toxicological studies, and other expenses necessary for the registration of a herbicide were estimated at over $2 million in 1966. This estimate increased to over $4 million by 1969 (7). As our understanding of the fundamental biochemistry of herbicidal activity is limited, a factor in increasing costs is the empirical nature of the approach which requires synthesis and screening of a large number of organic compounds. Other major cost items are toxicological, metabolism, and residue studies. The transformations of herbicides in plants, soils, and water yield a variety of products which must be examined to ensure that they do not produce harmful effects in man and animals. Specific and sensitive analytical methods are essential for this purpose.

The definition of a weed as "a plant growing where it is not wanted" is convenient although perhaps not scientific. It does focus upon one of the major problems of weed control, ie the selective killing of weeds without damage to the growing crop. Selectivity may often depend on the metabolism of a herbicide by the plant. Significant differences may exist between the metabolism of the chemical in the crop and in the weeds which are to be controlled. The time of application, placement, and formulation of the herbicide are factors which may influence the effectiveness of control. Increased effectiveness reduces associated residue problems by reducing the quantity of a herbicide used.

A prime source of information is the *Herbicide Handbook* (8) which provides a compact account of the physical, chemical, toxicological, and other properties of many herbicides in current use. Table 5 lists chemical names and common names approved by the Terminology Committee of the Weed Science Society of America.

Table 5. Common and Chemical Names of Herbicides

Common name or designation	Chemical name
acrolein	acrolein
ametryne	2-(ethylamino)-4-(isopropylamino)-6-(methylthio)-s-triazine
amiben	3-amino-2,5-dichlorobenzoic acid
amitrole	3-amino-s-triazole
AMS	ammonium sulfamate
atratone	2-(ethylamino)-4-(isopropyl-amino)-6-methoxy-s-triazine
atrazine	2-chloro-4-(ethylamino)-6-(isopropylamino)-s-triazine
barban	4-chloro-2-butynyl m-chlorocarbanilate 4-chloro-2-butynyl N-(m-chlorophenyl)-carbamate
benefin	N-butyl-N-ethyl-α,α,α-trifluoro-2,6-dinitro-p-toluidine
bromacil	5-bromo-3-sec-butyl-6-methyl-uracil
bromoxynil	3,5-dibromo-4-hydroxybenzonitrile, 2,6-dibromo-4-cyanophenol
buturon	3-(p-chlorophenyl)-1-methyl-1-(1-methyl-2-propynyl)urea
butylate	S-ethyl di(isobutyl)thiocarbamate
cacodylic acid	hydroxydimethylarsine oxide
CDAA	N,N-diallyl-2-chloroacetamide
CDEA	2-chloro-N,N-diethylacetamide
CDEC	2-chloroallyl diethyldithiocarbamate
chlorazine	2-chloro-4,6-bis(diethylamino)-s-triazine
chloroxuron	3-[p-(p-chlorophenoxy)phenyl]-1,1-dimethylurea
chlorpropham	isopropyl m-chlorocarbanilate, isopropyl N-(m-chlorophenyl)carbamate
cycloate	S-ethyl N-ethylthiocyclohexanecarbamate
cycluron	3-cyclooctyl-1,1-dimethylurea
cypromid	3′,4′-dichlorocyclopropanecarboxanilide
dalapon	2,2-dichloropropionic acid
DCPA	dimethyl tetrachloroterephthalate
dinoseb	2-sec-butyl-4,6-dinitrophenol
diallate	S-(2,3-dichloroallyl) di(isopropyl)-thiocarbamate
dicamba	3,6-dichloro-o-anisic acid, 3,6-dichloro-2-methoxybenzoic acid
dichlobenil	2,6-dichlorobenzonitrile
dichloroprop, 2,4-DP	2-(2,4-dichlorophenoxy)propionic acid
dicryl	3′,4′-dichloro-2-methylacrylanilide
diphenamid	N,N-dimethyl-2,2-diphenylacetamide
diquat	6,7-dihydrodipyrido[1,2-a;2′,1′-c]-pyrazinediium salts, 1,1′-ethylene-2,2′-bipyridinium salts
diuron	3-(3,4-dichlorophenyl)-1,1-dimethylurea
DNOC	4,6-dinitro-o-cresol
DSMA	disodium methanearsonate
EPTC	S-ethyl dipropylthiocarbamate
fenac	2,3,6-trichlorophenylacetic acid

(*continued*)

Table 5 (*continued*)

Common name or designation	Chemical name
fenuron	1,1-dimethyl-3-phenylurea
fluometuron	1,1-dimethyl-3-(α,α,α-trifluoro-*m*-tolyl)urea
ioxynil	4-hydroxy-3,5-diiodobenzonitrile
	4-cyano-2,6-diiodophenol
isocil	5-bromo-3-isopropyl-6-methyluracil
Lambast[a]	2,4-bis[(3-methoxypropyl)amino]-6-methylthio-*s*-triazine
Lasso[a] (alachlor)	2'-chloro-2,6-diethyl-*N*-(methoxymethyl)acetanilide
linuron	3-(3,4-dichlorophenyl)-1-methoxy-1-methylurea
MCPA	[(4-chloro-*o*-tolyl)oxy]acetic acid
MCPB	4-[(4-chloro-*o*-tolyl)oxy]-butyric acid
mecoprop	2-[(4-chloro-*o*-tolyl)oxy]-propionic acid
metham	sodium methyldithiocarbamate
metobromuron	3-(*p*-bromophenyl)-1-methoxy-1-methylurea
molinate	*S*-ethyl hexahydro-1*H*-azepine-1-carbothioate
monolinuron	3-(*p*-chlorophenyl)-1-methoxy-1-methylurea
monuron	3-(*p*-chlorophenyl)-1,1-dimethylurea
MSMA	disodium methanearsonate
naptalam, NPA	*N*-1-naphthylphthalamic acid
neburon	1-butyl-3-(3,4-dichlorophenyl)-1-methylurea
nitralin	4-(methylsulfonyl)-2,6-dinitro-*N,N*-dipropylaniline
nitrofen	2,4-dichlorophenyl *p*-nitrophenyl ether
norea	3-(hexahydro-4,7-methanoindan-5-yl)-1,1-dimethylurea
paraquat	1,1'-dimethyl-4,4'-bipyridinium salts
PCP	pentachlorophenol
pebulate	*S*-propyl butylethylthiocarbamate
picloram	4-amino-3,5,6-trichloropicolinic acid
prometone	2,4-bis(isopropylamino)-6-methoxy-*s*-triazine
prometryne	2,4-bis(isopropylamino)-6-(methylthio)-*s*-triazine
propachlor	2-chloro-*N*-isopropylacetanilide
propanil	3',4'-dichloropropionanilide
propazine	2-chloro-4,6-bis(isopropylamino)-*s*-triazine
propham, IPC	isopropyl carbanilate, isopropyl *N*-phenylcarbamate
pyrazon	5-amino-4-chloro-2-phenyl-3(2*H*)-pyridazinone
sesone	2-(2,4-dichlorophenoxy)ethyl sodium sulfate
siduron	1-(2-methylcyclohexyl)-3- phenylurea
silvex	2-(2,4,5-trichlorophenoxy) propionic acid
simazine	2-chloro-4,6-bis(ethylamino)-*s*-triazine
simetryne	2,4-bis(ethylamino)-6-(methylthio)-*s*-triazine
solan	3'-chloro-2-methyl-*p*-valerotoluidide
swep	methyl 3,4-dichlorocarbanilate

Table 5 (*continued*)

Common name or designation	Chemical name
terbacil	3-*tert*-butyl-5-chloro-6-methyluracil
TCA	trichloroacetic acid
triallate	S-(2,3,3-trichloroallyl) di(isopropyl)thiocarbamate
tricamba	3,5,6-trichloro-*o*-anisic acid, 3,5,6-trichloro-2-methoxybenzoic acid
trifluralin	α,α,α-trifluoro-2,6-dinitro-N,N-dipropyl-*p*-toluidine
2,3,6-TBA	2,3,6-trichlorobenzoic acid
2,4-D	(2,4-dichlorophenoxy)acetic acid
2,4-DB	4-(2,4-dichlorophenoxy)butyric acid
2,4-DEP	tris[2-(2,4-dichlorophenoxy)ethyl] phosphite
2,4,5-T	(2,4,5-trichlorophenoxy)acetic acid
vernolate	S-propyl dipropylthiocarbamate

[a] Registered trademark of Monsanto Co.

There is no general agreement on the best classification of the herbicides. Biochemical information is inadequate to permit classification based on mode of action, and fundamental relationships between structure and activity are not understood. For the chemist, the structural formula is the most useful basis for consideration of a herbicide as the structure most conveniently codifies a number of interrelationships with which he is well acquainted. Other methods of classification based on use and mode of application have been used. The variety of organic herbicides is increasing but the chlorinated aryloxy acids remain the most important class, historically and in terms of herbicide production. They were first introduced as plant-growth regulators in 1942 (9).

Chlorinated Aryloxyalkanoic Acids

Economically the herbicide market has been dominated by the phenoxyalkanoic acids, particularly 2,4-D(2,4-dichlorophenoxyacetic acid). Although their predominance has decreased markedly in the 1960s, together they represented 136.6 million lb of the total U.S. herbicide production of 402.8 million lb in 1968. Military use overseas has increased the demand for these herbicides.

Chlorinated phenoxyalkanoic acids are generally selective killers of broad-leaved weeds. 2,4-D and its derivatives have been used widely to control weeds in cereal grain crops and to a lesser extent in hay and pasture crops. The importance of 2,4,5-T derivatives is in noncrop uses. 2,4,5-T has a more potent effect on a number of woody species and is used for clearance of rangeland and rights-of-way. Crop uses have been primarily in hay, pasture, rice, and sugar cane. Silvex (2-(2,4,5-trichlorophenoxy)-propionic acid) is also effective against a number of woody plants.

On April 15, 1970, the registration of a number of 2,4,5-T containing products was suspended (10). Liquid formulations for use around the home or recreation areas and all formulations for use in lakes, ponds, or on ditch banks were included. Formulations intended for control of weeds, brush, and range, pasture-rights-of-way, forest and nonagricultural land are not included in this action. The suspension was based on the reports of birth defects produced by injection into mice at high dose rates.

Dioxins, including 2,3,7,8-tetrachlorodibenzo-*p*-dioxin, have been detected as contaminants in samples of 2,4,5-T and are also suspected of being able to produce birth defects. The manufacture of 2,4,5-trichlorophenol, the starting material for 2,4,5-T, requires the hydrolysis of tetrachlorobenzene under vigorous conditions and dioxins and other impurities may be formed at this stage.

Chemical Properties and Synthesis. The formulas,

$$\text{OCHR(CH}_2)_n\text{COOH}$$

names, melting points, and solubilities of the major chlorinated aryloxyalkanoic acids are given in Table 6 (9a). These compounds are used as the free acids, metal salts,

Table 6. Properties of Chlorinated Aryloxyalkanoic Acids

Name	R	X	Y	Z	n	Melting point, °C	water	acetone	ethanol	benzene
		Formula					Solubility, ppm[e]			
2,4-D[a]	H	Cl	Cl	H	0	140.5	0.07[25]	45[33]		1.07[28, f]
2,4-DB	H	Cl	Cl	H	2	117–119	46[25]			
2,4,5-T	H	Cl	Cl	Cl	0	158	238[80]		590[60]	
2,4,5-TB	H	Cl	Cl	Cl	2	120–121				
dichloroprop	CH₃	Cl	Cl	H	0	118	350[20]	59.5[28]		85[28, f]
MCPA[b]	H	CH₃	Cl	H	2	119	825		153[f]	
MCPB	H	CH₃	Cl	H	2	100	44			
mecoprop	CH₃	CH₃	Cl	H	0	94–95	600			0.472[25, f]
silvex[c]	CH₃	Cl	Cl	Cl	0	179–181	140[25]	18[25]		
sesone[d]	OCH₂CH₂OSO₃Na structure					170 (free acid) 245 (Na salt)	26.5[25]	0.64[25]		0.05[25, f]
2,4-DEP	[OCH₂CH₂O–]P structure ₃					200 (bp at 0.1 mmHg)				

[a] U.S. Pat. 2,390,941 (1945), to Amer. Chemical Paint Co.
[b] U.S. Pat. 2,740,810 (1956), to Diamond Alkali Co.
[c] U.S. Pat. 2,749,360 (1956), to The Dow Chemical Co.
[d] U.S. Pat. 2,573,769 (1951), to Union Carbide and Carbon Corp.
[e] Unless otherwise stated.
[f] Grams per 100 g solvent.

amine salts, ammonium salts, or esters. The method of synthesis of 2,4-dichlorophenoxyacetic acid (2,4-D) is generally applicable to the majority of the class. 2,4-Dichlorophenol and monochloroacetic acid are reacted in the presence of aqueous sodium hydroxide affording 2,4-D (10a). In the case of the butyric acids, γ-butyrolactone is used as a starting material in place of the chlorinated acid.

Phenoxyalkanoic acids are typical ω-substituted alkanoic acids and may be converted to salts and esters by standard techniques. The photochemical reactions of 2,4-D in aqueous solution have been investigated by Crosby and Tutass (11). Irradiation at 254 nm cleaved the ether linkage and replaced the chlorine by hydroxyl groups, giving a mixture of products. The ultimate product was 1,2,4-benzenetriol (**1**) which could be isolated as the acetate if the reaction was carried out in the presence of sodium bisulfite as an oxidation inhibitor. In the absence of an inhibitor, dark-colored polymeric material was formed by oxidation of benzenetriol. The photolysis of 2,4-D in solution on filter paper by sunlight gives rise to dark-colored material, although 2,4-D, 4-chloro-2-methylphenoxyacetic acid (MCPA), 2,4,5-trichlorophenoxyacetic acid (2,4,5-T), and 2-(2,4,5-trichlorophenoxy)-propionic acid (Silvex) are stable under dry conditions.

Analysis. Formulations (see under Application and selectivity below) of phenoxyalkanoic acids may be analyzed by the determination of total chlorine or acid equivalent content. Residue analysis presents greater difficulties and bioassay methods have been used (12). A color reaction was discovered by Freed (13) and adapted by Marquardt and Luce for determination of 2,4-D (14). This reaction involved heating phenoxyacetic acids with chromotropic acid (4,5-dihydroxy-2,7-naphthalenedisulfonic acid) in concentrated sulfuric acid. The reaction probably proceeds by liberation of formaldehyde which gives a color with chromotropic acid. This method suffers from interferences which produce high blanks and no reaction is obtained with phenoxypropionic acids. In an improved procedure the phenoxy acids are cleaved to the corresponding phenols by heating with pyridine hydrochloride. The phenols are then determined colorimetrically after reaction with 4-aminoantipyrine in basic solution (15). This method may present difficulties for analysis of extracts of green plant materials where misleading results have been reported. Infrared spectrometric analysis may be useful on a semimicro scale (16). Gas chromatography affords the most sensitive method for quantitative analysis of the acids, but requires their prior conversion to methyl esters by treatment with diazomethane or BF_3 in methanol. The method is highly specific and may be modified by the addition of internal standards to check losses through the clean-up procedure (17).

The presence of the phenoxyalkanoic acids in plant material as conjugates (see below) requires that complete analysis be conducted also on samples which have undergone acid hydrolysis. Other plant degradation products present specific problems of analysis which require individual consideration.

Metabolism. Selective inhibition of plant growth by herbicides may result from a combination of physiological and biochemical factors. Of greatest interest to the chemist, are studies which attempt to define biochemical sites of action and identify metabolic products and pathways. Considerable literature has accumulated relating to mode of action and metabolism of the phenoxyalkanoic acid herbicides (18).

Intensive investigations to define the primary site of action of the phenoxyalkanoic acids have been undertaken. The areas studied include synergism and antagonism with known auxins, effects on metabolism of nitrogen, respiration, carbohydrate metabolism, photosynthesis, enzymic systems, mineral uptake, and the ability of 2,4-D to chelate with metals. Examination of structure–activity relationships has shown that there are differences in selectivity according to ring substitution in the phenoxyacetic acids. The relative abilities of plant species to degrade phenoxyacetic acids obviously affect their susceptibility to the action of the herbicides. Conjugation or degradation by the plant may be associated with resistance, but it has recently been suggested that susceptibility of a plant may depend on its ability to degrade 2,4-D to a toxic metabolite, ie chloroacetic acid (22).

Phenoxyalkanoic acids are applied as weed killers in the form of salts or other derivatives, such as esters. Differential absorption may be an important factor in the selective response of a plant to a particular derivative, but conversion to the free acid within the plant is considered a prerequisite for herbicidal activity. Further reaction of the phenoxyacetic acids within the plant may follow one of several pathways: (a) physical or chemical binding to cellular constituents; (b) side-chain fission or degradation to yield the corresponding phenol; (c) ring hydroxylation; and (d) conjugate formation. Higher members of the homologous series of phenoxyalkanoic acids may be degraded by β-oxidation of the side-chain. Conversely detoxication on the surface of the plant leaf by conversion of 4-(2,4-dichlorophenoxy)butyric acid to higher homologs has been reported (19).

Conjugation with glucose after ring hydroxylation gives the corresponding phenolic glucosides. Thus, 2,4-D was hydroxylated in beans to 2,5-dichloro- and 2,3-dichloro-4-hydroxyphenoxyacetic acid which were isolated as glucosides (20). In oats, 2-chlorophenoxyacetic acid was converted to 2-chloro-4-hydroxyphenoxyacetic acid which could be obtained as the 4-O-β-D-glucoside (21). 2,4-D was converted to its β-D-glucose ester. 2,4-Dichlorophenoxyacetylaspartic acid was formed from 2,4-D in wheat, peas, and currants. "Protein complexes" of 2,4-D have been investigated, but their true nature remains unclear.

Many plant species are capable of degrading the side chain of 2,4-D to a limited extent but this pathway of 2,4-D metabolism plays a major role in only a few plants, such as red currants, certain apple varieties, strawberries, and garden lilac. The formation of 2,4-dichlorophenol from 2,4-D has been demonstrated in sunflower, corn, and barley. Investigation of side-chain degradation reveals that two separate mechanisms may exist. Some species may lose the side chain as a two-carbon unit, whereas others may effect stepwise degradation through a postulated one-carbon side-chain intermediate.

The aromatic ring of a number of phenoxyalkanoic acids is hydroxylated at the 4-position in oats, barley, corn, and wheat and pea tissue. In oats, phenoxyacetic acids unsubstituted at the 4-position were hydroxylated. Hydroxylation of 2,4-D in beans was accompanied by a shift of the Cl atom. 2,5-Dichloro-4-hydroxyphenoxyacetic acid was obtained as the major product and 3,5-dichloro-4-hydroxyphenoxyacetic acid as the minor.

Wain and his colleagues (23) have examined the metabolism of the higher homologs of 2,4-D and have demonstrated the important role of the enzymic system responsible for β-oxidation in plants. Degradation of carboxylic acids by a process involving oxidation at the β-position to the carboxylic acid group is well known in animals. Evidence for this process in flax seedlings was obtained using a homologous series of ω-phenoxy acids ($C_6H_5O(CH_2)_nCO_2H$, where $n = 1$–10). Only those acids with an even number of carbon atoms caused production of phenol in the plants, although from the unusual amount of phenol produced from 10-phenoxy-n-decanoic acid it was concluded that ω-oxidation was also a metabolic pathway (24). The presence of a β-oxidase enzyme system in wheat, tomatoes, and peas was inferred from growth-response observations. These and other related investigations led to the significant postulate that, as oxidation of homologs of phenoxyacetic acid was necessary for activity, only those plants possessing a β-oxidase system would be susceptible to the herbicidal activity of the phenoxyalkanoic acids. The β-oxidase system may differ in specificity from one plant to another and its functioning is influenced by the position and type of substitution in the aromatic ring (25). These differences also provide a basis for selective herbicidal activity.

A number of related herbicides are based on the phenoxyalkyl structure. The herbicidal activity is considered to depend on their conversion to phenoxyacetic acid by soil microorganisms. Amongst these are tris-(2,4-dichlorophenoxyethyl)phosphite (2,4-DEP), 2,4-dichlorophenoxyethyl sulfate (sesone) and a number of related phenoxyethylsulfate derivatives.

For the degradation of the phenoxyalkanoic acid herbicides by microorganisms in soils, see Soil chemistry of pesticides.

s-Triazines

$$\begin{array}{c} R_3 \\ \text{N} \diagup \diagdown \text{N} \\ R_1NH \diagdown \text{N} \diagup HNR_2 \end{array}$$

Chemical Properties and Synthesis. The herbicidal value of derivatives of s-triazine was discovered in 1952 (26). 2-Chloro-, 2-methylthio-, and 2-methoxy-4,6-bis-(alkylamino)-s-triazines are used as herbicides and some 2-azido substituted triazines have also been introduced (27). The 2-chloro derivatives are of major commercial importance. Properties of major herbicidal s-triazines are given in Table 7.

The starting material for the synthesis of the s-triazines is cyanuric chloride (28). It is obtained industrially by trimerization of cyanogen chloride and is also an intermediate in the manufacture of dyestuffs, optical brighteners, and plastics. The three chlorine atoms may be replaced by amines, phenols, alcohols, mercaptans, thiophenols, and related compounds. Two chlorine atoms are quite reactive; therefore replacement of a single chlorine atom demands controlled reaction conditions. Usually temperatures of -15 to $0°C$ are employed. An equivalent of base, such as sodium hydroxide or an amine is added to react with liberated hydrogen chloride. (See also Triazinetriol).

Table 7. Properties of *s*-Triazine Herbicides

Name	Formula R₁	R₂	R₃	Melting point, °C	Solubility, ppm water	acetone	ethanol	methanol	benzene	Vapor pressure at 20°C, mm Hg
ametryne[a]	C₂H₅	i-C₃H₇	SCH₃	84–86	185[20]					8.14×10^{-7}
atratone	C₂H₅	i-C₃H₇	OCH₃	94–96	70[27]			18,000[27]		3.0×10^{-7}
atrazine[b]	C₂H₅	i-C₃H₇	Cl	173–175	70[27]			18,000[27]		
simazine[c]	C₂H₅	C₂H₅	Cl	225–227	5[20]			400[20]		6.1×10^{-9}
simetone[a]	C₂H₅	C₂H₅	OCH₃							
simetryne	C₂H₅	C₂H₅	SCH₃							
prometone[a]	i-C₃H₇	C₂H₅	OCH₃	91–92	730[20]	>500,000[29]			>250,000	2.3×10^{-6}
prometryne[a]	i-C₃H₇	i-C₃H₇	SCH₃	118–120	48[29]					1.0×10^{-6}
propazine[d]	i-C₃H₇	i-C₃H₇	Cl	212–214						2.9×10^{-8}
Lambast[d,e]	NH(CH₂)₃OCH₃[e]		SCH₃	55	100	v sol	sl sol		v sol	
chlorazine[f]	N(C₂H₅)₂	N(C₂H₅)₂	Cl							

[a] U.S. Pat. 2,909,420 (1958), to J. R. Geigy Akt. Ges.
[b] U.S. Pat. 2,891,855 (1958), to J. R. Geigy Akt. Ges.
[c] Swiss Pat. 342,784 (1960), to J. R. Geigy Akt. Ges.
[d] Registered trademark Monsanto Co.
[e] For R₁ and R₂.
[f] Lambast and chlorazine have unsaturated rings.

The 2,4-dichloro-6-alkylamino-s-triazines are relatively more stable. A reason for this difference in reactivity may be the ability of the N—H bond in the alkylamino-s-triazine or the tautomeric imino structure to form a hydrogen bond with water. The attack on a neighboring carbon–chlorine bond, as in (2), would be facilitated (28).

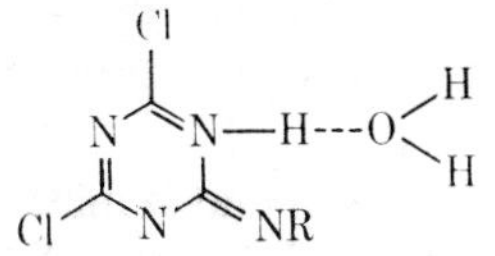

(2)

Replacement of two chlorine atoms by amines, alcohols, or thiols requires somewhat higher reaction temperatures, generally 20–50°C. Two equivalents of base are added to neutralize the liberated acid. Asymmetrically alkylated 2-chloro-4,6-bis-(alkylamino)-s-triazines can be prepared in good yield by reacting the two amines with cyanuric chloride.

The 2-O-methyl derivatives can be prepared by heating the 2-chloro derivatives with methanol in the presence of sodium methoxide. The 2-thioalkyl substituted triazines are obtained in an analogous manner. They may also be obtained by methylation of the 2-thiol substituted derivatives which exist in the —SH form.

Hydrolysis of the 2-chloro compounds by acid or base affords 2-hydroxy-4,6-bis-(alkylamino)-s-triazines. On methylation they do not, however, give O-methyl derivatives as they may exist in the tautomeric 2-oxo-4,6-bis(alkylamino)-1,2-dihydro-s-triazine form (29). Hydrolysis of the 2-alkylthio- or 2-alkyloxy-4,6-bis(alkylamino)-s-triazines by acid also gives the 2-hydroxy derivatives.

Oxidation of the 2-methylthio-4,6-bis(alkylamino)-s-triazines (3) affords the corresponding methylsulfinyl (4) and methylsulfonyl homologs (5) (30,31).

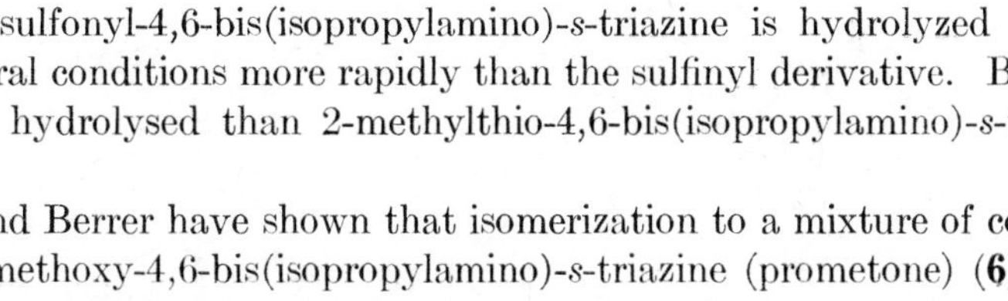

2-Methylsulfonyl-4,6-bis(isopropylamino)-s-triazine is hydrolyzed under acidic, basic, or neutral conditions more rapidly than the sulfinyl derivative. Both are much more rapidly hydrolysed than 2-methylthio-4,6-bis(isopropylamino)-s-triazine (prometryne).

Knüsli and Berrer have shown that isomerization to a mixture of compounds occurs when 2-methoxy-4,6-bis(isopropylamino)-s-triazine (prometone) (6) is heated at 260° (31).

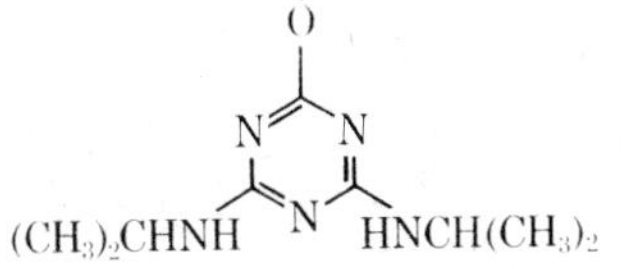

(6)

The amino groups of the 2-chloro-4,6-bis(amino)-s-triazines are generally un-reactive and cannot be acylated with normal acylating agents but ketene may be used.

N-Nitroso derivatives have been prepared and an N-nitroethylamino derivative has been obtained from 2-chloro-4,6-bis(ethylamino)-s-triazine (simazine).

Oxidative degradation of the alkylamino side chains of the triazines by hydroxyl radicals (Fenton's reagent) gives dealkylated products but the reaction is of little preparative value (32).

The amino groups of the 2-methoxy and 2-methylthio-4,6-bis(amino)-s-triazines are sufficiently basic to form mono or bis(acylamido) derivatives with the common acylating agents.

The heterocyclic ring of the s-triazine herbicides is very stable and is not cleaved except under drastic conditions.

The decomposition and loss of herbicidal activity of the triazines by light have been demonstrated. A major product of photolysis of simazine in methanol irradiated at 253.7 nm is 2,4-bis(ethylamino)-6-methoxy-s-triazine (simetone). Shorter wave-lengths effect replacement of chlorine or alkylamino groups by hydrogen and may cause ring cleavage (33). Irradiation of the crystalline 2-methylthio-4,6-bis(alkyl-amino)-s-triazines gives the corresponding 2H-4,6-bis(alkylamino)-s-triazines (34).

Analysis. The 2-chloro-bis(alkylamino)-s-triazines may be analyzed in herbicide formulations by reaction with morpholine, ie the reaction of a weak base with an active halogen atom. Chloride ion is liberated and determined by potentiometric titration. The ethylamino group of simazine may be titrated with perchloric acid. Other tri-azines with two amino groups interfere but 2,4-dichlorotriazines do not.

Residue analysis is based on the hydrolysis of the 2-chloro-s-triazine to a 2-hy-droxy-s-triazine by sulfuric acid, followed by absorbance measurements at 225, 240, and 255 nm for quantitative determination. Bioassay has been used for analysis of residues in soils. On heating a 2-chloro-s-triazine with pyridine on a steam bath for several hours a quaternary pyridinium halide (7) is produced which is hydrolyzed by water, to give a tertiary carbinol base (8). The pyridine ring is opened by a 1% sodium hydroxide solution to form a glutaconic aldehyde derivative (9). This compound has an absorption maximum at 436 nm. Maximal color development is produced by satu-rating the solution with glycine. As the color fades rapidly, measurements of absorbance are made immediately after reaction at one-minute intervals (35,35a).

Gas chromatography has been used for the determination of 2-chloro-s-triazines in soils (36).

The Zeisel method of methoxyl determination may be used for analysis of formulations of 2-methoxy-4,6-bis(alkylamino)-s-triazines. Titration with perchloric acid provides an alternative method. Hydrolysis by acid to the 2-hydroxy derivative is the basis of the method for residue analysis. The 2-methylthio derivatives can also be determined by acid hydrolysis to the hydroxy-s-triazine or, for formulation analysis, the liberated methylmercaptan can be absorbed and oxidized by iodine. Excess iodine is then backtitrated.

Infrared spectrometry has been used for quantitative analysis of the s-triazines.

Metabolism. Various investigations indicate that the ability of the triazines to interfere with photosynthesis is probably responsible for their biological activity. Simazine depletes carbohydrate by inhibiting the formation of sugars (37). The triazines inhibit the Hill reaction, ie the formation of oxygen by chloroplasts of certain plants in the presence of light and ferric salts (38). Interference with this reaction is an indication of herbicidal activity, but does not necessarily afford a direct measure.

The triazines are metabolized by plants. In addition to differences in uptake by and movement within the plant, chemical reactions have been examined as the proposed basis for their selective activity. The ability of corn to detoxify 2-chloro-4-ethylamino-6-isopropylamino-s-triazine (atrazine) and simazine (see above) by converting them to the nontoxic 2-hydroxy derivatives was correlated with the presence of an active principle in corn extracts. Wheat, which is susceptible, does not possess this compound identified as 2,4-dihydroxy-7-methoxy-1,4-benzoxazin-3-one (**10**) (37).

(**10**)

where R = glucose.

This compound is not present in all resistant plants and the conversion of 2-chloro-s-triazines to nontoxic 2-hydroxy-s-triazines can take place by alternative biological mechanisms. N-Dealkylation of the alkylamino side-chain of atrazine in pea plants gives 2-chloro-4-amino-6-isopropylamino-s-triazine. As this degradation pathway does not afford a completely nontoxic material, pea plants are moderately susceptible to the action of atrazine (39). In some plants simazine undergoes both N-dealkylation and conversion to the 2-hydroxy derivative. Additional pathways of s-triazine metabolism exist in plants. Radioactive labeling methods have shown the existence of metabolites, as yet unidentified, in polar and nonpolar solvent extracts.

Microorganisms can effect oxidation of side-chain alkyl groups to carbon dioxide. Partial deamination may also accompany hydrolysis and dealkylation in certain cases, as 2,4-dihydroxy-6-amino-s-triazine (ammelide) was identified as a product of simazine metabolism by *Aspergillus fumigatus Fres.* (40).

There is as yet no evidence for the identity of intermediate products formed by degradation of the s-triazine ring in plants or microorganisms. However, such processes have been reported to occur and certain microorganisms may utilize triazines as carbon sources (41).

Soils and crops, including sorghum, oats, soybeans, and cotton liberate ^{14}C-labeled carbon dioxide from ^{14}C-ring-labeled triazines. A biguanide structure has been proposed as the intermediate in corn metabolism of 2-chloro-4,6-bis(isopropylamino)-*s*-triazine (propazine), where the carbon atom at position 2 was removed as carbon dioxide (42).

Metabolism of triazines in animals gives many products. Conjugation of triazines and their metabolites takes place. Dealkylation and side-chain oxidation are important reactions and occur in both the alkylamino and methylthio substituents. Prometryne (11) in rats and rabbits is converted to both the thiol (12) and the corresponding disulfide (13), together with a number of other urinary metabolites (43).

Triazines have been used as selective herbicides in corn, cotton, sorghum, sugarcane, and other crops but they also command a significant nonagricultural market.

Urea Herbicides

Synthesis and Chemical Properties. The herbicidal activity of the substituted phenylureas was discovered in the late 1940s (44). Of many compounds screened, about twelve are commercially available; the properties of the most important are shown in Table 8.

The synthesis of 3-(*p*-chlorophenyl)-1,1-dimethylurea (monuron) (14) is typical of the general method used on a commercial scale and in the laboratory. Reaction of *p*-chloroaniline in dioxane or another inert solvent with anhydrous hydrogen chloride and phosgene at 70–75°C generates *p*-chlorophenyl isocyanate which reacts with dimethylamine at 25°C to give monuron.

monuron (14)

The ureas are generally insoluble in the inert solvent and precipitate out. A methoxyurea is obtained if dimethylamine is replaced by *O*-methylhydroxylamine. For example, 3,4-dichlorophenyl isocyanate and *O*-methylhydroxylamine give 3-(3,4-dichlorophenyl)-1-methoxy-1-methylurea (linuron).

Chemically, the ureas are quite stable. Hydrolysis by reflux with alkali generates the parent amine. This reaction is used as a basis for colorimetric analysis.

Table 8. Properties of Urea Herbicides

Name	Formula				Melting point, °C	Solubility, ppm				Vapor pressure	
	X	Y	R₁	R₂		water	acetone	ethanol	benzene	at °C	mm Hg
fenuron[a]	H	H	CH₃	CH₃	133–134	3850[25]				60	1.6×10^{-4}
monuron[b]	Cl	Cl	CH₃	CH₃	174–175	230[25]	52,000[27]		2,900[27]	25	5×10^{-7}
diuron[b]	Cl	Cl	CH₃	CH₃	158–159	42	53,000		1,200	30	0.31×10^{-6}
					180–190 (dec)						
fluometuron	H	CF₃	CH₃	CH₃	163–164	90[25]	3			100	5×10^{-5}
linuron	Cl	Cl	CH₃	OCH₃	93–94	75[25]	500,000[25]		150,000[25]		
metobromuron	Br	H	CH₃	OCH₃	95.5–96	330	sol	sol			
monolinuron	Cl	H	CH₃	OCH₃		930[20]				24	1.5×10^{-4}
neburon[b]	Cl	Cl	CH₃	C₄H₉	102–103	4.8[24]					
siduron	H	H	(2-methylcyclohexyl)	H	133–138	18		160,000			
chloroxuron	H	C₆H₄Cl	CH₃	CH₃	151–152	3.7[20]					
buturon	H	Cl	CH₃		145–146	30[20]					
chlorbromuron	Br	Cl	CH₃	OCH₃	94–96	50				20	4×10^{-7}
norea	(methanoindanyl)—NHCON(CH₃)₂				170–172	150[25]	sol	sol			
cycluron	(cyclooctyl)—NHCON(CH₃)₂				138	1200[20]					

[a] U.S. Pat. 2,655,447 (1953), to E. I. du Pont de Nemours & Co., Inc.
[b] U.S. Pat. 2,655,445 (1953), to E. I. du Pont de Nemours & Co., Inc.

Monuron, 3-(3,4-dichlorophenyl)-1,1-dimethylurea (diuron), 1-*n*-butyl-3-(3,4-dichlorophenyl)-1-methylurea (neburon), and 3-phenyl-1,1-dimethylurea (fenuron) were degraded by light (45). 3-(*p*-Chlorophenoxy)-phenyl-1,1-dimethylurea (chloroxuron) (15) was degraded in 90% yield in 13 hr by ultraviolet irradiation. Mono- and didemethylated products were identified and there was considerable loss of the urea carbonyl group as carbon dioxide (46).

When 3-(*p*-bromophenyl)-1-methoxy-1-methylurea (metobromuron) (16) is exposed to sunlight in aqueous solution, the bromine is replaced by a hydroxyl group. Preferential loss of the *N*-methoxyl group, followed by loss of the *N*-methyl group was also observed (47).

The 4-halogen substituent was replaced by hydroxyl when linuron and monuron were exposed to sunlight in aqueous solutions. Demethylation also occurred (48). Monuron (14), on irradiation in water under aerobic conditions, gives 3-(4-chloro-2-hydroxyphenyl)-1,1-dimethylurea together with 1,3-di(*p*-chlorophenyl)urea and a number of oxidized and polymeric compounds. 1,3-Di(*p*-chlorophenyl)urea may be formed by a reaction which proceeds through the isocyanate as a postulated intermediate. Dimethylamine is eliminated from monuron to form *p*-chlorophenyl isocyanate which could react with free *p*-chloroaniline (49).

polymeric and oxidized compounds

Analysis. A widely applicable method for analysis of formulations of urea herbicides involves hydrolysis with strong alkali. The liberated aliphatic amine is distilled and quantitatively absorbed in acid which is backtitrated. Alternatively, the free aniline may be extracted into an organic solvent and estimated.

Hydrolysis to the corresponding aniline followed by colorimetric estimation is the basis of a method for residue analysis. The aniline is diazotized and coupled with N-(1-naphthyl)ethylenediamine hydrochloride. Maximal absorbance is read at 560 nm. The azo dyes formed in this reaction have been used in identification and separation of the ureas by thin-layer chromatography.

The anilines may be determined by gas chromatography without further treatment or by diazotization and conversion to an iodo derivative. The latter method, which gives products suitable for measurement by the electron-capture detector, has been used for determination of residues in crops and soil (50,51).

Metabolism. A number of substituted ureas are specific and sensitive inhibitors of photosynthesis. Studies of structure–activity relationships show that monuron is one of the most potent inhibitors. In the light stages of photosynthesis, adenosine triphosphate (ATP) and reduced nicotinamide adenine dinucleotide phosphate (NADPH) are formed with liberation of oxygen. Carbon dioxide is then reduced in a reaction series which uses ATP and NADPH. The urea herbicides interfere with the mechanism of oxygen evolution. It is not known which enzyme is affected but the addition of riboflavin or flavine mononucleotide (FMN) reverses the effects of monuron when they are incubated together with an algal suspension. A complex may be formed between FMN and monuron in this reaction (52). There exists the possibility that a flavin-containing enzyme may be involved in the photosynthetic reaction sequence in which oxygen is produced. Blocking this enzyme by a urea would result in the death of the plant by starvation or as a result of the accumulation of a phytotoxic product.

Urea metabolism has been studied by radioactive tracer techniques. A fraction of the urea remained in the plant as a complex after extraction and a proteolytic enzyme was required to effect its release. Such complexes and conjugates may involve metabolites rather than the unchanged herbicides.

Dealkylation is an important pathway of metabolism and has been studied for metobromuron (**16**) in plants and cultures of soil bacteria by Geissbühler et al (53) who identified mono- and bis-dealkylated products. Stepwise N-dealkylation is a common pathway of degradation of urea herbicides in a variety of plant species. Differential rates of degradation may determine the resistance or susceptibility of a plant species (54). The anilines do not appear to accumulate in plants but 3,4-dichloronitrobenzene has been identified as a transformation product of diuron in corn. It was not certain however that this compound was a true metabolite (55). ^{14}C ring- and methyl-labeled compounds were applied to foliage and added to nutrient solutions; 3,4-dichloronitrobenzene was detected.

Dealkylation and hydroxylation of the aromatic ring of phenylureas take place during animal metabolism (56).

Carbamates

The biological activity of the carbamates is extremely varied. Suitable substitution gives a range of compounds used as herbicides, insecticides, medicinals, nematicides, miticides, or molluscides. The inhibition of plant-root growth by ethyl N-phenylcarbamate was discovered in 1929 (57). Isopropyl N-phenylcarbamate

Table 9. Properties of Carbamate Herbicides

Name	Formula	Melting point, °C	Solubility, ppm[e]			
			water	acetone	ethanol	benzene
propham (IPC)	NHCOOCH(CH₃)₂	87–88		v sol		
chlorpropham (CIPC)[a]	NHCOOCH(CH₃)₂ Cl	38–40		miscible	v so	
barban[b]	C₆H₅NHCOOCH₂C≡CCH₂Cl	75–76	0.0011[f]			37.0[f]
swep	Cl NHCOOCH₃ Cl	133–134				
Sirmate[c] UC 22463	CH₂OCONHCH₃ Cl Cl	52	170	sol		sol
Azak[d]	C(CH₃)₃ COONHCH₃ CH₃ C(CH₃)₃	200	7			

[a] U.S. Pat. 2,695,225 (1954), to Columbia Southern Chemical Corp.
[b] U.S. Pat. 2,906,614 (1959), to Spencer Chemical Co.
[c] Registered trademark Union Carbide Co.
[d] U.S. Pat. 3,140,167 (1964), to Hercules Powder Co.; registered trademark.
[e] Unless otherwise indicated.
[f] Grams per 100 g solvent.

(IPC) (**17**) was successfully used as a weed-control agent in 1945 and substituted derivatives, such as isopropyl N-(m-chlorophenyl) carbamate (CIPC) were introduced after 1950.

Chemical Properties and Synthesis. Table 9 gives the formulas and properties of the most widely used herbicidal carbamates. Two principal methods of synthesis are available. A substituted aniline is reacted with the corresponding chloroformate ester, or, alternatively, the carbamate is prepared by reaction of an isocyanate with an alcohol.

A modification of the former route used for the preparation of 4-chloro-2-butynyl N-*m*-chlorophenylcarbamate (barban) involves the reaction between *m*-chlorophenyl isocyanate and butyne-1,4-diol (1 mole) followed by reaction of the free hydroxyl group with thionyl chloride. Barban is decomposed by heating above its melting point (75°C) with the loss of HCl. Carbamates generally may be hydrolyzed to the aniline by acid or alkali and this reaction forms the basis for analytical determination. Barban in sodium hydroxide solution at 25°C very rapidly undergoes hydrolytic loss of the terminal chlorine atom but is more stable to acid. IPC (**17**) is relatively volatile and sublimes slowly at room temperature. The carbamates decompose on heating into an isocyanate and an alcohol. IPC photodecomposes in an analogous manner. At least two routes of photolysis are known (58), as shown below.

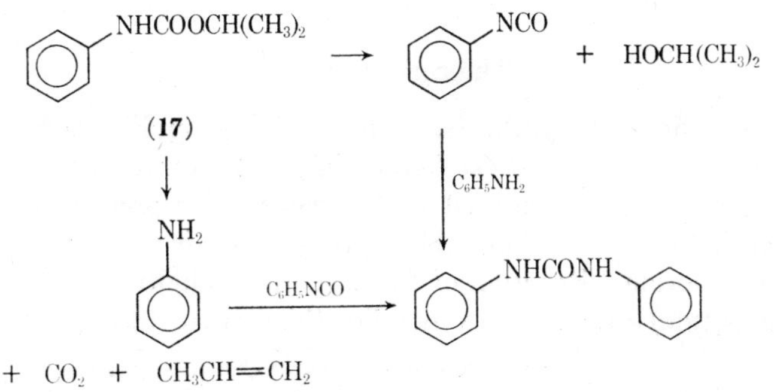

Analysis. Hydrolysis of carbamates by acid or base affords the corresponding amine. The mechanism of the reaction has been extensively investigated, particularly in relation to the activity of the insecticidal carbamates against cholinesterase (59). The reaction may be utilized for analyzing the carbamate herbicides in technical formulations or as residues in crops and soil. For analysis of the former, acid hydrolysis generates carbon dioxide which is trapped in standard sodium hydroxide solution. Excess base is backtitrated with standard acid. Alternative methods are total nitrogen analysis, total chlorine or, for barban, measurement of the uv absorption at 277.5 nm.

Residue analysis of phenylcarbamates depends on colorimetric estimation of the aniline generated by hydrolysis. As with the ureas, the aniline is diazotized and coupled with N-(1-naphthylethyl)amine hydrochloride and the absorbance of the solution measured at an appropriate wavelength. A color may also be developed from the aniline by reaction with calcium hypochlorite, phenol, and ammonia. The carbamate may be hydrolyzed by acid or base but acid hydrolysis does not permit estimation of aniline without extractive separation, as the accompanying decomposition and charring produces interfering substances. Infrared analytical techniques may also be used but are less sensitive than colorimetric methods (60).

Metabolism. The basic mechanism by which the carbamate herbicides exert their phytotoxic action has not been demonstrated, although the carbamates affect a number of physiological and biochemical processes. Structurally carbamates are related to the amides and the ureas. Like these, they affect photosynthetic processes and it has been suggested that the imino group, together with the adjacent carbonyl group, may be the required structural unit for binding to a catalytically active surface. The effects of carbamates on plant respiration and carbohydrate metabolism have been investigated (61).

There are several routes by which carbamates are biologically degraded. Hydrolysis of the carbamate function is a significant route but ring hydroxylation may also occur. Substituents (*O*- or *N*-alkyl) may be dealkylated through intermediate hydroxymethyl compounds. Such reactions occur in mammals and insects. Although many of these transformations are effected by mammalian microsomes, similar conversions may take place in plants. For example, the insecticidal carbamate, 1-naphthyl Δ′-methyl carbamate (carbaryl, see Vol. 11, p. 711) is converted to the corresponding hydroxymethyl carbamate. Hydroxylation of alkyl groups on the aromatic ring is followed by conjugation of the product with a sugar, amino acid, or peptide molecule. Little specific information is available on the products of carbamate metabolism in plants but a number of metabolites appear to retain the carbamate linkage intact and form water-soluble conjugates (62).

Thiocarbamates

Chemical Properties and Synthesis. *S*-Ethyl di-*n*-propylthiocarbamate (EPTC) was introduced as a herbicide in 1959 (63). Table 10 shows a number of thiocarbamates and related compounds also used as herbicides. The preparation of EPTC is typical of the synthetic methods (63). Di-*n*-propylamine is reacted with phosgene to give a carbamoyl chloride. Reaction of the carbamoyl chloride with a thiol (as the sodium alkylmercaptide) gives the corresponding thiocarbamate, as shown below:

$$(n\text{-}C_3H_7)_2NH + COCl_2 \rightarrow (n\text{-}C_3H_7)_2NCOCl + HCl$$
$$(n\text{-}C_3H_7)_2NCOCl + C_2H_5SNa \rightarrow (nC_3H_7)_2NCOSC_2H_5 + NaCl$$

Alternatively the amine is reacted with an alkyl chlorothiolformate in the presence of a proton acceptor:

$$RSCOCl + NHR'R'' \rightarrow RSCONR'R'' + HCl$$

2-Chloroallyl-*N*,*N*-diethyldithiocarbamate (CDEC) may be synthesized by reaction of allyl chloride with sodium diethyl dithiocarbamate:

$$(C_2H_5)_2NCSNa + ClCH_2CCl{=}CH_2 \rightarrow (C_2H_5)_2NCSSCH_2ClCCl{=}CH_2$$

Sodium dimethyldithiocarbamate (metham) possesses fungicidal and herbicidal activity in soil possibly due to the liberation of methyl isothiocyanate. It is prepared by the reaction of methylamine and carbon disulfide with sodium hydroxide.

Thiocarbamates are generally volatile compounds. EPTC is stable to hydrolysis by weak acids and bases at room temperature. CDEC is inactivated by ultraviolet irradiation (64).

Analysis. A variety of methods are available for the analysis of thiocarbamates. For macroquantities, determination of nitrogen (Kjeldahl), sulfur (Parr) or, where applicable, chloride (Parr) have been used. CDEC has been determined by total chlorine estimation or by measurement of ultraviolet absorption at 278 nm. Residue analysis is based on a general reaction of a number of thiocarbamates. The free amine is liberated by acid hydrolysis and distilled from an alkaline solution. The amine is reacted with cupric dithiocarbamate prepared by shaking carbon disulfide in chloroform with ammoniacal cupric sulfate solution. With amines the cupric dithiocarbamate reagent gives a color measured at 435 nm. The majority of the thiocarbamates are sufficiently volatile for gas chromatography to be a satisfactory method of analysis with microcolometric detection (for sulfur) or flame-photometric detection (65).

Table 10. Properties of Thiocarbamate Herbicides

Name	Formula	Boiling point, °C	Solubility, ppm[f]				Vapor pressure	
			water	acetone	ethanol	benzene	at °C,	mm Hg
EPTC[a]	$(C_3H_7)_2NCOSC_2H_5$	127_{20}	375[25]	miscible		v sol	24	1.97×10^{-2}
SMDC[b]	$CH_3NHCSSNa \cdot 2H_2O$		72.2[g]					
vernolate[a]	$(C_3H_7)_2NCOSC_3H_7$	140_{20}	109[24]	sol		sol	24	5.4×10^{-3}
CDEC[c]	$(C_2H_5)_2NCSSCH_2CCl{=}CH_2$	$128{-}130_1$	92				200	2.2×10^{-3}
pebulate[d]	$\begin{matrix}C_4H_9\\C_2H_5\end{matrix}NCOSC_3H_7$	142_{20}	30	miscible	miscible		25	4.8×10^{-3}
diallate	$\left(\begin{matrix}CH_3\\CH_3\end{matrix}CH\right)_2NCOSCH_2CCl{=}CHCl$	150_9	14	miscible				
triallate	$\left(\begin{matrix}CH_3\\CH_3\end{matrix}CH\right)_2NCOSCH_2CCl{=}CCl_2$	148_9	4	sol	sol			
butylate	$\left(\begin{matrix}CH_3\\CH_3\end{matrix}CH_2CH\right)_2NCOSC_2H_5$	137.5_{21}						
molinate	(ring) NCOSC$_2$H$_5$	137_{10}	1000[20]	miscible				
cycloate[e]	(ring) $\begin{matrix}C_2H_5\\|\end{matrix}NCOSC_2H_5$	146_{10}						

[a] U.S. Pat. 2,913,327 (1959), to Stauffer Chemical Co.
[b] U.S. Pat. 2,791,605 (1957), to Stauffer Chemical Co.
[c] U.S. Pat. 2,919,182 (1959), to Stauffer Chemical Co.
[d] U.S. Pat. 3,175,897 (1965), to Stauffer Chemical Co.
[e] U.S. Pat. 3,185,720 (1965), to Stauffer Chemical Co.
[f] Unless otherwise indicated.
[g] Grams per 100 ml.

Table 11. Properties of Amide Herbicides

| Name | Formula | Melting point,[e] °C | Solubility, g/100 g solvent[e] | | | | Vapor pressure at °C, mm Hg |
			water	acetone	ethanol	benzene	
solan[a]	NHCOCHC$_3$H$_7$ (ring with Cl, CH$_3$)	85–86	8–9[f]				
dicryl	NHCOC=CH$_2$ —CH$_3$ (ring with Cl, Cl)	127–128	insol	20			
propanil	NHCOC$_2$H$_5$ (ring with Cl, Cl)	93–94	500				
diphenamid[b]	CHCON(CH$_3$)$_2$	134–135	0.5[f]	15[f]			

	Structure							
propachlor[e]	C₆H₅–N(CH(CH₃)₂)(COCH₂Cl)	67–76	700^{20}[f]	30.9		50	110	0.3
Lasso	(2-CH₃-C₆H₄)–N(CH₂OCH₃)(COCH₂Cl)	$110^{9.03}$						
CDAA[c]	CH₂ClCON(CH₂CHCH₂)₂	$74^{0.3}$			50^{-13}		20	9.4×10^{-3}
CDA	CH₂ClCON(C₂H₅)₂							
naptalam (NPA)[d]	1-naphthyl–NHCO–(2-COOH-C₆H₄)	185	200	5000				
cypromid	(3,4-Cl₂-C₆H₃)–NHCO–(cyclopropyl)	131	<0.01			3		

[a] U.S. Pat. 3,020,142 (1962), to F.M.C. Corp.
[b] U.S. Pat. 3,120,434 (1964), to Eli Lilly and Co.
[c] U.S. Pat. 2,864,683 (1958), to Monsanto Chemical Co.
[d] U.S. Pats. 2,556,664 and 2,556,664 (1951), to U.S. Rubber Co.
[e] Unless otherwise stated.
[f] Ppm.
[g] Boiling point.

Metabolism. Thiocarbamates are generally applied to the soil surface. Their mode of action in the plant may involve interference with enzymic mechanisms by chelation of a metal, such as copper or iron. They are rapidly absorbed by plants and the sulfur atom appears to become part of the general metabolic pool. Experiments with ^{14}C-labeled compounds indicate that in tomatoes the propanethiol moiety of S-propyl butylethylthiocarbamate (pebulate) loses sulfur to give propanol, propionic acid, and pyruvic acid. Such compounds are readily assimilated into general metabolic pathways (66).

Amides

The amide herbicides include two major groups, herbicides based on substituted anilide structures and chloroacetamide derivatives. Both groups were introduced in 1956 (67). One of the earlier amide derivatives to be used was N-1-naphthylphthalamic acid (naptalam) which was reported in 1949 to be active as a herbicide (68).

Chemical Properties and Synthesis. A general method of synthesis is the reaction of an acid chloride with a suitable amine in the presence of an acid acceptor (69).

$$(C_6H_5)_2CHCO_2H \xrightarrow{SOCl_2} (C_6H_5)_2CHCOCl + HCl$$

$$(C_6H_5)_2CHCOCl + (CH_3)_2NH \rightarrow (C_6H_5)_2CHCON(CH_3)_2 + HCl$$

Naptalam is prepared by reaction of α-naphthylamine with phthalic anhydride. Properties of herbicidal amides are given in Table 11. The amides are generally stable crystalline compounds. However, naptalam is unstable in alkaline solutions. Other amides can be hydrolyzed under more vigorous conditions.

The photodecomposition of the amides has been investigated in few cases. N,N-Dimethyl-2,2-diphenylacetamide (diphenamid) was photolyzed in aqueous solution by ultraviolet irradiation to give N-methyl-2,2-diphenylacetamide as the principal product. Benzophenone, benzhydrol, and benzoic acid were also detected (70).

Analysis. Macroquantities of chlorine-containing amides may be determined by estimation of chloride. The anilide herbicides may be hydrolyzed to the aniline which is determined spectroscopically after diazotization and coupling. For analysis of microquantities naptalam hydrolysis with 30% sodium hydroxide gives naphthylamine which reacts with diazotized sulfanilic acid. The product has maximal absorbance at 534 nm. N,N-Diallyl 2-chloroacetamide (CDAA) can be determined by gas chromatography or by hydrolysis to diallylamine measured as its cupric dithiocarbamate complex. Gas chromatography has also been used for determination of 3′,4′-dichloropropionanilide (propanil), 3′-chloro-2-methyl-p-valerotoluidide (solan), 3′,4′-dichloro-2-methylacrylanilide (dicryl), and diphenamid (71,71a).

Metabolism. Propanil is a selective herbicide used on weeds commonly occurring in rice. The resistance of rice to the action of this herbicide is probably due to the ability of this plant to convert propanil to the less toxic 3,4-dichloroaniline (72). The acyl anilides have structural features in common with the ureas and carbamates suggesting that interference with the photosynthetic pathway may be the route by which herbicidal activity is effected. However, other amide herbicides, such as CDAA, appear to affect quite different metabolic systems and may react with sulfhydryl-containing enzymes (73).

A number of different pathways of microbial metabolism have been reported. Diphenamid undergoes stepwise demethylation (74). An interesting reaction is the conversion of propanil by soil microorganisms to 3,3′,4,4′-tetrachloroazobenzene. The conversion proceeds by way of the free aniline. Anilides which can be converted to anilines with a free ortho position undergo similar reactions. Condensation products involving more than two anilino residues have also been detected among the reaction products (75).

Chlorinated Aliphatic Acids

The use of trichloroacetic acid (TCA) as a herbicide was reported in 1944 (76). 2,2-Dichloropropionic acid (dalapon) and 2,2,3-trichloropropionic acid were introduced later (77) and the herbicidal properties of other halogenated aliphatic acids have since been investigated. Dalapon has been used to provide control of grass in a variety of crops and also in noncrop application, such as the banks of ditches and aquatic weeds.

Chemical Properties and Synthesis. The chlorinated aliphatic acids are quite strongly acidic, as shown below. Other properties are given in Table 12.

Acid	pK_a at 25°C
trichloroacetic acid	0.08
2,2-dichloropropionic acid	1.53
2,2,3-trichloropropionic acid	1.00

TCA may be prepared by nitric acid oxidation of chloral or by direct chlorination of acetic acid. Dalapon is similarly obtained by chlorination of propionic acid.

These acids are susceptible to hydrolysis in aqueous solution at room temperature and the rate of hydrolysis of the sodium salt of dalapon becomes rapid at 50°C. On hydrolysis TCA gives chloroform and carbon dioxide and dalapon gives pyruvic acid. At physiological pH an α-Cl atom may be replaced by a sulfhydryl group (RSH). 2,3,6-Trichlorophenylacetic acid (fenac) is for convenience included within this class of compounds. It is prepared by chlorination of toluene to give 2,3,6-trichlorobenzyl chloride. The chloride is reacted with sodium cyanide to yield trichlorobenzyl cyanide which can be hydrolyzed with sodium hydroxide to the acid. A mixture of chlorinated isomers is obtained and the commercial compound contains about 60% of the 2,3,6-trichloro isomer.

Table 12. Chlorinated Aliphatic Acid Herbicides

Name	Formula	mp, °C	bp, °C	Solubility, g/100 g solvent		
				water	acetone	benzene
TCA	CCl_3COOH	59	197.5[d]	1306[25]	850[25]	201[25]
	CCl_3COONa			833[25]	0.76[25]	0.007[25]
dalapon[a]	CH_3CCl_2COOH		185–190	v sol		
	CH_3CCl_2COONa	193–194		502	0.014	0.002
fenac[b,c]	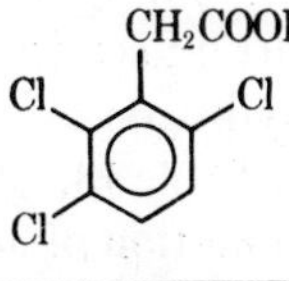	159–160		sl sol		

---

[a] U.S. Pat. 2,642,354 (1953), to The Dow Chemical Co.
[b] A mixture of isomers; data refer to the 2,3,6-trichloro isomer.
[c] U.S. Pat. 2,977,212 (1961), to Heyden Newport Chemical Corp.
[d] Vapor pressure at 76.99°C, 5.0 mm Hg.

Analysis. Chlorinated acids may be determined in macroamounts by standard techniques of volumetric analysis or analysis of the total chlorine content. Determination of residues of dalapon in crops may be carried out by purification of an extract by ion-exchange chromatography to remove interfering compounds. The acid is then hydrolyzed to pyruvic acid which is converted to the phenylhydrazone with 2,4-dinitrophenylhydrazine reagent. The absorbance of the solution is measured at 440 nm. TCA may be determined by a colorimetric procedure. The sample is heated with pyridine and sodium hydroxide solution. The colored solution has a uv absorption maximum at 570 nm (78).

Gas chromatographic methods have been developed for residues of dalapon and fenac (79).

Metabolism. Chlorination α to the carboxyl group appears to be a requirement for herbicidal activity. There is evidence that interference with pantothenate metabolism may be partly responsible for phytotoxicity but other biochemical systems must also be involved. Studies of metabolism in plants indicate that very little of the applied chemical is converted to metabolites and there is little evidence as to their nature. In animals, some dalapon was excreted unchanged in the urine, but it has been postulated that it is also converted to pyruvic acid. Pyruvic acid yields acetate and carbon dioxide which become part of the general metabolic pool (80). The degradation of chlorinated aliphatic acids by soils and isolated microorganisms has been extensively investigated. *Arthrobacter sp.* converts trichloracetate to carbon dioxide and chloride ion (81).

Chlorinated Benzoic Acids

The chlorinated benzoic acid herbicides are a heterogeneous group comprising a number of polychlorinated aromatic compounds with a variety of ring substituents. For convenience 2,6-dichlorobenzonitrile (dichlobenil) is discussed in this section, but the hydroxynitriles are discussed below under Phenols. 2,3,6-Trichlorobenzoic acid (2,3,6-TBA) is a rather nonselective herbicide first evaluated in 1948 and introduced experimentally in 1954 (82). It is a soil sterilant and is extremely phytotoxic. The synthesis of 2,3,6-TBA starts from a toluene derivative; 2,3,6-trichlorotoluene can be oxidized with nitric acid or 2,5-dichlorotoluene may be chlorinated and then oxidized. Mixtures of polychlorinated benzoic acids (2–5 chlorine atoms) are also formulated for herbicidal use.

Replacement of the 2-chloro-group of TBA by a methoxy group gives 3,6-dichloro-2-methoxybenzoic acid (dicamba) which is somewhat more selective in its action.   A related herbicide is the compound 3,5,6-trichloro-2-methoxybenzoic acid (tricamba).

2,3,6-TBA and dicamba are quite stable under oxidizing and hydrolytic conditions. The latter appears resistant to photodecomposition but on ultraviolet irradiation of 2,3,6-TBA in methanol, chloro substituents are replaced by hydrogen atoms. Properties of these and some other members of this group are given in Table 13.

Analysis. The method of analysis may be based on determination of acid or, in a few cases on determination of chlorine content. Ultraviolet and infrared spectra have also been employed for analysis. Residue analysis is most satisfactorily accomplished by conversion to the methyl ester (with diazomethane, or methanol and boron trifluoride), followed by gas chromatography with an electron-capture detector. Residues of 2,3,6-TBA and dicamba have been determined by this technique. Dimethyl

Table 13. Properties of Chlorinated Benzoic Acid

Name	Formula	Melting point, °C	Solubility, g/100 g solvent[f]				Vapor pressure at °C, mm Hg
			water	acetone	ethanol	benzene	
amiben[a]	NH$_2$, Cl, Cl, COOH	210	700[25,g]	23.27[24]	17.28[24]	200[24,g]	
DCPA[b]	COOCH$_3$, Cl, Cl, Cl, Cl, COOCH$_3$	156 (dec 360–370)	<0.5[g]	10		25	40 0.01
dicamba[c]	COOH, OCH$_3$, Cl, Cl	114–116	0.45		92.2	100	3.75 × 10^{-3}
tricamba	COOH, OCH$_3$, Cl, Cl, Cl	137–139	sl sol				
dichlobenil[d]	CN, Cl, Cl	139–145	18[20,g]			20	5.5 × 10^{-4}
2,3,6-TBA[e]	COOH, Cl, Cl, Cl	125–126	0.84	60.7		23.8	

[a] U.S. Pat. 3,014,063 (1958), to Amchem. Products Inc.
[b] U.S. Pat. 2,923,634 (1960), to Diamond Alkali Co.
[c] U.S. Pat. 3,013,054 (1961), to Velsicol Chem. Corp.
[d] U.S. Pat. 3,027,248 (1962), to North Amer. Philips Co., Inc.
[e] U.S. Pat. 2,848,470 (1956), to Heyden Newport Chem. Corp.
[f] Unless otherwise stated.
[g] ppm.

2,3,5,6-tetrachloro-terephthalate (DCPA) is sufficiently volatile for analysis by gas chromatography.

Metabolism. The metabolism of benzoic acids in plants generally follows the routes of hydroxylation or conjugation with cellular components. There is little evidence for degradation of the ring. Reports of the isolation of specific metabolites are few. When dicamba is metabolized by wheat and bluegrass the major product is 5-hydroxy-2-methoxy-3,5-dichlorobenzoic acid (as a conjugate). Conjugated 3,6-dichlorosalicylic acid was obtained in smaller quantity (83).

Amiben. A substituted benzoic acid of economic importance is 3-amino-2,5-dichlorobenzoic acid (amiben) which was discovered to be herbicidally active in 1958. It was first registered for this use in the United States in 1961.

Amiben is prepared by nitration of 2,5-dichlorobenzoic acid and the major product, 2,5-dichloro-3-nitrobenzoic acid, can be separated from the resultant mixture of isomers by fractional precipitation. It has similar herbicidal properties as amiben, but the latter is better tolerated by soybeans. The 3-nitro compound is then reduced to the 3-amino derivative. The purity of the final compound must be carefully controlled to remove isomeric aminodichlorobenzoic acids which reduce selectivity.

Amiben functions both as a benzoic acid and as an aromatic amine. In the solid state it is very stable but in aqueous solution it darkens if exposed to light. The photochemical decomposition of amiben appears to be a photooxidation process typical of an aromatic amine, ie polymerization through a photolytically generated NH radical. This is accomplished by the photolytic loss of chlorine with replacement by hydrogen (84). The fact that reductive dechlorination occurs in water is noteworthy (as water is not a good proton donor in such cases). Sodium bisulfite represses the photochemical oxidation of the amino group which could otherwise occur. The ability of the amino group of amiben to diazotize and couple is typical of an aromatic amine.

Analysis. Formulation analysis is based on diazotization of the amino group by titration with standard sodium nitrite solution. Gas chromatography of the methyl ester of amiben may be used for residue analysis. A hydrolytic step is probably required for the analysis of plant material, as an *N*-glycoside of amiben has been shown to be a metabolite in a number of treated plants (85). A colorimetric procedure is generally applicable to a variety of crops. Diazotization and coupling with *N*-(1-naphthyl)ethylenediamine give a colored solution which has an absorption maximum at 528 nm (86,86a).

Metabolism. Amiben was converted to *N*-(3-carboxy-2,5-dichlorophenyl)glucosylamine in the roots of treated soybeans, tomato plants, and other species (86,86a). The metabolite remained unchanged in the tissue and it has been suggested that this reaction affords a method of detoxication in tolerant plant species. In soybeans the related herbicide 3-nitro-2,5-dichlorobenzoic acid is converted to a metabolite which shows the same chromatographic behavior as the *N*-glucoside of amiben (87). Reduction of the nitro group is presumably followed by conjugation with glucose.

Dichlobenil. The herbicidal properties of 2,6-dichlorobenzonitrile (dichlobenil) (18) were first discovered in 1960 (88). It inhibits germination of weed seeds and has been used in the control of weeds in fruit trees, alfalfa, and dormant cranberries. It is also used to control certain weeds infesting waterways and irrigation canals, thereby reducing the flow of water.

Dichlobenil is a relatively volatile compound of low water solubility. It has been synthesized by reacting 2,6-dichlorotoluene with chlorine. Hydrolysis to the aldehyde, conversion to the oxime, and dehydration gives 2,6-dichlorobenzonitrile. It is hydrolyzed in alkaline solution to 2,6-dichlorobenzamide.

(18)

Dichlobenil is stable to sunlight. In methanol decomposition by ultraviolet irradiation takes place slowly with replacement of chlorine by hydrogen. Replacement of the second chlorine atom proceeds less readily. The compound 2,6-dichlorothiobenzamide (**19**) is also herbicidally active. It is converted to dichlobenil in soils and in plants.

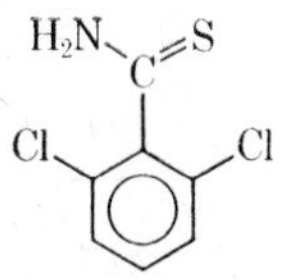

(**19**)

Phenols

Nitrophenols were used in Europe as selective herbicides in the 1930s (6). Initially 2-methyl-4,6-dinitrophenol (DNOC) was used and later, in 1946, 2-*sec*-butyl-4,6-dinitrophenol (dinoseb) was introduced (89). DNOC and dinoseb have been used as contact herbicides for many years. Dinoseb may be formulated as the ammonium or amine salts for selective use or in oil or oil–water emulsion for nonselective uses. Pentachlorophenol (PCP) has been available since 1936 as a wood preservative; its use as a herbicide was first recorded in 1940 (90). The benzonitrile derivatives, 4-cyano-2,6-dibromophenol (bromoxynil) and 4-cyano-2,6-diiodophenol (ioxynil), were discovered more recently and were first tested in 1960 (91). Bromoxynil has been used for postemergence control of broadleaf weeds in corn.

Chemical Properties and Synthesis. Table 14 gives the properties of the herbicidal phenols. PCP is manufactured by catalytic chlorination of phenol. Dinitrophenols are prepared by controlled nitration of an alkylated phenol.

The synthesis of bromoxynil and ioxynil starts with *p*-hydroxybenzaldehyde, as follows:

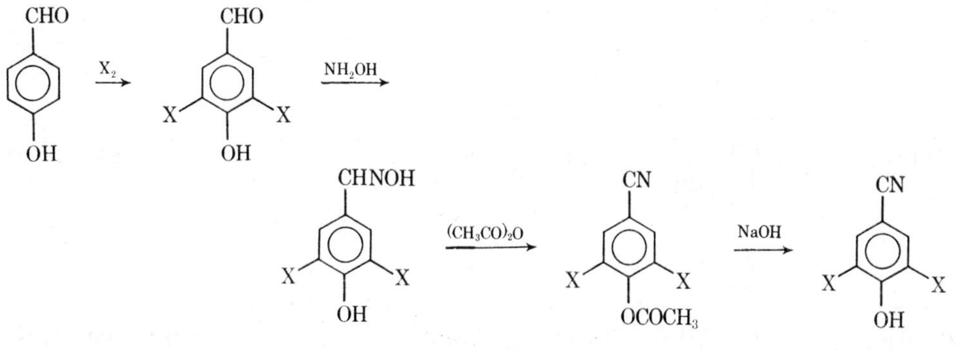

Nitrophenols may present explosive hazards if completely dry. They form water-soluble salts of alkali metals and may be readily reduced to the monoamino nitro derivatives. They form typical molecular compounds with amines, hydrocarbons and phenols. They are susceptible to decomposition by ultraviolet irradiation in alkaline solution. PCP decomposes in sunlight. Munakata and his co-workers identified monomeric and dimeric oxidation products (**20–24**) formed by photolysis of the sodium salt in water (92).

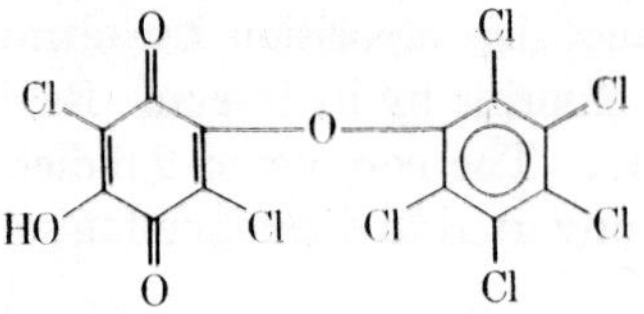

(20)

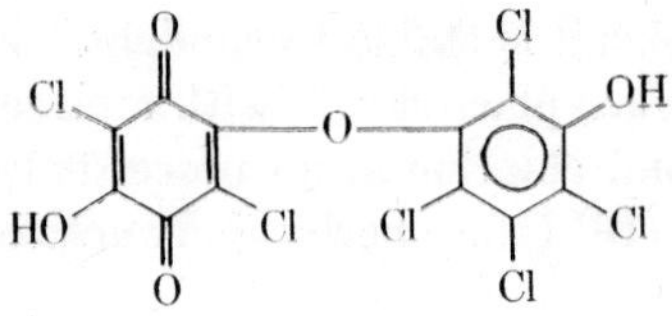

(21)

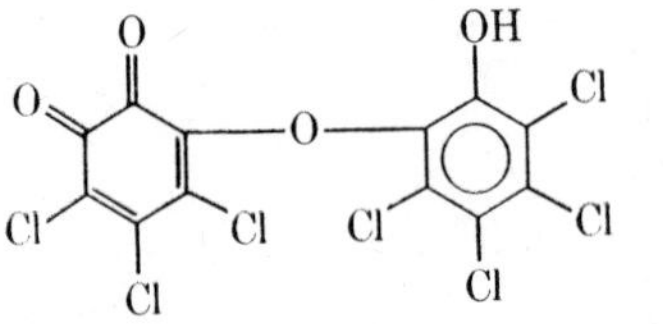

(22)

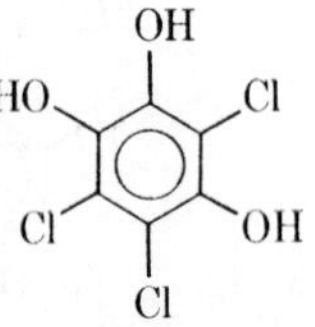

(23)

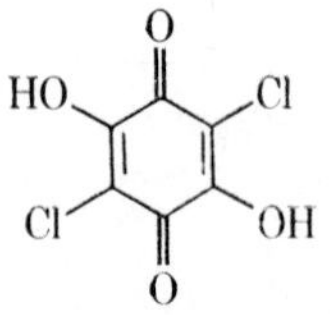

(24)

Table 14. Properties of Phenol Herbicides

Name	Formula	Melting point, °C	Solubility, g/100 g solvent[b]				Vapor pressure at °C, mm Hg	
			water	acetone	methanol	ethanol		
bromoxynil	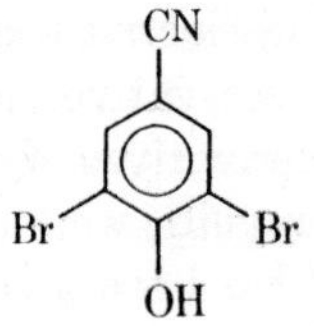	194–195	5[20,c]	17[25]	9[25]			
ioxynil	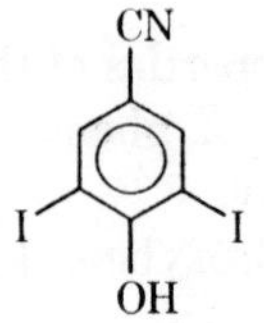	212–213.5	50[25,c]	7[25]	2[25]			
PCP[a]	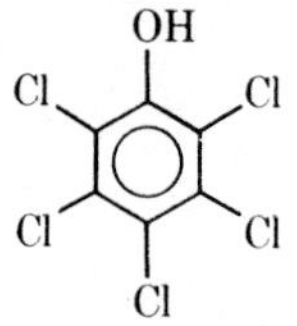	190–191	20–25[20,c]				100	0.12
DNOC	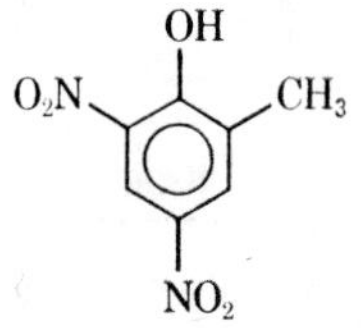	85.8	130[c]				25	105 × 10⁻⁶
dinosep (DNSP)	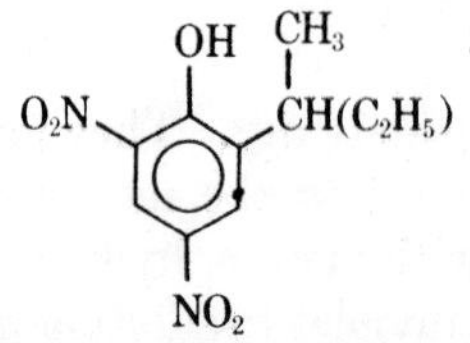	38–42	0.0052[25]			48[25]	151	1.0

[a] U.S. Pat. 2,131,259 (1938), to The Dow Chemical Co.
[b] Unless otherwise stated.
  Ppm.

Ioxynil undergoes photolysis in benzene solution on irradiation with light of 253.7 nm. The two iodo groups are replaced by phenyl groups in a free-radical reaction (93). Bromoxynil forms radicals on irradiation in aqueous solution and oxidative polymerization follows the replacement of bromine by hydroxyl groups (94).

Analysis. The earlier methods of analysis of residues of the nitrophenol derivatives relied on colorimetric methods by measuring the absorbance of the free phenol or of a dichromate oxidation product of the phenylenediamine obtained by reduction. A more sensitive method utilizes gas chromatography with an electron-capture detector for analysis of the ethers obtained by methylation with diazomethane. Macromethods of analysis may be based on titrimetric or gravimetric procedures.

For microestimation of PCP, methods are based on determination of chlorine content or on the measurement of the color developed with a variety of chromogenic reagents. In soil water and fish, residues of PCP have been measured by gas chromatography of the methyl ether (95).

Metabolism. Nitrophenols and chlorophenols act as uncoupling agents. They interfere with the phosphorylation process and prevent the generation of ATP. They are not translocated and act directly by contact with the plant causing cell necrosis. There is little information concerning products of metabolism in plants.

It has been suggested that bromoxynil and ioxynil may act by formation of free radicals which interact with sulfhydryl groups within the plant. There is no direct evidence for this reaction but iodide ions are produced in bean plants treated with ioxynil.

Substituted Dinitroanilines

One of the best known substituted dinitroaniline herbicides is α,α,α-trifluoro-2,6-dinitro-N,N-di-n-propyl-p-toluidine (**27**) (trifluralin). The herbicidal use of this compound was first reported in 1960 (96). A number of other anilines show herbicidal activity. Of these, the 2,6-dinitro-substituted compounds are most effective. Trifluralin and nitralin have been used for selective preemergence control of annual grasses in cotton, soybeans, and other situations.

Chemical Properties and Synthesis. The properties of compounds of commercial interest are listed in Table 15. They may be prepared by the reaction of a 4-chloro-3,5-dinitrotoluidine derivative with an appropriate amine, as follows:

where R and R$'$ = alkyl.

These products are typically substituted 2,6-nitroaniline derivatives and are deep orange in color. Trifluralin (di-n-propyl) is extremely insoluble in water but soluble in organic solvents; it is slightly volatile. Trifluralin is susceptible to photodecomposition in the solid state and in solution. Macmahon has studied the behavior of α,α,α-trifluoro-2,6-dinitro-N-methyl-p-toluidine (**25**) in n-heptane solution when irradiated with an ultraviolet source. Demethylation occurred and formaldehyde and α,α,α-trifluoro-6-nitro-2-nitroso-p-toluidine (**26**) were identified. The N-propyl analog (**28**)

Table 15. Properties of Substituted Dinitroaniline Herbicides

Name	Formula	mp, °C	bp, °C	Solubility[a] water, ppm	Vapor pressure at °C, mm Hg	
benefin	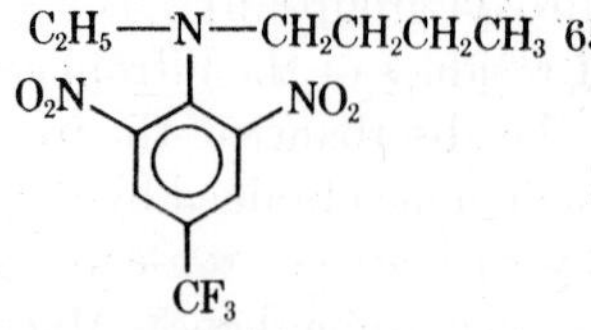 C_2H_5—N—$CH_2CH_2CH_2CH_3$ O_2N … NO_2 … CF_3	65–66.5	$121–122_{0.5}$ $148–149_7$	70^{25}	25	4×10^{-7}
trifluralin	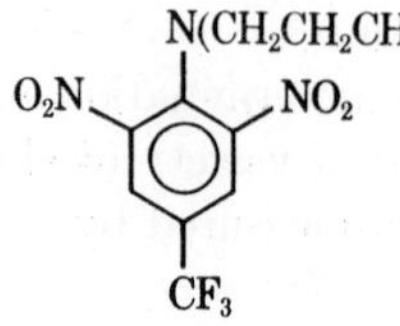 $N(CH_2CH_2CH_3)_2$ O_2N … NO_2 … CF_3	48.5–49	$96–97_{0.18}$	1^{27}	24.5	1.99×10^{-4}
nitralin	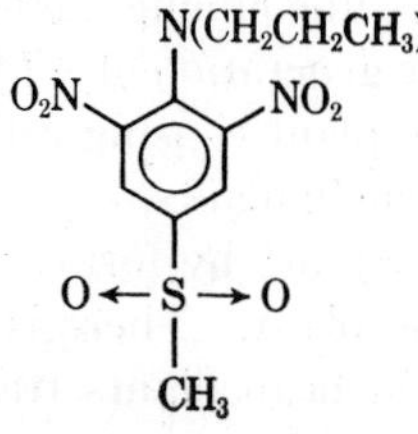 $N(CH_2CH_2CH_3)_2$ O_2N … NO_2 … $O \leftarrow S \rightarrow O$ … CH_3	151–152		0.6^{25}	25	1.5×10^{-6}

[a] All three compounds listed are soluble in acetone and slightly soluble in ethanol.

gave propionaldehyde. At least ten compounds were obtained by photolysis of trifluralin (**27**) in methanol and two of the products were identified as α,α,α-trifluoro-2,6-dinitro-N-propyl-p-toluidine (**28**) and α,α,α-trifluoro-2,6-dinitro-p-toluidine (**29**) (97). Postulated reaction sequences are shown below.

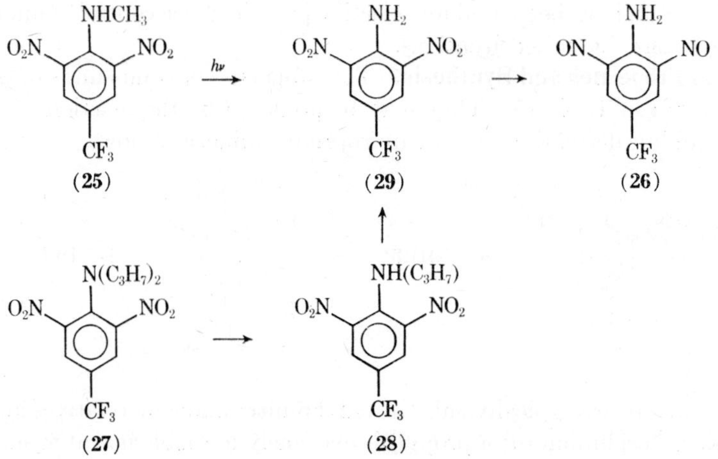

A sensitive method of the determination of trifluralin residues in soils is the gas chromatography of extracts after cleanup on a florisil column (98).

Metabolism. Trifluralin prevents development and growth of plant roots and shoots. Cell division is inhibited in trifluralin treated plants. The nuclei divide but the cell enlarges without separating into daughter cells. Schultz and Funderburk found that DNA, total RNA, and protein decreased in root tips of corn following trifluralin treatment but there is no clear explanation for the mode of action (99). Plants vary in their susceptibility to trifluralin but selectivity may be achieved by limiting the depth of soil incorporation so that the crop root remains outside of the region of chemical control.

The substituted dinitroanilines are degraded by soil, plants, and animals. There was some conversion of trifluralin into the monodealkylated product (**28**) in carrots and traces of α,α,α-trifluoro-5-nitro-N-propyltoluene-3,4-diamine and 4-(di-n-propylamino)-3,5-dinitrobenzoic acid were also identified (100). In peanuts, conversion to the monodealkylated derivative (**28**) occurred, whereas in sweet potatoes α,α,α-trifluoro-5-nitro-N-propyltoluene-3,4-diamine was the degradation product initially obtained (101).

In soils dealkylation and reduction of trifluralin and benefin occurs. One or both nitro groups may be reduced to amino functions, depending on whether anaerobic or aerobic conditions are maintained, but there is little evidence for breakdown by specific microorganisms.

Animal metabolism of trifluralin follows a similar pathway to that in other organisms. The major metabolites of trifluralin fed to rats were reduced ($NO_2{\rightarrow}NH_2$) and dealkylated (102). The presence of polar compounds as degradation products was reported in a number of these studies. These were compounds of low chromatographic mobility and it has been suggested that they are oxidation products.

Bipyridinium Herbicides

Chemical Properties and Synthesis. Bipyridinium diquaternary salts were prepared in the last century but the herbicidal activity of 1,1-ethylene-2,2'-bipyridinium (diquat) salts was first reported in 1958 (103). 1,1'-Dimethyl-4,4'-bipyridinium di-(methosulfate) (paraquat) is the other major herbicide of this group. The bipyridinium compounds are marked by their very strong adsorption to clay minerals. Consequently they may be sprayed between rows of crops to eliminate weeds. The herbicide is rapidly adsorbed by the soil and does not damage the growing crop.

Paraquat (U.S. Pat, 2,972,528 (1959), to I.C.I.), melts at 175–180°C (dec) and is soluble in water, but insoluble in organic solvents. Diquat (Brit. Pat. 785,732 (1957), to I.C.I.) melts at 335–340°C and decomposes at higher temperatures. It is very soluble in water, and slightly soluble in acetone. As strong electrolytes the bipyridinium salts are dissociated into ions and are practically insoluble in nonpolar solvents. They are stable to acids but diquat decomposes above pH 9 to give colored products. Paraquat is more stable but decomposes above pH 12. The spectrum of the blue solution produced by the action of sodium hydroxide on paraquat salts is identical with that of the radical produced by reduction.

Bipyridine may be synthesized by reduction of pyridine with sodium in liquid ammonia which gives an ion radical which is oxidized to 4,4'-bipyridine (**30**). Raney nickel reduction gives 2,2'-bipyridine (**31**). Quaternization with methyl bromide or ethylene bromide affords paraquat (**32**) or diquat (**33**), respectively. The syntheses of paraquat and diquat labeled in various positions with tritium, ^{14}C or ^{15}N have been described by Calderbank (104,104a).

Oxidation of paraquat with potassium ferricyanide gives a dipyridone (**34**) and a monopyridone (**35**).

Oxidation with hydrogen peroxide also gave the two pyridones but degradation of one pyridine ring took place to give a betaine (**36**) as major product. Another compound, probably 4-carboxyl-1-methyl-2-pyridone (**37**) was also obtained. Alkaline hydrogen peroxide gave mainly oxalic acid.

The property of diquaternary-4,4'-bipyridinium salts to form intensely colored solutions on reduction has been known for a long time (105). The appearance of color results from addition of a single electron to give a water-soluble free radical. The reaction is reversible and some of the quaternary salts have found use as redox indicators known as viologens.

The free radical is stabilized by delocalization of the odd electron over the extended conjugated system. ESR measurements indicate a high electron density on the nitrogen atoms and a lower and almost equal density on the ortho and meta carbon atoms. Reversible reduction of diquat is somewhat easier than that of paraquat. Further reduction of paraquat is not reversible and gives at first a dihydro derivative and ultimately a fully saturated compound.

Concentrated solutions of paraquat and diquat are corrosive to steel, tinplate, galvanized iron, and aluminum. Corrosion inhibitors are normally included in formulations.

Both paraquat and diquat are degraded by uv irradiation. If paraquat is adsorbed on a surface its maximum absorption is shifted from 257 to 290 nm and it may absorb sufficient solar radiation to cause degradation. The betaine (**36**) was formed by uv irradiation of an aqueous solution of paraquat in oxygen and was identical with the products formed by photochemical degradation from plants sprayed with paraquat (106). Methylamine was also isolated as a salt. On irradiation by uv, unfiltered, diquat affords a tetrahydropyridopyrazinium salt (**38**) as the major product of photolysis. The decomposition, under carefully controlled conditions of irradiation, is shown in Scheme 1.

Scheme 1.

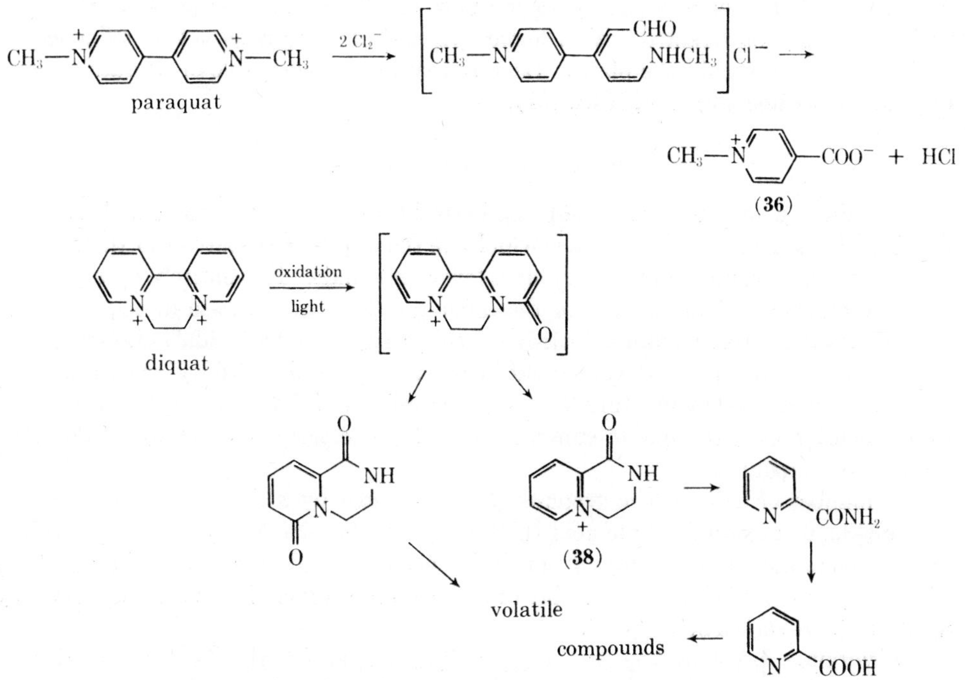

Analysis. The analysis of paraquat and diquat in formulations is based on the reduction of the solution with sodium dithionite. The herbicide can be determined by measuring the color intensity. Paraquat develops a color measured at 600 nm and diquat at 430 nm. The color fades in air but is stable in the presence of excess reducing agent.

Residue analysis presents difficulties as the bipyridinium salts are strongly adsorbed and prolonged boiling of samples with dilute acid may be necessary to liberate the herbicide. Solutions may be concentrated by ion-exchange resins and the quantity of herbicide determined by measurement of the color of the reduced salt.

Polarography is of value for the analysis of the bipyridinium herbicides and was used to determine diquat in food crops. The photolysis product, 4-carboxy-1-methyl pyridinium chloride, has also been estimated by polarographic methods (104,104a).

Metabolism. Examination of a variety of bipyridine derivatives led to the conclusion that the ability to form a free radical by one-electron reduction is necessary for herbicidal activity. All herbicidally active compounds are capable of reduction. From this it has been inferred that a similar process takes place within the plant to produce free radicals from the quaternary bipyridinium salts.

A plant treated with a bipyridinium herbicide dies rapidly in light but not in the dark. As light is required and green tissue is affected, the photosynthetic apparatus is probably involved. The action of the bipyridinium salts follows reduction during the photosynthetic process and spectroscopic evidence demonstrates that paraquat is reduced by illuminated fresh chloroplasts. Interference with the photosynthesis alone would result in the slow death of the plant but as death occurs rapidly, it is presumed that some other mechanism is responsible. The reduced form of the herbicide can be oxidized by molecular oxygen to regenerate the herbicide and oxygen is simultaneously reduced to hydrogen peroxide. Presumably radical intermediates are produced during oxidation and they or hydrogen peroxide may be the actual toxic agents. Paraquat and diquat are translocated within higher plants but there is as yet no evidence for their metabolism within the plant. Bacterial systems are capable of degrading paraquat to the betaine (**36**) (104, 104a, 107).

Miscellaneous Herbicides

In addition to the major groups of herbicides previously discussed, there are a number of compounds of importance which are the sole representatives of their classes and a number of groups containing only a few compounds. It is entirely possible that, as patterns of use change, some of these compounds may increase in economic importance. Some fundamental knowledge of the principles of herbicidal action would be of value in guiding future development but factors of cost, selectivity, and safety continue to be influential in affecting the choice of herbicides. The following section is not comprehensive but attempts to summarize some properties and reactions of the more significant compounds.

A number of heterocyclic systems are represented by a single herbicide, such as 4-amino-3,5,6-trichloropicolinic acid (picloram) and 3-amino-1,2,4-triazole (amitrole). These compounds and the uracil herbicides are discussed in the following section. Their properties and those of a number of compounds not included in classes previously mentioned, are shown in Table 16.

Picloram. 4-Amino-3,5,6-trichloropicolinic acid is a herbicide and growth regulator which has an activity similar to 2,3,6-TBA. Its biological activity was reported in 1963 (108). It is a white crystalline solid which is quite stable in aqueous solution and in the solid state. It is active at low concentrations and under certain conditions its biological activity in soil may be prolonged. It is used as a brush killer.

By heating the free acid alone or in solvent, the decarboxylated product, 4-amino-2,3,5-trichloropyridine, can be obtained in good yield. The photodecomposition of picloram in aqueous solution gives nonphytotoxic products (109). The rate of photol-

$$
\underset{\substack{\parallel \\ \text{NH}}}{\text{H}_2\text{NNHCNH}_2 \cdot \text{H}_2\text{CO}_3} + \text{HCOOH} \longrightarrow \underset{\substack{\parallel \\ \text{NH}}}{\text{H}_2\text{NNHCNH}_2 \cdot \text{HCOOH}} \xrightarrow[-\text{H}_2\text{O}]{120°\text{C}}
$$

$$
\underset{\substack{| \\ \text{H}}}{\underset{\text{OHC} \diagdown \text{N} \diagup \text{NH}}{\text{HN}{=}\text{C}{-}\text{NH}}} \xrightarrow[-\text{H}_2\text{O}]{120°\text{C}} \underset{\substack{| \\ \text{H}}}{\underset{\text{N}}{\overset{\text{N}-\text{N}}{\diagup}}{\diagdown}\text{NH}_2}
$$

(**39**)

Table 16. Properties of Miscellaneous Herbicides

Name	Formula	mp, °C	bp, °C	Solubility, g/100 g solvent		
				water	acetone	ethanol
amitrole[a]		159		28^{25}	insol	
picloram[b]		215 (dec)				
hexachloro-acetone[c]	CCl_3COCCl_3		204			
isocil						
bromacil		158–159		0.0815^{25}	16.7^{25}	13.4^{25}
terbacil						
acrolein[d]	$CH_2{=}CHCHO$		52		miscible	
acrylonitrile	$CH_2{=}CHCN$		77.3–77.5			
allyl alcohol[e]	$CH_2{=}CHCH_2OH$		96–99	miscible		
pyrazone		207		3.54^{20}		
nitrofen[f]		70–71		insol	sol	

[a] U.S. Pat. 2,670,282 (1954), to Amer. Paint Co.

[b] U.S. Pat. 3,285,925 (1966), to The Dow Chemical Co.

[c] U.S. Pat. 2,635,117 (1953), to Allied Chemical and Dye Corp.

[d] U.S. Pats. 2,959,476 (1960) and 2,987,475 (1961), to Shell Oil Co.

[e] U.S. Pat. 2,773,331 (1956), to Shell Development Co.

[f] U.S. Pat. 3,080,225 (1963), to Rohm and Haas Co.

ysis is relatively rapid at wavelengths below 250 nm. There is evidence that complete breakdown of the heterocyclic ring occurs. Photolysis in methanol results in some replacement of nuclear chlorine by hydrogen.

Amitrole. The herbicide 3-amino-1,2,4-triazole (amitrole) (**39**) can be synthesized by heating formic acid with aminoguanidine bicarbonate. The reaction proceeds by cyclodehydration of the intermediate N-formyl derivative (110).

Amitrole was introduced as a herbicide in 1952–1953 (111). It is a water-soluble, crystalline solid. The aqueous solution is neutral but amitrole acts as a weak base and forms salts with acids. The primary amino group reacts to give aldimines or ketimines with aldehydes or ketones and can be diazotized like a typical aromatic amine. The triazole ring confers stability on the compound. Amitrole forms chelate compounds with a number of metals including iron and copper. Although the heterocyclic ring is stable to a variety of reagents, it is cleaved by free-radical generating systems. Carbon dioxide was liberated by Fenton's reagent, and urea and cyanamide were identified as reaction products. Irradiation of an aqueous solution of amitrole at 220 nm gave the same reaction products. Daylight illumination of a solution of amitrole to which riboflavin has been added brings about decomposition to similar products. It was postulated that polymeric compounds are also produced in these reactions (112).

Analysis. Potentiometric titration with hydrochloric acid may be used for analysis of amitrole in technical products. Residue methods are based on color reactions. In the original method devised by Sund (113) a green color, developed by nitroprusside in basic solution, was measured. An improved colorimetric method depends on coupling the diazotized amine with N-(1-naphthyl)ethylene diamine (114).

Metabolism. The mode of action of amitrole has been the subject of intensive investigation. Amitrole interferes with synthesis of chlorophyll in plants. A metabolite, 3-(3-amino-1,2,4-triazol-1-yl)-2-aminopropionic acid (**40**), occurs in considerable amounts in tissue where destruction of chlorophyll is most severe, but this compound is probably not phytotoxic.

(40)

Amitrole reduces the protein of treated bean plants by disrupting the amino acid or nucleic acid metabolism. Indirect interference with purine metabolism by inhibition of histidine biosynthesis has been postulated as a mechanism of action. In yeast and *Salmonella typhimurium* the addition of amitrole inhibits the enzyme imidazoglycerol phosphate dehydratase responsible for converting imidazoleglycerol phosphate to imidazoleacetol phosphate. Imidazoleglycerol phosphate accumulates (115).

Several other enzymes, including catalase and fatty acid peroxidase, are also inhibited by amitrole. It has been suggested that the action of amitrole may lie in its ability to participate in one-electron oxidation processes with formation of a free radical which may attack enzyme systems causing inhibition. The role of amitrole as a radical scavenger has been demonstrated by its reactions with Fenton's reagent and this property probably underlies its effectiveness as a protective agent against high-energy radiation.

In a number of plants amitrole is converted to metabolites which are generally lower in phytotoxicity. The triazole ring is retained intact in these compounds and the *N*-glucosyl derivative has been isolated from a number of plants. As the glucose adduct forms readily from glucose and amitrole in vitro, the status of this compound remains in doubt. The structure and formation of the plant metabolite identified as 3-(3-amino-1,2,4-triazol-1-yl)-2-aminopropionic acid (**40**) has been investigated in some detail. It is probably formed by the condensation of serine and amitrole. Other unknown plant metabolites have been reported (116).

Uracils. The herbicidal activity of substituted pyrimidines was announced in 1962 (117); 5-Bromo-3-*sec*-butyl-6-methyluracil (bromacil) (**41**), 5-bromo-3-isopropyl-6-methyluracil (isocil) (**42**), and 3-*tert*-butyl-5-chloro-6-methyluracil (terbacil) (**43**) are all crystalline solids soluble in aqueous bases.

(**41**)   (**42**)   (**43**)

Uracil may be prepared by the condensation of acetoacetic ester with a substituted urea. The *N*-substituted methyl uracil is then halogenated. Bromacil and isocil are decomposed by ultraviolet irradiation (118). In acid solution the tertiary butyl group of terbacil may be lost.

Analysis. Infrared spectroscopy has been used for analysis of uracils. For analysis of residues in plants, soils, and animal tissues, samples were extracted with aqueous sodium hydroxide and determined by a gas chromatographic method employing a microcoulometric detector (119).

Metabolism. The uracils inhibit photosynthesis in plants (120). 5-Bromo-3-*sec*-butyl-6-hydroxymethyluracil was identified as the major metabolite of bromacil in soil and orange plants. The major metabolite of bromacil in the urine of rats is 5-bromo-3-*sec*-butyl-6-(hydroxymethyl)uracil which is excreted as a conjugate. Five other metabolites are present in lower concentration. They are formed by oxidation of the 3-methyl or 3-*sec*-butyl group, and by debromination of the 5-bromo group, or by a combination of these. In an analogous manner, terbacil is metabolized to 3-*tert*-butyl-5-chloro-6-(hydroxymethyl)uracil (**44**) excreted in the urine of dogs fed with a diet incorporating terbacil. A similar metabolic compound (**45**) is formed by elimination of water from the oxidized product (**46**) (121).

(**44**)   (**46**) $-H_2O$ (**45**)

A synthesis of uracil herbicides is shown below:

$$RNH_2 + KCNO + HCl \rightarrow RNHCONH_2 + KCl$$

$$RNHCONH_2 + CH_3COCH_2COOR' \xrightarrow{H^+} RNHCONHC{=}CHCOOR' \xrightarrow{NaOCH_3}$$
$$\underset{CH_3}{|}$$

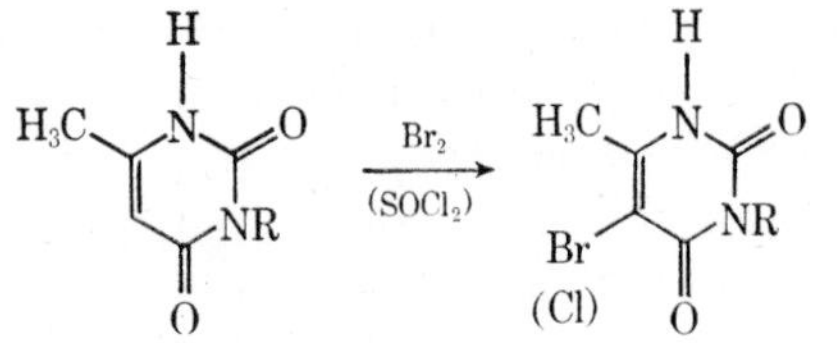

where R = sec-butyl, isopropyl, or *tert*-butyl and R' = alkyl.

Inorganic Compounds

A number of inorganic chemicals have been used as herbicides. One of the earliest to be used (in 1909) was *sulfuric acid* at a dilution of about 10%. Related compounds more recently introduced are sulfamic acid (in 1942) and ammonium sulfamate. Both compounds are relatively stable if stored in closed containers but are hygroscopic and undergo hydrolysis. Sulfamic acid hydrolyzes slowly in solution to ammonium hydrogen sulfate.

Various *boron compounds* are used as herbicides. Sodium tetraborate is used in various hydrated forms (borax is the decahydrate). It is a nonselective herbicide of relatively low toxicity to wildlife and fish. Borax may be formulated with sodium chloride, also a herbicide, to reduce the fire hazard of the chlorate. Sodium metaborate is similar in its activity to the tetraborate.

A number of *arsenical compounds* are used as weed killers. Sodium arsenite is a general herbicide and kills all vegetation. Its high mammalian toxicity renders its use extremely hazardous. Arsenic acid is somewhat less toxic than the arsenious acid salts. More recently organic arsenic compounds of much lower mammalian toxicity have been introduced, such as the sodium salts of methanearsonic acid and sodium cacodylate (monosodium dimethylarsinic acid).

Organic arsenicals are used extensively in cotton farming as selective postemergence contact herbicides, in noncrop areas, and for industrial applications.

Sodium chlorate is one of the most common defoliants and has also been used on railroad beds for seasonal weed control.

Application and Selectivity

The manufacture of a herbicide gives a technical product. The active ingredient is normally diluted with a carrier before application. To prepare such a formulation requires skilled technical knowledge.

Herbicides may be formulated as aqueous solutions, wettable powders, granules, dusts, or emulsifiable concentrates. If the chemical is to be applied as a spray the least expensive diluent is water. Most herbicidal chemicals can be formulated so that they may be diluted with water. Water-soluble herbicides may be sold as aqueous concentrates or as solids. Insoluble organic compounds may be dissolved in organic solvents and formulated with suitable emulsifying agents. Dispersion in water then gives a stable emulsion. Solids may be sold as wettable powders or pastes which can be dispersed as suspensions in water. Alternatively, a mineral oil or organic solvent may be used as a diluent.

Solid formulations are obtainable in powder or granular form. Talc or clays are used as inert carriers. Granular forms may be prepared so as to prolong the activity of the herbicide by releasing it slowly into the soil.

Emulsifiers and surface-active agents have an important function in spray formulations. Although biologically inert they potentiate the action of the spray by influencing penetration, retention, surface tension, and drop size. They also maintain a stable dispersion during storage and application. Composition of the formulation must be adapted to the geographical region where application is to be made. Local factors, such as the nature of the water to be used as diluent, require special consideration. Sequestering agents are generally added to prevent sedimentation in hard-water areas. Typical compositions may occasionally be found in the literature and examples of the formulation of urea herbicides have been published (122). The effectiveness of a herbicide is extremely dependent on its formulation.

Selectivity against weeds is achieved in a number of ways. The use of a spray against low-growing weeds in an area covered by bushes or trees illustrates the method of selective application. Paraquat and other herbicides which have no action through the soil may be used as contact herbicides on weeds before the crop emerges from the soil. If the crop and weeds are present together, selectivity must be based on physical or biochemical factors. If the spray is applied to the plant, it should preferentially penetrate or be retained by the weed. If the herbicide is applied to the soil there must be differences in uptake between weed and crop, or the roots of the crop must be located where uptake cannot occur. If both crop and weed take up herbicide, physiological and biochemical factors which control adsorption, translocation and metabolism must differ sufficiently so the weed species alone is affected.

Penetration is fundamentally affected by the formulation of the active ingredient. For example, the penetration of 2,4-D into the plant depends on the pH of the spray. In beans it is maximal at pH 2. Ester formulations of 2,4-D were found to be more effectively absorbed than salts, but lower alkyl esters were not effectively translocated in the plant and adjacent crops were damaged by vapor drift from the volatile esters formerly used as sprays. Tissue penetration appears to be a function of the partition of the active ingredient between polar and nonpolar phases and varies with the plant species.

Retention of spray by the plant depends on a number of factors including the morphology of the plant, adherence of the spray droplets, and the wettability of the leaf. The ultrastructure of the leaf surface and the nature of the leaf wax are probably of major importance. The surface tension of the formulation influences the amount of spray retained and is varied by addition of surface-active agents. If the differences in spray retention between the crop and the weed are sufficiently marked, selective control results (123).

The application and placement of soil-applied herbicides is extremely influential in determining effectiveness and selectivity. The behavior of herbicides in soils and the factors involved in retention of biological activity or persistence have been discussed previously.

Safety

Weed killers are generally less toxic to mammals than other classes of pesticide. There are exceptions to this generalization, such as acrolein, acrylonitrile, arsenical compounds and nitrophenol derivatives, and others, which are highly toxic compounds. The toxicity of a formulated product depends on the quantity of active ingredient and the type of formulation. General guides for the safe use and handling of agricultural and household pesticides are available (124). Precautions attached to selection, use, and storage of a pesticide are outlined in several publications (125). Such precautions include the use of correct protective clothing and respiratory devices during handling and application. The safe handling of pesticides and proper disposal of empty containers and unwanted pesticides is the responsibility of the user.

The labeling and marketing of pesticides sold or distributed in interstate commerce are regulated by the Federal Insecticide, Fungicide and Rodenticide Act of 1947 (as amended) which is administered by the U.S. Department of Agriculture. Most states have similar laws which regulate marketing within the state. Federal registration does not remove the requirement for state registration.

For the protection of the pesticide user, consumer pesticide products must comply with rigid standards before they can be registered. The Pesticides Regulation Division of the U.S. Department of Agriculture is responsible for reviewing the data provided by the manufacturer. Supporting evidence is required to prove that the product is effective and is not hazardous to humans, and that it is beneficial to animals or crops which may be exposed when the product is used in accordance with the directions on the label. Scientific findings substantiating the manufacturers' claims about the duration and amount of residues resulting from the use of a pesticide are examined.

If the use of any pesticide involves food or feed, the manufacturer must also petition the Food and Drug Administration to set a tolerance for his product before it can be registered. The FDA is responsible for protecting consumers from potentially harmful residues remaining on foods when marketed. A tolerance is the maximum amount of pesticide residue (usually expressed in ppm per weight) that may lawfully remain on food or feed when it is marketed or used. The Miller Pesticide Chemicals Amendment of 1954 to the Federal Food Drug and Cosmetic Act of 1938 (as amended), administered by the Food and Drug Administration, directed that the FDA establish safe tolerances on raw agricultural products after the U.S. Department of Agriculture approves the usefulness of such chemicals. The Food Additive amendment of 1958 provides for establishing safe limits of pesticide residues in processed foods.

Bibliography

"Weed killers" in *ECT* 1st ed., Vol. 15, pp. 18–24, by D. E. H. Frear, The Pennsylvania State University.

1. *The World Food Problem, A Report of the Presidents' Science Advisory Committee*, Vol. II, The White House, 1967, p. 206.
2. U.S. Tariff Commission, *United States Production and Sales of Pesticides and Related Products*, U.S. Govt. Printing Office, Washington, D.C., 1968.
3. "Chemical Week Report: Pesticides Part I" (April 12) and "Chemical Week Report: Pesticides Part II" (April 26), *Chem. Week* **104** (1969). (Provides a detailed list of major products and discussion of current trends in pesticide research and markets.)
3a. C. R. Wherry, *Chemical Economics Handbook*, 573.7420A, Stanford Research Institute, Stanford, Calif., 1968.

4. F. Kögl, A. J. Haagen-Smit, and H. Erxleben, *Z. Physiol. Chem.* **90**, 228 (1934).

5. C. D. Woods and J. M. Bartlett, *Maine Agr. Expt. Sta. Bull.* **167** (1909).

6. Brit. Pat. 424,295 (May 29, 1935), G. Truffaut and I. Pastac.

7. C. G. McWhorter and J. T. Holstun, Jr., "Science Against Weeds," in *Protecting Our Food*, U.S. Dept. of Agriculture Year Book, U.S. Govt. Printing Office, Washington, D.C., 1966, p. 79. *Chem. Eng. News* **47**, 22 (June 9, 1969).

8. *Herbicide Handbook*, Weed Science Society of America, 1967.

9. P. W. Zimmermann and A. E. Hitchcock, *Contrib. Boyce Thompson Inst.* **12**, 321 (1942).

9a. G. W. Bailey and J. L. White, *Residue Rev.* **10**, 97 (1965). H. Martin, *Guide to the Chemicals Used in Crop Protection*, 5th ed., Can. Dept. of Agriculture, Ottawa, 1968. See also references 3 and 8. (Most of the data listed in Tables 6–16 are compiled from these sources.)

10. *Chem. Eng. News* **48**, 60 (1970). J. Kimmig and K. H. Schulz, *Dermatologica* **115**, 540 (1957).

10a. R. Pokorny, *J. Am. Chem. Soc.* **63**, 1768 (1941).

11. D. G. Crosby and H. O. Tutass, *J. Agr. Food Chem.* **14**, 596 (1966).

12. C. P. Swanson, *Botan. Gaz.* **107**, 507 (1948).

13. V. H. Freed, *Science* **107**, 98 (1948).

14. R. P. Marquardt and E. N. Luce, *J. Agr. Food. Chem.* **3**, 51 (1955).

15. *Ibid.*, **9**, 266 (1961).

16. H. A. Glastonbury and M. D. Stevenson, *J. Sci. Food Agr.* **7**, 379 (1959).

17. A. Bevenue, G. Zweig, and N. L. Nash, *J. Assoc. Offic. Agr. Chem.* **45**, 990 (1962).

18. M. A. Loos, in P. C. Kearney and D. D. Kaufman, eds., *Degradation of Herbicides*, Marcel Dekker, New York, 1969, pp. 1–49; A. S. Crafts, *The Chemistry and Mode of Action of Herbicides*, Interscience Publishers, Inc., New York, 1961.

19. D. L. Linscott, R. D. Hagin, and J. E. Dawson, *J. Agr. Food Chem.* **16**, 844 (1968).

20. E. W. Thomas, B. C. Loughman, and R. G. Powell, *Nature* **204**, 884 (1964).

21. *Ibid.*, 286 (1964).

22. D. G. Crosby, *Abstr. No. 11, Am. Chem. Soc. Div. Agr. Food Chem., 155th Ann. Meet., San Francisco, 1968*.

23. R. L. Wain and F. Wightman, *Proc. Roy. Soc. (London) Ser. B.* **142**, 525 (1955). R. L. Wain, *Ann. Appl. Biol.* **42**, 151 (1955). R. L. Wain, *J. Agr. Food Chem.* **3**, 128 (1955).

24. C. H. Fawcett, J. M. A. Ingram, and R. L. Wain, *Proc. Roy. Soc. Ser. B* **142**, 60 (1954).

25. C. H. Fawcett, R. M. Pascal, M. B. Pybus, H. F. Taylor, R. L. Wain, and F. Wightman, *Proc. Roy. Soc. Ser. B* **150**, 95 (1959).

26. A. Gast, E. Knüsli, and H. Gysin, *Experientia* **11**, 107 (1955); *Ibid.*, **12**, 146 (1956).

27. H. Gysin and E. Knüsli, in R. L. Metcalf, ed., in *Advances in Pest Control Research*, Vol. III, Interscience Publishers, Inc., New York, 1960, p. 289.

28. J. T. Thurston et al., *J. Am. Chem. Soc.* **45**, 2981 (1951).

29. J-Y. T. Chen, *J. Assoc. Offic. Anal. Chem.* **50**, 595 (1967).

30. P. C. Kearney and D. D. Kaufman, eds., *Degradation of Herbicides*, Marcel Dekker, New York, 1969, p. 59.

31. E. Knüsli and D. Berrer, unpublished results, cited by E. Knüsli, D. Berrer, G. Dupuis, and H. Essner, in reference 30, p. 59.

32. J. R. Plimmer, P. C. Kearney, and J. R. Rowlands, *Abstr. 47, Am. Chem. Soc., Div. Agr. Food Chem., 156th Ann. Meeting, Atlantic City, 1968*.

33. B. E. Pape and M. J. Zabik, *Abstr. 23, Am. Chem. Soc., Div. Pesticide Chem., 158th Ann. Meet., New York, 1969;* J. R. Plimmer, *Residue Rev. (in press)*.

34. J. R. Plimmer, P. C. Kearney, and U. I. Klingebiel, *Tetrahedron Letters* **44**, 3891 (1969).

35. G. Zweig, ed., *Pesticides, Plant Growth Regulators and Food Additives*, Vol. IV, Academic Press, Inc., New York, 1964. (This volume also contains relevant articles on pp. 13, 27, 33, 171, 179, and 187.)

35a. E. Knüsli, H. Burchfield, and E. E. Storrs, in reference 35, p. 213.

36. R. C. Tindle, C. W. Gehrke, and W. A. Aue, *J. Offic. Assoc. Anal. Chem.* **52**, 682 (1969).

37. J. L. Hilton, L. L. Jansen, and H. M. Hull, *Ann. Rev. Plant Physiol.* **14**, 353 (1963).

38. R. Hill, *Nature* **139**, 881 (1937); *Ibid.*, **146**, 61 (1940).

39. R. H. Shimabukuro, *J. Agr. Food Chem.* **15**, 557 (1967).

40. D. D. Kaufman, P. C. Kearney, and T. J. Sheets, *J. Agr. Food Chem.* **13**, 238 (1965).

41. E. Knüsli, D. Berrer, G. Dupuis, and H. Esser, in reference 30, p. 63.

42. M. L. Montgomery and V. H. Freed, *J. Agr. Food Chem.* **12**, 11 (1964).

43. C. Böhme and F. Baer, *Food Cosmet. Toxicol.* **5**, 23 (1967).

44. H. E. Thompson, C. P. Swanson, and A. G. Norman, *Botan. Gaz.* **107**, 476 (1946).

45. L. S. Jordan, C. W. Coggins, B. E. Day, and W. A. Clerx, *Weeds* **12**, 1 (1964).

46. H. Geissbühler, C. Haselbach, H. Aebi, and L. Ebner, *Weed Res.* **3**, 277 (1963).

47. J. D. Rosen, R. F. Strusz, and C. C. Still, *J. Agr. Food Chem.* **15**, 568 (1968).

48. *Ibid.*, **17**, 206 (1969).

49. D. G. Crosby and C-S. Tang, *J. Agr. Food Chem.* **17**, 1041 (1969).

50. W. K. Lowen, W. E. Bleidner, J. J. Kirkland, and H. L. Pease, in reference 35, p. 16.

51. G. Yip, *J. Assoc. Offic. Anal. Chem.* **52**, 273 (1969).

52. P. B. Sweetser, *Biochem. Biophys. Acta.* **66**, 78 (1963).

53. H. Geissbühler, in reference 30, p. 98.

54. J. W. Smith and T. J. Sheets, *J. Agr. Food Chem.* **15**, 577 (1967).

55. J. H. Onley, G. Yip, and M. H. Aldridge, *J. Agr. Food Chem.* **16**, 426 (1968).

56. W. Ernst and C. Böhme, *Food Cosmet. Toxicol.* **3**, 789 (1965).   C. Böhme and W. Ernst, *Food Cosmet. Toxicol.* **3**, 797 (1965).

57. G. Friesen, *Planta* **8**, 666 (1929).

58. D. G. Crosby, in reference 30, p. 346.

59. M. J. Kolbezen, R. L. Metcalf, and T. R. Fukuto, *J. Agr. Food Chem.* **2**, 864 (1954).   L. W. Dittert and T. Higuchi, *J. Pharm. Sci.* **52**, 852 (1963).

60. T. R. Hopkins, in reference 35, p. 37.   See also pp. 49 and 139.

61. D. E. Moreland and K. L. Hill, *J. Agr. Food Chem.* **7**, 832 (1959).

62. R. J. Kuhr and J. E. Casida, *J. Agr. Food Chem.* **15**, 814 (1967).   See also reference 37, p. 358.

63. H. Tilles, *J. Am. Chem. Soc.* **81**, 714 (1959); U.S. Pat. 2,913,327 (Sept. 29, 1959), H. Tilles and J. Antognini (to Stauffer Chemical Co.).

64. R. B. Taylorson, *Weeds* **14**, 155 (1966).

65. G. G. Patchett, G. H. Batchelder, and J. J. Menn, in reference 35, p. 117.

66. S. Yamaguchi, *Weeds* **9**, 374 (1961); S. C. Fang and E. Fallin, in reference 30, p. 163.

67. P. C. Hamm and A. J. Speziale, *J. Agr. Food Chem.* **4**, 518 (1956).

68. O. L. Hoffmann and A. E. Smith, *Science* **109**, 588 (1949).

69. A. J. Speziale and P. C. Hamm, *J. Am. Chem. Soc.* **78**, 2556 (1956).

70. J. D. Rosen, *Bull. Environ. Contam. Toxicol.* **2**, 349 (1967).

71. L. J. Audus, ed., *The Physiology and Biochemistry of Herbicides*, Academic Press, Inc., New York, 1964.

71a. V. H. Freed, in reference 71, p. 62.

72. D. S. Frear and G. G. Still, *Phytochemistry* **7**, 913 (1968); R-Y. Yih, D. H. McRae, and H. F. Wilson, *Plant Physiol.* **43**, 1291 (1968).

73. E. G. Jaworski, *Science* **123**, 847 (1956); H. Lindley, *Biochem. J.* **82**, 418 (1962).

74. C. D. Kesner and S. K. Ries, *Science* **155**, 210 (1967).

75. R. Bartha and D. Pramer, *Science* **156**, 1617 (1967).

76. U.S. Pat. 2,393,086 (1944), E. W. Bousquet (to E. I. duPont de Nemours & Co., Inc.).

77. U.S. Pat. 2,642,354 (1951), K. C. Barrons (to the Dow Chemical Co.); U.S. Pat. 2,807,530 (1957), K. C. Barrons (to the Dow Chemical Co.).

78. G. N. Smith and E. Yonkers, in reference 35, p. 79; V. H. Freed, in reference 71, p. 54.

79. M. L. Beall, E. A. Woolson, T. J. Sheets, and C. I. Harris, *J. Agr. Food Chem.* **15**, 208 (1967).

80. J. K. Leasure, *J. Agr. Food Chem.* **12**, 40 (1964).

81. P. C. Kearney, C. I. Harris, D. D. Kaufman, and T. J. Sheets, *Advan. Pest Control Res.* **6**, 1 (1965).

82. R. C. Brian, in reference 71, p. 14.

83. N. A. Broadhurst, M. L. Montgomery, and V. H. Freed, *J. Agr. Food Chem.* **14**, 585 (1966).

84. J. R. Plimmer and B. E. Hummer, *J. Agr. Food Chem.* **17**, 83 (1969).

85. C. R. Swanson, R. E. Kadunce, R. H. Hodgson, and D. S. Frear, *Weeds* **14**, 319 (1966); C. R. Swanson, R. H. Hodgson, R. E. Kadunce, and H. Swanson, *Weeds* **14**, 323 (1966).

86. G. Zweig, ed., *Pesticides, Plant Growth Regulators and Food Additives*, Vol. V, Academic Press, Inc., New York, 1967.

86a. H. Segal, in reference 86, pp. 321–334.

87. S. R. Colby, *Weeds* **14**, 197 (1966).

88. H. Koopman and J. Daams, *Nature* **186,** 89 (1960).

89. A. S. Crafts, *Plant Physiol.* **21,** 348 (1946).

90. C. Chabrolin, *Compt. Rend.* **210,** 262 (1940).

91. A. R. Cooke, R. D. Hart, and N. E. Achuff, *Proc. Northeast Weed Control Conf.* **19,** 321 (1965).

92. M. Kuwahara, N. Kato, and K. Munakata, *Agr. Biol. Chem. (Tokyo)* **30,** 232, 239 (1966).

93. E. N. Ugochukwo and R. L. Wain, *Chem. Ind. (London)* **1965,** 35.

94. D. G. Crosby and M-Y. Li, in reference 30, p. 351.

95. G. Yip and S. F. Howard, *J. Assoc. Offic. Anal. Chem.* **51,** 24 (1968); A. Stark, *J. Agr. Food Chem.* **17,** 871 (1969).

96. E. F. Alder, W. L. Wright, and Q. F. Soper, *Proc. Northern Central Weed Control Conf.* **17,** 23 (1960).

97. R. E. McMahon, *Tetrahedron Letters* **21,** 2307 (1966). See also R. E. McMahan, in reference 30, p. 259.

98. J. B. Tepe and R. E. Scroggs, in reference 86, p. 527.

99. J. Hacskylo and V. A. Amato, *Weed Science* **16,** 573 (1968).

100. T. Golab, R. J. Herberg, S. J. Parka, and T. B. Tepe, *J. Agr. Food Chem.* **15,** 638 (1967).

101. P. K. Biswas and W. Hamilton, Jr., *Weed Science* **17,** 206 (1969).

102. J. L. Emmerson and R. C. Anderson, *Toxicol. Appl. Pharmacol.* **9,** 84 (1966).

103. R. C. Brian, R. F. Homer, J. Stubbs, and R. L. Jones, *Nature* **181,** 446 (1968).

104. R. L. Metcalf, ed., *Advances in Pest Control Research*, Vol. VIII, Interscience Publishers, a div. of John Wiley & Sons, Inc., New York, 1968.

104a. A. Calderbank, in reference 104, p. 127.

105. H. Weidel and M. Russo, *Monatsh. Chem.* **3,** 863 (1882).

106. A. Calderbank, in reference 104, p. 189.

107. H. H. Funderburk, Jr., in reference 30, p. 283.

108. J. W. Hamaker, H. Johnston, R. T. Martin, and C. T. Redemann, *Science* **141,** 363 (1963).

109. P. C. Kearney, E. A. Woolson, J. R. Plimmer, and A. R. Isensee, *Residue Rev.* **29,** 137 (1969).

110. C. Ainsworth, *Org. Syn.* **40,** 99 (1960).

111. R. Behrens, *Proc. N. Central Weed Control Conf.* **10,** 61 (1953).

112. P. Castelfranco and M. S. Brown, *Weeds* **11,** 116 (1963); J. R. Plimmer, P. C. Kearney, D. D. Kaufman, and F. S. Guardia, *J. Agr. Food Chem.* **15,** 996 (1967).

113. K. A. Sund, E. C. Putala, and H. M. Little, *J. Agr. Food Chem.* **8,** 210 (1960).

114. J. Burke and R. W. Storherr, *J. Assoc. Offic. Anal. Chem.* **44,** 196 (1961); G. L. Sutherland, in reference 35, p. 17.

115. J. L. Hilton and P. C. Kearney, *Weeds* **13,** 22 (1965).

116. M. Carter, in reference 30, p. 187.

117. R. W. Varner and C. W. Bingemann, *Proc. Southern Weed Control Conf.* **15,** 215 (1962); H. C. Bucha et al., *Science* **137,** 537 (1962).

118. L. S. Jordan, J. D. Mann, and B. E. Day, *Weeds* **13,** 43 (1965).

119. H. L. Pease, *J. Agr. Food Chem.* **16,** 54 (1968); W. H. Gutenmann and D. J. Lisk, *J. Assoc. Offic. Anal. Chem.* **51,** 688 (1968).

120. C. E. Hoffmann, J. W. McGahen, and P. B. Sweetser, *Nature* **202,** 577 (1964).

121. J. A. Gardiner, R. C. Rhodes, J. B. Adams, Jr., and E. J. Soboczenski, *J. Agr. Food Chem.* **17,** 980 (1969); R. C. Rhodes, R. W. Reiser, J. A. Gardiner, and H. Sherman, *J. Agr. Food Chem.* **17,** 974 (1969); J. A. Gardiner, R. W. Reiser, and H. Sherman, *J. Agr. Food Chem.* **17,** 967 (1969).

122. U.S. Pat. 2,655,447 (Oct. 13, 1953), C. W. Todd (to E. I. du Pont de Nemours & Co., Inc.).

123. K. Holly, in reference 71, p. 423.

124. U.S. Dept. of Agriculture, *Safe Use of Agricultural and Household Pesticides*, Agriculture Handbook No. 321, U.S. Govt. Printing Office, Washington, D.C. 1967; U.S. Dept. of Agriculture, *Summary of Registered Agricultural Pesticide Chemical Uses*, Vol. I, *Herbicides, Defoliants and Plant Growth Regulators*, 3rd. ed. and Suppl., U.S. Govt. Printing Office, Washington, D.C., 1968. *Safe Use of Pesticides, A Manual for Public Health Personnel*, The Am. Public Health Assoc., Inc., New York, 1967, 92 pp.

125. U.S. Dept. of Agriculture, *Suggested Guide for Weed Control, Uses of Herbicides*, Agriculture Handbook No. 332, U.S. Govt. Printing Office, Washington, D.C., 1968.

General References

References 3, 8, 9a, 18, 35, 71, and 86.

E. K. Woodford and S. A. Evans, eds., *Weed Control Handbook*, 4th ed., Blackwell Scientific Publications, Oxford, England, 1965.

R. L. Metcalf ed., *Advances in Pest Control Research*, Vols. I–VIII, Interscience Publishers, a div. of John Wiley & Sons, Inc., New York, 1957–1968.

F. Gunther, ed., *Residue Reviews*, Academic Press, Inc., New York (a periodical.)

JACK R. PLIMMER
U.S. Dept. of Agriculture

WEIGHING AND PROPORTIONING

Weighing is the operation of determining the weight of material in one or more objects or in a definite quantity of bulk material. The terms weight and weighing are used somewhat loosely in this connection, since what is desired actually is the mass of the material. The weight is really a measure of the gravitational force on a mass and so is dependent upon the location of the weighing station and upon its altitude. Since it is common practice to calibrate scales against dead weights at the location of the installation, for practical purposes the weighing operation actually determines the mass of material. Measuring by weight rather than by volume eliminates variations due to changes in specific gravity of liquids with temperature and variations in density of solids due to voids.

Proportioning is the weighing and controlling of two or more materials to a specific formula to make a definite blend of the materials for a mixed product or for a chemical process.

The four most common applications of weighing in industry are: the measuring of incoming material by weighing received shipments; controlling ingredients to the proper proportions; putting the product into packages of uniform weight (either by packaging directly on a scale or by using a scale to check the performance of other filling equipment); and measuring outgoing shipments for billing and transportation-charge purposes. In addition, for interdepartmental accounting, materials are frequently measured by weight when being transferred from one department to another. Also, in processing operations, the rate of material flow to various pieces of equipment, such as grinders or kilns, is very important to secure best efficiency of operation. See also Conveying.

Types of Equipment

A scale is a device or machine used to perform a weighing operation. There are many types of scales and several principles of operation used (1). The type of scale to be used depends upon the conditions of the operation and the performance required.

In mechanical scales, the load is measured either by comparing it with a known weight or by directly measuring the distortion of a spring caused by the load. With the even-balance type, the load is measured by direct comparison with known weights. Industrial scales of this type are most commonly used to measure material to predetermined quantities.

In beam-type scales such as shown in Figure 1, the force due to the load, reduced by the multiplication of the lever system, is measured by the position of a known weight (the poise) on a graduated beam. The poise is moved by hand until the beam is in

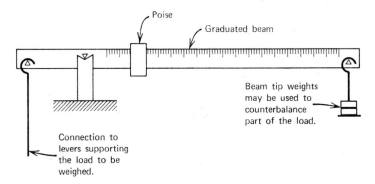

Fig. 1. Beam scale.

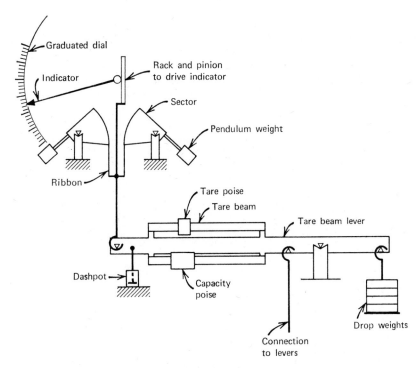

Fig. 2. Pendulum scale.

balance. Many beam scales also use fixed weights at the tip of the beam to offset definite amounts of load. For instance, a 1000-lb scale could have a beam graduated to 200 lb by 1 lb and four 200-lb tip weights. With this scale any load to 1000 lb can be weighed with 1-lb graduations. With the poise at 200 lb, tip weights would be added until the beam moved down. Then the poise would be moved back until the beam balanced. Both even-balance and beam-type scales are characterized by high sensitivity in relation to capacity, since motion of the scale is required only for the measurement of the final small percentage of the total weight.

Automatic indicating scales eliminate the manual operation of placing weights on a platter, for an even-balance scale, or positioning a poise on a graduated beam, for a beam-type scale. In pendulum-type dial or automatic indicating scales, as shown in

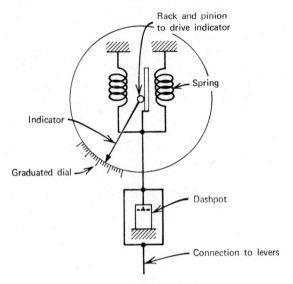

Fig. 3. Spring scale.

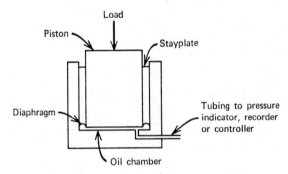

Fig. 4. Hydraulic scale.

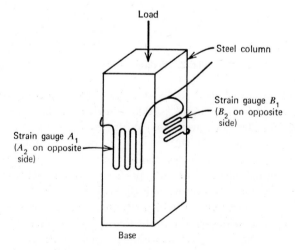

Fig. 5. Strain gauge load cell, column type.

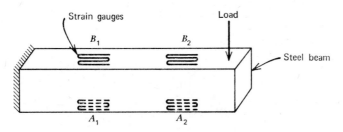

Fig. 6. Strain gauge load cell, beam type.

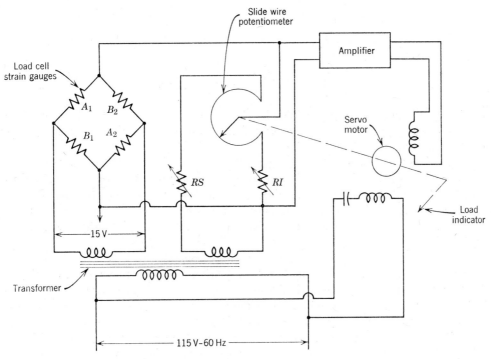

Fig. 7. Load cell scale, analog type.

Figure 2, the known weight (the pendulum) is moved by rotation of a sector or cam until it counterbalances the force from the load, and the amount of movement is measured and indicated by an indicator which is driven by a rack and pinion. The load is placed on a platform or in a hopper supported by levers which reduce the force due to the load by a factor called the multiplication of the lever system. Since it is frequently desired to offset the empty weight of the container, most automatic indicating scales are equipped with a tare beam which is mounted on a lever. A poise on this beam can be moved until the empty weight of the container is offset so the scale indicator is at zero. The capacity of the automatic indicating scale can be extended by use of a "capacity" poise on a beam, or this additional poise can be used to offset additional tare weight. Automatic indicating scales require a device such as a dashpot to dampen oscillations of the indicator. The dashpot normally consists of a plunger moving in a cylinder containing oil. Frequently, drop weights are added to the scale to extend its range. For instance, the dial might have a capacity of 1000 lb by 1 lb. By adding four

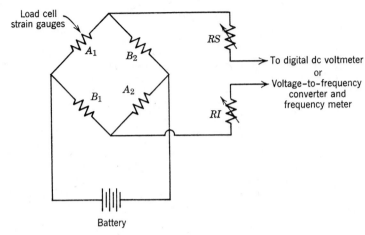

Fig. 8. Load cell scale, digital type.

drop weights, each equivalent to 1000 lb, the capacity of the scale can be extended to 5000 lb. Since each drop weight is equivalent to one dial, the drop weights are frequently called unit weights.

In spring-type automatic indicating scales, such as shown in Figure 3, the deflection of a spring is measured and indicated by an indicator. A dashpot is required to dampen the oscillations of this scale also. In some cases, the load is supported directly from the springs. Many spring scales, however, are connected to levers and may use a tare beam lever with poises and unit weights the same as a pendulum type automatic indicating scale.

A hydraulic scale measures weight by supporting the load on a piston in a hydraulic capsule, as shown in Figure 4, and measuring the hydraulic pressure developed by the use of a Bourdon tube or similar pressure-indicating device. See Pressure measurement. Since in many weighing operations it is required that the load be supported at three or more points, the individual hydraulic pressures developed by the various capsules are added by a hydraulic summing device, by means of a mechanical lever, where the total force can be measured by a mechanical scale, or by using suitable electrical sensing elements and adding the electrical signals. The hydraulic equipment thus performs the same function as a mechanical lever system in reducing the load to an easily measured force. Compressed air can be used to measure weight by measuring the air pressure required to lift a load which is supported by a diaphragm.

Weighing can also be done by electrical means. One method is to measure the change in length of a steel column, as shown in Figure 5, or the bending of a steel beam, as shown in Figure 6, by measuring the change in resistance of grids of wire or foil bonded to the steel member. These grids are called strain gages. Four strain gages are connected in the form of a Wheatstone bridge as shown in Figure 7. When the load is applied as shown in Figures 5 and 6, gages A_1 and A_2 are shortened so their resistance decreases and gages B_1 and B_2 are lengthened so their resistance increases. Referring to Figure 7, when a voltage such as 15 V at 60 Hz is applied across the bridge, the output from the bridge would be 30 mv for a load equivalent to the full rating of the load cell. This output voltage is applied to a resistor and slide-wire potentiometer circuit which is connected to a servo amplifier. Since there is normally some initial load on the load cell, RI is adjusted so that it balances the output of the cell with the initial load applied.

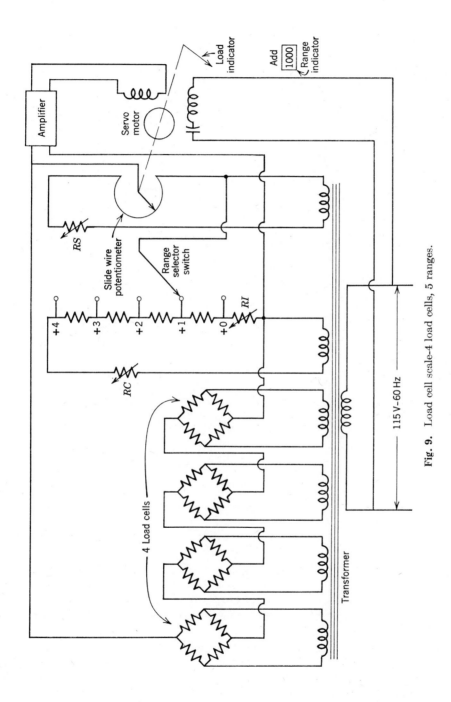

Fig. 9. Load cell scale–4 load cells, 5 ranges.

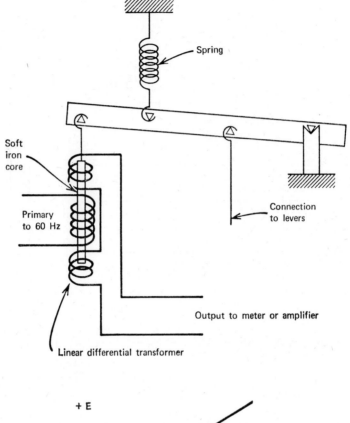

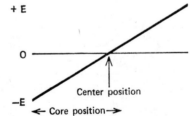

Fig. 10. Scale using linear differential transformer.

RS, or the span adjusting resistor, is adjusted until the full output from the potentiometer balances the output from the load cell for the desired capacity of the scale. When a load within the capacity of the scale is applied to the load cell, an output signal is applied to the potentiometer circuit. If the position of the sliding contact of the potentiometer does not correspond to the output from the load cell, a signal is amplified and applied to the servo motor to drive the potentiometer contact until the potentiometer balances the output from the load cell. An indicator, also driven by the servo motor, indicates the corresponding load on a dial. Figure 8 illustrates how a digital indication of a load can be obtained. By exciting the Wheatstone bridge with direct current, a direct current signal is developed by the load cell. *RI* is adjusted to offset the initial load, and *RS* is adjusted to provide the proper voltage, depending upon the capacity of the scale, to apply to a digital dc voltmeter or a voltage-to-frequency converter with a frequency meter to provide the digital indication. Figure 9 shows how the output from four load cells, which might be supporting a platform or a tank, can be combined to indicate the total load on the load cells, and also shows how range steps can be added to increase the resolution of the weight indication. For instance, with a dial equivalent to 1000 lb by 1 lb and four range steps, a load up to 5000 lb can be measured with a resolution of 1 lb. The calibration resistor, RC, is adjusted so the voltage across each range resistor equals the voltage across the slide wire potentiometer. When the range selector switch is moved, it moves a range indicator flag advising the operator to add weight corresponding to the range steps in use to the weight indicated on the dial.

Figure 10 illustrates another method of obtaining a direct electrical output from a scale. This uses a linear differential transformer consisting of a primary winding and two equal secondary windings spaced so that, when a soft iron core is centered in the transformer, the outputs of the two secondary windings balance each other, and the output of the transformer is zero. When the soft iron core is moved up, the output of the upper secondary increases and when the core is moved down, the output of the lower secondary increases. By proper design this output can be made linear. The output can be read directly on a meter or through an amplifier. The most common application for this type of scale is "over-and-under weighing," as for checkweighers.

Mass can be determined directly by measuring the absorption of γ or β rays due to the presence of the unknown quantity of material. See Radioisotopes.

Factors Affecting Weighing

Variations in the Force of Gravity (2). Weight is a measure of the force with which a mass is attracted to the earth. The basic force equation

$$F = ma$$

where F is force in newtons or poundals; m is mass in kg or lb; and a is acceleration, in cm/(sec²) or ft/(sec²). For weight rather than force,

$$W = mg$$

where g is the acceleration due to gravity. The value of g is not the same everywhere on the earth. The earth is rotating on its axis, and thus a body at the equator is subject to small centrifugal force, whereas a body at the poles is not. Also, the earth is not spherical, but bulges at the equator, so that a body at one of the poles is actually nearer to the center of the earth. For both of these reasons the acceleration due to gravity, and the weight of a given mass, increase with increasing latitude, from the equator to the poles. Over and above this, the acceleration due to gravity decreases with increasing altitude. The effect of these factors is shown in Helmert's equation for g in cgs units:

$$g = 980.616 - 2.5928 \cos 2\,\phi + 0.0069 \cos^2 2\phi - 3.086 \times 10^{-6}\,H$$

where ϕ is the latitude in degrees and H is the altitude in cm. Values from this equation can be used to compare the weight of a definite mass at different locations, such as the location where a scale was manufactured and originally calibrated as compared to a location where it will be installed:

$$W = \frac{mg}{g_c}$$

where W is force or scale reading, in kg or lb; m is mass, in kg or lb; g is g at point of installation, in cm/sec^2; and g_c is g at a point of calibration, in cm/sec^3. Weighing equipment which measures force, such as spring scales including the popular strain gage load cell and hydraulic or pneumatic scales, may have to be recalibrated at the point of installation.   As an example, Table 1 shows g for various locations (3), and the effect at the various locations on the reading of a spring scale calibrated at Toledo, Ohio.

Table 1. Readings for g and effects on a spring scale at various locations

Country	Acceleration due to gravity, g	Effect on spring scale calibrated at Toledo, Ohio Parts per 1000	
United States			
St. Michael, Alaska	982.192	high	1.99
Quiet Harbor, Alaska	981.624	high	1.41
Los Angeles, Calif.	979.595	low	.66
San Francisco, Calif.	979.965	low	.28
Denver, Colo.	979.609	low	.66
Hartford, Conn.	980.336	high	.01
Wash., D. C.	980.095	low	.15
Key West, Fla.	978.970	low	1.30
Miami, Fla.	979.053	low	1.21
Honolulu, Hawaii	978.591	low	1.68
St. Louis, Mo.	980.001	low	.24
New York, N. Y.	980.267	high	.03
Toledo, Ohio	980.241	—	0
El Paso, Texas	979.124	low	1.14
Seattle, Wash.	980.733	high	.50
Other Countries and Areas			
Arctic Red River, Canada	982.434	high	2.22
Vancouver, B.C., Canada	980.949	high	.72
Canal Zone, Panama	978.243	low	2.04
Greenwich, England	981.188	high	.96
Königsberg, Germany	981.477	high	1.26
Rome, Italy	980.367	high	.13
Monrovia, Liberia	978.165	low	2.11
Bergen, Norway	981.922	high	1.71
Stockholm, Sweden	981.843	high	1.63
Balia, Brazil	978.331	low	1.95
Kingston, Jamaica	978.591	low	1.68
North or South Pole, sea level	983.217	high	3.03
Equator, sea level	978.039	low	2.21
Equator, 5000 meters above sea level	976.496	low	3.82

Conventional balances, beam scales, and pendulum-type automatic indicating scales measure an unknown mass by comparing it with a known mass. Since a variation in the acceleration due to gravity has the same effect on the known mass as on the unknown, this factor does not affect the operation of these scales.

Buoyant Effect of the Air (2). Since most weighing operations are performed in air, a variation in density of the air between the point of calibration of a force-measuring scale and its point of use will affect the calibration (1). This effect is so small it can be ignored except for the most precise measurements. When weighing materials of very low density, there can be a measurable difference due to variation in air density between the point of origin and the destination.

Moisture Content. Materials subject to variable moisture content are frequently treated on a dry weight basis or on some standardized moisture content basis. In such cases, the actual moisture content of the material being weighed must be determined, and the actual weight converted to the desired basis.

$$W_D = W \ (1 - m_c/100)$$

where W_D is dry weight; W is actual wet weight; and m_c is moisture content as percent of wet weight; or

$$W_D = \frac{W}{1 + m_c/100}$$

where W_D is dry weight, w is actual wet weight; and m_c is moisture content as percent of dry weight.

Commercial Requirements

When the weight measurement is used as the basis of paying for material or for labor or services, the weighing equipment is subject to state or local weights and measures inspection and must meet certain specifications and tolerances (4). This requirement does not apply to equipment used only for control of processes or to measure yield or inventory.

Accessories are available for most types of scales. Printing attachments will record the weight as well as time, date, and other data on a ticket, sheet, or strip. Circular or strip recorders provide a continuous record, plotting weight vs time. Remote indication duplicates the scale indication at a remote point. Analog-to-digital converters are used for direct digital control or to transmit weight information to remote printers or to digital computers for record or control purposes. Control devices can be coupled directly to the scale. A paddle on the indicator which cuts a light beam to a photocell (in electrical circuitry) or which cuts a low pressure air jet (in pneumatic circuitry) can be used for simple or complex control functions. A potentiometer coupled to the indicator shaft or a pneumatic transducer coupled to the lever system provides analog electrical or pneumatic outputs for control. A good scale is a very accurate sensor, but extreme care must be taken to make sure the coupling of the control device to the scale does not reduce that accuracy by introducing friction or other variable forces. See also Instrumentation and control.

Choice of Scale

Since the error in a scale, as in many other instruments, is approximately the same throughout its indicating capacity, for best accuracy a scale should be chosen so that the majority of the weighings to be made on it are in the upper range of its indicating capacity. The capacity of a scale can be extended by the use of drop weights for a mechanical scale or—range steps—for a load cell type scale. This permits greater accuracy in weighing through a wide range of loads. As an example, a 1000-lb dial scale with four 1000-lb drop weights can be expected to weigh to within 1 lb for the first 1000 lb, and within 5 lb at its capacity of 5000 lb. A 5000-lb dial scale would be expected to weigh within 5 lb through the 5000-lb range.

As a guide to the selection of the most suitable type of scale, the relative advantages and disadvantages of the various types of weighing equipment are discussed below.

Mechanical Scales. Advantages: (a) high accuracy; (b) relatively inexpensive particularly in smaller capacities; (c) comparatively simple maintenance because of mechanical construction; (d) no electricity required; (e) lends itself to simple control equipment when a simple contact such as a magnetically operated mercury switch is used for the control element; very accurate primary element for use of other controls such as pneumatic or electronic controllers; and is capable of high overload capacity with relatively little sacrifice of accuracy.

Disadvantages: (a) relatively complicated and expensive installation for large sizes; (b) damage from such causes as corrosive fumes or liquids; (c) remote indication adds to cost; (d) excessive vibration may cause damage and difficulty of reading; and (e) motion of the platform or weighing container is relatively high, particularly for small-capacity scales.

Hydraulic Load Cell Scales. Advantages: (a) compact and self-contained unit when a single load cell suffices, such as on a crane or hanging scale; (b) inexpensive installation; (c) relatively easily protected from corrosion; (d) effects of vibration can be eliminated by damping; and (e) motion of platform or weighing container is relatively small.

Disadvantages: (a) additional equipment, either mechanical or electrical, is required if it is necessary to totalize the load on several load cells, such as for a large platform scale, and (b) expensive electrical equipment must be added to perform control operations.

Pneumatic Load Cell. Advantages: (a) compact unit; (b) inexpensive installation; (c) no electricity required; (d) readily adaptable to pneumatic control equipment; (e) simple construction lending to ease of maintenance; (f) vibration is no serious problem; and (g) motion of platform or weighing container is relatively small.

Disadvantages: (a) limited to single-cell application so that a large platform, hopper, or tank cannot be supported on several cells and the total forces added; (b) a supply of compressed air is required; and (c) speed of response is relatively low.

Electric Load Cell Scales. Advantages: (a) easily protected from corrosion (b) inexpensive installation; (c) relatively easy remote indication and use of one remote unit to read from several scales; (d) easily adapted to controls; and (e) motion of platform or weighing container is very small.

Disadvantages: (a) relatively expensive, particularly small-capacity units; (b) somewhat slower recovery from sudden changes in ambient temperature; (c) load cell mountings must be designed to minimize side forces; and (d) electricity is required.

Nuclear Scales (Measuring Mass Directly by Absorption of γ or β Rays) (5). Advantages: a simple installation, (b) not affected by extraneous forces on container such as connections to tank or hopper and tension or stiffness of belt conveyor, (c) no contact with material is required, (d) no moving parts subject to wear or corrosion, and (e) electrical output permits direct readout and adaptation to controls.

Disadvantages: (a) variations in geometry, as well as in the density of the material being measured will affect accuracy; (b) proper shielding and care are required to avoid exposure of personnel to radiation, (c) frequent calibration is required to compensate for loss of source emission with time, and (d) electricity is required.

Forms of Scales (6)

Scales of the various types are available in different forms to facilitate the weighing operation. Bench scales are small units, generally with a platform, on which small quantities can be placed by hand or moved across the scale on a roller conveyor installed on the platform. Portable scales generally have a platform near the floor, and the scale is on wheels so it can be moved about from place to place. Bench scales can be placed on a wheeled stand for similar use. Floor scales are platform scales installed in the floor. Motor truck scales for weighing highway vehicles may be installed in a pit or may be self contained so that they can be moved from one site to another.

Railroad track scales are installed in a pit and support the proper length of rail to weigh a railroad car. For static weighing of cars, the rails are generally long enough to support the complete car, although with two-draft-weighing the car can be weighed one end at a time. Some railroad weighing is done with the cars coupled and in motion. In this case the rails of the scale may be long enough to weigh the pair of axles at each end of the car, or short, weighing a single axle at a time. An adding machine or simple computer is used to obtain total car or train weight.

Where a product is conveyed by roller hangers from a monorail, scales weigh a section of the overhead rail. Hopper or tank scales support a hopper for dry materials or tank for liquids. Generally, the dead weight of the hopper or tank is offset on the scale so the scale reads the net weight of material.

Belt conveyor scales (7) weigh a section of the belt on which material is conveyed and transmit a signal proportional to the weight into a totalizing device. A measurement of the belt speed is also fed into the totalizing device which multiplies the two signals and provides the total weight of material that has passed over the belt between two consecutive readings. Controls can be provided to stop the flow after a predetermined amount of material has been delivered or to control the rate of feed. Because of the effect of the stiffness of the belt, such equipment is generally limited in accuracy. Under ideal conditions, accuracy in which the error is of the order of 0.5% can be obtained.

Weighing of definite objects is accomplished by placing the object on a suitable scale. Objects that are accounted for by count rather than by weight can be measured either by dividing the net weight by the known average weight of one object or by the use of scales equipped with special counting attachments. One type of attachment is the "fixed ratio scoop." This scale is equipped with one or more scoops mounted so that a piece in a scoop counterbalances a definite number of pieces on the platform. Thus, by counting relatively few pieces placed in the scoops, a total count is easily determined. Another type is the "variable ratio scoop." Here the scoop is moved along the lever connected to the scale. With 1, 2, 5, or 10 pieces in the scoop, it is moved along the lever until the scale balances. The count is indicated on a graduated bar.

Measuring Definite Quantities

For the measurement of bulk materials, which may be handled in anything from small packages to railroad cars, the weighing operation can consist of measuring the gross weight of material and container and subtracting the tare weight of the container. When the tare weight is known, its weight can be offset on the scale so that the net

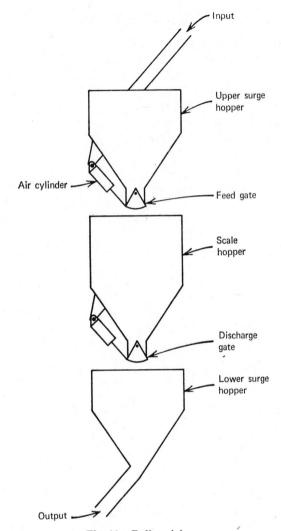

Fig. 11. Bulk weigher.

weight is indicated. The weighing operation can be performed by any one of the following three methods.

The material is brought to the scale, which is fixed in position, and placed on it. Bench scales are used for weighing small packages, floor scales for larger loads, and motor truck or railroad track scales can be used for weighing materials in trucks and cars. The scale location should be chosen so it is handy for the weighing operation but the scale does not get unnecessary abuse from excessive traffic that is not weighed.

The scale is brought to the load; scales of various types and capacities are available in portable form so that they can be moved about and temporarily located where the weighing operation is to take place.

Scales can be built into such material-handling devices as cranes, lift trucks, and conveyor belts so that the weighing operation can take place while the material is being transported. A hopper or tank scale can be installed so that it accumulates, weighs, and discharges the material as it is being transported. The quantity to be weighed can be measured in one draft or in multiple drafts. A hopper scale used for multiple draft weighing of material in transit is commonly called a "bulk-weigher," as shown in Figure 11. An accumulating or surge hopper with a feed gate is installed above the scale hopper, which has a discharge gate. A surge hopper is also installed below the scale hopper. The scale is generally equipped with an adjustable cutoff device to interrupt the flow of materials to the scale by closing the feed gate above the scale hopper, a signal device to close the scale discharge gate when the scale is almost empty, and a readout device to operate an adding machine. In operation, material flows into the scale until the feed gate is closed by the cutoff device. The actual load on the scale is transmitted to the adding machine, the discharge gate is opened, the scale discharges to a predetermined point, and the discharge gate is closed. The actual weight of material left on the scale is also transmitted to the adding machine which subtracts the two to arrive at the net weight of that draft. This operation is repeated until the quantity to be measured has passed through the scale. Since actual weights of repeated drafts are used, with this system it is possible to weigh a definite amount, such as contained in a railroad car or ship, with an error of less than 0.1% of net weight. With the input from storage and the output to a car or ship, the device is a shipping scale. With the input from a car or ship and the output to storage, the device is a receiving scale.

Since net weight of material is desired when measuring definite quantities, various approaches can be taken. For example, any one of the following methods of weighing the material in a railroad car can be used. The methods are arranged in approximate increasing order of accuracy: (1) weighing the loaded car on a coupled-in-motion track scale with the net weight determined by subtracting the stenciled weight of the car; (2) weighing the material, either as it is put into the car or removed from the car, on a belt conveyor scale; (3) weighing the gross weight of the car on a static railroad track scale and determining the actual empty weight of the car, either before the loading operation or after the unloading operation; (4) weighing the complete load either delivered to or received from the car as a single draft in a hopper or tank scale; and (5) using a "bulk-weigher" as described above to measure the material either being placed in or removed from the car.

Controlling Quantity

The most common application for the controlling of quantities of material is to fill containers to predetermined weights. One method is the net-weighing process, in which material is fed into a fixed container such as a hopper or tank permanently attached to the scale. The dead weight of this container can be counterbalanced, so that the scale indicates the weight of material to be added; or the weight of container plus the desired weight of material can be counterbalanced, so that the scale indicates the deviation from the desired weight. The filling operation may be performed manually by an operator observing the scale and stopping the flow of material at the proper time, or automatically by equipping the scale with a suitable control device to stop the feeding equipment at the desired weight. The hopper or tank is then discharged into the shipping container. In the gross-weighing method, the shipping container is placed directly on the scale and filled. The tare weight of the individual container is offset on the scale, and if this weight varies, the tare weight must be adjusted for each container to maintain a uniform net weight. In manual operation the tare is adjusted by the operator and the filling operation is controlled by the operator. Equipment for performing such operations automatically is available. The tare can be set and the feeding equipment can be controlled, automatically.   Figure 12 illustrates an oil drum filling machine on which the tare of the drum is offset automatically by means of a motor-driven poise.   The filling tube is then lowered into the

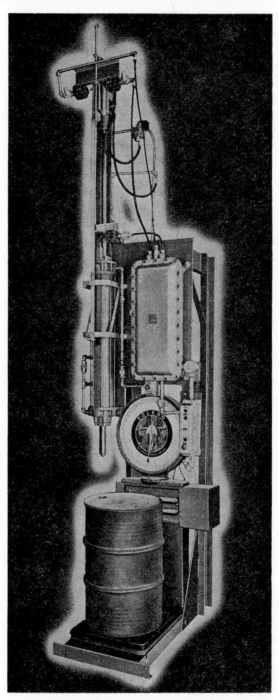

Fig. 12. Oil drum filling machine.

drum automatically. The feed valve, on the bottom of the filling tube, is opened and the tube automatically rises as material flows into the drum. To prevent foaming, the feed valve is kept below the surface of the oil in the drum. The feed valve is closed automatically by a control contact in the scale. The tare, and net, and gross weights can all be read on the scale and stencilled on the drum.

Continuous Weighing and Controlling

Sometimes it is required to determine the weight per linear foot or yard of material in sheet or strip form as it is being transported or in the process of manufacture. This weight information, with suitable controls, can be used to control moisture content, addition of sizing or other materials, or an extrusion process. The material can pass over a roll which is weighed or over a power driven belt-conveyor mounted on a scale. Direct mass measurement with a nuclear scale can be used for this application also.

Weighing and controlling bulk material being continuously conveyed is often necessary for optimizing the performance of such devices as grinders or pulverizers, for adding materials to a process at a continuous rate, or for controlling additives, such as to water supplies. A belt-conveyor scale can be installed in a belt-conveyor, or a short belt feeder can be mounted on a scale. If a belt-conveyor scale is mounted in the conveyor, the scale can be equipped with controls to maintain the feed rate within limits by controlling the operation of the device feeding the material to the conveyor. A short belt feeder mounted on a scale can have controls to regulate either the rate of feed to the scale or the speed of the belt on the scale to maintain a uniform flow. Direct mass measurement with a nuclear scale can also be used to measure and control such a continuous stream of material. Uniform continuous flow of solid or liquid materials can also be achieved by the "loss-in-weight" system. See Proportioning.

Proportioning

There are several methods of proportioning ingredients to obtain the desired formulation. Selection of the proper method depends upon the processing equipment and also upon the accuracy required. Where a definite quantity of a mixture is to be injected into a process, such as being placed in a batch mixer, or where the greatest accuracy is required, the batch system is generally used. In this system, definite amounts of each of the ingredients are weighed and delivered to the process. The various ingredients can be weighed consecutively in a single hopper scale, or each ingredient can be weighed on a separate scale. The latter method provides the greater accuracy because each scale can be of the proper capacity for its specific ingredient, and also any error in the weighing of one ingredient does not affect the accuracy of weighing of another. As a compromise between the two methods described, a sequence method of weighing can be used wherein one ingredient is weighed and then discharged and succeeding ingredients are weighed and discharged in the same manner all on one scale. If all ingredients weigh about the same amount, accuracies of quite high order can be obtained. A point to be considered, however, is the possibility of contamination in case the same equipment is used for different ingredients for different formulas. Another factor is the time involved. With a scale for each material, all ingredients are weighed simultaneously. The scale operation can be manual, wherein an operator controls the feed and discharge, or the operation can be automatic in varying degrees.

With proper control, automatic equipment can function in conjunction with material-handling and processing equipment so that the full process cycle proceeds automatically.

To summarize, the following factors should be considered in the choice of equipment for a particular batching operation: (1) accuracy to which each ingredient must be weighed; (2) frequency in changing the formula or proportion of ingredients; (3) number of formulas or products for which equipment is to be used; (4) importance of avoiding any contamination of ingredients; (5) speed of operation required; and (6) volume of storage of raw materials required.

Two examples out of many available will illustrate different approaches to different problems. The first example is the operating station for weighing one to five of a possible ten ingredients on one scale (contamination is not a problem, large capacity storage is not required, and speed and accuracy requirements can be met). Because the formula is changed quite frequently, controls that permit selecting the ingredients to be used and setting the amount of each at a point remote from the scale are chosen, and these controls are furnished in duplicate. Thus, while a series of formula A is being weighed, the operator can be setting up the next ingredients on the formula B controls. The second example is the operating station for a system where eighteen scales are used to weigh the ingredients (large volume storage is required and high accuracy in combination with high speed is necessary). Three different formulas are used, but the amount of each ingredient in a formula is rarely changed. The controls for each scale are located at the corresponding scale and the interlocking controls for the scales, as well as the control for delivering the batches to the two mixers and to the ultimate destination, are incorporated in the station.

When the manufacturing process is continuous, consideration can be given to special forms of weighing equipment that deliver a continuous controlled stream of material. Equipment to perform this operation may be of the *continuous feeder* type or of the loss-in-weight type. In the case of a continuous feeder for dry materials, a suitable belt conveyor is installed on a scale. The belt runs at a definite controlled rate and material is delivered to the belt by means of a feeder. The rate of delivery of material from the feeder is controlled by the scale. The scale is preset to the desired weight of material to be delivered per unit of time and corrections to the feed rate are made when the feed rate deviates from the desired rate. Accuracies on the order of 1% for short intervals of time are possible with equipment of this type.

For the *loss-in-weight system* of weighing, a hopper or tank containing a supply of the material is suspended from a scale. This hopper or tank is discharged by a controllable feeder. It the scale is of the beam type, a poise is moved automatically along the beam at a rate in accordance with the flow rate desired for the material. In the case of an automatic indicating scale, a flow rate controller is used. The scale then controls the discharge feeder so as to deliver material at the preset rate. Since correction is continually made back to the weight in the hopper, this system can be said to have a "closed-loop" control as compared to the "open-loop" control of the continuous feeder. Accuracies to the order of 0.5% can be achieved. The system has the disadvantage that to maintain continuous flow two units must be used for each material. One unit is being refilled while the other is discharging.

Both the continuous feeder and the loss-in-weight feeder can be equipped with remote-set controls which can be connected to equipment for measuring an uncontrolled flow so the controlled stream or streams can be definitely proportioned to the uncontrolled stream.

Scales and Computers (8–10)

Scales can be modified to perform some types of computing directly. The counting scales described above under "Forms of Scales" is one example. A strain gage load cell scale can be built with variable gain to act as a direct reading counting scale by adjusting the gain to read the correct count for a definite number of pieces and then putting the unknown quantity on the scale to read actual count. By providing an adjustable ratio lever system or an adjustable gain load cell scale, material with a known moisture content can be caused to read on a dry weight or a definite percentage moisture basis. Scales equipped with analog or digital readout devices can feed weight information directly into analog or digital computers. Small special purpose computers can be used for counting by dividing the lot weight by the weight per piece, for determining dry weight or weight on a specified moisture content basis, for certain proportioning, optimizing, and other applications. Weight information fed directly into a central computer can be used for inventory control, proportioning with corrections for actual weights delivered, optimizing production, and other requirements. With proper interface equipment weight information can be transmitted to distant computers.

Feeding Equipment Suitable for Weighing Operations. For best performance, careful consideration must be given to the choice of proper equipment for the feeding of materials to scales for such operations as filling containers or for proportioning ingredients to a definite formula. The feeder should deliver material at as uniform a rate as possible and should respond quickly to control signals from the scale. When the feeder stops, there is a column of material in the air that falls into the weigh hopper. The cut-off point can be preset to allow for the average weight of this column, but deviations from the average weight will result in errors. For best accuracy, the weight in the column after cut-off should be small in proportion to the total weight. When the weighing operation is to be performed relatively fast, such as in filling the container in less than a minute, or when the rate of flow will change from such causes as a variation in head of material above the feeder, the feeder should be capable of being slowed down before the desired weight is in the scale. For this operation the bulk of the material is added to the scale at a high rate of feed and the speed of flow is decreased shortly before the desired weight is in the scale. When the control element on the scale signals the completion of the weighing operation, the feeder should stop quickly. This means that motor-operated feeders may require brakes or plugging controls or the use of an auxiliary gate to stop the flow instantly.

The following are recommended types of feeders for various materials:

A. Free-flowing materials that do not tend to arch or "rat hole" and flood.
   1. Electric vibrating feeders.
   2. Belt feeders.
   3. Screw feeders.
   4. Power-operated gates.
B. Material predominantly in large lumps.
   1. Electric vibrating feeders.
   2. Belt or slat feeders.
   3. Screw feeders of proper design if lump size is not too large.
C. Finely pulverized materials that tend to flood.
   1. Screw feeders of sufficient length, preferably double-flight or half-pitch. An auxiliary gate may be required.

2. Rotary vane or star feeders. The compartment or pocket size should be small enough so the discharge of one pocket will not cause an unacceptable error. For some applications it may be desirable to use a rotary vane feeder in conjunction with a vibrating feeder. With this combination, flooding is prevented, and the desired smooth control of flow is achieved with the vibrating feeder.

3. Air slide feeders. If sufficient length cannot be provided to prevent the possibility of flooding, a rotary vane feeder may be desirable to deliver material to the air slide.

D. Liquids. Power-operated valves are generally used for control of the weighing of liquids. The valve must close quickly and should be of proper design and size for the material to be handled at the rate of flow desired. Pneumatically operated diaphragm valves are frequently used for larger sizes and solenoid operated valves for small sizes. Valves of special construction to provide two rates of feed are available, or two valves in tandem with an adjustable by-pass as shown in Figure 13 can be used. Both valves are open for the fast rate of feed. Valve *A* closes at the first cut-off when near the desired weight. Valve *B* closes at the desired weight for final cut-off. The hand valve in the by-pass is adjusted to give the desired slow or dribble rate of feed for the maximum accuracy. See also Conveying.

Discharging the Scale. When material is weighed into a hopper or tank on a scale, as in the net-weight method of filling containers or in a batching system, the hopper or tank can be emptied completely by means of a gate or valve. When a gate is used, very little control can be exercised over the rate of discharge. In some cases, particularly when the scales of a batching system discharge onto a conveyor belt, it is very desirable to control the rate of discharge. This can easily be done by using a suitable feeding device in place of the gate. By proper adjustment of the discharge rate, a certain amount of premixing can be done on the belt. Also, with a controllable feeder as the discharge device, it is not necessary to empty the hopper or tank completely. By

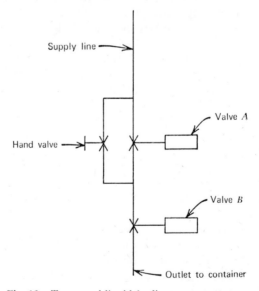

Fig. 13. Two-speed liquid feeding arrangement.

proper setting of the discharge control point, a definite amount of material can be left in the scale hopper or tank. This means the material is weighed out of, as well as into, the scale. With materials that cling to the sides of a hopper, this has a definite advantage in that every bit of material does not have to be removed. In the case of liquids, more rapid discharge of a definite amount can be accomplished because the last material does not have to be discharged with zero head. Of course this scheme cannot be used when different materials are accumulated in the scale hopper, because the composition of the material remaining in the hopper would vary.

Factors Affecting Scale Performance. Since weighing scales are actually highly accurate instruments, they should be protected from unnecessary abuse. Corrosive atmospheres and excessive dust will cause deterioration of weighing equipment unless it is properly protected. Some forms of load cells for electric, hydraulic, or pneumatic scales can be adequately protected by proper protective coatings. Exposed parts of mechanical scales can be coated with protective material if it does not interfere with the weighing operation. If mechanical scale lever systems are enclosed, they can be quite effectively protected from corrosion and dust by purging the enclosure with clean air.

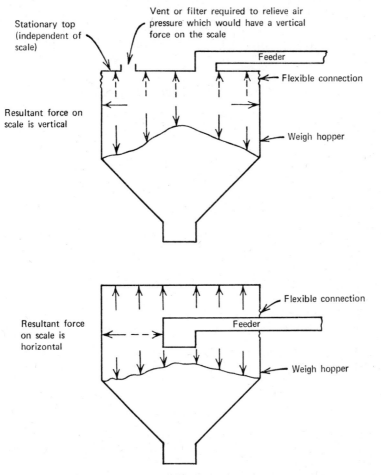

Fig. 14. Effect on scale of air pressure in weigh hopper (arrows indicate direction of air pressure. Solid arrows show force against scale, dotted arrows show force against stationary area).

Indicating elements of scales can be protected by locating away from the damaging atmospheres or by purging with clean air.

Certain precautions must be observed in the installation of weighing equipment and in the connection of other devices, such as feeding equipment to such elements of the scale as the hopper or tank.   Pipes conveying liquid to or from a weigh tank should be equipped with connections that are suitably flexible for the capacity of the scale and the movement of the weigh tank.   Precautions must be observed to see that any exterior forces such as might be caused by changes in temperature of the piping are applied to the scale in a horizontal rather than a vertical direction.   When weighing dusty materials, it is very desirable to keep the dust out of the surrounding air.   This is done by using dust seals or flexible connections where the material flows into and out of the scale hopper.   Since the filling of a hopper or tank displaces air in the hopper or tank, it is essential to make certain that either the displaced air is allowed to flow out through a filter or through a vent pipe, or that the increased pressure does not exert a vertical force on the scale.   This can be accomplished by designing the weigh hopper or tank so that material is fed in through the side of the hopper with the flexible connnection in the vertical plane as shown in Figure 14.   Horizontal forces applied to a weigh hopper can be overcome by suitable check links or stay rods without affecting the weighing accuracy of the scale.   Vertical forces of unknown magnitude or of magnitudes that cannot be definitely controlled may have a serious effect on the accuracy of any weighing equipment.

Bibliography

"Weighing and Proportioning" in *ECT* 1st ed., Vol. 15, pp. 25–34, by Donald B. Kendall, Toledo Scale Company.

1. D. B. Kendall, "Weight and Weight Rate of Flow," in D. M. Considine, ed., *Handbook of Applied Instrumentation*, McGraw-Hill Book Co., Inc., New York, 1964, pp. 5–41 to 5-54.
2. G. B. Anderson and R. C. Raybold, "Studies of Calibration Procedures for Load Cells and Proving Rings as Weighing Devices," *National Bureau of Standards Technical Note 436*, 1969.
3. C. D. Hodgman, ed., *Handbook of Chemistry and Physics*, 44th ed.,   The Chemical Rubber Publishing Co., Cleveland, Ohio, 1962, pp. 3480–3484.
4. "Specifications, Tolerances, and Other Technical Requirements for Commercial Weighing and Measuring Devices," *National Bureau of Standards Handbook*, 3d ed. **44**, 1965.
5. S. Rowe, "Nuclear Weighing," *Instr. Control Systems* **40**, 85–86, (Feb. 1967).
6. *Terms and Definitions for the Weighing Industry*, Scale Manufacturers' Association, Inc., Washington, 1964.
7. H. Colijn and P. W. Chase, "Continuous Weighing," *Instrumentation Technol.* **14**, 64–68, (May, 1967), 75–80, (June, 1967).
8. C. W. Hibscher, "Six Good Reasons for Computer/Scale Systems," *Mod. Mater. Handling* **23**, 82–85, (Nov. 1968).
9. C. W. Hibscher and R. J. Phillips, "Unique Computer Controlled Batching," *ASME Paper No. 61-WA-183*, 1961.
10. E. S. Savas, *Computer Control of Industrial Processes*, McGraw-Hill Book Co., Inc., New York, 1965.

General References

D. M. Considine, "Process Weighing," *Chem. Eng.* **71**, 113–132, (Aug. 1964), 199–206, (Sept. 1964).
M. W. Jensen and R. W. Smith, "The Examination of Weighing Equipment," *National Bureau of Standards Handbook* **94**, 1965.
D. B. Kendall, "Industrial Weighers" in D. M. Considine, ed., *Process Instruments and Controls Handbook*, McGraw-Hill Book Co., Inc., New York, 1957, pp. 7–8 to 7–39.

J. M. Logan and H. F. Williams, "Controlling Weight in Production," *Automation* **14**, 86–93, (Jan. 1967), 94–98, (Feb. 1967).

G. W. van Santen, *Electronic Weighing and Process Control*, Springer-Verlag New York Inc., New York, 1967.

DONALD B. KENDALL
Toledo Scale Company,
Division of Reliance Electric Company

WELDING

A weld is defined by the American Welding Society as "a localized coalescence of metal wherein coalescence is produced either by heating to suitable temperatures, with or without the application of pressure, or by the application of pressure alone, and with or without the use of filler metal. The filler metal either has a melting point approximately the same as the base metals or has a melting point below that of the base metals but above 800°F (427°C)."

The main divisions of welding are: arc welding, resistance welding, gas welding, brazing, solid-state welding and other processes (1). These main divisions are subdivided into the 34 different welding processes in commercial use today (Fig. 1). For the most part these processes complement each other, each having specific applications according to the metals to be joined, the nature of the structure of product being made, and the production facilities available. Of these 34 welding processes, the following are used most in the chemical industry: shielded metal-arc welding, submerged-arc welding, gas metal-arc welding, gas tungsten-arc welding and flux-cored arc welding.

Shielded Metal-Arc Welding

Shielded metal-arc welding is defined as "an arc-welding process wherein coalescence is produced by heating with an arc between a covered metal electrode and the work. Shielding is obtained from decomposition of the electrode covering. Pressure is not used and filler metal is obtained from the electrode" (Fig. 2). This welding process requires either an ac or dc source of electric power called a welding machine, which may be one of four types.

An *ac welding machine* is a transformer designed to have current characteristics that are specifically suitable for welding. All standard arc-welding machines, ac and dc, are built and rated in accordance with the *Arc-Welding Machine Standards* of the National Electrical Manufacturers Association (NEMA). They are available in varying sizes according to their deliverable current. Ac welding machines are built in capacities ranging from 150 to 1500 amperes. Two ac machines may be connected in parallel and operated together to obtain more current per arc than either will deliver separately.

Standard *dc motor-generator machines* are available in ratings of 150–600 A. Where there is a concentration of welders in a small area, multiple-operator machines, with capacities as high as 2000 A may be used. *Gas-engine-driven welding machines* are available in about the same range of ratings as the motor-generator sets. *Rectifier machines* are available in ratings of 200–800 A.

Filler metal for shielded metal-arc welding is in the form of covered electrodes. Such electrodes consist of a metal core and a covering of primarily nonmetallic material, which together provide a deposit of suitable chemical and mechanical properties.

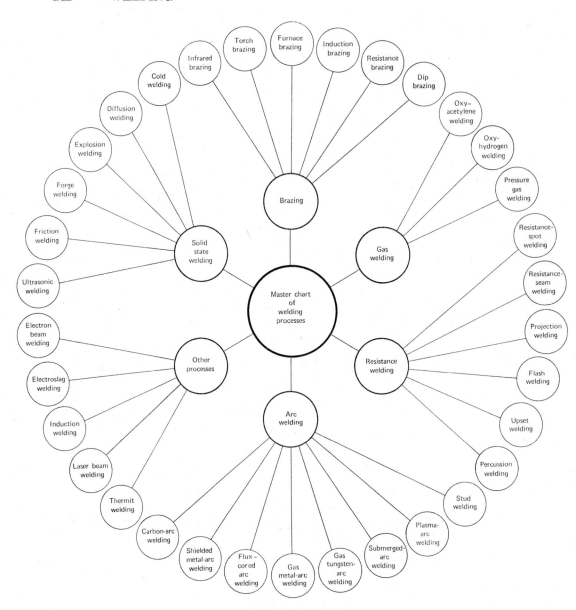

Fig. 1. Master chart of welding processes.

The covering is intended primarily to provide the "shielding" to prevent oxidation. It may also be used to introduce metallic elements into the deposited metal.

The formulation of electrode coverings is precise and complex. The covering constituents may serve as a deoxidizer, slag-former, arc-stabilizer, or binder. The usual binder is sodium silicate. Cellulose is a common deoxidizer. Ferromanganese is used as a deoxidizer and also as a slag-former. Aluminum silicate in the form of clay or feldspar, magnesium silicate in the form of asbestos or talc, silica, and manganese dioxide are used as slag-formers. Calcium carbonate, titanium dioxide, and iron oxide

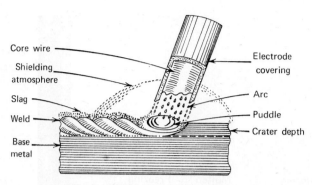

Fig. 2. Shielded metal-arc welding.

are used as slag-formers and arc-stabilizers. Not all these materials are used in any one covering.

Submerged-Arc Welding

Submerged-arc welding is defined as "an arc-welding process wherein coalescence is produced by heating with an arc or arcs between a bare metal electrode or electrodes and the work. The welding is shielded by a blanket of granular, fusible material on the work. Pressure is not used and filler metal is obtained from the electrode and sometimes from a supplementary welding rod" (Fig. 3). This process is automatic or semiautomatic.

The welding operation is usually started by striking an arc beneath the flux on the workpiece. The heat produced causes the surrounding flux to become molten, thus forming a conductive pad. The upper, visible portion of the flux is not melted. It may be automatic if the welding head is so mounted as to progress along the joint to be welded or the joint is made to move under the welding head. If the progression of the welding is manual, the process is said to be semiautomatic.

Submerged-arc welding is especially characterized by the use of high welding currents. This makes possible higher electrode melting rates and higher welding speeds for a given thickness of base metal. These characteristics also make possible easier welding of greater thicknesses. However, while welding speed is greater than for manual shielded metal-arc welding, time is required to set up the work for welding. Therefore, this process is most effectively used on long joints or on repetitive welding of shorter joints where jigs are used to align and hold the work.

Filler metal is in the form of bare wire electrodes and must be used in combination with the proper flux which functions in a manner similar to the covering of covered electrodes.

The work must be positioned for this process so that welding is done in the flat position because the flux is deposited over the joint and adheres to it by gravity.

Gas Metal-Arc Welding

Gas metal-arc welding is defined as "an arc-welding process wherein coalescence is produced by heating with an arc between a continuous filler metal (consumable) electrode and the work. Shielding is obtained entirely from an externally supplied gas or gas mixture. Some methods of this process are called MIG or CO_2 welding."

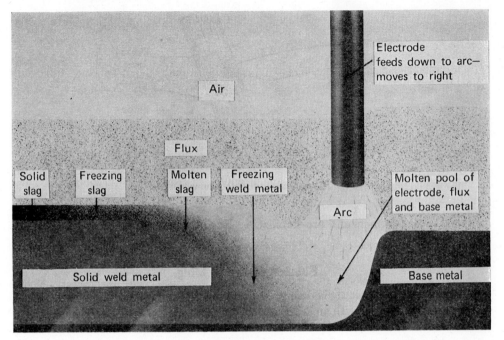

Fig. 3. Submerged-arc welding.

In this process, the filler metal is the electrode. Since the electrode has no covering, and no flux or additional filler metal is used, the electrode must contain all the ingredients necessary to meet the required material specifications. The shielding gases most commonly used are helium, argon, argon–helium mixtures or carbon dioxide. Dc is usually used as the power source for this process.

Gas Tungsten-Arc Welding

Gas tungsten-arc welding is defined as "an arc-welding process wherein coalescence is produced by heating with an arc between a single tungsten (nonconsumable) electrode and the work. Shielding is obtained from a gas or gas mixture. Pressure may or may not be used and filler metal may or may not be used. (This process has sometimes been called TIG welding)."

In this process, the tungsten electrode is not consumed but serves to maintain the arc between itself and the work. Filler metal, when used, is added in the form of a rod and it does not form part of the welding circuit. Argon or argon–helium mixtures are the shielding gases most commonly used. Either ac or dc power may be used with this process.

Flux-Cored Arc Welding

Flux-cored arc welding is defined as "an arc-welding process wherein coalescence is produced by heating with an arc, between a continuous filler metal (consumable) electrode and the work. Shielding is obtained from a flux contained within the electrode. Additional shielding may or may not be obtained from an externally supplied gas or gas mixture."

In this process, the filler metal is supplied by the electrode. The flux, contained within the hollow electrode, generates gas for shielding and may supply alloying elements. Additional shielding gas, usually CO_2, from an external supply may be used. Dc is used as the power source for this process.

Other Welding Processes

Resistance welding is a group of welding processes, rather than a single welding process, in each of which "coalescence is produced by the heat obtained from resistance of the work to electric current in a circuit of which the work is a part, and by application of pressure."

The resistance welding processes include *spot welding, seam welding,* and *projection welding* as well as three other processes (flash welding, upset welding, and percussion welding) that are less used in chemical industry. In resistance-spot welding, separate spot welds, or localized welds, are made in a joint, the several spots acting much like a series of rivets in a joint. In resistance-seam welding, the spots are spaced so close together as to overlap, providing an entirely gas- and watertight joint. In projection welding, one or both of the pieces of base metal to be joined are embossed by means of a die to obtain a "projection." The weld is made at the point of projection, thereby localizing the heat and reducing the dissipation of heat to adjacent base metal. This concentration of heat makes it possible to weld the same thickness with less current or greater thicknesses with the same current.

The resistance welding processes are usually used only on thin materials, up to about $\frac{1}{4}$ in., although greater thicknesses have been welded. The welding heat is obtained from resistance to electric current, hence a source of high current is required.

The rate of welding is extremely high. A spot weld may be made in as little as two cycles or one-thirtieth of a second. However, effective use of these processes necessitates a comparatively high capital outlay so that their use requires a high production rate or a large volume of production over a period of time.

Gas Welding. The principal difference between arc and gas welding is that in gas welding the source of welding heat is a gas flame produced by the combustion of a mixture of fuel gas and oxygen. The fuel gases used include acetylene, ethylene, butane, propane, natural gas, and a number of other petroleum derivatives. Acetylene is most used. However, the welding heat obtained with this process is lower, and hence the rate of welding is slower, than with arc welding. Therefore, gas welding is used in the chemical industry for incidental welding or for maintenance and repair welding where the extreme portability of the welding equipment is an advantage.

Brazing differs from the other welding processes in that coalescence is produced by heating to a suitable temperature and by using a filler metal having a liquidus above 800°F (427°C) and below the solidus of the base metals. The filler metal is distributed between the closely fitted surfaces of the joint by capillary attraction. When brazing is used there is no appreciable melting of the base metal, as is the case with the other welding processes. See Solders and brazing alloys.

Base Metals To Be Welded

The base metals used in the chemical industry include carbon steel; low-alloy steels; stainless steels; aluminum and nickel and high-nickel alloys; and lead.

Carbon steels are steels having no deliberate additions of alloying elements; they contain, in addition to iron, only carbon and manganese in appreciable quantities. Their principal use is for load-carrying at normal atmospheric temperatures. In the chemical industries, carbon steels are used for structural elements, such as beams and columns; for storage tanks intended to contain water or oil or a gas where pressure and strength are factors; and for piping in use at normal temperatures where corrosion is not a factor.

Of all the metals commonly joined by welding, the carbon steels are most easily welded. The heat of welding has a metallurgical effect on the base metal according to its composition. This effect may reduce the strength or the corrosion resistance of a metal or otherwise change its properties, but it is less in carbon steels than in other steels. Steels containing up to 0.30% carbon and 1.00% manganese can be readily welded in thicknesses up to about 1½ to 2 in. without special techniques. For greater thicknesses or where the carbon or manganese content is higher, it may be necessary to preheat the base metal adjacent to the joint to be welded. The amount of preheat will vary from 100°F to as high as 400°F according to the combination of thickness and chemical analysis involved. All the welding processes discussed are used for welding carbon steels. Filler metal is usually of a mild steel type.

Low-alloy steels are carbon steels to which other elements have been deliberately added to impart a particular property or properties to it. Alloying elements commonly used include nickel, chromium, molybdenum, vanadium, and silicon. Nickel is added to improve the mechanical properties of a steel at low temperatures. Chromium, molybdenum, and vanadium or combinations of these elements are added to improve the properties of a steel at elevated temperatures. Elements such as silicon are added to increase the mechanical properties of a steel at ordinary temperatures. These steels are not used where corrosion is a prime factor and are usually considered separately from the so-called "stainless steels." Low-alloy steels are used for pressure piping systems operating at high temperatures, for storing liquid nitrogen or oxygen and similar materials at low temperatures, or where greater strength with less mass is required at normal ambient temperatures.

Low-alloy steels, although less easily welded than the plain carbon steels, can be readily welded using preheat and suitable filler metal. Where the heat of welding impairs the metallurgical properties of the steel, it may be necessary to subject the welded joint or the entire weldment to heat treating in order to restore these properties. Where dimensional stability is a factor or where the welding operation is such as to create considerable restraint in the weldment, it is usual to follow welding by a stress-relief heat treatment. This consists of heating the weldment to a temperature of 1150–1350°F (according to its composition) and holding it at that temperature for one hour per inch of thickness or fraction of an inch, followed by slow cooling to about 500 or 600°F. The specific preheat and postheat treatments used would be a function of the amount of alloying elements present, the thickness of the parts assembled, and the intended service requirements. They are usually specified by codes such as the USAS *Code for Pressure Piping*, the ASME *Boiler Code*, etc. All of the welding processes previously mentioned may be used to weld the low-alloy steels. Low-alloy steel filler metal is used. Where high- or low-temperature properties are required, a filler metal of similar composition to the base metal is used.

Stainless Steels. The so-called stainless steels are of two general types: austenitic chromium–nickel steels and straight chromium steels. These steels have exceedingly wide use because of their resistance to corrosion by all kinds of acids, salts, and other materials. The subject of corrosion resistance is a complex one and will not be dealt with here. It is usual practice to select a steel suitable for the required corrosive medium or atmosphere involved and then establish a welding procedure that will provide a welded joint of corrosion resistance equal to that of the base metal.

Austenitic Chromium–Nickel Steels. There are about twenty different steels in this category according to classifications of the American Iron and Steel Institute. These include the well-known 18/8 variety as well as 19/9, 25/12 and 25/20 grades. These steels are identified by type numbers such as 308, 309, 310, and 347. They contain as the principal alloying elements, varying amounts of chromium and nickel. An 18/8 steel (Type 304) has a nominal composition of 18% chromium and 8% nickel. Other alloying elements, which are added to these steels in lesser quantities, are columbium, molybdenum, manganese, and silicon.

When welded, these steels go through a metallurgical change known as carbide precipitation. This reduces the corrosion resistance of the weld and any adjacent base metal affected by the welding heat. By use of the so-called stabilized grades, in which columbium or molybdenum has been added, or by use of the extra-low-carbon (ELC) grades it is possible to weld these steels without encountering carbide precipitation. Alternatively, the unstabilized grades may be welded and the weldment subjected to a heat treatment above 1900°F followed by very rapid cooling to below 800°F. This heat treatment will restore the precipitated carbides to solution and the weldment to its original corrosion resistance.

Stainless steels have higher mechanical properties than the carbon steels. Where they are used for strength alone, carbide precipitation is not a factor and the unstabilized grades are as suitable as the stabilized grades.

These steels are readily welded by all of the welding processes previously mentioned using filler metal selected to provide a weld deposit of the same general composition as the base metal.

Chromium Steels. The chromium steels, or "chromium irons" as they are sometimes called, contain only chromium in amounts ranging from 3% to about 27% or more. The metallurgy of these steels is rather complex. They harden to different degrees, being generally of higher hardenability as the chromium content increases. These steels are welded by submerged-arc, shielded metal-arc, gas metal-arc, gas tungsten-arc, and flux-cored arc welding, using carefully planned welding procedures including preheat and postheat treatment. The filler metal is either of a straight chromium analysis or of a chromium–nickel analysis, in which case the chromium and nickel must be high enough and so proportioned as to allow for dilution, which results from mixing with the base metal, so that the final analysis of the weld metal will contain an adequate amount of austenite.

Aluminum and aluminum alloys are used wherever lightness or atmospheric corrosion resistance are factors. These metals are also used where milder corrosive fluids are involved. Typical applications are railroad tank-car tanks, pressure vessels, and tanks for storing chemicals and dairy products.

There are two broad classifications or general types of aluminum alloys: nonheat-treatable and heat-treatable. Strengths of commercially pure aluminum and the non-heat-treatable alloys are developed by strain hardening and by alloying elements, of which magnesium, manganese, and silicon are typical examples. The alloying elements in heat-treatable aluminum alloys are dissolved in the pure aluminum at a high temperature by a process known as solution heat treatment. In addition, specific aluminum alloys are available in different tempers from hard to soft.

The heat incident to welding will produce an annealing effect in aluminum and the nonheat-treatable alloys at or near the weld, reducing the strength of the metal imparted by cold working. In the heat-treated alloys, the heat of welding produces a change in the metallurgical structure of that portion of the base metal affected by welding. The weldment may be heat-treated after welding, if of a heat-treatable grade, to restore its properties, provided a heat treatment is otherwise feasible. At any rate, use of gas metal-arc and gas tungsten-arc welding or resistance welding minimizes the deleterious effects. These processes use very high heats concentrated at the point of welding. In fact, the development of procedures for gas metal-arc and gas tungsten-arc welding has greatly expanded the use of welding for joining aluminum and its alloys. Shielded metal-arc welding and gas welding are still used to some extent.

On very thin sections, filler metal is not used, the base metal being flanged sometimes to serve as the filler metal. Filler metal, when used, is either pure aluminum or an aluminum–silicon alloy for shielded metal-arc welding, or it may be of an aluminum alloy of about the same composition as the base metal for gas metal-arc, gas tungsten-arc, or gas welding. Gas welding requires the use of a flux. Shielded metal-arc welding is done using covered electrodes. Gas metal-arc and gas tungsten-arc welding is done with bare filler metal. Shielding is obtained from either helium or argon and a flux is not required.

Nickel and Nickel Alloys. This group of metals includes commercially pure nickel and a variety of nickel alloys including Monel, Inconel, Incoloy, and Hastelloy. Commercially pure nickel has a nominal nickel content of 99.4%. It has a high corrosion resistance and is magnetic.

Nickel 200 is the standard grade of commercially pure nickel. Nickel 201 is the low-carbon version (.02% max) of nickel 200. It is used where maximum ductility is required at temperatures above 600°F. It has the same corrosion resistance as nickel 200.

The *Monel* alloys (International Nickel Co.) are nickel–copper alloys having a nominal composition of 67% nickel and 30% copper. Monel alloys are identified by numbers in the 400 and 500 series. Monel alloys have good strength, toughness and resistance to corrosion. They are slightly magnetic.

The *Inconel* alloys (International Nickel Co.) are nickel–chromium alloys having a nominal composition of 77% nickel and 16% chromium. Inconel alloys are identified by numbers in the 600 and 700 series. These alloys offer a high resistance to scaling and are used at elevated temperatures.

The *Incoloy* alloys (International Nickel Co.) are nickel–iron–chromium alloys having about 32% nickel, 21% chromium and 45% iron. Incoloy alloys are identified by numbers in the 800 series. These alloys have high oxidation and heat resistance.

The *Hastelloy* alloys (Haynes Stellite Co.) are nickel–molybdenum–iron alloys which are resistant to corrosive mineral acids and possess good mechanical properties. There are 9 grades; B, C, D, F, G, N, R-235, W, X, each intended for particular service conditions. All the above nickel alloys can be welded. The applicable welding processes are given below (2).

For shielded metal-arc welding of any of these metals and alloys, dc reverse polarity (electrode positive, work negative) is preferred. For submerged-arc welding, a suitable granular material and electrode of proper chemical analysis must be used. For gas metal-arc or gas tungsten-arc welding either helium or argon can be used as the shielding gas. Helium is preferred for the nickel alloys because it permits greater speeds and because sounder welds are obtained. Argon, however, is preferred for small parts or thin sections, because of the lower heat input with this gas. For welding the Hastelloys, argon is preferred. Dc straight polarity is recommended for gas tungsten-arc welding; dc reverse polarity is recommended for gas metal-arc welding.

Copper and Copper Alloys. The coppers are divided into oxygen-bearing coppers and oxygen-free coppers, which have been processed to make them essentially free of oxygen. In addition to copper itself, there are a number of copper alloys which are of commercial importance, including the alloys of copper and zinc (brasses); phosphor bronzes, which are alloys of copper and tin; silicon bronzes, alloys of copper and silicon; aluminum bronzes, alloys of copper and aluminum; copper–nickel alloys; nickel silvers, alloys of copper, zinc, and nickel; and beryllium copper.

All of these materials are weldable and have been welded commercially. When welding copper itself, the oxygen-free coppers are used if the joint strength is required to be equal to that of the base metal. Shielded metal-arc, submerged-arc, gas metal-arc and gas tungsten-arc welding may all be used. All copper alloys can be welded using shielded metal-arc welding. Submerged-arc welding has been used to weld copper, and to a limited extent to weld the brasses and bronzes. Gas metal-arc and gas tungsten-arc welding are the preferred processes for welding copper. The exception to all of the foregoing is nickel silver which is usually not welded by any of these processes but is readily joined by oxyacetylene braze welding or brazing.

Lead is used for tank linings and other equipment used to store, mix, or contain acids, such as sulfuric and chromic acids, and other chemical solutions. It is most commonly welded by the oxyacetylene, oxynatural gas, or oxyhydrogen flame, hydrogen being considered the most satisfactory fuel gas. Filler metal of the same composition as the lead to be welded is generally used. Cleanliness is especially important. The joint area to be welded must be cleaned to a bright shiny surface by shaving or wire brushing. This process although actually welding, is often improperly referred to as lead "burning."

Filler Metals

Filler metal is metal to be added in making a weld. For those arc-welding processes in which it forms a part of the electrical circuit, it is used in the form of an electrode. This is the case in shielded metal-arc, submerged-arc, gas metal-arc and flux-cored arc welding. For gas welding, gas tungsten-arc welding, and those arc-welding processes in which the filler metal does not form a part of the welding circuit (does not conduct current), it is used in the form of a welding rod. The same filler metal may sometimes be used as a welding rod or as an electrode depending upon the welding process being used.

Carbon Steels. Covered electrodes for shielded metal-arc welding of carbon steels, up to about 0.35% carbon are standardized in the *Specification for Mild Steel Covered Arc-Welding Electrodes*, AWS Designation A5.1; bare electrodes and fluxes for the submerged-arc welding of carbon steels are covered in the *Specification for Bare Mild Steel Electrodes and Fluxes for Submerged-Arc Welding*, AWS Designation A5.17; bare electrodes for the gas metal-arc welding of carbon steels are covered in the *Specification for Mild Steel Electrodes for Gas Metal-Arc Welding*, AWS Designation A5.18;

bare electrodes for the flux-cored arc welding of carbon steels are covered in the *Speci-fication for Mild Steel Electrodes for Flux-Cored Arc Welding*, AWS Designation A5.20.

Welding rods for gas welding mild steel are covered in the *Specification for Iron and Steel Gas-Welding Rods*, AWS Designation A5.2.

Low-Alloy Steels.   Covered electrodes for shielded metal-arc welding of low-alloy steels are standardized in the *Specification for Low-Alloy Steel Covered Arc-Welding Electrodes*, AWS Designation A5.5.

Weld metal deposited from these electrodes contains alloy elements to obtain the required properties.   The alloy elements may be added through the use of a low-alloy steel core wire or by adding them to the electrode covering.   Electrodes may be selected on the basis of the chemistry of the weld deposit as well as the mechanical properties of the weld metal.   The electrodes included, all of which are commercially available, are ½ Mo; ½ Cr-½Mo; 1¼ Cr-½ Mo; 2¼ Cr-1 Mo; ½ Cr-1 Mo; 2½ Ni; and 3½ Ni.

Specifications are currently (1970) being prepared for low-alloy steel electrodes for the submerged-arc, gas metal-arc and flux-cored arc welding of low alloy steels.

Chromium and Chromium–Nickel Steels.   Covered electrodes for shielded metal-arc welding of corrosion-resisting chromium and chromium–nickel steels are standardized in *Specification for Corrosion-Resisting Chromium and Chromium–Nickel Steel Covered Welding Electrodes*, AWS Designation A5.4.

Bare filler metal which may be used as an electrode for submerged-arc and gas metal-arc welding and as a welding rod for gas tungsten-arc welding is standardized in *Specification for Corrosion-Resisting Chromium and Chromium–Nickel Steel Welding Rods and Bare Electrodes*, AWS Designation A5.9.

Aluminum and Aluminum Alloys.   Covered electrodes for shielded metal-arc welding of aluminum and its alloys are standardized in *Specification for Aluminum and Aluminum-Alloy Arc-Welding Electrodes*, AWS Designation A5.3.   Bare aluminum and aluminum-alloy filler metal which may be used as an electrode with the submerged-arc and gas metal-arc welding processes and as a welding rod with the gas tungsten-arc welding process are covered in the *Specification for Aluminum and Aluminum-Alloy Welding Rods and Bare Electrodes*, AWS Designation A5.10.

Nickel and Nickel Alloys.   Electrodes for shielded metal-arc welding of nickel and nickel alloys are standardized in the *Specification for Nickel and Nickel-Alloy Covered Welding Electrodes*, AWS Designation A5.11.   Bare nickel and nickel-alloy filler metal which may be used as an electrode for the submerged-arc and gas metal-arc welding processes and as a welding rod for the gas tungsten-arc and oxyacetylene welding processes are covered in the *Specification for Nickel and Nickel-Alloy Bare Welding Rods and Electrodes*, AWS Designation A5.14.

Copper and Copper Alloys.   Bare and covered electrodes for the gas metal-arc, submerged-arc, and shielded metal-arc welding of copper and copper alloys are standardized in the *Specification for Copper and Copper-Alloy Arc-Welding Electrodes*, AWS Designation A5.6.   Welding rods for the gas tungsten-arc and oxyacetylene welding of copper and copper alloys are covered in the *Specification for Copper and Copper-Alloy Welding Rods*, AWS Designation A5.7.

Titanium and Titanium Alloys.   Bare titanium and titanium-alloy filler metal for use as an electrode with the gas metal-arc welding process and as a welding rod with the gas tungsten-arc welding process are covered in the *Specification for Titanium and Titanium-Alloy Bare Welding Rods and Electrodes*, AWS Designation A5.16.

Magnesium Alloys. Bare magnesium-alloy filler metal for use as an electrode with the gas metal-arc welding process and as a welding rod with the gas tungsten-arc and oxyacetylene welding processes is covered in the *Specification for Magnesium-Alloy Welding Rods and Bare Electrodes*, AWS Designation A5.19.

Cast Iron. Filler metal for welding cast iron, both welding rods and covered electrodes, is standardized in the *Specification for Welding Rods and Covered Electrodes for Welding Cast Iron*, AWS Designation A5.15.

Design for Welding

The design of a weldment is a function of the service conditions for which the completed structure is intended. Service conditions may include merely strength to resist static loads at normal atmospheric temperatures, or resistance to dynamic loading producing maximum stress many times repeated, or strength at elevated or subzero temperatures, or corrosion resistance to some specific atmosphere or material.

The basis for design for usual structures is prescribed by the governing code or specification. For structural purposes where only static unit stresses are involved, the allowable design unit stresses for the welded joint are usually the same as the corresponding allowable stresses for the base metal. Where resistance to dynamic loading is a factor, the allowable stresses for both the base metal and the welded joint are reduced according to the number of repetitions of loading to be expected. For storage tanks and pressure vessels, design is usually on the basis of a joint efficiency ascribed

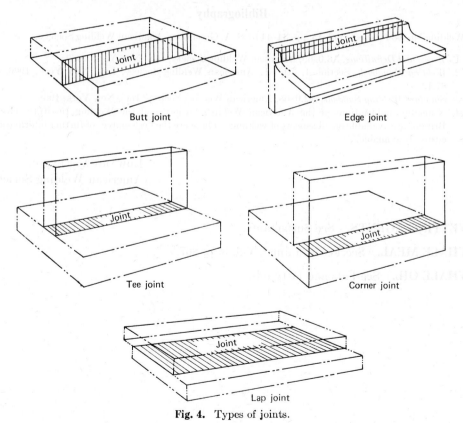

Fig. 4. Types of joints.

to a welded joint according to its location or the procedure under which it was made. Such joint efficiencies are usually expressed as a percentage of the allowable stress in the base metal. Design for corrosion-resistance requires the selection of a base metal and weld metal that are known by experience or test to provide the necessary protection for the expected life.

There are five basic types of joint that are used for welded connections. These are shown in Figure 4. Each of these can be used with the various types of weld indicated in Figure 4. The types of weld most usually made with the welding processes described here are the *fillet weld* and the various kinds of *groove welds*.

The specific details of a joint design as to the groove angle (the total included angle between the parts to be joined by a groove weld), the root opening (the separation between the members to be joined, at the root of the joint), the root face (that portion of the groove face adjacent to the root of the joint) and the size of weld vary not only from one metal to another but from one fabricating shop to another. Many companies have established standards of design which are used by them and which vary in details from the standards of other companies. These differences do not affect the quality of the welded joint but merely reflect different practices and techniques.

The identification of welded joints and the designation of geometry of such joints on design drawings has been standardized in *Standard Welding Symbols* of the American Welding Society (3). These symbols are equally applicable for all the welding processes and the base metals discussed here.

Bibliography

"Welding" in *ECT* 1st ed., Vol. 15, pp. 34–44 by S. A. Greenberg, American Welding Society.

1. *Terms and Definitions*, A3.0–69, American Welding Society, Inc., New York, 1969.
2. *Welding Handbook*, Section 4, 5th ed., American Welding Society, Inc., New York, 1966, p. 67.12.
3. *Standard Welding Symbols*, A2.0–68, American Welding Society, Inc., New York, 1968.
4. Numerous publications of the American Welding Society and the Welding Institute (Great Britain) are available on all aspects of welding. These are the best sources of further information currently available.

EDWARD A. FENTON
American Welding Society

WETTING AGENTS.   See Surfactants.

WHALE MEAL.   See Feeds, animal, Vol. 8, p. 867.

WHALE OIL.   See Fats and fatty oils.

WHEAT AND OTHER CEREAL GRAINS

Cereals comprise members of the grass family, Gramineae, that are grown for their edible grains. Most important are wheat, rye, barley, maize (corn), oats, rice, and grain sorghum. The cereals were grown by primitive peoples before any recorded history of man so that the wild forms from which they were evolved are often unknown and their origin is uncertain.

Although the ancestry of the common species of wheat grown today is problematical, bread wheats apparently originated by hybridization of various wild grasslike plants (1,2). Cultivation of wheat began in Asia Minor at least six thousand years ago. Wheat first reached Europe in the form of einkorn and emmer (primitive forms of wheat) about 3000 BC. Bread wheats spread over Europe from Southern Russia about 2000 BC, and were brought to America by early explorers.

Corn, or maize, originated in the western hemisphere; it was the only cereal cultivated by the American Indian, though some other grains were harvested in the wild state (3). Oats apparently were unknown to the ancient Chinese, Hebrews, or Hindus (4). The common oat (*A. sativa*) originated in Asia Minor. The first records on cultivated oats appear at the beginning of the Christian era. Oats were first brought to America with other small grains by Captain Gosnold in the early 17th century and cultivated on one of the Elizabeth Islands off the southern coast of Massachusetts. Ancient records show that cultivated barley was used by Neolithic cultures in Egypt between 5000 and 6000 BC (5). Major centers of barley, where cultivated varieties may have developed, include Abyssinia and the highlands of Sikkim and Southern Tibet. Cultivated rye seems to have originated from several types of rye-weeds growing in crops of barley and wheat (6). Perennial wild ryes could have given rise to annual wild ryes, possibly the forerunners of cultivated rye. Rye seems to have entered Europe from two sources in Asia, Transcaucasia and Turkestan-Afghanistan. Rye was introduced into the northeastern part of the United States during colonial times by English and Dutch settlers. Sorghum, thought to have originated in Africa, was brought to the western hemisphere in slave ships (7). Rice probably originated in southern Asia (8). Rice production in China dates back to about 2800 BC, and in India to 1000 BC. It was cultivated in the Euphrates Valley in 400 BC and was brought to southern Europe in medieval times by the Saracens. Rice culture in the United States began about 1685 in South Carolina.

Table 1. Botanical Classification of Common Grains (Family Gramineae) Grown in the United States

Common name	Tribe	Genus	Species	Subspecies
barley				
two-row	Hordeae	*Hordeum*	*distichon* L.	
six-row	Hordeae	*Hordeum*	*vulgare* L.	
corn or maize				
dent corn	Maydeae	*Zea*	*mays*	*indentata* Sturt.
flint corn	Maydeae	*Zea*	*mays*	*indurata* Sturt.
popcorn	Maydeae	*Zea*	*mays*	*everta* Sturt.
sweet corn	Maydeae	*Zea*	*mays*	*saccharata* Sturt.
oats				
white and yellow	Aveneae	*Avena*	*sativa* L.[a]	
red	Aveneae	*Avena*	*byzantina* L.	
rice	Oryzeae	*Oryza*	*sativa* L.	
rye	Hordeae	*Secale*	*cereale* L.	
sorghum, grain	Andropogoneae	*Andropogon*	*sorghum* Brot.	
wheat				
club wheat	Hordeae	*Triticum*	*sativum*	*compactum* Host
common wheat	Hordeae	*Triticum*	*sativum*	*vulgare* Vill.
durum wheat	Hordeae	*Triticum*	*sativum*	*durum* Desf.

[a] Includes *Avena orientalis* Schreb, or Tartary oats, which have one-sided panicles.

Purposeful plant breeding, which has been an important feature of American agriculture, introduced marked advances in the development of varieties better adapted to various cultural conditions or of higher quality for specific purposes than those previously grown. The botanical classification of the economically important cereals grown in the United States is given in Table 1.

Cereals are very important foods because they supply most calories per acre, can be stored safely for a long time, and can be processed into many foods acceptable throughout the world; cereals are adapted to widely varying soil and climatic conditions, can be cultivated with only a small amount of labor, and have high food value. Most developing countries rely heavily on cereals as food sources. More than two-thirds of the world's cultivated area is planted to grain crops. These crops provide more than half of the calories consumed for human energy, while in the developing areas of the world they probably provide two-thirds of the total food. They also contribute greatly to meat and livestock products which provide much of the remainder of the calories needed for the energy consumed. Cereals provide about 65% of the world's protein supply. The cereals contribute about 72 million metric tons of pro-

Table 2. Total Grain Consumption[a] as Food in the United States, 1960–1968, million bu[b]

Year	Wheat	Rye	Rice[c]	Corn	Oats	Barley
1960	489	4.6	10.8	151	42	5
1962	491	4.7	13.4	164	44	6
1964	501	4.8	13.2	187	46	6
1966	499	5.3	14.1	191	47	6
1967	507	5.4	14.6	195	47	6
1968	508	5.7	15.4	198	47	6

[a] Excludes quantities used for alcoholic beverages.
[b] One U.S. bushel = 35.23 liter.
[c] Million 100 lb; consumption for the year beginning August previous to the year stated.

teins, of which wheat provides 25 million, corn 20 million, and rice 14 million. The legumes supply about 8.5 million metric tons of protein; minor quantities of protein are supplied by tubers and nuts. The total animal protein supply is about 30 million metric tons, equivalent to about 25% of the world's protein resources. Civilian per-capita consumption of major food grain products in the United States is summarized in Table 2. See also Bakery processes; Starch.

Production

A summary of grain production and trade is given in Table 3 (9). Production of the chief grains in the major areas is summarized in Table 4 and the yields per acre in Table 5. Information on exports of grain showing the chief exporting regions is given in Table 6.

Table 3a. World Production[a] of Grain, Million Tons, av (9)

Grain	1946–1951	1952–1956	1957–1961	1962	1963	1964	1965	1966[b]	1967[b]
wheat	130.8	173.2	210.1	207.7	232.8	213.4	247.1	236.7	278.3
coarse grains[c]	250.6	273.5	320.0	331.6	344.5	348.0	351.7	362.2	385.4
rice[d]	66.1	79.5	93.8	104.5	104.2	111.4	118.0	108.6	110.0

Table 3b. World Exports[a] of Grain, Million Tons, av (9)

Grain	1946–1950	1951–1955	1956–1960	1961	1962	1963	1964	1965[b]	1966[b]
wheat[c]	22.8	27.4	34.3	45.2	41.3	48.9	55.9	54.9	55.6
coarse grains[c]	9.2	13.7	18.7	23.8	28.6	28.3	30.8	33.4	31.1
rice[d]	3.1	4.4	4.9	5.4	5.2	5.9	6.3	6.6	5.5

Courtesy, The Commonwealth Secretariat, London.
[a] Excluding China.   [d] Milled rice equivalent.
[b] Estimated.   [e] Including wheat flour in wheat equivalent.
[c] Maize, barley, oats, and rye.

Wheat is the chief cereal of temperate regions, where it continues to be a basic food, though higher incomes and increased meat, milk, and egg consumption have de-creased the per-capita consumption of wheat. In developing countries, wheat con-sumption is increasing and accompanies rising living standards. The trend is rein-forced by shortage of rice. In temperate regions, lower-quality wheat grades are marketed for animal feeding purposes. In addition, milling by-products, about a quarter of the weight of wheat milled for flour, are also utilized for animal feed. Wheat is the most widely cultivated grain crop in temperate areas, where it is the highest yielding crop and commands the highest price. Wheat may be grown in rotation with barley and oats to reduce loss of soil fertility. In areas with limited rain and short growing period, barley may replace wheat. Rye is better adapted than wheat to poor soil and is more winterhardy; oats thrive better in areas with cool summers. Corn and rice are particularly suited for cultivation in the tropics and the warmer temperate areas. Grain sorghum, like maize and rice, is a "tropical" cereal but is better suited than corn to cultivation under semiarid conditions (10).

Corn is now the most important grain used for animal feed in the temperate re-gions. In many tropical areas it is a basic human food. There are the following five types of corn: flint, dent, flour, pop, and sweet. Dent varieties, grown most widely

Table 4. Production[a] of the Chief Grains in 1966, Million Tons (9)

Country	Wheat	Maize	Barley	Oats	Rye	Rice[b]
Western Europe	44.6	8.8	29.4	11.4	5.2	
Eastern Europe[c,d]	21.4	17.0	7.5	4.5	11.5	
U.S.S.R.	58.7		20.0	6.1	16.0	
North America	54.9		13.2			
United States		102.1		13.2		
other America		33.0				
India						30.1
Pakistan						11.7
Japan						11.4
other Asia						42.9
America						7.7
others	57.1	35.7	16.4	9.0	2.3	4.8
China[c]						55.0
total[e]	236.7	196.6	86.4	44.2	35.0	108.6

Courtesy, The Commonwealth Secretariat, London.
[a] Grand total (excluding China) was 707.5 million tons.
[b] Milled rice equivalent.
[c] Estimate.
[d] Including Yugoslavia.
[e] Excluding China.

Table 5. Average Grain Yields in Various Countries, 1961–1966, 100 lb/acre (9)

Country	Wheat	Maize	Barley	Oats	Rye	Rice
Argentina	12.2	14.7	9.9	10.0	6.1	19.0
Australia[a]	9.7	15.1	9.0	6.9	3.3	33.4
Brazil		10.7			6.1	8.0
Burma		4.9[b]				8.7
Canada	11.0	37.4	13.2	13.4	9.2	
Denmark	32.9		30.8	29.6	23.3	
France	23.3	24.0	22.3	16.5	12.4	20.6
Great Britain	32.2		28.6	22.6	21.5	
India	6.7	7.9	6.9			7.8
Indonesia		7.3[b]				9.1
Italy		26.1			12.9	26.4
Japan		20.3				28.4
Pakistan	6.6	8.3	5.0			8.8
Rhodesia[c]		22.1[b]				
South Africa[c]		12.5[b]				
Sweden	26.1		21.4	20.1	18.8	
Thailand		15.4				8.0
Turkey		11.2			8.4	12.7
United States	13.5	33.2[d]	15.3	12.9	9.8	24.9
U.S.S.R.	7.7	17.7[d]	9.0	6.6	7.4	12.9
West Germany	26.4	28.0	24.0	23.0	21.2	
Yugoslavia		18.1			8.3	

Courtesy, The Commonwealth Secretariat, London.
[a] Per sown acre.
[b] Average for less than the full period shown.
[c] Excluding native production.
[d] Harvested for grain.

Table 6. Exports[a] of Grain from the Chief Exporting Regions in 1965, 1000 tons

Region	Wheat[b]	Rye	Maize	Barley	Oats	Rice
Argentina			2,804		345	
Australia	7,141				301	
Burma						1,326
Canada and U.S.	31,073	172	14,923	2,081	660	
South Africa			321			
Thailand						1,829
United States						1,502[c]
U.S.S.R. and Eastern Europe	2,120	62	1,267	2,141	12	
Western Europe	6,832	52		2,547		
others	7,738	166	4,079	1,122	339	1,909[d]
total	54,904	452	23,394	7,891	1,657	6,566

Courtesy, The Commonwealth Secretariat, London.

[a] Grand total of all grains was 94,864,000 tons.

[b] Including wheat flour in wheat equivalent.

[c] Excluding relief shipments.

[d] Excluding China, which exported 764,000 tons (figure taken from returns from importing countries).

in the United States and South Africa, tend to be higher yielding. Flint corn has a somewhat higher food value and is common in Europe, Asia, and South America. With the introduction of U.S. hybrid corn, dent varieties are grown more extensively in those areas. Both flint and dent corn have white and colored varieties differing little in feeding value except that yellow corn is rich in carotenoids.

Pop and sweet corns are important, though relatively small, food crops in some areas. Flour corns are grown in some parts of South America where their easy friability makes them valued for hand milling.

Genetic variants of corn have different kinds of starch. Ordinary starch has two components, one linear in molecular structure (amylose), the other branched (amylopectin). Waxy corn, so-called because of its appearance when the mature kernel is cut, contains only the branched starch molecules. Amylomaize (corn rich in amylose) contains unusually high percentages of linear starch molecules. Both types are grown commercially in the United States for starch production; the starches go into food and industrial uses for which they are suited. High-lysine corn is another genetic variant.

Barley is a winter-hardy and drought-resistant grain. It matures more rapidly than wheat, oats, or rye and is widely distributed. Its use for human foods has declined and it is used mainly as feed for livestock and in the milling and malting industries. The two main types of barley, depending on the arrangement of grains in the ear, are two-rowed and six-rowed. The former predominates in Europe and parts of Australia; the latter is more resistant to extreme temperatures and is grown in North America, India, and the Middle East. Both types can be malted. In addition, small amounts of a hull-less ("naked") barley that is more easily processed for food are grown in Japan and neighboring countries. See also Malt.

Oats are grown most successfully in cool, humid climates and on neutral to slightly acid soils. High-quality oats are ground into oatmeal, or are manufactured into rolled oats and proprietary breakfast foods. The bulk of the crop is consumed as animal feed (particularly by horses).

Rye is characterized by good resistance to cold, pests, and diseases, but is less profitable (especially on good soils and under improved cultivation practices) than wheat. It is used as bread grain in parts of Europe and in the United States and other countries in conjunction with wheat flour. Elsewhere, it is used predominantly as a feed grain; a small proportion is used in the distilling industry.

In 1965–1966 the wheat crop in the United States was 35,240,000 tons. Two botanical species of wheat, namely, *Triticum aestivum* L. (common wheat) and *T. durum* (amber durum wheat), and *T. compactum* (white and red club wheat), each representing many varieties, are grown in the United States but members of the common group represent about 92% of the total production (11). Common or vulgare wheats differ widely in character. For commercial purposes, they are classified into the following four major groups: hard red spring, hard red winter, soft red winter, and white wheats (winter and spring). Spring wheat is sown in the spring and harvested in the late summer, whereas winter wheat is sown in the fall and is harvested early the next summer. In 1964, 57% of the U.S. wheat crop was seeded to varieties of the hard red winter market class, 14% to soft red winter, 16% to hard red spring, 8% to white, and 5% to the amber durum class (11,11a).

Several rather sweeping changes have taken place in wheat breeding in the 1960s. Very high-yielding semidwarf varieties with straw 5 to 10 in. shorter than the older varieties, that have a good stand in the field and that respond to high rates of fertilizer and ample irrigation, have been developed. Some of these varieties are grown widely on the Pacific coast. High-yielding hybrid wheats have been developed and probably will soon be grown commercially. Small differences in protein between varieties and selections grown in a common environment for years have discouraged plant breeding efforts to increase protein content, as has the general negative correlation between yield and protein content. The report of Middleton et al. (12) that certain varieties can, with adequate fertilization, produce increased grain and protein yields, indicates that the genetic barrier to increased protein content can be broken. Another new development is the production of triticale, a cross between wheat and rye. Several of the crosses have proved high-yielding. Triticale appears to have high potentiality as a source of energy and some protein in foods and feeds.

The principal areas of hard red spring wheat in the United States are in the northern semiarid portion of the Great Plains, in Minnesota, North Dakota, South Dakota, and Montana. The hard red winter-wheat region of the Great Plains is south of the hard red spring-wheat area, with the greatest acreages in Kansas, Nebraska, Oklahoma, and Texas. Soft red winter wheat is chiefly grown east of the Great Plains and south of Wisconsin; Ohio and Indiana are the leading producers. White wheats are grown principally in the Pacific Coast states and in New York and Michigan. Durum wheats are adapted to regions of limited rainfall and most of the U.S. crop is produced in North Dakota, with small amounts in South Dakota and Minnesota. Red durums are not in demand by millers, and the amber varieties (used for macaroni products) are grown more extensively.

In Canada high-quality hard red spring wheat is grown mainly in the three prairie provinces (Manitoba, Saskatchewan, and Alberta). It is known in the trade under the name Manitoba wheat, though about 55% is from Saskatchewan, 30% from Alberta, and only 15% from Manitoba. The average yearly wheat production in those provinces is 440 million Imperial bushels, up to 70% of which is exported. (An Imperial bushel is 36.35 liter, ie 1.032 times the U.S. bushel.) About 8% of the wheat grown in Canada is of the durum type and most of it is exported. Canada produces and exports also some soft wheat grown mainly in the East.

Corn is the leading grain in the United States. In 1965, the corn crop was 114,-362,000 tons. The average yield of corn per acre was 4130 lb as compared to 2190 lb for wheat. Certain corn and wheat varieties, however, can produce much higher yields under irrigation. Eightfold increases were recorded in isolated cases (11).

Corn requires warm, moist growing conditions. Principal commercial production is in the so-called "corn belt," which includes all of Iowa and Illinois and parts of Indiana, Kansas, Kentucky, Michigan, Minnesota, Missouri, Nebraska, Ohio, South Dakota, and Wisconsin. Here, the dent type (*Zea mays*, var. *indentata*) is predominant; dent corn constitutes more than 90% of the entire U.S. production. Flint corn, which is early maturing, is grown principally in northern Pennsylvania, New York, the New England States, California, and the western and northern edges of the corn belt. Sweet corn (*Z. mays* var. *saccharata*), adapted to the cooler areas, is grown for canning purposes and used in the fresh state; the North Atlantic and Central States are important producers of sweet corn. Popcorn (*Z. mays* var. *everta*) is grown chiefly in the corn belt. Marked improvements in the yield and quality of corn for specific purposes have been made by the widespread culture of hybrid corn (the first generation hybrid between two or more inbred strains or between two first-generation hybrids); in the corn belt today hybrid corn is grown almost exclusively. Discovery of high-lysine corn is of great significance in improving the nutritional value of cereal grains.

Barley was grown on only 9,140,000 acres in 1965–1966 in the United States; the crop totaled 8,406,000 tons, yielding an average of 1840 lb per acre. Barley is grown principally in Western and the North Central States. Six-rowed barleys are grown much more extensively than two-rowed varieties; hybrid smooth-awned varieties are now extensively grown in Minnesota, North Dakota, and South Dakota, the leading barley-producing states of the Midwest.

The area under oats in the United States in 1965–1966 was 18,479,000 acres (more than twice that of barley), but the yield was 13,241,000 tons, or an average of 1430 lb per acre. Oats are grown throughout the United States, but the chief producing areas conform to the north central corn belt and extend further northward. Iowa, Minnesota, Illinois, and Wisconsin are the leading oat-producing states. White and yellow varieties are most commonly grown in the North, while red and gray oats predominate in the South.

Only 1,469,000 acres were under rye in 1965–1966 in the United States; the total yield was 831,000 tons (out of a world total of 35 million tons) and the average yield per acre was 1130 lb per acre. Winter varieties of rye are of major importance and the chief producing areas are North Dakota, South Dakota, and Minnesota.

Rice is the staple food of about half of the human race. Half of the 160 million tons consumed each year is eaten on farms where it is grown. Only about 5% enters the international trade. Rice is produced mainly in tropical and semitropical regions and over four-fifths of the world crop is grown and consumed in the Far East.

Rice production in the United States is confined to southwestern Louisiana, southeastern Texas, eastern Arkansas, and the Sacramento and San Joaquin valleys of California. Only lowland or irrigated varieties are grown and include long-grain, medium-grain, and short-grain (or Japanese) types. Short-grain varieties predominate in California, and medium-grain varieties in other areas. The crop in 1965–1966 was

2,457,000 tons from an acreage of 1,793,000 acres; the average yield in the United States per acre was 2720 lb (3080 lb per acre in California alone).

Sorghum has been used in many countries as human food. In the United States, it is considered a feed grain; only 3–5% of U.S. production is used for food and industrial purposes. Production of sorghum grain in the United States has increased rapidly from less than 2 million tons harvested from 3.4 million acres in 1930 to about 21 million tons from 13.3 million acres in 1965. The rapid increase has been attributed to the development of high-yielding hybrids that can be harvested mechanically, expanded need for starch and tapioca substitutes, the ability of sorghum to tolerate heat and drought, and certain favorable agroeconomic factors (10). Sorghum is now the third largest U.S. cereal crop; in Texas, sorghum is the number one grain crop, with an annual cash value exceeding $300 million.

Grain sorghum, one of four general classes (sorgo or sweet sorghum, broom corn, grass sorghum, grain sorghum) of this plant produced in the United States, thrives well under drought conditions and is grown principally in the southern sections of the Great Plains area and in parts of the Southwest. Several groups of grain sorghums (milo, kafir, feterita, durra, kaoliang, and shallu) have been grown in the United States but in recent years crosses of milo with kafir have become the leading varieties. Like certain varieties of corn and other cereals, some varieties of sorghum yield so-called waxy or glutinous starch, similar to cassava, the root starch from which tapioca is prepared. During World War II when imports of tapioca from The Netherlands Indies were cut off, the waxy varieties of sorghum were cultivated and this grain was first used for starch production in 1942. In many varieties of sorghum, the nucellar tissue contains pigments that complicate the production of white starch from this grain. Varieties, such as (waxy) cody, with no nucellar tissue persisting to maturity, can be processed by wet-milling methods closely resembling those used for cornstarch production. Sorghum is attracting attention as a source of starch, particularly for use in food products, and plant breeders are developing new and better adapted types of grain that yield starches of improved properties. Grain sorghums may become much more important as a source of starch than they are today.

Kernel Structure

The grain or kernel of a cereal is really a one-seeded fruit, with the fruit coat adherent to the seed, called a caryopsis. As the fruit ripens, the seed coat, or pericarp, becomes rather firmly attached to the wall of the seed proper and forms the outer tissue of the bran. The plantlet, or embryo, which is monocotyledonous and develops into a new plant upon germination, occupies only a small part of the seed. The bulk of the seed is taken up by the endosperm, which constitutes a food reservoir.

In the grass family, the floral envelopes (modified leaves), or chaffy parts, within which the caryopsis develops, persist to maturity. In rice and most varieties of oats and barley, chaffy structures envelop the caryopsis so closely that they remain attached to it when the grain is threshed; these structures constitute the hull of such grains (which are said to be covered). In the common wheats, rye, hull-less barleys, and the common varieties of corn, the caryopsis readily separates from the floral envelopes on threshing; these grains are said to be naked.

The structure of the wheat kernel is shown in Figure 1. The dorsal side of the wheat grain is rounded while the other side (the ventral side) has a deep groove or crease (cut along in the longitudinal section), which extends the entire length of the

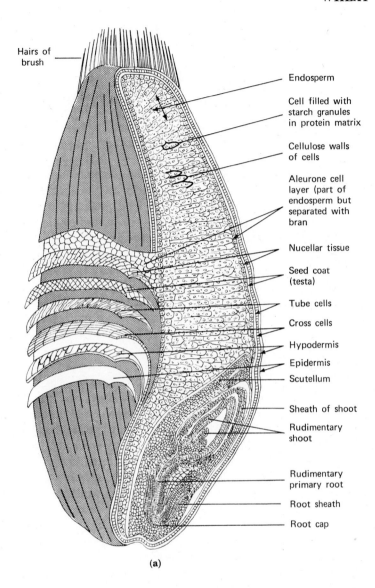

Hairs of brush

Endosperm

Cell filled with starch granules in protein matrix

Cellulose walls of cells

Aleurone cell layer (part of endosperm but separated with bran

Nucellar tissue

Seed coat (testa)

Tube cells

Cross cells

Hypodermis

Epidermis

Scutellum

Sheath of shoot

Rudimentary shoot

Rudimentary primary root

Root sheath

Root cap

(a)

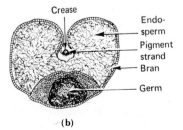

Crease

Endo-sperm

Pigment strand

Bran

Germ

(b)

Fig. 1. (a) Longitudinal and (b) cross section of wheat kernel (enlarged approx 35 times). (Courtesy, Wheat Flour Institute, Chicago, Ill.)

kernel. At the apex or small end (stigmatic end) of the grain is a cluster of short fine hairs known as brush hairs. The pericarp, or dry fruit coat, which corresponds roughly to the shell of a nut, consists of four outer bran layers (epidermis, hypodermis, cross cells, and tube cells). The remaining tissues, the inner bran (seed coat and nucellar tissue), endosperm, and embryo (germ) comprise the seed proper. The aleurone layer, sometimes erroneously called the gluten layer, consists of large rectangular, heavy-walled, starchfree cells; botanically it is the outer layer of the endosperm, but as it tends to remain attached to the outer coats during milling, it is shown in the diagram as the innermost bran layer.

The embryo (germ), or plantlet, consists of the plumule and radicle, which are connected by the hypocotyl; the latter is connected with the scutellum or cotyledon, which serves as a special organ for absorbing the food products from the endosperm and transferring them to the plantlet. The surface of the scutellum that adjoins the endosperm consists of columnar cells known as the epidermal layer or epithelial layer. These epithelial cells contain prominent nuclei and assume a glandular, or enzyme-secreting, function during germination of the seed. In a well-filled wheat kernel, the germ comprises about 2–3% of the kernel, the bran 13–17%, and the endosperm the remainder. The bran layers are high in protein, cellulose, hemicelluloses, and mineral constituents; biologically, they function as a protective coating and remain practically intact when the seed germinates. The germ is rich in proteins, lipids, sugar (chiefly sucrose), and ash constituents, while the endosperm consists largely of starch grains embedded in a matrix of protein.

Grains of other cereals are similar in structure to wheat. The corn grain is the largest of all cereals. The kernel is flattened, wedge-shaped, and broader at the apex than at the point of attachment to the cob. The aleurone cells contain largely protein and oil and also contain the pigments that make certain varieties blue, black, or purple. Beneath the aleurone layer of the corn endosperm, horny endosperm and floury endosperm are found, both containing a large percentage of starch, less protein, and practically no oil. In dent varieties, the horny endosperm is found on the sides and back of the kernel and bulges in toward the center at the sides while the floury endosperm fills the crown (upper part) of the kernel and extends downward to surround the germ. As dent corn matures, the soft floury endosperm at the top of the grain shrinks and causes kernel indentation. In dent corn, the hull comprises about 6%, the germ 11%, and the endosperm 83% of the kernel. In flint varieties of corn, the horny endosperm is proportionately much greater and only a small amount extends around the top of the kernel so that the floury portion of the endosperm is found around the embryo.

In the common varieties of oats, the fruit (or caryopsis) is enveloped in a hull comprising certain of the floral envelopes; naked or hull-less varieties are not extensively grown. The percentage hull in oats varies with the test weight per measured bushel. In light thin oats of low test weight, hulls may comprise as much as 45% of the grain, while in very heavy or plump oats, they may represent only 20%. While test weights of oats vary from about 20 to 50 lb per bushel, they average from 32 to 36 lb; the hull normally makes up about 30% of the grain. Kernels of oats, obtained by removing hulls, are called groats.

Classification and Standards

The United States Grain Standards Act was passed in 1916, so that grain might be marketed economically and advantageously under uniform standards. This Act provided for establishment of official grain standards for the common grains, and federal licensing and supervision of the work of grain inspectors (13,14). The standards have been revised and extended from time to time; those at present in force for all the cereal grains, except rice, are published in the *Official Grain Standards of the United States* (15). They provide market classifications for grains as to types and quality by division into classes and frequently subclasses, with a number of grades in each. Factors employed in establishing the standards for wheat serve to illustrate the principles involved.

Wheat is divided into the following seven classes, on the basis of type and color: hard red spring wheat, durum wheat, red durum wheat, hard red winter wheat, soft red winter wheat, white wheat, and mixed wheat. Several classes are further divided into subclasses, which differ in kernel density or geographical origin. For example, hard red spring wheat comprises three subclasses—dark northern spring, northern spring, and red spring—depending upon the percentage of vitreous kernels present. White wheat is also divided into subclasses: hard white, soft white, white club, and western white. Division of wheat into classes and subclasses effects a classification according to suitability of different wheats for specific purposes. Hard red spring, hard red winter, and hard white wheats differ in milling and baking properties and are valued for the production of flours used in making yeast-levened breads. Soft red winter, soft white, and white club wheat flours are especially suited for making chemically leavened baked products. Amber durum wheat is prized for the manufacture of semolina and flour used in producing alimentary pastes (macaroni, vermicelli, etc), while red durum is used principally as a poultry and livestock feed. Protein content of wheat determines suitability for these purposes. Kernel texture is useful in classification since the vitreousness or hardness of the kernels is generally associated with protein content.

The wheat in each subclass is sorted into a number of grades on the basis of quality and condition characteristics. In general, "quality" refers to the plumpness, soundness, and cleanliness of the grain and is measured by test weight per bushel, damaged kernels, and "dockage" or foreign material. "Condition" refers to the moisture content, odor, presence of smut (a fungus disease), garlic, and insects, and to the extent of weathering or artificial treatment. Quality requirements, summarized in Table 7, become less stringent as the grade number decreases. Since moisture content is related to stability in storage, as well as the quantity of dry matter, the standards specify a maximum moisture limit for all the numerical grades of wheats. When the moisture exceeds this limit but is below a certain higher maximum, the word "tough" is added to the grade. For the hard red spring, durum wheat, and mixed wheat classes, "tough" is applied when the moisture content falls between 14.5 and 16.0%; for the remaining classes the moisture limits are 14.0–15.5%. Special grades are provided for wheat that contains garlic, smutty grain, excess moisture, etc.

Other cereals are graded similarly. **Barley** is divided into three classes (barley, western barley, and mixed barley) and three subclasses (malting barley, blue malting barley, and barley) on the basis of production and color.

Corn is divided into the following three classes: yellow corn, white corn, and mixed corn. These class designations apply when the corn is of the dent type; the word "flint" is added when the corn consists of 95% or more of flint varieties. When the corn consists of more than 5% but less than 95% flint varieties, the designation "flint and dent" is added to the above class names. For corn (and also the grain

Table 7. Sample Grade, Numerical Grades, and Grade Requirements for All Classes of Wheat, Except Mixed Wheat (15)

Grade	Min test wt lb/bu		Max limits of							
			Defects						Grain of other classes[a]	
	Hard red spring wheat	All other classes	Heat-damaged kernels, %	Dam-aged kernels,[b] %	Foreign material, %	Shrunken and broken kernels, %	Total de-fects, %		Con-trasting classes, %	Total other classes,[b] %
1	58.0	60.0	0.1	2.0	0.5	3.0	3.0		0.5	3.0
2	57.0	58.0	0.2	4.0	1.0	5.0	5.0		1.0	5.0
3	55.0	56.0	0.5	7.0	2.0	8.0	8.0		2.0	10.0
4	53.0	54.0	1.0	10.0	3.0	12.0	12.0		10.0	10.0
5	50.0	51.0	3.0	15.0	5.0	20.0	20.0		10.0	10.0

[a] Red durum wheat of any grade may contain not more than 10.0% of wheat of other classes.
[b] Total.

NOTE: *Sample grade* shall be wheat which does not meet the requirements for any of the grades from No. 1 to No. 5, inclusive; or which contains stones; or which is musty, or sour, or heating; or which has any commercially objectionable foreign odor except of smut or garlic; or which contains a quantity of smut so great that any one or more of the grade requirements cannot be applied accurately; or which is otherwise of distinctly low quality.

sorghums), maximum moisture limits are specified. These cereals vary considerably in moisture content as they come to the market, and moisture determines their dry-matter values and their storage properties. In fact, moisture content is the principal factor determining the grade of market corn, especially at the beginning of harvest.

Oats are divided into the following classes on the basis of color: white oats (including yellow oats), red oats, gray oats, black oats, and mixed oats. Oats containing more than 14.5% but not more than 16% moisture have the word "tough" added to the grade designation. Special grades are provided for heavy, extra heavy, bright, thin, bleached, weevily, smutty, ergoty, and garlicky oats.

Rough rice is divided into three classes, long, medium, and short. There are standards for rough, brown, and milled rice. All rough rice is graded on the basis of milled-rice yields.

Rye is classified into five regular grades and a number of special grades, for example, plump, tough (14.0–16.0% moisture), smutty, garlicky, weevily, and ergoty.

Table 8. Requirements for Cereal Grains According to the Standards of the European Economic Community

	Soft wheat	Durum wheat	Rye	Barley	Maize	Oats	Canary-seed	Sorghum	Millet
Moisture, %	16		16	16	15	16	16	15	13
admixture[a] (Besatz), %	5	2	5	3	8	3	3	16	
test wt, hl/kg	75	78	71	67		49	70		

Courtesy, Commonwealth Secretariat, London.
[a] Damaged grains, grain of other cereals, etc, according to specification for each type of grain.

Grain sorghum is divided into the following four classes: yellow, white, brown, and mixed.

The three factors that determine standard quality of sound grain in the regulations of the European Economic Community are moisture content, Besatz (admixture of all parts of a sample that do not constitute faultless specimens of the basic grain), and test weight (expressed as kg/hl).   The quantitative requirements are listed in Table 8.

Storage Properties

Cereal grains have been important to civilization because of their excellent keeping qualities.   Problems in grain storage are discussed by Oxley (16) and reviewed comprehensively in the second monograph of the American Association of Cereal Chemists (17).   An annotated bibliography on grain storage is also available (18).   Large quantities of grain may be difficult to transport and store because of heating and other types of damage, such as loss of viability, development of acidity, loss of nutritive qualities and processing value, insect damage, molding, and rotting.   Moisture content is the major factor in determining the storage behavior of grain, which is also influenced by temperature, oxygen supply, inherent characteristics, history, and condition of the grain, length of storage, and biological factors (molds and insects).   Cereal grains, although stored in the dormant state, are living entities and continually respire and produce heat, water, and carbon dioxide.   Where hexose sugars are the food materials being utilized, the process may be summarized by the following equation:

$$C_6H_{12}O_6 + 6 O_2 \rightarrow 6 CO_2 + 6 H_2O + 677.2 \text{ kcal}$$

As the quantity of carbon dioxide respired is related to the heat production, measurement of carbon dioxide is a convenient laboratory method for investigating deterioration of stored grain; the small quantity of heat evolved per unit of time is difficult to measure accurately.

In dry grain, the respiratory rate is very low.   As the moisture content is raised, the respiration increases gradually until a certain critical moisture range is reached above which marked acceleration in the rate occurs and heating tendencies appear (19). This sharp increase in respiration is due to the germination and growth of certain molds (various species of *Aspergillus* and *Penicillium*) commonly found in soil and in previously used storage bins.   Molds are invariably present on the grain and within the seed coats, even though the grain is harvested under ideal conditions.   The molds that grow under storage conditions are usually acquired after harvest from trucks, railroad cars, and storage bins.   The most xerophytic molds begin to grow at a relative humidity of about 70%, and moisture contents of individual kinds of grain, which are in equilibrium with this humidity, have been found to be very close to the critical moistures.   If the grain is shriveled or immature, or has been frozen, weathered, or damaged during threshing, mold spores grow and germinate at somewhat lower humidity than on sound, mature seed because of the greater availability of the nutrients of the damaged seed.   Critical moisture content for various cereals lies around 14%.   Grain often can be stored at 14–15% moisture during the winter because growth and reproduction of molds are checked by low temperature.   If such grain is heated by the sun in farm bins or railroad cars, mold growth may become profuse within several days.

The rate of respiration at any subcritical moisture level (at constant temperature) is not only low but is also relatively constant from day to day.   However, at moistures and temperatures that provide an adequate water activity for mold growth, respiratory rate increases with time and closely parallels the increase in mold population. Adding chemicals effective in preventing mold growth on grain also lowers the marked acceleration in respiratory rate that normally occurs when the moisture content rises above the critical value.

Deterioration of grain and cereal products in storage is accompanied by increased acidity, including formation of free fatty acids, acid phosphates, and amino acids. During early stages of deterioration, fat acidity increases much faster than all other types of acidity combined. Consequently, fat acidity often is used as one of the best measures of grain damage. Using fat acidity as the sole criterion of damage can be misleading (20). Recent investigations have shown that breakdown of lipids by molds in flour stored in adverse conditions considerably impairs breadmaking potentialities (21). Detrimental effects on quality due to malt lipase are associated with a soapy taste. The rancid odor of badly dried green malt during kilning and mashing can be ascribed to lipoxidase.

Growing molds also produce toxic substances. The chemical nature and biological effects of some toxic factors are summarized in a book edited by Wogan (22). Bacteria are generally not considered primarily responsible for initiating spoilage, since moisture content must approach equilibrium with an atmosphere of about 95% relative humidity before they grow readily. Insects, when present, increase the tendency of grain to heat, since their respiratory rate is many times that of the grain. Once insects become established, the increase in temperature and moisture produced by respiration is favorable to their growth, so that damage to the grain proceeds rapidly.

In addition to moisture content, the temperature at which grain is stored has an important bearing on its propensity to heat. At low temperature, mold growth is inhibited and respiration of insects and of grain is decreased. Grain containing moisture in excess of safe limits has been successfully stored during winter months in the northern sections of the United States.

Heating of grain, chiefly attributable to microorganisms, usually commences in localized areas of high moisture content. Grain has a low specific heat and if heat is produced more rapidly than it is dissipated by the low rate of thermal conduction through the grain, by radiation, by conduction and by convection in the interseed air, and by evaporation of water, the temperature of the grain rises. The rise in temperature increases the rate of respiration so that a continually self-accelerating process occurs. Damage from bin-burning occurs if the temperature is allowed to exceed about 110°F.

In bulk storage of cereal grains, interseed air occupies about one-third of the space filled by the grain. The quantity of oxygen readily available for respiration is limited and tends to be depleted with accumulation of carbon dioxide, which decreases the respiratory rate. However, airtight storage of high-moisture grain to prevent spoilage is often not practical since anaerobic respiration occurs under these conditions with the production of alcohols, acids, and other substances, which ruin the grain for both seed and industrial purposes. The use of this method has increased, however, in recent years for the storage of corn as livestock feed.

Although cereal grains normally are considered safe for storage if the moisture content does not exceed about 14%, prolonged storage in bulk, where temperature differences occur, may result in spoilage. Temperature gradients across a bulk of grain, such as at the surfaces due to a change in the outside temperature, cause a transfer of moisture by convection from warmer to cooler areas. If the temperature gradients are sufficiently steep and prolonged, the humidity of the interseed air in the cooler portions may become sufficiently high to promote mold growth and heating.

Grain of a moisture content too high to store without serious danger of heating, is normally dried on farms and terminal elevators by forcing heated air through thin columns of the grain to lower the moisture to a safe level. Several commercial dryers consisting of drying and cooling sections have been devised for this purpose.

Chemical Composition

The chemical composition of the dry matter of different cereal grains, in common with other foods of plant origin, varies widely, as it is influenced by genetic, soil, and climatic factors. Variations are encountered in the relative amounts of proteins, lipids, carbohydrates, pigments, vitamins, and ash; mineral elements present also vary widely.

As a food group, cereals are characterized by relatively low protein and high carbohydrate contents; the carbohydrates consist essentially of starch (90% or more), dextrins, pentosans, and sugars.

The various components are not uniformly distributed in the different kernel structures. The hulls and bran are high in cellulose, pentosans, and ash; the germ is high in lipid content and rich in proteins, sugars, and ash constituents. The endosperm contains the starch and is lower in protein content than the germ and, in some cereals, than bran; it is also low in crude fat and ash constituents. As a group, cereals are low in nutritionally important calcium, and its concentration and that of other ash constituents is greatly reduced by the milling processes used to prepare refined foods. In these processes, hulls, germ, and bran, which are the structures rich in minerals and vitamins, are more or less completely removed. Oats are a partial exception since essentially only their hulls are removed.

All cereal grains contain vitamins of the B group but all are completely lacking in vitamin C (unless the grain is sprouted) and vitamin D. Yellow corn differs from white corn and the other cereal grains in containing carotenoid pigments (principally cryptoxanthin, with smaller quantities of carotenes), which are convertible in the body to vitamin A. Wheat also contains yellow pigments but they are almost entirely xanthophylls, which are not precursors of vitamin A. The oils of the embryos of cereal grains, particularly wheat, are rich sources of vitamin E. The relative distribution of different vitamins in various kernel structures is not uniform, although the endosperm invariably contains the least. See also Vitamins.

Protein contents of wheat and barley are important indexes of their quality for manufacture of various food products and are influenced by climate, weather, soil, and the variety of grain. The breadmaking potentialities of bread wheat are largely associated with the quantity and quality of its protein (23). The quantity of protein is influenced mainly by environmental factors, but its quality is heritable, and a high-quality variety produces good bread over a fairly wide range of protein percentages, whereas a low-quality variety produces relatively poor quality bread even when its protein content is high. The protein content of wheat depends on the relative amounts of carbohydrates and nitrogenous compounds made available to developing grains during the translocation of nutrients. The limiting factors in protein production appear to be the amount of available nitrogen in the soil, and environmental factors that determine yield. Total rainfall, seasonal distribution of rainfall, and temperature profoundly affect the amount of protein and, in some cases, the characteristics of the pro-

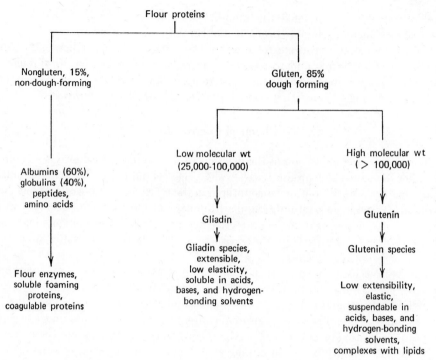

Fig. 2. Wheat flour proteins. Courtesy, J. Holme, Ogilvie Flour Mills, Montreal, Can.

teins produced. Differences in protein content among varieties of wheat grown under comparable conditions are small compared with differences due to environment.

Soils in areas where bread wheats are grown are generally high in nitrogen and produce high-protein grain. The amount and distribution of rainfall affect the protein content of wheat directly and indirectly. The indirect effect is from soils in areas of low rainfall, generally being much higher in organic matter, and therefore in nitrogen, than those of high rainfall. Rainfall affects the protein content of wheat directly because limited rainfall produces low yields of high-protein wheat. In wheat grown under humid conditions, a large part of the available nitrogen is used in the vegetative stage so that relatively little is available at the grain-forming stage. Relatively cool temperatures during kernel growth promote formation of carbohydrates, prolong maturation, and result in larger quantities of starch and, consequently, lower percentages of protein. Conversely, relatively high temperature and lower soil moisture after the crop has headed, tend to reduce carbohydrate synthesis and storage and to shorten the maturation period, thus resulting in lower grain yields of higher protein content. In general, higher yield per acre has been associated with lower protein content for different varieties grown under the same environment and for the same variety grown under different environments.

The cereal grains contain water-soluble proteins (albumins), salt-soluble proteins (globulins), alcohol-soluble proteins (prolamins or gliadins), and acid- and alkali-soluble proteins (glutelins). The prolamins are characteristic of the grass family and, together with the glutelins, comprise the bulk of the proteins of cereal grains. In corn, the prolamin called zein is the chief protein, whereas oats and rice are low in prolamins, and glutelins comprise the bulk of their proteins. The glutelin of oats is called avenin and the glutelin of rice, oryzenin. Wheat, rye, and barley are intermediate between corn, on the one hand, and oats and rice, on the other, in regard to the relative proportions of prolamins and glutelins.

The various proteins are not distributed uniformly in the kernel. Thus, the proteins fractionated from the inner endosperm of wheat consist chiefly of a prolamin (gliadin) and a glutelin (glutenin) apparently in approximately equal amounts. The embryo proteins consist of nucleoproteins, an albumin (leucosin), a globulin, and proteoses, while in wheat bran a prolamin predominates with smaller quantities of albumins and globulins. When water is added, the wheat endosperm proteins, gliadin and glutenin, form a tenacious colloidal complex, known as gluten (see Fig. 2). Gluten is responsible for the superiority of wheat over the other cereals for the manufacture of leavened products, since it makes possible the formation of a dough that retains the carbon dioxide produced by yeast or chemical leavening agents. The gluten proteins collectively contain about 17.55% nitrogen; hence, in estimating the crude protein content of wheat and wheat products from the determination of total nitrogen, the factor 5.7 is normally employed rather than the customary value of 6.25, which is based on the assumption that, on the average, proteins contain 16% nitrogen.

The adequacy of solubility and precipitation as bases for protein classification has been much discussed. Fractionation procedures based on differences in solubility are empirical. The fractions contain (sometimes overlapping) mixtures of proteins as shown by starch gel and polyacrylamide gel electrophoresis. Yet the fractionation procedures are used widely, as no more meaningful or accepted procedure has been developed.

As a class, cereal proteins are not so high in biological value as those of certain legumes, nuts, or animal products. Zein, the prolamin of corn, lacks lysine and is low in tryptophan. The limiting amino acid in wheat endosperm proteins is lysine. While biological values of the proteins of entire cereal grains are greater than those of the refined mill products, which consist chiefly of the endosperm, the American diets normally comprise various cereals, as well as animal products. Under those conditions, different proteins tend to supplement each other, and the cereals are important and valuable sources of amino acids for the synthesis of body proteins.

Technology

BARLEY

The major industrial use of barley is the production of malt. Smaller amounts are used to manufacture many other products. Agrotechnical aspects and barley processing are reviewed in a *U. S. Dept. of Agriculture Handbook* (24).

Milling. Most barley that goes directly into human food is consumed as *pot barley* or *pearl barley;* barley flour is a secondary product. Barley flour, as well as barley grits, is also made by a gradual reduction roller-milling process similar to that used in milling wheat to flour. Pot and pearl barley both are manufactured by gradually removing the hull and outer portion of the barley kernel by abrasive action; the pearling or decortication process is carried further in the manufacture of pearl barley. To produce the highest quality demanded by the consumer, pot and pearl barley must be made from barley free from discolored kernels and foreign matter. Coloring matter must be absent from the bran layers, and the endosperm must be as white as possible. Well-filled high-grade barley of uniform size is desired; uniformity in kernel size is important because the finished products must be produced in definite sizes. Two-rowed varieties, grown in the western United States, produce large plump kernels and are preferred for pot pearling.

One type of pearling machine consists of three to eight abrasive discs, coated with carborundum or emery, that are rapidly revolved within a perforated cylinder. The hull and outer portion of the kernel are gradually rasped off by rubbing against the discs and the perforated cylinder. From the first pearler, the mixture of offal and partially hulled barley is sent to a reel that removes the hulls. The barley is then as-pirated to remove fine particles and is subsequently transferred to a cooling bin where the heat developed by the attrition process is dissipated. That series of operations is repeated until a product of the desired size and purity is obtained. After the third pearling, the bran, together with most of the aleurone layer, is largely removed. At this stage, the product may be graded or classified and sold as pot barley. Grain sub-jected to five or six pearling operations gives pearl barley that is small, round, and white. After the last sequence of operations, the product is classified according to size by means of a grading reel. 100 lb of barley normally yields about 65 lb of pot barley or 35 lb of pearl barley. On the basis of the decreases in weight and the changes in chemical composition, it has been computed that six pearlings remove 74% of the protein, 85% of the fat, 97% of the fiber, and 88% of the mineral ingredients contained in the origi-nal barley (25).

Barley Flour and Grits. In the manufacture of pearl barley, some flour is produced as a by-product. The pearl barley may be ground and sifted to produce a granular prod-uct, barley grits, and/or barley flour. The highly refined flour is known as patent barley flour. Barley grits and a less highly refined barley flour may also be made by a process of roller milling, bolting, and purification, similar to that used in milling flour from wheat. In the larger mills, most hulls are removed before milling, a process accom-plished by passing the barley through a pearler or a hulling machine similar to an oat huller, and then sifting and aspirating.

Malting. For the biology, chemistry, and technology of barley and malting, see references 26–31. See also Malts and malting.

CORN

Agrotechnical aspects of corn production are discussed in a book edited by Sprague (32), the structure and reproduction of corn by Kiesselbach (33), and by Wolf et al. (34), wet milling by Goodwin (35), and dry milling by Larsen (36), Katz (37,37a), and Roberts (38). Component parts of mature dent corn kernels and their chemical composition are given in Table 9 (39). About three-fifths of the processed corn is used by the wet-milling industry to produce cornstarch and other carbohydrates, corn oil, and various feed by-products. The remainder is used to prepare various food products and alcoholic beverages. Corn is prepared in several ways as human food: (*1*) parched to be eaten whole; (*2*) ground to varying degrees of fineness to make hominy, corn meal, or corn flour; (*3*) treated with alkali to remove the pericarp and germ to make lye hominy; and (*4*) converted to a variety of breakfast foods by special processes. Yellow corn products are preferred in the Northern States; white corn products, in the South-ern States and Rhode Island; and blue, black, and red corn products in the Southwest where the Spanish influence dominates.

Dry milling of corn for human consumption is carried out by two general methods, which are designated as "old process" and "new process." "Old process" meal is also

Table 9. Component Parts of Mature Dent Corn Kernels and Their Chemical Compositions (39)

Part		Dry wt of whole kernel, %	Composition of kernel parts, %[a]				
			Starch	Fat	Protein	Ash	Sugar
germ	range	10.5–13.1	5.1–10.0	31.1–38.9	17.3–20.0	9.38–11.3	10.0–12.5
	mean	11.5	8.3	34.4	18.5	10.3	11.0
endosperm	range	80.3–83.5	83.9–88.9	0.7–1.1	6.7–11.1	0.22–0.46	0.47–0.82
	mean	82.3	86.6	0.86	8.6	0.31	0.61
tip-cap	range	0.8–1.1		3.7–3.9	9.1–10.7	1.4–2.0	
	mean	0.8	5.3[b]	3.8	9.7	1.7	1.5
pericarp[c]	range	4.4–6.2	3.5–10.4	0.7–1.2	2.9–3.9	0.29–1.0	0.19–0.52
	mean	5.3	7.3	0.98	3.5	0.67	0.34
whole kernels	range		67.8–74.0	3.9–5.8	8.1–11.5	1.27–1.52	1.61–2.22
	mean	100	72.4	4.7	9.6	1.43	1.94

Courtesy, *Cereal Chemistry*.
[a] Dry basis.
[b] Composite from nine different corn-belt hybrids.
[c] Also known as hull or bran.

known as "water-ground" because the mills making it were formerly operated by water-power.

Old-Process Corn Milling. In the old process, corn (preferably white dent) is ground to a coarse meal between millstones run slowly at a low temperature, with the meal, especially in many of the small mills, frequently not being bolted (sifted). In the larger mills, about 4–6% of the coarser particles of the hull are bolted out. Such meal is essentially a whole corn product and, owing to the presence of the germ, has a rich oily flavor but the meal deteriorates rapidly. The meal is softer and more flourlike than the more highly refined new-process meal, and is preferred by many in the Southern States. In many of the larger mills, corn is dried to 10–12% moisture before being ground; the kiln-drying permits more rapid grinding and improves the keeping qualities of the meal.

New-Process Corn Milling (40). The new-process corn milling is characteristic of the northern United States but is also used considerably today in the South. Steel rolls are used, as in the milling of wheat. The objective is to remove the bran and germ, and to recover the endosperm in the form of hominy or corn grits, coarse meal, fine meal, and corn flour. A schematic outline of dry corn processing is given in Figure 3. Corn grits and coarse meal consist largely of particles of flinty (sharp) endosperm, whereas fine meal and corn flour are derived chiefly from the starchy (soft) endosperm. Flint varieties of corn are considered too "sharp" for grinding to meal. Dent corn, both white and yellow, is almost invariably used. Since grits used in the manufacture of corn flakes are made mainly from white corn, large quantities are milled to make grits. Meal and flour are regarded as by-products. U.S. Grade No. 2 corn is the principal grade milled.

The corn is first thoroughly cleaned and is then usually passed through a scourer to remove the tip cap from the germ end of the kernel. The hilar layer under the tip is frequently black and causes black specks in the meal. In some areas, light-colored hilar layers are desired because the black specks are considered to be "dirt." In other areas the specks constitute the sign of a good corn meal. The corn is then tempered by two additions of water to a moisture content of 21–24%. Subsequently it is frequently passed through a corn degerminator, which frees the bran and germ, and breaks the endosperm

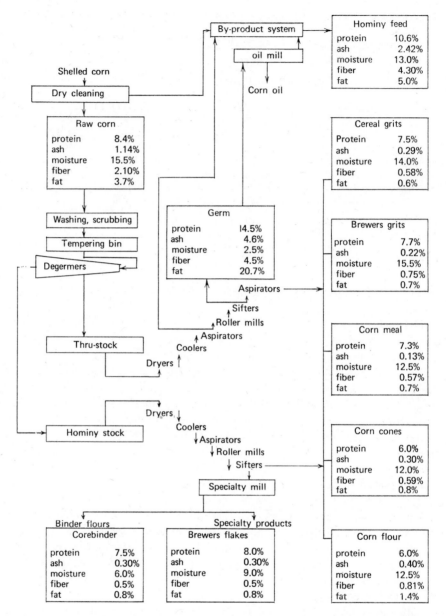

Fig. 3. Dry corn processing. Courtesy, A. B. Ward, Kansas State University, Manhattan, Kans.

to two or more pieces. Stock from the degerminator is dried to 14–16% moisture in revolving dryers equipped with steam coils, and then cooled in revolving or gravity-type coolers. The largest endosperm pieces are often reserved for making corn flakes. The stocks are next passed through a hominy separator, which consists of an intricate series of grading reels. It first separates the fine particles (feed) and then grades the larger fragments to four sizes and "polishes" them. The various grades of broken corn are passed through centrifugal-type aspirators to remove any loose bran from the endosperm fragments.

The various grades of aspirated corn fragments are reduced to coarse, medium, and fine grits by gradual reduction between corrugated rolls and subsequent sifting or bolting of the chop, as in milling wheat. The coarsest stock from the aspirators (unless made directly into commercial hominy) goes to the first break rolls; these are spaced farther apart, have coarser corrugations (5–8/in.), and operate at a lower speed differential than succeeding breaks. The coarsest grade of hominy is the one most highly contaminated with germ. The object is to free the germ and flatten it between the break rolls, with minimum endosperm grinding, so that the germ may be separated by bolting. The successive steps in the gradual reduction process for corn are analogous to those described for wheat. Modern corn mills are designed in such a way that grits of any particular size may be taken off, or they may be subjected to further grinding, sifting, and aspirating to produce meal and flour (a coarser product than wheat flour). The various grades of grits or meal are generally dried at 150°F for several minutes and cooled before packing. Tailings from the various grindings and aspirations enter the hominy feed, which is often reground in attrition or hammer mills.

The flattened germs, which are separated from the grits and hominy by the bolting process, are generally used to produce corn oil. The germ is dried to about 2–3% moisture, ground, tempered with steam, and passed through expellers (oil presses of the continuous or screw type). The germ cake from which most of the oil has been removed is frequently reground and may be extracted by a solvent before packaging.

Table 10. Yield and Particle-Size Range of Milled Corn Products (41)

Product	Particle size		Yield, %
	Mesh	In.	
grits	14–28	0.054–0.028	40
coarse meal	28–50	0.028–0.0145	20
fine meal	50–75	0.0145–0.0095	10
flour	through 75	below 0.0095	5
germ	3–30	0.292–0.0268	14
hominy feed			11

Courtesy, *Bulletin of the Association of Operative Millers.*

Hominy or grits intended for industrial uses, such as brewing and manufacture of wallpaper paste, are flaked. The grits are steamed and passed between heavy-duty heated iron rolls, and the resulting flakes dried (but not toasted). As a result, the starch is partially gelatinized.

Yields. Relative yields of various mill products depend on whether the main objective is to produce grits or meal and, in the latter instance, whether the corn was degermed before being subjected to grinding. Typical yields and particle sizes of milled corn products are given in Table 10 (41). In milling corn for grits and meal by the degerminating process, the following average yields are typical: grits, 52%; meal and flour, 8%; hominy feed, 35%; and crude corn oil, 1.0%. Shrinkage is about 4%, largely from differences between moisture contents of corn and its various mill products. When corn is not degermed before grinding, a yield of about 20% feed and 72% corn meal is produced with a shrinkage of about 7%. Of the total meal produced, approximately two-thirds is highly refined, containing about 1.4% fat, and one-third is standard meal, containing about 4.7% fat.

OATS

Agrotechnical aspects of oat production are discussed by Findlay (42) and in a book edited by Coffman (43); the histology and development of the oat plant by Bonnett (44), and the processing and technological uses by Kent (45) and by Carroll and Richards (46).

Oat products are milled in England to provide numerous foods, such as oatmeal for porridge and oatcake, rolled oats for porridge, and oat flour for baby foods, and to manufacture ready-to-eat breakfast cereals, and "white groats" to make "black puddings," a popular dish in the English Midlands (45). Human consumption of oats in England is substantially higher than in the United States. Oat milling is a major cereal industry for the production of breakfast cereals. Rolled oats and oatmeal are high in protein, fat, and energy value and are the richest sources of calcium, phosphorus, iron, and thiamine among cereal foods. Their high nutrient value is due to the fact that they are made from oat groats, which are obtained by removing the fibrous hull and adhering portions from the oat grain. Groats correspond to the caryopsis of wheat and corn; the bulk of the bran, the aleurone layer, and the germ, which are rich in proteins, vitamins, and minerals, remain with the portion used as human food. Hence, rolled oats and oatmeal, like brown rice, are essentially whole-grain products from the nutritive standpoint.

Oat hulls are an important by-product because on acid-steam digestion they yield furfural (qv) by conversion of the pentosans in the hulls. Furfural is used extensively in the manufacture of phenolic resins and as a solvent. A diagram of an oat milling process is given in Figure 4.

For milling purposes, plumpness, soundness, and freedom from heat damage, foreign odors, wild seed, smut, must, and molds are important, and only high-grade oats (No. 1 and No. 2) are employed. The milling machinery in each producing area is adapted to the efficient milling of the highest test weight oats that are regularly available. In the North Central States white and yellow oats are milled. Since red varieties are the class principally grown in the South Central States, red oats are milled in this area. The color of groat of red oats is similar to that of other classes.

The initial step in the oat-milling process involves thorough cleaning to remove foreign matter and light oats, after which the milling oats are dried, or slowly roasted, to reduce the moisture content to about 6%, thereby increasing the brittleness of the hulls and facilitating their later removal. The milling process proper comprises the following principal steps: grading or sizing, hulling, separating hulls and unhulled oats from the groats, steaming the groats, rolling the groats into flakes, and packaging. The parched oats from the storage bins are passed through graders, which sort them into five or six groups, according to length. Each stream is sent to separate hulling stones, the groats from the large oats yielding the choicest grade of rolled oats. The hullers are similar to the machines formerly employed in hulling rough rice and consist of two "stones"; the lower one is flat and stationary; the upper one is slightly conical and rotates rapidly. The space between the stones is carefully adjusted to less than the length of the grain, but slightly more than the length of the groat. As the oats are carried from the center to the periphery of the stones, they are turned endwise so that pressure exerted on the ends of the grain shatters the hull and releases the groat.

From the hulling stones, the mixture of unhulled oats, hulls, whole and broken groats, and flour, is first passed through a horizontal sifting reel to remove oat flour and then through a succession of air separators to take out the hulls. The remaining mix-

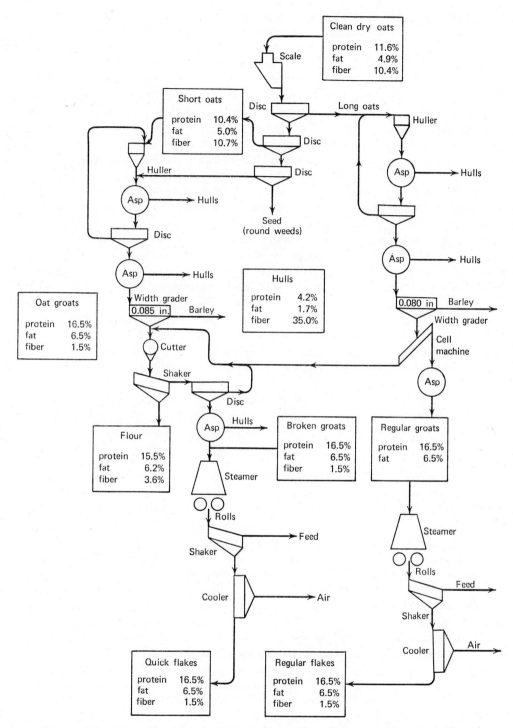

Fig. 4. Oat milling. Courtesy, A. B. Ward, Kansas State University, Manhattan, Kans.

ture of unhulled oats and groats is graded to take out broken groats, and then is passed through a series of apron machines. These machines are provided with pockets of the correct size to receive the groats and carry them over the machine while the unhulled oats slide off the apron at the bottom. The groats thus obtained are further processed in several ways depending on the product desired.

In the manufacture of the standard type of slower-cooking choice rolled oats, whole groats are steamed directly with live steam at atmospheric pressure. This treatment sterilizes and partially cooks the groat; it also increases the moisture content, which is essential to avoid excessive production of fine particles or flour in the subsequent rolling process. If, on the other hand, the so-called quick-cooking rolled oats are desired, the groats are steel-cut, by means of rotary-type cutters, to pieces of varying size, often classified as A (fine), B (medium), and C (coarse) steel-cut oats. Before rolled oats became prominent, steel-cut or Scotch oats were the chief form consumed as human food. The B or medium steel-cut oats are the size normally used to produce quick-cooking rolled oats; each fragment represents about one-third of a groat and makes one flake. After cutting, the groats are heat-treated in the same manner as the whole groats used to produce large flakes of standard rolled oats; the broken groats employed in the manufacture of bulk-rolled oats also are steamed in the same manner.

The final step in the manufacturing process is flaking, which is accomplished by feeding steamed groats between two large steel rolls. In producing quick-cooking rolled oats, rolls are adjusted to produce a thinner flake than that of the standard slower-cooking type. As heat treatments of groats for the two types of rolled oats are similar, the quicker cooking properties are attained by decreasing the size and thickness of the flake.

Medium-quality Grade No. 2 white oats yield about 42% good-quality rolled oats, 30% hulls, and 28% other products (including oat shorts, oat middlings, cereal grains, weed seeds, and other material removed in the cleaning process).

RYE

Manufacture of rye flour is described by Schulerud (47) and by Schmalz (48). Rye is the main bread grain of several European countries. The total annual consumption of rye in Germany is about 3.5 million tons compared with only 0.6 million tons in the United States. In addition, whereas about 40% of the rye consumed is used for human food in Germany, the percentage is only about 20% in the United States. Rye is sometimes attacked by a fungus, ergot. Other cereal grains are also attacked but comparatively rarely. Ergot is a poisonous contaminant and should be removed almost completely prior to milling. Ergot grain that contains above 0.3% attacked kernels presents also problems in wheat and rye cleaning and in feed where ergot tends to concentrate. It is impossible to wash gluten from doughs made entirely of rye flour and hence rye flour is inferior to wheat flour in baking properties. In the United States most so-called rye bread is baked from mixtures of rye and wheat flour (usually a first clear grade). Most of the major mills market rye blends made from a mixture of strong spring or hard winter wheat and rye. They roughly follow the pattern of 80% clear and 20% dark rye, 70% clear and 30% medium rye, and 50% clear and 50% white rye (48). In the manufacture of breads made largely from rye flour, a proper level of α-amylase activity is the major factor in determining the quality of the bread. When there is insufficient α-amylase, the bread is characterized by a dry brittle crumb and the crust becomes cracked and torn on cooling; whereas excess

α-amylase produces bread with a wet, soggy crumb, which frequently pulls away from the crust and leaves large hollow spaces in the bread. Level of α-amylase activity is normally assessed indirectly by determining hot-paste viscosity, which is generally determined by continuously recording changes in viscosity of a flour suspension as it is heated at a uniform rate of $1.5°C/min$ for $30°C$ to about $95°C$.

Rye is milled into flour by a process similar to that described for wheat; but, because the bran adheres tenaciously to the endosperm, it is not practical to make clean "middlings" or to purify them by aspiration. Moreover, rye is a "tough" grain. The middlings are not so friable as those of wheat and, if ground between smooth rolls, tend to flake or flatten rather than pulverize. Accordingly, the reduction rolls are finely corrugated (40–50 corrugations per in.). Since little advantage is to be gained by producing middlings, the objective in milling is to produce flour during the breaking process. The break rolls are set relatively close and have somewhat finer corrugations than those used for corresponding breaks in milling wheat. Rye grain requires either little or no tempering (modifying the physical condition of the kernel prior to milling by adding water, followed by a rest period to improve milling and separation of bran from the endosperm). The particle size of white rye flour is 24–35 μ, as compared with 50–70 in bread wheat flour, and 45–60 in dark rye flour. White rye flour is normally bleached with benzoyl peroxide or with chlorine to remove color.

The highest grade of flour is produced by the first break rolls. As the purity of the flour decreases, it becomes darker and has a more pronounced rye flavor. In American milling three main grades of rye flour are produced: white or light, medium, and dark. The white (light) rye flour represents 50–65% of the grain. About 80% of the rye flour used in breadmaking in the United States is of the white type. Medium rye flour corresponds to straight-grade wheat flour while dark rye flour corresponds to the "clears," ie the dark flour represents that portion of the total flour that remains after the white or patent rye flour is removed. In addition to those three main grades, "cut" and "stuffed" straights are sometimes produced; the former is a medium rye flour from which a small percentage of white rye flour has been removed, whereas a stuffed straight is a medium rye flour to which a small percentage of dark has been added. For baking pumpernickel rye bread, rye millers produce coarse unbolted rye meal known as pumpernickel flour. The milling of rye normally yields 65% light or patent rye flour, 15–20% dark rye flour, and 15–20% offal.

RICE

Agrotechnical aspects of rice production are discussed by Grist (49) and in a handbook of the U.S. Dept. of Agriculture (49a). Harrel and Dirks (50), Kester (51), Deobald (52), and Kent (45) describe processing of rice; and Hampel (53) outlines uses of rice in Europe.

Rice is a covered cereal; in the threshed grain (or rough rice), the kernel is enclosed in a tough siliceous hull, which renders it unsuitable for human consumption. When this hull is removed, the kernel (or caryopsis), comprising the pericarp (outer bran) and the seed proper (inner bran, endosperm, and germ), is known as brown rice or sometimes as unpolished rice. Brown rice is in little demand as a food. Unless stored under very favorable conditions, it tends to become rancid and is more subject to insect infestation than the various forms of white rice. When brown rice is subjected to further milling processes, the bran, aleurone layer, and germ are removed and the purified endosperms are marketed as white rice or polished rice, which is classified according to size as head rice (whole endosperm), and various classes of broken rice, known as second head, screenings, and brewers' rice, in order of decreasing size.

The U.S. production of rice is only about 1% of the world total; the annual per capita consumption in the United States is about 8 lb of rice, compared with 200–500 lb in some Asiatic countries. About 75% of the domestic rice production is in the South, the rest in California. The more popular long- and medium-grain rice varieties are grown in the South, mostly short-grain varieties in the West. Very limited amounts of a short-grain, glutinous (waxy) rice are grown in California. The starch in the latter is amylopectin. The starch of long-grain rice may contain up to 25% amylose; of short-grain rice 14% or less. The vitreous long-grain rices are used for cooking and processing (canned soups). The soft short-grain rices are used to make puffed rice. Both types are manufactured into dry breakfast cereals and are parboiled. Short-grain rice is more difficult to cook, but is preferred by the brewing industry. The cooking and processing behavior of rice improve with storage.

Milling (54). Excellent descriptions of rice milling and processing equipment have been published by the Food and Agriculture Organization of the United Nations (55). The objective of rice milling is to remove the hull, bran, and germ with minimum breakage of the endosperm. The rough rice, or paddy, is first thoroughly cleaned by passing it through several mechanical devices and then conveying it to shelling machines that loosen the hulls. Conventional shellers consist of two steel plates, 4 or 5 ft in diameter, mounted horizontally; the inner surfaces are coated with a mixture of cement and coarse carborundum. One plate is stationary and the other, set at the proper distance to permit rice grains to assume a somewhat vertical position, is rotated. As the plate revolves, the pressure on the ends of the upturned grains disengages the hulls. The hulls are removed by strong aspiration and the remaining hulled and un-hulled grains are separated in an ingenious device known as a paddy machine. It consists of a large box shaker fitted with vertical, smooth steel plates set on a slight incline to form zigzag ducts. The plates and shaking action cause the less dense, unhulled rice, or paddy grains, to move upward while the heavier, hulled grains move downward. The unhulled grain is conveyed to auxiliary shellers, which are set to smaller clearances than those used in the first shelling operation. Rough rice may also be shelled with rubber rolls or with a rubber belt operating against a ribbed steel roll. The process causes less mechanical damage, and thus improves stability against rancidity. Hulled or brown rice is sent to machines, erroneously called hullers, which consist essentially of grooved, tapering cylinders that revolve rapidly in stationary uniformly perforated cylinders composed of upper and lower halves with abutting edges. In operation, the entire machine is filled with grain and the packing force is regulated by a blade that protrudes between the upper and lower halves of the perforated cylinder. The outside bran layers and the germ are removed by the scouring action of the rice grains moving against themselves near the surface of the perforated cylinder, and re-tarded by the blade. After passing through a succession of hullers, the rice, some of which is broken, is practically free from germ and outer bran. Scouring is usually com-pleted by polishing in a brush machine, which consists of a large, vertical wooden cyl-inder (dressed with overlapping soft leather strips) which revolves at high speed within an outer stationary cylinder of wire screen. As the rice passes downward, it is polished between the leather strips and the screen wire. The polished rice contains whole endosperms and broken particles of various sizes; these are separated into head rice (whole endosperm), second head rice (one-half and three-quarter grains), screenings rice (one-quarter to one-half grains), and brewers' rice (less than one-quarter grain size) by grading reels or disc separators.

The yield of white rice normally varies between 66 and 70%, based on the weight of the rough rice. As head rice is the most valuable product, its yield determines the milling quality of rough rice. The price obtained for the various classes of broken rice decreases with size.

A solvent-extraction process was developed to increase the yield of whole grain rice (56). Dehulled brown rice is softened with rice oil, to improve bran removal. Fully milled rice is sometimes treated with a talc and glucose solution to improve its appearance. After the coating is evenly distributed on the kernels and dried with warm air, the rice emerges from the equipment with a smooth glistening luster and is known as coated rice.

Chemical Composition. Production of brown rice from rough rice increases protein, fat, and starch contents since the hulls are low in these constituents; conversely, there is a decrease in the crude fiber and ash contents. Conversion of brown rice to white (or polished rice) removes about 10% of the protein, 85% of the fat, and 70% of the minerals. Rough and brown rice differ little in vitamin content but conversion of the latter to white rice decreases the vitamin values markedly. Thus, head rice contains only about 20% as much thiamine, 45% as much riboflavin, and 35% as much nicotinic acid (niacin) as brown rice. These losses have created much interest in the development of practical methods for retaining more of the B vitamins in the milled rice. Studies in this field have been summarized by Kik and Williams (57).

The problem of improving the vitamin content of processed rice has been approached in two ways: (*1*) by removing less of the bran layers and germ in milling, ie by the production of so-called undermilled (or unpolished) rice; and (*2*) by processing the rice before milling to diffuse the vitamins and other water-soluble nutrients in the outer portions of the grain into the endosperm. The first method is rather impractical, since it is difficult to control the degree of undermilling and the product is less stable in storage than polished rice. Processing the rough rice to increase vitamin retention involves parboiling or some modification thereof. For parboiling, rough rice is soaked in water, drained, steamed, and finally dried. In 1940, a process for the manufacture of so-called converted rice was developed and patented in England (45). The cleaned rough rice is exposed to a vacuum, treated with hot water under pressure, and then steamed, after which the grain is dried and milled. The converted rice process is particularly effective for the retention of vitamins, as shown in Table 11, for twenty-five commercially milled samples (57).

The modern trend in processed foods is toward convenience items. Quick-cooking rices, that may be prepared for serving in 2–15 min, are available. They are manufactured by precooking in water and drying under special and closely controlled conditions or by application of dry heat (51). Other convenience items include canned and frozen cooked rice.

Table 11. Vitamin Retention in Converted Rice (57), γ/g rice

	Thiamine	Riboflavin	Niacin
rough rice	3.25	0.61	58.70
converted rice	3.00	0.43	45.50
vitamins retained, %	92.20	70.80	77.60

SORGHUM

Agronomic characteristics and cultural production practices of sorghum are reviewed by Ross and Webster (58), by Quinby et al. (59), and by Kramer (7). Information on chemical composition and processing is reviewed by Martin and Mac-Masters (60), by Watson (61), by Werler (62), and by Rooney and Clark (10).

Component parts of mature sorghum (milo) kernels and their chemical compositions are given in Table 12, the proximate analysis in Table 13. Sorghum is similar in composition to corn. However, it contains slightly more starch and protein, but less

Table 12. Component Parts of Mature Milo (Sorghum) Kernels and Their Chemical Composition

Component	Germ	Endosperm	Bran	Whole kernel
parts dry wt[a]				
range	7.8–21.1	80.0–84.6	7.3–9.3	
mean	9.8	82.3	7.9	
starch				
range		81.3–83.0		72.3–75.1
mean	13.4[b]	82.5	34.6[b]	73.8
composition of kernel parts, %[c]				
protein				
range	18.0–19.2	11.2–13.0	5.2–7.6	11.5–13.2
mean	18.9	12.3	6.7	12.3
fat				
range	26.0–30.6	0.4–0.8	3.7–6.0	3.2–3.9[c]
mean	28.1	0.6	4.9	3.6
ash				
range		0.30–0.44		1.57–1.68
mean	10.36[d]	0.37	2.02[c]	1.65

Courtesy, *Cereal Chemistry.*
[a] Percent of whole kernel.
[b] Composite.
[c] Dry basis.
[d] Includes five varieties.
[e] Includes wax.   Content of whole kernel (dry basis): range, 0.29–0.44%; mean, 0.32%.

Table 13. Proximate Analysis of Sorghum Grain (61,64)

Constituent	Range,[a] %	Average,[a] %
water,[b] %	8–20	15.5
starch	60–77	74.1
protein (N × 6.25)	6.6–16	11.2
fat, CCl₄ extract[c]	1.4–6.1	3.7
ash	1.2–7.1	1.5
fiber (crude)	0.4–13.4	2.6
pentoglycans	1.8–4.9	2.5
sugars (as dextrose)	0.5–2.5	1.8
tannin	0.003–0.17	0.1
wax	0.2–0.5	0.3

Courtesy, *Cereal Science Today.*
[a] Dry basis.
[b] Wet basis.
[c] Includes wax.

lipids than the latter. Corn contains yellow pigments, of which most sorghum grains contain only small quantities. But yellow-endosperm sorghums are being developed by plant breeders. Sorghum grain is slightly higher than corn in niacin and ribo-flavin contents; otherwise the vitamin content of the two grains is similar. Sorghum is unique among cereal grains in that the coating contains relatively large amounts (0.2–0.5% of kernel weight) of a wax with properties much like those of carnauba wax.

Dry Milling (62). The conventional roller-milling process can be employed for manufacture of whole or refined sorghum products. Sorghums are harder to grind than wheat, barley, or oats, and slightly easier than corn. During cracking, grits and a break flour are obtained. The grits are used for feeding purposes and are comparable in composition to the whole grain; part of the fibrous materials can be removed by aspiration. The break flour (10–15% yield) contains little protein (4.0–4.5%, as com-pared to 8–9% in the whole grain). Roller-milled flours are ground and sieved to yield products varying in extraction and composition. A product obtained in 70% yield contains 0.5% ash and 0.8% fiber, and is reasonably free from objectionable specks. For the production of high-extraction flours (eg 90%) impact grinding is preferable, as it requires less space and equipment than a comparable capacity roller-mill system. Quality products from grain sorghums require loosening and separation of the hull and germ from the rest of the kernel prior to milling. Because the sorghum kernel is round, the bran can be removed mechanically by abrasion with a barley pearler. Without tempering, a 75–80% yield of almost completely dehulled berries (except for the portion left on the germ) can be obtained. Proper tempering or other treatments to assist in loosening the hull can increase the yield to 85%. The germ can be removed by passing the dehulled kernel through cracking rolls or impact machines, or by corn-milling techniques. Germ separation can be accomplished by flattening and sifting, and air classification or gravity separation. The final product is milled to a white flour with low ash and fiber contents, and a minimum of objectionable specks. The yield is 58–65%.

Almost all dry-milled sorghum grain products are for industrial uses many of which overlap those of dry-milled corn products. Properly dehulled, degermed, and ground grits are used by the fermentation industry.

Wet Milling. The sorghum wet-milling process is similar to that of corn, although finer mesh screens are necessary for efficient operation (61). Chemically and micro-scopically, regular sorghum starch is very similar to corn starch, which it also resembles in gelatinization characteristics. They can be used interchangeably in most food and industrial applications. Milo starch is more bland in flavor and does not develop rancidity, so it is more suitable for use in edible fillings (65). Other products of wet milling are oil and gluten feeds.

WHEAT

The agrotechnical aspects of wheat production and improvement are covered in a monograph of the American Society of Agronomy (11). Various aspects of wheat chemistry, technology, and utilization are reviewed in a monograph of the American Association of Cereal Chemists (66), and in two excellent books published in Germany (67,68). Wheat is discussed in the classical text on cereal chemistry by Kent-Jones and Amos (69), in the book on the chemistry and technology of cereals as food and feed edited by Matz (1), and in the book on technology of cereals by Kent (45). Flour-milling technology is the subject of several books (70–72) and so are the history and chemistry (73) and technology of breadmaking (74–75). The chemistry of wheat starch and gluten and their conversion products (77), and of enzymes and their role in wheat technology (78) were each the subject of a text. Conditioning of wheat for milling (79), and reviews on methods for determining the quality of wheat and flour for breadmaking (80), on the relation between chemical composition and breadmaking potentialities of wheat flour (23), and on quality control in processing cereals (81) were presented in comprehensive bulletins or as chapters in books. A series of articles dealt with the structure of the mature wheat kernel (82). Nearly entire issues of *Baker's Digest* (82a), *Cereal Science Today* (82b), and *Proceedings of the Nutrition Society* (82c), deal with the chemistry, processing, utilization, and nutrition of wheat and wheat products. In addition to several scientific, technical, and popular journals dealing largely with wheat, abstracts on industrial utilization of wheat are published periodically (83).

Approximately two-thirds of the annual U.S. harvest of wheat is processed for food, the principal use of wheat. The limited use for industrial purposes is due mainly to its high price in relation to other grains. The main use of wheat for food is the manufacture of flour for making bread, biscuits, and pastry products. A small portion is converted into breakfast cereals and raw materials for the macaroni industry. Large quantities of flour are not sold in the form in which they come from the mill, but are utilized for blended and prepared flours.

Industrial uses of wheat include the manufacture of malt, potable spirits, starch, gluten, pastes, and core binders. Because of its relatively high price, wheat malt has not been used to any extent in the brewing and distilling industries; it is used chiefly by the flour-milling industry to increase the amylase activity of high-grade flours. Small quantities of wheat flour (chiefly low-grade) are used to manufacture wheat starch and gluten; the latter is a raw material for the manufacture of monosodium glutamate, which is used to accentuate the natural flavors of foods. Low-grade flours are used to some extent in the manufacture of pastes for bookbinding and paper hanging, in the manufacture of plywood adhesives, and in iron foundries as a core binder in the preparation of molds for castings.

Wheat and Flour Quality. In wheat and flour technology, the term quality is purely relative and denotes the suitability of the material for some particular purpose; it has no reference to nutritive values. Thus, hard wheat flour is of high breadmaking quality but is inferior to soft wheat flours for chemically leavened products, such as biscuits, cakes, and pastry.

The miller, naturally, desires a wheat that mills easily and gives a high flour yield. Wheat kernels should be plump and uniformly large to permit ready separation of foreign material without loss of millable wheat. The wheat should produce a high yield of flour with maximum and clean separation from the bran and germ without undue consumption of power. Since the endosperm is denser than the bran and the germ, the higher-density wheat (for a given moisture content) yields more flour. The test weight (weight of a bushel of clean wheat) is affected by the kernel shape, the moisture content, wetting and subsequent drying, and even the handling, because these characteristics and operations affect the grain packing.

Thus, under certain conditions, the relationship between test weight and flour yield may lose significance. Above 57 lb per bu, the test weight has relatively little influence on flour-milling yield. At lower test weights, the milling yield usually falls off rather rapidly with decreasing test weight. Average test weight per bushel of U.S. wheat is about 60 lb and may be as high as 65 lb.   Badly shriveled kernels have test weights of 45 lb or less.   Certain varieties and also wheats weathered by exposure to intermittent rains during harvest give flour yields that are out of line with their test weights. Weathering lowers the test weight per bushel by swelling kernels but the proportion of endosperm remains the same.   Environmental and various other factors also influence the ease of milling; bran of weathered and frosted wheats tends to pulverize so it is difficult to secure clean separation of flour from bran.

The endosperm proteins of wheat possess the unique and distinctive property of forming gluten when wetted and mixed with water.   Gluten imparts physical properties that differ from those of doughs made from any other cereal grain.   It is gluten formation, rather than any distinctive nutritive property, that gives wheat its prominence in the diet.   When water is added to wheat flour and mixed, the water-insoluble proteins hydrate and form gluten, a complex coherent mass, in which starch, added yeast, and other dough components are embedded.   Thus, the gluten is, in reality, the skeleton or framework of wheat-flour dough and is responsible for gas retention, which makes production of lightly leavened products possible. A schematic outline of wheat-flour proteins is given in Figure 2.   The percentage of dry gluten is closely related to the percentage of protein.   Before the Kjeldahl method for determining nitrogen came into general use, gluten determinations were widely used as an index of the value of the flour for baking purposes.   Many European cereal chemists continue to determine gluten to estimate protein contents.   Gluten determinations offer several advantages over the conventional Kjeldahl-protein analysis.   The physical properties of the cohesive gluten ball can be tested by an experienced operator.   Large differences in protein quality of various varieties or advanced stages of deterioration, which cannot be detected by the Kjeldahl test, are brought out simply by washing out a gluten ball.   However, the gluten test is used little in the United States because: (1) it is not precise; (2) gluten can be washed out easily from flour but not from wheat; and (3) the test is not suited for large-scale routine determinations.

The importance of gluten in breadmaking is demonstrated by comparing loaves baked from an untreated flour and from gluten–starch mixtures varying in protein content.   Loaf volume increases with increasing protein content.   In addition to gluten proteins, lipids are essential for maximum performance.   Bread baked from defatted flour or gluten–starch mixtures is consistently lower in volume than bread baked at comparable protein content from flour that has not been defatted (84).

The value of a wheat flour for a particular baking purpose depends upon the quantity and quality of the protein present and the rheological characteristics of the gluten that is formed upon making a dough.   On the basis of their suitability for the manufacture of yeast-leavened bread, common, or vulgare, wheats and the flours milled therefrom are classified broadly into two groups: hard or strong, and soft or weak. Hard (strong) wheat flours contain a relatively high percentage of proteins that form a tenacious, elastic gluten of good gas-retaining properties and are capable of being baked into well-risen shapely loaves possessing good crumb grain and texture.   They require considerable water to make a dough of proper consistency to give a high yield of bread. The doughs have excellent handling qualities, and are not critical in their mixing and fermentation requirements; for this reason, they yield good bread over a wide range of baking conditions and have good fermentation tolerance.   In contrast, soft (weak) flours have a relatively low protein content and form a soft, weak, relatively nonelastic gluten of poor gas-retaining properties.   They have relatively low water-absorbing

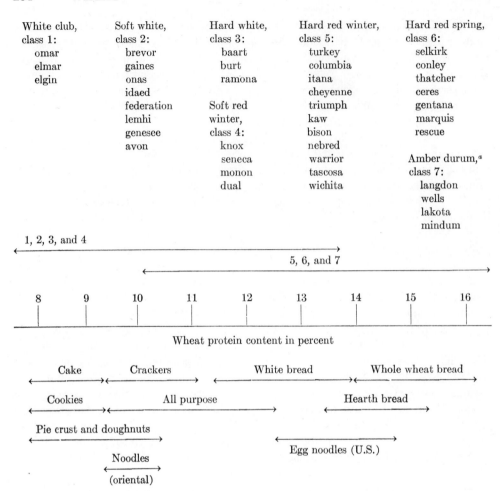

White club,   Soft white,   Hard white,   Hard red winter,   Hard red spring,
class 1:   class 2:   class 3:   class 5:   class 6:
   omar   brevor   baart   turkey   selkirk
   elmar   gaines   burt   columbia   conley
   elgin   onas   ramona   itana   thatcher
                      idaed   cheyenne   ceres
                      federation   Soft red   triumph   gentana
                      lemhi   winter,   kaw   marquis
                      genesee   class 4:   bison   rescue
                      avon   knox   nebred
                                            seneca   warrior   Amber durum,ᵃ
                                            monon   tascosa   class 7:
                                            dual   wichita   langdon
                                                                                       wells
                                                                                       lakota
                                                                                       mindum

Fig. 5. Range of protein contents of varieties of wheat grown in the United States and uses of the protein levels of market classes (85). LEGEND: 1, 2, and 4, pastry; 3, 5, and 6, bread; **7, macaroni.**

capacity, yield doughs of inferior handling quality (which give trouble in machine baking), and are very critical in their mixing and fermentation requirements so that they are more likely to fail in baking. Weak flours require less mixing and fermentation than strong flours to give optimum baking results. The protein test mentioned above, while not included as a grading factor in the U.S. standards for wheat, is accepted as a marketing factor. Customarily, data on protein content are made available to buyers of wheat and flour. The relation between protein contents and potential use of wheat is illustrated in Figure 5 (85).

To determine the quality of the protein and to evaluate breadmading potentialities (strength) and performance characteristics under mechanized conditions, physical dough testing devices are used, which evaluate and predict plant breeder samples, provide quality control during milling and breadmaking, and assist in basic rheological studies.

American bakery practice and quality demands of the U.S. consumer dictate the use of strong relatively high-protein flours for breadmaking. Of the commercial classes, the hard red spring and hard red winter wheats are excellent for yeast-leavened bread.

Soft wheats are better for chemically leavened products. There is no fermentation to "mellow" or "ripen" the gluten, and the low protein content and the soft, mellow characteristics of the gluten formed by soft or weak flours produce a lighter, more tender product than do strong flours. Varying degrees of strength or weakness are required by the bakery trade. Thus, the general strength of bread flours sold for the family trade is greater for commercial hearth bread than for commercial pan bread. Home baking involves rather mild treatment (hand or very slow-speed mixing and gentle fermentation), so that good results are obtained with a flour of lower protein content and more easily conditioned gluten than would be satisfactory for commercial bakeries. For the manufacture of pan bread, a medium-strong flour is required to withstand high-speed mixing and produce a dough possessing the physical characteristics that permit machine manipulation. Bread baked on the hearth of the oven without pans requires a flour of still higher protein content to yield a strong dough that does not flatten unduly under its own weight. Other types of bread flours must be supplied by the flour miller to suit different markets.

With soft wheat flours, the protein content and gluten quality desired also vary widely. For example, in soda cracker manufacture, flours—depending upon whether they are used in the sponge (see Vol. 3, p. 46) or as doughing flours—varying in protein content from about 9 to 10% are required; for cake making, very weak flours, with protein contents ranging from about 7 to 9% and yielding batters that have a pH of 5.1–5.3, give the best results. Baking quality depends upon so many factors, which vary from flour to flour, that no single test has yet been found to be a completely reliable index of the behavior of a flour under any particular set of baking conditions. This has made it necessary to devise standard experimental baking tests for bread, biscuits, cakes, pies, etc. The results of such tests must be interpreted in terms of the conditions under which the flour is to be used, and this requires no little skill and experience.

Quality factors other than those related to protein content and gluten quality must be considered by the flour miller; for example, with bread flours, he must ensure that there is adequate amylase activity for optimum starch modification. The flour miller maintains the level of amylase activity by wheat selection and by the use of small quantities of malted wheat flour or malted barley flour.

The miller must also control the yellow pigment content of the flour, since the majority of the American bread-eating public demands bread with a relatively white crumb. This consumer requirement is met by the use of very small quantities of bleaching agents, which convert the carotenoid pigments of the flour to colorless compounds. Enrichment of flour with certain vitamins and minerals has introduced an additional set of variables that the miller must control. Scientific and technological advances have required the miller to exercise greater scientific control to produce flour that meets the rigid and highly specialized requirements of the baking trades.

Roller Milling Process for Producing Bread Flours. Milling grain as food for man has been traced back more than 8000 years. Flour milling has progressed from a laborious household task to a vast industry essential to modern civilization. In the manufacture of white flour, the aim is to separate the endosperm of the grain from the bran and germ, followed by pulverization. A partial mechanical separation of these

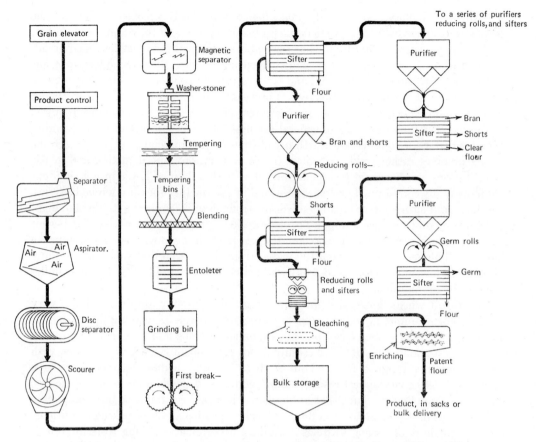

Fig. 6. Milling process of flour.   Courtesy, Wheat Flour Institute, Chicago, Ill.

closely adhering particles is possible because of differences in their physical properties. Bran is tough because of its high fiber content, while the starchy endosperm is friable. The germ, because of its high oil content, readily flakes when passed between smooth rolls.   Because these particles also differ in density it is possible to utilize air currents to separate them.   The differences in the friability of the bran and endosperm are accentuated by a process known as tempering or conditioning, which consists of adding water several hours before the wheat goes to the rolls in order to toughen the bran and to mellow the endosperm.   The milling process proper comprises a gradual reduction in particle size, first between corrugated or break rolls and, later in the process, between smooth or reduction rolls.   The separation is not quantitative, however.   Some of the endosperm is lost in the offals and some bran and germ are present in certain flour streams.   These variations in purity give rise to different grades of flour.

Steps involved in wheat-flour production are wheat selection and blending, cleaning, conditioning or tempering, breaking, bolting or sieving, purification, reduction, and bleaching.   Added, when necessary, are enzymes for "diastating" (conversion of starch into sugar), and vitamins and minerals for enrichment.   An outline of wheat milling is shown in Figure 6.

Wheat Selection and Blending.   Since the miller must produce a flour of definite characteristics for a particular market and maintain uniform quality, selection of wheats and binning according to quality are essential phases of modern milling. An adequate supply of wheat, sorted according to quality characteristics, permits the miller to blend or "build a mill mix," which provides uniform flour over long periods.

Cleaning. Commercial wheat, as received at the mills, contains such impurities as stinking smut (a fungus disease), weed seeds, other cereal grains, and soil. Many special machines have been developed for removing the various types of impurities. Preliminary cleaning involves the use of sieves, air blasts, and disc separators. This is followed by dry scouring in which the wheat is forced against a perforated iron casing by beaters fixed to a rapidly revolving drum. This severe treatment removes foreign materials lodged in the creases of the kernels, as well as in the brush hairs. Many American mills are equipped with wheat washers in which the wheat is scrubbed under a flowing stream of water. The washed wheat is passed through a "whizzer" (centrifuge), which removes the bulk of the water.

Conditioning or Tempering. This process immediately follows the final cleaning. When wheat is washed, part of the water required for tempering is added in the washers. In tempering, a suitable addition of water is made and the wheat allowed to stand long enough to secure maximum toughening of the bran with optimum mellowing of the endosperm. Both the quantity of water added and the tempering time must be varied with different wheats to bring the grain to the optimum condition for milling. The quantity of water required to be added increases with decreasing moisture content of the wheat, with increasing vitreousness, and with increasing plumpness. Normally, hard wheat after tempering contains 15–16% moisture. Heat is frequently applied to accelerate the tempering.

In the customary tempering method the wheat is frequently scoured again, after it has been held in the tempering bins for several hours. A second small addition of water is made to raise the moisture content an additional 0.5% about 20 min to 1 hr before the wheat goes to the rolls. A number of new developments have taken place in the 1960s (86), many still in the experimental stage. Ultrasonic treatment of wheat is reported to improve milling and baking results. With mills increasing their capacity and using the same tempering bins, it has been necessary to have shorter tempering times. With pretempering in the grain elevator, and the use of steam for regular tempering, some mills have reduced the tempering time to $3\frac{1}{2}$ hr. Regular cold tempering is still the most common method.

Claims have been made that chlorine dioxide added to wheat during tempering aids in the separation of the endosperm and in the production of sterile flour.

Breaking. The first part of the grinding process is carried out on corrugated rolls (break rolls), usually 24–30 in. long and 9 in. in diameter and turning in opposite directions at a differential in speed of about 2.5:1; each stand has two pairs of rolls. In the first break rolls, there are usually ten to twelve corrugations per in. This number increases throughout the break system, to as high as twenty-six to twenty-eight corrugations per in. on the fifth break roll. The corrugations run the length of the roll with a spiral cut, which is augmented with an increase in the number of corrugations. As the rolls turn rapidly toward each other, the edges of the corrugations of the fast roll cut across those of the slow roll, producing a shearing as well as a crushing action on the wheat, which falls in a rapid stream between them. The first break rolls are spaced so that the wheat is lightly crushed and only a small quantity of fine material or flour is produced. After bolting (sieving), as described below, the coarsest material is conveyed to the second break rolls; these are set a little closer together than the first break rolls, so that the material is crushed a little finer and fragments of endosperm are released.   This grinding and sifting process is repeated until the wheat passes through four to six breaks.   The material going to each succeeding break contains less and less endosperm. After the fifth or sixth break, the largest fragments consist of flakes of the outer covering of the wheat; they are passed through a special machine (called a bran duster), which removes a small quantity of low-grade flour, and are then packed as bran.

Bolting or Sieving.   After each grinding on the break rolls, the crushed material, called stock or chop, is conveyed to a sifter or bolter, which is a large box fitted with a series of sloping sieves.   The break sifters have a relatively coarse wire sieve at the top and progressively finer silk sieves arranged below, ending with a fine flour-silk at the bottom.   The sifter is given a gyratory motion so that the finer particles of stock pass through the sieves from the top (head) to the bottom (tail).   Particles too coarse to pass through any particular sieve tail over it and are removed at once from the sifter box. Three classes of material are separated: (*1*) coarse fragments, which are fed to the next succeeding break until only bran remains; (*2*) fine particles, or flour (first break flour, second break flour, etc), which pass through the finest or flour sieve; and (*3*) intermediate granular particles which are called middlings.   The latter are quite different from the cattle feed known as middlings.

Purification.   The middlings consist of fragments of endosperm, small pieces of bran, and the released embryos.   Several sizes are separated from each of the break chops; individual streams of similar size and degree of refinement resulting from the bolting of different break chops are combined.   The next step is to remove, as far as possible, the branny material from the graded middlings.   This is accomplished in machines known as purifiers which, in addition, produce a further classification of the middlings according to size, and hence complete the work of the sifters.   In the purifier, the shallow stream of middlings is made to travel over a large sieve, shaken rapidly backward and forward.   The sieve consists of tightly stretched bolting silk, or grits gauze, progressively coarser from the head to the tail end of the machine.   An upward air current through the sieve draws off very light material to dust collectors and tends to hold branny particles on the surface of the moving middlings so that they drift over the tail of the sieve.   Electrostatic separators are used to a limited extent to purify middlings.   In some of the modern mills, purifiers have been eliminated.

Reduction.   The purified and classified middlings are gradually ground to flour between smooth rolls, called reduction rolls, which revolve at a differential of about 1.5:1.   The space between the rolls is carefully adjusted to suit the granulation of the middlings.   The endosperm fragments passing through the rolls are subjected to a crushing and rubbing action which reduces them to finer middlings and flour; any remaining fibrous fragments of bran tend to be flaked or flattened.   After each reduction, the resulting chop is bolted through its own sifter, as in the case of the break stock. Most of the bran fragments are removed on the top sieve while flour passes through the finest or bottom sieve.   The remaining middlings are classified according to size by the intermediate sieves, are moved to their respective purifiers, and then pass to other reduction rolls.   These steps are repeated until most of the endosperm has been converted to flour and most of the bran chips have been removed as offal by the reduction sifters. All that remains is a mixture of very fine middlings and bran with a little germ, that is called feed middlings.   Impact mills have been used for some time in reduction grinding, especially with soft wheats.   Very close grinding using clean middlings on reduction rolls, followed by some type of flake disrupter in the form of a pin mill, detacher, etc. increases the amount of flour made on a particular reduction step. This process has been adapted more for soft than for hard wheats.

The embryos are largely released by the break system and appear as lemon-yellow particles in certain of the coarser middlings streams. These streams are known as sizings. In reduction of the sizings, the embryos are flattened and separated as flakes upon bolting the chop. All the germ thus obtained was formerly mixed with the shorts; subsequently, on account of its richness in certain vitamins, special uses were found for it. Germ may also be separated without reduction of the sizings by gravity and regular air currents.

Flour Grades. Each grinding and bolting operation results in a stream of flour. In addition to the various breaks (first, second, third, etc) and middlings flours (first, second, third, etc), a small quantity of flour originates from special units, such as dust collectors and bran and shorts dusters. Each flour stream has its individual characteristics and properties, but the first few middlings separations are the most highly refined. With each successive reduction, the flour contains more and more pulverized bran and germ. The flour from the last reduction, commonly called "red dog," is dark in color and high in components characteristic of bran and germ, such as protein, ash, fiber, pentosans, lipids, sucrose, and vitamins. Such flour bakes into inferior, dark-colored, coarse-grained bread, and is generally sold as feed flour. For information on millfeed products, see reference 86a.

In a large mill, there may be thirty or more streams of varying composition and purity, which must be collected and merchandised as flour. Various flour streams obviously may be grouped together in a number of ways. If all streams are combined, the resulting flour is called straight flour. But a straight flour of 100% does not mean a whole wheat flour, but 72% of the kernel weight (mainly endosperm), because milled wheat yields 72% flour and 28% feed products. Frequently the more highly refined streams—ca 60% of the straight flour—are taken off and sold separately as patent flours, while the remaining streams, which contain more bran and germ, are known as clear flours. The percentage of the total flour merchandised as patent flour varies widely; the more streams of successively decreasing refinement are included in the patent flour, the lower the quality of the remaining (clear) flours. A diagram of flours and mill products is given in Figure 7.

The clear flours are too dark in color and too low in baking quality to make acceptable white bread. Some of the better clear flours are used in mixtures with rye or whole wheat flour; the lower grades are used in the manufacture of gluten, starch, monosodium glutamate, and pet foods. The blended flours are always rebolted through a centrifugal reel or high-speed sifter before packing.

Yield of Mill Products. The plump wheat grain consists of approx 84% endosperm, 14% bran, and 2% germ, but these three structures are not completely separated by the milling process. Normally, the yield of total flour varies from about 70 to 74% instead of the theoretical 84%, and the flour obtained contains some pulverized bran and germ. In ordinary milling processes only about 0.2% of germ is recovered as such. Bran recovered in a relatively pure condition and sold as bran normally varies from about 12 to 16% of the wheat milled. Remaining by-products are approximately equal in quantity to the bran and, when bulked together (with at times the omission of a small percentage of feed flour), constitute shorts. The low-grade flour and feed middlings may be sold separately, giving rise to a series of other feed by-products. The more efficient the milling process, the greater the yield of flour and other products for human consumption.   The lower the moisture content and the higher the test weight of the wheat, the greater, in general, is the potential flour yield.   The relation between test weight and flour yield is not linear.   It seems that the decrease of each pound in test weight results in a decrease of about 0.6–0.8% in flour yield.   Wheats of equal test weights may, however, vary as much as 2% in flour yield, depending upon the size and density of the kernel as influenced by the variety and by environmental conditions.

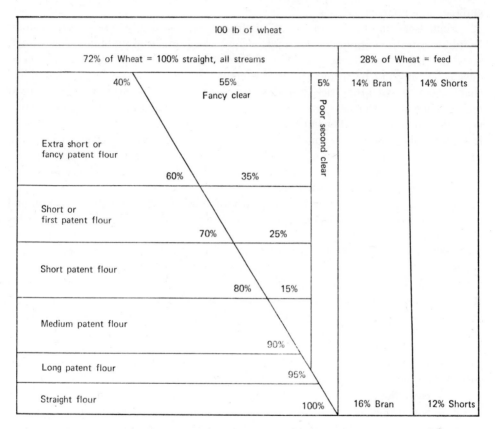

Fig. 7. Grades of flour (85).   Courtesy Burgess Publishing Co., Minneapolis, Minn.

Flour Fractionation by Fine Grinding and Air Classification.   Wheat flour produced by conventional roller milling contains particles of different sizes (from 1 to 150 μ), such as large endosperm chunks, small particles of free protein, free starch granules, and also small chunks of protein still attached to starch granules.   The regular flour can be ground relatively fine to free the high protein material from the starch granules.   The reground flour is passed through an air-classifier.   A fine fraction, made up of particles about 40 μ and smaller, is removed and passed through another air-classifier.   Particles about 20 μ and smaller are separated there, called the "fines" or fine fraction.   This fraction has a high protein content (15–22%) and the weight is from 15 to 5%, respectively, of the original flour.   This high-protein flour is used for fortifying and blending with other flours.

Air classification is a new process, which has created considerable interest in the milling industry. It is relatively inexpensive, and its advantages are numerous, such as manufacture of more uniform flours from varying wheats; increase of protein content of bread flours and decrease of protein in cake and cookie flours; controlled particle size and chemical composition; and production of special flours for specific uses.

A number of equipment and process patents on fine grinding and separation have been issued (86). The technology of the process is well known, yet its benefits and potential have not been fully utilized. Results of air-classification milling for wheat and other cereal flours have also been reported (88,89).

Soft Wheat Milling. Soft wheats are milled by the method of gradual reduction described above for the hard or bread wheats. As the flours are used for a variety of purposes, each with special quality requirements, the soft-wheat miller carefully selects his wheat and mills to produce the type of flour desired for particular trades. Thus, patent flours containing from 7 to 9% protein milled from soft red winter wheat are especially suitable for making chemically leavened biscuits and hot breads so widely consumed in the South. Special mixtures of soft wheats, including white wheats, are converted into well-known brands of cake flours for use in pastry and cake making. Such flours usually contain 8% protein or less and are milled to very short patents (commonly about 30%) having low ash content. Treatment with rather heavy dosages of chlorine to bring the pH to about 5.1 to 5.3 weakens the gluten and facilitates the production of short pastry. Cake flours are bolted through silk of finer mesh than that used for biscuit or bread flours, the last having the coarsest granulation.

Manufacture of low-protein flours by air classification is also practiced. Specifications for cookie, pastry, cake, cracker, biscuit, and cracker flours are summarized by Larsen (36). Cookie and cracker technology is discussed by Matz (90). Soft wheat flour products were reviewed by Gilles and Shuey (91) and Finney and Yamazaki (92).

Durum Wheat Milling. Semolina (middlings from durum) is the preferred raw material for the manufacture of alimentary pastes, such as macaroni, spaghetti, and noodles. In durum milling, the object is the production of a maximum yield of highly purified semolina. Although the same sequence of operations (ie cleaning, tempering, breaking, sizing, purification, and reduction) is employed in the production of semolina and flour, the milling systems differ materially in design. In milling flour, differences in the friability of the endosperm, bran, and germ on the smooth reduction rolls aid in separating these products. In semolina manufacture, however, impurities and the mill offals must be removed by the cleaning and purifying systems. Durum wheat milling involves thorough cleaning and proper conditioning of the grain, light and careful grinding, and thorough purification. The cleaning, breaking, sizing, and purifying systems are, therefore, much more elaborate and extensive than in flour mills. On the other hand, the reduction system is much shorter in durum mills because the primary product is removed and finished in the granular condition and the reduction rolls are required only for the reduction of the tailings from the various parts of the middlings system.

To secure the maximum yield of large endosperm particles, break rolls with U-cut corrugations are employed. The break system is very extensive, to permit lighter and more gradual grinding than in flour mills. Durum wheat of good milling quality normally yields about 62% semolina, 16% clear flour, and 22% feeds. Highest yields of best semolina are produced from clean, large, vitreous, and uniform kernels. The wheat should be high in protein, and have medium strong gluten, high concentration of carotenoid pigments, and low lipoxidase activity. Particle-size distribution and granulation of semolina are highly important in the manufacture of macaroni. Processing and milling specifications of durum wheat are given by Larsen (36).

Flour Bleaching and Maturing. The use of flour bleaching and maturing agents is well established. Bleaching of flour was introduced as early as 1879 in Britain, and around 1900 in America (93). In the earliest days flour was treated with nitrogen peroxide. Subsequently other methods came into use that not only made the flour whiter, but simultaneously induced changes in the flour quality to improve the baking characteristics. In fact, the treated flour possessed baking properties similar to those of flour that had been stored, ie naturally aged.

Today, practically all bread and cake flours in the United States are bleached. In addition, maturing agents are used to obtain maximum baking performance. Flour improvers are used in Great Britain, Canada, and many other countries. In France and West Germany only ascorbic acid may be used legally as a flour improver. In still other countries, for example, Italy no flour improver is allowed (94).

In the early days of the federal food laws, treatment of flour with nitrogen "peroxide" (dioxide) was questioned. It was ruled that all flour undergoing treatment must be labeled "bleached." However, the significance of the term "bleached" has been extended to cover the products obtained by treatments by other methods. All methods now in use make flour whiter to a greater or lesser degree, and some methods also increase the baking quality.

Carotenoids. Many varieties of wheat are used in the manufacture of flour. These varieties differ in color of the flour they produce, but such flours are always pale yellow or cream-colored. This color is due to a group of pigments known as carotenoids present in unbleached flour in small amounts, about 1–3 ppm. In the earlier days of flour bleaching, it was thought that the carotenoid pigment carotene was present in flour. Carotene is a precursor of vitamin A; hence if carotene were destroyed by bleaching, it might be significant from a nutritional standpoint. Later investigations of the wheat pigments have shown only traces of carotene. Zechmeister and Escue (95) obtained from American whole wheat flour 1.5–2.0 mg/kg of total pigments, which on chromatographic fractionation produced only 0.01–0.04 mg/kg of carotene and 1.0–1.4 mg/kg of xanthophylls, which are not precursors of Vitamin A and have no nutritional value. Munsey (96) investigated this subject and concluded that xanthophyll is the main pigment in wheat flour and that carotene is present only in mere traces, if at all. This indicates that the bleaching of flour, so far as color removal is concerned, is due to conversion of xanthophyll into a colorless compound. However, the nutritional aspects of flour improvers cannot be completely disregarded (94). For example, chlorine dioxide can destroy tocopherol, vitamin E, accelerating the oxidation of unsaturated fatty acids in flour lipids. The bleaching of flour destroys 50–90% of its vitamin E content; a similar decrease is observed during storage of flour for six months. Oxidation of essential polyunsaturated fatty acids by bleaching agents is insignificant. A very large amount of the flour manufactured is enriched with added B vitamins. Permitted bleaching agents have little destructive effect on these vitamins, even though the flour is bleached after enrichment. Even in continuous breadmaking, the high levels of oxidants used have little effect on the yeast activity and the panary fermentation.

Aging. When unbleached white flour is stored, reduction in its color occurs through the oxidation of xanthophyll. During the storage of unbleached flour, other important changes take place which improve the baking quality to a rather limited extent. Freshly milled flour, frequently referred to as "green" flour, usually possesses poor baking quality. The dough from such flour is sticky, has inadequate oven-spring, and produces baked products of inferior texture, volume, and crumb quality. As natural aging progresses, the flour matures and becomes better suited for making acceptable products and the dough is easier to handle. These storage changes continue slowly until the baking quality of the stored flour reaches a peak value. Further storage causes a deterioration in desirable baking qualities. It is impossible to set a time when this peak condition may be expected. This depends on storage conditions, such as temperature, humidity, and ready access of air to the flour. When proper chemical improvers are carefully applied, the treated flour is ready for immediate use, and the quality of the oven products is comparable and often superior to those obtained from untreated flour after adequate storage. In other words, the storage of flour, the cost of which is high, becomes unnecessary. During storage periods, market fluctuations may take place; furthermore, insect infestation may occur. Chemical treatment obviates both these disadvantages.

For years cereal chemists have attempted to explain the rather dramatic effects of trace amounts of chemical improvers (oxidants) on breadmaking. Wheat-flour components assumed to participate, directly or indirectly, in such mechanisms include water-soluble pentosans, lipids, enzymes, and proteins (97). The most prevalent theory is that the effects involve the participation and possibly modification of protein sulfhydryl groups.

Agents which have maturing action but little or no bleaching action include bromates, iodates, peroxysulfates, peroxyborates, calcium peroxide, and ascorbic acid (which is enzymically converted to dehydroascorbic acid, an oxidizing agent). Agents which have both bleaching and maturing effects include oxygen, ozone, chlorine, chlorine dioxide, and nitrogen trichloride. However, nitrogen trichloride was removed by the Food and Drug Administration from the list of permissible improvers, because dogs and certain other animals fed large amounts of flour and bread containing this compound showed harmful effects.

Two new improvers, azodicarbonamide and acetone peroxide (93,98), have been approved by the Food and Drug Administration for inclusion with the Standards of Identity for flour as bleaching and maturing agents. Acetone peroxide (see Peroxides) performs a dual function of bleaching and maturing. Azodicarbonamide, $H_2NCON=NCONH_2$, is reduced to hydrazodicarbonamide (biurea), $H_2NCONHNHCONH_2$. It has a maturing action, only. Benzoyl peroxide is added primarily as a bleaching agent. Additional agents, used less commonly for bleaching, include nitrogen peroxide, fatty acid peroxides, hydrogen peroxides, and certain preparations (for example from untreated soyflour) containing the enzyme lipoxidase.

Flour Bleaching Methods. Flour standards were authorized by the Food, Drug and Cosmetic Act (1948) and are administered by the Food and Drug Administration. Improving agents cannot be used if their addition conceals damage or inferiority or makes the flour appear better or of greater value than it is. With these limitations, the flour standards permit the use of one or any combination of optional ingredients, and they may not be added "in a quantity more than sufficient for bleaching, or, in case such an ingredient has an artificial aging effect, in a quantity not more than sufficient for bleaching and such artificial aging effect." From the practical standpoint, intentional overtreatment is unlikely as it damages the handling properties of the dough and impairs the quality of the baked product. The tolerance of flour to over-treatment varies with different improvers. Flour can tolerate overtreatment with ascorbic acid much better than with other improvers. It is practically impossible to overtreat with ascorbic acid, but a small excess of potassium iodate may be detrimental.

Various other additives have been recommended, approved, and used. They include yeast foods (sometimes mixed with maturing oxidants), and numerous dough conditioners and improvers such as calcium stearyl-2-lactylate and sodium stearyl-fumarate. Little is known about their improving action.

Quantitative requirement for oxidation by flours depends on several factors. Generally, as the protein content increases, the requirement for oxidants increases. Correlation between protein contents and oxidation requirement is, however, low because of the variations in mixing time within a wheat variety and the interrelation of mixing time and oxidation level (99). Mixing time and oxidation levels compensate each other to some extent, though they are not completely interchangeable. With inadequate oxidants, overmixing improves the bread. With optimum or near optimum oxidant, either undermixing or overmixing reduces loaf volumes and bread crumb texture, the decrease being proportional to the departure from the optimum mixing time. At a fixed protein content of about 12.5% oxidation requirement decreases materially with mixing time and reaches a minimum at about 5 min mixing time, beyond which the oxidation requirement is approximately constant. Similarly, as the degree of milling refinement or flour grade is lowered, oxidation requirements increase, because protein sulfhydryl groups susceptible to oxidation are found in highest concentration in the aleurone layer and germ. Low-grade flours have more of these tissues than patent flours. In addition, the required oxidation level depends on growth conditions, on the wheat variety (or mixture) from which the flour was milled, the milling systems used in flour manufacture, and the breadmaking process.

In general, the level of potassium bromate, iodate, acetone peroxide, or azodicarbonamide added at the mill varies from 5 to 20 ppm of flour; ascorbic acid is added at about 75 ppm. Chlorine dioxide (see Vol. 5, p. 35) containing about 20% free chlorine is used at the rate of about 15 ppm (94). For the continuous breadmaking process, the amount of improver used ranges from 50 to 75 ppm—up to five times higher than that used in conventional breadmaking. The difference in oxidant is supplemented by the baker. Generally, a combination of oxidants, ie of bromate and iodate (about 4:1) is used; the combination provides for a complementary action of a slow and a rapid agent. Oxidants vary in their oxidation potential and oxidation rate. Atmospheric oxygen is a slow oxidant; chlorine dioxide functions rapidly. Iodates, acetone peroxide, and azodicarbonamide act much more rapidly than bromates. Oxidation by bromate is materially enhanced by raising the hydrogen ion concentration (93).

Flour improvers have increased in importance and need with the introduction of continuous breadmaking processes. In all such processes, a chemical improver (or a combination of maturing improvers) is essential and is used at a much higher level than in conventional breadmaking.

Chemical Composition of Mill Streams. In progressing from the head to the tail end of the mill, the flour streams contain progressively more of the components characteristic of the bran and germ. These structures are richer in protein than the endosperm; also the protein content of the endosperm steadily decreases in progressing from the outer to the central zone. The ash content of the bran is from twenty to twenty-five times, and that of the germ from ten to fifteen times, that of the endosperm.

The ash content of a flour is, therefore, useful to the miller as a measure of the "purity" of the various flour streams. As the purity decreases, the ash content rises much more sharply than the protein content, since the protein content of the bran and germ exceeds that of the endosperm only by about 1.4 and 2.0 times, respectively. Table 14 shows the composition of various parts of the wheat kernel as compared with the composition of flours varying in extraction rate. Gross chemical composition of wheat and mill streams, based on analyses of flours and mill feeds, from nine wheat mixes are summarized in Table 15.

Table 14. Weight, Ash, Protein, Lipid, and Crude Fiber Contents of the Main Parts of the Wheat Kernel and of Flours Varying in Milling Extraction Rate (Yield of Total Kernel) (100)

Parameter	Wheat kernel fractions				Milling extraction		
	Pericarp	Aleurone layer	Starchy endosperm	Germ			
weight, %	9	8	80	3	75	85	100
ash, %	3	16	0.5	5	0.5	1	1.5
protein, %	5	18	10	26	11	12	12
lipid, %	1	9	1	10	1	1.5	2
crude fiber, %	21	7	>0.5	3	>0.5	0.5	2

Courtesy, *Baker's Digest.*

Table 15. Gross Chemical Composition[a] of Wheat and Mill Streams (100)

	Yield, %	Ash, %	Protein (N × 5.7),[b] %	Crude fat, %	Crude fiber, %	Starch, %
wheat	100	1.2–1.7	9.2–13.8	1.1–1.9	1.7–2.6	54.1–61.8
flour	72.6–77.0	0.35–0.42	8.6–13.0	0.8–1.0		64.3–73.7
germ	0.6–1.1	3.5–4.3	21.7–24.5	6.3–10.6	2.8–4.0	14.0–23.9
red dog	1.4–4.7	1.5–2.7	12.7–15.2	2.3–4.7	1.2–3.2	37.0–47.8
shorts	6.6–8.9	3.1–4.3	13.8–16.5	3.6–6.3	5.6–7.2	15.9–21.7
bran	12.5–16.9	4.7–7.1	12.1–15.4	3.0–4.2	9.2–11.6	4.6–7.2

Courtesy, *Baker's Digest.*

[a] Based on a 14%-moisture content.

[b] This multiplier is used for wheat.

The amino acid composition, vitamin content, and mineral composition of wheat, flour, bread, and certain mill streams are compared in Tables 16–18.

Supplementation of Cereal Products with Nutrients. Investigations of the diets of the American people and others, before World War II, established that serious and widespread deficiencies existed in all age groups in intake of thiamine, riboflavin, nicotinic acid, and iron. A deficiency of calcium was found in certain population groups, while vitamin D deficiency was quite prevalent among infants and children. Millers, bakers, and many nutrition authorities in the United States realized that a rapid improvement in general nutrition could not be achieved by increasing the consumption of long-extraction or whole wheat flours, which contain aleurone, bran, and germ, because of the strong preference of the public for white flour and white bread. Less than 5% of the bread in the United States is made with whole wheat flour. It was recognized that the controlled addition of commercially prepared vitamins and minerals to white flour and other types of refined food products, which had long met with

Table 16a. Amino Acids of Wheat, Flour, and Bread, g per 16 g nitrogen (100 g protein) (101)

Amino acid	Wheat	Flour	Bread
alanine	3.25	2.78	2.93
arginine	4.69	3.80	3.56
aspartic acid	5.09	4.14	4.60
cystine	1.97	2.11	1.88
glutamic acid	28.5	34.5	31.7
glycine	3.88	3.22	3.21
histidine	1.92	1.88	1.89
isoleucine	3.90	4.26	4.32
leucine	6.48	6.98	7.11
lysine	2.74	2.08	2.48
methionine	1.76	1.73	1.90
phenylalanine	4.42	4.92	4.80
proline	9.85	11.7	11.1
serine	5.06	5.44	5.45
threonine	3.02	2.82	3.01
tryptophan	1.09	1.02	0.97
tyrosine	3.10	3.25	3.32
valine	4.50	4.54	4.68

Table 16b. Amino Acids of Certain Mill Products, g per 16 g nitrogen (100 g protein) (101)

Amino acid	Low grade	Red dog	Bran	Shorts	Germ
alanine	3.10	4.70	4.65	4.74	5.23
arginine	4.68	6.84	6.60	6.85	6.88
aspartic acid	4.50	6.76	6.64	6.95	7.48
cystine	1.67	1.40	1.52	1.36	1.04
glutamic acid	29.6	17.9	16.2	16.6	14.0
glycine	3.70	4.98	5.12	5.33	5.22
histidine	2.14	2.22	2.22	2.20	2.26
isoleucine	3.72	3.42	3.29	3.31	3.48
leucine	6.33	5.77	5.51	5.64	5.75
lysine	2.54	4.13	3.77	4.18	5.28
methionine	1.67	1.70	1.48	1.62	1.91
phenylalanine	4.61	3.55	3.58	3.44	3.38
proline	10.16	6.30	6.11	6.03	5.03
serine	5.12	4.85	4.58	4.69	4.60
threonine	2.76	3.11	2.86	3.03	3.42
tryptophan	1.01	1.25	1.58	1.29	0.98
tyrosine	3.20	2.85	2.82	2.82	2.85
valine	4.45	4.91	4.00	4.84	4.90

Courtesy, *Baker's Digest.*

public acceptance, would assure the best distribution of thiamine, riboflavin, niacin, and iron.

The four following approaches are possible in dealing with the improvement of the nutritional value of cereals (and other products) (102): (*1*) *restoration,* ie addition of one or more nutrients to a processed food to restore them to a preprocessed (or natural) level; (*2*) *enrichment,* ie addition of specific amounts of selected nutrients to a processed food in accordance with official regulations; (3) *fortification,* ie addition of nutrients to foods and food products at levels that may exceed natural levels; and (*4*) *supplementation,* ie addition of nutrients.

Table 17a. Vitamin Content of Wheat, Flour, and Bread, mg/100 g dry weight (101)

Vitamin	Wheat	Flour	Bread
thiamine	0.40	0.104	0.46
riboflavin	0.16	0.035	0.29
niacin	6.95	1.38	4.39
biotin	0.016	0.0021	0.0029
choline	216.0	280.0	202.0
pantothenic acid	1.37	0.59	0.69
folic acid	0.049	0.011	0.040
inositol	370.0	47.0	53.0
p-aminobenzoic acid	0.51	0.050	0.092

Table 17b. Vitamin Content of Certain Mill Products, mg/100 g dry weight (101)

Vitamin	Low grade	Red dog	Bran	Shorts	Germ
thiamine	1.25	3.25	0.730	1.55	1.56
riboflavin	0.144	0.374	0.387	0.403	0.565
niacin	4.48	9.27	26.2	18.6	5.25
biotin	0.013	0.029	0.051	0.041	0.020
choline	171.0	202.0	179.0	204.0	307.0
pantothenic acid	1.17	2.11	4.54	3.09	1.21
folic acid	0.049	0.139	0.102	0.157	0.238
inositol	395.0	937.0	1554.0	1253.0	988.0
p-aminobenzoic acid	0.342	0.906	1.72	1.46	0.429

Courtesy, *Baker's Digest.*

Table 18a. Mineral Composition of Wheat, Flour, and Bread (101)

Mineral	Wheat	Flour	Bread
potassium,[a] %	0.454	0.105	0.191
phosphorus,[a] %	0.433	0.126	0.183
magnesium,[a] %	0.183	0.028	0.034
calcium,[a] %	0.045	0.018	0.127
sodium, ppm	45	9.8	0.858[b]
zinc, ppm	35	7.8	9.7
iron, ppm	43	10.5	27.3
manganese, ppm	46	6.5	5.9
copper, ppm	5.3	1.7	2.3
molybdenum, ppm	0.48	0.25	0.32
cobalt, ppm	0.026	0.003	0.022

Enrichment of white flour and bread in the United States is so regulated that six slices of enriched bread in the average daily diet would add sufficient amounts of thiamine, riboflavin, niacin, and iron to protect against deficiencies.

Flour standards in 1941 under the Federal Food, Drug and Cosmetic Act of 1938 included provisions for the enrichment of flours. In a series of publications, the Committee on Cereals, Food and Nutrition of the National Academy of Sciences (103) kept pace with developments in the cereal-enrichment program from its inception in 1941 until 1958. New information led to the establishment of revised standards which are listed in Table 19. Flour may also be enriched with Vitamin D and calcium. In en-

Table 18b. Mineral Composition of Certain Mill Products

Mineral	Low grade	Red dog	Bran	Shorts	Germ
potassium,[a] %	0.434	0.903	1.67	1.29	0.889
phosphorus,[a] %	0.294	0.781	1.57	1.31	0.923
magnesium,[a] %	0.156	0.342	0.688	0.541	0.268
calcium,[a] %	0.046	0.110	0.128	0.133	0.048
sodium, ppm	8.5	30.5	30.6	37.2	23.2
zinc, ppm	44.5	105.3	99.4	100.1	100.8
iron, ppm	43.6	131.4	141.3	145.7	66.6
manganese, ppm	35.8	121.4	136.5	164.7	137.4
copper, ppm	5.5	14.2	15.2	13.3	7.4
molybdenum, ppm	0.39	0.70	0.83	0.79	0.67
cobalt, ppm	0.023	0.074	0.109	0.099	0.017

Courtesy, *Baker's Digest.*
[a] Dry basis.
[b] Percent.

riched self-rising flour the addition of calcium is required at 500–1500 mg/lb. The standards state that rice, after a rinsing test, must contain at least 85% of the minimum vitamin level. The Governments of Puerto Rico and the Philippines also require the rinsing test. If the method of enrichment does not permit the rinsing requirement to be met, consumer-size packages must bear the statement "Do not rinse before or drain after cooking." The South Carolina law does not require a rinsing test on packages less than 50 lb, as rice in small packages is presumed to be clean.

The early history of enrichment, and development in relation to the world food situation, were reviewed in several papers included in the symposium on enrichment of the American Association of Cereal Chemists (104).

Table 19. Enrichment Requirements[a] for Cereal Grain Foods in the United States,[b] mg/lb

Enriched product	Thiamine (B_1) Min	Thiamine (B_1) Max	Riboflavin (B_2) Min	Riboflavin (B_2) Max	Niacin Min	Niacin Max	Iron Min	Iron Max
bread or other baked products	1.1	1.8	0.7	1.6	10.0	15.0	8.0	12.5
flour	2.0	2.5	1.2	1.5	16.0	20.0	13.0	16.5
farina	2.0	2.5	1.2	1.5	16.0	20.0	13.0	none[c]
macaroni and noodle products[d]	4.0	5.0	1.7	2.2	27.0	34.0	13.0	16.5
corn meal	2.0	3.0	1.2	1.8	16.0	24.0	13.0	26.0
corn grits[e]	2.0	3.0	1.2	1.8	16.0	24.0	13.0	26.0
milled white rice	2.0	4.0	1.2[f]	2.4[f]	16.0	32.0	13.0	26.0

Courtesy, Hoffmann-La Roche, Inc., Nutley, N.J.
[a] The maximim and minimum levels shown are in accordance with the Federal Standards of Identity or state laws. The Government of Puerto Rico requires the use of enriched flour in all products made wholly or in part of flour, such as crackers and pretzels.
[b] Collected from Food and Drug Administration regulations.
[c] No maximum level established.
[d] Levels allow for 30–50% losses in kitchen procedure.
[e] Levels must not fall below 85% of minimum figures after a specific test described in the Federal Standards of Identity.
[f] The requirement for vitamin B_2 is optional, pending further study and public hearings, because of certain technical difficulties encountered in the application of this vitamin.

Flour is enriched at the mill by metering a vitamin–mineral premix mixture by means of a mechanical feeder into a flowing flour stream or by adding a weighed quantity of the enrichment mixture to flour in a batch-mixing operation. The miller is relieved of the troublesome task of making premixes and assured of more uniform enrichment by the use of commercially available, accurately compounded, finely milled enrichment mixtures. In the United States little interest has been shown in adding the optional ingredients, calcium and vitamin D, to flour at the mill.

Flour enriched at the mill finds little application in commercial baking operations. Bakers use regular flour and add wafers or tablets of the necessary ingredients. The tablets, consisting of the amounts of ingredients required for units of 50 or 100 lb of flour with a rapidly swelling starch as carrier, are allowed to disintegrate in a portion of the water used to make the dough, and the resulting suspension is added during dough mixing. Two types of tablets are available, one for bread made with less and one for bread with more than 3% milk powder. Milk is a good source of riboflavin, and bread made from enriched flour and milk contains more riboflavin than required by the Standard of Identity for Enriched Bread.

Enrichment of corn grits presents several problems. To avoid leaching losses during washing of grits, somewhat costly rinse-resistant premixes are used. Regular premixes can be used if no rinsing is anticipated.

Enriched macaroni products are made from enriched flours. Losses of about 50% of the added thiamine, 40% of the niacin, and 30% of the riboflavin in cooking and draining must be allowed for.

Enrichment of rice poses several problems, ie leaching during washing and discoloration of the surface by riboflavin. Several methods of coating, penetration, and simple admixture are available, depending on mode of use. Riboflavin addition to rice is, thus far, optional.

In 1967 the cost of enriching 57 lb of bread was one cent. The enriched products are nutritionally improved; yet the color, texture, aroma, and overall appearance are unaffected. Nearly 97% of the white bread in the United States is enriched, even though only thirty states have laws which require enrichment of white bread. Enrichment regulations do not apply to cakes, pies, pastries, and similar sweet goods.

Enrichment of cereals with vitamin E and pyridoxine has been suggested (104). Fortification of bread with proteins in the form of nonfat dry milk solids is widely accepted by the baking trade in the United States. This supplement not only increases the total protein content, but improves the overall nutritional value; nonfat milk solids contain proteins of high biological value and are rich in amino acids (notably lysine) that are deficient in wheat products. Supplementing cereal proteins with limiting amino acids to improve biological value has been studied intensively, as has been the blending of protein sources to achieve a balanced amino acid profile. In the United States, soyflours are used to a limited extent. Fortification with pure synthetic amino acids has met with little acceptance. Fortification of staple cereals with protein concentrates or pure amino acids may improve their nutritional value and aid in alleviating malnutrition in developing countries (105). In addition to soyflour, proteins from peanuts, sesame, or cottonseed can be utilized. Fish proteins have high nutritive value and are potentially the cheapest animal protein supplement. Single-cell protein, grown on petroleum or other low-cost substrates, has potential for animal and human feeding. Various sources of proteins are reviewed in a publication of the American Chemical Society (106). See also Proteins.

Special Types of Flours.   *Bromated Flours.*   For certain markets strong high-protein flours containing potassium bromate (to improve rheological dough properties and breadmaking performance) are best suited for the type of bread desired and the particular baking methods employed.   For these special market requirements, the original federal standard for bromated white flour permitted the addition of up to 75 ppm of potassium bromate to flours that contained at least 15% protein.   In 1948 the definition and standard of identity for bromated flour were revised at the request of the Millers' National Federation to permit the bromation of white flour regardless of its protein content to a maximum of 50 ppm of potassium bromate.   This revision provided an alternative means of flour improvement to treatment with nitrogen trichloride, which has been discontinued as a permissible improver of flour.

Whole-Wheat or Graham Flours.   Whole-wheat flours (entire wheat flour or graham flour) are used for whole-wheat bread; they must be made from hard or bread wheats ground to a fairly fine flour.   Whole-wheat flours are frequently made by recombining in the proper proportions the mill streams obtained in the gradual reduction roller-milling process.   The flour streams may be bleached in the usual manner and malted wheat or malted barley flour may be added where necessary to raise the amylase activity to a satisfactory level.   Whole-wheat flours may be bromated, if so labeled, at levels not exceeding 75 ppm to improve their baking properties.   In addition, crushed wheat and coarse-ground wheat are produced by crushing cleaned wheat other than durum or red durum.   It contains, in natural proportions, all the constituents of the cleaned wheat.   Cracked wheat is prepared by cracking or cutting cleaned wheat, other than durum or red durum, into angular fragments.

Phosphated Flours.   Phosphated flours represent a very large percentage of the so-called family flours, which are marketed in the southeastern states, and comprise bleached soft-wheat flours to which monocalcium phosphate, $CaH_4(PO_4)_2$, has been added.   These flours are used with baking soda (sodium bicarbonate, $NaHCO_3$) and sour milk to make sour-milk biscuits.   When made with plain flour, such biscuits frequently have a yellow color and soda taste because of insufficient acidity in the milk to react with all the soda; addition of monocalcium phosphate provides an acid-reacting material to neutralize the excess soda.   Proportions recommended in the trade are 0.5 lb monocalcium phosphate monohydrate, $CaH_4(PO_4)_2$. H_2O (or 0.375 lb of the anhydrous salt), to each 100 lb flour.

Self-Rising Flours.   In the South self-rising flours are in large demand.   They are made from bleached, soft-wheat patent flours and contain salt, sodium bicarbonate, and an acid-reacting ingredient (monocalcium phosphate) in the proper proportions to produce chemically leavened biscuits.   Effective leavening requires a minimum carbon dioxide production of 0.5% of the weight of the flour.   To avoid a yellow color and a disagreeable "soda taste" (somewhat characteristic of soap), sufficient acid ingredient must be added to react with all the soda.   A common formula for the production of self-rising flours contains 1.5 parts sodium bicarbonate, 1.875 parts monocalcium phosphate monohydrate, and 1.75 parts by wt of salt to each 100 parts by wt of flour.

Ready-Mixes.   Ready-mixed flours for biscuits, cakes, doughnuts, pancakes, etc, contain all the necessary ingredients other than the liquid (water or milk) in the proportions required to make the product.   The various ingredients in such mixes perform definite functions and the formula must be properly balanced to produce baked products of optimum quality (65).

Ready-prepared mixes containing shortening must have a fat that does not readily become rancid; shortenings prepared from vegetable oils of high stability are generally used. The incorporation of the shortening with the other finely divided "dry" ingredients requires special mixers; some manufacturers purchase a powdered preparation of shortening in dried milk. In ready-mixed preparations the acid-reacting ingredient for leavening is frequently sodium acid pyrophosphate, $Na_2H_2P_2O_7$; for many products this is said to give superior volume and texture to those obtained with monocalcium phosphate. Since, however, the pyrophosphate leaves a distinctly alkaline residue, monocalcium phosphate is frequently also added in the preparation of mixes yielding white baked products. The addition of the monophosphate results in baked goods of lower pH and thereby decreases the tendency toward the development of a yellowish-white crumb. Properly balanced formulas require considerable experimental work and technical control. Convenience foods in the form of frozen and refrigerated ready-mixes are finding increasing use in the United States.

Durum Wheat Products. Amber durum wheat is frequently called macaroni wheat from the principal use of *semolina*, the coarsely ground, highly purified middlings obtained by milling durum wheat. The quality of durum wheat for semolina milling is associated with its ability to produce a high yield of semolina that is free of visible specks, and which produces macaroni of the desired brilliant yellow color and cooking characteristics. When boiled for 10 min, a good macaroni swells to at least twice its original size and retains its tubular shape and its firmness without becoming pasty.

Macaroni products consist basically of dough made with semolina, farina (the purified middling of wheat other than durum), or flour (and sometimes salt), and water, which has been formed under pressure into a variety of sizes and shapes and then dried. This entire group of foods is usually called "alimentary pastes" (which is a translation of the French term *pâtes alimentaires* and the Italian *paste alimentari*), and includes macaroni itself, spaghetti, noodles, vermicelli, alphabets, rings, stars, sea shells, and many other special forms. From the standpoint of manufacture, the various macaroni products are classified as "long-goods" and "short-goods"; the long-goods are dried over sticks while the short-goods are dried on screen trays. For the production of long-goods, such as macaroni, spaghetti, and vermicelli, semolina is considered the prime raw material to avoid manufacturing troubles from undue stretching and breaking during drying or curing. For short-goods, such as alphabets, elbows, bow knots, and shells, the use of semolina is not so important, since these products are dried on trays. For manufacturing noodles (which contain egg solids), durum or common wheat flour is invariably used; the egg proteins act as a binder and impart a pleasing yellow color.

The steps in macaroni manufacture are mixing, kneading, pressing, and curing or drying. Throughout the process, every effort is made to develop and retain the maximum intensity of yellow pigmentation. Definitions and standards were established in 1941 under the Federal Food, Drug and Cosmetic Act for semolina, durum flour, and whole durum flour. Standards for macaroni products have also been established. Macaroni products are defined as the class of food prepared from semolina, durum flour, farina, or flour with or without the addition of egg white, disodium phosphate, onions, celery, garlic, bay leaf, and wheat gluten. These optional ingredients must be specified. Standards are also provided for macaroni products (macaroni, spaghetti, and

Table 20. Proximate Composition, Fuel Value, Mineral, and Vitamin Contents of Cereal Products Prepared for Human Consumption (112)

	Water, %	Food energy, kcal[a]	Protein, %	Fat, %	Carbohydrate		Ash, %	Calcium, mg[a]	Phosphorus, mg[a]	Iron, mg[a]	Thiamin, mg[a]	Riboflavin, mg[a]	Niacin, mg[a]
					Total, %	Fiber, %							
barley, pearled													
light	11.1	349	8.2	1.0	78.8	0.5	0.9	16	189	2.0	0.12	0.05	3.1
pot	10.8	348	9.6	1.1	77.2	0.9	1.3	34	290	2.7	0.21	0.07	3.7
biscuits, made with													
enriched flour	27.4	369	7.4	17.0	45.8	0.2	2.4	121	175	1.6	0.21	0.21	1.8
unenriched flour	27.4	369	7.4	17.0	45.8	0.2	2.4	121	175	0.5	0.04	0.10	0.5
bread													
cracked-wheat	34.9	263	8.7	2.2	52.1	0.5	2.1	88	128	1.1	0.12	0.09	1.3
French or Vienna	30.6	290	9.1	3.0	55.4	0.2	1.9	43	85	2.2	0.28	0.22	2.5
rye-American	35.5	243	9.1	1.1	52.1	0.4	2.2	75	147	1.6	0.18	0.07	1.4
pumpernickel	34.0	246	9.1	1.2	53.1	1.1	2.6	84	229	2.4	0.23	0.14	1.2
white, 3–4% NFDMS[b]	35.6	270	8.7	3.2	50.5	0.2	2.0	84	97	2.5	0.25	0.21	2.4
cornbread	53.9	207	7.4	7.2	29.1	0.5	2.4	120	211	1.1	0.13	0.19	0.6
corn flakes, enriched	3.8	386	7.9	0.4	85.3	0.7	2.6	17	45	1.4	0.43	0.08	2.1
corn flour	12.0	368	7.8	2.6	76.8	0.7	0.8	6	164	1.8	0.20	0.06	1.4
corn grits, degermed	12.0	362	8.7	0.8	78.1	0.4	0.4	4	73	1.0	0.13	0.04	1.2

doughnuts	23.7	391	4.6	18.6	51.4	0.1	1.7	40	190	1.4	0.16	0.16	1.2
oats, meal or rolled	8.3	390	14.2	7.4	68.2	1.2	1.9	53	405	4.5	0.60	0.14	1.0
popcorn, unpopped	9.8	362	11.9	4.7	72.1	2.1	1.5	10	264	2.5	0.39	0.11	2.1
rice													
brown	12.0	360	7.5	1.9	77.4	0.9	1.2	32	221	1.6	0.34	0.05	4.7
white enriched	12.0	363	6.7	0.4	80.4	0.3	0.5	24	94	2.9	0.44		3.5
rye													
whole grain	11.0	334	12.1	1.7	73.4	2.0	1.8	38	376	3.7	0.43	0.22	1.6
flour													
light	11.0	357	9.4	1.0	77.9	0.4	0.7	22	185	1.1	0.15	0.07	0.6
dark	11.0	327	16.3	2.6	68.1	2.4	2.0	54	536	4.5	0.61	0.22	2.7
sorghum, grain	11.0	332	11.0	3.3	73.0	1.7	1.7	28	287	4.4	0.38	0.15	3.9
wheat													
bran	11.5	213	16.0	4.6	61.9	9.1	6.0	119	1276	14.9	0.72	0.35	21.0
flour,													
80% extraction	12.0	365	12.0	1.3	74.1	0.5	0.65	24	191	1.3	0.26	0.07	2.0
patent	12.0	364	10.5	1.0	76.1	0.3	0.43	16	87	2.9	0.44	0.26	3.5
cake	12.0	364	7.5	0.8	79.4	0.2	0.31	17	73	0.5	0.03	0.03	0.7
germ	11.5	363	26.6	10.9	46.7	2.5	4.3	72	1118	9.4	2.01	0.68	4.2
whole grain	12.5	330	12.3	1.8	71.7	2.3	1.7	46	354	3.4	0.52	0.12	4.3

[a] Per 100 g edible portion.
[b] Nonfat dry-milk solids.

vermicelli), which contain milk, soy, or vegetables. Noodles must contain at least 5.5% by wt of egg solids or egg yolk, and they may also contain one or more of the optional ingredients. The quality of macaroni products is determined by the following factors (107): color and appearance; mechanical properties of the uncooked product (of special significance in packaging and transportation); and behavior during cooking (cooking loss and texture of cooked product).

The annual per-capita consumption of alimentary pastes is highest in Italy; in the United States less than one-tenth of that amount is consumed, and in England even less.

The chemistry and physics of macaroni products are reviewed by Radley (108), processing of durum products by Irvine (109), manufacture by Hoskins and Hoskins (110), and evaluation of durum wheats and their products by Holliger (107) and Fabriani (111).

Chemical Composition of Cereal Products

The proximate composition, fuel value, vitamin B contents, and percentages of certain minerals, as prepared for human consumption, are given in Table 20 (112).

Bibliography

"Cereals" in *ECT* 1st ed., Vol. 3, pp. 591–634, by W. F. Geddes, University of Minnesota, and F. L. Dunlap, Wallace & Tiernan Co., Inc.

1. S. A. Matz, ed., *The Chemistry and Technology of Cereals as Food and Feed*, Avi Publishing Co., Westport, Conn., 1959.
2. J. A. Shellenberger, *Wheat*, in reference 1, Chap. 1.
3. S. A. Matz, *Corn*, in reference 1, Chap. 2.
4. T. R. Stanton, *Oats*, in reference 1, Chap. 3.
5. A. D. Dickson, *Barley*, in reference 1, Chap. 4.
6. H. L. Shands, *Rye*, in reference 1, Chap. 5.
7. N. W. Kramer, *Sorghum*, in reference 1, Chap. 6.
8. M. H. Beachell, *Rice*, in reference 1, Chap. 7.
8a. *Agricultural Statistics 1969*, U.S. Dept. of Agriculture, Washington, D.C., 1969, p. 49.
9. *Grain Crops—A Review*, The Commonwealth Secretariat, Her Majesty's Stationary Office, London, 1967.
10. L. W. Rooney and L. E. Clark, *Cereal Sci. Today* **13,** 258 (1968).
11. K. S. Quisenberry and L. P. Reitz, eds., *Wheat and Wheat Improvement*, Monograph No. 13, American Society of Agronomy Inc., Madison, Wis., 1967.
11a. L. P. Reitz, *World Distribution and Importance of Wheat*, in reference 11, Chap. 1.
12. G. K. Middleton, C. E. Bode, and B. B. Bayles, *Agron. J.* **46,** 500 (1954).
13. W. P. Carroll and A. B. Combs, *U.S. Dept. Agr. Misc. Publ.* **325** (1938).
14. *U.S. Dept. Agr. Misc. Publ.* **740** (1957).
15. *Official Grain Standards of the United States*, SRA-C and MS-177, U.S. Dept. of Agriculture, Washington, D.C., 1964.
16. T. A. Oxley, *The Principles of Grain Storage*, Northern Publishing Co., Liverpool, England, 1948.
17. J. A. Anderson and A. W. Alcock, eds., *Storage of Cereal Grains and Their Products*, American Association of Cereal Chemists, Minneapolis, Minn., 1954.
18. E. James, *U.S. Dept. Agr. Crops Res. Div. Publ.* 34-15-1 (1961) and 34-15-2 (1963).
19. C. H. Bailey, *Plant Physiol.* **15,** 257 (1940).
20. Y. Pomeranz and J. A. Shellenberger, *Am. Miller Processor* **94** (6), 9 (1966).
21. Y. Pomeranz, R. D. Daftary, M. D. Shogren, R. C. Hoseney, and K. F. Finney, *Agr. Food Chem.* **16,** 92 (1968).
22. G. N. Wogan, ed., *Mycotoxins in Foodstuffs*, The M.I.T. Press, Cambridge, Mass., 1965.

23. Y. Pomeranz, *Advan. Food Res.* **16,** 335 (1968).
24. G. A. Wiebe, ed., *U.S. Dept. Agr. Handbook* No. 338, U.S. Dept. of Agriculture, Washington, D.C., 1968.
25. J. A. LeClerc and C. D. Garby, *Ind. Eng. Chem.* **12,** 451 (1920).
26. A. H. Cook, ed., *Barley and Malt,* Academic Press, Inc., New York, 1962.
27. P. R. Witt, *Malting,* in reference 1, Chap. 18.
28. A. D. Dickson, *Cereal Sci. Today* **10,** 284 (1965).
29. S. Laufer, *Am. Brewer* **97** (5), 75 (1964).
30. H. A. Cook and G. Harris, *Brewers Digest* **39** (7), 70 (1964).
31. J. A. Brussman, W. A. Hardwick, and J. Schwaiger, *Cereal Sci. Today* **10,** 496 (1965).
32. G. F. Sprague, ed., *Corn and Corn Improvement,* Monograph No. 5, American Society of Agronomy, Inc., Madison, Wis., 1955.
33. T. A. Kiesselbach, *The Structure and Reproduction of Corn,* Bull. 161, University of Nebraska, Lincoln, Nebr., 1949.
34. M. J. Wolf, C. L. Buzan, M. M. MacMasters, and C. E. Rist, *Cereal Chem.* **29,** 321 (1952).
35. J. T. Goodwin, *Wet-milling,* in reference 1, Chap. 13.
36. R. A. Larsen, *Milling,* in reference 1, Chap. 9.
37. *Proc. Eighth Symp. Central States Section, American Association of Cereal Chemists, St. Louis, Mo., February 17–18, 1967.*
37a. W. W. Katz, *Manufacture of Dry Corn Mill Products,* in reference 37.
38. H. J. Roberts, *Usage of Corn Flour Products,* in reference 37.
39. F. R. Earle, J. J. Curtis, and J. E. Hubbard, *Cereal Chem.* **23,** 504 (1946).
40. E. P. Stimmel, *Am. Miller* **69,** 30 (1941).
41. T. S. Stiver, Jr., *Bull. Assoc. Operative Millers* **2168** (1955).
42. W. M. Findlay, *Oats,* Oliver and Boyd, Edinburgh, Scotland, 1956.
43. F. A. Coffman, ed., *Oats and Oat Improvement,* Monograph No. 8, American Society of Agronomy Inc., Madison, Wis., 1961.
44. O. T. Bonnett, *The Oat Plant; Its Histology and Development,* Bull. 672, University of Illinois, Urbana, Ill., 1961.
45. N. L. Kent, *Technology of Cereals, With Special Reference to Wheat,* Pergamon Press, Ltd., Oxford, 1966.
46. L. P. Carroll and A. W. Richards, *Manufacture and Usage of Oat Products,* in reference 37.
47. A. Schulerud, *Das Roggenmehl,* 2nd ed., Schäfer Verlag, Berlin, 1957.
48. F. D. Schmalz, *Manufacture and Usage of Rye Flour,* in reference 37.
49. D. H. Grist, *Rice,* 3rd ed., Longmans, London, 1959.
49a. *U.S. Dept. Agr. Handbook No. 289,* U.S. Dept. of Agriculture, Washington, D.C., 1966.
50. C. G. Harrel and B. M. Dirks, "Cereals and Cereal Products," in F. C. Blank, ed., *Handbook of Food and Agriculture,* Reinhold Publishing Co., New York, 1955.
51. E. B. Kester, *Rice Processing,* in reference 1, Chap. 16.
52. H. J. Deobald, *Manufacture and Usage of Rice Flour,* in reference 37.
53. G. Hampel, *Cereal Sci. Today* **13,** 64 (1968).
54. T. B. Wayne, *Food Ind.* **2,** 492 (1930).
55. A. Aten and A. D. Faunce, *Equipment for the Processing of Rice,* Development Paper No. 27, Food and Agriculture Organization of the United Nations, Rome, 1953.
56. Anonymous, *Cereal Sci. Today* **12,** 218 (1967).
57. M. C. Kik and R. R. Williams, *Bull. Natl. Res. Council (U.S.)* **112** (1945).
58. W. M. Ross and O. J. Webster, *Agr. Inf. Bull. No. 218,* U.S. Government Printing Office, Washington, D.C., 1960.
59. J. R. Quinby, N. W. Kramer, J. C. Stephens, K. R. Lahr, and R. E. Karper, *Texas Agr. Expt. Station, Bull.* **912,** 1958.
60. J. H. Martin and M. M. MacMasters, "Industrial Uses of Grain Sorghum," in A. Stefferud, ed., *Crops in Peace and War,* Yearbook of Agriculture, 1950–1951, U.S. Dept. Agr., Washington, D.C., 1951.
61. S. A. Watson, "Manufacture of Corn and Milo Starches," in R. L. Whistler and E. P. Paschall, eds., *Starch Chemistry and Technology,* Vol. II, *Industrial Aspects,* Academic Press, Inc., New York, 1967, Chap. 1.
62. P. F. Werler, *Manufacture and Usage of Sorghum Flour,* in reference 37.
63. J. E. Hubbard, H. H. Hall, and F. R. Earle, *Cereal Chem.* **27,** 415 (1950).

64. D. F. Miller, *Natl. Acad. Sci. Natl. Res. Council Publ.* **585** (1958).

65. W. F. Geddes, *Agr. Food Chem.* **9,** 605 (1959).

66. I. Hlynka, ed., *Wheat Chemistry and Technology*, American Association of Cereal Chemists, Minneapolis, Minn., 1964.

67. M. P. Neumann and P. F. Pelshenke, *Brotgetreide und Brot*, 5th ed., Verlag Paul Parey, Berlin, 1954.

68. M. Rohrlich and G. Bruckner, *Das Getreide*, Vol. I, *Das Getreide und Seine Verarbeitung*, Verlag Paul Parey, Berlin, 1966.

69. D. W. Kent-Jones and A. J. Amos, *Modern Cereal Chemistry*, 6th ed., Food Trade Press, Ltd., London, 1967.

70. L. Smith, *Flour Milling Technology*, 3rd ed., Northern Publishing Co., Liverpool, 1944.

71. J. A. Scott, *Flour Milling Processes*, 2nd ed., Chapman and Hall, Ltd., London, 1951.

72. J. F. Lockwood, *Flour Milling*, 4th ed., Northern Publishing Co., Liverpool, 1960.

73. T. Horder, E. C. Dodds, and T. Moran, *Bread*, Constable, London, 1954.

74. E. J. Pyler, *Baking Science and Technology*, 2 Vols., Siebel Publishing Co., Chicago, Ill., 1952.

75. E. B. Bennion, *Breadmaking*, Oxford University Press, England, 1954.

76. S. A. Matz, ed., *Bakery Technology and Engineering*, Avi Publishing Co., Westport, Connecticut, 1960.

77. J. W. Knight, *The Chemistry of Wheat Starch and Gluten and Their Conversion Products*, Leonard Hill, London, 1965.

78. J. A. Anderson, ed., *Enzymes and Their Role in Wheat Technology*, Interscience Publishers Inc., New York, 1946.

79. D. Bradbury, J. E. Hubbard, M. M. MacMasters, and F. R. Senti, "Conditioning Wheat for Milling," *U.S. Dept. Agr. Misc. Publ.* **824** (1960).

80. B. S. Miller and J. A. Johnson, *A Review of Methods for Determining the Quality of Wheat and Flour for Breadmaking*, Bull. 76, Kansas State University, Manhattan, Kans., 1954.

81. Y. Pomeranz, "Analytical Methods, Cereals," in A. Kramer and B. Twigg, eds., *Fundamentals of Quality Control for the Food Industry*, Vol. II, Avi Publishing Co., Westport, Conn., 1969, Chap. 6.

82. D. Bradbury, I. M. Cull, and M. M. MacMasters, *Cereal Chem.* **33,** 329, 342, 361, 373 (1956).

82a. *Baker's Digest* **41** (5) (1967).

82b. *Cereal Sci. Today* **10** (6) (1965).

82c. *Proc. Nutr. Soc. (Engl. Sect.)* **17** (7) (1958).

83. M. F. Adams, ed., *Abstracts for the Advancement of Industrial Utilization of Wheat*, Washington State University, Pullman, Wash.,

84. Chien-Mei Chiu, Y. Pomeranz, M. D. Shogren, and K. F. Finney, *Food Technol.* **22,** 1157 (1968).

85. L. P. Reitz, *Qualitas Plant. Mater. Vegetabiles* **11,** 1 (1964).

86. A. B. Ward, *Recent Manufacturing Techniques and Applications of Wheat Flour*, in reference 37.

86a. *Millfeed Manual*, The Millers National Federation, Chicago, Ill., 1967.

87. C. O. Swanson, *Wheat and Flour Quality*, Burgess Publishing Co., Minneapolis, Minn., 1938.

88. A. C. Stringfellow, V. F. Pfeifer, and E. L. Griffin, Jr., *Baker's Digest* **36** (4), 38 (1962).

89. A. C. Stringfellow and A. J. Peplinski, *Cereal Sci. Today* **11,** 438 (1966).

90. A. S. Matz, *Cookie and Cracker Technology*, Avi Publishing Co., Westport, Conn., 1968.

91. K. A. Gilles and W. C. Shuey, *Soft Flour Products*, in reference 66, Chap. 13.

92. K. F. Finney and W. T. Yamazaki, *Quality of Hard, Soft, and Durum Wheats*, in reference 11, Chap. 14.

93. J. A. Johnson, D. Miller, and H. C. Fryer, *Baker's Digest* **36** (6), 50 (1962).

94. C. C. Tsen and I. Hlynka, *Baker's Digest* **41** (5), 58 (1967).

95. L. Zechmeister and R. B. Escue, *Proc. Natl. Acad. Sci. U.S.* **27,** 528 (1941).

96. V. E. Munsey, *J. Assoc. Offic. Agr. Chem.* **21,** 331 (1938).

97. Y. Pomeranz, *Baker's Digest* **42** (3), 30 (1968).

98. F. D. Vidal, H. Parker, and R. R. Jointer, *Baker's Digest* **37** (6), 69 (1963).

99. K. F. Finney and M. A. Barmore, *Cereal Chem.* **22,** 244 (1945).

100. Y. Pomeranz and M. M. MacMasters, *Baker's Digest* **42** (4), 24 (1968).

101. W. B. Bradley, *Baker's Digest* **41** (5), 66 (1967).

102. R. S. Harris, *Agr. Food Chem.* **16,** 149 (1968).

103. C. L. Brooke, *Agr. Food Chem.* **16,** 163 (1968).

104. *Cereal Sci. Today* **11** (6) (June 1966).
105. G. K. Parman, *Agr. Food Chem.* **16,** 168 (1968).
106. "World Protein Resources," *Advan. Chem. Ser.* **57** (1966).
107. A. Holliger, *Cereal Chem.* **40,** 235 (1963).
108. J. A. Radley, *Food Manuf.* **27,** 329, 369, 406, 436, 481 (1952); *ibid.* **28,** 11 (1953).
109. G. N. Irvine, *Durum Wheat and Paste Products*, in reference 66, Chap. 12.
110. C. M. Hoskins and W. G. Hoskins, *Macaroni Production*, in reference 1, Chap. 11.
111. G. Fabriani, *Cereal Sci. Today* **11,** 339 (1966).
112. B. K. Watt and A. L. Merrill, "Composition of Foods; Raw, Processed, Prepared," *U.S. Dept. Agr. Handbook No. 8*, U.S. Dept. of Agriculture, Washington, D.C., 1963.

Y. POMERANZ
U. S. Dept. of Agriculture
and M. M. MacMASTERS
Kansas State University

WHEAT GERM OIL.   See Fats and fatty oils.

WHEY.   See Feeds, animal, Vol. 8, p. 865; Milk and milk products.

WHISKEY.   See Alcoholic beverages, distilled, Vol. 1, p. 510.

WHITE LEAD.   See Pigments, inorganic, Vol. 15, p. 500.

WHITENING AGENTS.   See Brighteners, optical.

WHITING, $CaCO_3$.   See Pigments, inorganic, Vol. 15, p. 508.

WINE

The term wine was probably first applied to the fermented juice of grapes; however, the fermented juices of many fruits are now called wine, and the term is also sometimes incorrectly applied to the alcoholic fermented juice of various plant materials containing sugars, either di- or monosaccharides. It is thus used in contrast to fermented liquids made from starch-containing materials, such as beer (see Beer and brewing) and related beverages. When used alone, the term "wine" applies only to wine produced from fermented grapes. In addition to grape wine, orange, peach, cherry, blackberry, loganberry, currant, apple, strawberry, and other fruit wines are produced commercially in the United States. There is also a small home industry that produces wines from various herbs and vegetables by adding sugar—dandelion, bean, rhubarb, rose, and similar wines—or from fermented honey (mead). See also Alcoholic beverages, distilled.

An important aspect of wine is its intimate association with man's artistic, cultural and religious activities throughout history. Thus at the beginning of recorded history, wines were described or their production portrayed and their properties critically evaluated and praised. The Egyptians, for example, by 2500 B.C. had evolved hieroglyphics that described various types of wines and their place of production. Noah's experience in planting a vineyard and making wine may indicate the early development of winemaking. The Old Testament contains many references to wines and their properties. Greek literature also has many reports on various types of wines and grapes. One special feature of Greek wines was the frequent addition of herbs, perfumes, and other odorous materials. This practice still survives with the modern Greek wine containing resin (retsina) and in various herb-containing wines produced elsewhere, such as vermouth. The origin of this practice is not known, but it is logical to assume that under primitive conditions, with poor storage facilities and a warm climate, many Greek and other Middle-East wines bacterially spoiled or acetified. The addition of the herbs might then be accounted for as an attempt to mask the spoiled aroma or hopefully (and usually unsuccessfully) to prevent its development. The addition of honey probably arose from the same conditions, the high sugar content masking the spoiled flavor. However, the added honey may have resulted in a sufficiently high alcohol content to prevent spoilage. Some herbs were probably added because of their real or supposed medicinal or aphrodisiac value.

The religious and allegorical significance of wine was developed by the Greeks and has been utilized by many other religions, including Christianity. It is quite unlikely that grape juice was more than a vintage season beverage before the 19th century. Its ready fermentability, the general ignorance of sterilization procedures, and the lack of sealable storage containers (at least on a cheap, sure, and large scale) are all reasons for this conclusion.

With the Romans, wine production became increasingly highly organized and specialized. The writings of Columella and Pliny, for example, indicate that a great amount of organized information on the cultivation of grapes and the production of wine was available. The various varieties of grapes and different methods of producing wines were described. Clay amphora (of Greek origin) and later wood casks were available for aging and a small volume of wine was stored for many years in glass. Crude methods of clarification, preventing spoilage, and treating spoiled wines were developed by the Romans. Although many of their wines must have been very poor by modern standards, it is clear that the Romans had a cultivated taste for the beverage, lavished much care on its production, and gave it both literary and artistic praise. The writings of such Roman poets as Horace and Virgil on wines are well known.

During the Middle Ages, wine production continued in the Mediterranean countries and also in France and Germany. The need of wine for religious use and the large number of monastic groups led to the special production of wines by monasteries. Many important modern European vineyards owe their origin to Cistercian and other monastic orders. As the cooper's art improved, as bottlemaking became less expensive, and particularly after corks became available, wines could be stored safely and for longer periods. The use of caves for storage was an early and sound development because only in caves could constant and relatively cool storage conditions be easily obtained: both of these are essential for the proper aging of wines.

Finally, in the 19th century the work of Pasteur and others not only showed the role of yeasts in fermenting grape juice but indicated as well the microorganisms responsible for spoilage and methods for their control. Wine production then gradually developed from an uncertain art to a scientific industry. Winemaking is still something of an art, but commercial production based on technology is producing more and more of the world's wine. The producer who allows a certain percentage of his wine to spoil does not survive.

Definitions

Wine is the fermented juice of the fruit of one of the several species of *Vitis*, usually of cultivars of *Vitis vinifera*, with or without the addition of sugar, grape concentrate, or reduced must (boiled-down grape juice), herbs, or alcohol. The present legal definition in the United States is:

United States Internal Revenue Code. Sec. 3044. *Definitions.* (a) Natural Wine. Natural wine within the meaning of this subchapter shall be deemed to be the product made from the normal alcoholic fermentation of the juice of sound, ripe grapes, without addition or abstraction, except such as may occur in the usual cellar treatment of clarifying and aging.

(b) Wine. The product made from the juice of sound, ripe grapes by complete fermentation of the must under proper cellar treatment and corrected by the addition (under the supervision of a storekeeper-gauger) of a solution of water and pure cane, beet, or dextrose sugar (containing respectively, not less than 95 per centum of actual sugar, calculated on a dry basis) to the must or to the wine, to correct natural deficiencies, when such addition shall not increase the volume of the resultant product more than 35 per centum, and the resultant product does not contain less than five parts per thousand of acid before fermentation and not more than 13 per centum of alcohol after complete fermentation or, if sweetened, after complete fermentation and sweetening, shall be deemed to be wine within the meaning of this subchapter, and may be labeled, transported, and sold as "wine," qualified by the name of the locality where produced, and may be further qualified by the name of its own particular type of variety.

(c) Sweet Wine. Wine as defined in this section may be sweetened with cane sugar or beet sugar or pure condensed grape must and fortified under the provisions of this subchapter, and wines so sweetened or fortified shall be considered sweet wine within the meaning of this subchapter.

(d) Pure Sweet Wine. This is defined in Section 3036 (a). It is a wine which consists of fermented or partially fermented grape juice only but which may be fortified with wine spirits and may contain condensed grape must or may be sweetened with sugar. ["Wine spirits" is alcohol distilled from wine. It is usually 190 proof, or 95% alcohol.]

This definition applies to wines in interstate commerce, but under the 21st amendment to the U.S. Constitution, the several states control the production and sale of wine within the state. The California definition of wine is also important since about 85% of the wine produced in the U.S. is from that state. It is:

California Alcoholic Beverage Control Act. Sec. 23007. *"Wine."* "Wine" means the product obtained from normal alcoholic fermentation of the juice of sound, ripe grapes or other agricultural products containing natural or added sugar or any such alcoholic beverage to which is added grape brandy, fruit brandy, or spirits of wine, which is distilled from the particular agricultural product or products of which the wine is made and other rectified wine products and by whatever name and which does not contain more than 15 per cent added flavoring, coloring, and blending material and which contains not more than 24 per cent of alcohol by volume, and includes vermouth and sake, known as Japanese rice wine.

Various states and countries have slightly different definitions based on local needs.

Fruit wines indicate the fruit of origin, such as orange wine, blackberry wine, etc. They are legally defined in the United States Internal Revenue Code as follows:

Sec. 3045. *Application of Natural Wine Provisions to Citrus-Fruit Wines and Other Like Wines.*   The provisions of the internal revenue laws applicable to natural wine shall apply in the same manner and to the extent to citrus-fruit wines, peach wines, cherry wines, berry wines, apricot wines, prune wines, plum wines, pear wines, pawpaw wines, papaya wines, pineapple wines, cantaloup wines, and apple wines, which are the products, respectively, of normal alcoholic fermentation of the juice of sound ripe (1) citrus-fruit (except lemons and limes), (2) peaches, (3) cherries, (4) berries, (5) apricots, (6) prunes, (7) plums, (8) pears, (9) pawpaws, (10) papayas, (11) pineapples, (12) cantaloupes, (13) apples, with or without the addition of dry cane, beet, or dextrose sugar (containing, respectively, not less than 95 per centum of actual sugar, calculated on a dry basis) for the purpose of perfecting the product according to standards, but without the addition or abstraction of other substances, except as may occur in the usual cellar treatment of clarifying or aging: *Provided,* That in the case of wines produced from loganberries, currants, or gooseberries, respectively, having a normal acidity of twenty parts or more per thousand, the volume of the resultant product may be increased more than 35 per centum but not more than 60 per centum by the addition of sugar and water solution under such regulations as the Commissioner of Internal Revenue may prescribe.

Standards

There are two types of standards for wines in the United States and in most other countries. First, there are those based on taxes. In the U.S. the federal tax paid for wines varies according to the alcohol content: less than 14%, 17¢/gal; 14–20%, 67¢/gal; 21–24%, $2.40/gal. Very few wines of 21% or more are produced in this

Table 1. Federal and California Standards for Wines

Agency	Type	Alcohol, %	Maximum volatile acidity, % acetic[a]	Minimum fixed acidity, % tartaric	Maximum sulfur dioxide, mg/l	Minimum extract,[b] g/100 ml
federal	red table	not over 14	0.140	none[c]	350[d]	none
federal	white table	not over 14	0.120	none[c]	350[d]	none
federal	dessert	17–21	0.120	none[c]	350[d]	none
California	red table	10.5–14	0.120	0.40	350	1.8
California	white table	10–14	0.110	0.30	350	1.7
California	dessert	19.5–21	0.110	0.25	350	none[e]

[a] Exclusive of sulfur dioxide.
[b] Minimum soluble solids content.
[c] No specification given.
[d] Not more than 70 of which may be "free."
[e] Either minimum degree Balling or minimum percent reducing sugar is required, varying with the type.

country or abroad. The federal government and some states have also set up elementary standards for spoiled wines to prevent these wines from reaching the consumer. Poor quality wines, it may be noted, might reduce the potential sale by the producer, and thus lessens the tax receipts. The most important federal and California standards are given in Table 1.

The federal and state public health authorities also frequently set up standards for wines on the basis of chlorides, sulfates, lead, arsenic, and other constituents. It is seldom necessary to determine these in the U.S., since they are not known to be present in U.S. wines in excessive amounts.

It is worth while noting that salicylic, benzoic, and monochloracetic acids may not be used in American wines and that permissible practices such as agents used for clarification and the addition of sulfur dioxide or citric acid are specifically listed, the presumption being that all others are prohibited. Other countries likewise generally prohibit certain practices, permit others, and those not specifically listed may not generally be employed until permission for their use is granted.

Classification

Wines may be classified in many ways: place of production, color, alcohol content, method of production, or variety of fruit or grape. The following classification combines several of these standards. Wines of below 14% alcohol have been defined as table wines since they are normally consumed with meals. Wines of over 14% are called dessert or aperitif wines since they are usually consumed after or before meals. This system is faulty because some table wines have a high sugar content and are preferably (and usually) consumed after meals. It is true, however, that wines of over 14 or 15% alcohol are seldom consumed with meals.

I. Wines with added herbs or plant materials containing alkaloids or other flavoring materials
- A. With a red color.
  1. Proprietary types—includes certain wines containing quinine (Byrrh, Dubonnet, Campari, etc)
  2. Medicinal or home-produced types, such as iron- or herb-containing.
- B. Without a red color.
  1. Nearly dry—dry or French-type vermouth.
  2. Sweet (usually with a muscat flavor)—sweet or Italian-type vermouth.
  3. Proprietary types of herb-flavored dessert wine. (Thunderbird, Silver Satin, etc)
  4. Medicinal or home-produced types, such as gentian, rhubarb, dandelion, etc (see also A, 2 above).

II. Wines without added herbs or plant materials.
- A. Wines with excess carbon dioxide.
  1. From fermentation of added sugar. Usually 2–6 atm of pressure.
    a. Containing anthocyanin and related (red) pigments.
      (1) Pink—pink sparkling types.
      (2) Red—sparkling burgundy or champagne rouge.
    b. Not containing red pigments.
      (1) With muscat flavor—sparkling muscat and muscato spumante.
      (2) Without muscat flavor—California (etc) champagne, bulk process, Champagne, Sekt, spumante, espumante, shampanski, etc.
        (a) Below 1% sugar—brut type.
        (b) Above 1% sugar—sec (dry), demi sec, and doux types.
  2. Wines with excess carbon dioxide, not from added sugar. Usually 0.2–2 atm of pressure.

   a. Gassiness from fermentation of residual grape sugar. Includes occasional wines in Germany, France (some Vouvray), Italy, and the muscato amabile type of California.
   b. Gassiness from malo-lactic fermentation—vinhos verdes wines (white and red) from northern Portugal and some Italian wines.
  3. Wines with added carbon dioxide.
   a. Containing anthocyanin and related (red) pigments—carbonated burgundy.
   b. Not containing red pigments—carbonated moselle, some Swiss and Californian types.
B. Wines without obvious excess carbon dioxide.
  1. Wines 8–14% alcohol.
   a. Wines with anthocyanin and related (red) pigments.
    (1) Pink wines.
      (a) Dry—pink or rose types and varietal types, such as Gamay, Grenache, Tavel and Grignolino.
      (b) Sweet—Aleatico (see, however, 3a (1) below).
    (2) Full red color.
      (a) Dry.
        -1- With distinguishable (usually) varietal aromas.
          -a- With high acidity—Barbera.
          -b- With moderate acidity—Barolo, Beaujolais, Bordeaux (Médoc, St.-Émilion, etc), Burgundy (French), Cabernet Sauvignon or C. franc, Châteauneuf-du-Pape, Chianti (Italian), Fresia, Gamay, Hermitage, Merlot, Pinot noir, Rioja, Zinfandel, etc.
        -2- Without (normally) distinguishable varietal aromas—California (etc) burgundy, California (etc) claret, California (etc) dry red table, California (etc) red chianti, Charbono, Durif, Malvoisie, Mourestel, Nebbiolo, Red Pinot, Valdepeñas.
      (b) Sweet. Proprietary types with or without Concord aroma (see also 2a below), California (etc) red table, and California sweet red table.
   b. Wines without anthocyanin and related (red) pigments.
    (1) Wines usually with distinguishable varietal aromas.
      (a) Containing sugar—German Auslese, etc, types, and many non-Auslese German wines from the Moselle, Rhine, etc, regions, Hungarian Tokay, light muscat, light sweet muscat, some Loire wines, Sauternes, sweet Catawba, sweet Sauvignon blanc, sweet Sémillon, and various proprietary wines.
      (b) Not containing sugar—Catawba, Chablis (French), Chardonnay, Delaware, Folle blanche, Gewürztraminer, Graves, some Grey Riesling, some Loire wines, Moselle (Germany), Pinot blanc, Riesling, Rhine (Germany), Sauvignon blanc, Sémillon, White Pinot, and White Riesling, etc.
    (2) Wines usually without distinguishable varietal aromas.
      (a) Containing sugar—chateau types, California (etc) sweet white table, sauterne, sweet sauterne, and various proprietary labeled types. (See also 2b below.)

      (b) Not containing sugar—California (etc) dry sauterne, California (etc) rhine, California (etc) white chianti, California (etc) white table, French Colombard, some Grey Riesling, Sylvaner, Ugni blanc (also called Trebbiano or St.-Émilion).

2. Wines with 14–17% alcohol.
  a. Containing anthocyanin or related (red) pigments—miscellaneous sweet red types—mainly with proprietary names.
  b. Not containing red pigments—miscellaneous sweet white types—mainly with proprietary names.
  c. Special types
    (1) Blending—foreign types, usually to increase the alcohol or to impart a special flavor to other wines or even other products, ie, whiskey.
    (2) Ecclesiastical, usually sweet—vino santo—and various wines with dessert-type names but specially produced for special markets.
    (3) Fino and manzanilla types in Spain.

3. Wines with 17–21% alcohol.
  a. Containing anthocyanin and related (red) pigments.
    (1) With a muscat flavor.
      (a) Pink—Aleatico (see also B1a(1) above).
      (b) Red—red or black muscatel.
    (2) Without a muscat flavor.
      (a) With a baked odor—California tokay.
      (b) Without a baked odor.
        -1- Brownish red—tawny port.
        -2- Red—port (including ruby port).
  b. Not containing red pigments.
    (1) With a muscat aroma—Muscat blanc (ie, Muscat Frontignan and Muscato Canelli), Samos, Setúbal, Sitges, etc.
    (2) Without a muscat aroma.
      (a) With a special odor due to treatment or aging.
        -1- With a raisin, cooked, reduced-must, or rancio odor.
          -a- Raisin odor—Malaga.
          -b- Baked odor—California (etc) sherry (dry, medium, or sweet), Madeira.
          -c- Reduced-must or burnt odor—Marsala.
          -d- Rancio (aged, slightly oxidized) odor—Banyuls (may have tawny color), Tarragona (such as Priorato).
        -2- With a film yeast odor.
          -a- Dry—amontillado (Spanish), California dry flor sherry, Château Chalon (France), Spanish fino-type sherry outside Spain.
          -b- Sweet—California medium or sweet flor sherry, some sweet Spanish sherry types.
      (b) Without a special odor due to treatment or aging.
        -1- With amber color—Angelica and some oloroso (Spanish) types
        -2- Without amber color—white port.

NOTE: Only the more important types have been included in this classification with particular emphasis on California wines. Proprietary wines are wines having a name that is owned by a producer. All wine types are in lower case except varietal types, proprietary types, and those having a specific regional significance. Thus, California burgundy is not capitalized as there is no geographical significance to the name as employed here. On the other hand, Burgundy means a wine from the region of that name in France. The "etc" following California usually represents other U.S. wines, such as New York or Ohio wines.

Legal Restrictions

Not only is the composition of the wines offered for sale subject to legal restrictions, but every producer of wines in the U.S. (except those produced in limited quantities in the home) must secure a federal permit and take out a bond before beginning operations. Since this basic permit may be canceled for willful violation of federal laws or regulations, the federal government possesses a powerful tool to deter violation of the regulations. The regulations are frequently changed and will not, therefore, be given in full here. The regional office of the Alcohol and Tobacco Tax Division of the Internal Revenue Service in the Department of the Treasury should be consulted by anyone contemplating going into the wine business as to the necessary forms, bonds, drawings, etc, which must be submitted. The bonded premises must be properly posted, protected by locks, etc, and possess certain equipment (such as scales, ebulliometer, etc) to facilitate inspection by government agents (commonly called gaugers). The capacity of all tanks must be determined and accurate records of production and movement, not only of the wines but also of permissible additives, must be maintained. Where sugar is needed there are strict federal limits on the amount of sugar that may be added. (Sugar may not be used in the production of most types of wines in California.) State laws, as well as those of other countries, on establishing and operating a winery are not so strict or detailed. However, the regulations of many countries on production and the labeling of wines are often detailed and required expert interpretation.

The sale of wines in this country (and in other countries) is subject to a tax. The tax varies from state to state (or country) and mainly depends on the alcoholic content or type of wine. Wines imported into this (or into most other countries) must pay *in addition* a special customs tax. In some cases special tax stamps are purchased and pasted on the bottle or container. Special state taxes for state-sponsored advertising and research may also be assessed, eg, by the Wine Advisory Board in California.

Special regulations apply to the addition of alcohol (fortification) and the production of vermouth and sparkling wines. The amounts of each of the herbs used for vermouth must be given to the authorities. The formula of the sweetening agents used in production of sparkling wines must likewise be stated.

Production Methods

GENERAL

For the details of alcohol fermentation see Fermentation. The special character of each of the various types of wines depends to a considerable extent on the composition of the fruit juice fermented.

Composition of the Raw Material. The most important component is, of course, the sugars, since the alcohol of the wine is derived from them. Grapes contain about 15–25% sugar, but partially dried grapes containing as much as 30–40% sugar are sometimes fermented. The percentage of sugar in the grapes, the extent of the fermentation, and the losses or additions of alcohol during treatment and storage determine the percentage of alcohol in the finished product. Since at least 9% alcohol (by volume) is usually necessary to prevent rapid acetification or spoilage, sugar must sometimes be added to permit fermentation to reach at least this value. The amount of sugar necessary is approximately calculated from the Gay-Lussac equation for fermentation $(C_6H_{12}O_6 \rightarrow 2\ C_2H_5OH + 2\ CO_2)$. Roughly each per cent of sugar fermented yields 0.55% alcohol (by volume). A minimum sugar content of about 16.4% is therefore necessary to produce a wine of 9% alcohol. The soluble solids content of mature grape juice is over 90% sugar. It is common, therefore, to float a hydrometer in grape juice and to determine roughly, the percentage of sugar. The Balling or Brix hydrometer is used in the U.S. This reads grams of sugar per 100 grams of liquid. In Europe the Baumé or Oechsle hydrometers or other types are employed. The refractometer can also be used to determine the approximate percentage of sugar. For fruits, the hydrometer or refractometer is also usually employed. See Sugar analysis.

The practice of adding sugar is necessary in the U.S. only in the midwestern and eastern states. In Europe it occurs most frequently in Germany and occasionally in Switzerland and parts of France. Elsewhere grapes nearly always attain a high enough sugar content to produce a wine of at least 9% alcohol. Addition of sugar is prohibited in California, but in the very cool year of 1948 some grape concentration (about 70% sugar) was added to a few musts (grape juice) from the cooler regions to insure wines of over 10% alcohol. Other fruits and vegetables contain lesser amounts of sugar and the addition of sugar is usually essential. Sucrose or invert sugar is commonly used. The amount of sugar added to fruit is greater than that added to grapes, since the finished

Table 2. Sugar and Acid Content of Certain Fruits

Fruit	Sugar, %	Acid, %[a]	pH
apple	11.1	0.47M	2.9–3.3
apricot, fresh (edible portion, flesh, or flesh and skin)	10.4	1.19M	3.6–4.0
blackberry	6.1	0.91C	3.2–3.6
cherry, fresh			
sour, sweet, and hybrid	14.0		
sour	9.5	1.38M	
sweet	11.6	0.68M	
currant	5.7	2.30C	
fig, dried	55.0	0.60M	
grapefruit, all (edible portion)	6.5		3.0–3.3
california-grown	6.6	2.23C	
florida-grown	6.5	1.16C	
loganberry	6.0	2.18C	
orange	8.8	0.68C	3.0–4.0
peach	8.8	0.64M	3.4–3.6
pear	8.9	0.29C	3.6–4.0
pomegranate	13.3	1.05C	
strawberry	5.3	1.09C	3.1–3.5

[a] M = malic acid; C = citric acid.

wine is usually not fermented dry but is allowed to remain sweet. The sugar and acid contents of various fruits are summarized in Table 2.

The organic acids are the second most important constituent of the fruit. For grapes these are almost entirely tartaric and malic acids. The percentage present varies depending on the state of maturity (decreasing as the grapes ripen), the variety, and the climatic conditions of the season or region (less in warmer regions). It is always higher under cool conditions and may sometimes be so high as to yield an unpleasantly tart beverage. This occurs in eastern U.S., Canada, Switzerland, and Germany in some years, and regulations permit the addition of water or of calcium carbonate to reduce the acidity. The pH of normal grape juice is between 3.0 and 3.6 and the titratable acidity 0.5–1% (calculated as tartaric acid). In this range most spoilage organisms grow slowly or not at all, thus allowing free growth of the desirable yeast. The high titratable acidity of musts also aids in the extraction of color from the skins and in wine clarification. In other fruits the acidity is due to malic, citric, oxalic, and isocitric acids in various proportions. Their acidity is usually sufficient to permit a disease-free fermentation and a stable product. Some fruit and berry wines must be sweetened to mask their excessive acidity.

Only a small amount of nitrogenous material is found in grapes, 0.3–1.0%. But the nitrogen fractions are of considerable significance for yeast nutrition as well as bacterial stability, largely because of the amino acids present. The most important amino acids have now been reported in grape juices or wines: alanine, arginine, aspartic acid, cystine, glutamic acid, glycine, histidine, isoleucine, leucine, lysine, methionine, phenylalanine, proline, serine, theonine, tryptophan, tryrosine, and valine. Smaller amounts of a number of other amino acids have also been found. During fermentation the total amino acid content decreases, although individual acids may be higher in the finished wine than in the must because of their release by autolysis of yeast cells. In other fruits, the low nitrogen content is sometimes the limiting factor in yeast propagation, and nitrogen is then added to stimulate yeast growth. Apple and pear juices, for example, may ferment slowly for this reason.

The pigments of grapes and fruits are usually located in the cells of the epidermis. During alcoholic fermentation the cells are killed and release these pigments. By separating the skin of red grapes from the juice before fermentation, it is possible to produce a white or nearly white wine. Not only are the anthocyanin and related red pigments of grapes responsible for the color but they probably also aid in clarification. Red wine (from red grapes) also contains considerable tannin material, which is important in the taste and color of the wine and also influences its oxidation-reduction potential.

The pectins of some fruits and grapes are a source of difficulty in the clarification of their juices. However, pectins are rather insoluble in alcohol, and more or less complete precipitation occurs during alcoholic fermentation.

The inorganic constituents are not of critical importance as they are usually present in sufficient amounts for yeast or enzyme functioning, but excess iron or copper may cause turbidity.

The average composition of a grape must is given in Table 3.

Microorganisms. Wines are normally produced by fermentation with the yeast *Saccharomyces cerevisiae* var. *ellipsoideus.* This and other yeasts are found on the grapes or fruit and they multiply rapidly in the sweet juice, eventually causing fermentation. While this system is adequate for grapes under most conditions, it may be inadequate

Table 3. Composition of a Grape Must[a]

Content	Range, %	Content	Range, %
water	70–85	polyphenol and related	
carbohydrates	15–25	compounds	
dextrose	8–13	anthocyans	T
levulose	7–12	chlorophyll	T
pentoses	0.08–0.20	xanthophyl	T
arabinose	0.05–0.15	carotene	T
rhamnose	0.02–0.04	flavonol	
xylose	T	quercetin	T
pectin	0.01–0.10	quercetrin	T
inositol	0.02–0.08	rutin	?
alcohols and related		tannins	0.01–0.10
compounds		catechin	T
ethyl	T	gallocatechin	T
methyl	0.0	epicatechin gallate	T
higher	0.0	gallic acid	T
2,3-butylene glycol	0.0	ellagic acid	T
acetoin	0.0	chlorogenic acid	T
glycerol[b]	0	isochlorogenic acid	T
sorbitol	T	caffeic acid	T
diacetyl	0.0	p-coumarylquinic acid	T
aldehyde	T	nitrogenous compounds	
organic acids	0.3–1.5	total	0.03–0.17
tartaric	0.2–1.0	protein	0.001–0.01
malic	0.1–0.8	amino	0.017–0.110
citric	0.01–0.05	humin	0.001–0.002
succinic	T	amide	0.001–0.004
lactic	0	ammonia	0.001–0.012
acetic	0.00–0.02	residual	0.01–0.02
formic	0	mineral compounds	0.3–0.5
propionic	0	potassium	0.05–0.25
butyric	0	magnesium	0.01–0.025
gluconic	?	calcium	0.004–0.025
glucuronic	?[b]	sodium	T–0.020
glyceric	?	iron	T–0.003
glyoxylic	?	aluminum	T–0.003
α-ketoglutaric	T	manganese	T–0.0051
mesoxalic	?	copper	T–0.0003
mucic	T[b]	boron	T–0.007
pyruvic	T	rubidium	T–0.0001
saccharic	?[b]	phosphate	0.02–0.05
amino	0.01–0.08	sulfate	0.003–0.035
pantothenic	T	silicic acid	0.0002–0.005
quinic	0	chloride	0.001–0.010
p-coumaric	T	fluoride	T
shikimic	T	iodide	T
sulfurous	0	carbon dioxide	T
carbonic	T	oxygen	T

[a] Source: M. A. Amerine and M. A. Joslyn, *Table Wines, the Technology of Their Production,* Univ. California Press, Berkeley, 1970.
[b] Except more for botrytised grapes.

for fruits. When the climatic conditions are unfavorable and the grapes are in poor condition, it may also be unsatisfactory for grapes. For this reason it is customary to add a pure culture of yeast. Various strains of *S. cerevisiae* var. *ellipsoideus* are available but only small differences in composition or quality of the resulting wines have been noted when different strains are used on identical samples of must and fermented under the same conditions. However, in Spain a special film-forming strain is naturally present in the sherry district. There have been suggestions that the restricted activity of non-*Saccharomyces* species of yeast might contribute desirable or special odors to certain wines, but no critical experiments with adequate sensory controls appear to have been made. Where pure cultures of yeasts are used, it is customary to add about 1–3%. The actively fermenting culture is usually grown in sterilized must. Recently, pressed wine yeasts grown in non-grape media have been used. These are cheaper than can be produced on a grape media and are easy to add. See also Yeasts.

To prevent growth and competition of undesirable organisms, 50–200 mg/l sulfur dioxide is usually added about two hours before the pure yeast culture. The sulfur dioxide acts as a selective antiseptic and thus permits more or less unrestricted growth of the added yeasts. Originally a piece of sulfur (in the form of a wick) was introduced into the cask and burned. When the must was introduced as much as 25 mg or more per liter of sulfur dioxide might be absorbed from the air in the cask or from sulfur dioxide absorbed on the moist wooden walls of the cask. This practice is objectionable as the amount of sulfur dioxide introduced is not easily controlled, because elemental sulfur sublimes onto the walls of the cask, or because pieces of the sulfur wick drop onto the bottom of the cask. During fermentation this free sulfur is reduced to hydrogen sulfide, giving the new wine a very unpleasant odor. For this reason, salts that yield sulfur dioxide, such as potassium metabisulfite (pyrosulfite), aqueous solutions of sulfur dioxide, or the liquefied gas, are commonly used. The warmer the must and the poorer its quality, the more sulfur dioxide used. No more than 150 mg/l is usually necessary. The sulfur dioxide kills or inhibits the growth and activity of undesirable bacteria and yeasts, increases the extraction of color and soluble material from the skins, and acts as an antioxidant. The resulting wines are thus of higher alcohol, extract, and total acid content, are lower in volatile acidity, and are lighter in color in the case of white wines, somewhat darker in the case of red wines, than those produced without added sulfur dioxide. Any necessary additions, such as sugar, water, acid, or nitrogen, are made at the same time.

Equipment. Wines can be produced with little specialized equipment, but for large-scale operation much specialized equipment has been developed for greater labor efficiency and better results.

To crush the grapes, combined stemmers and crushers are used. These remove the stems first (by centrifugal force) and then crush the detached berries. The actual crushing may be done by centrifugal force or by passing the fruit through rollers or by both. Must pumps are used to transfer the crushed grapes (*must*) to the fermenters or presses.

Fermentation tanks may be of wood, concrete, lined iron (epoxy or their resins or a thin layer of stainless steel) or stainless steel and may be open or closed. Open wooden or concrete tanks are usually used for red musts and closed containers for white musts. The tendency in California and elsewhere is to use stainless steel or lined iron tanks for both fermentation and storage. Large open tanks should have coils, preferably of stainless steel, for cooling or heating the must.

The problem of transferring the residue (stems and skins, often called *pomace* or *marc*) after red-wine fermentation from the fermenter to the press is always difficult. In some wineries, electric elevators are lowered into the tank and the pomace is raised to the top of the tank. There it is either dumped directly into the press or into a trough with a continuous belt or chain to carry it to the press. In some cases, the fermentation tanks are raised above the floor and have a slanting floor level. The pomace is flushed out the bottom of the tank into the press. Sometimes the conveyor arrangement runs along the floor and the pomace is flushed out of the tank and carried to the press. This is especially favored where the wash material is not desired for wine but can be used for distillation.

Two types of presses are used for grape wines and one of these or a third for fruit wines. The oldest types of press still used is the hydraulic basket press. Usually these are vertically operated, either from above or below, but horizontally operated hydraulic presses have recently been used. The hydraulic press is more expensive to operate but produces a clearer juice, whether used on red pomaces or white musts, than the continuous screw-type press. The continuous press seldom operates well on fresh must but is the cheapest to operate with fermented pomace and gives a high yield of liquid (albeit cloudy) with this type of pomace. Where the press-wine can be used for distillation the cloudiness is of no importance.

For some fruits, and occasionally for high-pectin grape musts, the rack-and-cloth press is used. This consists of putting the must in cloth sacks which are placed between wooden racks. A pile of these is then pressed in a hydraulic press. The yield of clear juice is better than with either of the other types, but the cost of operation is greater and the press cloths are difficult to clean. Recently the Willmes press has been widely used. It has an inflatable rubber bag inside a cylinder with perforated holes. The cylinder is filled with crushed grapes. When the bag is inflated the must in the cylinder is pressed. Belt presses which crush and press are also available.

WHITE TABLE WINES

These are defined as white wines of less than 14% alcohol. They include two distinct types, dry and sweet wines, and since they are produced by different procedures, they will be discussed separately. They are usually made from white grapes. A white wine is occasionally made from red grapes (as in the Champagne district of France) by separating the skins from the juice immediately after crushing.

Dry. The basic problem in the production of the dry types is to produce a wine that retains a light color and a fresh, unoxidized flavor. The winemaker's skill is thus directed to rapid transportation of the grapes from the vineyard to the winery. Crushing, stemming, and pressing should follow immediately after picking. The free-run juice is darker in color and higher in tannin. The juice may be "settled" before fermentation. This practice consists of placing the juice in a tank, adding 50–150 mg/l of sulfur dioxide and allowing it to settle for 24–36 hr. It is desirable that the juice be cool (50°F) during this time. Modern wineries often cool the juice. During this settling period pieces of skin and other solid material fall to the bottom. The clear supernatant liquid is then drawn off and used for the fermentation. Settling is carried out most satisfactorily in smaller (less than 2000-gal) containers. It is most useful with varieties that yield a pulpy must and where one desires to obtain a new wine that will clarify rapidly.

The juice, settled or not, is inoculated with a pure culture of yeast. Fermentation should be conducted in closed containers and at not over 60°F. Many white wines are fermented at 40–50°F, but more time is required for fermentation at these temperatures. When fermentation is conducted in tanks of over 2000-gal capacity, the volume of carbon dioxide given off by fermentation is such as to prevent oxygen reaching the wine. For smaller containers fermentation traps permitting outflow of gas but preventing inflow are often employed.

The temperature and percentage of sugar (degree Brix or Balling with the hydrometer) should be determined daily during the fermentation. If the temperature rises much above 60°F, it may be necessary to cool the must. This is usually done by pumping the most through a tubular heat exchanger or into a sump having cold pipes. If the fermentation stops, refermentation may be necessary.

As the fermentation nears its end (the degree Brix or Balling falls towards 0) the tank should be filled with other wine of the same type, or the bung closed with a trap or covered with a thick cloth to prevent air reaching the wine. When the Brix or Balling value reaches about 1° below 0°, the wine is *racked* (drawn off) into another container. This container should be filled completely and the bung tapped in gently. Each day thereafter until fermentation ceases the bung can be loosened to release gas pressure. If the cellar where the wine is stored is cool (about 50°F) the wine will normally settle clear in about six weeks. Settling consists in the precipitation of yeast cells, mucilaginous material, and cream of tartar (potassium acid tartrate). (See Tartaric acid.) The settling is slower the warmer the cellar and the larger the container. Whether the settling is complete or not, the wine must be racked from the sediment in about six weeks. This is particularly important if the wine is stored in a warm cellar since autolysis of the yeast cells may then occur and hydrogen sulfide and other undesirable compounds will be produced. In transferring white wines from one tank to the other as little oxygen as possible should be introduced into the wine (unless the wine has a yeasty or hydrogen sulfide odor, in which case some aeration is helpful). To rack out of contact with air the new tank may be filled with carbon dioxide gas, or the wine may be pumped from the old tank to the bottom of the new tank, or a sulfur wick may be burned in the new tank to remove the oxygen.

The present-day tendency in the U.S. as well as abroad is to bottle this type of wine during the first year. The wine thus receives only two or three rackings before it is ready to be bottled. Special stabilization procedures are necessary to insure that wines bottled this young will remain clear in the bottle.

The general procedure is to rough-filter the wine through a filter press about the first of the year. Later in the spring the wine is chilled to 25°F and held at this temperature for several days to precipitate the excess tartrates. Alternatively, the wine, or a portion of it may be ion-exchanged in a cation-exchanger where the portion of the potassium is exchanged for sodium or hydrogen. This will increase the sodium content. In some countries, Germany for example, the process is not applicable as the sodium content will exceed the legal limit (60 mg/l). It is then racked and clarified (fined) with a slurry of bentonite. The wine may then be closed-filtered through a pad filter directly into the bottle. The more modern and careful wineries usually make a trial bottling which they subject to heat, cold, and sunlight to determine whether the wine will remain brilliant after bottling. If the trial bottling is unsuccessful, the wine may require longer chilling, a heavier fining, or a closer filtration. Just before bottling, the sulfur dioxide content should be adjusted. American practice is to add enough to make the total sulfur dioxide about 150 mg/l. The more rational precedure is to add enough sulfur dioxide to secure about 15–35 mg/l of free sulfur dioxide.

Sweet. This class of wines is difficult to produce and even more difficult to stabilize because the wines contain only 13 or 14% alcohol and may contain from 0.5 to 20% sugar.

The production problem is twofold. First, one must secure a must that is sufficiently high in sugar to produce 13% alcohol and yet have residual sugar. Second, the fermentation must be stopped before the alcohol reaches about 14%, and this is not always easy. In Europe some wines of this types contain slightly over 14% alcohol, but American tax regulations make it essential to avoid this.

In the U.S. a high sugar content in the grapes is secured by delaying the harvest of suitable varieties as much as possible. The harvest cannot be too long delayed however, or the grapes will sunburn or turn to raisins on the vines. The raisin flavor and dark color are undesirable in white table wines. It is thus usually necessary to add grape juice, dessert wine, or grape concentrate to the finished wine in order to secure the necessary sugar content. In Europe the high humidity permits the growth of the fungus *Botrytis cinerea* on the ripe grapes. This fungus extracts water from the grapes, which then shrivel and increase in percentage of sugar, sometimes to as much as 40%. The Sauternes of France, the sweeter Tokays of Hungary, and the sweetest German (Auslese) wines are all produced from this type of grape. Another system is to dry the grapes between straw, by hanging them on strings or laying them in shallow boxes. After two or three months the grapes are crushed and pressed. The first procedure is used to a limited extent in France while the other two are often employed in central and northern Italy, especially for sacramental wines.

The grapes are crushed, stemmed, and pressed as for dry white wines. The pressing may be somewhat stronger in order to get the maximum yield of juice from the shriveled grapes. Settling is definitely advantageous with this type of must. The fermentation must also be conducted at a low temperature, 40–45°F, if possible. Various methods of stopping the fermentation are employed. Frequent rackings during fermentation keep the yeast population low, the successive propagation of the yeast cells reduces the nitrogen content, and the increasing alcohol content in the presence of sugar restricts further yeast growth. Some operators rack off the yeast and add a large amount of sulfur dioxide. If the wine is further cooled at the same time it is helpful. It may even be filtered to remove as much of the yeast as possible.

The sweet new wine must be watched constantly for signs of referementation. Cooling, racking, filtration, and addition of sulfur dioxide or other antiseptic, such as diethyl pyrocarbonate (DEPC), may all be employed to prevent refermentation. As the wine ages it has less and less tendency to referment. Sweet table wines are frequently held in the cask for three or even four years in order to secure stability.

The wine should be bottled only after a trial bottling and a stability test. Various methods for insuring bottle stability are employed. Least desirable is a very high sulfur dioxide content, which injures the sensory quality of the wine. Pasteurization is also used. The wine is first heat-stabilized to a temperature slightly higher than the final pasteurization temperature, then cooled and filtered to remove heat-precipitatable solids. Finally, it is pasteurized hot into the bottles. Screwcap closures are used for sealing in this case. More recently sterile filtration through pad or membrane filters has been employed. This requires a low filter pressure, which reduces the volume, and careful control of the prefiltration sterilization. Use of DEPC is also possible.

While the fermentation-produced and fermentation-stable sweet table wines are generally considered the best, many wines of this types are produced in the U.S. by adding a sweetening agent to a dry table wine. In the eastern U.S., sucrose is often used, with both white and red wine. The wine is usually a blend of California table wine and eastern-Concord-flavored wine. The final sugar content is often 13–14%. Such wine always requires pasteurization or germ-proof filtration. Producers who require a lower percentage of sugar may blend in a sweet dessert wine such as Angelica or port. A number of white, red, and rosé table wines of about 1.5 to 2.5% sugar are produced in California by the use of either dessert wine or concentrate. Occasionally grape juice is preserved with sulfur dioxide (called *muté*) and this is added to make a white sweet table wine. The high sulfur dioxide content is a disadvantage.

RED TABLE WINES

A large percentage of the world's wine is red table wine of less than 14% alcohol. In France and Italy, particularly, it constitutes an important item in the daily caloric intake. Red wines are relatively easy to produce compared with white table wines, and they are less subject to spoilage during aging.

The harvesting should be done when the grapes are sufficiently ripe to produce 11–13% alcohol, if climatic conditions permit. A sugar content of about $21\frac{1}{2}$–$23\frac{1}{2}°$ is optimum. The grapes should be transported from the vineyard to the crusher without delay to prevent development of spoilage bacteria. Prompt crushing and destemming are essential. The must is usually pumped to open fermenters.

Fermentation. The primary problem in red-wine production is the management of the "cap," the floating mass of skins that rises above the liquid during fermentation. If the cap is allowed to become dry it will usually acetify. Furthermore, if the cap is not submerged into the liquid too little color will be extracted from the skins. When small fermenters are used, the cap can be submerged manually. With very large (over 2000-gal) containers, the cap is too heavy to be forced down with wooden paddles. The juice is then pumped from the bottom of the container and sprayed over the cap. Instead of allowing the cap to float "free," various submerging systems have been devised. The oldest was a wooden latticework that fitted into the tank. The must was introduced under the framework so that when fermentation started the cap was retained under the lattice while the juice rose and covered it. In other systems a permanent cover was constructed with a narrow opening into a basin. Sufficient must was introduced into the tank so that when the fermentation started, the cap pressed against the cover and the juice rose through the narrow opening into the basin. Both these systems insured good color and flavor extraction by intimate contact of the cap and the liquid. The disadvantage of the submerged-cap systems is the problem of management of the wooden latticework, the possibility of too much oxidation with the free-floating surface of fermenting must, and an excessively rapid fermentation.

New methods of making red wines include fermentations under pressure, continuous systems, and automatic procedures. To ferment under pressure it is necessary to use a metal tank lined with some inert material. Usually a pressure of about 3 atm is maintained. This does not prolong the fermentation unduly and increases the color extraction. The cap, being under a carbon dioxide atmosphere, cannot acetify. The high cost of the equipment is one disadvantage. Continuous systems of fermentation have been used in Argentina, Italy, the Soviet Union, France and elsewhere. A tall tank is used and the sulfited must introduced about midway. The pomace is continuously taken off the top of the tank and the partially fermented wine is removed from the bottom. The possibility of the fermentation becoming contaminated is one disadvantage of the system. Another is the difficulty of providing grapes of the same variety for a sufficiently long period to make the system truly continuous. Automatic procedures have also been employed in Algeria. The most ingenious of these provides automatic circulation of the must. These methods use the carbon dioxide pressure to raise a portion of the must which then flows back over the cap. Construction costs are high and the quality of the resulting wine is not the best.

Red wines are normally fermented in contact with the skins until 70–90% of the sugar has fermented. Since the color pigments are extracted more rapidly than the tannin, it is better to separate the skins and juice as soon as color extraction has reached its maximum, usually when about 50 to 70% of the sugar has fermented. Moreover, where the grapes contain many shriveled or raisined berries, the longer contact of the skins with the juice increases the extraction of sugar from the high-sugar fruit and may result in wines that are too alcoholic.

Hydraulic basket presses are used for red wines, and if not operated at too high a pressure are very satisfactory. At high pressures too much tannin, bitter-tasting material, and solids are extracted. The same is true when continuous presses are employed with too much pressure. However, if the press juice is used only for distilling, the extra solid material is not important, and continuous presses are less expensive to operate.

After pressing, fermentation of the residual sugar will require from one to six weeks if the cellar is not too cold. It is best to complete the fermentation rather than leave residual sugar in the wine. A low-alcohol sweet wine is very subject to spoilage by a variety of microorganisms, particularly if the pH is above 3.6.

Aging. As soon as the sugar has fermented out, the wine should be separated from the yeast deposit. Red wines clarify themselves more easily than white, and usually a simple racking is sufficient. However, a rough filtration may also be necessary. In warm climates, where the total acidity is low, an early racking is very desirable. In cold climates, where the total acidity is high, the sediment (called *lees*) is kept longer in contact with the wine. The reason is that the wine often contains microorganisms (bacteria) capable of decarboxylating malic to lactic acid. This reduces the titratable acidity, and, since lactic is a weaker acid than malic, leads to a higher pH. The longer contact of the sediment with the wine also results in yeast autolysis which releases amino acids. The autolysis of the yeast provides the amino acids essential for the growth of these bacteria.

Because of their high tannin content, many red wines must be stabilized before bottling. The best red wines are aged for two or three years in wooden cooperage. The excess tannins are then gradually oxidized or combine with aldehydes and precipitate. Tartrate stability is also accomplished by this aging. When red wines are bottled young, a special chilling to remove excess tartrates is necessary. In addition, clarification (fining) with gelatin to remove excess tannin may be required. Just before bottling a close (pad) filtration is used to bring the wine to a perfect state of clarity. Even so, red wines frequently throw a slight deposit when aged in the bottle for several years.

SPARKLING WINES

Wines containing a permanent visible excess of carbon dioxide are called sparkling wines. The nomenclature of the sparkling wines is most complicated. The most famous name is champagne, originally produced only in the region of that name in

France.   In that country this appellation may be used only for wines produced and fermented in that region.   However, the name is used in other countries as well, although local names are used in some countries, for example Sekt in Germany and spumante in Italy.   In the U.S., the term champagne is used for all sparkling wines produced by a secondary fermentation of sugar in closed containers, although for those fermented in tanks it must be so stated on the label, to distinguish them from those fermented in the bottle.   They are made by at least five different processes all of which will be described here, although only three are commonly used in the U.S.

In Europe many wines are bottled with a slight residual sugar content.   During aging this sugar may ferment and the wines become slight gassy.   The slightly gassy wines of Alsace, the Loire, and Switzerland are frequently of this type.   Naturally, a certain amount of yeast growth is necessary to produce this gassiness, but the yeast deposit is often surprisingly small.   This type of wine often does not travel well because if it becomes warm during transit, it loses the gassiness that is one of its chief attractions.   Quite another type of gassiness results when the malic acid of the wine ferments, yielding lactic acid and carbon dioxide.   Wines of high malic acid content are necessary.   The vinho verde wines of Portugal are of this type, but some gassy northern Italian wines also owe their gassiness to such a fermentation.   In the U.S., sparkling wines are produced either by a secondary fermentation or by carbonation.   The former is by far the more important and will be described in detail.

Fermentation-Produced.   Wines intended for a secondary fermentation must be specially produced.   The requirements for white sparkling wines are: 10–11½% alcohol, a pH of 3.0–3.4, a low (below 0.04%) volatile acidity, a fresh fruity taste, a light unoxidized color, and freedom from all undesirable odors or tastes.   For red wines, the color must be a medium red without brown or violet color, and the tannin content must be low.

Production of a wine of these characteristics requires careful control of the harvesting, crushing, pressing, and fermentation.   The harvest must be timed to obtain a sugar content adequate for at least 10% alcohol.   However, it must not be delayed too long or the acidity will decrease too much.   The crushing should be rapid and complete, and pressing and settling of the must should follow immediately.   Where red grapes are used to produce white musts (as in the Champagne region of France), crushing may be omitted and the grapes sent straight to the presses.   This gives a clearer and lighter colored must.

The fermentation should be accomplished at not over 60°F and should be complete, that is, no sugar should remain.   The use of pure yeast cultures is highly desirable for this fermentation.   The new wines are clarified as rapidly as possible.   Cooling to 40°F for several weeks or chilling to 25°F for several days is common.   The wines are fined and filtered to produce a brilliant young wine.

Since no single wine usually has all of the desired qualities, blending is common.   Thus wines of higher and lower alcohol and acidity are used.   However, all the wines employed must be especially low involatile acidity and free from off-odors or off-tastes.   Sometimes it is necessary to correct the total acidity by adding small amounts of citric acid.   The total sulfur dioxide content of the wine should not be greater than about 50 mg/l at this time lest it interfere with the secondary fermentation and contribute an undesirable odor tone to the wine.   The blended wine should be filtered.   It is then placed in a tank with stirrers.   It is analyzed in the laboratory for its sugar, iron, copper and protein.   Recommended maximums are 200, 5, 0.3 and 25 mg per liter respectively.

The total sugar content is then brought to 24 g/l. This is sufficient to give a pressure of 6 atm in equilibrium with the wine if the carbon dioxide is not allowed to escape. Sucrose is not employed, commercial invert sugar solutions being used in the U.S. Rock candy is also sometimes dissolved in wine containing about 0.5% citric acid and heated so as to give a 50% (by volume) solution of invert sugar. An actively fermenting culture of *Saccharomyces cerevisiae* var. *ellipsoideus* is then added. The so-called "champagne" strain is usually employed since it is an "agglutinating" type, about 1% of the yeast culture in active fermentation being required. To aid in later clarification some producers also add a small amount of clarifying agent, such as bentonite, to the wine. At this stage the wine may either be transferred to bottles or to a pressure tank for fermentation. Both procedures are used in the U.S. The tank method is cheaper and less hazardous as the temperature is easily controlled, excess pressure can be allowed to escape, and the wine can be stabilized in the tank. It has the disadvantage that air comes in contact with the wine during the transfer from tank to tank since no counterpressure of carbon dioxide may be used in the U.S. The traditional procedure of fermenting in bottles is considerably more costly, requiring many hand operations and much equipment and time. Greater skill is also needed to conduct the whole operation successfully. It does produce wines of lower aldehyde content and greater bouquet (presumably due to greater yeast autolysis during the longer aging period in bottles).

In the bottle operation the blended, sweetened, yeasted wine is continuously agitated during bottling. A special bottle capable of withstanding a pressure of at least 8 atm is used. A large and special cork is then inserted into the bottle and an iron clamp placed over it. Crown caps are often used as they are cheaper and more convenient to attach and remove. The bottles are then stacked in a constant-temperature room. In the U.S. a fermentation temperature of 60–70°F is sometimes used to insure rapid and complete fermentation. However, the better producers believe quality is enhanced by use of a lower temperature (50–55°F) and there is less breakage of bottles. Bottle breakage is mainly due to defective bottles, but it also occurs when the blend was not properly stirred during bottling so that some bottles have received too much of the sugar. Less than 1% of the bottles should break. The fermentation takes from three weeks to six months, depending on the temperature and the wine. Wines of higher alcohol (11½% or over), lower nitrogen content and at low temperatures (below 50°F) may ferment more slowly. The wine is aged in the bottles for one to three years, since its quality improves during this aging in contact with the yeast.

During fermentation a yeast deposit will form on the lower side of the bottle. Before the wine is to be sold this yeast deposit must be removed. This is done in the bottle process by two operations, both requiring much hand labor. First the bottles are placed upside down in special "A" racks to permit shaking the yeast deposit gradually down onto the cork. This is called riddling and takes about three to six weeks if all goes well. Then the deposit must be removed. The wine is under a pressure of 6 or more atmospheres. To minimize decreases in gas pressure and losses of wine when the cork is removed, the bottles are chilled to about 35–40°F. The neck of the bottle is frozen solid by submerging it in an ice-salt mixture or in special freezers. At this temperature the pressure is reduced considerably, and when the cork is removed the solid plug from the neck of the bottle is ejected, carrying with it the yeast deposit. This is called disgorging. Immediately after the plug is ejected the bottle is placed in a special machine where the bottle is temporarily closed and loss of pressure is thus prevented. In this machine a small measured amount of sweetening agent is added to the bottle and enough wine of the same kind to fill the bottle. The sweetening agent is usually a mixture of brandy, wine, and sugar, the formula being different for each company. One blend is 50% sugar, 35% wine and 15% or more brandy. American producers use little or no brandy. The best aged sparkling wines receive only enough to bring the sugar content to about 1%. This is sold as "brut" or "nature." The so-called "dry or "sec" wines are much sweeter (2–3% sugar), and the "demi-sec" may have 5% or more sugar. Very little "sweet" or "doux" sparkling wine is prepared; as much as 10% sugar is used for this type. A new closure (cork or plastic) is then inserted and a special wire netting placed over it to hold it in. The bottle is then ready for labeling and shipment. Losses during aging and disgorging reduce the pressure within the bottle of the finished wine to about 3–4 atm.

Instead of individually disgorging bottle-fermented wines they are sometimes disgorged into a tank and then filtered into a bottle. No riddling is necessary. This is called the transfer system. To counteract oxygen pick-up some sulfur dioxide is usually added.

In the tank system two or more tanks are required. The first tank has a double wall so that the temperature may be controlled during fermentation and clarification by circulating warm or cold water in the jacket. The fermentation is carried out at about 60–65°F since the equipment is costly and the main advantage of the tank system is its rapid turnover of wine. Some producers carry out fermentation at lower temperatures, 45–50°F, since the quality is better. Fermentation is thus complete in 1 to 4 weeks. The wine is then chilled and filtered to a second tank. Here the requisite sugar is added and the wine bottle under pressure. Occasionally the sugar solution is placed directly in the bottle and the wine then added. The whole process may be completed in four to six weeks. Since there is considerable contact with air during the transfer operation, the wine may become oxidized if no sulfur dioxide is employed as an antioxidant. The sulfur dioxide also helps to prevent fermentation of the added sugar. Unfortunately the odor of the sulfur dioxide is easily discernible when the wine is served, as the release of the gas from the wine serves to release the sulfur dioxide. In the tank process there is little time for the yeast cells to die and a few may pass the filter and get into the wine. Hence the addition of sulfur dioxide is also advantageous in inhibiting yeast growth. In the bottle process, when properly conducted, most of the yeast cells are dead after two or three years' storage.

Sparkling burgundy and other red or pink sparkling wines are produced by exactly the same procedures. Sparkling burgundy and other red types, because of their higher tannin content, are frequently sweetened to 5% to mask the tannin flavor.

Carbonation. Instead of using costly fermentation processes to secure an excess of carbon dioxide in the wine, the gas may be added. The main problem is to secure a good impregnation of the gas in the wine. Most of the old French carbonators employed a counterflow of wine and carbon dioxide gas over glass balls. The modern spray water carbonation techniques have not been applied to wines as much as they should. These provide a finer and more persistent carbonation than the older flow procedures. Carbonated wines are traditionally sold cheaper than fermentation-produced wines. Producers, therefore, sometimes do not exercise as much care as would be desired in selecting high quality wines for carbonation. Somewhat older wines may be used for carbonation than for fermentation, but they should still retain a fruity flavor, have a light unoxidized color, and be free of cask or other off-odors. For the best flavor the wines should have a relatively low redox potential: small amounts of sulfur dioxide and ascorbic acid may be desirable additions. Before carbonation, the wine is clarified and the requisite amount of sugar added, about 1–4%. Carbonation is usually to a pressure of about 4–5 atm. The same closure (cork or plastic) is employed as for other sparkling wines.

DESSERT WINES

About 60% of all the wine produced for sale in California is fortified or dessert wine of 17–21% alcohol, in contrast to other areas where table wines predominate. Both red and white dessert wines are produced, and the white may be either dry or sweet.

White Types. Muscatel, Angelica, white port, and sweet sherry are the important sweet types of dessert wines produced in the U.S.; dry sherry is the only dry type and it is more often than not at least slightly sweet. Abroad, Malaga, Marsala, and Madeira are additional sweet types.

The most important requisite in producing muscatel and related types is a must of high sugar content, 24° or over. However, the fruit should not be raisined. The crushing is as for dry white table wines. But pressing is frequently dispensed with in California. The free-run juice only is drawn off and used for wine. The sweet pomace is then mixed with water and used for producing wine for distillation. This is economically sound as a very large amount of alcohol is required in the production of dessert wines. Since only alcohol from wines may be used, a huge quantity of distilling wine is required. In the production of muscatel the must and skins usually remain in contact for a day or two in order to increase the extraction of muscat flavor from the skins.

Normally about 100 mg/l of sulfur dioxide is added after crushing, but except at the beginning of the season pure yeast cultures are seldom used in dessert wine production. This is because the fermentation period is very short, usually only two or three days. For example, a must of 25° will be fermented only to about 15° in order to produce a finished fortified wine of 12% sugar. At 15° there will be less than 5% alcohol. Since dessert wines are sold at 17 to 20% alcohol, about 12 to 15% of alcohol must be added. In the U.S., this addition of alcohol is called fortification and is carried on under strict control and supervision of the agents of the U.S. Internal Revenue Service, called storekeeper-gaugers. The fortifying brandy (or alcohol), about 95% or 190 proof, is kept in locked tanks. The wine to be fortified is pumped to another tank. There the gauger (or the winemaker) determines the volume of wine and its percentage of alcohol. The correct amount of brandy to add to bring the percentage of alcohol to over 17% (but less than 21%) is then calculated and weighed out in a weighing tank. The fortifying brandy is then pumped into the fortifying tank and mixed. The mixing is usually accomplished with compressed air. Fortifications of 25,000–100,000 gal are commonly made in California.

The new wine is then pumped to storage tanks where the yeast is allowed to settle. The first racking and rough filtration follow in a few weeks. Dessert wines can be brought to brilliancy earlier than table wines because they are not so sensitive to the undesirable oxidative changes that can occur with repeated manipulations of table wines. Chilling, pasteurization, heavy fining, and repeated filtrations are all used to secure a brilliant wine. Many dessert wines reach the market within six months of the vintage and are stored only in lined metal or concrete tanks.

However, dessert wines do profit by aging in wooden casks for several years and, where the original quality justifies it, this aging period is desirable. Because of their higher alcohol content, dessert wines are not so subject to acetification and thus can remain in the cask at higher temperatures for longer periods than table wines.

California white port is very young Angelica which has been partially decolorized with large amounts of charcoal or made from very light-colored free-run musts. It is seldom aged. Portuguese white port is not decolorized. California white port, Angelica, and muscatel are marketed with about 12–14% sugar.

California sherry is a fortified wine of low sugar content which acquires its characteristic odor by being baked at about 135°F for three of four months. It therefore resembles the wines of Madeira much more than those of the Spanish sherry district, which are produced by an entirely different process. This illustrates one of the disadvantages of using foreign appellations for native wines. The must is fermented as for sweet dessert wines except the fortification is postponed until only 1–7% sugar remains. The fortified wine is clarified and then placed in large redwood, concrete or lined metal tanks. The wine is brought to the desired temperature by circulating hot water or steam in pipes placed in the tanks. After the wine reaches the desired temperature it can be maintained there with only a small additional amount of heat, especially if the tanks are insulated or are placed in insulated rooms. Very little wine is baked by placing it in barrels in the sun, but this is expensive as there is a large loss of wine by evaporation. After the heating, the wine is clarified and marketed, although it improves by aging in oak cooperage—50–2500 gal in size.

Sweet California sherry is usually a blend containing some slightly sweet sherry produced as above plus some very sweet California Angelica. If a very sweet wine is baked at 140°F, a burnt taste may result, hence blending is necessary. A wine type called California Tokay is also produced by blending about equal proportions of sherry, port, and Angelica. This has a brown-pink color and some of the slightly burnt flavor of sherry.

In Spain for sherry, after the primary fermentation to about 15% alcohol, a surface film stage of the yeast occurs. A thick film forms on the surface of the wine which is kept in barrels that are filled only about three-fourths full. Various oxidation-reduction changes occur in the wine during the growth of the film, including increase in acetal and acetaldehyde and decrease in acetic acid, which give the wine a characteristic and much appreciated bouquet. The same process is now employed on a large scale in South Africa and to a lesser degree in Australia and California.

Recently the film-type yeasts have been successfully grown in submerged culture. The recommended procedure is to use a suitable dry white wine of 15.5% alcohol in a lined tank equipped with a stirrer. A rapidly fermenting culture of the film-type yeast is added. An air pressure of about 1 atmosphere is maintained and a small amount of air bubbled through the wine. Periodically the stirrer is operated to prevent the yeast from settling to the bottom of the tank. Under these conditions the aldehyde content will increase rapidly, often reaching 500–700 mg/l in 2 or 3 weeks. This is more aldehyde than is needed in commercial wine so it is blended to about 150–200 mg/l for commercial sale. The process is being used commercially in Canada, the Soviet Union, and California and perhaps elsewhere.

Red Type. Port and port-type wines are the main red dessert wines in the U.S. although a small amount of red muscatel is produced. The primary problem in the production of red dessert wines is to secure an adequate extraction of color during fermentation. It is made more difficult by the limited period of fermentation and the normally low color of grapes grown under warm climatic conditions. Most of these wines are produced in warm regions in order to secure grapes of high sugar content. The grapes are harvested at a 24° or over and crushed and stemmed as for red table wine. In order to secure a good extraction of color from the skins one of two methods is used: heating or special management of the cap. The juice and skins are heated to release the color pigments from the skins. In the best procedures the juice and skins are heated very rapidly to about 180°F, held for one or two minutes at this temperature, and then cooled. In the older system the mass of grapes was gradually heated to 140–160°F and then cooled usually with hot and cold pipes in the tank. This takes more time and may give a cooked taste. The cooled must is then fermented to the proper Brix or Balling.

Where no heating is used the cap must be pressed down into the fermenting liquid frequently. This is difficult with large tanks, and pumping over of the juice is then employed. The treading of grapes in shallow tanks in Portugal has the same objective —extraction of the maximum color from the skins during the short fermentation period. Treading, however, is disappearing in Portugal for more modern procedures. Whatever the system, pressing or drawing-off occurs in about two to four days depending on the temperature of the grapes and other factors.

Ports are relatively simple to clarify and age. Some wines made from heated musts contain too much pectin and remain cloudy. In this case a pectin-splitting enzyme may be used to aid clarification. Two types of ports are commonly made, one in which the normal red color is retained and the other, a tawny type, which may require three to five years to develop the proper tint. Alternatively, early-maturing tawny ports may be produced by using tawny-colored grapes. Imitation tawnys are sometimes produced by heating for three of four weeks at 140°F. A very limited amount of the finest Portuguese and ports of other countries are bottled after two years' aging. After 10 to 20 years in bottle this wine develops a special bouquet which is much appreciated. This is called vintage port.

FRUIT WINE

The production of fruit wines is similar to that of grape wines except the use of sugar is always necessary because of the low sugar content of the berries or fruit.

The crushing problem is not difficult as most fruits are soft and easily crushed. The juice is separated from the pulp of apples and pears, but most other fruits, particularly berries, are fermented with the pulp. Sufficient sugar is added to make a finished wine of 11–13% alcohol. Usually less sugar is used for European fruit wines, only enough to produce 9–11% alcohol. If fermented with the pulp, the juice is separated when color extraction is complete. The new wine is clarified in the usual way, but most fruit wines in the U.S. are sweetened to 10% or more sugar before sale, although apple and pear wines are occasionally sold without sweetening. A few wines, peach and apricot, have been fortified to 18% alcohol as well as sweetened.

Grape and berry wines may not be mixed unless sold and labeled as a mixture. All berry and fruit wines are best when consumed young. After aging, the color fades and the characteristic odor of the fruit often disappears or is greatly reduced. Fortunately, these wines are easily stabilized (the tartrate precipitation problem does not exist) and they can safely be marketed soon after production. They are almost always bottle-pasteurized to prevent fermentation in the bottle.

VERMOUTH

Vermouths are nearly dry or sweet fortified wines to which herbs or herb extracts are added. The nearly dry or French type is used straight or for making martini cocktails. The sweet or Italian type is used for Manhattan cocktails but in Europe is more often drunk as a dessert wine or even as an aperitif. The herbs in vermouth should be easily detectable, but the odor of no single herb should be allowed to predominate.

Dry. To make dry vermouth, a light-colored wine of moderate total acidity is fortified to about 17–18% alcohol. In the U.S., dry table wines and dessert wines, such as low-sugar sherry material, are sometimes blended to yield a wine of the proper alcohol content and 1–4% sugar. To this base a mixture of herbs is added. This includes wormwood, gentian, orris, marjoram, centuary, bitter orange peel, pomegranate root, anise, nutmeg, vanilla, cinnamon, and others. The herbs and the proportions of each used are the commercial secret of the various producers. Typical formulas are given in many texts, *Dessert, Appetizer and Other Flavored Wines* is one.

The herb mixture is placed in sacks that are submerged in the wine, or it may be mixed directly with the wine. The extraction lasts from a few days to a fortnight, depending on the amount of vermouth flavor desired in the finished wine. Too long a period is undesirable since tannins and other bitter substances are also extracted. In some wineries herb extracts are used.

The wine is then separated from the herbs, filtered, and stabilized. Some aging is desirable but the color should be kept light. Much of the vermouth sold in the U.S. today is partially decolorized with charcoal. This is because bartenders prefer a very light-colored vermouth in making martinis, the consumer then believing that he is getting a four to one gin to vermouth ratio, whereas the bartender can use a two or three to one ratio with the cheaper, light-colored vermouth. In addition the demand is now for a colorless martini.

Sweet. The Italian-type vermouth was originally produced with a muscatel wine as a base. At present any sweet white dessert wine is used and grape concentrate or sucrose may be employed for sweetening. Furthermore, some caramel or dark-colored sherry is frequently added. Herb extracts rather than the herbs are employed. The appropriate mixture of herbs is extracted with alcohol and the herb extract filtered. Then the proper amount is added to the wine base. The sweet vermouth herb mixtures often contain some vanilla. Clarification of sweet vermouths may be difficult if the herb extract was not clear or if the herbs were placed directly in the wine. Chilling and heating and heavy clarification with bentonite are then used to secure clarity.

FINISHING

The aging, blending, clarification, and bottling practices in American wineries are carefully controlled by laboratory tests. Trained chemists taste and analyze the wines and prescribe the proper treatments to secure stability and the best quality. The clarification practices commonly used are filtration, fining refrigeration, and pasteurization.

Filtration. This operation consists of passing the wine through a pad or layer of inert material with a fine pore size. For removing large particles the filter press with coarse pads is used. This has a large filtration surface and the plates are relatively large. In order to prolong the length of the filtration period, a filter aid of diatomaceous earth powder is mixed with the wine as it passes to the filter. Other filters consist of fine screens on which a precoat of porous inert material is placed before the wine and filter aid are passed through. Smaller filters with pads of small pore size are employed for polishing filtration. These are also used for prebottling filtration. For sweet table wines sterile filtration may be required. The pore size of the pads is sufficiently small to remove yeast cells. The entire bottling line must be sterile and the filtration is conducted at a low, steady pressure. Membrane filters of small pore size but a large per cent of pores per unit area are now being used for this purpose.

Fining. Adding to the wine a material that mechanically or chemically clarifies it is a practice of ancient origin. A solution of gelatin combines with the tannin in the wine, and the precipitate mechanically fines or clarifies the wine. A solution of egg white or isinglass may also be used. More recently the organic fining agents have largely been supplanted by bentonite, a montmorillonite clay of tremendous swelling properties and capable of some adsorption of material. Its fining effectiveness depends both on this and the mechanical clarification produced as its particles settle. It may be used on hot or cold wines and addition of an excess will not lead to cloudiness, as it may with the organic agents. It is especially effective in removing proteins. One objection is the large flocculant deposit which bentonite produces.

Refrigeration. Under natural conditions in Europe, wines lose their excess cream of tartar because of the cool winters, the small containers (greater surface to volume), and the longer period of aging. In California the mild winters, the huge containers, and the relatively shorter aging period necessitate artificial chilling to remove tartrates. Table wines are chilled to about 25°F and dessert wines to below 20°F. Both chilled rooms and insulated tanks with circulating systems are found. The process requires about two weeks and the wine should be filtered cold off the sediment to prevent resolution of tartrate.

Ion-exchange resins are now used by many wineries, partially replacing potassium by sodium and/or hydrogen. These must be used with caution. At least 500 mg/l of potassium should remain in the finished wine.

Pasteurization. Pasteurization is used both to stop progress of disease and to aid in clarification. The latter is by far the more important in the U.S. Contrary to common belief, pasteurization does not prevent reinfection of wines; in fact, pasteurized wines may be more sensitive to infection. Pasteurization, however, when properly used does kill undesirable bacteria and when the bacteria activity has not progressed too far is a useful practice. Temperatures of above 150°F are needed. When heating to aid clarification, even higher temperatures but short periods are used. The wine is then cooled and filtered.

Marketing

Automatic bottling, corking or capping, and labeling lines are employed in the larger American wineries. These require less labor and are more satisfactory than hand procedures. The bottles are cased in cardboard boxes and are then ready for shipment. Most wines are sold in $\frac{1}{5}$-gallon bottles but $\frac{1}{10}$-, $\frac{1}{2}$-, and 1-gallon containers are also standards of fill.

Corks are employed for wines to be aged in the bottle, that is, for the better red and white table wines. Plastic or metal screw caps with inert liners are used for other table wines and most dessert wines.

Diseases. Wines are subject to few diseases since the alcohol content and their low pH make them relatively immune to spoilage. Table wines exposed to the air may acetify.   Keeping the containers full is the solution to this problem, or pasteurization if acetification has proceeded too far.   Low-acid wines may have one of several types of spoilage from various types of lactic acid bacteria; low concentrations of sulfur dioxide prevent these from being a problem.

Excess iron and copper occasionally cause cloudiness in white table wines.   The excess copper and iron do not come from the grapes but from pickup from copper and iron filters, pipes, pasteurizers, etc.   The best procedure is prevention by installing stainless steel or other inert equipment.   Iron cloudiness is partially inhibited by use of small amounts of citric acid, which forms a complex with the iron and thus prevents the formation of ferric phosphate.   Copper cloudiness is also partially inhibited by fining the wine while hot with bentonite.

Economic Aspects

The production and consumption of wines in the U.S. is only a small fraction of the world total.   In 1967, the estimated world production was 4,852,000,000 gal, of which France and Algeria produced 1,616,676,000 gal; Italy, 924,000,000 gal; Spain, 528,360,000 gal; and the U.S., 120,816,000 gal.   Per capita consumption per year was 36.4 gal in France, 25.2 in Italy, and about 0.88 in the U.S.   The wine industry of the U.S. has been actively engaged in various types of advertising to increase the low per capita consumption in the U.S.

Table 4.   U.S. Production of Wines, 1959–68 Crop Year Average[a]

Type	California, 1000 gal	Rest of U.S., 1000 gal	Total, 1000 gal
gross	160,045	30,903	190,948
sparkling	3,705	3,315	7,020
vermouth	3,071	1,883	4,953
other special natural wines	14,721	697	15,418

[a] Gross wine production is quantity removed from fermenters plus increase after fermentation by amelioration, sweeting and addition of wine spirits, less withdrawals for distillation.   Includes wines subsequently used in producing sparkling wines, vermouth, or other special natural wines.   Prepared by Wine Institute from reports of U.S. Treasury Department, Internal Revenue Service.

Table 5.   Wine Entering Distribution Channels in the United States, 1960–69 Calendar Year Average[a]

Type	U.S.-produced, 1000 gal	Imported, 1000 gal	Total, 1000 gal
table, still	65,748	9,581	75,329
table, sparkling	6,686	1,478	8,164
dessert	81,245	1,551	82,795
vermouth	4,636	4,130	8,766
other special natural wines	14,761		14,761
Total	173,076	16,740	189,816

[a] Prepared by Wine Institute from Reports of U.S. Treasury Department, Internal Revenue Service and U.S. Department of Commerce, Bureau of the Census.

While world production consists predominantly of red and white table wines, the U.S. produces about equal amounts of table and dessert wines. However the proportion of table wine is increasing.

Table 4 indicates the dominance of the California wine industry in the U.S. Actually some California grape concentrate and wine spirits are shipped to eastern markets and then appear as eastern production. California produces more sparkling wine and vermouth than the rest of the country. California wine also is shipped east and is there converted into sparkling wine or vermouth.

Consumption of the various types of wines, both domestic and imported, is given in Table 5.

The U.S. market is thus largely for domestic wines except for sparkling wines and vermouth, which are still imported to a considerable extent.

The prices of California wines have fluctuated very widely over the years. Many factors other than grape supply appear to have influenced this wide fluctuation. In 1967 bulk dessert wines were selling at 55¢ per gallon f.o.b. the winery, and red and white table wines at 35–45¢ per gallon.

Bibliography

"Wine" in *ECT* 1st ed., Vol. 15, pp. 48–72, by Maynard A. Amerine, University of California.

Am. Soc. Enologists, Proc. **1950–1953** (Davis, Calif., 4 vols.).

M. A. Amerine, *Wines & Vines* **34** (12), 25–28 (1953); **35** (1), 29–30, **35** (2), 27–30 (1954).

M. A. Amerine, *Advances in Food Research* **5**, 353–516 (1954).

M. A. Amerine and M. A. Joslyn, *Table Wines, the Technology of Their Production*, Univ. California Press, Berkeley, 1970.

M. A. Amerine and R. E. Kunkee, *Ann. Rev. Microbiol* **22**, 323–358 (1968).

M. A. Amerine and L. Wheeler, *A Check List of Books and Pamphlets on Grapes and Wine and Related Subjects*, 1938–1948, Univ. California Press, Berkeley, 1951.

M. A. Amerine, H. W. Berg, and W. V. Cruess, *The Technology of Wine Making*, 2nd ed., Avi Publishing Company, Westport, Conn., 1967.

Association of Official Analytical Chemists, *Methods of Analysis*, 10th ed., Washington, D.C., 1965.

F. von Basserman-Jordan, *Geschichte des Weinbaues*, 2nd ed., Frankfurter Verlag-Anstalt, Frankfurt-am-Main, 1923, 3 vols.

H. Berg, *Proc. Am. Soc. Enologists*, **1951**, 90–147 (1951).

L. Benvegnin, E. Capt, and G. Piguet, *Traité de Vinification*, 2nd ed., Librarie Payot, Lausanne, 1951.

G. Chappaz, *Le Vignoble et le Vin de Champagne*, Louis Larmat, Paris, 1948.

C. Chatfield and G. Adams, "Proximate Composition of American Food Materials," *U.S. Dept. Agr. Circ.* **549**, Washington, D.C., 1940.

M. Ferrarese, *Enologia Practica Moderna*, 3rd ed., U. Hoepli, Milano, 1951.

J. C. M. Fornachon, *Bacterial Spoilage of Fortified Wines*, Australian Wine Board, Adelaide, 1943.

J. C. M. Fornachon, *Studies on the Sherry Flor*, Australian Wine Board, Adelaide, 1953.

L. Genevois and J. Ribereau-Gayon, *Le Vin*, Hermann & Cie, Paris, 1947.

M. Gonzalez Gordon, *Jerez-Xeres-Scheris*, Jerez de la Frontera, Spain, 1948.

C. von der Heide and F. Schmitthenner, *Der Wein, Weinbau und Weinbereitung*, Vieweg, Braunschweig, 1922.

K. Hennig, *Deut. Wein-Ztg.* **87**, 551–56 (1951); **88**, 258–60, 281–83, 569–70 (1952); **89**, 29–30, 372–76, 387–92 (1953); **90**, 384–86, 388 (1954).

W. Q. Hull, W. E. Kite, and R. C. Auerbach, *Ind. Eng. Chem.* **43**, 2180 (1951).

P. Jaulmes, *Analyse des Vins*, 2nd ed., Librairie Coulet, Dubois et Poulain, Montpellier, 1951.

M. A. Joslyn and M. A. Amerine, *Dessert, Appetizer and Other Flavored Wines*, Division of Agricultural Sciences, Berkeley, 1964.

J. Marcilla Arrazola, *Tratado Prático de Viticultura y Enologia Españolas*, SAET, Madrid, 1950, 2 vols.

A. Marescalchi and G. Dalmasso, *Storia della Vite e del Vino in Italia*, Presso Arti Grafiche Enrico Gualdoni, Milano, 1930–1937, 3 vols.

P. A. Maveroff, *Enologia*, J. Best, Mendoza, Argentina, 1949.

G. L. Mehren and S. W. Shear, *Economic Situation and Market Organization in the California Grape Industries*, Giannini Foundation of Agr. Econ., Mimeo. Report 107, 1950.

J. Ribéreau-Gayon and E. Peynaud, *Analyse et Contrôle des Vins*, Librairie Polytechnique Ch. Béranger, Paris and Liège, 1947.

J. Ribéreau-Gayon and E. Peynaud, *Traité d'Oenologie*, Librairie Polytechnique Ch. Béranger, Paris, Vol. 1, 1960; Vol. 2, 1961.

J. L. St. John, *Food. Inds.*, **13** (12), 65–67, 103 (1941).

F. Schoonmaker and T. Marvel, *The Complete Wine Book*, Simon and Schuster, 1934.

G. Troost, *Die Technologie des Weines*, 3rd ed., E. Ulmer Verlag, Stuttgart, 1961.

U.S. Internal Revenue Service, *Federal Wine Regulations*, 26CFE (*1954*) Part 240, Commerce Clearing House, Inc., Chicago, 1954, pp. 16001–16180; 16201–16240.

M. A. AMERINE
University of California
Davis, California

WINGSTAY. See Rubber chemicals.

WINTERGREEN OIL, methyl salicylate, $C_6H_4(OH)COOCH_3$. See Oils, essential, Vol. 14, p. 214; Salicylic acid, Vol. 17, p. 730.

WITHERITE, $BaCO_3$. See Vol. 3, p. 82.

WOLFRAM AND WOLFRAM ALLOYS

Wolfram, W, at. no. 74, is a transition metal in group VI of the periodic table, below molybdenum, and between tantalum and rhenium horizontally. The free atom has the electronic configuration (2) (8) (18) (18) ($5s^2, 5p^6, 5d^4$) $6s^2$, resulting in chemical valences of $+2, +3, +4, +5,$ and $+6$, the last being the most stable. Its atomic weight is 183.85 (based on carbon-12, International Union of Pure and Applied Chemistry, 1961). There are five stable naturally occurring isotopes: 180 (0.14%), 182 (26.41%), 183 (14.40%), 184 (30.64%), and 186 (28.41%) (1). Eight artificially radioactive isotopes have been produced. [188]W has the longest half-life, 65 days.

Wolfram has a higher melting point, 3387°C, than any element and all but a few compounds. As a consequence, it can be used for components and structural members at very high temperatures. Its use as filaments in incandescent lamps is the most widely appreciated, but consumption in high-speed tool steels and in wolfram carbide account for the largest tonnages. Modern weaponry has expanded its applications in unusual ways. There are very few wolfram-based alloys in the usual sense; they are discussed under Uses below.

The origin of the word wolfram has been variously explained. Agricola in 1556 called wolfram ore *lupi spuma* (foam of the wolf), and the explanation is said to be that the mineral is generally discovered in conjunction with tin, which it "eats up as a wolf eats up sheep." *Lupus*, the Latin for wolf, has been translated in the works of the metallurgists of Germany into their own *wolf* and *ram*, a middle-high-German equivalent of *Rahm* meaning soot or dirt, presumably because of the black color and friability of wolframite ore to which the word wolfram referred up to the nineteenth century. In English-speaking countries, the metal is commonly called tungsten, especially commercially. The name is derived from the Swedish words *tung*, meaning heavy or ponderous, and *sten*, meaning stone. Its application to wolfram derives from the very early work of the famous Swedish chemist, Scheele. In 1781, he established that the mineral, later named in his honor, yielded lime and a yellow acidic powder, which he called tungstic acid. Despite the efforts of the Fifteenth Conference of the International Union of Pure and Applied Chemistry in 1959 to have the term "wolfram" adopted universally, the British and American usage "tungsten" remains in competition with wolfram and both are officially sanctioned by the major scientific societies in these countries.

Two Spanish brothers with the name de Elhuyar (variously spelled, they themselves did not always use the same spelling) showed in 1783 that wolfram and wolfram ores yielded Scheele's acid. The same year they produced, by igniting an intimate mixture of wolframic acid and powdered charcoal, the first metallic wolfram.

In 1785 the German Raspe demonstrated that wolfram hardens steel. The greatly improved cutting qualities of steels containing 2–6% wolfram were demonstrated in the mid-nineteenth century. The use of wolfram for this purpose constitutes a major consumption. It is added, often as ferrowolfram, to give steels with up to 20% wolfram.

The filaments of incandescent lamps developed from carbon to osmium to tantalum and finally to wolfram. Technical efforts to draw the refractory metals began in Germany in 1902. The initial processes started by extruding pastes of tantalum or wolfram powders together with binders, such as dextrin, starch, and sugar, through a diamond die. The discovery leading to modern processes for manufacturing wolfram filaments was made by W. D. Coolidge in the United States in 1909. He demonstrated the workability of the metal at high temperatures. Pressed and sintered powders were worked to rods and drawn into wire which, by virtue of its fibrous structure, retained its ductility at room temperature. That powder-metallurgical process is in principle the one used for almost all massive wolfram available today as sheet, rod, wire, tubes, and forged shapes. Although melted and extruded metal is available, it is a costly product not in general use.

The next development leading to further widespread use of wolfram was the hard-metal-carbide industry dating from perhaps 1914 when Voigtländer and Lohman filed the first of a series of German patents involving the manufacture of wolfram carbides and evolving into the modern cemented carbides containing principally wolfram carbide and cobalt, with or without tallium and titanium carbide.

Physical Properties

Massive wolfram exists primarily as a silvery gray or tin-white body-centered-cubic (bcc) metal, the so-called α-wolfram; the physical properties are shown in Table 1. Many of the most reliable data are still the result of the very extensive work done at the time of the First World War and summarized in 1925 (2) by two outstanding investigators, W. E. Forsythe and A. G. Worthing of the General Electric Co. in the United States. An up-to-date and critical review of properties is given by Rieck (3). The thermochemical values have also been critically reviewed by Kubaschewski, Evans, and Alcock (4).

Polemic discussion has gone on for years concerning β-wolfram, a cubic structure with a lattice parameter of 5.046Å, a shortest interatomic distance of 2.52Å, and eight atoms per unit cell instead of the two associated with the bcc structure. Some workers have maintained that it is a lower oxide, W_3O, but it is now generally believed to be an

Table 1. Physical Properties

Property	Value
lattice constant, bcc, at 25°C, Å	3.16522 ± 0.00009
shortest interatomic distance, Å	2.741
atomic radius, Å	1.37
mp, °C	3387^a
bp, °C	5900^b
density, d_4^{25} for monocrystals, g/cm³	19.254
density, d_4^{25} for wrought wire, g/cm³	19.17^c
specfic heat C_p (25–2727°C), cal/(°C)(mole)	$5.90 + 0.78 \times 10^{-3}t$
latent heat of fusion, kcal/mole	8.4^d
latent heat of vaporization at bp, kcal/mole	197.0 ± 5.0
heat content (25–2727°C), H(t)–H(25°C), cal/mole	$5.90t + 0.39 \times 10^{-3} t^2 - 147.7$
coefficient of thermal expansion [$(1/l_0) (dl/dt)$] × 10⁶ (0–2027°C), (°C)$^{-1}$	$4.438 + 5.44 \times 10^{-5}t + 6.60 \times 10^{-7}t^2$
linear expansion from 0 to t°C (0–2026°C), $(l - l_0)/l_0 \times 10^6$	$4.438t + 2.72 \times 10^{-5}t^2 + 2.20 \times 10^{-7}t^3$
logarithm of the vapor pressure p, $\log_{10} p$ (25–3387°C), torr	$-44000/T + 0.50 \log_{10} T + 8.76$
electrical resistivity (−200 to 750°C), microohm-cm	$5.0(1 + 4.7 \times 10^{-3}t + 1.4 \times 10^{-6}t^2)$
electrical resistivity at 25°C, microohm-cm	5.6^e
temperature coefficient of electrical resistivity ρ_0 ($d\rho/dt$), where $\rho_0 = \rho$ at 0° t (−200 to 750°C), (°C)$^{-1}$	$4.7 \times 10^{-3} + 2.8 \times 10^{-6}t$
thermal conductivity (0–2200°C), cal/(cm)² (sec) (°C/cm)	$0.361 - 1.17 \times 10^{-4}t + 2.32 \times 10^{-8}t^2$

^{a} The mp of wolfram is a secondary and the highest reference standard of the 1968 International Practical Temperature Scale (5).

^{b} Calculated from rates of evaporation.

^{c} American Society for Testing and Materials, Standard Specification F290.

^{d} Estimated.

^{e} The value varies from below 5.0 to almost 8.0 microohm cm as one goes from the best monocrystals to highly worked wire. The temperature coefficient, $(l/\rho_0)/(d\rho/dt)$, for annealed wolfram applies fairly accurately to all unalloyed wolfram of reasonable purity.

allotrope of wolfram, which, although it can form only in the presence of oxygen, has no fixed stoichiometry. The maximum oxygen which it can retain corresponds to W_3O. It is unstable above about 630°C.

Chemical Properties

Wolfram's resistance to oxidation varies considerably with conditions. As a very fine powder, ie, with a specific surface area of 5 m²/g or more, it is pyrophoric, but as massive metal, the oxidation rate is very low at room temperature. Brown and dark-blue lower oxides form at low temperatures. They tend to be more protective than the nonadherent yellow trioxide forming at higher temperatures. Appreciable oxidation begins to occur around 400°C and any high-temperature use of the metal requires a neutral or reducing atmosphere or an oxidation-resistant coating. See also Wolfram compounds. Wolfram resists water and water vapor at room temperature, but is oxidized rapidly at red heat.

Dry chlorine, free from air, attacks wolfram at about 250–300°C, forming the hexachloride; but if air or moisture is present, the oxychlorides are formed. Fluorine reacts with wolfram at ordinary temperatures. The pentabromide is formed at red heat in the absence of moisture. The diiodide is formed when wolfram is treated with iodine at red heat.

Wolfram is not attacked by hydrofluoric acid. Warm nitric acid easily oxidizes it to the yellow trioxide, WO_3. Hot concentrated hydrochloric acid and hot concentrated sulfuric acid react only slightly with the metal. A mixture of nitric and hydrofluoric acids acts on wolfram with the formation of the trioxide. Aqua regia causes superficial oxidation at room temperature. Table 2 gives the chemical resistance of sintered wolfram to common acids.

Table 2. Chemical Resistance of Sintered Wolfram

Corrosive agent, 10% soln	Time, hr	Wt loss, g/dm²
nitric acid, cold	24	0.00
hot	1	0.06
sulfuric acid, cold	24	0.00
hot	1	0.01
hydrochloric acid, cold	24	0.00
hot	1	0.01

Wolfram is resistant to molten sodium hydroxide, but the presence of alkali-metal nitrate, nitrite, or chlorate, lead dioxide, or other oxidants in the hydroxide will cause rapid corrosion. Molten nitrates, nitrites, and peroxides react violently. Aqueous alkaline solutions, including ammonia, do not corrode wolfram in the absence of oxygen.

The gases, carbon dioxide, carbon disulfide, sulfur, carbon monoxide, nitric oxide, and nitrogen dioxide react with wolfram at high temperatures. Molten sulfur and phosphorus attack the metal slowly, while the vapors of these elements react with it vigorously. Carbon, boron, and silicon form compounds with wolfram at high temperatures. Nitrogen does not attack it up to 1500°C, but the nitride W_2N is formed when wolfram is reacted with ammonia at 700–800°C. The wolfram metal is not attacked by hydrogen.

Occurrence

Although wolfram is one of the rarer elements, comprising only approximately $5 \times 10^{-4}\%$ of the earth's crust, some fairly rich ores are available around the world. The largest deposits are in China and North Korea. The commercially important ores are of two general types, scheelite and wolframite.

Scheelite is calcium wolframate, $CaWO_4$, which contains 80.6% WO_3 when pure, the commonest impurity being MoO_3. Molybdenum is not an objectionable impurity for the manufacture of alloy steels, but is highly undesirable for making incandescent-lamp filaments. The scheelite deposits are crystalline, white, yellow, or reddish in color, but never dark. The crystals are soft and easily scratched with a knife. The hardness is 4.5, and the specific gravity is about 6, making mechanical separation easy.

Wolframite is essentially a wolframate of Fe(II) and Mn(II), $(Fe,Mn)WO_4$, of varying proportions. Technically, samples of ore which contain more than 20% FeO are called ferberite, while those containing more than 20% MnO are known as hübnerite, and intermediate materials are called wolframite. The term wolframite, however, is often used to cover all materials in the series. The mineral is either crystalline or granular in form, always dark in color, with a tendency toward brown in hübnerite and shining black in ferberite. The mineral is soft enough to be scratched with a knife, is very brittle, and shows perfect cleavage, breaking into thin, lustrous flakes. The hardness is 5.5 and the specific gravity is about 7.5, equal to that of galena and greater than that of cassiterite, with which it is commonly associated in nature.

Wolframite generally occurs in veins in granitic rocks and in metamorphic rocks close to granitic intrusives. Quartz is the commonest gang mineral; cassiterite is usually present, and native bismuth, bismuthinite, molybdenite, and scheelite are occasionally present. Scheelite is mostly mined from contact metamorphic deposits in limestone near granitic intrusives; garnet, vesuvianite, and fluorite are commonly present.

Wolfram ore in the United States occurs underground in low concentrations, 0.5–3%. The grade is generally less than 1%; ore running 3% is considered high-grade and is rare.

Manufacture

Concentration of Ore. Since the ores contain at best only 2–3% wolfram, a concentrate has to be produced first. That is done by crushing the ore to a powder and by various gravity concentrations (qv) and/or by flotation (qv). The objective is to produce a concentrate containing at least 60% WO_3. Since cassiterite, a nonmagnetic mineral of tin oxide, is frequently found associated with tungsten ores, magnetic separators have proved especially useful in beneficiation of the feebly magnetic wolframite. The method of magnetic separation, however, is not applicable to the separation of tin from scheelite concentrates since, in that case, both minerals are nonmagnetic.

Preparation of Oxide. All uses for wolfram, except the use for alloying of steels, require high to very high purity. Interstitial impurities greatly influence the workability of the metal and substitutional impurities are generally deleterious in high-temperature applications. The chemical refinement of ores is consequently somewhat complicated. Three procedures are in common use.

In the case of scheelite, the comminuted refined ore concentrate is digested in hot strong hydrochloric acid. The insoluble wolframic acid formed is filtered from the soluble calcium chloride and washed. The acid is then dissolved in ammonium hydroxide to free it principally from silica. Acidification precipitates a pure ammonium parawolframate, $(NH_4)_{10}[H_{10}W_{12}O_{46}]$, which is readily ignited to the yellow tungstic oxide, WO_3.

In the case of wolframite ores, the classic procedure is digestion in strong caustic to form soluble sodium wolframate, insoluble silica, and iron and manganese hydroxides. After filtration, a pure dihydrate of sodium wolframate is fractionally crystallized, converted to wolframic acid with hydrochloric acid, and washed free of sodium. The acid is then dissolved, as in the case of scheelite, with ammonium hydroxide, and ammonium parawolframate is precipitated by acidification or by evaporation.

A third and new commercial procedure is coming into use for wolframite ores. An acidified aqueous sodium wolframate solution is equilibrated in a multistage process with a tertiary amine sulfate in an organic medium, usually kerosene. The wolframate ion exchanges with the sulfate ion, passing into the organic phase, which is next stripped of wolframate by equilibration with ammonium hydroxide. The result is a very pure aqueous solution from which ammonium parawolframate is recovered.

None of the above processes removes molybdenum. When very low molybdenum contents are required, generally less than 50 ppm, in incandescent-lamp filaments the producer either selects ores with very low molybdenum, or he removes it by precipitation of molybdenum disulfide with sodium hydrosulfide in a slightly acid solution.

Reduction. Although a large quantity of wolfram is prepared for alloying by a carbon reduction or by a thermite process to yield ferrowolfram, reduction with hydrogen is used to produce all of the high grades of metal powder for conversion to metal-cutting grades of cemented wolfram carbides, and for processes leading to sheet, rod, wire, and forgings. The hydrogen reduction can be controlled to give a powder varying from a fine black pyrophoric material to coarse crystals up to 500 mμ in diameter. Powders which are to be consolidated into workable slabs, rods, or forging billets are fine, with average particle sizes from 2 to 8 mμ.

The reduction furnaces are generally tubular with an adjustable temperature gradient from about 500 to 800°C. The fineness of the powder is controlled by the rate and temperature at which the reactions are carried out. The different oxides are produced consecutively. In the case of powder produced for wire, one portion of the wolfram trioxide is often reduced to a lower so-called blue oxide and the second portion to the still lower brown dioxide. Small quantities of "dopants," potassium silicate and an aluminum compound, are added to the blue oxide via an aqueous solution. The two oxides are then blended and reduced to elemental wolfram containing the dopants, which are ultimately responsible for a secondary recrystallization to an elongated interlocking structure in the finished wire. Other dopants for specialized products are added either to the original trioxide or to one of the intermediates. The most common one is thorium dioxide for either shock-resistant incandescent-lamp filaments or TIG (tungsten–inert-gas) welding electrodes. The TIG process is also called GTA (gas-tungsten-arc process).

Powder Metallurgy (qv). The high melting point of wolfram has had the result that practically all commercially available materials originate from worked sintered shapes. The fine powders are consolidated by mechanical or isostatic pressing and sintered in hydrogen at temperatures of 1800°C and above. For manufacture into wire, the powders, doped with aluminum, potassium, and silicon, are mechanically pressed into bars and sintered by an electric current (joule heating) in bell jars water-cooled to about 2800°C. The high temperature vaporizes much of the residual impurity, leaving only about 100 ppm of the combined dopants.

For manufacture into rolling or forging billets, undoped powder is pressed isostatically and is loaded into a flexible mold, sealed tightly, and immersed in water within the press. The press, which is often a converted naval gun barrel, is closed and pumped up to a pressure of 15,000 psi or more. The irregular shape and fine size of suitable wolfram powder result in sound strong pieces. Only thin plates and long rods are difficult to process in that way.

An interesting exception to the above process is one in which wolfram hexafluoride is reduced in a fluidized bed to give spherical particles ranging from 50 to 400 mμ. They are consolidated by hot isostatic pressing to a completely dense billet, which is, however, difficult to work. Mill forms, which are worked from the material, have abnormally high recrystallization temperatures. The process is not a commercial success because of its high cost.

Metalworking. The primary metalworking of wolfram is carried out at very high temperatures, usually about 1500°C. The temperature is lowered as a worked grain structure develops in the metal. In the case of metal destined for incandescent-lamp filaments, the processing must be done in such a way that the worked grain structure in the finished filament can be "flashed" by rapid heating to a structure with long interlocking grains whose diameters are approximately that of the filament itself. It is now believed (6,7) that the role of the dopants is to collect in minute cavities during the hot consolidation of the powder. The cavities are not obliterated during hot working because of the contained dopants, but become long thin regions, which are converted to submicroscopic "stringers" of spherical voids during annealing at high temperature. The stringers, which can be seen in the electron microscope, are believed to be responsible for the unusual secondary recrystallization leading to the coarse elongated interlocking structure formed during the flashing of the filaments. That structure is responsible for the resistance to creep in high-temperature incandescent filaments. Wolfram, being bcc and the most refractory of all metals, is highly brittle and notch-sensitive below its ductile-to-brittle transition temperature, which varies widely with the degree of cold-work and purity. Only monocrystals are ductile at room temperature in the unworked state. This means also that wolfram is very difficult to machine. There is now one method which looks promising for deforming at low temperatures, hydrostatic extrusion. In this scheme, wolfram billets, rod, or wire is put in a high-pressure vessel and extruded through a die with hydrodynamic lubrication. If the receiving chamber is also under high pressure, sound material seems possible from relatively unworked starting stock. Typical pressures would be 400,000 psi into a receiver at 200,000 psi. In the case of drawn wire, further reduction is possible near room temperature by drawing it through a die from a high-pressure chamber. The technique is exciting considerable interest with the advent of reasonably priced high-pressure vessels.

All commercial metal is still processed at high temperatures. Rods are rolled, swaged, or drawn. The lubricant is usually graphite, although glasses are also useful. In wire drawing, the lowest temperatures are used in the drawing of the finest wire, about 500°C for diameters around 0.4 mils (10 μm). Forming at a high rate of energy can also be practiced; again, it is always done hot.

Melting and casting have been developed despite the very high melting point. Vacuum arc melting is done by pressing and sintering hydrogen-reduced powder into electrodes, which are melted by striking an arc between the electrode and a wolfram stool in a water-cooled copper mold. As molten wolfram drips into the mold, the electrode is retrieved so as to maintain a constant arc between the electrode and a molten pool on the solidified ingot. Such material has a very coarse grain structure which is normally refined by extrusion. Ingots can also be made by replacing the arc by an electron beam, which plays locally on the end of an electrode.

Wolfram is exceptionally easily purified and converted to monocrystals for research purposes by electron-beam zone refining. A rod from 0.060 to 0.250 in. in diameter is disposed vertically in a high-vacuum system and a narrow electron beam from a circumscribing cathode is allowed to melt a zone in the rod. The zone is retained by surface tension. Although many metals are now purified by sweeping impurities ahead of a molten zone in a number of passes, wolfram appears to be purified mainly by evaporation of the impurities. Exceedingly high purities are obtained, and single crystals form readily whether or not the molten zone is seeded. See also Zone refining.

Machining is difficult because of wolfram's high modulus of rigidity and very high hardness. It requires extraordinarily stiff machines and tooling. Grinding is usually used as the finishing operation. Ways to limit the difficulties are to machine consolidated shapes before sintering or to use mechanical pressing with the required tolerances. Unlike sintering of pressed iron parts, however, sintering of wolfram parts involves high shrinkage which makes the holding of close tolerances difficult. Doping with thorium oxide at levels of 1–5% renders the metal more machinable, but the greatest machinability comes with the use of 5–10% copper, iron, and/or nickel in varied combinations. In such materials, the wolfram forms a discrete phase. Such combinations have, of course, completely different properties. Their uses are discussed below.

Economic Aspects

As ore concentrate, wolfram is priced on the basis of the "short ton unit" of contained wolfram trioxide. (The short ton unit (stu) is 1% of 2000 lb, or 20 lb.) Over the years, the price has ranged widely. For some years after World War II, the U.S. government was supporting the price by buying large quantities for the strategic stockpile at $65/stu, which is equivalent to $4.10/lb wolfram. The buying kept many marginal mines open in the United States. Since 1964, however, the government has been stabilizing the price by selling it at $43/stu, equivalent to $2.70/lb wolfram; consequently there were only four domestic mines producing in 1969. The largest was

Table 3. 1968 World Production of Wolfram Ore and Concentrate[a] (18)

Country	Contained wolfram, 1000 lb
North America	
Canada	2,855
Mexico	586
United States	10,188
South America	
Bolivia	4,000
Brazil	958
Peru	1,120
Europe	
Portugal	2,855
USSR[b]	13,600
Africa	
Rwanda	708
Asia	
Burma	307
China, mainland[b]	17,700
Japan	1,165
North Korea[b]	4,720
South Korea	4,615
Thailand	988
Oceania	
Australia	2,530
other	1,018
total	69,813

[a] For countries exceeding 250,000 lb.
[b] Estimate.

the Bishop scheelite mine in California. The source second in importance was the molybdenite mine in Colorado, of Climax Molybdenum Co., which produces wolfram as a by-product.

The importation of ore, which is subject to duty ($6.34/stu in June 1969), has been declining in the United States. U.S. consumption is now primarily from domestic production and releases from the government stockpile. Table 3 gives the 1968 world production of ore and concentrate by countries whose production exceeded 250,000 lb of contained wolfram. The importance, judged by consumption, of the uses to which

Table 4. 1968 U.S. Consumption by End Uses, 1000 lb of Contained Wolfram (8)

Use	Ferro-wolfram[a]	Metal powder[b]	Carbide powder	Other wolfram materials[c]	Total[d]
steel (ingots and castings)					
high-speed and tool	952			1018	1970
stainless	140			61	201
alloy (excluding stainless)	200	unknown[e]	unknown[e]	122	322
other	unknown[e]			97	97
cast iron	<0.500			63	63
cutting and wear-resistant materials					
cemented or sintered		834	4012	215	5060
other	<0.500	38	31	35	104
welding and hard facing					
rods and materials	8	391	534	223	1155
nonferrous alloys	160	305		213	678
electrical material	<0.500	104	unknown[e]	27	131
chemical and ceramic uses					
pigments				141	141
other		unknown[e]		221	221
miscellaneous and un-specified uses	91	1941[f]	672	259	2963
total[d]	1552	3612	5249	2695	13108

[a] Includes metal base and metal pellets.
[b] Includes carbon- and hydrogen-reduced wolfram powder.
[c] Includes wolfram chemicals, natural and synthetic scheelite, wolfram scrap, and other.
[d] Totals may not tally since the figures are rounded off.
[e] Withheld as confidential industrial information.
[f] Includes wolfram consumed as filaments in lamps, electron tubes, etc.

wolfram is put is shown in Table 4. It displays the end uses and the form of the wolfram going into those uses. Production and consumption of wolfram are expected to increase about 5% per year in the 1970s, according to unpublished industrial projections.

Prices for various forms of wolfram, eg, powder, mill forms, and wire, depend, of course, on the form. Hydrogen-reduced powder for the best carbide grades ranges from $5.43 to $6.36 (July 1970), depending on the particle size. Forging stock starts at about $9.00/lb. The highest price is for very fine wire. At 0.5-mil diameter there are about 185,000 m/lb, making the price per pound somewhat over $700.

Specifications and Standards

The American Society for Testing and Materials has issued the following (9):

B297-65T	Tentative Specification for Tungsten Arc-Welding Electrodes
B410-68	Standard Specification for Unalloyed Sintered Tungsten Billets, Bars, Rods, and Preforms for Forging
B459-67	Standard Specification for Tungsten-Base, High-Density Metal (often referred to as heavy alloys or heavy metals)
B482-68	Recommended Practice for Preparation of Tungsten and Tungsten Alloys for Electroplating
F73-66T	Tentative Specification for Tungsten–Rhenium Alloy Wire for Electron Devices and Lamps
F288-66T	Tentative Specification for Tungsten Wire for Electron Devices and Lamps
F290-68	Standard Specification for Round Wire for Winding Electron Tube Grid Laterals

Analysis

In qualitative analysis, wolfram is usually converted to a wolframate. A white amorphous precipitate of hydrated tungstic acid is obtained by treating the wolframate solution with nitric acid; upon boiling, the yellow anhydrous acid is obtained. Care is needed when arsenic and phosphoric acids are present as they solubilize large amounts of wolfram as heteropoly acids. If the solution of an alkali wolframate is treated with hydrochloric acid and zinc or aluminum or tin, it turns a beautiful blue. Ferrous sulfate gives a yellowish-brown precipitate which is not turned blue by acid. Lead acetate, barium chloride, and silver nitrate also precipitate white wolframates.

In quantitative analysis, the most satisfactory method for determining wolfram is based upon its separation as the slightly soluble wolframic acid. This is accomplished by digesting the wolframate solution with hydrochloric acid or nitric acid. Since there is some residual wolfram left in the filtrate, a second precipitation, generally in the presence of cinchonine hydrochloride, is necessary. The cinchonine hydrochloride forms a complex with the wolframate ion, which is insoluble in cold dilute hydrochloric acid. Its presence also prevents the formation of the acid wolframate and of colloidal solutions upon washing. The two precipitates are then combined and carefully ignited to the trioxide, WO_3, which is weighed. Since the trioxide is volatile above 750°C, the ignition should not be prolonged or carried out at too high temperatures. If niobium, tantalum, and titanium are present, the weight of the ignited oxide must be corrected. That is done by redissolving in caustic and precipitating the three impurity elements with cupferon after complexing the wolfram with tartaric acid. That precipitate is ignited and the original weight corrected.

Health and Safety Factors

There are no documented cases of wolfram poisoning in humans. The thresholds (10) set for airborne dusts are 1 mg/m³ for soluble wolfram compounds and 5 mg/m³ for insoluble forms, based on studies with animals. The very high densities of wolfram and wolfram carbides make control of dusts easy. Obviously, dusts of wolfram–thoria need special precautions owing to the radioactivity of thorium; however, years of industrial use of wolfram–2%-thoria-welding electrodes have resulted in no problems. The pyrophoricity of very fine wolfram powders dictates caution.

Uses

Innumerable ferrous alloys in wide usage contain up to 20% wolfram as an addition. The wolfram enables the steel to resist the effects of high temperature and forms hard, wear-resistant carbides in the steel. Prominent among these wolfram tool steels is the so-called Grade T1 or 18–4–1 high-speed steel whose composition is Fe–18W–4Cr–1V. (See also Tool materials.) The percentage of vanadium may vary up to 3% with or without the addition of cobalt. The high-speed steels (18% W) are the most widely used wolfram-containing steels. Their main characteristic is the maintenance of a sharp cutting edge at working temperatures far above those that ruin carbon-steel tools, a property generally known as red-hardness. The wolfram imparts a cutting or lathe efficiency approximately five times that of ordinary carbon steel. Hot-work steels (10% W) are used in operations where a minimum loss of hardening is desired when the steel is used at about the tempering temperature. Wolfram hot-work steels generally retain considerable hardness up to about 600°C, while ordinary chrome steels have a temperature limit of about 400°C. Finishing steels (3.5% W) possess extreme wear resistance or abrasion resistance, whereby a keen cutting edge can be retained. They, however, are not designed for high-temperature operations and a service temperature greater than 150°C is not recommended. Wolfram is added to other steels for various reasons, for example, creep resistance.

The other high-volume use of wolfram is in the carbide industry, where large quantities go into WC for cutting tools and wear-resistant surfaces. See also Carbides. It is also used in tire studs to improve automotive traction on ice. The extreme wear resistance of WC allows the studs to withstand the abrasion of road surfaces.

Wolfram carbide is usually made by intimate mixing—ball-milling for 5 days is common—of pure grades of wolfram and carbon and reacting in covered graphite pots heated by induction to 1500–1600°C. The exothermic reaction makes temperature control important, since very high temperatures result in particle growth having the ultimate effect of producing a softer cemented carbide. The resulting carbide is crushed and mixed with cobalt powder, which acts as a cementing agent when the product is pressed into shapes and sintered. The result is a cemented wolfram carbide that approaches the diamond for hardness. Because of its high hardness and transverse strength, wolfram carbide is a component of hard materials used for cutting tools, dies, and wear-resistant parts, and has become important as a material for insert bits used in rock drilling. Various materials containing wolfram carbide are available. The straight wolfram carbide with cobalt binder is used in tools for cutting metals and also many nonmetallic materials, in dies for drawing wire, bar, and tubing, in dies for blanking, and in dies for mechanical pressing, as well as for machine parts where resistance to wear is important. The steel-cutting grades consist largely of wolfram carbide with additions of titanium carbide or tantalum carbide or both. The added constituents produce a material that resists cratering better than do any of the straight wolfram carbide grades. Solid solutions of WC in TiC are also becoming more popular as a component in cemented carbides.

The most pervasive use of wolfram is as the filaments in incandescent lamps. Every household light bulb has, as its heart, a small coil, often a coiled coil or double helix, of fine wolfram wire. Much research and engineering is still going into increasing the luminous efficiency and life of that household appliance.

The heaters for electronic tubes are also wolfram and sometimes a W–3Re alloy. Rhenium has a ductilizing effect on the group VIB elements. It is the only element which has commercial application as a substitutional alloying element in a wolfram lattice. It is used in thermocouple wire also, where W–25Re and W–5Re form a couple useful to 2400°C, a considerably higher temperature than that reached with any other couple.

The application requiring the greatest resistance to heat and erosion is the throat in solid-fuel rocket engines. In the United States, the three-stage Minuteman intercontinental ballistic missiles use forged and machined pure wolfram throats. In some military rocket engines, the throats are machined from porous wolfram preforms whose pores have been infiltrated with silver. The silver evaporates during the short firing time, cooling the nozzle through which burning fuel at over 3500°C delivers the enormous thrust propelling the rocket.

Other high-temperature uses are as vacuum-metalizing filaments, heat shields, heating elements, and x-ray targets. In the last two particularly, the very low vapor pressure of wolfram is critical. Induction furnaces operating above 2000°C in hydrogen or especially in vacuum usually employ wolfram susceptors. Tantalum is generally a competitor of wolfram in high-temperature furnaces. It is far less brittle after recrystallization, but it suffers from a vapor pressure which is 5–15 times higher in the 1800–3000°C range, a disastrous tendency to absorb hydrogen to form a hydride, and a higher material cost.

Wolfram is also used in electrical contacts and spark gaps, either pure, as in automotive points in distributors, with thoria in arc-discharge lamps, or with barium aluminate in spark igniters. Its use in automobile distributors is the second most common application of wolfram in the world today.

Welding electrodes in the so-called TIG (tungsten–inert-gas) welding technique use mostly W–2ThO$_2$, and also some pure wolfram and W–ZrO$_2$. Plasma-arc cutting is a further procedure involving wolfram as one electrode in an arc. Electrical contacts in switch gear often employ W–Cu or W–Ag, prepared either by sintering mixtures of powder or by infiltrating porous wolfram shapes. The mutual insolubility of the two components results in two-phase materials in which the high conductivity of the copper or silver is combined with the arc-pitting resistance of the wolfram. Such materials are also readily machinable.

So-called dispenser cathodes are made from porous wolfram bodies. In this case, the bodies are infiltrated, generally with copper, but sometimes with a thermoplastic. The stock is then machined to the proper configuration, the infiltrating phase is expelled in a vacuum or by dissolving with a solvent, and the resulting body is infiltrated with an electron-emitting material, such as alkaline-earth carbonates or thoria. After activation, the body serves as a cathode in power tubes.

Wolfram–copper and wolfram–silver also serve as tools in EDM (electric-discharge machining) and ECM (electrochemical machining), two processes in which electrodes, as tools, with shaped cross section accurately produce configurations of the same shape.

There is a group of two-phase alloys in which wolfram particles are surrounded by a continuous matrix, generally of nickel–copper or nickel–iron, up to about 17%. The group is always made by pressing mixtures of elemental or alloy powders and sintering at 1200–1500°C. They are readily machinable. They are used in applications requiring high density, ie, radiation shielding, gyroscopes, and counterweights.

One of the newer applications is in the manufacture of boron filaments for the strengthening phase in boron–epoxy or boron–aluminum composites. A 0.5-mil wolfram wire serves as the substrate for the chemical vapor deposition of boron from a gaseous mixture of boron trichloride and hydrogen. The wolfram wire is heated to about 1300°C via mercury contacts as it passes into the reaction chamber. The final boron filament has a 4-mil diameter and a wolfram boride core. It has exceptionally high stiffness and strength-to-weight ratio.

Bibliography

"Tungsten and Tungsten Alloys" in *ECT* 1st ed., Vol. 14, pp. 353–362, by Bernard Kopelman, Sylvania Electric Products, Inc.

1. *Chart of the Nuclides*, 9th ed., Knolls Atomic Power Laboratory, General Electric Co., under direction of Naval Reactors, U. S. Atomic Energy Commission, Schenectady, July, 1966.
2. W. E. Forsythe and A. G. Worthing, *Astrophys. J.* **61**, 146–185 (1925).
3. G. D. Rieck, *Tungsten and Its Compounds*, Pergamon Press, Oxford, 1967.
4. O. Kubaschewski, E. L. Evans, and C. B. Alcock, *Metallurgical Thermochemistry*, 4th ed., Pergamon Press, Oxford, 1967.
5. Comité International des Poids et Mesures, *Metrologia* **5** (2), 35–44 (1969).
6. R. C. Koo, "Evidence for Voids in Annealed Doped Tungsten," *Trans. Met. Soc. AIME* **239**, 1996–1997 (1967).
7. G. Das and S. V. Radcliffe, "Internal Void Formation in Powder-Metallurgy Tungsten," *Trans. Met. Soc. AIME* **242**, 2191–2198 (1968).
8. *1968 Minerals Yearbook*, Vols. 1–2, U.S. Dept. of Interior, Bureau of Mines, U. S. Govt. Printing Office, Washington, D.C., 1969.
9. *Book of Standards*, Parts 7–8, American Society for Testing and Materials, Philadelphia, Pa., 1968–1969.
10. *Documentation of Threshold Limit Values*, Committee on Threshold Limit Values, American Conference of Governmental Industrial Hygienists, Cincinnati, Ohio, 1966, Appendix C.

BERNARD KOPELMAN AND
JAMES S. SMITH
Sylvania Electric Products Inc.

WOLFRAM COMPOUNDS

Like molybdenum, wolfram forms compounds in which it has the valence states of 0, +2, +3, +4, +5, or +6. However, wolfram alone has not been observed as a cation. The most stable, and therefore the most common, valence state of wolfram is +6. Wolfram has a wide variety of stereochemistries in addition to the variety of oxidation states, and its chemistry is among the most complex of the transition elements. Complex formation is exemplified by the large number of polywolframates. Some simple wolfram compounds, such as the halides, are also known.

Wolfram Hexacarbonyl. Wolfram hexacarbonyl, $W(CO)_6$, is a colorless to white solid which decomposes without melting at approximately 150°C although it sublimes in vacuo. It is a zero-valent monomeric compound with a relatively low vapor pressure (0.1 mm at 20°C and 1.2 mm at 67°C). It is fairly stable in air, water, or acid, but it is decomposed by strongly basic solutions and is attacked by halogens. Wolfram carbonyl is somewhat soluble in organic solvents but insoluble in water. It can be purified by sublimation or steam distillation. It may be prepared, in yields over 90%, by the aluminum reduction of wolfram hexachloride in anhydrous ether under a pressure of 100 atmospheres of carbon monoxide at 70°C.

It is reported that wolfram hexacarbonyl and its organic derivatives are being investigated in various applications such as lubricant additives, dyes, pigments, and catalysts. Wolfram can be deposited from wolfram hexacarbonyl (1,2) but carbide formation and gas-phase nucleation present serious problems. As a result, wolfram halides are considered to be the preferred approach.

Wolfram Halides and Oxyhalides. Wolfram forms binary halides for all oxidation states between $+2$ and $+6$, while oxyhalides are only known for oxidation states of $+5$ and $+6$. In general, they are reactive toward water and oxygen in the air and must therefore be handled in an inert atmosphere. They are all solid colored compounds at room temperature, except the fluorides, and many decompose on heating before they melt. The hexachloride and hexafluoride are commercially available and are particularly suitable starting materials for the chemical vapor deposition of wolfram. Wolfram prepared by that technique is of growing importance in the making of coatings and free-standing parts such as thin-walled tubing. Usually the resulting structure is columnar but recently a method has been described for obtaining a fine-grained noncolumnar wolfram structure (3).

A list of the known halides and oxyhalides is presented in Table 1.

Table 1. Halogen Compounds of Wolfram

Fluorides	Chlorides	Bromides	Iodides
WF_6	WCl_6	WBr_6	
	WCl_5	WBr_5	
WF_4	WCl_4	WBr_4	WI_4
		WBr_3	WI_3
	WCl_2	WBr_2	WI_2
WOF_4	$WOCl_4$	$WOBr_4$	
WO_2F_2	WO_2Cl_2	WO_2Br_2	WO_2I_2
	$WOCl_3$		

Wolfram hexafluoride, WF_6, is a colorless gas at room temperature, sp gr 12.9, with respect to air. At 17.5°C, the hexafluoride condenses into a pale yellow liquid, and at 2.5°C a white solid is formed. It may be prepared by treating either hydrogen fluoride, arsenic trifluoride, or antimony pentafluoride with wolfram hexachloride or the direct fluorination of wolfram:

$$WCl_6 + 6\ HF \rightarrow WF_6 + 6\ HCl$$

$$WCl_6 + 2\ AsF_3 \rightarrow WF_6 + 2\ AsCl_3$$

$$WCl_6 + 3\ SbF_5 \rightarrow WF_6 + 3\ SbF_3Cl_2$$

$$W + 3\ F_2 \rightarrow WF_6$$

Direct fluorination of pure wolfram in a flow system at atmospheric pressure, and at a temperature of 350–400°C, is the most convenient procedure (4). Wolfram hexafluoride is extremely unstable in the presence of moisture, hydrolyzing completely into wolframic acid:

$$WF_6 + 4\ H_2O \rightarrow H_2WO_4 + 6\ HF$$

When wolfram hexafluoride is dissolved in organic solvents such as benzene or cyclohexane, the solution becomes bright red, while a pale red is obtained in dioxane, and a violet-brown in ether.

Wolfram tetrafluoride, WF_4, is a reddish-brown solid which is nonvolatile and hygroscopic. It has been prepared in low yields by the reduction of the hexafluoride with phosphorus trifluoride in the presence of liquid anhydrous hydrogen fluoride at room temperature (5).

Wolfram oxytetrafluoride, WOF_4, exists in the form of colorless plates, mp 110°C, bp 187.5°C. It may be prepared by the action of an oxygen–fluorine mixture on the metal at elevated temperatures (6). This compound is extremely hygroscopic, decomposing into wolframic acid in the presence of water.

Wolfram dioxydifluoride, WO_2F_2, is a white solid which may be prepared by the careful hydrolysis of WOF_4 (7). Its chemistry has not been investigated.

Wolfram hexachloride, WCl_6, is a blue-black crystalline solid, mp 275°C, bp 346.7°C. It is prepared by the direct chlorination of pure wolfram in a flow system at atmospheric pressure and at a temperature of 600°C. Solidification usually occurs without incident. Further cooling may result in a violent, explosion-like expansion of the solid wolfram hexachloride mass at about 168–170°C. The phenomenon may be associated with the $\alpha_2 \rightarrow \alpha_1$ transition. However, it may be safely cooled if the wolfram hexachloride occupies not more than one-half of the containing vessel. In the presence of moisture or oxygen, some $WOCl_4$ is formed as an impurity. It is very soluble in carbon disulfide but decomposes in water, forming wolframic acid. It is easily reduced by hydrogen to the lower halides, and finally to the metal itself (2).

Wolfram pentachloride, WCl_5, is a black crystalline deliquescent solid, mp 248°C, bp 275.6°C. It is very slightly soluble in carbon disulfide and decomposes in water to the blue oxide, $W_{20}O_{58}$. Magnetic properties suggest that it may contain trinuclear clusters in the solid state, but its structure has not been defined. It may be prepared by the reduction of the hexachloride with red phosphorus (8).

Wolfram tetrachloride, WCl_4, is obtained as a crystalline gray solid which decomposes upon heating. It is diamagnetic and may be prepared by the reduction of WCl_6 with Al by the thermal-gradient method (9). *Wolfram dichloride*, WCl_2, is an amorphous powder. It is a cluster compound and may be prepared by the thermal disproportionation of wolfram tetrachloride (8). *Wolfram oxytetrachloride*, $WOCl_4$, is a red crystalline solid, mp 211°C, bp 327°C. It is soluble in carbon disulfide and benzene, and is decomposed to wolframic acid by water. It may be prepared by refluxing sulfurous oxychloride (thionyl chloride, $SOCl_2$) on wolfram trioxide (10), and purified after evaporation by sublimation. *Wolfram dioxydichloride*, WO_2Cl_2 is a pale-yellow crystalline solid, mp 266°C. It is soluble in cold water and in alkaline solution although it is partly decomposed by hot water. It may be prepared by the action of carbon tetrachloride on wolfram dioxide at 250°C in a bomb (11). *Wolfram oxytrichloride*, $WOCl_3$, a green solid, may be prepared by the aluminum reduction of $WOCl_4$ in a sealed tube at 100–140°C (12).

Wolfram hexabromide, WBr_6, bluish-black crystals, mp 232°C, is formed by gently heating wolfram in bromine vapor in the absence of air or moisture. *Wolfram pentabromide*, WBr_5, violet-brown crystals, mp 276°C, bp 333°C, is extremely sensitive to moisture. It may be prepared by the action of bromine vapor on wolfram at 450–500°C (13). *Wolfram tetrabromide*, WBr_4, black orthorhombic crystals, is formed by the reduction of WBr_5 with aluminum by the thermal-gradient method similarly to WCl_4 (9). *Wolfram tribromide*, WBr_3, is prepared by the action of bromine on WBr_2 in a sealed tube at 50°C (14). It is a thermally unstable black powder which is insoluble in water. *Wolfram dibromide*, WBr_2, formed by the partial reduction of the pentabromide by hydrogen, is a black powder decomposing at 400°C. *Wolfram oxytetrabromide*, $WOBr_4$, black deliquescent needles, mp 277°C, bp 327°C, is formed by the action of carbon tetrabromide on wolfram dioxide at 250°C (11). *Wolfram dioxydibromide*, WO_2Br_2, light red crystals, is formed by passing a mixture of oxygen and bromine over wolfram at 300°C.

Wolfram tetraiodide, WI_4, is a black powder, which is decomposed by air. It may be prepared by the action of concentrated hydriodic acid on wolfram hexachloride at 100°C. *Wolfram triiodide*, WI_3, may be prepared by the action of iodine on wolfram hexacarbonyl in a sealed tube at 120°C (15). *Wolfram diiodide*, WI_2, is a brown powder, sp gr 6.79. It may be prepared by the action of anhydrous hydrogen iodide on wolfram hexachloride at 400–500°C. Research by other investigators, however, failed to show the existence of the diiodide, but a stable wolfram dioxydiiodide, WO_2I_2, is described (16).

Oxides, Acids, Salts, and Sulfides

Wolfram forms a series of oxides. These oxides form well defined ordered phases which can be assigned precise stoichiometric formulae. The simple ones are WO_3, and WO_2. The known oxides have been studied extensively (17–24) and are presented in Table 2.

The composition of the wolfram oxides may vary over a fixed range without change in crystalline structure. Thus, the homogeniety ranges are represented by the formulae $WO_{2.95-3.0}$, $WO_{2.88-2.92}$, $WO_{2.664-2.766}$, and $W_{1.99-2.02}$. Each wolfram atom is octahedrally surrounded by six oxygen atoms. In WO_3, these WO_6 units are joined through sharing of corner oxygen atoms only; but as the oxygen to wolfram ratio decreases, the WO_6 units become more intricately joined in combinations of corners, edges, and faces to form chains and slabs. The loss of each oxygen atom from the oxide lattice means that two electrons are added to the conduction band of the lattice and it is meaningless to speak of pentavalent and tetravalent wolfram atoms in such a lattice.

Wolfram trioxide, WO_3, is a yellow powder. However, the smallest diminution of oxygen brings about a change in color. It is pseudorhombic at room temperature, but above 700°C, it transforms to a tetragonal form. Normally WO_3 is prepared from wolframic acid or wolframates. It is the most important of the wolfram oxides and is generally used as the starting material for the production of wolfram powder. Wolfram trioxide is reduced to metal by carbon above 1050°C, and by hydrogen as low as 650°C. At lower temperatures, intermediate oxides are formed. Wolfram trioxide is insoluble in water and in acid solutions (except hydrofluoric) but reacts with strong alkali forming the wolframate:

$$2NaOH + WO_3 = Na_2WO_4 + H_2O$$

Table 2. Wolfram Oxides

Phase	Average O/W	Formula	Color	Theoretical density, g cm^{-3}
α	3.00	WO_3	yellow	7.29
β	2.90	$W_{20}O_{58}$	blue-violet	7.16
γ	2.72	$W_{18}O_{49}$	reddish-violet	7.78
δ	2.00	WO_2	brown	10.82
(β-W)	0.33	W_3O	gray	14.4

When it is heated in a hydrogen chloride atmosphere, it is completely volatilized at about 500°C, forming the oxydichloride, WO_2Cl_2.

Wolfram dioxide, WO_2, is a brown powder. It is formed by the reduction of WO_3 with hydrogen at 575°C to 600°C. Generally, this oxide is obtained as an intermediate product in the hydrogen reduction of the trioxide to the metal. When WO_3 is reduced, it forms first the "blue oxide" and then the "brown oxide" (WO_2). The composition of the blue oxide was in doubt for a long time. However, it now has been resolved that $W_{20}O_{58}$ and $W_{18}O_{49}$ are formed as intermediates. They may also be prepared by the reaction of wolfram with WO_3.

The substance W_3O has been regarded as an oxide and as a metal phase. It is gray in color and has a density of 14.4 g cm^{-3}. It may be prepared by the electrolysis of fused mixtures of WO_3 and alkali metal phosphates. It decomposes into W and WO_2 at about 700°C. It seems now clear that β-Wolfram is W_3O.

Wolfram Bronzes. Wolfram bronzes (25,26) are a series of well-defined non-stoichiometric compounds having the general formula $M_{1-x}WO_3$ where x is a variable falling in the range $0 < x < 1$ and M is some other metal. In this series, the cation is generally an alkali metal, although many other metals can be substituted.

The system most extensively investigated, however, is the sodium wolfram bronze. These compounds are intensely colored, ranging from golden-yellow to bluish-black depending on the value of x, and, in crystalline form, exhibit a metallic sheen. They have a positive temperature coefficient of resistance for Na:WO_3 ratios greater than 0.3 and a negative temperature coefficient of resistance at lower ratios. The sodium wolfram bronzes are inert to chemical attack by most acids but may be dissolved by basic reagents. Recently, it has been reported that the sodium wolfram bronzes serve as promoters for the catalytic oxidation of carbon monoxide and reformer gas in fuel cells (27). In general, these bronzes form cubic or tetragonal crystals, the lattice constants increasing with sodium concentration. They may be prepared by electrolytic reduction, vapor-phase deposition, fusion, or, solid-state reaction. The latter method is the most versatile in which the reagents are finely ground and heated at 500–850°C, in vacuo, for prolonged periods of time, to yield a product via solid-state reaction.

Wolfram Blue. The mild reduction, for example, by Sn(II), of acidified solutions of wolframates, wolfram trioxide, or wolframic acid in solutions yields intense blue products, which are referred to by the general name of "wolfram blues." Thus they resemble molybdenum blues in many respects. Wolfram trioxide will acquire a bluish tint merely on exposure to ultraviolet light, under water. If hydrogen is produced in a wolframate solution by means of zinc and hydrochloric acid, a blue precipitate is formed which is quite stable in air. It is believed that these are hydrogen analogs of the wolfram bronzes. These "hydrogen bronzes," $H_{1-x}WO_3$, blue in color, can be prepared by the wet reduction of wolframic acid and have been shown to be structurally related to the alkali wolfram bronzes (28,29). The wolfram blues have a strong tendency to form colloids.

Acids and Salts. *Wolframic acid*, H_2WO_4 or $WO_3 \cdot H_2O$, forms an amorphous yellow powder, which is practically insoluble in water or acid solution, but dissolves readily in a strongly alkaline medium. It may be precipitated from hot normal wolframate solutions by the action of strong acids. However, if the wolframate solution is acidified in the cold, a white voluminous precipitate is formed. This white precipitate, hydrated wolframic acid, has been reported to have the formula, $WO_3 \cdot xH_2O$,

Table 3. Properties of Normal Wolframates

Compound	Color, mp, crystal structure	Sp gr	Soly, g solute per 100 cm³ water, (°C)
$BaWO_4$	colorless, tetragonal, $a = 5.64$, $c = 12.70$	5.04	very slightly soluble
$CdWO_4$	yellow rhombic		0.05
$CaWO_4$	white, tetragonal, $a = 5.24$, $c = 11.38$, n_D^{29} 1.9263	6.06	0.00064[15]
$Ce_2(WO_4)_3$	yellow monoclinic, $a = 11.51$, $b = 11.72$, $c = 7.82$, $\beta = 109°$ 48', mp 1089°C.	6.77	insoluble
$PbWO_4$	colorless, monoclinic, mp 1123°C	8.46	0.03
$K_2WO_4.2H_2O$	monoclinic	3.113	51.5
Ag_2WO_4	pale yellow		0.05[15]
Na_2WO_4	white rhombic, mp = 698°C	4.179	57.50
$Na_2WO_4.2H_2O$	white rhombic; loses $2H_2O$ at 100°C	3.245	410
$SrWO_4$	white, tetragonal, $a = 5.40$, $c = 11.90$	6.187	0.14[15]

where x is approximately 2. It may be converted to the yellow form by boiling in an acid medium. Both the yellow and white forms have a great tendency to become colloidal on washing. Wolframic acid forms a series of stable normal salts of the types $M_2^{1+}WO_4$, $M^{2+}WO_4$, and $M_2^{3+}(WO_4)_3$. Some of these normal salts also exist in the hydrated form. Except for wolframates of the alkali metals and magnesium, these salts are generally sparingly soluble in water. The action of hot, mineral acids (except phosphoric) decomposes them, precipitating the yellow wolframic acid. The insoluble wolframates can be obtained by adding a sodium wolframate solution to a solution of the appropriate salt. Some properties of these wolframates are listed in Table 3.

Normal ammonium wolframate, $(NH_4)_2 WO_4$, cannot be obtained from an aqueous solution since it decomposes when such a solution is concentrated. However, this salt may be obtained by the addition of hydrated wolframic acid to liquid ammonia.

Sodium wolframate, Na_2WO_4, may be obtained anhydrous by fusing wolfram trioxide, WO_3, in the proper proportion with either sodium hydroxide or sodium carbonate:

$$WO_3 + 2\ NaOH = Na_2WO_4 + H_2O$$

$$WO_3 + Na_2CO_3 = Na_2WO_4 + CO_2$$

On crystallization from solution, the dihydrate, $Na_2WO_4.2H_2O$, is generally obtained; the decahydrate, $Na_2WO_4.10H_2O$, is formed below 6°C.

The normal wolframates are of particular interest in electronic and optical applications. They are used for ceramics, catalysts, pigments, corrosion, and fire inhibitors, etc.

An important and characteristic feature of the wolframates is the formation of condensed complex ions of isopolywolframates in acid solution. As the acidity increases, the molecular weight of the isopolyanions increases, until wolframic acid precipitates. They have probably been investigated more than most systems, but our understanding of them has been limited by the inability to obtain well-defined solid derivatives. The isopolywolframates have been reviewed in detail by Kepert (30).

The chemistry of wolfram in solution has been studied by polarography, chromatography, turbidity, light absorption, molecular weight, ultracentrifugation, Raman spectroscopy, infrared and absorption spectroscopy, radiochemistry, solvent extraction, and precipitation. Much of the reported work concerns the existence of wolframate species in acid solutions with particular reference to the molar ratio of soluble wolframate species.

If one considers polywolframates as formed by the addition of acid to WO_4^{2-}, then a series of isopolywolframates appear in which the degree of aggregation in solution increases as the pH is lowered. The relationships of the species, in order of increasing ratio of H_3O+/WO_4^{2-}, are listed in Table 4 (31).

Table 4. Polywolframates in Order of Increasing Ratio of H_3O^+/WO_4^{2-} Based on $12WO_4^{2-}$ Ions in the Equation: $aH_3O^+ + 12WO_4^{2-} = bW_xO_yH_z^{n-} + wH_2O$

$H_3O^+/$ WO_4^{2-}, $a/12$	Polywolframate	Common name
0.33	$W_{12}O_{46}^{20-}$	para Z
0.667	$W_3O_{11}^{4-}$	triwolframate
	$H_4W_3O_{13}^{4-}$	
1.167	$H_{10}W_{12}O_{46}^{10-}$	para B
	$H_2W_6O_{21}^{5-}$	para A
1.33	$W_{12}O_{40}^{8-}$	
1.50	$H_2W_{12}O_{40}^{6-}$	meta
	$H_3W_6O_{21}^{3-}$	pseudo meta
2.00	$WO_3 \cdot H_2O$	wolframic acid

Metawolframates of the alkali, alkaline earth metals, rare-earth and transition metals have been reported. However, classical synthesis rarely results in high yields of the pure compounds. The rare-earth wolframates, $Ln_2(H_2W_{12}O_{40}) \cdot xH_2O$, may be prepared by the action of lanthanon carbonates on metawolframic acid, $H_6(H_2W_{12}O_{40})$. Other salts may be prepared by the action of carbonates or sulfates of the corresponding metal on metawolframic acid or metawolframates. Typical examples are: Ni_3-$(H_2W_{12}O_{40}) \cdot 22H_2O$, $Zn_3(H_2W_{12}O_{40}) \cdot 20H_2O$, and $Ag_6(H_2W_{12}O_{40}) \cdot 5H_2O$. Generally these compounds are heat sensitive and should be recovered by freeze-drying. Alkali metal and ammonium metawolframates, $M_6(H_2W_{12}O_{40}) \cdot xH_2O$, may be prepared by the digestion of hydrated wolfram trioxide with the corresponding base. All these salts are generally known for their high solubility in water.

The *parawolframates* generally are crystallized from slightly acid solutions. By far the most important salt is *ammonium parawolframate*, $(NH_4)_{10}$ $(H_{10}W_{12}O_{46})$, which is usually known as the "heavy" form of commercial ammonium parawolframate. It is usually formed directly by crystallization from a boiling solution. However, if crystallization is allowed to take place slowly at room temperature, a hydrate, $(NH_4)_{10}$ $(H_{10}W_{12}O_{46}) \cdot 6H_2O$, is formed. This hydrate is known as the "light" form of ammonium parawolframate. Both forms are insoluble in water and can be decomposed by acid or alkali. Reduction to metal occurs by heating in a hydrogen atmosphere. It is also widely used as a catalyst. *Peroxywolframic acid*, $H_2WO_5 \cdot H_2O$ or $WO_3 \cdot H_2O \cdot H_2O$, is a deep orange amorphous compound obtained by treating wolfram trioxide with a solution of hydrogen peroxide. Similarly, *peroxywolframates* are known.

Heteropolyanions are closely related to the isopolyanions and over thirty elements have been reported as capable of acting as the heteroatom with many stoichiometric ratios between the heteroatom and the anion. Both the acids and salts are known and are usually hydrated when crystallized from aqueous solutions. As a class, heteropoly compounds are characterized by a number of properties independent of the heteroatom and metallic component. (See Vol. 13, p. 651). Typically, heteropoly wolfram compounds show the following:

(*1*) high molecular weight, usually greater than 3000. (*2*) a high degree of hydration. (*3*) unusually high solubility in water and some organic solvents. (*4*) strong oxidizing action in aqueous solution. (*5*) strong acidity in free acid form. (*6*) decomposition in strongly basic aqueous solutions to give normal wolframate solutions. (*7*) highly colored anions or colored reaction products.

Heteropoly anions may be classified according to the ratio of the number of central atoms to wolfram, as shown in Table 5.

Table 5. Principal Species of Heteropolywolframates

Ratio of hetero atoms to W atoms	Principal central atoms	Typical formulae	Structure by x ray
1:12	P^{+5}, As^{+5}, Si^{+4}, Ge^{+4}, Ti^{+4}, Co^{+3}, Fe^{+3}, Al^{+3}, Cr^{+3}, Ga^{+3}, Te^{+4}, B^{+3}	$[X^{+n}W_{12}O_{40}]^{-(8-n)}$	known
1:10	Si^{+4}, Pt^{+4}	$(X^{+n}W_{10}O_x]^{-(2x-60-n)}$	unknown
1:9	Be^{+2}	$[X^{+2}W_9O_{31}]^{-6}$	unknown
1:6	series A: Te^{+6}, I^{+7}	$[X^{+n}W_6O_{24}]^{-(12-n)}$	isomorphous with -6-molybdates
	series B: Ni^{+2}, Ga^{+3}	$[X^{+n}W_6O_{24}H_6]^{-(6-n)}$	known
2:18	P^{+5}, As^{+5}	$[X_2^{+n}W_{18}O_{62}]^{-(12-n)}$	known
2:17	P^{+5}, As^{+5}	$[X_2^{+n}W_{17}O_x]^{-(2x-102-2n)}$	unknown
1m:6m (m unknown)	As^{+3}, P^{+3}	$[X^{+n}W_6O_x]_m^{-m(2x-36-n)}$	unknown

X ray structural determinations have been made on heteropolywolframate and isopolywolframate compounds. The simplest way to represent the anion structures is by polyhedra which share corners and edges with one another. Each W is at the center of an octahedron, and an O atom is located in each vertex of the octahedron. The central atom is similarly located at the center of an XO_4 tetrahedron or XO_6 octahedron. Each such polyhedron containing the central atom is generally surrounded by WO_6 octahedra which share corners or edges (or both) with it and with one another so that the correct total number of oxygen atoms is utilized. Each WO_6 octahedron is directly attached to a central atom through a shared oxygen atom. In the actual structures the octahedra are frequently distorted. The oxygens are relatively large spheres, and practically all of the space within the anion structure is taken up by the bulky oxygens which are either close-packed or nearly so.

When the large heteropolywolframate anions pack together as units in a crystal, the interstices between the anions are very large compared either to water molecules or to most simple cations. In most compounds there is apparently no direct linkage between the individual heteropoly anions as shown in the structure of $K_5(CoW_{12}O_{40})$. $20H_2O$, and $K_6(P_2W_{18}O_{62}) \cdot 14H_2O$. Instead, the complexes are joined by hydrogen bonding through some molecules of water of hydration. These principles are illustrated in the crystal structure of 12-wolframophosphoric acid hydrate, as determined by X ray diffraction (31a,31b).

Heteropoly salts of larger cations, such as cesium, frequently crystallize as acid salts no matter what the ratio of cations to anions is in the mother liquor. Furthermore, salts of these cations are frequently less highly hydrated than salts of smaller cations. Apparently the larger cations take up so much of the space between the heteropoly anions that there is less room left for water. In fact, there is often not enough room for all the large cations required to form a normal salt. Instead, (solvated) hydrogen ions fill in to balance the negative charge of the anions and a crystalline acid salt results.

Commercially, heteropolywolframates, particularly the 12-heteropolywolframates, are produced in large quantities as precipitants for basic dyes, with which they form colored lakes or toners (see Vol. 15, p. 555). They are also used in catalysis, passivation of steel, etc.

Sulfides. Wolfram and sulfur form two binary compounds, wolfram disulfide, WS_2, and wolfram trisulfide, WS_3. Although WS_2 can be made by heating wolfram powder with sulfur at 900°C, crystalline WS_3 is prepared only by treating diethylamine thiowolframate with hydrochloric acid (32).

Wolfram disulfide, WS_2, although found in nature, is usually prepared from the elements. It is a soft, grayish-black powder, and has sp gr 7.5. Wolfram disulfide is relatively inert and unreactive. It is insoluble in water, hydrochloric acid, alkali, and organic solvents, or oils. It decomposes in hot, strong oxidizing agents such as aqua regia, concentrated sulfuric acid, and nitric acid. Heating in air or in the presence of oxygen oxidizes WS_2 to WO_3. However, it has a thermal stability advantage over MoS_2 of approximately 90°C in air.

Wolfram disulfide has the ability to form adherent, soft, continuous films on a variety of surfaces and exhibits good lubricating properties.

The fundamental mechanism of lubrication is similar to that of molybdenum disulfide and graphite (33). It is also reported to be a semiconductor.

Wolfram trisulfide, WS_3, is a chocolate brown powder, slightly soluble in cold water, but readily forming a colloidal solution in hot water. It is generally prepared by treating an alkali-metal thiowolframate with excess acid. The compound is soluble in alkali carbonates and hydroxides.

Wolfram also forms a series of *thiowolframates* corresponding to the normal salts but with one, two, three, or all of the oxygen atoms replaced by sulfur. The *potassium thiowolframates* have the following formulae: $K_2WO_3S \cdot H_2O$, $K_2WO_2S_2$, $K_2WOS_3 \cdot H_2O$, K_2WS_4, and similar compounds of other metals are known.

These compounds are formed when solutions of the alkali or alkaline earth wolframates are saturated with hydrogen sulfide. These salts vary in color from pale yellow to yellowish-brown and, in general, crystallize well. Upon acidifying a solution of these salts, however, wolfram trisulfide is obtained. *Potassium tetrathiowolframate*, K_2WS_4, forms yellow rhombic crystals which are quite soluble in water. *Ammonium tetrathiowolframate*, $(NH_4)_2WS_4$, forms bright orange crystals having a metallic iridescence which are stable in dry air and are quite soluble in water. It is generally prepared by treating a solution of wolframic acid with excess ammonia and saturating with hydrogen sulfide. It is readily decomposed in a nonoxidizing atmosphere to WS_2 making it a convenient source of WS_2.

Interstitial Compounds

Wolfram forms hard, refractory and chemically stable interstitial compounds with nonmetals, particularly C, N, B, and Si. Their properties make them useful for cutting tools, structural elements of kilns, gas turbines, jet engines, sandblast nozzles, protective coatings, etc. Wolfram and carbon form two binary compounds, wolfram carbide, WC, sp gr 15.63, and diwolfram carbide, W_2C, sp gr 17.15, both of which are prepared by heating together wolfram and carbon at high temperatures. The presence of hydrogen or a hydrocarbon gas promotes the reaction. The relative quantities of the constituents used and the temperature of the reaction determines the phase formed. They may also be prepared from oxygen-containing compounds of wolfram, but because of the tendency to form oxycarbides, a final heating in vacuo above 1500°C is necessary. Both compounds melt at approximately 2800°C, and have a hardness approaching that of diamond. These compounds are insoluble in water, but they are readily attacked by a mixture of HNO_3–HF.

"Hard metals" (see Vol. 13, p. 814) are the most important commercial application of WC and W_2C. These carbides themselves are brittle, but in combination with, for example, cobalt, the properties of that binding metal decrease the brittleness.

The nitrides of wolfram are quite similar to the carbides. Although nitrogen does not react directly with wolfram, they can be prepared by heating wolfram in ammonia. The two phases, W_2N and WN have been extensively studied.

Wolfram borides, W_2B, and WB, prepared by hot pressing wolfram and boron, and W_2B_5, prepared by heating wolfram trioxide, graphite and boron carbide in vacuo, are extremely hard and exhibit almost metallic electrical conductivity. However, the existence of the phase, WB_2, has been questioned (34). Recently, the formation of wolfram–boride phases in the manufacture of boron filaments for composites, which are useful, or potentially so, as structural materials in space vehicles and aircraft have become extremely important.

Wolfram silicides are of interest because of the protective oxide layer they form over wolfram to prevent its catastrophic oxidation at elevated temperature. The layer fails to protect at lower temperatures (a behavior referred to as "disilicide pest"). The pest failure can be explained by the silicon being initially oxidized to SiO_2 at the surface and depleting the surface of Si, forming W_5Si_3. At high temperatures a uniform layer of W_5Si_3 is formed, but at "pest" temperatures, the attack is nonuniform and seems to follow grain boundaries or subgrain boundaries in the disilicide. The next stage in the pest is the rapid growth and penetration of the complex oxide into the disilicide layer. This process ultimately consumes the disilicide, causing oxidation of the wolfram substrate. The existence of W_3Si_2, and probably W_2Si, has been reported recently (35).

Diwolfram trisilicide, W_2Si_3, gray, sp gr 10.9, is insoluble in water, acid, or alkaline solutions. It is readily attacked by a mixture of nitric and hydrofluoric acids and by fused alkali metal carbonates and hydroxides.

Wolfram disilicide, WSi_2, forms bluish-grey tetragonal crystals with a $= 3.212$ and c $= 7.880$. It is insoluble in water and melts at 2160°C. The compound is attacked by fluorine, chlorine, fused alkalies, and HNO_3–HF mixtures. It may be used for high-temperature thermocouples in combination with $MoSi_2$, in an oxidizing atmosphere.

Anionic Complexes

Compounds of wolfram with acid anions, other than halides and oxyhalides, are relatively few in number, and are known only in the form of complex salts. A number of salts containing hexavalent wolfram are known. The compound K_2WF_8 can be prepared by the action of KI on $W(CO)_6$ in an IF_5 medium. The addition of wolframates to aqueous hydrofluoric acid gives salts which are mostly of the type $M_2^{1+}(WO^{2-}F_4)$. Similarly, double salts of wolfram oxydichloride are known.

Salts containing pentavalent wolfram may be obtained by the reduction of alkali wolframates in concentrated hydrochloric acid. Salts of the type $M_2^{1+}(WOCl_5)$ (green), $M^{1+}(WOCl_4)$ (brown-yellow), and $M^{1+}(WOCl_4 . H_2O)$ (blue), have been isolated. Thiocyanato and bromo salts are also known.

Salts containing tetravalent wolfram have been prepared by various methods. The most important are the octacyanides, $M_4^{1+}(W(CN)_8)$. They form yellow crystals and are very stable. They can be isolated as either salts or as free acids such as H_4-$(W(CN)_8).6H_2O$. The compounds can be oxidized by $KMnO_4$ in H_2SO_4 to compounds containing pentavalent wolfram, $M_3^{1+}(W(CN)_8)$ (yellow).

The only known trivalent wolfram complex is of the type $M_3^{1+}(W_2Cl_9)$. It can be prepared by the reduction of strong hydrochloric acid solutions of K_2WO_4 with tin. If the reduction is not sufficient, a compound containing tetravalent wolfram, K_2-$(WCl_5(OH))$, is formed (36).

Toxicity

A considerable difference in the toxicity of soluble and insoluble compounds of wolfram has been reported. For soluble sodium wolframate, $Na_2WO_4 . 2H_2O$, injected subcutaneously in adult rats, LD_{50} was 140–160 mg W/kg. Death is ascribable to generalized cellular asphyxiation. Guinea pigs treated orally or intravenously with $Na_2WO_4 . 2H_2O$ suffered anorexia, colic, incoordination of movement, trembling, and dyspnea.

Orally in rats the toxicity of sodium wolframate was highest, wolfram trioxide was intermediate, and ammonia parawolframate least. In view of the degree of systemic toxicity of soluble compounds of wolfram, a threshhold limit of 1 mg of wolfram per m³ of air is recommended. A threshhold limit of 5 mg of wolfram per m³ of air is recommended for insoluble compounds (37).

Uses. Wolfram compounds, especially the oxides, sulfides, and heteropoly complexes, form stable catalysts for a variety of commercial chemical processes. Petroleum processing is a major field of application for wolfram compounds (38). The processes involved are hydrocracking, hydrotreating, dehydrogenation, isomerization, reforming, and polymerization. See Catalysts; Hydroprocesses; Petroleum. The wolfram compounds may function as principal catalysts or promoters of other catalysts. Wolfram compounds, particulary the blue oxide, $W_{20}O_{58}$, are important in industrial chemical synthesis involving hydration, dehydration, hydroxylation, and epoxidation (39). It is expected that the use of wolfram catalysts will increase greatly since many wolfram compounds have recently become commercially available.

Wolfram disulfide has the ability to form adherent, soft, continuous films on a variety of substrates and exhibits superior lubrication under extreme conditions of temperature, load, and vacuum. Applied as a dry powder, suspension, bonded film, or aerosol it can be an effective lubricant in wire drawing, metal forming, valves, gears, bearings, packing materials, etc. Oil soluble wolfram compounds, such as the amine salts of wolframate or tetrathiowolframate, are reported to be effective lubricating oil additives. See Lubrication and lubricants.

Heteropoly wolframic acids, particularly the 12-heteropolys, are useful as reagents in analytical chemistry and biochemistry; in atomic energy work as precipitants and inorganic ion-exchangers; in photographic processes as fixing agents, oxidizing agents; in plating processes as additives; in plastics, adhesives, and cements for imparting water resistance and in plastics and plastic films as curing or drying agents. An important use for 12-phosphowolframic acid and its sodium salts is in the manufacture of organic pigments (see Vol. 15, p. 579). These compounds are extensively used for "carroting" or surface treating furs. In the textile industry, the salts are useful as antistatic agents. The acids are used in diverse applications, such as, printing inks, paper coloring, nontoxic paints, and wax pigmentation.

The wolframates and molybdates are good corrosion inhibitors and have been used for some time in antifreeze solutions. In addition, they are used as laser host materials, phosphors, flameproofing of textiles, etc.

The carbides (qv) are widely used in the manufacturing of the hard carbides, which are used for high-speed machining tools, wire-drawing dies, etc.

The halides, particularly the hexachloride and hexafluoride, are used for the chemical vapor-deposition of wolfram. High-purity wolfram in the form of crucibles, tubing, rod, etc, can readily be prepared. Complex parts, such as, honeycomb structures, can also be clad with a protective layer of wolfram by the reduction of the hexahalides.

Bibliography

"Tungsten Compounds" in *ECT* 1st ed., Vol. 14, pp. 363–372, by B. Kopelman, Sylvania Electric Products, Inc.

1. J. J. Lander and L. H. Germer, *Am. Inst. Mining Met. Engrs., Inst. Metals Div., Metals Technol.* **14** (6), *Tech. Publ. 2259* (1947).
2. C. F. Powell, J. H. Oxley, and J. M. Blocher, Jr., *Vapor Deposition,* John Wiley and Sons, New York (1966).
3. R. L. Landingham and J. H. Austin, *J. Less-Common Metals* **18** (3), 229 (1969).
4. E. J. Barber and G. H. Cady, *J. Phys. Chem.* **60,** 505 (1956).
5. T. A. O'Donnell and D. F. Stewart, *Inorg. Chem.* **5,** 1434 (1966).
6. G. H. Cady and G. B. Hargreaves, *J. Chem. Soc.* **1961,** 1568.
7. O. Ruff, F. Eisner, and W. Heller, *Z. Anorg. Allgem. Chem.* **52,** 256 (1907).
8. G. I. Novikov, N. Y. Andreeva, and O. G. Polyachenok, *Russ. J. Inorg. Chem.* **6,** 1019 (1961).
9. R. E. McCarley and T. M. Brown, *Inorg. Chem.* **3,** 1232 (1964).
10. R. Colton and I. B. Tomkins, *Australian J. Chem.* **18,** 447 (1965).
11. E. R. Epperson and H. Frye, *Inorg. Nucl. Chem. Letters* **2,** 223 (1966).
12. G. W. A. Fowles and J. L. Frost, *Chem. Commun.* 252 (1966).
13. R. Colton and I. B. Tomkins, *Australian J. Chem.* **19,** 759 (1966).
14. R. E. McCarley and T. M. Brown, *J. Am. Chem. Soc.* **84,** 3216 (1962).
15. C. Djordjevic, R. S. Nyholm, C. S. Pande, and M. H. B. Stiddard, *J. Chem. Soc. A,* **16** (1966).
16. J. Tillack, P. Eckerlin, and J. H. Dettingmeijer, *Angew. Chem.* **78,** 451 (1966).
17. E. Gebert and R. J. Ackermann, *Inorg. Chem.* **5** (1), (Jan 1966).
18. J. Neugebauer, T. Miller, and L. Imre Tungsram Techn. Mitteil. No. 2 (March 1961).
19. G. Hagg and A. Magneli, *Rev. Pure Appl. Chem.* **4,** 235 (1954).

20. O. Glemser and H. Sauer, *Z. anorg. Chem.* **252**, 144 (1943).
21. L. L. Y. Chang and B. Phillips, *J. Amer. Cer. Soc.* **52** (10), 527 (1969).
22. G. Hagg and N. Schonberg, *Acta. Cryst.* **7**, 351 (1954).
23. A. Magneli, *Arkiv. Kemi* **1**, 513 (1950).
24. A. Magneli, *J. Inorg. Nucl. Chem.* **2**, 330 (1956).
25. P. G. Dickens and M. S. Whittingham, *Quart. Rev.* **22** (1), 30 (1968).
26. M. J. Sienko, *Advan. Chem. Ser.* **39**, 224 (1963).
27. L. W. Niedrach and H. I. Zeliger, *J. Electrochem. Soc.* **116** (1), 152 (1969).
28. O. Glemser and C. Naumann, *Z. anorg. Chem.* **265**, 288 (1951).
29. P. G. Dickens and R. J. Hurditch, *Nature* **215**, 1266 (1967).
30. T. K. Kim, R. W. Mooney, and V. Chiola, *Separation Sci.*, **3** (5), 467 (1968).
31. D. L. Kepert, *Progr. Inorg. Chem.* **4**, 199 (1962).
31a. A. J. Bradley and J. W. Illingworth, *Proc. Roy. Soc. (London)*, **A157**, 113 (1936).
31b. R. Signer and H. Gross, *Helv. Chim. Acta.* **17**, 1076 (1934).
32. O. Glemser, H. Saver and P. Konig, *Z. Inorg. Chem.* **257**, 241 (1948).
33. V. R. Johnson, M. T. Lavik, and E. E. Vaughn, *J. App. Phys.* **28**, 821 (1957).
34. B. Post, F. W. Glaser, and D. Moskowitz, *Acta Met.* **2**, 20 (1954).
35. N. N. Matynshenko, L. N. Efimenko, and D. N. Solonikin, *Fiz. Met. i. Metalloved.* **8**, 878 (1959).
36. E. Konig, *Inorg. Chem.* **2**, 1238 (1963).
37. *Documentation of TLV, Appendix C*, American Conference of Industrial Hygienists, Cincinnati, Ohio (1966).
38. C. H. Kline and V. Kollonitsch, *Ind. Eng. Chem.* **57** (7), 53 (1965).
39. C. H. Kline and V. Kollonitsch, *Ind. Eng. Chem.* **57** (9), 53 (1965).

General References

G. D. Rieck, *Tungsten and its Compounds*, Pergamon Press, London, 1967.
K. C. Li and C. Y. Wang, *Tungsten*, Reinhold Publishing Corp., New York, 1955.
C. J. Smithells, *Tungsten*, Chapman and Hall Ltd., London, 1952.
J. H. Canterford and R. Colton, *Halides of the Transition Elements*, John Wiley & Sons, Inc., New York, 1968.

M. MacInnis
Sylvania Electric Products Inc.

WOOD

Wood is one of man's most important natural resources. It supplies him with material for items necessary to modern living, such as homes, furniture, bridges, poles, and railroad ties. Wood yields fiber for pulp, paper, and fiberboard and provides material for plywood, particleboard, and pallets and also for the valuable textile, rayon. In many localities wood serves as a fuel. It is a source of important industrial chemicals and such products as charcoal (see Carbon) and naval stores (see Rosin; Turpentine).

Statistics on the total production of wood products and residues (1) are difficult to interpret. This is due in part to the variety of units of measurement in common use.

The *cord*, commonly used to measure log volume, refers to a pile 8 ft long by 4 ft by 4 ft. The weight of wood in such a pile varies from 1600 to 3300 lb depending on the density of the wood and the bark and free space in the pile. Sawn lumber is commonly measured in *board feet*, the unit being a board 1 in. thick by 12 in. by 12 in.

The annual production of lumber in the United States has remained essentially constant for years at about 37 billion board feet. The production of paper and paperboard, however, has been increasing steadily. Production was 21 million tons in 1947, 31 million tons in 1957, and rose to 46 million tons in 1967 (2).

Considerable quantities of residue are produced in the forest-based industries. In 1962 about 3 billion ft³ of residues were generated, of which roughly half was used for fuel and conversion to particleboard and pulp fiber. The other half was not used. Because this represents an economic waste and a source of pollution, a challenge is presented to the chemical industry for developing uses for these residues.

Wood Structure

Wood is a complex substance, both anatomically and chemically (3). The anatomical structure affects strength properties, appearance, resistance to penetration by water and chemicals, resistance to decay, pulp quality, and the chemical reactivity of wood. To use wood most effectively in chemical technology requires not only a knowledge of the amounts of various substances that make up wood, but also of how they are distributed in the cell walls.

Trees native to the United States are divided into the following two classes: hardwoods, which have broad leaves, and softwoods, which have scalelike or needlelike leaves. Hardwoods, except in the warmest regions, are deciduous and shed their leaves at the end of each growing season. Native softwoods, except bald cypress, tamarack, and western larch, are evergreens. Softwoods are known also as conifers, because all native species of softwoods bear cones of one kind or another. The terms hardwood and softwood have no direct application to the hardness or softness of the wood of the two classes. In fact, such hardwood trees as cottonwood and aspen have softer wood than the western white pines and true firs; certain softwoods, such as longleaf pine and douglas fir, produce wood that is as hard as that of basswood or yellow poplar.

The cells that make up the structural elements of wood are of various sizes and shapes and are firmly bonded together (4). Dry wood cells may be empty or partly filled with deposits, such as gums or resins, or with tyloses, intrusive growths from one kind of cell into another. Long and pointed cells are known as fibers or tracheids; they vary greatly in length within a tree and among species. Hardwood fibers average about 1 mm in length; softwood fibers range from about 3 to 8 mm in length.

Many mechanical properties of wood, such as bending strength, crushing strength, and hardness, depend upon the density of wood; the heavier woods are generally stronger. Wood density is determined largely by the relative thickness of the cell wall and by the proportions of thick-walled and thin-walled cells present.

In temperate climates trees produce well-marked growth rings each year composed of thin-walled cells formed first in the growing season, called earlywood (springwood) and thicker-walled cells formed later, called latewood (summerwood). The earlywood and latewood bands are inconspicuous in some species and plainly evident in others. Changes in cell-wall thickness from earlywood to latewood may be gradual or abrupt. In certain abnormal types of wood (compression wood and tension wood, which develop when a tree is not growing vertically) the microstructure of the cell wall affects warping in lumber and veneer by increasing shrinkage along the grain of the wood.

The cells just under the bark of a tree, that take an active part in life processes, form the outer zone of the sapwood layer, which may vary in thickness and in the number of growth rings it contains. As the inner sapwood cells become inactive, they become heartwood. The heartwood of many species is less porous and more resistant

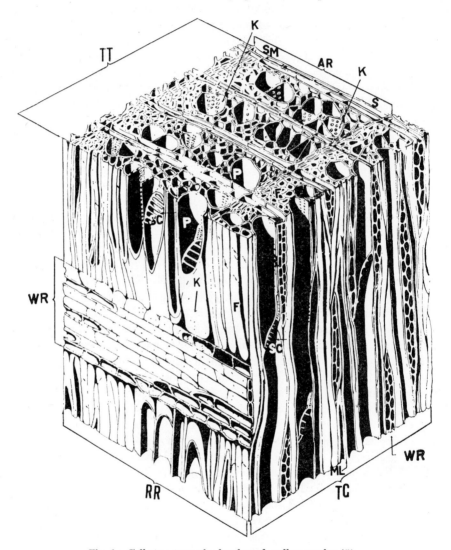

Fig. 1. Cell structure of a hardwood, yellow poplar (5).

to decay than the sapwood, primarily because of materials that are deposited within the cells. These factors are important in wood preservation.

Because of the extreme structural variations in wood, there are many possibilities for selecting a species for a specific purpose. Some species, like spruce, combine light weight with relatively high stiffness and bending strength. Very heavy woods, like lignum vitae, are extremely hard and resistant to abrasion. A very light wood, like balsa, has high thermal insulation value; a heavy one, like hickory, has extremely high shock resistance.

Structure of a Typical Hardwood. Figure 1 illustrates the cell structure of a $1/_{32}$-in. cube of hardwood (5). The horizontal plane TT, of the block corresponds to a minute portion of the top surface of a stump or end surface of a log. The vertical plane, RR, corresponds to a surface cut parallel to the radius (the center of the tree is to the right), and the vertical plane, TG, corresponds to a surface cut at right angles to the radius, or tangentially within the log. In hardwoods these three major planes along which wood may be cut are known commonly as end grain, TT, quarter sawed (edge grain), RR, and plain sawed (flat grain), TG, surfaces. In some hardwoods, such as the oaks, the quartered surface with its large lustrous "flakes" formed by wood rays, WR, entering and leaving the plane of the saw is valued for use in furniture.

The hardwoods have specialized vessels, or pores, P, for conducting sap, in contrast to the softwoods in which the sap is transferred through the tracheids or fibers.

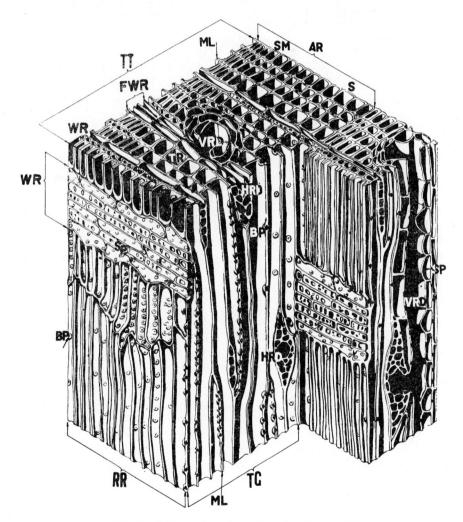

Fig. 2. Cell structure of a softwood white pine (5).

Table 1. Proximate Analyses of Representative Woods (9)

| Kind of wood | Ash | Soluble in | | | | Acetic acid | Meth- oxyl | Pento- san |
		Cold water	Hot water	Ether	1% NaOH			
longleaf pine	0.37	6.20	7.15	6.32	22.36	0.76	5.05	7.46
douglas fir	0.38	3.54	6.50	1.02	16.11	1.04	4.95	6.02
western larch	0.23	10.61	12.59	0.81	22.14	0.71	5.03	10.80
white spruce	0.31	1.12	2.14	1.36	11.57	1.59	5.30	10.39
western white pine	0.20	3.16	4.49	4.26	14.78	1.03	4.56	6.97
western yellow pine	0.46	4.09	5.05	8.52	20.30	1.09	4.49	7.35
yellow cedar	0.43	2.47	3.11	2.55	13.41	1.59	5.25	7.87
incense cedar	0.34	3.64	5.38	4.31	17.69	0.91	6.24	10.65
redwood	0.21	7.36	9.86	1.07	20.00	1.08	5.21	7.80
tanbark oak	0.83	4.10	5.60	0.80	23.96	5.23	5.74	19.59
mesquite	0.54	12.62	15.09	2.30	28.52	2.03	5.55	13.96
balsa	2.12	1.77	2.79	1.23	20.37	5.80	5.68	17.65
hickory	0.69	4.78	5.57	0.63	19.04	2.51	5.63	18.82
basswood	0.86	2.12	4.07	1.96	23.76	5.79	6.00	19.93
yellow birch	0.52	2.67	3.97	0.60	19.85	4.30	6.07	24.63
sugar maple	0.44	2.65	4.36	0.25	17.64	4.46	7.25	21.71
eucalyptus	0.24	4.67	6.98	0.56	18.57	1.85	6.73	20.09

Courtesy *Industrial and Engineering Chemistry.*

The vessels are made up of relatively large cells with open ends set one above the other and continuing as open passages for relatively long distances. In the heartwood and inner sapwood of many hardwoods the pores are filled with a frothlike ingrowth (tyloses) from the neighboring cells. The pores of hardwoods vary considerably in size, being visible without a magnifying glass in some species but not in others. In most hardwoods the ends of the individual cells of the pores are entirely open, whereas in others they are separated by a grating, as indicated at SC.

Most of the smaller cells seen in the cross section are wood fibers, F, which are the strength-giving elements of hardwoods. They usually have small cavities and relatively thick walls. Thin places or pits, K, in the walls of the wood fibers and vessels allow the passage of sap from one cavity to another. The wood rays, WR, are strips of short horizontal cells that extend in a radial direction. In the sapwood these rays store food, such as starch, and distribute it horizontally; in the heartwood they contain extraneous material or extractives. These short cells are a common source of the "fines" in pulp. In Figure 1 most of the rays shown in the surface TG are pictured as being two cells wide, but the width varies in different species of hardwoods from one to over fifty cells.

In temperate climates, the growth of one year, AR, commonly known as the annual ring, is usually sharply defined from that of the previous or following year. As a rule, wood formed in the early part of the growing season, the earlywood or springwood, S, is more porous than that formed later in the year, the latewood or summerwood, SM.

All the cells in wood, including pores, fibers, and ray cells, are firmly cemented together by a thin layer, the middle lamella, ML. This layer can be dissolved by certain chemicals, thus permitting the fibers to be separated as is done in manufacturing pulp from wood.

Table 1. (*continued*)

| Methyl-pen-tosan | Cellulose | Lignin | In the cellulose, % | | | | |
			Pen-tosan	Methyl-pen-tosan	α-Cellu-lose	β-Cellu-lose	γ-Cellu-lose
3.60	58.48		7.71	1.16			
4.41	61.47		5.34	1.20			
2.81	57.80		8.94	1.19			
3.55	61.85		9.63	0.72			
3.22	59.71	26.44	5.33	1.95	64.61	16.32	19.06
1.62	57.41	26.65	6.82	1.98	62.10	10.56	30.13
3.42	53.86	31.32	7.30	1.78	62.68	11.06	26.25
1.35	41.60	37.68	9.08	1.99	46.92	11.67	41.06
2.75	48.45	34.21	7.40	2.09	78.81	2.95	18.24
	58.03	24.85	22.82		56.77	19.92	23.03
0.70	45.48	30.47	17.75	0.81	76.48	2.35	21.17
0.86	54.15	26.50	19.99	1.35	75.64	0.27	24.08
0.80	56.22	23.44	21.89	1.41	76.32	2.82	20.35
3.73	61.24		24.28	1.54			
2.69	61.31		28.30	1.16			
2.39	60.78		24.48	0.96			
2.33	57.62	25.07	20.96	2.46	68.86	0.70	31.10

Structure of a Typical Softwood. Figure 2 illustrates the cell structure of a section of a $1/_{32}$-in. white-pine cube (5). The three planes—TT, RR, and TG—represent the same three planes described for hardwoods. In softwoods these planes are known commonly as end grain, TT, edge grain, RR, and plain sawed (flat grain), TG.

The rectangular units that make up the top surface, TT, are sections through vertical cells, mostly tracheids or water carriers, TR, the walls of which form the bulk of the wood substance. Between the various cell units is a cementing layer called the middle lamella, ML. Early wood or springwood cells, S, are distinguishable by their greater size. Latewood or summerwood cells, SM, are formed during the later part of the annual growing period. Earlywood growth is more rapid than latewood growth. Together the earlywood and latewood cells make up the annual ring, AR.

The wood rays, WR, store and distribute horizontally the food material of the tree. These rays—including the fusiform wood rays, FWR, or rays having horizontal resin ducts, HRD, at their centers—are found on the end surfaces as fine white lines running radially across the rings. The large hole, VRD, in the center of the top surface is a vertical resin duct.

The symbol SP indicates a simple pit, an unthickened portion of the cell wall through which sap passes from ray cells to fibers or vice versa. The bordered pits, BP, seen in section of surface TG have their margins overhung by the surrounding cell walls.

Chemical Composition

Figures 1 and 2 show that wood consists mainly of vertically oriented tubular fiber units or cells cemented together by the middle lamella. The various chemical components (6–8) are distributed in an interlocking network through this structure. The

Table 2. Carbohydrate Composition of Representative North American Woods, % of Extractive Free Wood[a]

Species	Glucan	Mannan	Galactan	Xylan	Arabinan	Uronic anhydride
hardwood						
trembling aspen						
(*Populus tremuloides*)	57.3	2.3	0.8	16.0	0.4	3.3
beech						
(*Fagus grandifolia*)	47.5	2.1	1.2	17.5	0.5	4.8
white birch						
(*Betula papyrifera*)	44.7	1.5	0.6	24.6	0.5	4.6
yellow birch						
(*Betula lutea*)	46.7	3.6	0.9	20.1	0.6	4.2
red maple						
(*Acer rubrum*)	46.6	3.5	0.6	17.3	0.5	3.5
sugar maple						
(*Acer saccharum*)	51.7	2.3	<0.1	14.8	0.8	4.4
sweet gum						
(*Liquidambar*						
styraciflua)	39.4	3.1	0.8	17.5	0.3	
white elm						
(*Ulmus americana*)	53.2	2.4	0.9	11.5	0.6	3.6
southern red oak						
(*Quercus falcata*)	40.6	2.0	1.2	19.2	0.4	4.5
softwood						
balsam fir						
(*Abies balsamea*)	46.8	12.4	1.0	4.8	0.5	3.4
eastern white-cedar						
(*Thuja occidentalis*)	45.2	8.3	1.5	7.5	1.3	4.2
eastern hemlock						
(*Tsuga canadensis*)	45.3	11.2	1.2	4.0	0.6	3.3
jack pine						
(*Pinus banksiana*)	45.6	10.6	1.4	7.1	1.4	3.9
white pine						
(*Pinus strobus*)	44.5	1.06	2.5	6.3	1.2	4.0
loblolly pine						
(*Pinus taeda*)	45.0	11.2	2.3	6.8	1.7	3.8
douglas fir						
(*Pseudotsuga taxifolia*)	43.5	10.8	4.7	2.8	2.7	2.8
black spruce						
(*Picea mariana*)	47.9[b]	10.5[c]		8.0		4.1
white spruce						
(*Picea glauca*)	46.5	11.6	1.2	6.8	1.6	3.6
tamarack						
(*Larix laricina*)	46.1	13.1	2.3	4.3	1.0	2.9

[a] Data from Forest Products Laboratory U.S. Dept. of Agriculture, Madison, Wis., and B. L. Browning (10).

[b] Including galactan.

[c] Including arabian.

approximate elementary composition of many species of woods is as follows: carbon, $49 \pm 1\%$; hydrogen $6.1 \pm 0.1\%$; nitrogen, 0.1–0.3%. The ash in most domestic woods is within the limits of 0.2–1.0%. Some exotic woods contain up to 8% and more ash. Oxygen, of course, constitutes the greater part of the balance. Table 1 gives proximate analyses of representative woods (9).

In describing the chemical components of wood, it is common to differentiate between cell-wall components and extraneous materials. The components of the cell wall are lignin and the total carbohydrate fraction, consisting of cellulose and hemicellulose.

The extraneous materials consist of substances that can be removed by extraction with nonreactive solvents, protein residues from the protoplasm of the growing cell, and the mineral constituents, some of which are difficult to remove. The distinction between cell-wall constituents and extraneous materials is arbitrary. There is extreme variation in the character of extraneous materials in different woods and the amount varies from about 4% to over 20%. The extractives, or the lack of them, affect such important factors as color, durability, taste (for wood used in contact with food), flammability, and pulping characteristics (75).

Cell-Wall Carbohydrates. The carbohydrates are the most prominent component of the cell wall, amounting in most cases to 65–75% of the weight of the wood. Hydrolysis of the total carbohydrate fraction yields mainly simple sugars, primarily glucose. The carbohydrates in representative softwoods and hardwoods are shown in Table 2.

Cellulose. The main constituent of the cell-wall carbohydrates is cellulose (qv), which in most respects is chemically similar to purified cotton cellulose. It consists of glucose residues joined through 1,4-β-glucosidic linkages. The resistant portion of wood cellulose shows a rate of hydrolysis in dilute acid nearly twice that of cotton cellulose. X-ray diffraction studies show that the crystallite is larger in cotton than in wood cellulose. No wood cellulose has been prepared that does not contain traces of mannan and xylan. Highly purified softwood celluloses are likely to contain 1% of both xylan and mannan. It is not known whether these traces of noncellulosic carbohydrates are physically trapped within the solid cellulose structure, or whether they are an integral part of the wood cellulose chains.

Hemicellulose. The cell-wall carbohydrates contain hemicellulose in addition to true cellulose. Historically, hemicellulose was defined as the easily hydrolyzed portion of cellulose. Structurally the material is amorphous and includes as components mannose, galactose, arabinose, xylose, uronic acids, and, in some cases, rhamnose. The distinction between "true" cellulose and hemicellulose is mainly of academic interest. No analytical method or preparative scheme succeeds in making a clear-cut separation.

Analyses of wood celluloses made for technical purposes frequently express the composition in terms of its α-, β-, and γ-cellulose content. The distinction is based on solubility in alkali. In general terms, α-cellulose is insoluble in 17.5% sodium hydroxide, β-cellulose is the soluble portion that is precipitated by acidification, and γ-cellulose is the soluble portion not precipitated by acid. α-Cellulose is sometimes mistakenly thought of as "true" cellulose. Actually, it may contain over 10% of carbohydrates other than glucosan.

Lignin (11). The noncarbohydrate component of the cell wall, lignin, is a polymer based primarily on a substituted phenylpropane unit. It usually comprises from 18 to 28% of the wood and it is present in the middle lamella and the layered cell wall. Lignin (qv) is difficult to characterize because in the process of isolation it undergoes changes that vary according to the methods used. For this reason the term "lignin" has no specific reference apart from a complete description of the methods used to prepare it. To make matters even more difficult, hardwood and softwood lignins differ, and it is probable that the ultimate composition of lignin may vary from species to species. Routine analyses of lignin are ordinarily carried out by dissolving the carbohydrate from wood using a primary hydrolysis with 72% sulfuric acid, followed by a secondary hydrolysis with dilute acid.

Extraneous Components (6,8,12–14). The mineral constituents of wood vary greatly between species, between trees within a species, and even within one tree. The most prominent cations found are calcium, potassium, and magnesium. Carbonates, phosphates, silicates, and sulfates are the common anions.

The organic extraneous materials in wood consist of many classes of compounds, such as aliphatic and aromatic hydrocarbons, terpenes, aliphatic and aromatic acids and their salts, alcohols, phenols, aldehydes, ketones, quinones, esters, and ethers. Certain woods contain essential oils and alkaloids. Others contain tannins, coloring matter, water-soluble polysaccharides, gums, and proteins.

Wood–Liquid Relationship

Adsorption. Dry wood substance is very hygroscopic. The amount of moisture adsorbed depends mainly on the relative humidity and temperature, as shown in Figure 3. Exceptions occur with species having high extractives contents (eg redwood, cedar, and teak). The equilibrium moisture contents of such woods are generally somewhat lower than those given in Figure 3.

In green wood, the cell walls are saturated while the cell cavities may be incompletely or completely full of water. Moisture in the cell walls is called bound, hygroscopic, or adsorbed moisture. Moisture in the cell cavities is called "free" or capillary water. The distinction is made because under ordinary conditions the removal of the free water has little or no effect on many properties of the wood. In contrast, the removal of the cell-wall moisture has a pronounced effect.

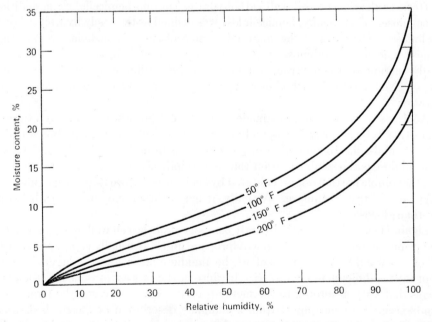

Fig. 3. Relationship between the moisture content of wood (% of dry wood) and relative humidity at different temperatures.

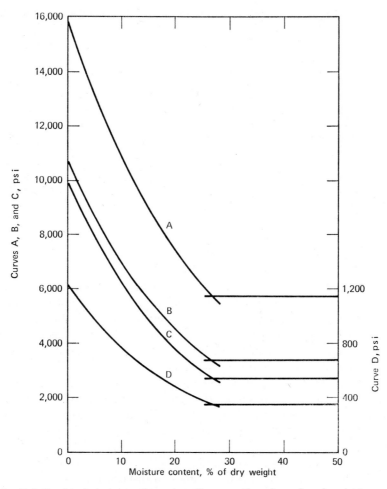

Fig. 4. Relationship between various strength properties of wood and moisture content: A, modulus of rupture; B, fiber stress at elastic limit in static vending; C, maximum crushing strength parallel to the grain; D, fiber stress at elastic limit in compression perpendicular to the grain.

In equilibrium with all relative humidities below 100%, the moisture in seasoned wood is present primarily in the cell walls. The cell cavities are essentially empty. The moisture content at which the cell walls would be saturated and the cell cavities empty is called the fiber-saturation point. Actually, such distribution is impossible. Beginning at about 90% relative humidity, some condensation may occur in the very small capillaries. The determination of the fiber-saturation point is based on the fact that certain properties of wood (for example, strength and volume) change uniformly at first with increasing moisture content and then become independent of the moisture content, as shown in Figure 4. The equilibrium-moisture content (usually determined by extrapolation), at which the property becomes constant, is the fiber-saturation point. The value usually varies between 25 and 35%.

The density of wood substance is about 1.50 g/ml. A species of a density of 0.5 g/ml has a void volume of 67%. Each gram of the wood occupies 2 ml and contains 0.67 ml of wood substance and 1.33 ml of void. Such wood floats in water. Waterlogg-

Table 3. Specific Gravity of Some Common Woods Grown in the United States

Wood	Sp gr[a]
aspen	0.40
birch, yellow	0.67
cottonwood, eastern	0.42
douglas fir, coast type	0.51
fir, balsam	0.38
hemlock, western	0.44
maple, sugar	0.68
oak, white	0.73
pine	
lodgepole	0.43
ponderosa	0.42
longleaf	0.62
spruce, engelmann	0.36
walnut, black	0.59

[a] Based on oven-dry weight and volume at 12% moisture content.

ing of sizable specimens is very slow. In the waterlogged condition, the cell walls adsorb about 0.28 ml of moisture; the cell cavities contain about 1.33 ml of water.

The average specific gravity of different species is given in Table 3 (4). The conventional way for expressing the specific gravity of wood is in terms of the oven-dry weight and volume at 12% moisture content. The specific gravity based on the volume of the oven-dry wood is about 6% higher.

At low relative humidities, adsorption is due to interaction of water with accessible hydroxyl groups. These are present on the lignin residues and on the carbohydrates in the noncrystalline or poorly crystalline regions. The high differential heat of adsorption by dry wood—about 260 cal/g of water—reflects a very high affinity for moisture (15–17).

At high relative humidities, adsorption is believed to occur in response to a tendency for cellulose chains and lignin residues to disperse themselves (solution tendency). Complete dispersion (dissolution) is prevented because of the strong interchain or interpolymer bonding at certain sites or regions. The differential heats of adsorption are much smaller than at low relative humidities.

Because the relative humidity of the atmosphere changes, the moisture content of wood undergoes corresponding changes. Effective protection against fluctuating atmospheric conditions is furnished by a surface coating of certain finishes, provided the coating is applied to all surfaces of wood through which moisture might gain access. However, no coating is absolutely moistureproof; coatings simply retard the rate at which moisture is taken up from or given off to the atmosphere. This means that coatings cannot be relied upon to keep moisture out of wood that is exposed to dampness constantly or for prolonged periods. Coatings vary markedly in their moisture-retarding efficiencies.

For some uses, it is important to protect wood against wetting by water. This is true for doors, windows, door and window frames, and the lap and butt joints in wood siding. Water repellents and water-repellent preservatives, long used in the millwork industry, provide protection from wetting. They are designed to penetrate into wood, but they leave a very thin coating of wax, resin, and oil on the surface of the wood which repels water. However, they are not as effective in resisting moisture vapor.

Neither coatings nor water repellents alter the equilibrium-moisture content or equilibrium swelling of wood. This can only be accomplished by depositing bulking agents that block normal shrinkage within the cell walls, chemically replacing the hygroscopic hydroxyl groups of cellulose and lignin with less hygroscopic groups, or forming chemical cross bridges between the structural units of wood (see below under Modified wood).

Shrinking and Swelling. The adsorption and desorption of water in seasoned wood is accompanied by external volume changes. At moisture contents below the fiber-saturation point, the relationship may be a simple one, merely because the adsorbed water adds its normal volume to, or the desorbed water subtracts its volume from, that of the wood. The relationship may be complicated by the development of stresses. Theoretically, above the fiber-saturation point, no volume change should occur with a change in the moisture content. Actually, owing to the development of stresses, changes in volume or shape may occur. The magnitude of such stresses is minimized by drying wood under carefully controlled and empirically established conditions (16).

In the absence of drying stress (ie with small specimens and extremely slow drying) the percentage of volumetric swelling from the green to oven-dry condition is, as a first approximation, proportional to the specific gravity of the wood. The value of the slope of the linear relationship is equal to the average fiber-saturation point of the wood. Serious deviations from the linear relationship may occur with species high in extractives content.

Swelling or shrinking of wood is highly anisotropic. Tangential swelling (occurring tangent to the rings) is 1.5–3.5 times greater than radial swelling (occurring along a radius of the rings). Longitudinal swelling (occurring in the direction of tree growth) is usually very small. In certain abnormal woods, however, such as compression or tension wood, longitudinal swelling or shrinking may be relatively high (as much as 1–2%) (17).

Permeability. Although wood is a porous material (60–70% void volume), its permeability (ie flow of liquids under pressure) is extremely variable. This is due to the highly anisotropic shape and arrangement of the component cells and to the variable condition of the microscopic channels between cells. In the longitudinal direction, the permeability is 50–100 times greater than in the transverse direction (16). The sapwood is considerably more permeable than the heartwood. In many instances (ie

Table 4. Relative Permeability of the Heartwood of Some Common Species, Decreasing from Group 1 to Group 4

Group 1	Group 2	Group 3	Group 4
ponderosa pine	coastal douglas fir	eastern hemlock	alpine fir
basswood	jack pine	engelmann spruce	interior douglas fir
red oaks	loblolly pine	lodgepole pine	tamarack
slippery elm	longleaf pine	noble fir	western red cedar
tupelo gum	western hemlock	sitka spruce	black locust
white ash	cottonwood	western larch	red beech
	aspen	white fir	red gum
	silver maple	white spruce	white oaks
	sugar maple	rock elm	
	yellow birch	sycamore	

interior-grown douglas fir) the permeability of the heartwood is practically zero. A rough comparison, however, may be made on the basis of heartwood permeability, as shown in Table 4.

Transport Phenomena. Physically, wood is composed of a complex capillary network through which transport occurs by capillarity, pressure permeability, and diffusion. A detailed study of the effect of capillary structure on each of the three transport mechanisms has been made by Stamm (16).

Drying Methods. The living tree holds many gallons of water in the cells of its wood. A 16-ft southern pine log 15 in. in diam, for example, may weigh as much as 1250 lb and contain about 47% or 70 gal, of water. Most of this water must be removed by seasoning to make lumber and other wood products.

There are a number of important reasons for seasoning: (1) it reduces the likelihood of stain, mildew, or decay developing in transit, storage, or use; (2) the shrinkage that accompanies seasoning can take place before the wood is put to use; (3) wood increases in most of its strength properties as it dries below the fiber-saturation point (30% moisture content); (4) the strength of joints made with common fasteners, such as nails and screws, is greater in seasoned wood than in green wood seasoned after assembly; (5) the electrical resistance of wood increases greatly as it dries; (6) dry wood is a better thermal insulating material than wet wood; (7) the appreciable reduction in weight that accompanies seasoning is important in reducing shipping costs.

Ideally the temperature and relative humidity during drying should be controlled; if wood dries too rapidly it is likely to split, check, warp, or honeycomb because of shrinkage stresses.

Air-drying is a process of piling sawmill products outdoors to dry. Control of drying rates is limited and great care must be used to avoid degrading the wood. Drying time is a function of climatic condition; in damp coastal areas wood dries slowly, whereas in the arid regions of the Southwest it dries rapidly. Typical air-drying times for 1-in. lumber of various species are shown in Table 5. For more details, see reference 18.

Kiln-drying is usually considered a controlled drying process. It is widely used for drying both hardwoods and softwoods. Dry-bulb temperatures seldom exceed 200°F and then only at the end of the drying schedule. In the initial stages of drying the

Table 5. Approximate Air-Drying and Kiln-Drying Periods for 1-in. Lumber

	Days required to	
Species	Air-dry to 20%	Kiln-dry to 6%[a]
bald cypress	100–300	10–20
hickory	70–200	7–15
magnolia	60–150	10–15
oak, red	100–300	16–28
oak, white	150–300	20–30
pine, southern	40–150	3–5
sweet gum	70–300	10–25
sycamore	70–200	6–12
tupelo	70–200	6–12
yellow poplar	60–150	6–10

[a] From 20% moisture content.

relative humidity is maintained at a high level to control the moisture gradient in the wood and thus prevent splitting and checking. Modern kiln installations use forced-air circulation, instrumented to give good control of both dry- and wet-bulb temperatures (see Vol. 2, p. 692). Most are equipped with automatic control instruments. The kilns are vented to exhaust the moisture evaporated from the wood. Most installations are steam-heated, although furnace-type kilns using gas or oil for fuel are now being built. Many species, especially hardwoods, are first air-dried to about 20% moisture content, then kiln-dried to the moisture content at which they will be used. Table 5 shows the approximate air-drying and kiln-drying periods for one-in. lumber. A typical time schedule for kiln drying a softwood is shown in Table 6. Schedules for drying hardwoods are generally more complex (19).

Table 6. Typical Softwood Kiln-Drying Time Schedules for 1-in. Ponderosa Common Pine

Hours in kiln	Dry-bulb temperature, °F	Wet-bulb temperature, °F	Relative humidity, %
heartwood[a]			
1–8	130	110	52
8–16	136	112	48
16 until dry	140	112	41
sapwood[b]			
1–12	130	110	52
12–24	136	112	48
24 until dry	140	112	41

[a] Kiln dried to average 12% moisture content in 24–36 hr.
[b] Kiln dried to average 12% moisture content in 48–72 hr.

Special drying methods, such as superheated-steam drying, solvent seasoning, vapor drying, vacuum drying, infrared radiation, and high-frequency dielectric and microwave heating, are occasionally used when accelerated drying is desired and the species being dried can withstand more severe drying conditions without damage. None of these methods are of significant commercial importance at the present time.

Wood as a Structural Material in the Chemical Industry

Strength and Related Properties. In the framing of a building or the construction of special industrial units in the chemical field, where wood may be used because of one or more of its unique physical properties, strength is a primary requirement. Different kinds or species of wood have different strength values (4). Strength is related to the amount of wood substance per unit volume, ie its specific gravity. Thus, a heavy wood, such as oak, is stronger than a lightweight wood, such as spruce. The strength of a piece of lumber depends also upon its grade. The strength values of a grade depend upon the size and number of such characteristics as knots, cross grain, shakes, splits, and wane (20). Wood free from these defects is known as "clear" wood.

Most strength properties of clear wood increase markedly as moisture content is decreased below about 30%, based on the oven-dry weight (see Fig. 4). There is no important difference in the strength properties of wood that is properly kiln-dried as compared with wood that is carefully air-dried.

In general, when the temperature of wood is raised above normal, the wood tends to become weaker in most strength properties; when the temperature is lowered, the

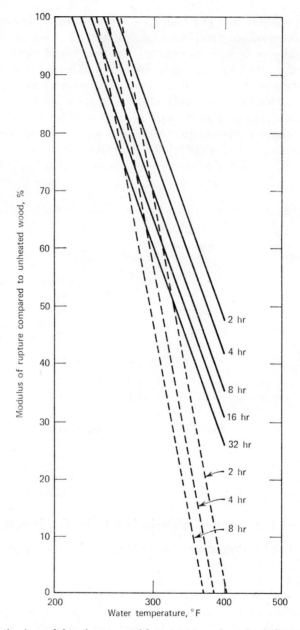

Fig. 5. Variation in modulus of rupture with temperature for softwoods and hardwoods soaked in hot water for various periods. Strength tests were made at ambient temperature and 12% moisture content. LEGEND: —— softwood; - - - - - - hardwood.

wood becomes stronger. This effect is immediate and its magnitude depends upon the moisture content of the wood and, when the temperature is raised, upon duration of exposure. Experiments indicate that air-dried wood can probably be exposed to temperatures up to nearly 150°F for one year or more without an important permanent loss in most strength properties. However, its strength at the elevated temperature will be temporarily reduced as compared to the strength at normal temperature (21–24).

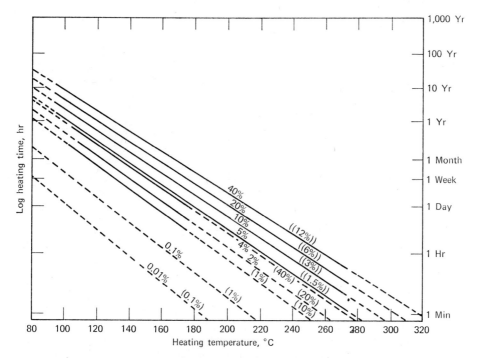

Fig. 6. Logarithm of heating time vs temperature to attain various degrees of degradation of wood (32). LEGEND: no parenthesis, weight loss on oven heating; single parenthesis, modulus-of-rupture loss on oven heating; double parenthesis, weight loss on heating beneath surface of molten metal. Courtesy, Industrial and Engineering Chemistry.

When wood is exposed to temperatures of 150°F or more for extended periods of time, as it may be in tanks used by the chemical industry, it is permanently weakened (25–27), even though the temperature is subsequently reduced and the wood is used at normal temperatures (see Fig. 5).

The effect of absorption of various liquids upon the strength properties of wood is largely dependent upon the chemical nature and reactivity of the absorbed liquid. In general, neutral, nonswelling liquids have little if any effect upon the strength properties (28). Any liquid which causes wood to swell causes a reduction in strength. This effect may be temporary, existing only while the liquid remains in the wood. It may also be permanent, as in the case of chemically reactive liquids (strong acids and bases) and of a magnitude dependent upon time and temperature of exposure and concentration of the solution (see p. 377).

Coal-tar creosote, water-gas creosote, creosote–tar mixtures, and creosote–petroleum mixtures are practically inert to wood and have no chemical influence that would affect its strength. (See Tar and pitch.) The 2–5% solutions of zinc chloride commonly used in preservative treatment apparently have no important effect.

Although wood preservatives are not harmful in themselves, the treatment used in injecting them into the wood can result in considerable loss of strength because of temperature effects. Under approved treating methods, however, the permanent loss of strength is not considered serious.

Resistance to Heat and Fire. Like any organic material, wood is subject to deterioration of its physical and chemical properties with age. The rate and extent of

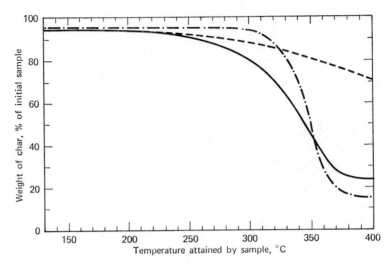

Fig. 7. Portion of dynamic thermogravimetric curves between 130 and 400°C, with temperature rising 6°C/min, for wood, lignin, and α-cellulose. LEGEND: —— wood; ------- lignin; –.–.– α-cellulose.

property loss is governed by the interdependent factors of temperature, time, and moisture. In locations not conducive to decay or insect attack, wood is extremely stable at ordinary temperatures. Thus, wood excavated from Egyptian tombs and Indian mounds after thousands of years in dry desert locations has shown no indication of change. However, as the temperature of exposure is increased, there is a degradation of surface layers of the wood which progresses into the interior layers, dependent upon the time of exposure. In prolonged heating at temperatures as low as 95°C, surface layers may become charred.

In rapid heating at temperatures up to 200°C, the principal products of decomposition of wood are water vapor and traces of carbon dioxide, formic and acetic acid, and glyoxal (29). The gases evolved are not usually ignitable, although there is some evidence that after 14–30 min at 180°C, a pilot flame can ignite small samples under certain conditions (30). Exothermic oxidation reactions may occur at temperatures near 200°C and, under conditions in which the heat is conserved, self-ignition may occur. Self-ignition of sound wood has not been obtained experimentally within this temperature range. However, many fire protection authorities (31) conservatively recommend exposure temperatures no higher than 77°C for wood in prolonged exposure near heating devices.

The degradation of wood at elevated temperatures is accompanied by a reduction in weight. The loss in weight varies not only with temperature, but also with the duration of heating (33), as shown in Figure 6. Thermogravimetric analysis (6°C/min) of wood (34), α-cellulose, and lignin (sulfuric acid type), as shown in Figure 7 (30), indicates that a slow initial weight loss for lignin and wood begins close to 220°C.

Differential thermal analyses of wood indicate that the thermal degradation reactions in an inert atmosphere release less than 5% of the heat released during combustion in air. The differential thermal analysis data for wood, cellulose, and lignin in an oxygen atmosphere, shown in Figure 8 (35), indicate there are two major exothermic peaks for wood, one near 320°C, which represents the flaming reaction, and the other near

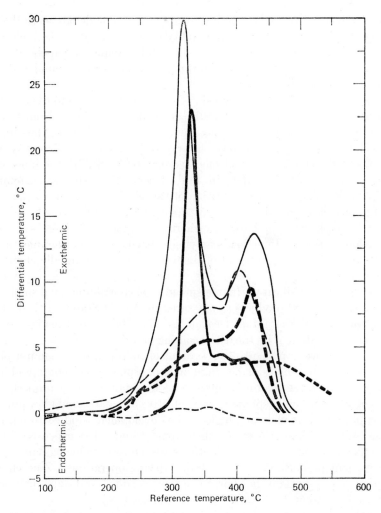

Fig. 8. Differential thermal analysis of the pyrolysis and combustion of wood and its components at a heating rate of 12°C/min and a gas-flow rate of 30 ml/min. Sample weight: wood, 40 mg; α-cellulose, 20 mg; lignin, 10 mg; charcoal, 12 mg. Treated wood sample contains 9% by wt of commercial fire retardant. LEGEND: ——— untreated wood in O_2; —·—·— untreated wood charcoal in O_2; ········ untreated wood in N_2; ▬▬ untreated α-cellulose in O_2; ▬ ▬ ▬ untreated sulfate lignin in O_2; ▬·▬·▬·▬ fire-retardant treated wood.

440°C, which represents the glowing reaction. The major exothermic peak for α-cellulose is near the flaming peak and the major exothermic peak for lignin is near the glowing peak. New techniques of thermogravimetric, differential thermal, and chemical analysis are being employed in basic studies of the pyrolysis and combustion reactions for wood and its components and the influence of chemical additives (34–40). This research will develop better knowledge of how the combustion reactions for wood may be controlled. (See also Thermal analysis.)

Fire performance codes for materials and constructions are principally concerned with noncombustibility (ASTM Standard E 136) (41), fire endurance (ASTM E 119 (42) and ASTM E 152 (43)), and fire hazard classification (ASTM Standard E 84) (44). Wood, even in its treated form, does not meet the requirements of ASTM Standard E

136 for noncombustibility. There is, however, considerable debate as to the safety of fireproof construction, when the combustibility of the contents of the buildings is not regulated. Some codes also permit materials with low flame-spread characteristics to be classified as noncombustible, and under this definition, certain treated wood products can be classified as noncombustible.

Wood in its untreated form has good resistance or endurance to the penetration of fire when used in thick sections for walls, doors, floors, ceilings, beams, and roofs. This endurance exists because wood has low thermal conductivity, thus limiting the rate at which heat is transmitted to the interior of the member. Typically, when fire temperatures at the surface of wood members are 1600–1800°F, the inner char zone temperature is about 550°F, and $\frac{1}{4}$ in. inward beyond the char zone, temperatures are 350°F or less. The rate of penetration of this char line is approx $\frac{1}{40}$ in. per min, dependent upon species, density, and moisture content (45,46). Due to this slow rate of penetration of the char and the low thermal conductivity, large wood members retain a major portion of their load-carrying capacity for considerable time during exposure to fire. Chemical treatment of the wood does not materially change this fire endurance.

The fire hazard classification (flame spread, fuel contributed, and smoke development) for wood and wood products as measured by ASTM Standard E 84 (44) can be reduced by the use of fire-retardant treatments (47). These treatments may be either by chemical impregnation or by the use of coating products.

Some of the more important fire-retardant chemicals are ammonium phosphate, ammonium sulfate, zinc chloride, dicyandiamide-phosphoric acid, borax, and boric acid. These chemicals are often used in combinations to obtain the best properties of each. For example, ammonium phosphate is very effective as a flame and glow inhibitor, but contributes to premature charring, high smoke development, and corrosion problems. Ammonium sulfate reduces the smoke development and is less expensive. Combinations of borax and boric acid give moderate effectiveness in reducing flame spread and afterglow without premature charring during drying operations. Zinc chloride is a very effective flame retardant, but boric acid is often added to the zinc chloride formulation to retard afterglow. (See also Fire resistant textiles.)

Solutions of these fire-retardant formulations are impregnated into wood under full-cell pressure treatment to obtain dry chemical retentions of 4–6 lb/ft³. This type of treatment greatly reduces flame spread and afterglow. There are many theories as to the reactions involved in retarding the flaming of wood. A principal theory is that the effective fire-retardant chemicals decompose at temperatures lower than the normal initial decomposition temperature for wood and release materials capable of quickly converting cellulose to charcoal and water, thus preventing the decomposition at higher temperatures to l-glucosan and volatile flammable products. The flaming characteristics of wood are thus reduced.

Some of the same chemicals mentioned and others, such as chlorinated rubber or paraffin, antimony trioxide, calcium carbonate, calcium borate, and pentaerythrithol, may be used to formulate fire-retardant paint coatings. Many of these paints are formulated so that the films intumesce (expand) when exposed to fire, thus insulating the wood surface from thermal exposure. The use of fire-retardant paints has been largely limited to existing construction because of the difficulty in job inspections, ie of seeing that the proper film thicknesses have been applied.

The use of these fire-retardant treatments is increasing because of their effectiveness in reducing flame spread, but costs, poor resistance to leaching and weather, and the effect on some of the related performance characteristics of wood have limited the use. Improved formulations are being developed with better resistance to leaching and weather and the effects on related performance characteristics are being minimized.

Treatment of wood by a suitable fire-retardant may prevent ignition by a small source of fire or delay the development of a large fire. However, in a fully developed fire, fire-retardant treatments are of minor effectiveness. The gases generated by combustible contents in a building can flash and spread fire to remote areas even when the enclosing surfaces are of low flammability or are noncombustible.

Fire-retardant treatments are used primarily for wood framing, roof-supporting members, and decking, when permitted in noncombustible buildings, also for wall paneling in corridor and assembly areas of public buildings, and for edge members in wood fire doors.

Resistance to Chemicals. Wood is widely used as a structural material in the chemical industry because it offers certain special advantages in addition to the well-known advantages that are responsible for its wide use in other industries (48–51). Foremost is its resistance to mild acids which is far superior to that of common steel. However it is not as good as some of the more expensive acid-resistant alloys. A well-constructed wood tank used to store cold, dilute acid generally lasts a long time. A redwood tank, for example, can be used for ten to twenty years depending on the acid concentration. But with increasing concentration and—even, more— with increasing temperature wood deteriorates rapidly.

Because traces of iron reduce the brilliance of many dyes, wood tanks have long been used in the manufacture of dyes. Similarly, wood tanks are used in processing vinegar and sour foodstuffs because common metals impart a metallic taste to these products. Ease of fabrication may be the deciding factor in the use of wood tanks in less accessible areas to which ready-made tanks of other materials can not be easily moved.

In the pickling of steel prior to applying a protective coating, such as zinc or enamel, the steel is commonly immersed in hot sulfuric acid (approx 8%) in large wood tanks. The deteriorating effect of the acid is kept to a minimum by the use of heart-wood of species that resist penetration by liquids. The thick timbers needed for strength in these large tanks offer more resistance than thin members. The service life of such tanks varies from about three to eight years.

Nitric acid attacks wood faster than the other common mineral acids, although wood is frequently used in contact with dilute nitric acid. Wood shows excellent resistance to organic acids, which gives it a distinct advantage over steel, concrete, rubber, and some plastics.

Alkaline solutions attack wood more rapidly than do acids of equivalent concentration; strong oxidizing chemicals are harmful to wood. Wood is seldom used where resistance to chlorine and hypochlorite solutions is required. Wood tanks are, however, satisfactory for holding hydrogen peroxide solutions and wood gives good service in contact with strong brine. Solutions of iron salts attack wood. In contact with iron under damp conditions, wood may show severe deterioration within a few years (52). Species that are high in acidic extractives seem especially prone to such attack.

Different species of wood vary in their resistance to chemical attack. The significant properties are believed to be in the structure, which governs the rate of ingress of the chemical, and the composition of the cell wall, which affects the rate of action at the point of contact (53).

The relative value of species for a specific use cannot be measured accurately, but arbitrary tests have been used to obtain indicative data (54,55).

Softwoods are generally more resistant to acids than are hardwoods. This has been explained as being due largely to a higher lignin content and lower hemicellulose content. The heartwood of certain conifers, notably cypress, dense southern pine, douglas fir, and redwood, has been widely used by the chemical industry, and several other coniferous species have been used in lesser amounts.

With the growth of the glued laminated-wood industry during the 1950s large-size wood beams and arches are now available in nearly any size and shape. Their use as structural members for warehouses for bulk fertilizer and other chemical storage warehouses, and for manufacturing-plant enclosures, is increasing. The large cross-section wood members provide good resistance to chemical attack by corrosive dusts and fumes.

Certain measures are taken to protect wood against chemicals in industry. There has been a steady increase in the use of wood tanks lined with materials, such as lead, rubber, ceramics, or plastics. Structural members have been coated with a variety of protective materials and, in some cases, the wood has been impregnated under pressure with bituminous materials or with resin-forming mixtures (56–59).

Wood impregnated with phenolic resin is produced commercially for items such as tanks, filter-press plates and frames, and large exhaust ducts, where resistance to acid is important and replacement cost is high.

Resistance to Microbiological Degradation. Wood may be subjected to attack by fungi or termites. Fungi break down the various wood components enzymically to forms which are readily assimilated by them. Wood which has undergone such attack is said to be decayed, rotted, or doty. Wood susceptible to fungal attack must be at a moisture content slightly above its fiber-saturation point; ie about 30% moisture content, based upon the oven-dry weight of the wood. Conversely, if wood is too wet, ie saturated, the oxygen supply in the wood is then too limited for fungal growth. Decay can be prevented by keeping wood either too dry or too wet for fungal development, by using naturally decay-resistant species, or by treating with fungicides or wood preservatives. Termite attack can be prevented, or lessened, through use of naturally resistant wood, or by treating with preservatives. In the case of the subterranean termite, which requires contact with the ground in order to survive, poisoning the soil around the wood structure is another means of preventing termite infestation. The nonsubterranean termite flies directly to the wood into which it bores. Barriers against their access to the wood, such as paint or screens, are further means of preventing infestation by termites. Despite the great differences between fungi and termites, chemicals that inhibit one usually inhibit the other.

For practical purposes, the sapwood of all species may be considered to be susceptible to decay. The heartwood of some species contains toxic extractives that protect it against fungal attack. Among the native species that have decay-resistant or highly decay-resistant heartwood are bald cypress, redwood, the cedars, white oak, black locust, and black walnut (60). Douglas fir, several of the pines, the larches, and honey locust are of intermediate decay resistance. Species low in decay resistance include the remainder of the pines, the spruces, true firs, hemlocks, ashes, aspens, birches, maples, hickories, red and black oaks, tupelo, and yellow poplar.

Wood treated with preservative is used in many chemical industries; in fact, wood treating is a major industry in itself. Oil-type preservatives, such as creosote or petroleum solutions of pentachlorophenol, are used most commonly where wood is in direct contact with the ground. However, copper–chrome–arsenate and ammoniacal-copper–arsenite waterborne salt preservatives are becoming more important in the treatment of woods to be used under conditions of severe exposure. When cleanliness and paintability are very important, wood treated with water-soluble preservatives rather than with colored oil-type preservatives is often used. Pentachlorophenol in light petroleum solvents or in a volatile solvent is also important in the treatment of wood where cleanliness is of paramount importance. Treating specifications pertaining to use of these preservatives are available (61).

Redwood used for cooling towers sometimes undergoes premature deterioration in which the attack may be either chemical or biological, or a combination of the two (52). It is generally accepted that fungi play the more important role in such deterioration; chemicals in the water render redwood more susceptible to decay by accelerating the loss of fungicidal extractives (62). While some cooling towers have been constructed of pressure-treated wood, many were built of untreated wood prior to the 1960s.

The most widely used in-place treatment for towers of untreated wood has been the double diffusion process (62). In this technique a toxic precipitate is formed within the wood by treatment first with a solution of one chemical, then with a solution of another (62a). The two chemicals diffuse separately into the wood, then react to form a relatively insoluble toxic precipitate. Copper, zinc, and nickel salts can be used for the first stage, and sodium chromate, arsenate, fluoride, borate, and phosphate for the second.

Confidence in the use of wood as a structural material in the chemical industry is well established. Properly treated with preservative, wood can be used under conditions where decay is usually severe. Treated posts instead of conventional foundations have been recommended for low-cost homes (63).

Modified Woods

One purpose of modifying wood other than to preserve it or to protect it from fire is to reduce shrinking and swelling under conditions of fluctuating relative humidity. It is well known that certain species with high extractives content, especially in the cell walls (for example, redwood and teak) have greater dimensional stability than species with low extractives content. This gives a clue to a means of obtaining still greater reductions in swelling and shrinking; that is, by deliberately adding large amounts of "extractives" or bulking agents to the cell walls.

Resin impregnation is a successful method of adding bulking agents provided the resin can permeate the lumens and penetrate the cell walls. To assure this, it is necessary to use resins of a very low degree of polymerization and to polymerize them after penetration has occurred. Thermosetting resins, such as the phenolics, have been used successfully. However the rate-determining step for the successful treatment is that of the approach of the resin to the cell wall. With green wood, this depends on the diffusion rate of the resin in the lumen-trapped water. With dry wood, pressure treatment can be used. Then the rate depends on the permeability of the wood. For both processes, wood size is very important. Treating times can be especially long with heartwood. The sapwood of some of the pines is sufficiently permeable to admit resin fairly uniformly and in a reasonable time, into 1-in. lumber several feet long. In general, however, thin veneers are treated. The veneers are pressure-impregnated (up to 200 psi) with a 30% aqueous solution of resin. (Water solubility indicates that the resin is sufficiently small in molecular size.) Following impregnation, the wetted wood is slowly dried and then heated at about 150°C for 20 min to set the resin. Laminates are built up by gluing the individual sheets together. The product is called *impreg*. Its density is about 20% greater than that of the original wood. Its color is that of the original wood or slightly darker.

Figure 9 shows the relationship between resin content and dimensional stability. Antishrink efficiency, defined as

$$\left(1 - \frac{\% \text{ swelling of treated specimen}}{\% \text{ swelling of control}}\right) \times 100,$$

is a measure of the extent to which the swelling and shrinking tendency has been reduced. It increases with increasing content of phenolic resin and then tends to level off at about 65% when the content reaches 30–35%, based on original wood. Impreg generally contains 25–35% resin.

Most of the mechanical properties are hardly affected by resin impregnation. The only property that is adversely affected is toughness. Resin impregnation reduces the Izod value for impact strength by over 60% (64). The treatment of wood with synthetic resins increases the decay resistance markedly. This may be due to the fact that the cell walls of the treated wood cannot take up sufficient moisture to support decay. Furthermore, residual unpolymerized resin would be toxic to organisms. Resin treatment increases the electrical resistivity (volume) of wood, especially under high relative humidity conditions. At 30% relative humidity the resistivity is increased tenfold; at 90% it is increased 1000-fold (65). Increasing the resin content from 30 to 70% has a negligible effect in changing the volume resistivity. At low humidities, ie below 40% relative humidity, surface resistivity is very high and therefore the insulating properties are determined solely by the volume resistivity, but above 40% relative humidity surface resistivity becomes an important factor.

Resistance to chemical attack is generally improved by resin impregnation. Impreg made with phenolic resin has an especially high resistance to acids. Impreg made with furfural resin has a high resistance to alkalies. Treated pine sapwood of 1-in. thickness is used for making chemical vats. The heat resistance of wood is improved markedly by resin impregnation. A block of impreg, subjected to forty-five 1-hr exposures at 400°F, showed no apparent loss in properties, although untreated wood showed signs of deterioration (66) after three 1-hr exposures.

Impreg is used to make die models for automobile body parts and patterns for shell-molding (66). In the latter procedure, the pattern is used to make a mold of resin-bonded sand. Molten metal is then poured into the mold. In such applications, advantage is taken both of the dimensional stability and heat resistance of the material. Such specialty uses for high-cost end products may stimulate the large-scale commercial production of impreg.

Compreg. If pressure is applied to dry, resin-treated veneers while they are being heat cured, a densified product (1.35 g/ml) is obtained. This material, called compreg, retains most of the advantages of impreg. In addition, owing to the two- to threefold

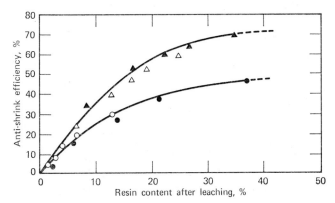

Fig. 9. Variation in antishrink efficiency for sitka spruce and sugar maple treated with urea and phenolic resins. LEGEND: urea: ● sitka spruce; ○ sugar maple; phenolic: ▲ sitka spruce; △ sugar maple.

increase in density, the mechanical properties are appreciably better than those of the original uncompressed wood. The strength properties of compreg are increased over those of untreated wood about in proportion to the degree of compression; only the toughness is lower than that of untreated wood but more than that of impreg.

Because of the plasticizing action of the resin-forming materials, the wood can be compressed under considerably lower pressures than dry, untreated wood. For example, treated spruce, cottonwood, and aspen veneer, dried to a moisture content of about 6% but not cured, are compressed, when subjected to a pressure of only 250 psi at 300°F, to about half the original thickness and a specific gravity of about 1.0 (67).

Compreg has a pleasing brown color and takes a high polish on buffing. The material is made commercially in small quantities and is used for items such as die stock (from which patterns or die models are carved), picker sticks (used in the textile industry), knife handles, gears, and certain woodwind instruments.

Staypak. In general, if wood is densified, then wetted, it springs back to its original dimensions and then swells beyond this to the normal swollen dimension. Under certain conditions (6–8% moisture, 160°C, 2000 psi), a densified material called staypak is obtained which does not spring back (68) and its normal swelling tendency is not affected. However, owing to the high density (1.35 g/ml), swelling occurs extremely slowly; in fact, the only manifestation of ordinary weathering is surface checking. The mechanical properties are much higher than those of the original wood; the modulus of rupture (with the grain) for a spruce-derived staypak is 37,000 psi in contrast to 12,000 psi for the uncompressed wood. Staypak is still an experimental material.

Wood-plastic combination is still another type of resin-treated product which is becoming commercially available. Wood is treated with a nonswelling vinyl-type monomer and cured by γ radiation (69) or by heat and a catalyst (70). The hygroscopic characteristics of the wood substance are not altered because little, if any, resin has penetrated the cell walls. However, because of the high resin content (70–100%), the normally high void volume of wood is greatly reduced. With the elimination of this very important pathway for vapor diffusion, the response of the wood substance to changes in relative humidity becomes very slow; in fact, the wood has a dimensional pseudostability. The important improvement in mechanical properties is the appreciable increase in hardness.

Wood bending is another wood-treating process. At temperatures above 175°F, green wood becomes readily deformable. On cooling to room temperature and drying under restraint, the new shape persists. This is the basis for commercial bending of wood to various shapes. The deformation, however, is not a fully plastic one. An elastic component persists and produces some strain recovery at high or cyclic humidification. A new process for plasticizing wood has been developed in 1964 (71). The wood member is immersed in liquid ammonia for a period of time, depending on the dimensions. Very appreciable plasticity is imparted to the wood. The ammonia is allowed to evaporate from the deformed wood and very little strain recovery occurs on humidification. The process is still in the experimental stage.

Wood As a Chemical Raw Material

The principal use of wood as a chemical raw material is in the pulp and paper industry (see Pulp; Paper). Pulp and other fiber-products industries consumed 57 million cords of wood in the United States in 1964. Another million cords were used in the production of charcoal. In addition, extractives produced by various tree species are important commercially.

Wood residues from logging and primary manufacturing operations represent a neglected source of organic raw material. Unlike mineral resources, this material is part of a renewable crop. In the United States over 100 million tons are available annually on a dry basis, a quantity comparable to the annual U.S. steel production.

The chemical utilization of significant amounts of wood residue has been limited mainly to the production of fiber products and charcoal, and, in Europe during World War II, to the production of yeast and alcohol from wood hydrolyzates (72). To some extent, this limitation has been due to a lack of integration of mechanical wood-processing plants with chemical plants to use wood to the best advantage. There is little chance, however, that the wood-chemical industry will grow rapidly as long as adequate supplies of organic chemicals are available at low cost from petroleum, natural gas, or coal.

Extractives. These are substances in trees or in their exudates which are soluble in water or in neutral organic solvents. Extractives, unlike lignin, the hemicelluloses, and cellulose, are not part of the cellular structure of wood. Extractives from wood or bark include a variety of compounds that differ widely, depending on the species and the part of the tree from which they are obtained. They include such products as carbohydrates, organic acids, dyes, tannins, polyphenols, fats, hydrocarbons, waxes, resins, terpenes, essential oils, sterols, and alkaloids. Some of these products are important commercially, but, with a few exceptions, the volume is small.

In the United States, the naval-stores industry is the largest of those based on wood extractives. In 1967, U.S. production of rosin (qv) from all sources was estimated at 1.9 million drums (of 520 lb) and of turpentine (qv) at 630,000 barrels (of 50 gal) (73). Tall oil (qv), a by-product of kraft pulping of pines, is composed of a mixture of rosin and fatty acids. In 1963 the total value of tall oil produced in the United States was about $58 million (74).

Tannin (see Leather) is obtained by extraction, but the bulk of the 100,000 tons of tannin used in the United States is imported. Latex, from which natural rubber is obtained, is an exudate. Exudates differ from the tree extractives in that they are formed in response to injuries to the tree.

Other extractives, such as the arabogalactan from western larch, and oils, such as cedarwood, sweet birch, and sassafras, are produced in minor amounts. An excellent source of information on wood extractives is the book by Hillis (75).

Hydrolysis. In the hydrolysis process, wood is treated with strong hydrochloric acid or with a dilute solution of sulfuric acid, forming a lignin or lignocellulose residue, and a liquor containing sugars, organic acids, furfural, and other chemicals. One of the many ways of utilizing this liquor is to ferment the sugar to alcohol, but there are many other possibilities. The hydrolysis process is adaptable to all species and all forms of wood waste.

Hardwood residues contain up to 6% acetyl, up to 6% methoxyl, more than 20% pentosan, and more than 20% lignin. The rest is predominantly hexosan. Softwoods contain less acetyl, less methoxyl, less pentosan, and more lignin. A differential hydrolysis can separate acetic acid and pentoses, and a second stage can remove hexoses and leave a lignin residue. Pentoses can be converted to furfural or yeast. Hexoses are useful as the crude sugar, or, by continuing the hydrolytic process, they can be converted to hydroxymethyl furfural and finally to levulinic and formic acids. A small part of the methoxyl appears as methyl alcohol during hydrolysis; the rest remains with the lignin. The lignin residue from hydrolysis can be made to react with hydrogen gas at elevated temperatures and pressures in the presence of a catalyst to form neutral oils, cyclic alcohols, phenols, and a high-boiling tarlike residue (76). It can also be used in plastic molding compounds or be converted to activated charcoal, possibly by a fluidized-bed process. So complete and integrated a utilization has not yet been attempted.

A number of commercial plants using the Scholler process (72) have been built in Germany, Switzerland, Japan, and Russia. Of these only the Russian plants are still in operation. In general, these plants were built to produce sugars for fermentation to ethyl alcohol. Except under special circumstances such a process has proved to be uneconomical. In the Scholler process, a dilute solution of sulfuric acid is percolated through a charge of chips and sawdust in several cycles under steam pressures of 150 to 200 psi. Sugar yield is 45–50%, and the sulfuric acid requirement is 10% based on the dry wood. The concentration of the resulting sugar solution is about 4%. Yields per 100 lb of dry wood are 2.4–2.5 gal of 100% ethyl alcohol and 4–4.5 lb of dry yeast.

A percolation process that required less time than the Scholler method was developed at the U.S. Forest Products Laboratory, Madison, Wis., during World War II (77,78). On concentrating the dilute sugar solution, a wood-sugar molasses is obtained containing about 50% reducing sugars (79–81). Wood-sugar molasses is comparable to blackstrap molasses (79) as livestock feed.

Although the original objective of hydrolyzing the carbohydrates in wood sugar was to produce alcohol by fermentation, a number of other products can be obtained. Wood sugar has been shown to be suitable for hydrogenation, oxidation, and fermentation to a wide variety of organic materials, including glycerol, oxalic acid, 2,3-butylene glycol, furfural, and lactic, butyric, and acetic acids (76). Alternatively, the main product from the fermentation might be the yeast itself.

Furfural can be produced from a variety of wood-processing by-products. The potential sources include spent sulfite liquor, liquors from the prehydrolysis of wood for kraft pulping, and hardwood wastes. Current production of furfural is from agricultural residues. Increasing furfural demand is expected to bring some of the alternative raw materials into the picture.

Although the hydrolysis of wood to produce simple sugars has not proved to be economically feasible, by-product wood-sugars obtained from sulfite pulping are being used to produce ethyl alcohol and feed yeast (82). Furthermore, a hemicellulose molasses, obtained as a by-product in hardboard manufacture, has found acceptance as a replacement for blackstrap molasses in cattle feeds (83).

Charcoal

Charcoal is produced by heating wood under conditions that severely limit the amount of oxygen available for combustion. When wood is heated slowly to about 280°C, an exothermic reaction occurs. Wood is carbonized by prolonged heating to final temperatures of 400–500°C in the absence of air. If the resulting charcoal is subjected to further heating, it undergoes marked chemical and physical changes (see Carbon).

The raw materials for wood charcoal are log and mill residues. Charcoal is made by the following three methods; heating in (*1*) kilns of one- to 100-cord capacity; (*2*) vertical batch retorts of about one-ton capacity; and (*3*) continuous multiple hearth retorts (about one ton of charcoal per hour). Yields range from 20 to 30%, based on the weight of wood.

Annual production dropped from 500,000 tons in 1909 to about 210,000 tons in 1947. Following this, production increased to 265,000 tons in 1956, 238,000 tons in 1961, and 500,000 tons in 1968 (84). The growth in production reflected the use of charcoal briquets for home barbecue cooking. Although the main raw material for such briquets was hardwood charcoal, charcoal made from softwoods, bark, and nut-shells was also used. Before about 1950 carbon disulfide was made from charcoal and sulfur, but now it is made from methane and sulfur. (See Carbon disulfide).

In the early 1900s, the destructive distillation of wood was also an important source of industrial acetic acid, methanol, and acetone. Since 1916 more economical processes for the production of these chemicals have been developed and the recovery of these chemicals in charcoal production has diminished in importance. In 1920, nearly one hundred recovery plants were in operation (85–87). In 1969, the last plant in the United States ceased operations.

Bibliography

"Wood" in *ECT* 1st ed., Vol. 15, pp. 72–102, by E. G. Locke, R. H. Baechler, E. Beglinger, H. D. Bruce, J. T. Drow, K. G. Johnson, D. G. Laughnan, B. H. Paul, R. C. Rietz, J. F. Saeman, and H. Tarkow, U.S. Dept. of Agriculture.

1. U.S. Dept. Agr. *Timber Trends in the U.S. Forest Service*, Forest Resource Rept. No. 17, 1965.
2. U.S. Dept. Comm., *Statistical Abstract of the United States*, Bureau of the Census, 1968.
3. A. J. Panshin, C. deZeeuw, and H. P. Brown, *Structure, Identification, Uses, and Properties of the Commercial Woods of the United States*, Vol. 1 of *Textbook of Wood Technology*, 2nd ed. McGraw-Hill Book Co., Inc., New York, 1964.
4. U.S. Forest Products Laboratory, *Wood Handbook*, U.S. Dept. Agr. Handbook No. 72, 1955. (Available from Superintendent of Documents, U.S. Govt. Printing Office, Washington, D.C.)
5. U.S. Forest Products Laboratory, *The Structure of Wood*, U.S. Forest Serv. Res. Note FPL-04, Madison, Wis., 1963.
6. E. Hägglund, *Chemistry of Wood*, Academic Press, Inc. New York, 1951.
7. A. J. Stamm and E. E. Harris, *Chemical Processing of Wood*, Chemical Publishing Co., Inc., New York, 1953.

8. L. E. Wise and E. C. Jahn, eds., *Wood Chemistry*, 2nd ed., Reinhold Publishing Corp., New York, 1952.

9. G. J. Ritter and L. C. Fleck, *Ind. Eng. Chem.* **14**, 1050 (1922).

10. B. L. Browning, *The Chemistry of Wood*, Interscience Publishers, a div. of John Wiley & Sons, Inc., New York, 1963, pp. 70–71.

11. F. E. Brauns, *The Chemistry of Lignin*, Academic Press, Inc., New York, 1952.

12. I. H. Isenberg, M. A. Buchanan, and L. E. Wise, *Paper Ind. Paper World* **28** (8), 816–821 (1946).

13. C. Wehmer, *Die Pflanzenstoffe*, Verlag Gustav Fischer, Jena, Vol. 1, 1929; Vol. 2, 1931; Suppl. Vol., 1935.

14. C. J. West, ed., *Nature of the Chemical Components of Wood*, Monograph No. 6, Technical Association, Pulp and Paper Industry, New York, 1948.

15. C. Skaar and W. T. Simpson, *Forest Prod. J.* **18** (7), 49–58 (1968).

16. A. J. Stamm, *Wood and Cellulose Science*, Ronald Press Co., New York, 1964.

17. F. F. P. Kollmann and W. A. Côté, Jr., *Solid Wood*, Vol. 1 of *Principles of Wood Science and Technology*, Springer-Verlag New York, Inc., New York, 1968.

18. E. C. Peck, *Air Drying of Lumber*, revised, U.S. Forest Prod. Lab Rept. No. 1657, Madison, Wis., 1961.

19. E. F. Rasmussen, *Dry Kiln Operators Manual*, U.S. Dept. Agr. Handbook No. 188, 1961. (Available from Superintendent of Documents, U.S. Govt. Printing Office, Washington, D.C.)

20. *National Design Specification for Stress-Grade Lumber and Its Fastenings*, revised, National Lumber Manufacturers' Association, Washington, D.C., 1951.

21. H. O. George, *Effect of Low Temperatures on the Strength of Wood*, N.Y. State College of Forestry, Syracuse Univ. Tech. Publ. 43, Syracuse, 1933.

22. F. F. P. Kollmann, *Mechanical Characteristics of Various Moist Woods Within Temperature Gradients from* −*200*° *to* +*200*°*C.*, (transl.), U.S. Natl. Advisory Comm. for Aeronautics Tech. Memo 984, Washington, D.C., 1941.

23. P. H. Sulzberger, *The Effect of Temperature on the Strength of Wood, Plywood, and Glued Joints*, Aeronautical Research Consultative Comm. (Australia) Rept. ACA-46, 1953.

24. B. Thunell, *Effect of Temperatures on the Bending Strength of Swedish Pine Wood*, Govt. Testing Inst. (Sweden), med. 80, 1940, Engl. summary, pp. 35–38.

25. J. D. MacLean, *Am. Wood-Preservers' Assoc. Proc.* **47**, 155–168 (1951).

26. *Ibid.*, **49**, 88–112 (1953).

27. J. D. MacLean, *Effect of Heating in Water on the Strength Properties of Wood*, Am. Wood-Preservers' Assoc., Washington, D.C., 1954.

28. H. D. Erickson and L. W. Rees, *J. Agr. Res.* **60**, 593–603 (1940).

29. F. L. Browne, *Theories of the Combustion of Wood and Its Control*, U.S. Forest Prod. Lab. Rept. No. 2136, Madison, Wis., 1958. (Reviewed and reaffirmed in 1963).

30. U.S. Forest Products Laboratory *Ignition and Charring Temperature of Wood*, revised, U.S. Forest Prod. Lab. Rept. No. 1464, Madison, Wis., 1958.

31. A. F. Matson, R. E. Dufour, and J. F. Breen, *Survey of Available Information on Ignition of Wood Exposed to Moderately Elevated Temperatures*, Bull. of Res. 51, Part II, Underwriters Laboratories, Inc., 1959.

32. A. J. Stamm, *Ind. Eng. Chem.* **48** (3), 413–417 (1956).

33. M. A. Millett, L. J. Western, and J. J. Booth, *Tappi* **50** (11), 74A–80A (1967).

34. F. L. Browne and W. K. Tang, *Fire Res. Abstrs. Rev.* **4** (1–2), 76–90 (1962).

35. H. W. Eickner and W. K. Tang, *Forest Prod. Intern. Union Forest Res. Organ., Proc., Sect.* **41**. (1963).

36. W. K. Tang and W. K. Neill, *J. Polymer Sci. C* **6**, 65–81 (1964).

37. F. L. Browne and J. S. Brenden, *Heat of Combustion of Volatile Pyrolysis Products of Fire-Retardant Treated Ponderosa Pine*, U.S. Forest Serv. Res. Paper FPL 19, Forest Products Laboratory, Madison, Wis., 1964.

38. W. K. Tang and H. W. Eickner, *Effect of Inorganic Salts on Pyrolysis of Wood, Alpha-cellulose, and Lignin Determined by Dynamic Thermogravimetry*, U.S. Forest Serv. Res. Paper FPL 71, Forest Products Laboratory, Madison, Wis., 1967.

39. W. K. Tang and H. W. Eickner, *Effect of Inorganic Salts on Pyrolysis of Wood, Cellulose, and Lignin Determined by Differential Thermal Analysis*, U.S. Forest Serv. Res. Paper FPL 82, Forest Products Laboratory, Madison, Wis., 1968.

40. J. J. Brenden, *Effect of Fire Retardants and Other Inorganic Salts on Pyrolysis Products of Ponderosa Pine at 250° and 350°C.*, U.S. Forest Serv. Res. Paper FPL 80, Forest Products Laboratory, Madison, Wis., 1967.

41. *Standard Method of Test for Determining Noncombustibility of Elementary Materials, ASTM Standard E 136-165*, American Society for Testing and Materials, Philadelphia, 1965.

42. *Standard Methods of Fire Tests of Building Construction and Materials, ASTM Standard E 119-167*, American Society for Testing and Materials, Philadelphia, 1967.

43. *Standard Methods of Fire Tests of Door Assemblies, ASTM Standard E 152-166*, American Society for Testing and Materials, Philadelphia, 1966.

44. *Standard Method of Test for Surface Burning Characteristics of Building Materials, ASTM Standard E 84-68*, American Society for Testing and Materials, Philadelphia, 1968.

45. E. L. Schaffer, *Review of Information Relating to the Charring Rate of Wood*, U.S. Forest Serv. Res. Note FPL-0145, Forest Products Laboratory, Madison, Wis., 1966.

46. E. L. Schaffer, *Charring Rate of Selected Woods—Transverse to Grain*, U.S. Forest Serv. Res. Paper FPL 69, Forest Prod. Lab. Madison, Wis., 1967.

47. H. W. Eickner, *ASTM J. Mater.* **1** (3), 625–644 (1966).

48. M. J. Brophy, *Wooden Tanks in Industry*, Dept. Mines Resources, Canada Dominion Forest Serv. Circ. 55, 1939.

49. *Redwood Tanks for Chemical Solutions*, California Redwood Association, San Francisco, 1946.

50. W. G. Campbell, *Wood* **9** (5), 99–101 (1944).

51. H. Poetter, *Chem. Tech. (Berlin)* **16**, 81–82, (1953).

52. R. H. Baechler and C. A. Richards, *Trans. Am. Soc. Mech. Engrs.* **73**, 1055–1059 (1951).

53. R. H. Baechler, *J. Forest Prod. Res. Soc.* **4** (5), 332–336 (1954).

54. J. D. Ross, *Forest Prod. J.* **6** (1), 34–37 (1956).

55. F. F. Wangaard, *Forest Prod. J.* **16** (2), 53–64 (1966).

56. R. H. Baechler, *Forest Prod. Res. Soc. Proc.* **1**, 120–123 (1947).

57. L. Perez et al., *Proc. Am. Wood-Preservers' Assoc.* **42**, 255–263 (1946).

58. A. J. Stamm, *Ind. Eng. Chem.* **39**, 1256–1261 (1947).

59. B. V. Volkening, *J. Forest Prod. Res. Soc.* **3** (5), 72–77 (1953).

60. U.S. Forest Products Laboratory, *Comparative Decay Resistance of Heartwood of Native Species*, U.S. Forest Serv. Res. Note FPL-0153, Madison, Wis., 1967.

61. *Wood Preservation: Treating Practices*, Fed Spec TT-W- 5711, U.S. General Service Administration, Washington, D.C.

62. R. H. Baechler, J. O. Blew, and C. G. Duncan, "*Causes and Prevention of Decay of Wood in Cooling Towers*," *Am. Soc. Mech. Engrs. Paper* **16**, part 5 (1961).

62a. R. H. Baechler, *J. Forest Prod. Res. Soc.* **3** (5), 170–176 (1953).

63. L. O. Anderson, *Low-Cost Wood Homes for Rural America—Construction Manual*, U.S. Dept. Agr. Handbook No. 364, 1969. (Available from Superintendent of Documents, U.S. Govt. Printing Office, Washington, D.C.)

64. E. C. O. Erickson, *Mechanical Properties of Laminated Modified Wood*, U.S. Forest Prod. Lab. Rept. No. 1639, Madison, Wis. 1952. (Reviewed and reaffirmed in 1965.)

65. R. C. Weatherwax and A. J. Stamm, *Trans. Soc. Elect. Engrs.* **64**, 833 (1945).

66. R. M. Seborg and A. E. Vallier, *J. Forest Prod. Res. Soc.* **4** (5), 305–312 (1954).

67. A. J. Stamm and R. M. Seborg, *Resin-treated, Laminated, Compressed Wood (Compreg)*, revised, U.S. Forest Prod. Lab. Rept. No. 1381, Madison, Wis., 1951.

68. A. J. Stamm, *Heat-Stabilized Compressed Wood (Staypak)*, U.S. Forest Prod. Lab. Rept. No. 1580, Madison, Wis., 1944. (Revised 1962).

69. Anonymous, *Nucleonics* **20** (3), 94 (1962).

70. F. C. Beall, J. A. Meyer, and C. Skaar, *Forest Prod. J.* **16** (9), 99–106 (1966).

71. C. Schuerch, *Forest Prod. J.* **14** (9), 377–381 (1964).

72. J. F. Saeman, E. G. Locke, and G. K. Dickerman, "Production of Wood Sugar in Germany and Its Conversion to Yeast and Alcohol," *U.S. Dept. Comm. Office Tech. Serv. PB Rept. 7736* (1945).

73. D. Hair and A. H. Ulrech, "The Demand and Price Situation for Forest Products," *U.S. Dept. Agr. Misc. Publ.* **1066** (1967). (Available from Superintendent of Documents, U.S. Government Printing Office, Washington, D.C.)

74. A. B. Anderson, *Econ. Botany* **21**, 15–30 (1967).

75. W. E. Hillis, *Wood Extractives and their Significance to the Pulp and Paper Industries*, Academic Press, Inc., New York, 1962.

76. E. G. Locke and K. G. Johnson, *Ind. Eng. Chem.* **46**, 478–483 (1954).

77. E. E. Harris, "Wood Saccharification," in W. W. Pigman and M. L. Wolfram, eds., *Advances in Carbohydrate Chemistry*, Vol. 4, Academic Press, Inc., New York, 1949, pp. 154–188.

78. R. A. Lloyd and J. F. Harris, *Wood Hydrolysis for Sugar Production*, U.S. Forest Prod. Lab. Rept. No. 2029, Madison, Wis., 1955.

79. N. Gilbert, I. A. Hobbs, and J. D. Levine, *Ind. Eng. Chem.* **44**, 1712–1720(1952).

80. N. Gilbert, I. A. Hobbs, and W. D. Sandberg, *J. Forest Prod. Res. Soc.* **2**, 43–49 (1952).

81. A. W. Goos and A. A. Reiter, *Ind. Eng. Chem.* **38**, 132–135 (1946).

82. L. A. Underkofler and R. J. Hickey, *Industrial Fermentation*, Chemical Publishing Co., Inc., New York, 1954.

83. H. D. Turner, *Forest Prod. J.* **14** (7), 282–284 (1964).

84. U.S. Dept. Agr., *Charcoal and Charcoal Briquette Production in the United States, 1961*, Forest Serv., Div. of Forest Econ. and Marketing Res., 1963.

85. L. F. Hawley, *Wood Distillation*, Chemical Catalog Co., New York, 1923.

86. H. K. Benson, *Chemical Utilization of Wood*, U.S. Dept. of Commerce, Washington, D.C., 1932.

87. U.S. Forest Products Laboratory, *Charcoal Production, Marketing, and Use*, Forest Prod. Lab. Rept. No. 2213, Madison, Wis., 1961.

HAROLD TARKOW   R. A. HANN
A. J. BAKER   R. C. KOEPPEN
H. W. EICKNER   M. A. MILLET
W. E. ESLYN   AND
G. J. HAJNY   W. E. MOORE
U.S. Department of Agriculture

WOOD PULP. See Pulp.

WOOL

Among the fibers used through the ages to clothe man, wool occupies a unique position; it is not only the oldest, but also it is one which has been continuously and universally used. Considerable research is in progress on the structure, processing, and chemical modification of wool to help keep it abreast of modern requirements in the textile industry.

Wool belongs to a family of proteins, the keratins, which also includes hair and other types of animal protective tissues such as horn, nails, feathers, and the outer skin layers. In the textile trade, the term "wool" is applied to the fibers from the fleece of the sheep (1).

Table 1 shows wool production and sheep numbers in the world's major producing countries.

Australia, which is the world's largest wool producer, has a high production per animal (10.6 lb). This is due to selective breeding practices coupled with the application of research results. About 75% of Australia's sheep are Merino which is a fine wool breed (average fiber diameter usually less than 26 microns). Merinos are also present in large numbers in South Africa and Argentina and in increasing numbers in the U.S.S.R. Medium wool breeds include types of English origin (Southdown, Hampshire, Dorset, Cheviot) as well as crossbreds from interbreeding Merino and long wool types (Columbia, Targhee, Corriedale, Polwarth). Long wool breeds produce coarse long wool and are bred chiefly for mutton (Lincoln, Cotswald, Leicester). The natural or unimproved breeds are prevalent throughout Asia, producing carpet type wools. Also in this class are certain British breeds (Scottish Blackface, Welsh Mountain) and Navajo in the United States.

Apart from wool, the fleece contains other constituents which can vary in content according to breed, nutrition, environment, and position on the sheep. The ether-

Table 1. World Sheep and Wool Production 1967–1968[a]

Country	Wool production, million lb (greasy)	Sheep, million
Australia	1774	166.9
New Zealand	728	60.5
United Kingdom	128	28.1
India	72	39.4
Pakistan	45	10.3
Lesotho	10	1.5
Canada	5	0.9
British Commonwealth, other	10	2.3
Argentina	428	49.0
South Africa	307	36.3
United States	227	22.1
Uruguay	186	21.0
Turkey	101	35.9
Spain	82	18.6
Brazil	62	23.1
France	48	9.5
Chile	51	7.0
Iran	43	22.4
Morocco	40	15.0
Iraq	28	8.5
Yugoslavia	31	9.7
Italy	31	8.6
Portugal	31	
Peru	28	16.0
Irish Republic	24	4.1
Greece	18	7.9
West Germany	10	
Asia, other	50	31.4
Africa, other	41	18.7
America, other	35	17.5
Western Europe, other	23	10.9
Soviet Union	871	138.0
Eastern Europe	192	35.2
China	175	59.0
world total	5935[b]	935.3

[a] Source: Commonwealth Secretariat, London.
[b] Clean equivalent is 3436 million lb.

soluble fraction of the fleece is called *wool-wax*. The water-soluble fraction is termed *suint*. Also present are dirt and vegetable matter derived from pastures. The term "yolk" is applied to all the natural components of the fleece except the fiber. It therefore consists of two main fractions, wool-wax and suint. Table 2 gives some

Table 2. Fleece Composition of Australian Wool

Wool	Oven-dry wool, %			Wax, %			Suint, %			Dirt, %			Moisture, %		
	Max	Min	Av	Max	Min	Av	Max	Min	Av	Max	Min	Av	Max	Min	Av
sheep															
merino	66.8	29.4	48.9	25.4	10.0	16.1	13.0	2.0	6.1	43.8	6.3	19.6	12.6	8.1	9.6
crossbred	72.2	49.3	61.0	19.3	5.3	10.6	13.6	4.4	8.2	23.7	4.3	8.4	14.2	9.5	12.0
skin[a]	70.0	50.6	63.0	20.0	9.9	15.8	1.4	0.1	0.6	21.5	6.4	11.2	9.6	7.2	8.8
lambs															
merino and crossbred	68.0	54.0	60.5	23.5	6.4	16.0	7.2	3.4	5.4	10.0	3.8	6.4	14.2	9.0	11.2

[a] Skin wool is the term applied to fellmongered wool usually removed from skins by the Sweating process involving bacterial action or by action of lime-sulfide depilatory.

Table 3. U.S.A. Official Standards for Grades of Wool and Wool Tops (in microns)

Grades	80s	70s	64s	62s	60s	58s	56s	54s	50s	48s	46s	44s	40s	36s
average diameter range														
min	18.1	19.6	21.1	22.6	24.1	25.6	27.1	28.6	30.1	31.8	33.5	35.2	37.1	39.0
max	19.5	21.0	22.5	24.0	25.5	27.0	28.5	30.0	31.7	33.4	35.1	37.0	38.9	41.2
fineness distribution (percentage)														
10–25, min	91	83												
10–30, min			92	86	80	72	62	54	44					
10–40, min										75	68	62	54	44
above 25.1, max	9	17												
above 30.1, max	1	3	8	14	20	28	38	46	56					
above 40.1, max			1	1	2	1	1	2	2	25	32	38	46	56
above 50.1, max										1	1	2	3	
above 60.1, max														4
min number of fibers to be measured	400			600	800		1000					1600		

figures for percentage of fleece constituents in Australian wool (2,3). These figures refer to what is termed free wools which contain little or no vegetable material. In certain areas, the latter can be as high as 20% of the fleece weight but is usually under 2%.

The yield of a greasy wool sample is its percentage content of pure fiber at a standard moisture content (regain). In the past, this has been estimated visually by wool valuers but there is now a marked movement to yield testing on samples usually taken as cores from the bales of raw wool. The samples are washed in detergent followed by weighing of the clean dry product. Development is taking place of rapid yield testing equipment which will cope with large numbers of samples (4).

Wool can be processed on two major systems of yarn manufacture; woolen and worsted. The *woolen* system is the simpler involving only two operations after scouring, namely, carding and spinning. *Worsted* (pronounced "woosted," "oo" as in "foot") processing is more complex requiring at least six operations between scouring and spinning. In a worsted yarn, the fibers lie parallel to the direction of the yarn due to combing prior to spinning. There are two major types of comb; French (rectilinear) and Noble (Bradford), which is used mainly in Britain. It has a higher production rate than the French, but is unsuitable for shorter wools and not as effective in removing vegetable matter. Woolcombing gives rise to two products—*top* and *noil*. The top contains the longer fibers in the form of a thick untwisted strand (known as sliver) with the fibers more or less parallel to its length. It is usually wound into large balls and is bought and sold in this form as the raw material of the worsted spinning sector of the trade. The noil is a less valuable by-product consisting mainly of the shorter fibers plus vegetable matter. It is used as raw material in the woolen sector or for making felts. The ratio of top to noil is referred to as the *tear*. There is no combing operation on the woolen system and the fibers are randomly disposed in the yarn. Worsted yarns are widely used in woven fabric such as suitings, knitwear, and hand knitting yarns. Woolen yarns are used to produce woven fabric such as tweeds, dress materials, and blankets and for certain types of knitwear, such as lambswool twinsets. Usually the shorter types of wool (locks, crutchings, bellies) go into the woolen system while the longer (or fleece wools) are processed on the worsted system. For fine wools, the division is roughly as follows: combing or staple (over $2\frac{1}{2}$ in. long), for the Bradford worsted system; French combing ($1\frac{1}{2}$ to $2\frac{1}{2}$ in.), for the French worsted system; and clothing (under $1\frac{1}{2}$ in.), wools too short to be combed and destined for the woolen method of processing.

Wool Quality. Commercial evaluation of raw (greasy) wool depends on the quantity and quality of pure fiber present. The main criteria of quality are fiber fineness and length but other characteristics such as strength, crimp, and color also need to be considered. Wool fineness has been traditionally expressed as *grade* or *count* which, in the worsted trade, represents the number of hanks, each 560 yards long, of the finest possible yarn that might be spun satisfactorily from a pound of the wool concerned. With finer fibers it is possible to spin out a given weight to finer counts of yarn due to the greater number of fibers per unit yarn cross section. Thus, the finer the wool the higher will be the count. Recently, there has been a greater trend in industry to use a simpler system based on fiber diameter. The relationships between the two are given in Table 3. Fiber fineness is determined by microscopic measurement on representative samples (5) or, more rapidly, by an air flow technique (6).

Fiber Growth and Morphology

Each fiber is produced in a special tubelike structure in the skin known as the wool follicle. There are two kinds of wool follicle; one is the primary which develops first in the outer skin of the unborn lamb and the other, which is late developing, is the secondary. Fig. 1 depicts the structure of the primary follicle. Wool wax is derived from the sebaceous glands while suint comes from the sudoriferous or sweat glands. The secondary follicles lack sudoriferous glands. There are about 200 million follicles in the skin of a Merino sheep. Studies on fiber formation in the follicle have helped in understanding the structure of the fully formed fiber in the fleece. The individual cells retain their separate identities from the time they first appear in the follicle until they end up in the keratinized fiber.

Microscopic examination shows the fiber to consist mainly or an inner cortex surrounded by an outer scale layer or cuticle. The protruding tips of the scales point towards the fiber tip. The cortex is made up of spindle shaped cortical cells about 100μ long. It is bilateral in structure with an orthocortex and paracortex which differ in reactivity (7). The disposition of the two segments varies with the breed of sheep. In a crimped fiber, (one with a pronounced waviness) such as Merino, the orthocortex lies on the outside of the crimp wave and the paracortex on the inner side. Fig. 2 shows this relationship. Dyes penetrate the orthocortex more readily than the paracortex. The orthocortex also shows greater reactivity toward other agents, which is partly due to its less densely packed cell structure. There are also differences in chemical composition between the two segments (7). Fig. 3 is a schematic diagram of a fiber, showing the relationships of the various histological components (8).

Cortical cells can be divided lengthwise into fibrils (9,10) which are composed of still smaller filaments called microfibrils embedded in an amorphous matrix (11). There is some doubt about the existence of a further substructural unit that has been

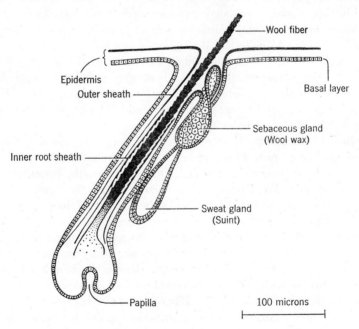

Fig. 1. The primary wool follicle showing production of fiber, wax, and suint.

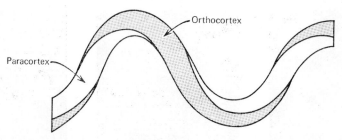

Fig. 2. The relationship between the crimp and the bilateral structure of the cortex in a Merino wool fiber.

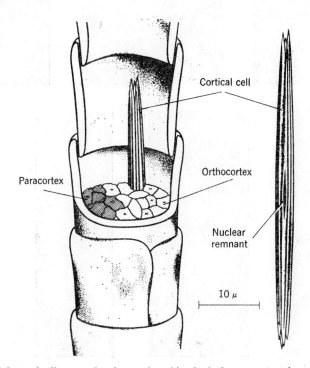

Fig. 3. Schematic diagram showing various histological components in a fiber (8).

called the protofibril. X-ray studies suggest that these might be present but electron microscopy has not produced positive corroboration. The x-ray evidence points to a fine structure of 2 or 3 α helical protein chains coiled around each other in the more organized portions of the microfibrils (12) (see under Chemical Structure below).

Coarse fibers, such as hair, sometimes contain a central core or medulla. This is not present in the better types of wool in which the cortex constitutes about 90% of the fiber mass. Kemp is a coarse fiber with a high content of medulla cells and lacking crimp. It frequently occurs in certain breeds of sheep producing coarse wools such as for the carpet industry.

The cuticle cells or scales can vary greatly in size and arrangement between breeds and animals and even along the length of one fiber. Fig. 4 shows some scale structures of various wool fibers as seen under the scanning electron microscope. Scales can be removed from individual fibers by mechanical means, by certain en-

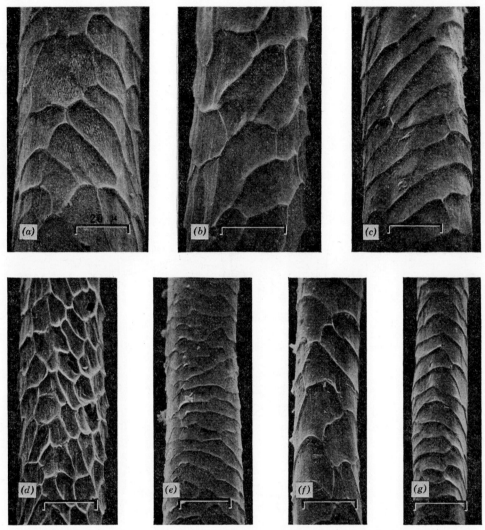

Fig. 4. Scanning electron micrographs of surface scales on various wool types. (Courtesy of Dr. C. A. Anderson.) (a) 36s Lincoln; (b) 46s Border Leicester; (c) 48s Corriedale; (d) 48s Cheviot; (e) 56s Dorst Horn; (f) 64s Super Comeback Polwarth; (g) 70s Merino. The bench mark on the micrographs measures 20 microns.

zymes, or by ultrasonics. They do not constitute a continuous layer along the fiber length but are separate from each other and overlap like the tiles on a roof. Each scale is considered to consist of a soft reactive outer component called the exocuticle and an inner more resistant layer called the endocuticle. The former is digested by enzymes (13,14). There is also a thin external semipermeable membrane over the exocuticle of the intact fiber called the epicuticle (15). It is less than 0.1% of the fiber and resists many reagents which normally react with other parts of the fiber. The epicuticle is easily removed or broken by mechanical means as when wool is commercially processed. Although it inhibits the diffusion of dyes into an intact fiber, this does not cause any difficulty in practice as the epicuticle has already been modified by the normal mechanical processing prior to dyeing.

Table 4. The Principal Physical Properties of Wool Fibers at 25°C (77°F)[a]

Regain, %	0	2	5	7	10	15	20	25	30	33
r.h., % absorption	0	2	14.5	25.5	42	68	85	94	98.5	100
desorption	0	1	8	17	31.5	57.5	79	91.5	98	100
specific gravity, d_{25}^{25}	1.304	1.3095	1.3135	1.3150	1.3150	1.3125	1.304	1.2915	1.2765	1.268
volume swelling, %	0	1.57	4.24	6.10	9.07	14.25	20.0	26.2	32.8	36.8
length swelling, %	0	0.23	0.55	0.75	0.93	1.08	1.15	1.17	1.18	1.19
radial swelling, %	0	0.66	1.82	2.62	4.00	6.32	8.88	11.69	14.57	16.26
heat of complete wetting (cal/g wool)	24.1	20.3	15.4	12.6	9.1	4.9	2.4	0.97	0.27	0
heat of complete absorption of liquid water (cal/g water)	204	180	149	129	103	66	34	24	10	8
relative Young's modulus	1.00	0.99	0.96	0.93	0.87	0.76	0.66	0.56	0.44	0.38
rigidity modulus (torsion), 10^{10} dynes/cm^2	1.76	1.71	1.60	1.48	1.26	0.90	0.50	0.28	0.16	0.11
electrical resistivity, ohm-cm $\times$ 10^6				3×10^6	4×10^4	800	40	6		
dielectric constant (at 10^4 Hz)	4.6	4.7	5.1	5.4	6.2	8.3	12.8			

[a] Reproduced from *Wool Research*, Vol. 2, by courtesy Wool Industries Research Association, Leeds, England.

Physical Properties

Table 4 summarizes the principal physical properties of wool.

Specific Gravity and Refractive Index. The exact value of specific gravity obtained for medulla-free wool depends on the displacement liquid used, and on the moisture content of the wool. For dry wool in benzene at 25°C the value is 1.304, related to water at 4°C. Wool is slightly birefringent, the refractive index perpendicular to the fiber axis (ω) being 1.542–1.546, and that parallel to the fiber (ϵ) 1.553–1.555 (17).

Moisture Relations. Wool is the most hygroscopic of the common textile fibers. Table 5 gives the relative moisture contents of various fibers. The regain (moisture

Table 5. Moisture Regain Values for Wool and Other Fibers[a]

Fiber	Regain, at 70°F, %	
	at 65% rh	at 95% rh
Acrilan acrylic	1.5	5
Orlon acrylic	1.5	4
nylon	2.8–5	3.5–8.5
polypropylene	0.01–0.1	0.01–0.1
Dacron polyester	0.4–0.8	
viscose rayon	13	27
acetate	6.5	14
cotton	7	
wool	16	22.5 (90% rh, 77°F)

[a] Based on data from *Textile World* (18).

content expressed on the dry weight of the fiber) is an important property in the processing of wool. It affects such operations as carding, combing, spinning, and weaving and, in fact, the relative humidity of the atmosphere needs to be controlled in many mill areas to ensure optimum processing. Although regain is mainly dependent on relative humidity, it may be affected by other factors operating during the previous history of the wool such as temperature of drying, possible damage to the fiber, and the presence of acid or alkali in the fiber.

The relationship between regain and relative humidity is shown in Fig. 5 (19). There is a hysteresis effect on drying, the desorption curve showing higher regains at each value of relative humidity. The lower adsorption curves represents the behavior of dry wool when exposed to atmospheres of higher relative humidity; the desorption curve shows the effect of drying wool of a given moisture content. In the normal range of relative humidities there is a difference of about 2% between the adsorption and desorption isotherms. To ensure the highest possible regain for spinning, wool tops should be stored. Apart from major effects in mechanical processing, the regain has important bearings on commercial transactions where wool is sold by weight. It has been necessary, therefore, to employ standard regains for such purposes, and these are given in Table 6. Unfortunately, there is no generally accepted value for use on a world basis due to historical obstacles where certain accepted procedures have been adopted over the years in different countries.

As seen from Table 4, most physical properties of the fiber are affected in one way or another by the regain. Due to the oriented structure of the fiber, there is marked radial swelling as the moisture content increases. At high regains, fibers are

Table 6. American and British Standard Regains

Form	Moisture regain, %	
	American[a]	British[b]
scoured wool	13.63	16
tops		
oil, combed	15	19
dry, combed	15	18.25
noils		
oil, combed		14
dry, combed		16
scoured and carbonized		16
yarn		
woolen	13	18.25
worsted, oil spun	13	18.25
worsted, dry spun	15	18.25
cloth		
woolen and worsted		16

[a] Listed by von Bergen (20).   [b] British Standard Specification 1051.

easier to bend and twist, and are much better conductors of electricity and less liable to generate static electricity.

Mechanical Properties. The mechanical properties of wool have been studied more than any other physical property of the fiber. Wool is characterized by a high extensibility and relatively low breaking strength. It has very unusual elastic properties particularly when the fiber is wet. Fig. 6 shows the stress/strain curves of wool at different relative humidities. The typical curve shows a rapid uncrimping of the fiber

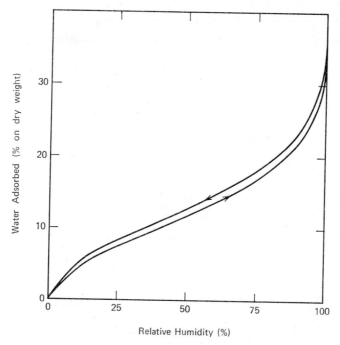

Fig. 5. The moisture adsorption isotherm of 50s Merino wool at 22°C.

which is followed by a steep section covering 2–3% extension which is known as the Hookean slope (where the fiber obeys Hooke's Law); following this, there is rapid extension with little increase in force which is known as the yield region, and the curve then steepens after about 30% extension continuing unchanged until the fiber breaks. If, up to 30% extension, the fiber is released in water, it will return to its original length. There is a hysteresis in the unloading curve as shown in Fig. 7 which represents a

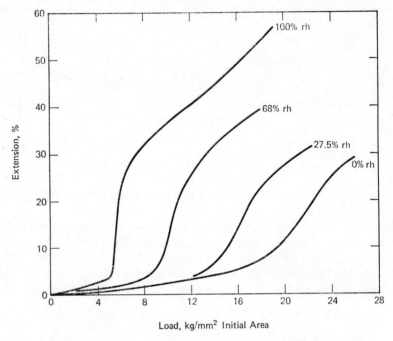

Fig. 6. Load/extension curves for a wool fiber at different relative humidities.

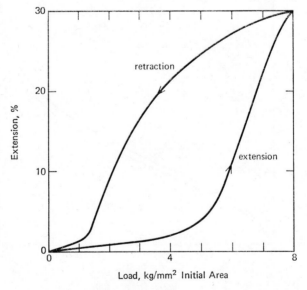

Fig. 7. Load/extension curve for a wool fiber in water.

typical stress-strain curve for a wool fiber in water. If the fiber is rested overnight in water at room temperature, the cycle can be repeated. This remarkable property has been extensively used in mechano-chemical techniques to study the effects on single fibers of treatments such as crosslinking, pH changes, polymer deposition, and disulfide breakdown (21).

Surface Friction. The scale structure gives wool a unique directional friction, the coefficient being greater when the fiber is moved in the root-tip direction. Due to this, a fiber tends to move in the root direction when rubbed along its axis. This is the basis of *felting* (22) which occurs when a mass of wool is subjected to mechanical action, usually in soap or detergent solutions. Felting can be an advantage in producing certain types of fabric such as blankets and billiard cloth but can be a disadvantage by causing shrinkage when wool garments are washed (see below). Various methods have been devised for measuring the surface friction of wool fibers (23). Considerable research has gone into the relations between surface friction and felting, particularly in explaining the action of various agents used to shrinkproof wool.

Chemical Structure

Wool has the most complex molecular structure of the major textile fibers. Elucidation of this problem has involved many workers in different laboratories throughout the world. Apart from the fact that keratin is a complex protein, there is the additional difficulty of chemical differences between the various morphological components. Results of analysis of the whole fiber therefore give only a general indication of its chemical structure. Nevertheless, such figures can be useful in considering possible chemical reactions of the fiber.

Unlike other proteins, keratin is characterized by a sulfur content which can vary with diet (24) but is usually about 3.0% (25). The sulfur is mainly contained in the amino acid, cystine, and occurs as disulfide linkages between neighboring polypeptide chains in the protein structure. There is also evidence that cystine may bridge amino acids in the same chain in certain parts of the fiber.

Using various methods of extraction, it has been found that there are two main groups of protein of low and high sulfur content respectively. It is generally agreed that the low-sulfur group is from the microfibrils whereas the high-sulfur group is from the matrix (26). Supporting this is the fact that sheep given cystine supplements will produce wool with up to 50% more high-sulfur protein together with a large increase in the amount of matrix (27). The low-sulfur proteins are fibrous in structure whereas the high sulfur proteins are amorphous. The latter give molecular weights on the order of 25,000 (28,29). There is less agreement on molecular weight of the low-sulfur proteins due to differing degrees of aggregation in various solvents. Figures upwards of 45,000 have been reported varying with the solvent and method of measurement (26).

Typical figures for the amino acid composition of Merino wool are given in Table 7.

These figures can differ with the breed of sheep (26). There are also differences in composition between cortex and cuticle (30). Reported differences in amino acid composition between ortho and para cortical cells (7) need to be accepted with reserve until more reliable experimental techniques are found. The composition can also be affected by changes in diet; for example, copper deficiency which produces "steely" wool by eliminating fiber crimp (31) can also decrease cystine content (32). Intro-

Table 7. The Amino Acid Composition of Merino Wool

Type of side chain	Acid	Mole, %
hydrocarbon (or hydrogen)	glycine	8.8
	alanine	5.5
	valine	5.6
	leucine	7.6
	iso-leucine	3.1
	phenylalanine	2.8
hydroxy	serine	11.5
	threonine	7.0
	tyrosine	4.0
acidic	aspartic acid	6.2
	glutamic acid	12.2
basic	lysine	2.6
	arginine	7.1
	histidine	0.8
sulfur-containing	cystine	6.3
	methionine	0.5
heterocyclic	proline	7.5
	tryptophan	0.9

duction of cystine, methionine, or casein into the abomasum or fourth stomach of the sheep increases wool production and sulfur content of the wool (33). Further research is in progress on this observation which could have important practical applications (34).

Despite the complex structure of wool, there is an underlying molecular pattern which can be used to interpret its high degree of reactivity. This generalized structure which was first postulated by Astbury and Speakman (35) in 1934 has provided a basis for much of the work undertaken since then. The molecular grid of keratin consists essentially of polypeptide chains bound together by; salt linkages between acid and basic side chains; covalent —S—S— bonds formed by cystine; and hydrogen bonds between —CO— and —NH— groups of neighboring peptides:

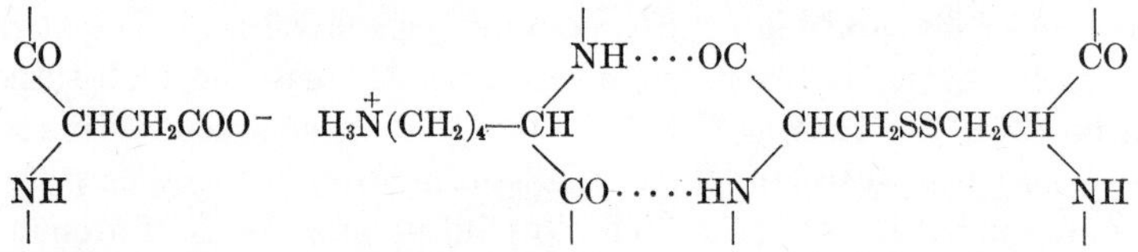

$$
\begin{array}{l}
\mathrm{CO} \\
\quad\diagdown \\
\qquad \mathrm{CHCH_2COO^-} \\
\quad\diagup \\
\mathrm{NH}
\end{array}
\qquad
\mathrm{H_3\overset{+}{N}(CH_2)_4{-}CH}
\begin{array}{l}
\mathrm{NH}\cdots\cdots\mathrm{OC} \\
\qquad\qquad\quad\diagdown \\
\qquad\qquad\qquad \mathrm{CHCH_2SSCH_2CH} \\
\qquad\qquad\quad\diagup \\
\mathrm{CO}\cdots\cdots\mathrm{HN}
\end{array}
\begin{array}{l}
\mathrm{CO} \\
\diagup \\
\mathrm{NH}
\end{array}
$$

salt linkage between　　　　　　　　　　　　sulfur linkage in
aspartic acid and　　　　　　　　　　　　　cystine
lysine residues

The chemical reactions of keratin (other than degradation by peptide bond rupture) involve the reactive side chains, particularly the crosslinkages. Wool is amphoteric, and it combines reversibly with acids, the salt linkages being ruptured:

$$-\mathrm{COO^-H_3\overset{+}{N}{-}} + \mathrm{HCl} \rightarrow -\mathrm{COOH} + \mathrm{H_3\overset{+}{N}{-}Cl^-}$$

Reaction with bases takes place in a corresponding way but is complicated by the fact that the disulfide bond is very readily degraded by alkalies. The complete acid–base titration curve is shown in Fig. 8. Attempts by various workers to determine the isoelectric point have given differing results. In view of the heterogeneous composition

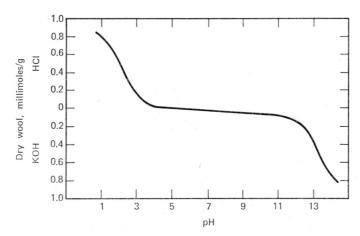

Fig. 8. The combination of wool with hydrochloric acid and potassium hydroxide as a function of pH at 0°C.

of wool, it is more likely that there is an isoelectric region rather than an isoelectric point (37).

The principal reaction under mildly alkaline conditions is the conversion of the combined cystine into combined lanthionine.

$$\text{>CHCH}_2\text{SSCH}_2\text{(CH)<} \rightarrow \text{>CHCH}_2\text{SCH}_2\text{(CH)<}$$

cystine linkage   lanthionine linkage

More severe conditions lead to complete elimination of sulfur from part of the combined cystine, probably by the formation of combined α-aminoacrylic acid, —CO(C= CH₂)—NH—. Possible mechanisms of these reactions are reviewed elsewhere (26).

The reactivity of cystine in wool has been studied extensively and is of great importance in processes such as bleaching, setting, scouring, and shrinkproofing; it is well reviewed in several publications (26,38).

Reduction of disulfide bonds is the basis of certain setting treatments for wool. Thioglycollate and other thiols reduce the —S—S— to —SH and there is evidence for the formation of mixed disulfides between the wool and reducing agent (39). Another reaction which has been closely studied is that between wool and sulfite:

$$\text{RSSR} + \text{SO}_3{}^{2-} \rightleftharpoons \text{RS}^- + \text{RSSO}_3{}^-$$

This is of importance in such processes as reductive bleaching and certain fabric setting procedures.

Oxidation, as in peroxide bleaching or certain shrinkproofing treatments (see below), involves reaction of the disulfide bonds. The final product is cysteic acid but various intermediate sulfoxides and sulfones can be produced (40,26). However, the whitening of wool which occurs in bleaching is not due to reaction of the disulfides (see below).

Chemical Processing

Scouring and Carbonizing. The first stage of wool processing is removal of fleece impurities, usually by scouring in warm, aqueous detergent solutions but sometimes by solvent extraction.

Early investigations on scouring in alkaline soap focused on control of pH in the liquor and established optimum conditions for minimum wool damage and consumption of chemicals (41). Later, the undesirable effects of felting and entanglement due to excessive mechanical agitation in the scouring liquor were recognized (42). Soap has now been replaced in many countries by synthetic detergents, particularly nonionics (see Surfactants), in neutral or slightly alkaline solutions (43). Grease is probably removed from the fibers by a distinct mechanism in which cryoscopic forces between wax, detergent, and water are the important factors, the principal action of detergent being to cause swelling of the grease and suint mixture on the wool (44,45). See also Vol. 6, p. 882. When the scour liquor is forced past the wool, as in squeezing or jetting, the swollen complex is washed off.

Conventionally, raw wool scouring is carried out by moving the freely-immersed wool through a sequence of four or five large bowls using rakes. Despite improvements in efficiency of the liquors, a major disadvantage remains since the wool still becomes entangled. One way of preventing entanglement is to force jets of aqueous

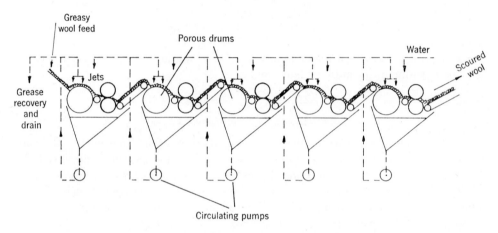

Fig. 9. C.S.I.R.O. jet scouring machine for raw wool.

detergent past the wool which is supported on a porous conveyor; jet-scoured wool is less entangled and therefore gives much less fiber breakage in subsequent processing than wool scoured in the usual way (46). Fig. 9 shows the CSIRO Jet Scouring machine developed for this purpose.

Other methods of producing scoured wool free from felting and entanglement rely on organic solvents to clean the wool, because it does not felt in most organic solvents. Another advantage of solvent degreasing is that yields of wool grease of at least 80% are obtained and there is no effluent problem. Yields above 40% are rare with aqueous systems, the usual centrifugal techniques being unable to recover additional wax. This is due to the emulsified wax in the scour liquor being separated by the centrifuge into oxidized (unrecovered) and unoxidized (recovered) wool wax (47). Recovery depends on the extent of oxidation of the wax, and is greater from base portions of fleece than from the tips (48) where atmospheric oxidation has occurred.

Carbonizing is a process often used to remove excessive contents of vegetable matter from wool. The usual procedure is to pass the scoured wool through aqueous sulfuric acid followed by baking. This converts the vegetable matter to a brittle black (carbonized) form which is then crushed to facilitate removal. The wool is finally neutralized in alkali.

Surface-active agents reduce fiber damage in the carbonizing process, as shown by fiber strength, yield (49,50), and amounts of protein (51) found in neutralizing and washing liquors. In Australia, most carbonizers add nonionic surfactants to the acid bath with higher yields of wool and less fiber breakage. This is particularly applicable under conditions of high acid concentration and rapid through-put which are more commonly used in Australian mills than in those of other countries. The detergents were first thought to act chemically in protecting the wool, but later work indicated that a major factor could be the more efficient removal of liquid in the squeezing operation when detergent is present (52).

Dyeing. Traditional batch procedures in which wool is dyed by boiling for at least 1 hr in solutions of dye are still the most commonly used. However, considerable progress has recently been made with methods for more rapid dyeing, particularly continuous dyeing. Dyes have also been developed which will react to form covalent linkages with wool and thus have outstanding fastness properties.

The chemistry of wool dyeing is most complex (53). Among the factors involved are pH and the temperature of the dye solution, nature and concentration of electrolytes in the solution, association of dye molecules, effect of dyeing auxiliaries in the solution, hydration of dye and auxiliaries, effects of contaminants and pretreatments on the fiber surface, affinity of dye for wool, and mode and depth of penetration of the dye into the fibers (54). An undesirable effect of dyeing is that wool can be damaged by boiling in aqueous acid solutions, the extent depending greatly on the pH of the solution (55).

There are three main groups of dye used in the wool industry. The *level-dyeing acid dyes*, which are easily applied at about pH 3, have excellent leveling properties but fastness to washing is low. This depends on such factors as electrovalent links to charged groups in the wool, hydrogen bonding, van der Waals attractive forces and aggregation of the dye molecules. See Dyes—application and evaluation.

The usual procedure to increase wash-fastness is to use dyes which are sorbed by a similar mechanism but can then be rendered less soluble by treatment with metal salts, usually those of chromium. These *chrome dyes* are now supplemented by *metal-complex dyes* of very high affinity for wool. They contain one metal atom for each one or two dye molecules, the metal being either chromium or cobalt. See Azo dyes. When first introduced, many metal-complex dyes could only be applied at pH 6–7 but nonionic dyeing assistants are now available which allow application at pH 5.0–5.5, thus decreasing fiber damage.

Auxiliaries, most of which are nonionic or cationic detergents, are often used in dye solutions to produce level dyeings. Surfactants are often used in dye solutions to produce level dyeings; in this context they are called auxiliaries and are mostly nonionic or cationic. The nonionic assist by dispersing dye aggregates, thus allowing better migration, and can form complexes with the dye in solution (56). These complexes aid in level dyeing by maintaining a low equilibrium concentration of free dye in solution during sorption by the wool; the net effect is to lower the rate of decrease of dye concentration compared with that in the absence of auxiliary (56).

Dyeing is hindered by a surface barrier (57,58) probably relating to the nature of the epicuticle (14,59). Removal of cuticle material by abrasion (60) or by chemical attack with alcoholic potassium hydroxide or chlorine (58,61,62), changes the fiber surface from weakly polar and hydrophobic to more hydrophilic, thus greatly increasing the rate of dyeing. This concept of the role of surface properties in dyeing has helped to overcome difficulties sometimes found with dyes that are more hydrophilic than normal. Cationic auxiliaries are used to form complexes with the dyes so that the complex is more hydrophobic than the dye itself, allowing easier penetration of the hydrophobic surface of normal wool fibers (62).

Sometimes, mixtures of nonionic and cationic auxiliaries give more level dyeings than either type alone (63). Presumably, both of the described mechanisms could then be operating simultaneously. Although level dyeings in which dyes have thoroughly penetrated the fibers are usually the aim of industrial dyers, some attention has been given to methods of producing several shades from one dyebath. One way is to modify some of the fibers before the fabric is woven to increase their rate of dyeing (eg, chlorination) or to decrease it (eg, acetylation). Some methods use mixtures of dyes with compounds of similar affinity for wool but having no chromophoric groups (64), which results in mixture effects similar to that obtained by mixing undyed and dyed fibers before spinning; others involve mixtures of dyes under conditions where some will dye level, forming an even background color, and others will dye unlevel under the given conditions, thus giving a nonuniform effect. Recently there has been greater use of the reactive dyes on wool. These form covalent links with the fiber producing bright shades of good wash- and light-fastness. They have enabled the production of bright shades on machine-washable wool garments which were not previously possible. See Dyes, reactive.

Vat dyeing, where a soluble, reduced form of a dye is applied to fibers at 50–60°C, the color being subsequently developed by oxidation, is used on cotton but rarely on wool. One reason is that the alkaline solutions often used with cotton may cause severe damage to wool. However, with careful control, this method can produce dyed wool less damaged than that by conventional methods (65). Small amounts of solvent have been used to increase rates of dyeing in aqueous dye solutions (66). Industrial methods based on such procedures would reduce wool damage and produce savings in time and steam since dyeing can be carried out in about 30 min at 60–80°C (67). Such methods would be very suitable for dyeing chemically modified wool, for example, wool shrinkproofed by oxidation methods which has increased susceptibility to damage in boiling solutions. So far, high cost of solvents has limited application, although there is some use of solvents in printing pastes (68). The most promising solvent appears to be benzyl alcohol.

Another approach to dyeing below the boil has been by using certain nonionic surfactants in the dyebath with careful control of pH (69). This method, being cheaper, has been more widely used than the solvent technique. Recently there has been renewed interest in continuous dyeing processes for wool. Tops can be dyed, with certain types of dye, by the Cibaphasol process (70), being run continuously through the liquor at pH 9 at about 60°C, steamed continuously for 10–20 min, and washed off in a normal backwashing unit. The dye liquor is called an aqueous coacervation system. The term "coacervation" is used to denote the separation of a colloidal system into two liquid isotropic phases. The concentrated phase is known as the "coacervate" and the more dilute phase as the "equilibrium liquid." As opposed to ordinary two-phase mixtures, the two phases are solutions in the same solvent, eg, water. The volume ratio of the two phases depends on the concentration

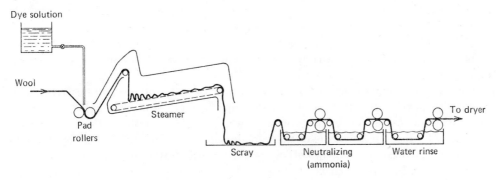

Fig. 10. C.S.I.R.O. Continuous dyeing machine for wool tops.

of surfactant, the agencies bringing about coacervation (eg, electrolytes), the temperature, and other physical factors.

A truly cold continuous technique is based on the use of solutions of dye in concentrated formic acid. There is no damage to the wool during this extremely rapid process. The high rate of dyeing is due to a number of reasons, including the rapid and thorough wetting obtained, the high degree of swelling in formic acid which assists rapid diffusion into the fibers, and the forces of attraction between dye and basic groups in wool which are at a maximum in formic acid (71). However, on economic grounds the process has not extended. A recent continuous technique for wool tops uses dye from a concentrated aqueous solution of urea plus suitable surfactant followed by brief steaming (72). This method which is now extending widely in industry has the advantages of an aqueous solution and low capital cost of equipment. The plant is shown in Fig. 10.

Bleaching. The origin of the yellowish tinge of normal wool is not known and research is in progress to elucidate this. The most widely used procedure is to treat with hydrogen peroxide. Reducing agents will also bleach wool but the bleach is not as stable as that with peroxide. Peroxide bleaching is simple and versatile (73–77). Wool is treated at about pH 9, often hydrogen peroxide (0.3–1.5% concentration), in the presence of pyrophosphate or silicate to stabilize the peroxide. Temperature of treatment can be 20–80°C and time 1–24 hr, depending on temperature, peroxide concentration, and required degree of bleaching. It may also be carried out continuously by impregnating the wool with peroxide followed either by drying at high temperatures or by standing at room temperature for 24 hr to allow the peroxide to react.

Although sodium bisulfite, $NaHSO_3$ will bleach wool, sodium hydrosulfite, $NaHSO_2$ gives a better result. A common industrial procedure is to follow a peroxide bleach with a reduction bleach with sodium hydrosulfite.

Wool that is dark colored due to the presence of the pigment melanin can be bleached with peroxide using iron as a catalyst. Impregnation of such wool with ferrous sulfate before peroxide treatment (76) results in a much higher degree of bleaching than in the absence of metal. However, in the normal bleaching of wool, metals such as iron will act as catalysts causing serious damage to the fiber (see below).

Peroxide harshens the handle of wool, greatly increases the solubility in alkaline solutions, weakens fibers, and causes a weight loss; considerable attention has been given to ways of preventing this damage. It is accelerated by the presence of traces of metals such as iron and copper and great care is required in preventing such contaminants from entering commercial bleaching systems based on peroxide. Formaldehyde pretreatment is sometimes used for protection, especially when bleaching pigmented fibers, but, while it certainly helps keep alkali solubility low, it has no effect on the strength loss in subsequent bleaching (78).

Application of fluorescent whitening agents (see Brighteners, optical) to wool was in the past of doubtful benefit to the consumer (79,80). The light-fastness of many of the fluorescent whiteners on wool was poor, so that their effectiveness was quickly lost. Further yellowing of the wool during use also occurred, possibly through photochemical decomposition of both wool and whitener, giving deep yellow and brown colors (79,81,82). Research is in progress to develop more satisfactory whitening agents for wool.

Both normal and bleached wool gradually become yellow in daylight which can dull white wool as well as many pastel-dyed shades. The question of yellowing has been reviewed (81,83) and, as yet, the mechanism is not completely understood, but evidence points to residues of tryptophan, cystine, and tyrosine being involved.

Investigations on preventing yellowing are now leading to practical results. One such treatment consists of a reductive bleach followed by treatment with a stilbene-type fluorobrightener in the presence of thiourea and formaldehyde (84).

Insectproofing. Waterhouse and Day (85) have studied the mechanism of digestion of wool by larvae of the clothes moth and various beetles. The preliminary stage of digestion is reduction of cystine crosslinks by enzyme systems in the midgut (86). Thus, it should be possible to mothproof wool by modifying the cystine structure so that enzyme digestion cannot occur. This has, in fact, been done in several ways (87,88) but such methods are difficult and costly to apply in practice. It is simpler if a mothproofing chemical can be added to the solution used for dyeing or scouring, rather than undertake a separate treatment.

Satisfactory methods for permanent mothproofing have been developed based on colorless, chlorinated aromatic sulfonic acids applied during dyeing. Compounds such as Mitin FF (Geigy) and Eulan CN (Bayer) have been marketed for some time, but cost has retarded their extension in industry. In view of this, investigations have been undertaken to produce cheaper methods of treatment. Many anionic surfactants can mothproof wool, the effectiveness of treatment depending on configuration of the alkyl chain and on the chain length (89). Although fastness to drycleaning is satisfactory, wash-fastness and light-fastness are not. Similar results were obtained with certain cationic surfactants (90). Dieldrin (see Insecticides) in extremely low concentration (91) is highly toxic to clothes moth and carpet beetle larvae, and most unexpectedly the mothproofing effect of aqueous emulsions of dieldrin is fast to drycleaning and washing (91,92). This is because a small amount of dieldrin enters the fiber structure from the emulsions and the sorbed dieldrin is quite resistant to subsequent drycleaning or washing (93). The method has been widely used. For most purposes, 0.05% of dieldrin on the weight of wool gives satisfactory protection. The fact that the dieldrin is located within the fiber and not on its surface is a major reason why there have been no reported harmful effects on wearers since the method was introduced 14 years ago. However, it may contribute to stream pollution due to traces in dyehouse effluents and its use in certain areas has recently been prohibited. Research is in progress to find alternative compounds which do not contain chlorine.

Insecticides such as dieldrin rely on a toxic effect when contacted or ingested by the insects. An entirely different approach is to starve insects by incorporating anti-metabolites in their diet (94). These are compounds which function as antagonists to dietary components of similar structure and so interrupt the metabolic cycle. Wool-digesting insects require certain factors of the vitamin-B complex for survival (95); in particular, nicotinic acid (see under Vitamins), and pantothenic acid are essential. Sulfanilamide and related sulfonamides (see Sulfonamides), antagonists of nicotinic acid, have therefore been found to act as mothproofing agents (94). However, this approach has not yet yielded a viable industrial process. Inferior fastness to such factors as atmospheric exposure, washing, and drycleaning has prevented extension of many treatments which have given satisfactory results in initial laboratory tests.

Setting. The setting of wool fabrics to give desired shape and dimensions is a traditional process. Twist is set into yarns by steaming, and certain fabrics are set after weaving to prevent undue distortion during wet finishing operations, a process generally known as *crabbing*. This is done by winding the fabric tightly on a roller and treating in water at 60–100°C. Fabrics are often given a setting treatment to impart a certain amount of dimensional stability. This is usually done by steaming, again, while the fabric is tightly wound on a roller, and is known as *blowing* or *decatizing*. Another traditional process is potting or roll-boiling to produce a soft-handle and a smooth glossy surface on a fabric such as billiard cloth. This can be quite lengthy, involving up to several days treatment of the rolled fabric in hot water at about 60°C. The effectiveness of these simple setting treatments depends on the tension, the pH of the wool (optimum being about pH 8.5), the moisture content of wool in steam setting, the temperature, and the time (96).

The mechanism of setting has been studied in experiments with single fibers which are held strained, usually at about 40% extension. After setting, the slack fibers may be treated for 1 hr in boiling water and the increase in length stable to this boiling is known as permanent set (97).

Setting with water or steam involves an initial attack of OH^- ions on cystine crosslinks (97) [which, however, is probably not a simple hydrolysis of disulfide bonds (38)]. Set imparted at any given temperature can be removed by treating in water at a higher temperature (98). Setting with boiling solutions of reducing agents is much faster and gives a set more stable to higher temperatures than setting with water alone (98). This work has proved of great industrial significance (99), and has led to the simple concept of setting, from the chemical point of view, as a breaking of disulfide bonds which allows peptide chains to be rearranged by boiling water in a process analogous to the denaturation of soluble proteins (98). In some cases, at least, the stability of the set is due mainly to the hydrogen bonds rearranging into very stable positions in the protein molecules (98,100). Thiol-disulfide interchange reactions must also occur (101,102). Simultaneously, a variety of covalent crosslinks form spontaneously and these also contribute to stability of set provided they come about after sufficient H— and —S—S— bond rearrangements have occurred (103).

Recognition of the role of boiling water in the setting of wool with aqueous solutions of reducing agents led to development of the Si-Ro-Set process (99). Here, durable creases and pleats are formed by spraying a solution of reducing agent (thioglycollate, monoethanolamine sulfite, or sodium bisulfite) onto the garment and steaming for a few seconds in the pressed state while the fabric still contains at least 40% by weight of water. Another application of setting with reducing agents is in a process known as *flat-setting* of fabrics (104). The fabric, containing 40% by weight of a sodium bisulfite solution, is steamed for approximately five min and acquires the ability to emerge flat and nonwrinkled from thorough wetting.

Hydrogen-bonded protein structures can be rearranged at low temperatures by concentrated solutions of compounds such as urea. Thus, solutions of urea and bisulfite rapidly impart set to strained wool fibers at 30°C under conditions where neither reagent alone has any effect (98). A technique for setting wool fabrics at room temperature has been developed from this observation (105) but the steaming method is most commonly used. If setting is applied to previously shrinkproofed fabrics, washable wool requiring only a minimum of ironing can be produced. "Drip-dry" garments obtained in this way have been commercially produced for some years. "Permanent-press" garments from wool that can be machine-washed and tumble-dried without ironing are also being developed. For this, the final garments require an extra treatment that stabilizes the set by preventing reversal during washing of rearranged H— and —S—S— bonds (106).

Shrinkproofing. There are two main types of shrinkage that can occur when wool garments are washed, namely, relaxation and felting. Relaxation shrinkage results from the release on wetting of strains in a fabric (introduced during finishing and drying). It is difficult to avoid such strains on the fabric in the conventional machines used for drying wool since the wet dimensions of cloths are greater than their dry dimensions. This means that a cloth dried at its exact wet dimensions is actually dried under tension and will later show some relaxation shrinkage.

Felting shrinkage and its elimination have been studied more extensively than any other aspect of the chemical processing of wool. A comprehensive review by Sule (107) refers to over 450 processes for reducing felting shrinkage in wool. Early workers developed theories of felting based on the interlocking of scales, but it is now generally accepted that the scales are responsible for a directional frictional effect (DFE), that causes felting by favoring fiber movement in the rootward direction. A third, minor, type of shrinkage is termed "consolidation shrinkage." It can occur after relaxation shrinkage and results from release of further strains in the fabric by gentle mechanical agitation in the wet state. The initial stages of machine-washing or tumble drying are sufficient to cause such shrinkage. It is difficult to distinguish sharply between the completion of consolidation shrinkage and the commencement of felting shrinkage. Methods for reducing felting are of two major types; chemical degradative treatments, and resin deposition, which either coats fibers or welds fibers together. Other techniques which are considered to alter solely the elastic properties of the fibers have so far been only of academic interest.

The earliest techniques were oxidation treatments with hypochlorite solutions. Various refinements were subsequently made such as bromination, chlorination under mild acid conditions, and chlorination in the presence of resins. Major disadvantages of these earlier methods were lack of uniformity of treatment and excessive fiber damage. A gaseous treatment was developed to overcome these difficulties. The wool is placed in an evacuated chamber and the required quantity of chlorine is then allowed to enter (108). To control the reaction, the moisture content of the wool has to be reduced prior to treatment, usually by allowing the wool to condition at low relative humidity. This process was used extensively during World War 2 in treating under-clothes and socks for the armed forces in Britain. Since then, its use has decreased with a preference for controlled aqueous treatments.

Other oxidizing agents, such as permanganate and bromate, have been used (109) as well as combinations of oxidizing agents, such as peroxyacetic acid with hypochlorite, and permanganate with hypochlorite. The latter has been widely adopted and, with careful chemical control, has given good results. A widely used method is based on oxidation with peroxymonosulfuric acid followed by reduction with sulfite (110). Aqueous oxidative treatments often weaken the fiber.

Ethanolic alkali (111) was used to treat goods for the U.S. armed forces in Australia during World War 2 (112). Uniformity of treatment by this method is not as difficult to achieve as with aqueous treatments because alcoholic solutions wet wool more easily. Further advantages are that there are no losses in weight or in fiber strength. The use of an inflammable organic solvent is however a disadvantage.

To control fiber damage, oxidizing agents have been applied from concentrated salt solutions (109). This tends to confine reaction to the fiber surface by inhibiting swelling. Treatment with potassium permanganate in concentrated salt solution has been used in several countries.

Surface deposits of polymer on the fiber have been used to reduce felting. These include melamines (113), polyamides (114), polyamides mixed with epoxy resins (115), acrylics (116), silicones (117,118), urea–formaldehyde polymers (119), and proteins (120). Polymerization within the fiber has also been studied but large quantities are needed for a significant reduction in felting (121,122). Surface deposits of polyglycine and polyamide resins have been synthesized on fibers by treatment in monomer solutions (123–126). Polyglycine, formed in this way from anhydrocarboxyglycine in an organic solvent, produces a scale-masking. This treatment is unique in giving a uniform thin layer of polymer on the surface so that large quantities are not necessary. Deposits on the order of 1% on the weight of wool will give very good nonfelting results.

Deposition of polyamide on the surface of wool fibers by interfacial polymerization has also been developed (127). The wool is treated consecutively in an aqueous solution of a diamine (eg, hexamethylenediamine) and a solution of a diacid chloride (eg, sebacoyl chloride) in organic solvent. Polymerization takes place very rapidly without need for heating or curing. More recently, reactive resins have been applied to knitwear from organic solvents using standard dry-cleaning equipment; this technique is now extending (128).

Some years ago, it was observed that shrinkproofing with resins could be more effective after the wool had been treated with ethanol to remove any soluble materials (129). This is believed to be due to removal from the fiber surface of impurities such as residual soaps and oils which may affect adherence between polymer and wool. Thus, it was shown that as much as 10% polyamide was only partially effective in shrinkproofing wool which had not been previously extracted with ethanol. However, 2% was found very effective on wool after ethanol extraction.

Other methods of cleaning the surface of the fiber were also investigated. A mild pretreatment with ethanolic alkali was most effective and, in this instance, the polyamide and the alkali were applied to the wool simultaneously from solution in ethanol (130). Pretreatment of wool with hydrogen peroxide appears to clean the surface of the fibers and also facilitates adherence of resin (129). With wool pretreated in this way, small quantities of polymer were again effective in preventing felting.

Proteins such as casein or gelatin were found to shrinkproof wool when applied with formaldehyde as a crosslinking agent (120). Again, if the wool were pretreated

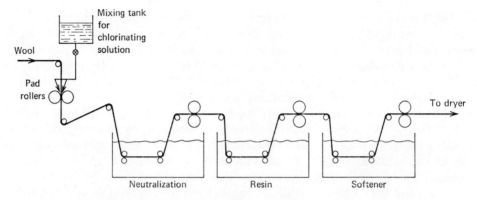

Fig. 11. Plant for continuous shrinkproofing of wool tops.

either by extraction with ethanol or bleaching with hydrogen peroxide, small quantities of protein were able to produce satisfactory results; without such pretreatments, excessive amounts of protein were required.

All the above methods of pretreatment have disadvantages which inhibit the application of resin treatment for shrinkproofing wool. The problem has been overcome by using a mild aqueous prechlorination (131). As reaction time is a matter of seconds, the treatment can be carried out continuously. This is the basis of a continous resin process for shrinkproofing wool tops which is now being used in several countries (132). The plant is shown in Fig. 11. Fig. 12 contains scanning electron micrographs showing the changes which occur at the fiber surface during this treatment. In common with other surface resin treatments, this process produces complete machine-washability. Garments subsequently made from treated wool can be washed in any type of washing machine without restriction. In contrast, the older chemical treatments usually produced results which were suitable only for handwashing or mild machine-washing.

The pretreatment required for resin application has also been found of value with other types of shrinkproofing treatments. Wool extracted with ethanol can be made nonfelting by subsequent treatment in a reducing agent, such as sodium bisulfite, which, on its own, produces no significant effect (133). Ethanol extraction possibly exposes certain sites on the fiber surface so that the reagents can react more readily with it. It might appear anomalous that sulfuryl chloride, SO_2Cl_2, will make wool nonfelting provided it has *not* been extracted with ethanol (134). In this case, however, residual oleic acid from soap remaining in the wool after scouring is considered to be essential in activating the sulfuryl chloride to give free Cl radicals which can then attack the wool.

Mechanisms. The chemical degradative reagents probably act mainly by altering the surface frictional properties of the fibers. With some treatments, the friction against the scales is reduced to approach the with-scale friction. In this way, the D.F.E. is reduced and fiber migration made more difficult. In other instances, such as with ethanolic alkali, there is an increase in both coefficients of friction so that the absolute values of friction are high and the fibers are limited in their movement. In older methods of wet chlorination, the scales were removed completely or else came away during the first washing treatment. Modern aqueous chemical treatments, however, do not have a visible effect on the scales (109) although sometimes there is a

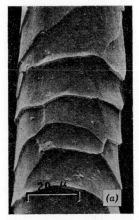  

Fig. 12. Scanning electron micrographs of fibers treated in the chlorine-resin shrinkproofing process. (a) untreated; (b) after chlorination stage; (c) after resin stage.

rounding of the edges which may be visible under the electron microscope but not under the ordinary light microscope. With these treatments (eg, permanganate/salt, peroxymonosulfuric acid), the reduction of the DFE is not nearly so great as with acid or gaseous chlorinations (135,136).

Speakman postulated that disulfide bonds in cystine are the only point of chemical attack in the fiber during shrinkproofing (137–139). This was based on indirect evidence such as pretreating wool with alkali which is known to convert cystine to lanthionine. Wool treated in this way could not be shrinkproofed with reagents such as chlorine, sulfuryl chloride, or alcoholic alkali. As the object is to limit the effect of the chemicals to the surface scales of the fibers, analysis of whole fibers may not show chemical changes which have taken place in the scales only. With techniques for removing the scales from treated and untreated wool fibers (140), it has been possible to undertake microanalysis of scale material before and after various treatments. The results confirm that disulfide attack is the major reaction in most treatments. The mechanism of resin shrinkproofing can be by scale-masking, fiber-bonding, or by increasing surface friction. Pretreatment with reagents, such as chlorine, is considered to increase the critical surface tension of the fiber, causing the polymer to spread more easily over it and giving better adhesion (141).

Wool Grease

When wool is scoured, the wool lipids remain in emulsion in the scouring liquors. This emulsion also contains the suint in the aqueous phase. Recovery of the lipid material initially produces a grease contaminated with detergent and suint and this product is called wool grease. It is usually designated according to the way in which it has been recovered such as centrifuged wool grease, acid cracked wool grease, neutralized wool grease etc.

Lanolin is wool grease that has been refined to lighten its color, reduce the odor and the free fatty acid content. *Wool wax* is the pure lipid material of the fleece, extractable with the usual fat solvents such as ether and chloroform. Because of its heterogeneous nature, different solvents may extract different components. Lanolin was first introduced as an article of commerce in Germany in 1885. The name lanolin was apparently first applied by Braun and Liebreich to the purified product obtained by a centrifugal recovery process in 1883 (142).

Early this century, Arlington Mills in Lawrence, Mass., and Erben and Harding in Philadelphia, adopted a batch solvent degreasing system which had originated in Belgium (143). Practically all of the grease was recovered as a crude product called *common degras*. Erben and Harding discontinued operations in the 1930s, and Arlington Mills closed down their solvent plant in 1950. The original Belgian plant has continued to operate in Verviers, although in recent years a modern continuous process based on newer principles has been developed and is now in production on the same site (144).

At the beginning of World War 1 in Europe, commercial centrifugal recovery of wool grease and production of highly purified lanolin was begun in the U.S. Subsequently, stream pollution in certain areas caused by discarded wool-scouring liquors led to the installation of acid-cracking grease recovery systems. The primary purpose of these is to produce a satisfactory sewage effluent, but grease recovery is essential to economical operation. The crude grease produced is called *acid degras*. A well-known example is the pioneer establishment at Bradford, England, where a large acid-cracking recovery system has been in use for many years on the city's effluent.

In the 1920s Duhamel introduced in Europe a system of aqueous wool scouring in which no soda and little soap was used (145). The principal detergent was suint liquor recovered from previously used wool scouring liquors. The system depended upon the removal of dirt and grease from these liquors by a series of centrifuges but difficulty was experienced with accumulation of heavy solids in the centrifuges. Adams introduced the first centrifuge which overcame these difficulties making practicable the continuous separation of dirt, grease, and clean soap solution (146,147). Continuous centrifugal recovery of wool grease is now widely used. It gives a much better quality product than acid cracking but the yields are lower.

Although wool grease is a mixture of compounds which are classed as waxes (qv), it does not have the physical characteristics usually displayed by waxes. It is soft with a greasy appearance and is slightly sticky. Some centrifugally recovered greases are light buff or ivory in color and practically odorless. Those recovered by

Table 8. Some Physical and Chemical Data for Wool Wax (148)

color	yellow to pale brown
density (at 15°C)	0.94–0.97
refractive index (40°C)	1.48
melting point	35–40°
free acid content	4–10%
free alcohol content	1–3%
iodine value (Wijs)	15–30
saponification value	95–120
molecular weight (Rast method, in phenyl salicylate)	790–880
fatty acids	50–55%
alcohols	50–45%
acids: melting point	40–45°
iodine value (Wijs)	10–20
mean molecular weight	330
alcohols: melting point	55–65°
iodine value (Wijs)	40–50
mean molecular weight	370

other methods contain dark-colored and odorous impurities as well as free fatty acids. Acid-cracked grease can be used in lanolin production, but the centrifugally recovered products are more suited to the alkali refining, bleaching, and deodorizing required in preparing pharmaceutical or cosmetic grade lanolin. Table 8 gives some physical and chemical data for wool wax.

Chemical Composition. Wool wax is a complex mixture of esters of water-insoluble alcohols and higher fatty acids. The alcohol and acid fractions obtained in approximately equal amounts by hydrolysis have been examined by many workers over the past 100 years. About 85% of the acids and 75% of the alcohols have now been identified. The acid fraction is unusual and contains very little of the straight chain fatty acids. Branched chain and hydroxy acids preponderate. Table 9 shows the approximate percentages of the various types present in the acid fraction.

Table 9. Composition of Wool Wax Acid Fraction[a]

Acid type	Formula	Percent of total
normal	$CH_3CH_2CH_2(CH_2)_nCH_2COOH$	7
iso	$CH_3CHCH_2(CH_2)_nCH_2COOH$ $\quad\ \ \ |$ $\quad\ \ CH_3$	22
anteiso	$CH_3CH_2CH(CH_2)_nCH_2COOH$ $\qquad\ \ |$ $\qquad\ CH_3$	29
α-hydroxy	$CH_3CH_2CH_2(CH_2)_nCHCOOH$ $\qquad\qquad\ \ |$ $\qquad\qquad\ OH$	25
α-hydroxy-iso	$CH_3CHCH_2(CH_2)_nCHCOOH$ $\quad\ \ \ | \qquad\quad\ \ |$ $\quad\ \ CH_3 \qquad\ \ OH$	3
unsaturated		probably the remainder[b]

[a] A total of 36 acids have so far been identified.

[b] About 14% from iodine values.

There is little, if any, free fatty acid in wool wax that has been extracted from the basal portions of the fleece (149). As the fleece grows, the composition of the wax changes due to atmospheric oxidation producing free fatty acid. This mainly accounts for the free acid contents shown in Table 8 which are figures for wax obtained from the whole fleece.

The alcohol fraction is likewise a very complex mixture. Cholesterol (about 20% of the fraction) was first identified as a component in 1868 (150) although its structure was then unknown. Four years later, another alcohol of similar molecular weight, but dextrorotary, was isolated (151). This was named isocholesterol (about 25% of the fraction). It was subsequently found to consist of a mixture of triterpene alcohols and their derivatives, the main component being designated "lanosterol" and a second component "agnosterol" (152). Aliphatic alcohols (about 20% of the fraction) are also present corresponding to the five different types of acid shown in Table 9. About 6% of the weight of wax has been isolated as free cholesterol (153). Traces of low-melting unidentified hydrocarbons have also been found, and about 22% of the alcohol fraction is still unidentified.

Methods of Recovery. As solvent degreasing is only used to a limited extent, recovery from this system will not be considered in great detail. The grease remains when the solvent is recovered by distillation of the clarified solution. In earlier systems the crude grease from the still was not further refined. In more recent processes refined greases have been produced (144). Centrifugal recovery from aqueous scour liquors is the most commonly used process. Liquor is usually withdrawn from the second bowl of a five-bowl scouring plant, heated to about 190°F, and passed into a large settling tank to remove the heaviest dirt particles. Next the hot emulsion is passed through a centrifuge designed to remove suspended solids, reheated, and run through a centrifuge of the cream-separator type. This separates most of the grease remaining in the soap solution, and the latter is cooled to 120°F and returned to the scouring train at an appropriate point or held in a reserve tank for later re-use. Latest developments in this field are well covered in a recent review (154).

The grease produced by this method contains up to 30% moisture and is sometimes sold in this condition. However, it is often processed immediately in the same plant by mixing it with a large volume of water at 190–210°F, and recentrifuging it in high-speed machines to a moisture content of 1 or 2%. Centrifugally recovered wool grease is the principal raw material for cosmetic and U.S.P. grades of lanolin. The refining process is primarily intended to remove free fatty acids, odor, and color. It consists essentially of treatment with hot aqueous alkali to convert free fatty acids to soaps followed by bleaching, usually with hydrogen peroxide. Before acid cracking, the liquors from the scour are stored in sedimentation tanks where coarse sand separates. They are then passed through a coarse filter to the cracking tanks. Sulfuric acid is added to give pH 3–4 and the solution is well agitated. The emulsion breaks, heavy sludge settles to the bottom, and light scum floats to the top leaving a third layer of almost clear liquor in between. As much of the clear liquid as possible is decanted and rejected. The remaining magma is heated with steam and passed through pressure filters while hot. The grease layer of the filtrate is separated from the water layer and dried by heating, sometimes followed by centrifuging. Increasing use of synthetic detergents which do not respond to acid cracking has produced difficulties in effluent treatment. However, it has been shown that nonionic scouring liquors can be acid cracked (155) and a satisfactory technique should therefore be feasible. Other chemical treatments, apart from acid cracking have been studied but there has been little application of results. Calcium hypochlorite has, however, been applied to a limited extent. It produces an initial cracking by forming calcium soaps. The scum and sludge are then treated with sulfuric acid and the wax recovered. This has the advantage of a cleaner and more sterile effluent (smaller BOD, biochemical oxygen demand) and a lighter colored grease due to oxidation by the acid hypochlorite (156).

Aeration processes have had limited application. These processes depend upon the accumulation of the grease in froth resulting from aeration by paddling manually or

Table 10. Properties and Uses of Wool Grease

Property	Resultant uses
stability to oxidation (no rancidity)	cosmetics; toilet soaps
adheres to most surfaces	rust preventives; adhesive tapes; sheep branding fluids; lubricants
does not dry (remains plastic)	belt and fur dressings; adhesive tapes
emulsifies readily	cosmetics and pharmaceutical bases; sheep branding fluids; cutting oils
does not crystallize and hinders crystallization	cosmetics; printing inks

mechanically (157,158) or by jets of compressed air (159,160). The resulting froth is washed with water, heated, and separated into grease and water layers. A later version (161) produces a froth concentrate and washes it continuously in a bank of modern subaeration flotation machines. The washed froth is dispersed in an alkaline medium from which high-grade lanolin is obtained by centrifuging.

Uses of Wool Grease. Wool grease or lanolin has several outstanding properties which account for its use in a wide range of different products. Table 10 lists these properties showing some of the uses resulting from the presence of one or several of them.

The above uses are by no means exhaustive but would cover the major outlets. In the United Kingdom, there is a wool grease distillation industry which produces lubricants used mainly in the textile industry and known as *wool oleine*. Another fraction is *wool stearine* which is used chiefly in the form of calcium salts as a lubricant in heavy duty ball bearings and steel rollers. It is also used in metal and floor polishes. In U.S.A. a big use is in metal lubricants (162). In most countries, cosmetics and pharmaceuticals are major outlets.

Bibliography

"Wool" in *ECT* 1st ed., Vol. 15, pp. 103–134, by John Menkart, Textile Research Institute.

1. *Textile Terms and Definitions, 5th ed.*, Textile Institute, Manchester, 1963.
2. M. Lipson and U. A. F. Black, *J. Roy. Soc. N.S.W.* **78,** 84 (1945).
3. R. B. Sweetten, *J. Textile Inst. Trans.* **40,** T727 (1949).
4. J. G. Downes, B. H. Mackay, A. B. Dehlsen and K. D. Sinclair, *Report No. 11, International Wool Textile Organization*, Technical Committee, Montreux, May 1968.
5. *A.S.T.M. Standard D2130–69*, American Society for Testing and Materials, Philadelphia, 1969.
6. *British Standard 3183:1959*, British Standards Institution, London (1959).
7. G. V. Chapman and J. H. Bradbury, *Arch. Biochem. Biophys.* **127,** 157 (1968).
8. R. D. B. Fraser and F. G. Lennox, *Textile J. of Australia* **37,** 120 (1962).
9. C. W. Hock, R. G. Ramsay, and M. Harris, *J. Res. Natl. Bur. Std.* **27,** 181 (1941).
10. *Ibid.*, 234 (1941).
11. J. L. Farrant, A. L. G. Rees, and E. H. Mercer, *Nature* **159,** 535 (1947).
12. R. D. B. Fraser, T. P. MacRae, and A. Miller, *J. Mol. Biol.* **14,** 432 (1965).
13. E. H. Mercer and A. L. G. Rees, *Australian J. Exptl. Biol. Med. Sci.* **24,** 147, 175 (1946).
14. J. Lindberg, E. H. Mercer, B. Philip, and N. Gralen, *Textile Res. J.* **19,** 673 (1949).
15. J. Lindberg, B. Philip, and N. Gralen, *Nature* **162,** 458 (1948).
16. W. J. Onions, *Wool*, Ernest Benn Ltd., London, 1962, p. 48.
17. C. W. Bunn, in J. M. Preston, ed., *Fibre Science*, Textile Institute, Manchester, 1949, Chap. 10.
18. *Textile World, Man-Made Fiber Chart*, McGraw-Hill Book Co., New York, 1968.
19. J. B. Speakman and C. A. Cooper, *J. Textile Inst. Trans.* **27,** T183 (1936).
20. W. von Bergen, in H. R. Mauersberger, ed., *Matthews' Textile Fibers*, 6th ed., John Wiley & Sons, Inc., New York, 1954, Chap. 11.
21. J. B. Speakman, *J. Textile Inst. Trans.* **38,** T102 (1947).
22. K. R. Makinson, *Wool Science Review* **24, 34**, International Wool Secretariat, London, 1964.
23. E. R. Kaswell, *Textile Fibers, Yarns, and Fabrics*, Reinhold Publishing Corporation, New York, 1953, p. 69.
24. P. J. Reis and P. G. Schinckel, *Australian J. Biol. Sci.* **16,** 225 (1963).
25. C. Earland, *Textile Res. J.* **31,** 492 (1961).
26. W. G. Crewther, R. D. B. Fraser, F. G. Lennox, and H. Lindley, *Advan. Protein Chem.* **20,** 191–346 (1965).
27. J. M. Gillespie, P. J. Reis, and P. G. Schinckel, *Australian J. Biol. Sci.* **17,** 584 (1964).
28. B. S. Harrap, *Australian J. Biol. Sci.* **15,** 596 (1962).
29. J. M. Gillespie and B. S. Harrap, *Australian J. Biol. Sci.* **16,** 252 (1963).

30. J. H. Bradbury and G. V. Chapman, *Australian J. Biol. Sci.* **17,** 960 (1964).

31. H. R. Marston, Papers, Fibrous Proteins Symposium, Society of Dyers and Colourists, May 1946, 207.

32. R. W. Burley and F. W. A. Horden, *Textile Res. J.* **30,** 484 (1960).

33. P. J. Reis and P. G. Schinckel, *Australian J. Biol. Sci.* **16,** 226 (1963).

34. K. A. Ferguson, J. A. Hemsley, and P. J. Reis, *Australian J. Sci.* **30,** 215 (1967).

35. W. T. Astbury and J. B. Speakman, *J. Soc. Dyers Colourists,* Jubilee Issue, 1934, pp. 24–45.

36. J. Steinhardt, C. H. Fugitt, and M. Harris, *J. Res. Natl. Bur. Std.* **25,** 519 (1940).

37. J. B. Speakman and E. Stott, *Trans. Faraday Soc.* **30,** 539 (1934).

38. R. Cecil and J. R. McPhee, *Advan. Protein Chem.* **14,** 255 (1959).

39. J. P. E. Human and P. H. Springell, *Australian J. Chem.* **12,** 508 (1959).

40. W. E. Savige and J. A. Maclaren, in N. Kharasch and C. Y. Meyers, eds., *The Chemistry of Organic Sulfur Compounds,* Vol. 2, Pergamon Press, Oxford, 1966, Chap. 15.

41. H. M. Phillips, *J. Textile Inst. Proc.* **27,** P208 (1936).

42. F. O. Howitt, *Proc. Intern. Wool Textile Res. Conf.,* Australia, 1955, E, 315.

43. V. A. Williams, *Textile Res. J.* **31,** 472 (1961).

44. A. S. C. Lawrence, *Nature* **183,** 1491 (1959).

45. R. P. Harker, *J. Textile Inst. Trans.* **50,** T189 (1959).

46. C. A. Anderson, M. Lipson, J. F. Sinclair, and G. F. Wood, *Intern. Wool Textile Res. Conf.,* 3rd, Paris, 1965, 3, 146.

47. C. A. Anderson and G. F. Wood, *Nature* **193,** 742 (1962).

48. C. A. Anderson, *J. Textile Inst. Trans.* **53,** T401 (1962).

49. W. G. Crewther, *Proc. Intern. Wool Textile Res. Conf.,* Australia, 1955, E, 408.

50. W. G. Crewther and T. A. Pressley, *Textile Res. J.* **28,** 67 (1958).

51. V. A. Williams, Unpublished data.

52. A. E. Davis, A. J. Johnson, and L. R. Mizell, *Textile Res. J.* **31,** 825 (1961).

53. J. M. Preston, *Wool Science Review* **30,** 1 (1966); **31,** 32 (1967); **32,** 36 (1967).

54. T. Vickerstaff, *The Physical Chemistry of Dyeing,* 2nd ed., Oliver and Boyd, London, 1954.

55. R. V. Peryman, *Proc. Intern. Wool Textile Res. Conf.,* Australia, 1955, E, 17.

56. W. Luck, *J. Soc. Dyers Colourists* **74,** 221 (1958).

57. J. A. Medley and M. W. Andrews, *Textile Res. J.* **29,** 398 (1959).

58. V. Køpke and B. Nilssen, *J. Textile Inst. Trans.* **51,** T1398 (1960).

59. H. E. Millson, *Am. Dyestuff Reptr.* **44,** P417 (1955).

60. H. R. Hadfield and D. R. Lemin, *J. Soc. Dyers Colourists* **77,** 715 (1961).

61. E. M. Kärrholm and J. Lindberg, *Textile Res. J.* **26,** 528 (1956).

62. H. R. Hadfield and D. R. Lemin, *J. Soc. Dyers Colourists* **77,** 97 (1961).

63. H. E. Millson and L. H. Turl, *Textile Res. J.* **21,** 685 (1951).

64. Brit. Pat. 740,003 (1955), A. J. Harding (to Maifoss Ltd.).

65. H. Luttringhaus, J. E. Flint, and A. Arcus, *Am. Dyestuff Reptr.* **39,** P2 (1950).

66. L. Peters and C. B. Stevens, *Dyer* **115,** 327 (1956).

67. W. Beal, K. Dickinson, and E. Bellhouse, *J. Soc. Dyers Colourists* **76,** 333 (1960).

68. J. Delmenico, *Textile Res. J.* **27,** 899 (1957).

69. R. J. Hine and J. R. McPhee, *International Dyer* **132** (7), 523 (1964).

70. R. Casty, *Dyer* **125,** 100 (1961).

71. B. S. Harrap, *J. Soc. Dyers Colourists* **75,** 106 (1959).

72. I. B. Angliss, P. R. Brady, J. Delmenico, and R. J. Hine, *Textile J. Australia* **43,** 17, (April 1968).

73. I. E. Weber, *J. Textile Inst. Proc.* **24,** P178 (1933).

74. W. S. Wood and K. W. Richmond, *J. Soc. Dyers Colourists* **68,** 337 (1952).

75. N. F. Crowder and W. A. S. White, *J. Soc. Dyers Colourists* **71,** 764 (1955).

76. L. Chesner and G. C. Woodford, *J. Soc. Dyers Colourists* **74,** 531 (1958).

77. B. K. Easton, *Can. Textile J.* **78** (18), 32 (1961).

78. J. E. Moore and R. A. O'Connell, *Textile Res. J.* **28,** 687 (1958).

79. R. S. Higginbotham and F. W. Thomas, *Nature* **181,** 1437 (1958).

80. Delaware Valley Section, A.A.T.C.C., *Am. Dyestuff Reptr.* **49,** P565 (1960).

81. D. R. Graham and K. W. Statham, *J. Soc. Dyers Colourists* **72,** 434 (1956).

82. A. Kling and J. Kurz, *Textil-Praxis* **17,** 250 (1962).

83. F. G. Lennox, *J. Textile Inst. Trans.* **51,** T1193 (1960).

84. W. E. Savige, *Textile Res. J.* **38,** 101 (1968).

85. D. F. Waterhouse, *Ann. Rev. Entomol.* **2,** 1 (1957).

86. K. Linderstrom-Lang and F. Duspiva, *Z. Physiol. Chem.* **237,** 131 (1935).

87. W. B. Geiger, F. F. Kobayashi, and M. Harris, *J. Res. Natl. Bur. Std.* **29,** 281 (1942).

88. J. R. McPhee and M. Lipson, *Australian J. Chem.* **7,** 387 (1954).

89. M. Lipson, *Proc. Intern. Wool Textile Res. Conf.*, Australia, 1955, E, 514.

90. G. N. Freeland and V. A. Williams, *Textile Res. J.* **37,** 408 (1967).

91. M. Lipson and R. J. Hope, *Nature* **175,** 599 (1955).

92. M. Lipson and R. J. Hope, *Proc. Intern. Wool Textile Res. Conf.*, Australia, 1955, E, 523.

93. M. Lipson and J. R. McPhee, *Textile Res. J.* **28,** 679 (1958).

94. R. J. Pence, *Soap Chem. Specialties* **35** (8), 65 (1959).

95. G. Fraenkel and M. Blewett, *J. Exp. Biol.* **22,** 156 (1946).

96. C. S. Whewell, *J. Textile Inst. Proc.* **47,** P851 (1956).

97. J. B. Speakman, *J. Soc. Dyers Colourists* **52,** 335 (1936).

98. A. J. Farnworth, *Textile Res. J.* **27,** 632 (1957).

99. Trade Circular No. 4, CSIRO. Wool Textile Research Laboratory, Geelong, Victoria, Australia, 1958.

100. M. Feughelman, A. R. Haly, and T. W. Mitchell, *Textile Res. J.* **28,** 655 (1958).

101. W. G. Crewther, *J. Soc. Dyers Colourists* **82,** 54 (1966).

102. J. B. Caldwell, S. J. Leach, and B. Milligan, *Textile Res. J.* **35,** 245 (1965).

103. J. R. Cook and J. Delmenico, *J. Textile Inst.* In press, (1970).

104. A. J. Farnworth, M. Lipson, and J. R. McPhee, *Textile Res. J.* **30,** 11 (1960).

105. M. Cednas and E. M. Karrholm, *Textile Res. J.* **34,** 973, 989 (1964).

106. I. B. Angliss, J. R. Cook, J. Delmenico, H. D. Feldtman, B. E. Fleischfresser, F. W. Jones, and M. A. White, *Proc. Intern. Wool Textile Res. Conf.*, 4th, San Francisco, 1970. In press.

107. A. D. Sule, *Wool and Woollens of India* **3** (9), 24 (1967).

108. H. Phillips, *Dyer* **96,** 299 (1946).

109. J. R. McPhee, *Textile Res. J.* **30,** 349 (1960).

110. Brit. Pat. 716,806 (1954), E. T. Fell (to Stevensons Dyers Ltd.).

111. M. R. Freney and M. Lipson, *Council Sci. Ind. Res.*, Australia, Pamphlet 94 (1940).

112. M. Lipson, *J. Textile Inst. Proc.* **38,** P279 (1947).

113. J. R. Dudley and J. E. Lynn, Fibrous Proteins Symposium, Society of Dyers and Colourists, May 1946, p. 215.

114. D. L. C. Jackson and M. Lipson, *Textile Res. J.* **21,** 156 (1951).

115. C. E. Pardo and R. A. O'Connell, *Am. Dyestuff Reptr.* **47,** P333 (1958).

116. F. H. Steiger, *Am. Dyestuff Reptr.* **50,** P97 (1961).

117. W. J. Neish and J. B. Speakman, *Nature* **156,** 176 (1945).

118. P. Alexander, D. Carter, and C. Earland, *J. Soc. Dyers Colourists* **65,** 107 (1949).

119. P. Alexander, *J. Soc. Dyers Colourists* **66,** 349 (1950).

120. D. L. C. Jackson and A. R. A. Backwell, *Australian J. Appl. Sci.* **6,** 244 (1955).

121. M. Lipson and J. B. Speakman, *Nature* **157,** 590 (1946).

122. M. Lipson and J. B. Speakman, *J. Soc. Dyers Colourists* **65,** 390 (1949).

123. A. W. Baldwin, T. Barr, and J. B. Speakman, *J. Soc. Dyers Colourists* **62,** 4 (1946).

124. P. Alexander, J. L. Bailey, and D. Carter, *Textile Res. J.* **20,** 385 (1950).

125. J. H. Bradbury and J D. Leeder, *Textile Res. J.* **30,** 118 (1960).

126. J. H. Bradbury and D. C. Shaw, *Textile Res. J.* **30,** 976 (1960).

127. R. E. Whitfield, L. A. Miller, and W. L. Wasley, *Textile Res. J.* **31,** 704 (1961).

128. International Wool Secretariat, *Am. Dyestuff Reptr.* **58** (16), 22 (1969).

129. D. L. C. Jackson, *Textile Res. J.* **23,** 616 (1953).

130. J. Delmenico, D. L. C. Jackson, and M. Lipson, *Textile Res. J.* **24,** 828 (1954).

131. Australian Pat. 284,572 (1966) (to CSIRO).

132. H. D. Feldtman, J. R. McPhee, and W. V. Morgan, *Textile Mfr.* **93,** 122 (1967).

133. A. J. Farnworth, *J. Soc. Dyers Colourists* **77,** 483 (1961).

134. A. J. Farnworth and J. B. Speakman, *J. Soc. Dyers Colourists* **65,** 162 (1949).

135. J. R. McPhee, *Textile Res. J.* **31,** 770 (1961).

136. J. R. McPhee, *Textile Res. J.* **32,** 14 (1962).

137. J. B. Speakman, B. Nilssen, and G. H. Elliott, *Nature* **142,** 1035 (1938).

138. J. B. Speakman, *J. Textile Inst. Trans.* **32,** T83 (1941).

139. A. J. Farnworth, W. J. P. Neish, and J. B. Speakman, *J. Soc. Dyers Colourists* **65,** 447 (1949).
140. J. H. Bradbury, *J. Textile Inst. Trans.* **51,** T1226 (1960).
141. H. D. Feldtman and J. R. McPhee, *Textile Res. J.* **34,** 634 (1964).
142. U.S. Pat. 271,192 (1883), Otto Braun and Oscar Liebreich.
143. U.S. Pat. 545,899 (1895), Emile Maertens.
144. J. Brach and A. Delforge, *Ann. Sci. Textiles Belges,* **7** (3) (Sept. 1967).
145. C. E. Mullin, *Textile Colorist* **54,** 381, 446, 529, 562 (1932).
146. Fr. Pat. 687,317 (1930), James W. Adams.
147. Ger. Pat. 578,856 (1933), James W. Adams.
148. E. V. Truter, *Wool Wax,* Cleaver-Hume Press Ltd., London, 1956, p. 32.
149. M. R. Freney, *Council Sci. Ind. Res.,* Australia, Bull. No. 130 (1940).
150. Hartmann, *Ueber den Fettschweiss der Schafwolle.* Inaug. Diss., Göttingen (1868).
151. E. Schulze, *Chem. Ber.* **5,** 1075 (1872); **6,** 251 (1873); **31,** 1200 (1898).
152. A. Windaus and R. Tschesche, *Z. Physiol. Chem.* **190,** 51 (1930).
153. M. Lipson, *Council Sci. Ind. Res.,* Australia, *J.* **13,** 273 (1940).
154. C. A. Anderson, *Wool Science Review* **37,** 23 (Oct. 1969).
155. C. A. Anderson, *Textile J. Australia* **40,** (4), 11 (1965).
156. D. H. S. Horn and F. W. Hougen, *J. Chem. Soc.* 3533 (1953).
157. Brit. Pat. 273,642 (1928), Eugene Mertens.
158. U.S. Pat. 1,853,871 (1932), Eugene Mertens.
159. U.S. Pat. 1,830,633 (1931), T. W. Barber.
160. U.S. Pat. 1,846,577 (1932), T. W. Barber.
161. L. F. Evans and W. E. Ewers, *Australian J. Appl. Sci.* **4,** 552 (1953).
162. R. S. Raymond and S. L. Mandell, U.S. Department of Agriculture, *Marketing Research Report* No. 89, 1955.

General References

W. von Bergen, ed., *Wool Handbook,* Vol. 1, 3rd ed., Interscience Publishers, a division of John Wiley & Sons, Inc., New York, 1963. Vol. II, 3rd ed., New York, 1969.

P. Alexander and R. F. Hudson, in C. Earland, ed., *Wool: Its Chemistry and Physics,* 2nd ed., Chapman & Hall, Ltd., London, 1963.

W. J. Onions, *Wool: An Introduction to its Properties, Varieties, Uses and Production,* Ernest Benn Ltd., London, 1962.

W. G. Crewther, R. D. B. Fraser, F. G. Lennox, and H. Lindley, "The Chemistry of Keratins" in *Advan. Protein Chem.,* **20,** 191–346 (1965).

M. Lipson, "Recent Developments in Wool Textile Processing" in J. E. Lynn and J. J. Press, eds., *Advances in Textile Processing,* Vol. 1, Textile Book Publishers, Inc., New York, 1961, pp. 193–223.

E. V. Truter, *Wool Wax: Chemistry and Technology,* Cleaver-Hume Press Ltd., London, 1956.

E. R. Kaswell, *Textile Fibers Yarns, and Fabrics: A Comparative Survey of their Behavior with Special Reference to Wool,* Reinhold Publishing Corp., New York, 1953.

Papers, Intern. Wool Textile Res. Conf. 1st Australia, 1955, CSIRO, Melbourne.

Intern. Wool Textile Res. Conf. 2nd Harrogate, 1960, *J. Textile Inst., Trans.* **51,** T489, 1960.

Papers, Intern. Wool Textile Res. Conf. 3rd Paris, 1965. L'Institut Textile de France, Paris.

Wool Science Review, International Wool Secretariat, London.

M. Lipson
Division of Textile Industry,
CSIRO,
Geelong, Victoria, Australia

WOONCOPON.   See Surfactants.

WORMWOOD.   See Oils essential, Vol. 14, p. 214.

WORT.   See Beer and brewing.

WYTOX.   See Rubber chemicals.

X

XANTHATES

The salts and *S*-esters of the mono-*O*-esters, ROCSSH, of dithiocarbonic acid (thionothiolcarbonic acid) are known as xanthates. The free acids decompose on standing. A salt, potassium ethyl xanthate, C_2H_5OCSSK, was first prepared in 1822 at the University of Copenhagen by W. C. Zeize from potassium hydroxide, carbon disulfide, and ethyl alcohol. Most alcohols, including cellulose, undergo this reaction to form xanthates (see Rayon), but normally phenols do not. Potassium phenyl xanthate was finally prepared in 1960 from potassium phenoxide and carbon disulfide in dimethylformamide (1). Xanthates remained a laboratory curiosity until the turn of the twentieth century when the rubber industry found a use for them in connection with the curing and vulcanization of rubber. A much larger use developed after 1925 when C. H. Keller patented his discovery that xanthates were excellent flotation reagents for the recovery of heavy-metal sulfides from ores. This is the principal use today for the noncellulose xanthates, and several of the alkali metal xanthates are commercially available.

Nomenclature. The names "xanthogenic acid" and "xanthogen" were applied by Zeize to the acid C_2H_5OCSSH and the radical C_2H_5OCSS—, respectively. "Xanthogenic" is still used to some extent instead of "xanthic" and "xanthogenate" for xanthate, eg sodium methyl xanthogenate for sodium methyl xanthate. The term xanthogen is sometimes used for the radical HOCSS—, as methylxanthogen acetic acid, $CH_3OCSSCH_2COOH$.

Commencing with Volume 66 of *Chemical Abstracts* (*CA*), entries are no longer listed under xanthic acid and xanthates, but are indexed under carbonic acid, dithio esters. Examples of the new nomenclature are given in Table 1, along with the corresponding common usage. The latter will be used in this article. Prior to this

Table 1. Nomenclature of Some Xanthic Acids and Related Compounds

Compound	CA usage	Common usage
CH_3OCSSH	*O*-methyl dithiocarbonic acid[a]	methyl xanthic acid
$CH_3OCSSNa$	sodium *O*-methyl dithiocarbonate[a]	sodium methyl xanthate
$CH_3OCSSC_2H_5$	*S*-ethyl *O*-methyl dithiocarbonate	ethyl methyl xanthate
$(CH_3OCS)_2S$	*O,O*-dimethyl dithiocarbonic acid anhydrosulfide	dimethyl xanthogen monosulfide
$(CH_3OCS)_2S_2$	*O,O*-dimethyl dithiobis (thionoformate)	dimethyl dixanthogen
$CH_3OCSSCOOC_2H_5$	*O*-methyl dithiocarbonic acid anhydrosulfide with *O*-ethyl thiocarbonic acid	methyl xanthogen ethyl formate
CH_3OCSCl	*O*-methyl chlorothioformate	methyl chlorothionoformate
$CH_3OCSNHC_2H_5$	*O*-methyl ethylthiocarbamate[a]	methyl ethylthionocarbamate
$CH_3OCSNHC_6H_5$	*O*-methyl thiocarbanilate[a]	methyl thionocarbanilate

[a] Conforms to IUPAC nomenclature.

change CA would call CH_3OCSSH, methyl xanthic acid and $CH_3OCSSC_2H_5$, ethyl methyl xanthate.

Physical and Chemical Properties

The relatively unimportant xanthic acids are unstable, colorless or yellow oils, and have been known to decompose with explosive violence. They are soluble in the common organic solvents, and slightly soluble in water (methyl xanthic acid at 0°C, 0.05 mole/liter; ethyl xanthic acid, 0.02 mole/liter). Values for the dissociation constant for ethyl xanthic acid have been reported in the range of 2.0 to 3.0 $\times$ 10^{-2} (2). Potentiometric determinations (3,4) were reported for C_{1-8} xanthic acids and showed a decreasing acid strength with increasing molecular weight. The values for the ethyl derivative were in the range of 1.82 to 3.4 $\times$ 10^{-3}. Similar values, by the same method, were reported for a series run in dimethylformamide (5).

The alkali metal salts, in contrast to the free acids, are relatively stable solids, pale yellow when pure, with a disagreeable odor. Pyrolysis studies were made on the potassium salts of ethyl, isopropyl, n-butyl, sec-butyl, and n-amyl xanthic acids, and the gaseous products were separated by gas chromatography (6). All salts produced carbonyl sulfide, carbon disulfide, thiols, alcohols, sulfides, disulfides, and some aldehydes. A comparison of literature values, T_{lit}, of xanthate melting points (7) and the temperature interval, T_{DTA}, taken from the start to the end of the differential thermal analysis interval (6) are given in Table 2.

Table 2. Decomposition Temperatures of Xanthates

Compound	T_{DTA} °C	T_{lit}
CH_3OCS_2K	165–185	182–186
$C_2H_5OCS_2K$	210–225	225–226
n-$C_3H_7OCS_2K$	220–240	233–239
i-$C_3H_7OCS_2K$	230–275	278–282
n-$C_4H_9OCS_2K$	235–255	255–256
i-$C_4H_9OCS_2K$	250–265	260–270
sec-$C_4H_9OCS_2K$	220–260	

The sodium salts, when exposed to air, tend to take up moisture and form dihydrates. The alkali metal xanthates are soluble in water, alcohols, the lower ketones, pyridine, and acetonitrile. They are not particularly soluble in the nonpolar solvents, such as ether or ligroin. The solubilities of a number of these salts have been determined (see Table 3) (8). Potassium isopropyl xanthate is reported to be soluble in acetone to about 6%, whereas the corresponding methyl, ethyl, n-propyl, n-butyl, isobutyl, isoamyl, and benzyl xanthates are soluble to more than 10% (7).

The heavy-metal salts, which vary in color depending upon the metallic portion, are relatively insoluble in water, but are soluble in many organic solvents. The solubilities of the heavy-metal xanthates in water have been placed in the following orders of increasing solubility by two different authors:

$$Hg^{2+} \ Ag^+ \ Cu^+ \ Co^{3+} \ As^{3+} \ Pb^{2+} \ Tl^+ \ Cd^{2+} \ Ni^{2+} \ Zn^{2+} \qquad (9)$$

$$Hg^{2+} \ Hg_2^{2+} \ Au^{3+} \ Ag^+ \ Cu^+ \ Bi^{3+} \ Pb^{2+} \ Cd^{2+} \ Ni^{2+} \ Fe^{2+} \ Zn^{2+} \qquad (10)$$

The solubility products of some of these salts at 20°C have been reported (see Table 4) (11). The value for silver ethyl xanthate has been reported elsewhere (10).

Table 3. Solubility of Some Alkali Metal Xanthates

Xanthate	Solvent	Solubility, g/100 g soln	
		0°C	35°C
potassium n-propyl	water	43.0	58.0
sodium n-propyl	water	17.6	43.3
potassium isopropyl	water	16.64	37.15
sodium isopropyl	water	12.1	37.9
potassium n-butyl	water	32.4	47.9
sodium n-butyl	water	20.0	76.2
potassium isobutyl	water	10.7	47.67
sodium isobutyl	water	11.2	33.37
potassium isoamyl	water	28.4	53.3
sodium isoamyl	water	24.7	43.5
potassium n-propyl	n-propyl alcohol	1.9	8.9
sodium n-propyl	n-propyl alcohol	10.16	22.5
potassium isopropyl	isopropyl alcohol		2.0
sodium isopropyl	isopropyl alcohol		19.0
potassium n-butyl	n-butyl alcohol		36.5
sodium n-butyl	n-butyl alcohol		39.2
potassium isobutyl	isobutyl alcohol	1.6	6.2
sodium isobutyl	isobutyl alcohol	1.2	20.5
potassium isoamyl	isoamyl alcohol	2.0	6.5
sodium isoamyl	isoamyl alcohol	10.9	15.5

The value for ferrous ethyl xanthate is not consistent with the place of Fe^{2+} in the second series above. The surface tension depression increases with the increase in molecular weight (12).

Table 4. Solubility Products of Heavy-Metal Xanthates

Xanthate	Zn	Ni	Cu	Fe(ous)	Ag
methyl	2.85×10^{-5}	1.15×10^{-14}	4.6×10^{-14}		
ethyl	1.31×10^{-6}	1.37×10^{-12}	5.6×10^{-15}	4.04×10^{-20}	3.5×10^{-17}
n-propyl		1.92×10^{-17}			
n-butyl	3.76×10^{-7}	4.51×10^{-18}	3.4×10^{-17}		

Literature reports indicate that there is some stabilization of xanthate solutions by alkalis. One report (13) indicates that decomposition is accelerated at pH values above 10 or below 9 and that between these limits no significant decomposition takes place during a day. This was in agreement with a report (14) that over an eight-day period 75% decomposition took place at pH 6.5 and only 25% at pH 10.8. Other studies over a wide pH range showed a minimum of decomposition rates at a pH of 10–13 (15). Decomposition studies using 1–4 N sodium or potassium hydroxide showed that the decomposition rate decreased with increased molecular weight of the xanthate and in going from a primary to a secondary alkyl xanthate (16). The decomposition rate of *tert*-butyl xanthate is more rapid than that for methyl xanthate.

Some of the prior publications on the well-studied decomposition of xanthates in acidic solution are discussed in reference 17. With a lower alkyl xanthate, the rate passed through a maximum at a low pH and then decreased smoothly as the acidity was increased. The decrease was not observed with tertiary butyl xanthate. Aqueous solutions of the alkali salts are fairly stable at room temperature (see Table 5) (18).

Table 5. Hydrolysis of Xanthates (0.01 M at 25°C)

Xanthate	Number of days							
	1	2	3	4	5	6	7	8
	Percent hydrolyzed							
sodium ethyl	0.0	0.6	1.1		2.0	2.3	2.4	
potassium ethyl	0.0	0.7	0.9	1.2	2.1	2.8		
potassium isobutyl	0.0	0.4	0.7	1.1			1.4	1.7
potassium isoamyl	0.0	0.7	0.8	1.2			1.7	2.2

REACTIONS

The chemistry of the xanthates is essentially that of the dithio acids. The free xanthic acid readily decomposes at room temperature into carbon disulfide and the corresponding alcohol:

$$ROCSS^- + H^+ \rightleftharpoons ROCSSH \rightarrow ROH + CS_2$$

The initial hydrolysis of the xanthates in aqueous solutions at room temperature is reported to give the following reaction, as shown for potassium ethyl xanthate:

$$6\ C_2H_5OCSSK + 3\ H_2O \rightarrow 6\ C_2H_5OH + 2\ K_2CS_3 + K_2CO_3 + 3\ CS_2$$

Further hydrolysis of the carbon disulfide and the trithiocarbonate produces hydrogen sulfide, etc (19).

The alkali metal xanthates react readily with the various alkylating reagents to form the *S*-esters:

$$C_2H_5OCSSNa + CH_3I \xrightarrow{\text{alcohol}} C_2H_5OCSSCH_3 + NaI$$

$$C_2H_5OCSSNa + ClCH_2COONa \xrightarrow{\text{water}} C_2H_5OCSSCH_2COONa + NaCl$$

The reactions are exothermic, and cooling may be required.

The alkyl esters react readily with ammonia or alkylamines to give the corresponding thionocarbamates and thiols. In general, heating is not required. In the following reaction, the thionocarbamate is the only water-insoluble material and separates as an oil:

$$C_2H_5OCSSCH_2COONa + C_2H_5NH_2 \rightarrow C_2H_5OCSNHC_2H_5 + HSCH_2COONa$$

The Chugaev reaction, or thermal decomposition of the *S*-methyl esters of the xanthates, gives olefins without rearrangements (20,21), for example:

$$iso\text{-}C_5H_{11}OCSSCH_3 \xrightarrow{\Delta} (CH_3)_2CHCH{=}CH_2 + COS + CH_3SH$$

Esters of benzyl xanthic acid yield stilbenes on heating (22).

Double bonds in, or dialkylamino groups on, the alkyl group of the *S*-methyl ester may facilitate isomerization to the dithiol ester (23), for example:

$$(CH_3)_2C{=}CHCH_2OCSSCH_3 \xrightarrow{\Delta} CH_2{=}CHC(CH_3)_2SCOSCH_3$$

In a relatively low-temperature procedure olefins are readily obtained from certain classes of xanthate esters (24):

$$C_6H_{11}OCSSC_2H_4CN \xrightarrow[\text{10 min}]{\text{BF}_3} \underset{\substack{\text{cyclo-}\\\text{hexene}}}{C_6H_{10}} + \text{other products}$$

The reaction of phosgene with an aqueous solution of a xanthate and ether at 15–20°C gives the anhydrosulfide in good yield:

$$2 \text{ ROCSSNa} + \text{COCl}_2 \rightarrow (\text{ROCS})_2\text{S} + \text{COS} + 2 \text{ NaCl}$$

The anhydrosulfide is obtained in the ether layer and is thus readily separated. The anhydrosulfides are useful in the preparation of thionocarbanilates by heating with an alcoholic solution of an arylamine, for example:

$$(\text{C}_2\text{H}_5\text{OCS})_2\text{S} + \text{C}_6\text{H}_5\text{NH}_2 \rightarrow \text{C}_2\text{H}_5\text{OCSNHC}_6\text{H}_5 + \text{C}_2\text{H}_5\text{OH} + \text{CS}_2$$

This method avoids the preparation of the intermediate aryl isothiocyanate.

Many oxidizing agents convert the alkali metal xanthates to the corresponding dixanthogen:

$$2 \text{ ROCSSK} + \text{KI}_3 \rightarrow (\text{ROCS})_2\text{S}_2 + 3 \text{ KI}$$

$$2 \text{ ROCSSK} + \text{K}_2\text{S}_2\text{O}_8 \rightarrow (\text{ROCS})_2\text{S}_2 + 2 \text{ K}_2\text{SO}_4$$

The reaction is generally carried out in water, and the resulting dixanthogen separates as a solid or oil. Cooling during these reactions may or may not be necessary. Copper salts are also capable of effecting the oxidation:

$$4 \text{ ROCSSNa} + 2 \text{ CuSO}_4 \rightarrow (\text{ROCS})_2\text{S}_2 + (\text{ROCSS})_2\text{Cu}_2 + 2 \text{ Na}_2\text{SO}_4$$

Both the dixanthogen and cuprous xanthate separate out of solution together. The dixanthogen can be separated by means of its solubility in ether. Older samples of alkali metal xanthates contain some dixanthogen believed to be formed by the following reaction (19):

$$2 \text{ C}_2\text{H}_5\text{OCSSK} + \tfrac{1}{2}\text{O}_2 + \text{CO}_2 \rightarrow (\text{C}_2\text{H}_5\text{OCS})_2\text{S}_2 + \text{K}_2\text{CO}_3$$

The sulfur chlorides give higher sulfides of the xanthates. For example, the following product is obtained from sulfur monochloride and an ether suspension of the xanthate at room temperature:

$$2 \text{ ROCSSNa} + \text{S}_2\text{Cl}_2 \rightarrow (\text{ROCS})_2\text{S}_4 + 2 \text{ NaCl}$$

Mixed anhydrosulfides can be readily obtained by adding an acetone solution of the xanthate to an acetone solution of the acyl halide at −35°C (25):

$$\text{C}_2\text{H}_5\text{OCSSK} + \text{RCOCl} \rightarrow \text{C}_2\text{H}_5\text{OCSSCOR} + \text{KCl}$$

The stability of these products varies (22). When R is methyl the compound decomposes below room temperature, but when R is phenyl it is stable up to 40–45°C. Increasing the molecular weight of the alkyl group and introducing an electronegative group on the phenyl increases the stability. The aliphatic mixed anhydrosulfide can be decomposed by two different routes (25):

$$\text{C}_2\text{H}_5\text{OCSSCOR} \xrightarrow{\ \ \ } \text{RCOOC}_2\text{H}_5 + \text{CS}_2$$
$$\downarrow \Delta$$
$$\text{C}_2\text{H}_5\text{OCSSR} + \text{CO}$$

The alkyl chloroformates, ROCOCl, react with cold ethereal dispersions of the xanthates to give the fairly stable xanthogen formates:

$$(\text{CH}_3)_2\text{CHOCSSNa} + \text{C}_2\text{H}_5\text{OCOCl} \rightarrow (\text{CH}_3)_2\text{CHOCSSCOOC}_2\text{H}_5 + \text{NaCl}$$

In the reaction between xanthates and sulfonyl chlorides, the xanthates are converted to dixanthogens, while the sulfonyl chlorides are reduced to sulfinic acids and other compounds (22):

$$2 \text{ C}_2\text{H}_5\text{OCSSK} + \text{C}_6\text{H}_5\text{SO}_2\text{Cl} \rightarrow (\text{C}_2\text{H}_5\text{OCS})_2\text{S}_2 + \text{C}_6\text{H}_5\text{SO}_2\text{K} + \text{KCl}$$

The acid chlorides of the xanthic acids can be prepared by the reaction of chlorine with a dixanthogen (26):

$$(C_2H_5OCS)_2S_2 + Cl_2 \rightarrow 2\ C_2H_5OCSCl + 2\ S$$

In the Leuckart thiophenol synthesis, the reaction of xanthates with diazonium compounds may be violent. The reaction can be controlled, however, by thermal decomposition of the intermediate azo compound as it forms, and thus provides a convenient and easy method of preparing the aryl esters, which may, in turn, be saponified to the thiophenols. The crude reaction mixture has been shown to be a more complex mixture than just the aryl alkyl xanthate (27):

$$C_2H_5OCSSK + C_6H_5N_2Cl \rightarrow C_2H_5OCSS\text{-aryl} + (C_6H_5\ S)_2CO + N_2 + KCl + \text{other products}$$

Both of these esters give the thiophenols on saponification and subsequent acidification.

The reaction of the xanthate with ethylenimine leads to the thermally unstable 2-alkoxy-2-mercaptothiazolidine which on heating gives 2-mercaptothiazoline (28):

Xanthates have been added to activated double bonds (29,30):

$$ROCSSH + CH_2{=}CHCN \rightarrow ROCSSCH_2CH_2CN$$

Preparation and Manufacture

The alkali metal xanthates are generally prepared by the reaction of sodium or potassium hydroxide, an alcohol, and carbon disulfide. The initial reaction is the formation of the alkoxide, which then reacts with the carbon disulfide to give the xanthate:

$$ROH + NaOH \rightleftharpoons RONa + H_2O$$
$$RONa + CS_2 \rightarrow ROCSSNa$$

When no water is present, xanthates of very high purity are obtained (31). The presence of water favors the principal side reaction:

$$6\ NaOH + 3\ CS_2 \rightarrow 2\ Na_2CS_3 + Na_2CO_3 + 3\ H_2O$$

It has been demonstrated that ethyl alcohol reacts slowly with sodium trithiocarbonate to produce sodium ethyl xanthate (32):

$$C_2H_5OH + Na_2CS_3 \rightarrow C_2H_5OCSSNa + NaHS$$

The nature of the inorganic by-products in technical xanthates varies as the result of exposure to air. Thus, Na_2S, Na_2SO_4, and $Na_2S_2O_3$ may be found in older samples exposed to air (33).

With tertiary and some of the more complex alcohols the use of alkali hydroxides is not feasible, and it is necessary to use reagents such as sodium hydride, sodium amide, or the alkali metal to form the alkoxide:

$$2\ ROH + 2\ K \rightarrow 2\ ROK + H_2$$

Various xanthates can be readily prepared in the laboratory from alcohols (up to C_{16}), potassium hydroxide, and carbon disulfide (34).

Another and quite different method that has been reported is the preparation of xanthates from ethers (35):

$$C_2H_5OC_2H_5 + 2 CS_2 + 2 NaOH \rightarrow 2 C_2H_5OCSSNa + H_2O$$

The heavy-metal xanthates have been prepared from aqueous solutions of alkali metal xanthate and the water-soluble compound of the heavy metal desired:

$$2 C_2H_5OCSSNa + ZnCl_2 \rightarrow (C_2H_5OCSS)_2Zn + 2 NaCl$$

These salts are often soluble in organic solvents, such as acetone or chloroform.

MANUFACTURE OF ALKALI METAL XANTHATES

The commercially available xanthates are prepared from various primary or secondary alcohols. The alkyl group varies from C_2 to C_6, and the alkali metal may be potassium or sodium. Not all of the commercially available alcohols in the C_{2-6} range are available as their xanthates, but most could be if there were sufficient demand for them.

Except for a few foreign articles and reports of the Office of Technical Services, most information on the manufacture of xanthates will be found in the patent literature. The patents are numerous and cover the temperature of reaction, relative amounts of reactants, use of diluents, etc. Although the important patent on the flotation use was taken out in 1925, there are still articles and patents appearing on the manufacture of xanthates, in particular from countries of the Eastern Bloc.

Iron has generally been reported to be suitable as a material of construction for the apparatus used in the manufacture of xanthates. Cooling is required to minimize side reactions.

In one procedure the reaction is carried out in an inert diluent, such as petroleum ether (36–38). One German plant employed this method for the preparation of sodium amyl and sodium hexyl xanthate. The kettles were made of copper and equipped with a stirrer and a means for cooling. A petroleum ether slurry was made of equimolar quantities of powdered sodium hydroxide and the alcohol. Carbon disulfide (5% excess) was then added and the temperature maintained around 35–40°C. After 3 hr the product was separated by filtration, washed with petroleum ether, and dried.

For the manufacture of potassium ethyl xanthate, one German plant (39) used a 400% excess of alcohol and equimolar quantities of carbon disulfide and 50% aqueous potassium hydroxide. The reaction temperature was maintained at 40°C by cooling. This step took a half hour. The product was dried in a vacuum drum dryer, and water was removed from the recovered alcohol by distillation. The xanthate was obtained in an almost quantitative yield and assayed at 95%. It is claimed that potassium amyl xanthate can be made using almost the same ratio of reactants, water being added to solid potassium hydroxide to make it about 60% caustic potash (40).

A third procedure consists in using nearly equimolar quantities of reactants and no diluent. In one application of this method (41), a 10% excess of alcohol and carbon disulfide is used. The initial reaction of the alcohol and caustic can be carried out in conventional iron mixing equipment fitted for cooling. The next step with carbon disulfide is carried out in an iron kettle also equipped for cooling. A strong agitator is needed. The temperature is maintained around 40°C during the addition of the carbon disulfide. The reaction product is removed from the kettle, air-dried, and then may be dried in an oven or other suitable drying device. The combined reaction times vary from approximately 3 hr for the ethyl xanthates to 5 hr for the amyl derivatives.

A more recent patent covers a continuous process using an aqueous alkali metal hydroxide, carbon disulfide, and an alcohol. The reported reaction time was 0.5–10 min before the mixture was fed to the dryer (42).

A Russian study reports the use of the water–alcohol azeotrope for water removal from isobutyl or isoamyl alcohol and the appropriate alkali hydroxide to form the alkoxide prior to the addition of carbon disulfide (43).

Due to hydrate formation the sodium salts tend to be difficult to dry. Excess water over that of hydration is believed to accelerate the decomposition of the xanthate salts. The effect of heat on the drying of sodium ethyl xanthate has been studied at 50°C (44):

Time of heating, hr	Xanthate content, %
0	79.32
40	74.38
64	65.66
88	65.06
112	63.85
136	60.41

Economic Aspects

There is a lack of information on production figures for xanthates. In addition to the regular commercial facilities for the manufacture of xanthates, there are some captive production units at a few of the mines, and at least one producer of flotation reagents makes his own xanthates as intermediates for his final products. The consumption of xanthates during 1965 by the mining industry in the United States has been estimated at 7.6 million lb (45).

The prices of xanthates have reflected, to some extent, the inflationary effects on the cost of labor and raw materials. The carload, freight allowed price of the xanthates ranges from 25 to 34¢/lb.

Standards

The alkali metal xanthates are obtainable in the technical grade only, and vary in particle size from powder to pellets, the color being various shades of yellow, and the odor of varying degrees of unpleasantness. There are no published specifications for the xanthates, but each manufacturer has his own standards. The assay on the commercial materials runs about 75% xanthate content for the potassium hexyl xanthate and around 90–95% for the others.

The assay does not necessarily indicate the collector ability of a xanthate (see Flotation). Although all xanthates decompose to some extent on prolonged storage, such decomposition does not necessarily result in corresponding reduction in mineral-collecting power. This lack of correlation between xanthate content and flotation-collecting power has been noted with material produced by different methods. Apparently, some of the xanthate oxidation or decomposition products are effective as flotation agents.

The shipping regulations call for packaging numbers 10, 27, and 34 of the Consolidated Freight Classification. The material is shipped in steel drums. No placards are required.

Analysis

Although there are several analytical procedures published, many are either tedious or inaccurate. For example, when assayed by the standard KI_3 method, by the copper sulfate method (46), or by the aqueous acid method (47), certain of the inorganic impurities assay as xanthate, giving high results. When specifications are drawn up for xanthate content, the method of assay should be given.

In a satisfactory procedure developed by The Dow Chemical Company, the xanthate content is determined by dissolving the sample in acetone, removing the inorganic solids by filtration, decomposing the xanthate (filtrate) with an excess of standard acid, and back-titrating with a standard base. The components of the inorganic solid left on the filter can be determined by any standard procedure, if desired.

The water content can be determined by the Karl Fischer method (48). The nonxanthate organic material can be determined by extraction with ether.

The ultraviolet absorption spectra of xanthates have been extensively used for the assay of dilute solutions of the latter (16,17). To avoid the interference of trithiocarbonates, a nickel xanthate can be formed, extracted into heptane, and the absorbance thereof at 425 nm measured (49).

Health and Safety

The alkali metal xanthates are relatively safe to handle. The standard precautions of rubber gloves, dust mask, and goggles are sufficient when handling the solid or solution. If xanthates come into contact with the body, they may produce a dermatitis, although susceptibility varies from person to person. Contaminated clothing should be laundered before reusing. Internally, whether from dust or fumes, the xanthates act in a manner similar to carbon disulfide (see Vol. 4, p. 383).

Under regulations for the enforcement of the Federal Insecticide, Fungicide, and Rodenticide Act, products containing over 50% sodium isopropyl xanthate must bear the label "Caution. Irritating Dust. Avoid Breathing Dust, Avoid Contact With Skin and Eyes" (50). Rubber goods in repeated contact with food may contain diethyl xanthogen disulfide not to exceed 5% by weight of the rubber products (51).

Xanthate drums should be kept as cool and dry as possible. Protection from moisture is the most important factor. A combination of moisture and hot weather has been known to cause sodium ethyl xanthate to ignite spontaneously.

Uses

Outside of the importance of cellulose xanthates in the manufacture of rayon (qv) and cellophane (see Vol. 9, p. 228) the primary use for the alkali metal xanthates is as collectors in the flotation (qv) of metallic sulfide ores. Allyl amyl xanthate, a relatively new product, is used for the flotation of copper–molybdenum sulfide ores. Nickel isopropyl xanthate, KPNi, is available as an antiozonant for rubber (see Vol. 17, p. 528). Other uses are very minor. This is not so much due to the lack of activity of performance of these compounds as it is to the fact that superior products are available. Xanthates have been used or recommended for vulcanization of rubber, as herbicides, insecticides, fungicides, high-pressure lubricant additives, and for analytical procedures.

Derivatives

The derivatives of principal interest fall into three classes. The dixanthogens, $(ROCS)_2S_2$ were produced on a small scale at one time as an additive for polymers, but are apparently no longer available. The mixed anhydrosulfides, the xanthogen formates, are used chiefly in acid pulp flotation of sulfide ores; the diethyl compound, $C_2H_5OCSSCOOC_2H_5$, is the one of most importance. Dialkyl thionocarbamates are used in the flotation of copper and zinc sulfides, in particular where the rejection of iron pyrites is of importance. The two principal products are isopropyl ethylthiocarbamate and ethyl butylthionocarbamate.

Bibliography

"Xanthic Acids and Xanthates" in *ECT* 1st ed., Vol. 15, pp. 150–157, by G. H. Harris, The Dow Chemical Company.

1. A. F. McKay, D. L. Garmaise, G. Y. Paris, S. Gelblum, and R. J. Ranz, *Can. J. Chem.* **38,** 2042–2052 (1960).
2. I. Iwasaki and S. R. B. Cooke, *J. Phys. Chem.* **63,** 1321–1322 (1959).
3. V. Hejl and F. Pechar, *Chem. Zvesti* **21,** 261–266 (1967); *Chem. Abstr.* **67,** 85514b (1967).
4. A. P. Gavrish, *Ukr. Khim. Zhur.* **29,** 900–904 (1963); *Chem. Abstr.* **60,** 6264 (1964).
5. A. P. Sanzharova, *Ukr. Khim. Zhur.* **35**(1), 91–93 (1969); *Chem. Abstr.* **70,** 105824v (1969).
6. I. Tydén, *Talanta* **13,** 1353–1360 (1966).
7. I. S. Shupe, *J. Assoc. Offic. Agr. Chem.* **25,** 495–498 (1942).
8. I. Yu. Keskyula, S. B. Faerman, Ch. I. Kondrat'ev, E. L. Goncharova, R. M. Sorokina, and K. K. Chevychalova, *Trans. State Inst. Appl. Chem. (USSR)* **30,** 68–94 (1936).
9. L. Malatesta, *Chim. Ind. (Milan)* **23,** 319 (1941).
10. A. T. Pilipenko, *Zhur. Anal. Khim.* **4,** 227 (1949).
11. A. T. Pilipenko, T. P. Varchenko, E. S. Kudelya, and A. P. Kostyshina, *Zhur. Anal. Khim.* **12,** 457–461 (1957).
12. C. C. DeWitt, R. F. Makens, and A. W. Helz, *J. Am. Chem. Soc.* **57,** 796–801 (1935).
13. P. J. C. Fierens, J. Adam, and E. Royers, *Compt. Rend. Congr. Intern. Chimie Ind. 31, Liège 1958* (Published as *Ind. Chim. Belge, Suppl. 1959*) **2,** 777–780 (1959); *Chem. Abstr.* **54,** 3055 (1960).
14. I. Iwasaki and S. R. B. Cooke, *J. Am. Chem. Soc.* **80,** 285–288 (1958).
15. B. Philipp and C. Fichte, *Faserforsch. Textiltech.* **11,** 118–124, 172–179 (1960); *Chem. Abstr.* **54,** 16139 (1960).
16. R. Joedodibroto, Ph. D. Thesis, Syracuse University, Syracuse, N.Y., 1963.
17. J. Dyer and L. H. Phifer, *Macromolecules* **1969,** 111–117.
18. K. Schaum, P. Siedler, and E. Wagner, *Kolloid-Z.* **58,** 341–348 (1932).
19. A. Pomianowski and J. Leja, *Can. J. Chem.* **41,** 2219–2230 (1963).
20. G. L. O'Connor and H. R. Nace, *J. Am. Chem. Soc.* **74,** 5454–5459 (1952); **75,** 2118–2123 (1953).
21. H. R. Nace, "The Preparation of Olefins by the Pyrolysis of Xanthates. The Chugaev Reaction," in A. C. Cope, ed., *Organic Reactions,* Vol. 12, John Wiley & Sons, Inc., New York, 1962.
22. G. Bulmer and F. G. Mann, *J. Chem. Soc.* **1945,** 666, 677, 680.
23. T. Taguchi, Y. Kawazoe, K. Yoshihira, H. Kanayama, M. Mori, K. Tabata, and K. Harano, *Tetrahedron Letters* **1965,** 2717–2722.
24. *Chem. Eng. News* **47,** 41 (Sept. 22, 1969).
25. D. H. R. Barton, M. V. George, and M. Tomoeda, *J. Chem. Soc.* **1962,** 1967–1974.
26. D. Martin and W. Mucke, *Chem. Ber.* **98,** 2059–2062 (1965).
27. J. R. Cox, Jr., C. L. Gladys, L. Field, and D. E. Pearson, *J. Org. Chem.* **25,** 1083–1092 (1960).

28. F. Drawert and K. Reuther, *Chem. Ber.* **93**, 3066–3070 (1960).
29. O. Bayer, *Angew. Chem.* **61**, 229–241 (1949).
30. J. L. Garraway, *J. Chem. Soc.* **1962**, 4072–4076.
31. U.S. Pat. 2,024,923 (Dec. 17, 1935), W. Hirschkind (to Great Western Electro-Chemical Co.).
32. G. Ingram and B. A. Toms, *J. Chem. Soc.* **1957**, 4328–4344.
33. S. N. Danilov, N. M. Grad, and E. I. Geĭne, *Zhur. Obshch. Khim.* **19**, 826–842 (1949); *Chem. Abstr.* **43**, 9439 (1949).
34. L. S. Foster, *Technical Paper No. 2*, Dept. of Mining and Metallurgy Research, University of Utah, Salt Lake City, 1928.
35. U.S. Pat. 2,534,085 (Dec. 12, 1950), B. M. Vanderbilt and J. P. Thorn (to Standard Oil Development Co.).
36. W. Hensinger, *FIAT Microfilm Reel M 31*, Frames 980, 988, 1022, 1675; *U.S. Dept. Comm. Office Tech. Serv. PB Rept. 74736*, 1934–1936.
37. U.S. Pat. 1,559,504 (Oct. 27, 1925), R. B. Crowell and G. F. Breckenridge (to Western Industries Co.).
38. U.S. Pat. 2,107,065 (Feb. 1, 1938), A. J. Van Peski (to Shell Development Co.).
39. W. A. M. Edwards and J. H. Clayton, *U.S. Dept. Comm. Office Tech. Serv. PB Rept. 34023*, 28 (1945).
40. Ger. Pat. 765,196 (July 28, 1955), H. Gunther (to Deutsche Gold- und Silber-Scheideanstalt vorm Roessler).
41. I. Yu. Keskyula, S. B. Faerman, Ch. I. Kondrat'ev, and E. L. Goncharova, *Trans. State Inst. Appl. Chem. (USSR)* **30**, 29–67 (1936), *Chem. Abstr.* **33**, 8573 (1939).
42. Czech. Pat. 97,929 (Jan. 15, 1961), O. Leminger, S. Novak, J. Chalupa, Z. Kubek, L. Kraus, F. Blechta, M. Morak, and J. Malek, *Chem. Abstr.* **56**, 333 (1962).
43. N. I. Gel'perin, E. M. Idel'son, A. K. Livshits, A. T. Borisenko, L. I. Gabrielova, and V. I. Zil'berg, *S. B. Nauchn. Tr. Gos. Nauchn. Issled. Inst. Isvestn. Metal,* **1959**, 170–179; *Chem. Abstr.* **57**, 12313 (1962).
44. U.S. Pat. 1,724,549 (Aug. 13, 1929), T. W. Bartram and W. C. Weltman (to Rubber Service Laboratories Co.).
45. *Minerals Yearbook*, Vol. 1, U.S. Dept. of Interior, Bureau of Mines, U.S. Govt. Printing Office, Washington, D.C., 1965, p. 75.
46. W. S. Calcott, F. L. English, and F. B. Downing, *Eng. Mining-J. Press* **118**, 980 (1924).
47. W. Hirschkind, *Eng. Mining-J. Press* **119**, 968 (1925).
48. A. L. Linch, *Anal. Chem.* **23**, 293–296 (1951).
49. D. Kyriacou, private communication to author, 1969.
50. *Federal Register* **27**, 2267–2277 (March 9, 1962).
51. *Ibid.*, **33**(109), 8338 (June 5, 1968).

G. H. HARRIS
The Dow Chemical Company

XANTHENE DYES

The dyes of the xanthene class date from 1871, when Baeyer synthesized fluorescein (Acid Yellow 73, CI 45350) from phthalic anhydride and resorcinol (1). They are based on the xanthene (**1**) or dibenzo-γ-pyran nucleus.

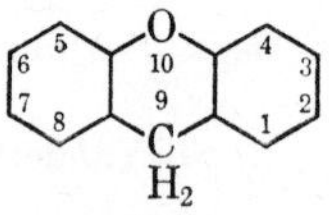

(**1**)
xanthene

The xanthenes are characterized by pure, brilliant hues and by fluorescence. They range in shade from yellow to violet.

Although the xanthenes comprise a relatively small class—the *Colour Index* (2) lists only sixty-eight xanthene dyes, in contrast to over a thousand entries for the largest class, the azo dyes—they have a broad spectrum of applications. In the textile field they are used for dyeing wool and silk directly and for dyeing cotton on a tannin mordant. Nontextile uses include the dyeing of wood, paper, and leather; the coloring of food, drugs, and cosmetics (see Colors for food, drugs, and cosmetics); biological staining; and various applications in the printing industry (see Inks) (3,4). Because of their fluorescence xanthene dyes are utilized in a number of specialty applications, such as fluorescent signs. Their high tinctorial power makes them useful as water-flow tracers (5). Many downed pilots owe their lives to the use of fluorescein as a sea marker. Xanthene lakes are used in paint, varnish, and printing applications.

Classification

In general xanthene dyes are cationic (basic) dyes. The class, however, comprises acid- and spirit-soluble dyes and mordant colors and lakes. The fluoresceins, which contain carboxyl and phenolic hydroxyl groups, are acid dyes. Acid xanthene dyes are obtained also by the introduction of the sulfonic group. Esterification of the carboxyl group yields spirit-soluble dyes. Introduction of *o*-dihydroxybenzene or salicylic acid moieties gives mordant dyes. Lakes are obtained by precipitation of hydroxyl- or carboxyl-containing dyes with salts of metals, such as lead or aluminum.

Xanthene dyes can be conveniently classified into three main groups: those containing amino groups, eg, Rhodamine B (**2**); those containing both amino and hydroxy groups, eg, Chromogen Red B (**3**); and those containing hydroxyl groups, eg, fluorescein (**4**).

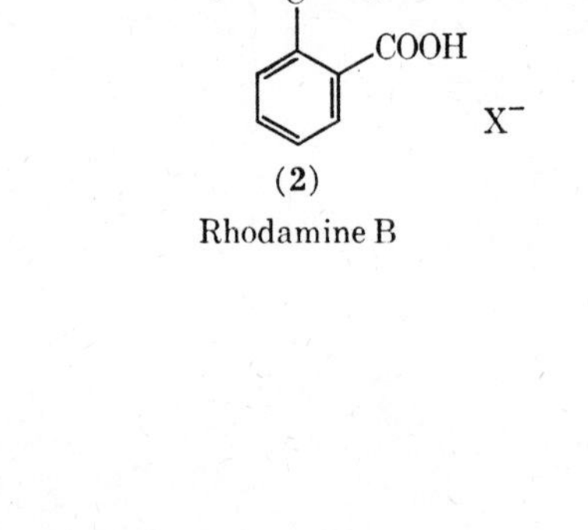

(**2**)
Rhodamine B

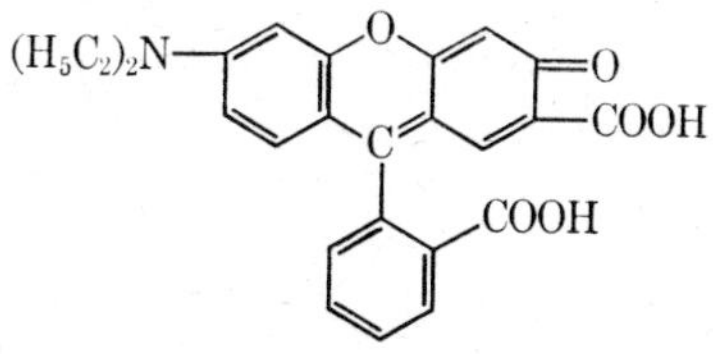

(3)
Chromagen Red B

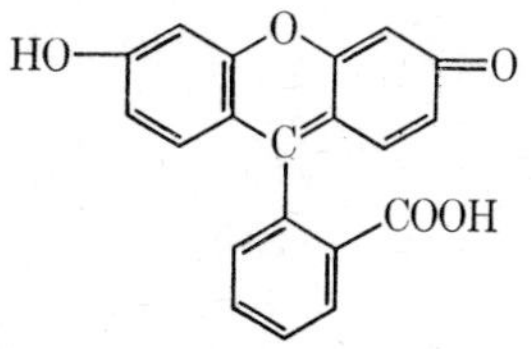

(4)
fluorescein
(neutral quinonoid form)

They are further divided by the *Colour Index* into subgroups:

1. Amino derivatives (fluorenes), ie, pyronines, succineins, sacchareins, rosamines, and rhodamines.
2. Aminohydroxy derivatives (rhodols).
3. Hydroxy derivatives (fluorones), ie, hydroxyphthaleins and anthrahydroxy-phthaleins.
4. Miscellaneous derivatives.

Color and Constitution

Xanthene dyes, like their close relatives the triphenylmethanes, are cationic resonance hybrids.

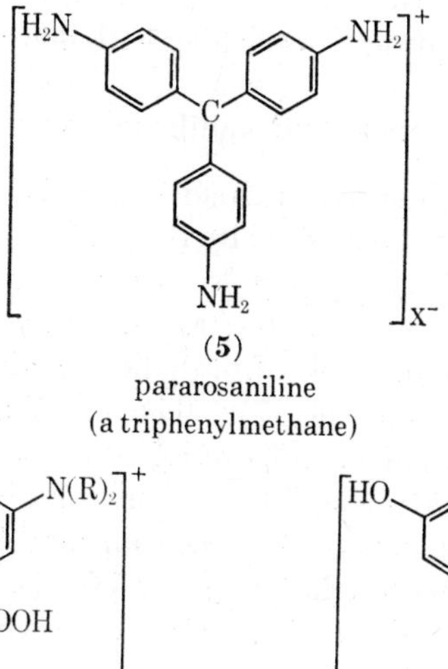

(5)
pararosaniline
(a triphenylmethane)

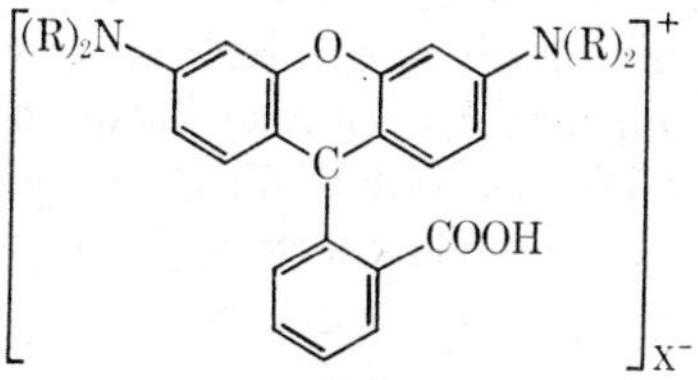

(6)
a rhodamine
(an aminoxanthene)

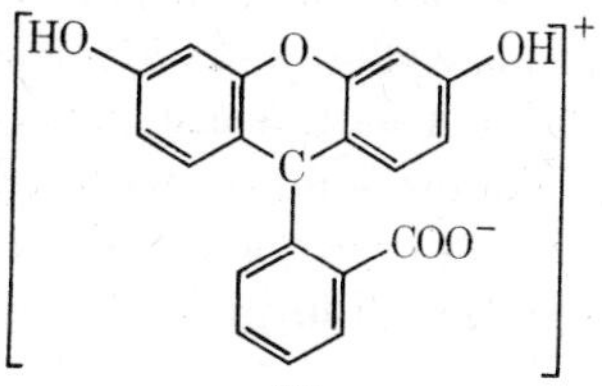

(7)
fluorescein
(a hydroxyxanthene)

The major contributing structures have the positive charge on the oxygen, nitrogen, or methane carbon atoms and are frequently drawn as such.

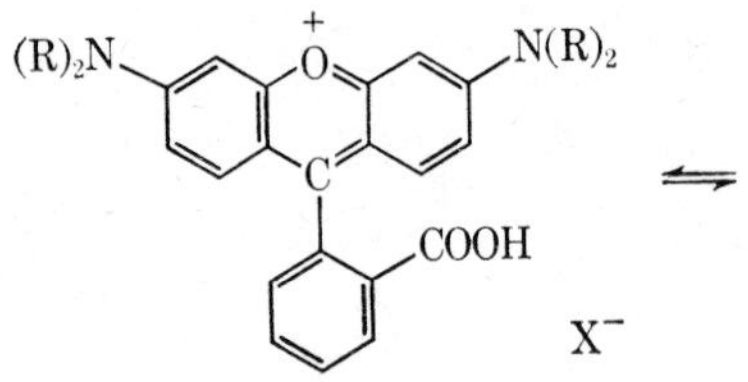

(**8**)
oxonium form

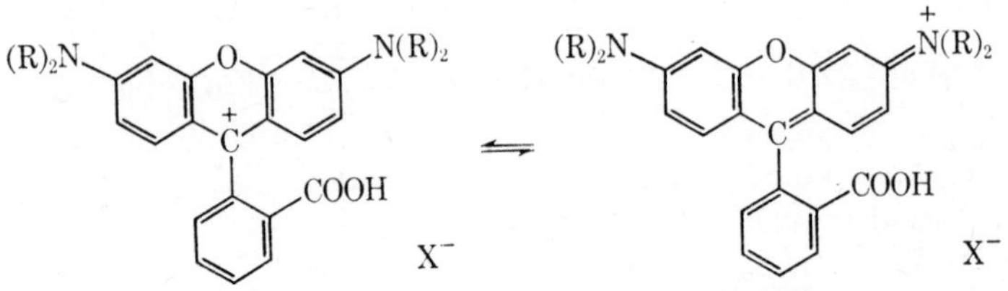

(**9**)
carbonium form

(**10**)
ammonium form

Hydroxyxanthenes can also exist in uncharged forms (see structures (**4**) and (**11**), fluorescein).

Most xanthene dyes are fluorescent. The oxygen atom, in heterocyclic systems, and ortho-para directing groups are known to contribute to fluorescence (see Fluorescent pigments). Generally they are also fugitive to light, as are many cationic dyes. Although no direct correlation has been established between fluorescence and lightfastness, many fluorescent dyes have poor lightfastness. However, a connection has been demonstrated by Rath and Brielmaier (6) who increased the lightfastness of fluorescent dyeings by pretreatment of the fiber with fluorescent quenching substances.

Important Xanthene Dyes

Hydroxy Derivatives. Fluorescein (Acid Yellow 73, CI 45350) is an acidic xanthene dye and one of the first synthesized by Baeyer in 1871. It is no longer used as a textile dye, but because of its strong fluorescence, which is detectable in high dilution (1:40,000,000), it is useful as a water-flow tracer. It may be prepared by condensing two moles of resorcinol with one mole of phthalic anhydride in the presence of a condensing agent such as zinc chloride (7). The disodium and dipotassium salts of fluorescein are known as uranine. Fluorescein has been isolated in at least two isomeric forms, yellow and red. The yellow form is obtained by precipitation of a cold alkaline solution with acetic acid. The red form may be obtained by precipitating a boiling alkaline solution with acid or by heating the yellow form. Orndorff and Hemmer (8) assigned the lactoid structure (**11**) to the yellow form and the quinonoid structure to the red form (**4**).

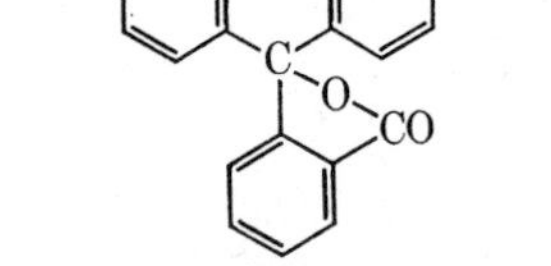

(**11**)
lactoid form

However, one would expect the lactoid form to be colorless. The exact constitution of these two forms is still in question and conflicting data exist in the literature. For example, Sklyar and Mikhailov (9), on the basis of infrared spectral data, ruled out the existence of the lactoid or quinonoid structures in favor of the zwitterion form (7), while Davies and Jones (10) concluded from their infrared evidence that the most likely structure is the lactoid form (11). Likewise, absorption spectra analyses of neutral alcohol solutions have led to two different conclusions: (1) that only the p-quinonoid form is present (11), and (2) that the two forms are in equilibrium—a colored, fluorescent quinonoid form and a colorless, nonfluorescent form (12).

Nagase et al. (13) isolated, besides the yellow and red forms, an orange modification. They assigned to these the lactoid, o-quinonoid, and p-quinonoid structures, respectively. On the basis of absorption spectral analysis in dioxane at varying pHs Zanker and Peter (14) obtained evidence for the existence of the four possible prototropic forms of fluorescein: the neutral lactone (11), the cationic form (12), and the mono- (13) and dianions (14).

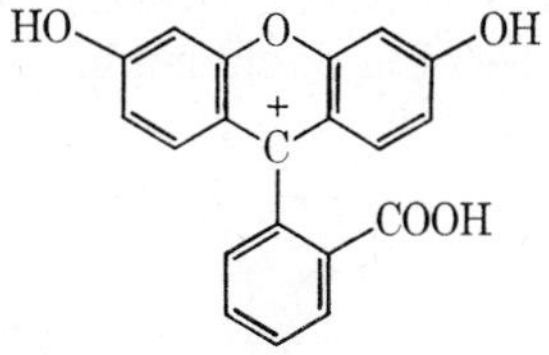

(12)   (13)   (14)

cation   monoanion   dianion

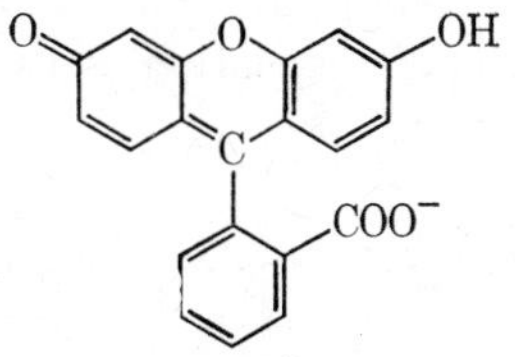

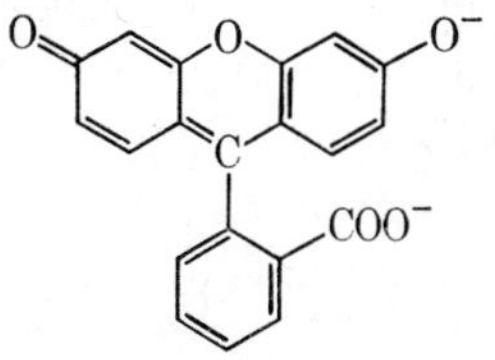

They concluded that the yellow form is the lactone and that it is converted to the quinone (red form) at high temperatures via lactone ring-opening and proton shift. The yellow color of the lactone, however, is still unexplained.

Several important dyes are obtained from fluorescein by halogenation. Eosine (15) (Acid Red 87; CI 45380) is the disodium salt of 2,4,5,7-tetrabromofluorescein. It may be obtained by treating fluorescein, in acidic aqueous alcohol, with bromine water and sodium chlorate solution, followed by conversion to the sodium salt (15). A colorless lactoid form of eosine has been isolated (8).

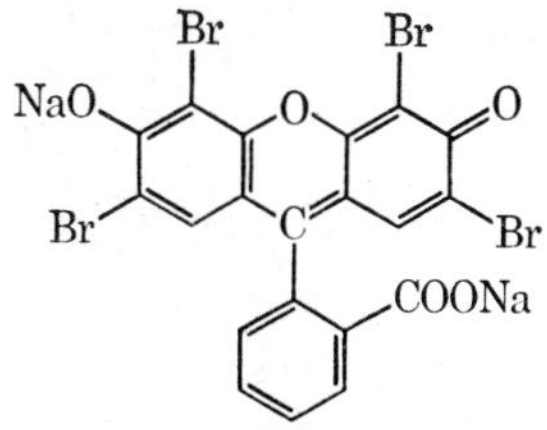

(15)

eosine

Eosine is used in the production of red lakes and toners (see Inks). These pigments are used extensively in printing inks. The lead lake of eosine, CI Pigment Red 90, is prepared by precipitation of eosine with a soluble lead salt.

Rose bengal (16) (Acid Red 95, CI 45440) is the disodium salt of dichloro-2,4,5,7-tetraiodofluorescein. It may be made by tetraiodination of tetrachlorofluorescein (16).

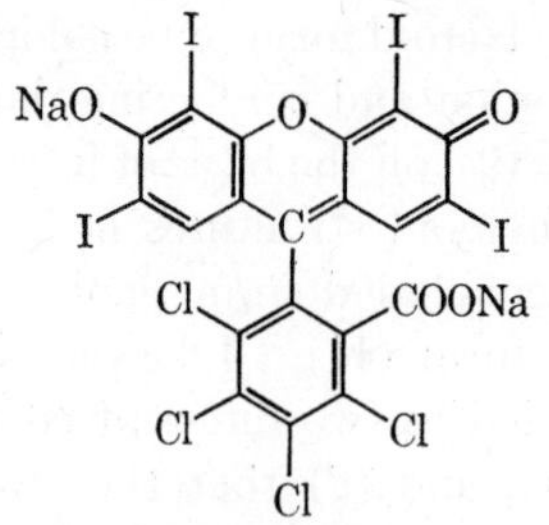

(16)
rose bengal

Mercurochrome, although not a dyestuff, is a widely used but not very effective mercurial antiseptic (see Antiseptics and disinfectants). It is the sodium salt of 4-hydroxy-2,7-dibromofluorescein.

Amino Derivatives. The most important amino-substituted xanthene dyes are the rhodamines. They are basic dyes, but acid rhodamines may be obtained by introduction of the sulfonic acid group. Rhodamine B **(17)** (Basic Violet 10, CI 45170) is prepared commercially by the condensation of two moles of m-diethylaminophenol with phthalic anhydride (17).

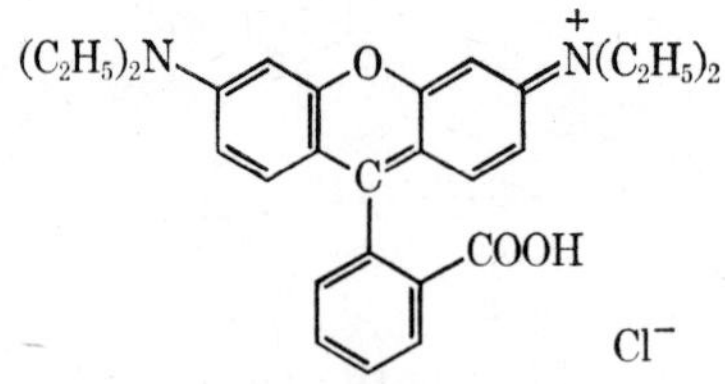

(17)
Rhodamine B

An alternative route is via the action of diethylamine on 3',6'-dichlorofluoran under pressure. The free base is Rhodamine B Base (Solvent Red 49, CI 45170B).

Rhodamine 6G **(18)** (Basic Red 1, CI 45160) is an esterified rhodamine. It is prepared by condensing 3-ethylamino-p-cresol with phthalic anhydride and esterifying the product with ethyl chloride under pressure (18).

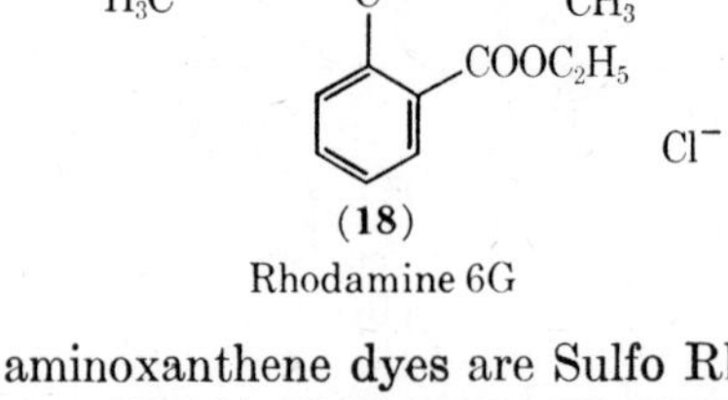

(18)
Rhodamine 6G

Two important acidic aminoxanthene dyes are Sulfo Rhodamine B **(19)** (Acid Red 52, CI 45100) and violamine **(20)** (Acid Violet 9, CI 45190). Sulfo Rhodamine B is obtained, not by sulfonating the rhodamine base, but by condensing m-diethylaminophenol with benzaldehyde-2,4-disulfonic acid, ring-closing in strong sulfuric acid, and oxidizing (19). Violamine is prepared by condensing o-toluidine with 3',6'-dichlorofluoran, sulfonating the base obtained, and converting it to the sodium salt (20,21). It is used mainly for dyeing wool and coloring antifreeze.

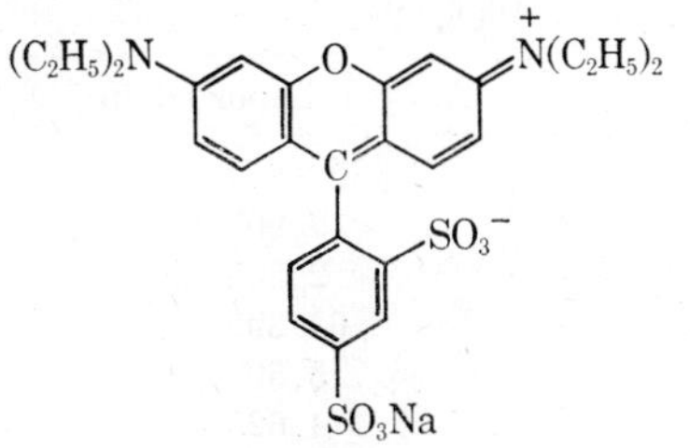

(19)
Sulfo Rhodamine B

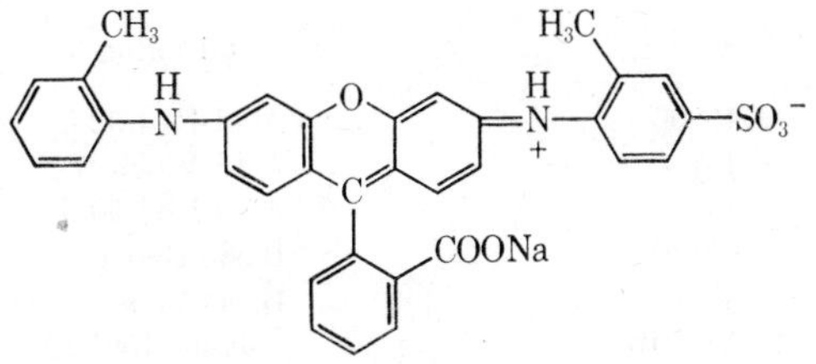

(20)
violamine

Aminohydroxy Derivatives. Aminohydroxyxanthenes, rhodols, are currently not of commercial importance in the United States. An example is Chromogen Red B (**3**) (Mordant Red 15; CI 45305). It is obtained by condensing *m*-diethylamino-phenol with phthalic anhydride and condensing the product with 2,4-dihydroxy-benzoic acid (22).

Economic Aspects

Xanthene dyes represent a small but important segment of the total dyes production and sales. Currently over 1500 synthetic dyestuffs are manufactured in the United States. Of these, about ten are xanthene dyes. U.S. dye sales in 1967 amounted to $332,000,000. Xanthene dyes accounted for $4,502,000 or about 1.4% (23). Table 1 lists the xanthene dyes manufactured in the United States in 1967 (23). Table 2 lists the xanthene dye imports of 1000 lb or more in 1968 (24).

Table 1. Xanthene Dyes Manufactured in the United States in 1967 (23)

CI no.	Cl name	Classical name	Manufacturers[a]
45100	Acid Red 52	Sulfo Rhodamine B	GAF
45160	Basic Red 1	Rhodamine 6G	BASF, DUP, GAF
45170	Basic Violet 10	Rhodamine B	ACCO, DUP, GAF
45170	Pigment Violet 1	Rhodamine B	ACCO, DUP, GAF
45170B	Solvent Red 49	Rhodamine B Base	ACCO, DUP, GAF, DSC
45350	Acid Yellow 73	fluorescein	ACS, GAF, NYC, SDH
45350	Acid Yellow 73	uranine	SDH, NYC
45380	Acid Red 87	eosine	AMS, NYC, SDH
45380	Pigment Red 90	eosine (lead lake)	SDH
45440	Acid Red 94	rose bengal	NYC

[a] *Manufacturers Code:*
ACCO American Cyanamid Company
ACS Allied Chemical Corporation, Specialty Chemicals Division
AMS Martin-Marietta Corporation, Ridgeway Color & Chemical Division
BASF BASF Corporation
DUP E. I. du Pont de Nemours & Company, Inc.
DSC Dye Specialties, Inc.
GAF GAF Corporation
NYC Tenneco Chemicals, Inc., New York Color Division
SDH Sterling Drug, Inc., Hilton Davis Chemical Company Division

Uses

Table 3 lists the principal and uses of xanthene dyes which were manufactured in the United States in 1967.

Table 2. U.S. Imports of Xanthene Dyes in 1968, (24)

CI no.	CI name	Amount imported, lb
45100	Acid Red 52	22,827
45190	Acid Violet 9	7,907
45220	Acid Red 50	2,508
45160	Basic Red 1	64,395
45175	Basic Violet 11	5,590
45170B	Solvent Red 49	4,625

Table 3. Uses of Xanthene Dyes in 1967

CI no.	CI name	Classical name	Uses
45100	Acid Red 52	Sulfo Rhodamine B	mainly a wool dye for bright baby-pink shade on baby clothing and to brighten up other reds used as base on wool; tracing water currents; nylon jackets and fluorescent parachutes
45160	Basic Red 1	Rhodamine 6G	paper dye for pinks and tinting whites; for manufacturing lake Pigment Red 81; for making daylight fluorescent pigments with resin; leather dyeing, to some extent; bright printing inks; tracing current in water pollution studies
45170	Basic Violet 10	Rhodamine B	paper dyeing; leather dyeing, to some extent; fluorescent pigments with resin; dyeing waxes; dyeing antifreeze; tracing water currents in antipollution studies
45170B	Solvent Red 49	Rhodamine B Base	ball-point pen inks; fluorescent pigments with resins; plastics coloring; shoe polish and waxes
45350	Acid Yellow 73	fluorescein	dye markers for locating downed aviators in ocean; tracing water leaks in sewers and water systems
45380	Acid Red 87	eosine	making D&C Red for cosmetics; making lakes (Pigment Red 90); coloring fertilizer; dyeing bright shades on wool
45440	Acid Red 94	rose bengal	dyeing wool; small amount for dyeing paper and leather

Table 4. Xanthene Dyes Certified for Use in the Food, Drug, and Cosmetic Industry

Official FDA name	CI no.	CI name
FD&C Red No. 3	45430	Food Red 14
D&C Red No. 3	45430	Lake of FD&C No. 3
D&C Orange No. 5	45370	Acid Orange 11
D&C Orange No. 10	45425A	Solvent Red 73
D&C Orange No. 11	45425	Acid Red 95
D&C Red No. 19	45170	Basic Violet 10
D&C Red No. 21	45380A	Solvent Red B
D&C Red No. 22	45380	Acid Red 87
D&C Red No. 27	45410A	Solvent Red 48
D&C Red No. 28	45510	Acid Red 92
D&C Red No. 37	45170B	Solvent Red 49
D&C Yellow No. 7	45350	Acid Yellow 73 (fluorescein)
D&C Yellow No. 8	45350	Acid Yellow 73 (uranine)

Several xanthene dyes have been certified for use in the food, drug, and cosmetic industry. These are listed in Table 4. See Colors for foods, drugs, and cosmetics.

Bibliography

"Xanthene Dyes" in *ECT* 1st ed., Vol. 15, pp. 136–149, by W. G. Huey and Shirley K. Morse, General Aniline & Film Corporation.

1. A. Baeyer, *Chem. Ber.* **4**, 555, 658 (1871).
2. *Colour Index*, 2nd ed., The Society of Dyers and Colourists, Bradford, England, and The American Association of Textile Chemists and Colorists, Lowell, Mass., 1956.
3. U.S. Pat. 3,244,728 (April 5, 1966), J. R. Johnson and E. J. Gosnell (to Burroughs Corp.).
4. Belg. Pat. 613,013 (Feb. 15, 1962), P. L. Foris (to National Cash Register).
5. U.S. Pat. 3,367,946 (Feb. 6, 1968), H. I. Stryker (to E. I. du Pont de Nemours & Co., Inc.).
6. H. Rath and H. J. Brielmaier, *Melliand Textilber.* **42**, 911 (1961).
7. F. O. Robitschek, "Dyestuffs Manufacturing Processes of the I. G. Farbenindustrie," *U.S. Dept. Comm. Office Tech. Serv. PB Rept. 70135, Reel 18G* (1943), Frames 1125–1135.
8. W. R. Orndorff and A. J. Hemmer, *J. Am. Chem. Soc.* **49**, 1272 (1927).
9. Y. E. Sklyar and G. I. Mikhailov, *Zh. Org. Khim.* (in Russian) **2** (5), 899–901 (1966); *Chem. Abstr.* **65**, 10556e (1966).
10. M. Davies and R. L. Jones, *J. Chem. Soc.* **1954**, 120.
11. W. R. Orndorff, R. C. Gibbs, and C. V. Shapiro, *J. Am. Chem. Soc.* **50**, 819 (1928).
12. R. Lucas, *Compt. Rend.* **205**, 864 (1937).
13. Y. Nagase, T. Ohno, and T. Goto, *J. Pharm. Soc. Japan* **73**, 1033–1039 (1953); *Chem. Abstr.* **48**, 9983d (1954).
14. V. Zanker and W. Peter, *Chem. Ber.* **91**, 572–580 (1958).
15. Ref. 7, Frames 1078–1087.
16. Ref. 7, Frames 1062–1072.
17. *BIOS (British Intelligence Objectives Subcommittee) Report 959; PB 63858* (1945), pp. 32–36.
18. Ref. 7, Frames 894–903.
19. Ref. 17, pp. 66–70.
20. Ger. Pat. 49047 (Aug. 21, 1889); *Frdl.* **2**, 79 (1887–1890).
21. Ref. 7, Frames 1148–1150.
22. Ref. 7, Frames 1277–1283.
23. *Synthetic Organic Chemicals—U.S. Production and Sales, Dyes 1967, TC Publ. 295*, U.S. Tariff Commission, U.S. Govt. Printing Office, Washington, D.C., 1969.
24. *Imports of Benzenoid Chemicals and Products, 1968, TC Publ. 290*, U.S. Tariff Commission, U.S. Govt. Printing Office, Washington, D.C., 1969.

General References

K. Venkataraman, *The Chemistry of Synthetic Dyes*, Vol. 2, Academic Press, Inc., New York, 1955, Chap. 24.

H. A. Lubs, *The Chemistry of Synthetic Dyes and Pigments*, Reinhold Publishing Corp., New York, 1955, pp. 291–301.

R. D. Elderfield, *Heterocyclic Compounds*, John Wiley & Sons, Inc., New York, 1951, Chap. 13.

U.S. Pats: Classes 260/335 and 336.

FRANK F. CESARK
American Cyanamid Co.

XENON, Xe.   See Helium group gases.

XENOTIME.   See Rare earth elements, Vol. 17, p. 147.

XERODUPLICATING.   See Reprography, Vol. 17, p. 339.

XEROGRAPHY.   See Reprography, Vol. 17, p. 342.

X-RAY ANALYSIS

The principal applications of x rays in chemistry are based on the study of the pattern of scattered (diffracted) rays produced when an x-ray beam strikes a specimen. The wave lengths of x rays are about the same as the distances between atoms in solids or liquids, and for this reason the diffracted rays form a pattern which is determined by the arrangement of atoms in the specimen. The pattern of diffracted rays may be used empirically as a "finger-print" for identification of the substance or substances in the specimen; or, interpreted in much more detail, it may be used to give information on the structure of the material—the arrangement of the atoms, the stereochemistry of molecules, and the interatomic distances. This is the application generally known as x-ray analysis. Another application of x rays, quite different in principle, is the identification and quantitative analysis of elements by their x-ray fluorescent emission and absorption spectra (see p. 456).

For the x-ray examination of engineering materials see Radioisotopes.

X-ray Diffraction: Principles and Experimental Methods

X rays are electromagnetic waves of very short wave length, about 1 Å (10^{-8} cm). When an x-ray beam passes through matter, a small fraction of it is scattered by the electrons of each atom. One portion is scattered with a change of wave length, and the other without a change of wave length. The portion scattered with a change of wave length (Compton scattering) does not concern us; since there are no phase relationships between the rays scattered by different atoms, this portion is said to be incoherent, and it tells us nothing about the structure of the scattering material; it forms part of the background intensity, fortunately a small part, and can be ignored (although its intensity has to be corrected for in some quantitative applications). It is the portion scattered without change of wave length, the coherent portion, that is our concern: the phase relations between rays scattered by different atoms result in destructive interference in certain directions and reinforcement in others, and the resulting pattern of diffracted rays depends on the spatial arrangement of the atoms as well as on the wave length of the x rays. The pattern is simplest to interpret when the x-ray beam is monochromatic,; but, in practice it need not be strictly monochromatic, and the radiation emitted when a metallic target is bombarded by electrons (radiation containing a high intensity of a narrowly defined dominant wave length together with some "white" radiation with a wide range of wave lengths) is suitable. Copper K radiation is most frequently used; the dominant wave length, 1.54 Å (Kα), is actually a very close doublet, but for most purposes it can be treated as a single wave length; the only other sharply defined wave length, Kβ, is removed by a nickel foil filter 0.02 mm thick. A collimator tube giving a beam 0.5 mm in diameter is suitable for most purposes, and the pattern of diffracted x rays produced by a specimen of about the same diameter is either recorded on photographic film or explored with a Geiger counter linked to a chart recorder.

The most detailed and sharply defined diffraction patterns are given by crystalline solids. In a crystal the atoms are precisely arranged in a three-dimensional repeating pattern based inevitably on straight lines and planes, and the diffracted rays may be regarded as reflections by these planes; but, owing to the three-dimensional character of the pattern, a diffracted ray is produced (see Fig. 1) only when a set of parallel reflecting planes is at a precisely defined angle θ to the incident beam, given by the Bragg

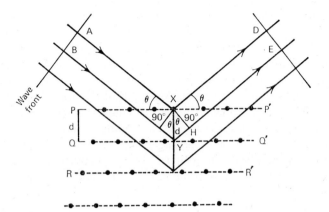

Fig. 1. Reflection of x rays by a crystal (1). Waves reflected by any one plane QQ' must be exactly one wave length (or a whole number of wave lengths) behind those reflected by the next plane above (PP') and the same amount in front of those from the next plane below (RR'). Path difference $= GY + YH = 2d \sin \theta$ which must equal $n\lambda$.

equation: $\lambda = 2d \sin \theta$, where λ is the x-ray wave length, and d the spacing of a set of planes; they are not produced at *any* angle of incidence, as they are by a two-dimensional optical diffraction grating. Consequently, if a single crystal is turned about in a monochromatic x-ray beam, in most positions it gives no diffracted rays; but occasionally, when one of the sets of planes is at the appropriate angle θ to the incident beam, a diffracted ray flashes out at an angle 2θ to the undiffracted transmitted beam. It is for this reason that even a crystalline powder in which the little crystals are oriented at random gives sharply defined diffracted beams: they form a set of cones around the incident beam, each cone being the combined reflections from many little crystals all oriented so that one particular crystal plane makes the correct angle with the incident beam. Each different type of crystal plane requires a different angle of incidence θ, and therefore gives a diffracted or reflected ray at a different angle 2θ to the undiffracted transmitted beam; the set of cones thus defines the set of reflecting planes in the crystal species. The cones of diffracted rays are either recorded simultaneously on a cylindrical strip of film around the specimen (which thus shows arcs where the cones intersect it), or successively by a Geiger counter on a rotating arm. The record gives a set of values of 2θ, which is readily converted by the Bragg equation to a set of plane-spacings d which is characteristic of the crystal species.

A rotating single crystal gives diffracted spots, not arcs. The usual arrangement in an x-ray goniometer is to rotate the crystal around one of its principal axes which is also at right angles to the incident x-ray beam; the spots produced on a cylindrical photographic film lie on lines, and their positions not only define the angles of diffraction but also partially define the orientation of the crystal planes responsible for them. There are also various types of moving-film goniometers in which a part of the pattern is selected by screening, and coupled movements of the film and the rotating crystal have the effect of sorting out the spots still further, completely defining the relative orientations of the crystal planes responsible for the diffracted beams. By rotation successively about two or three different crystal axes, a complete knowledge of the diffracting characteristics of a crystal (complete at any rate within the limitations imposed by the wave length of the x rays and the geometry of the camera) is obtained (2).

Noncrystalline substances, gases, liquids, and amorphous and glasslike solids, give diffuse diffraction patterns, because there is no regular repeating pattern of atoms in them. Nevertheless the distribution of diffraction intensity with angle of diffraction is usually far from featureless and often shows marked maxima and minima. The reason is that, although there is no precise repeating pattern of atoms in such substances, nevertheless the limitations of atomic size ensure that the arrangement of atoms is not random: atoms joined by chemical bonds always have their centers 1–2 Å apart, while atoms in neighboring molecules are usually 2.5–4 Å apart; moreover the stereochemical limitations ensure local precision of arrangement. It is these limitations which are responsible for the maxima and minima in the diffraction patterns, although it must not be assumed that maxima in the diffraction intensity at particular angles correspond directly to interatomic distances.

Identification and Analysis

Each different crystalline species gives its own characteristic powder photograph, consisting of a number of lines of different intensity; the scale of the pattern taken in a particular camera depends on the wave length of the x rays (longer wave lengths giving more spread-out patterns), but the plane spacings derived from the measured pattern by the Bragg equation are constant for the crystal, and the relative intensities of the lines, although not absolutely independent of wave length, are sufficiently nearly constant to be recognizably characteristic. Measurement with a millimeter scale is usually sufficient for identification purposes; and relative intensities can be estimated by eye. Geiger counter records provide more accurate information. Direct comparison with the patterns of known substances is the simplest method of identification; but the scope of the method is greatly enlarged by comparison of data with published information classified in a card index issued by the American Society for Testing and Materials. In this index there is one card for each substance, filed according to the magnitude of the spacing of the most intense line. In each group of cards corresponding to a particular spacing the cards are sorted according to the spacing of the second strongest line, etc. Cards are also available which make use of I.B.M. machines and sorters of the Key-Sort type. In practice it is often sufficient to look up the spacing of the strongest line; several substances may have this spacing, but the rest of the spacings and the relative intensities, which are listed on the cards, usually eliminate all but one of them.

True mixtures, in which the substances are present in separate crystals, give powder patterns in which all the lines of all the different crystal species appear together; identification of several constituents can usually be effected by successive elimination of sets of lines. But mixed crystals containing varying amounts of different substances in the same crystal give patterns in which the lines are in intermediate positions; there is no direct, straightforward method of dealing with such cases if there is no previous information on the crystal species present.

The x-ray method of identification is nondestructive, it needs a very small amount of materials, and, as compared with chemical methods, it identifies crystal species, not elements or groups; in mixtures, therefore, it tells which elements or groups are linked together in crystals.

The proportions of different substances in mixtures may be estimated from the relative intensities of their lines in the composite powder patterns; but there is no simple relation between the relative intensities of particular lines and the proportions of the substances; it is necessary to take patterns of standard mixtures of known proportions of the constituents to establish the relations empirically. This method of analysis is particularly useful when ordinary chemical methods are inadequate or not applicable, as for complex mixtures where the different elements or groups may be associated in various ways all compatible with the same chemical composition, and for mixtures of polymorphous forms of one and the same substance. The composition of mixed crystals (such as certain alloys) can be estimated from the positions of the lines, provided again the relations between line position and composition have already been established. In such cases it may be preferable to interpret the pattern as far as the determination of the dimensions of the unit cell (see below).

The powder method of identification has been much used for alloys, minerals, and inorganic substances generally. Its use for organic substances is less developed, partly owing to lack of data on the enormous numbers of substances. When the data become available, it seems desirable to subdivide the index into sections for the different classes of substances to avoid too unwieldy an index.

When crystals large enough to be handled individually are available, it may be preferable to use the x-ray diffraction patterns of single crystals, rather than to reduce the substance to a fine powder. But a systematic index of single crystal diffraction patterns, on the empirical basis used for powder photographs, is hardly feasible, and for this reason it will usually be necessary to interpret the patterns as far as the determination of unit cell dimensions, as described below. An index is available in which substances are arranged in order of their unit cell dimensions and crystal symmetry (3). Direct comparison of a diffraction pattern with those of authentic specimens of known substances is, of course, the most certain method of identification, for single crystals no less than for powders.

Structure Analysis

From the positions and relative intensities of the diffracted x-ray beams, it is often possible to deduce the distribution of electron density within a crystal, and thus to arrive at a very complete knowledge of its chemical structure. Each atom is a spherical concentration of electron density; the total number of electrons in a particular spherical concentration identifies the atom, while the position of maximum electron density pinpoints its center. Thus, not only the chemical constitution, but also a quantitative account of the stereochemistry is obtained. In some circumstances, which will be defined later, the experimental data can be combined in a direct way to give the three-dimensional distribution of electron density; in these circumstances, the x-ray method can take the place of much chemical work on the constitution. But direct structure determination is not always possible, and in these circumstances it is necessary to use trial methods, in which the intensities of diffracted beams are calculated for various postulated structures and compared with measured intensities, the criterion of correctness being the closest agreement of calculated and measured intensities of all the reflections. The complexity of structures soluble by such methods is limited; for complex organic compounds it is usually necessary to know the chemical constitution (the scheme of primary bonds linking the atoms) in order to solve the crystal structure, and so most of the specific contributions of x-ray analysis to chemical knowledge are confined to stereochemistry, to the precise distances between atoms and the spatial distribution of the bonds. In some cases, however, it has proved possible to settle certain details of chemical constitution by such methods; and, with the rapid advance in

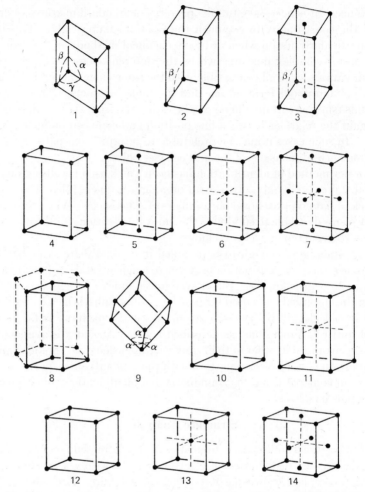

Fig. 2. The seven unit cell types and fourteen space lattices (1). 1, Triclinic (*P*); 2–3, monoclinic (*P* and *C*); 4–7, orthorhombic (*P*, *C*, *I*, and *F*); 8, hexagonal (*C*); 9, rhombohedral (*R*); 10–11, tetragonal (*P* and *I*); 12–14, cubic (*P*, *I*, and *F*). (The capital letters denote lattice type: *P* for primitive, *I* for inner- or body-centered, *C* for *C*-face centered, *F* for centered on all faces.)

methods of interpretation, ever more complex structures are being solved, sometimes on the basis of an incomplete initial knowledge of the chemical constitution.

Even if detailed interpretation in this sense cannot be achieved, partial interpretation often yields useful information; the process of interpretation falls into well-marked stages, each of which provides a different type of information. These stages, and the information obtainable at each, will be considered in turn.

The Unit Cell (see also Crystals). A crystal is a repeating pattern of atoms in space: the same grouping of atoms is repeated over and over again in all directions and always in straight lines (since only straight-line arrangements are capable of infinite exact repetition). The space pattern may be divided by sets of parallel planes into units of pattern each of which contains the smallest group of atoms from which, by exact contiguous repetition, the whole structure can be built up. Just as, for a plane pattern such as a wallpaper design, two sets of parallel lines give a unit of pattern which is always a parallelogram, so for a space pattern the unit is always a parallelepiped, a "box" bounded by three pairs of parallel planes. The shape and dimensions of the box, that is, the lengths of its three different sorts of edges and the angles between them, are different for each different crystal species; in some crystals the box is a cube, in others it is rectangular with unequal edges, in others one of the angles is not a right angle, and so on.

There are many ways of dividing up a crystal structure in this manner; but in practice it is obviously convenient to accept a *unit cell* which conforms to the symmetry of the crystal: thus, for a crystal having cubic symmetry, it is obviously better to use a cubic unit cell than one of the nonrectangular ones that could be drawn. The recognition of the paramount importance of crystal symmetry has a further result: for some crystals the accepted unit cell is one that contains more than one unit of pattern, simply because no cell containing only one unit of pattern conforms to the crystal symmetry. If each unit of pattern is symbolized by a *lattice point* (it need not be the center of any one atom of the pattern unit, but any convenient reference point), then the whole crystal structure can be represented by an array of points, which for simple structures of one lattice point per cell can be located at the corners of all the unit cells; for compound unit cells containing more than one lattice point per cell, the additional points are at either the face centers or the body center of the cell. The seven types of unit cell and the fourteen space lattices (first recognized by Bravais) are shown in Figure 2. The relations between the lengths of the unit cell edges (axes) and the angles between them, in the seven different types, are shown in Table 1 (α = angle bc, $\beta = ac$ $\gamma = ab$).

Table 1. Types of Unit Cells

triclinic	axes a, b, c all unequal	angles α, β, γ not $90°$
monoclinic	axes a, b, c all unequal	β not $90°$; $\alpha = \gamma = 90°$
orthorhombic	axes a, b, c all unequal	$\alpha = \beta = \gamma = 90°$
hexagonal	$a = b$; c (hexagonal axis) different	$\gamma = 120°$; $\alpha = \beta = 90°$
rhombohedral	$a = b = c$	$\alpha = \beta = \gamma = $ not $90°$
tetragonal	$a = b$; c (tetragonal axis) different	$\alpha = \beta = \gamma = 90°$
cubic	$a = b = c$	$\alpha = \beta = \gamma = 90°$

The different types of unit cells correspond to the different crystal "systems" which were originally distinguished on the basis of morphological measurements. The bounding faces of crystals are planes which are related in a simple way to the unit cell, so that from the measured interfacial angles it is possible to deduce a possible unit cell shape (although the morphologists did not actually use the term "unit cell" but spoke simply of relative axial lengths and interaxial angles). The shape of the unit cell deduced from morphological measurements is often identical with that of the true unit cell which was afterwards deduced by x-ray measurements; but, even when the two are not identical, there is always a simple relation between them, one of the morphological axes being perhaps twice as long or half as long as it should be in relation to the others.

Diffracted x-ray beams may be regarded as reflections by the various sets of planes which can be put through the lattice points. The spacings of the various sets of planes are calculated directly by the Bragg equation, and the first objective in the interpretation of x-ray diffraction photographs is to deduce (perhaps from the set of spacings alone or, better, from this and any information on the orientation of the reflecting plane

given by single crystal patterns) the unit cell implied by this information. Note that
the unit cell dimensions are deduced from the positions of the x-ray reflections without
reference to their intensities.

The nomenclature used in referring to the various planes is illustrated in Figure 3:
for any one set of planes, proceed along each axial direction in turn, and count the num-

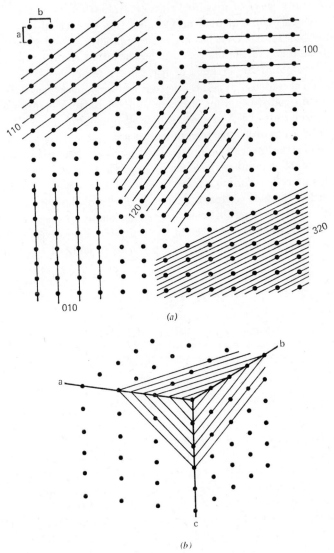

Fig. 3. Nomenclature of crystal planes (1). (*a*) Some planes of type *hk*0 seen in projection;   (*b*) a
general plane 312.

ber of planes crossed between one lattice point and the next; these numbers are the
index numbers, which are given in order of the axes *a,b,c*. The relations between plane-
spacings *d*, indexes *hkl*, and the unit cell dimensions *a*, *b*, and *c* are, for the more symmetrical crystals, fairly simple expressions:

cubic $\qquad d = \left(\dfrac{h^2 + k^2 + l^2}{a^2} \right)^{-\frac{1}{2}}$

tetragonal $\qquad d = \left(\dfrac{h^2 + k^2}{a^2} + \dfrac{l^2}{c^2} \right)^{-\frac{1}{2}}$

hexagonal $\qquad d = \left(\dfrac{4}{3} \dfrac{h^2 + k^2 + hk}{a^2} + \dfrac{l^2}{c^2} \right)^{-\frac{1}{2}}$

Reflections from planes with large spacings and small indexes are at small angles to the incident beam, while those from planes with small spacings and large indexes are at large angles.

Even if only a powder photograph is available, so that the only information is a set of plane-spacings, the problem can usually be solved for such crystals. It is of course simplest for cubic crystals, for which the plane-spacings are all in simple ratios (the squares of the reciprocal spacings are in the ratios $1:2:3:4:5:6:8 \ldots$). For tetragonal and hexagonal crystals, there are two variables, a and c, so that plane graphical methods are appropriate; charts (1,4) can be used which show, on a logarithmic scale, the relative spacings for a wide range of axial ratios c/a. For orthorhombic crystals the equation

$$d = \left(\dfrac{h^2}{a^2} + \dfrac{k^2}{b^2} + \dfrac{l^2}{c^2} \right)^{-\frac{1}{2}}$$

involves the three variables a, b, and c, so that plane chart methods are no longer applicable; trial methods, in which simple indexes are postulated for the first few reflections, are used, and there are ingenious methods for systematizing the process (5). For the less symmetrical monoclinic and triclinic crystals, the number of variables is still greater and the problem correspondingly more complex, but it is possible, with a sufficiently accurate set of plane-spacings, to arrive at a solution by a very ingenious procedure due to Ito (6). As soon as indexes have been assigned to the reflections, the unit cell dimensions can be calculated from the appropriate equations.

The determination of unit cell dimensions for the less symmetrical crystals is greatly simplified if single-crystal diffraction patterns are available. To begin with, a diffraction photograph, taken on a rotation goniometer, of a crystal rotated around a principal axis shows spots arranged in "layer lines," the distances between which give directly the length of the axis. For the rest, indexing of the spots is accomplished by the use of a concept known as the *reciprocal lattice*. Although an adequate treatment of this subject is beyond the scope of this article, the concept is of such importance in the interpretation of x-ray diffraction data that a definition and some comments are called for. From a point of origin, imagine lines drawn outwards perpendicular to the lattice planes; along these lines points are marked at distances inversely proportional to the spacings of the lattice planes; the points thus obtained form a lattice, that is, they fall on sets of parallel planes. This is the reciprocal lattice; its appropriateness as a theoretical tool for the interpretation of x-ray diffraction patterns lies in the fact that reciprocal lattice points representing planes with high indexes and small spacings lie far from the origin, while those representing planes with small indexes lie close to the origin, just as in the diffraction photograph itself. From any diffraction photograph it is a simple matter to derive, by graphical methods, the reciprocal lattice coordinates for all the reflections, and, once the reciprocal lattice is defined, the real lattice follows straightforwardly. There are, indeed, special diffraction cameras which, by appropriate mechanical means, register reflections directly as a layer of the reciprocal lattice in undistorted form (7).

Whenever a single crystal large enough to be handled and mounted in a goniometer camera is available (crystals as small as 0.1 mm in diameter, or even less, can be used), the unit cell can always be determined unequivocally. Arrival at this stage, whether by powder or by single crystal diffraction patterns, opens the way to several chemical applications. Identification by unit cell dimensions has already been mentioned. The determination of the composition of mixed crystals can also be achieved, provided the relation between composition and unit cell dimensions has already been established; this method has been much used for alloy systems. The determination of molecular weights is another important application.

Determination of Molecular Weight. The unit cell of a crystal contains a small whole number of molecules, sometimes one, more frequently two or four, more rarely six or eight or more; the weight of matter in the unit cell, which is obtained by multiplying the volume by the density, is therefore either equal to the molecular weight or a small multiple of it. If the approximate molecular weight is known, and this knowledge is usually available through chemical evidence, this will decide the number of molecules in the unit cell; it will then be possible to arrive at an accurate value for the molecular weight by dividing the unit cell weight by this number. An accuracy of about 1% is usually attainable, and this may be improved if the density can be measured sufficiently accurately. Often the x-ray method is the most accurate way of determining the molecular weight; it is particularly valuable if for any reason the standard physicochemical methods cannot be used, for instance for very insoluble substances which do not permit the use of boiling-point or freezing-point methods.

Molecular Dimensions. The size and shape of the unit cell may give a general idea of the overall dimensions of the molecule. Some caution is necessary here, because, even if there is only one molecule in the unit cell, it is possible to pack molecules of varying dimensions to give the same unit cell dimensions; however, if the one-molecule cell is very long or very flat, there will be little doubt that the molecules have the same general shape. When there is more than one molecule in the unit cell, it is more difficult to draw definite conclusions; but scale models of alternative possible chemical formulas may be made to see if they are compatible with the unit cell dimensions.

In the crystalline regions of specimens of long-chain polymers, the unit cell contains only sections of molecules: a small group of atoms, often only one or two monomer units, is repeated over and over again along the chain, and the side-by-side packing of the chains gives rise to the crystalline pattern of atoms. The chain molecules thread their way through the unit cell, and therefore the length of the cell in the chain direction is the repeat distance of the molecule itself. The magnitude of this repeat distance may lead directly to a knowledge of the geometry of the chain molecule; for instance, if the chemical constitution of the molecule is known, models can be made or calculated with the usual interatomic distances and bond angles, and, if the x-ray repeat distance is equal to that of the fully extended model (a plane zigzag chain in the simpler cases), there is no doubt of the chain configuration in the crystal. If, however, it is less than that of the fully extended model, the chain is evidently shortened by rotation around the chain bonds, there is usually more than one way of shortening, and further x-ray evidence will be necessary to decide which is correct; but models or calculations will indicate the possibilities.

The Symmetries of Atomic Arrangements. The number of different symmetry elements is remarkably small—the center of symmetry $(\bar{1})$, the plane of symmetry (m), two-, three-, four-, and sixfold axes of symmetry $(2,3,4,6)$, and the inversion axes $(\bar{3},\bar{4}, \text{and } \bar{6})$; by all the possible combinations of them, the 32 point groups are obtained (see Vol. 6, p. 519). In the grouping of atoms associated with each lattice point in a crystal, the same 32 point groups may occur, in space lattices having the appropriate symmetries. The point-group symmetries are, however, those of isolated objects, crystal shapes themselves or limited groups of atoms; and the symmetry elements involved in them are those which, by continued repetition, always bring us back to the

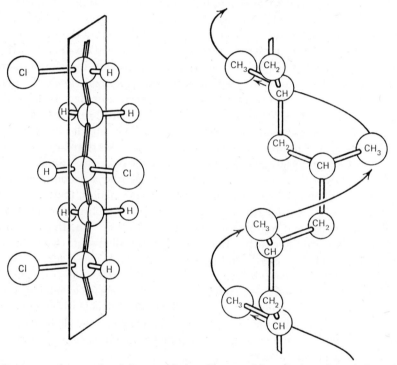

Fig. 4. Symmetry elements involving translation, illustrated by a hydrocarbon chain molecule (1). Left, glide plane; right, twofold screw axis.

crystal face, or the atom, from which we started. In repeating patterns in space, two additional types of symmetry elements may be discerned, elements which involve translation and therefore do not occur in point groups or crystal shapes: by continued repetition of them, we do not arrive back at the atom from which we started; we arrive at the corresponding atom associated with another lattice point, and then another, and so on throughout the crystal. These elements (see Fig. 4) are the glide plane, which involves simultaneous reflection and translation, and the screw axis, which involves simultaneous rotation and translation; they are represented by symbols similar to those used for the nontranslational elements, a, b, or c for glide planes with translation halfway along the a, b, or c axis respectively, n for a diagonal halfway glide, 2_1 for a twofold screw axis with halfway translation combined with halfway rotation, 3_1 for a threefold screw axis involving a one-third translation combined with one-third rotation, and so on. The various combinations of these, in association with the appropriate space lattices, give rise to the 230 space groups (8). (For the nomenclature for lattice types, see Fig. 2.)

In the x-ray analysis of a crystal, before attempting to locate the atoms in a complex structure it is desirable to determine as far as possible the space-group symmetry. The morphology, the optical properties, and the unit cell type determined from the positions of the reflections will have made clear to which of the seven systems the crystal belongs. A further narrowing down of the range of possible space groups, sometimes to one particular space group, is effected by examining the indexes of the x-ray reflections; it is usually found that particular types of reflections are missing, and it is these which identify symmetry elements involving translation. A diffracted x-ray beam is a reflection by a set of parallel planes, and the rays reflected by any one plane of atoms are exactly one wave length behind those reflected by the plane above it and one wave length in front of those reflected by the plane beneath it (Fig. 1). If there are translational symmetry elements in the structure, involving translation across these particular planes, this means that there are equivalent atoms *between* the planes, and the waves diffracted by these will be out of phase with those first mentioned; for instance, if there is a twofold screw axis in a particular direction, this means that reference planes of atoms perpendicular to this direction are interleaved at exactly halfway by exactly equivalent atoms; the phase of the waves scattered by the inverleaving atoms will be exactly opposite to those scattered by the reference atoms, and so the net intensity is zero. On the other hand, for the second-order reflection, which means a two-wave length difference between successive reference planes, the waves from interleaving atoms are one wave length behind or in front of those from the reference atoms, so that all are in phase and combine to produce a reflection. For this reason the even orders of diffraction are present, but the odd orders are all absent. Compound lattices (face-centered or body-centered) produce similar effects, because again there is interleaving; for a lattice face-centered on c, for instance, all reflections for which $h + k$ is odd are absent, and for a body-centered lattice all reflections for which $h + k + l$ is odd are absent. A survey of the entire range of reflections to decide which are absent thus settles the lattice type and shows which translational symmetry elements are present; this sometimes identifies the space group symmetry unequivocally, or else narrows it down to a very few possibilities.

Arrival at this stage may lead to definite conclusions on molecular symmetry. For each space group, it takes a specific number of asymmetric units to make up the complete symmetry, so that, if the number of molecules in the unit cell is, say, half this, each molecule must have twofold symmetry of some kind, and it may be possible to say what this symmetry is: a plane of symmetry, a twofold axis, or a center of symmetry (1).

Determination of Atomic Coordinates. The actual positions of the atoms in the unit cell are deduced from the intensities of the reflections; a set of relative intensities, measured photometrically or by Geiger counter methods, is usually sufficient, although absolute intensities relative to that of the incident beam have sometimes been measured, as they confer certain advantages. It is, however, possible to estimate absolute intensities from a complete set of relative intensities (9). The influence of atomic positions on the relative intensities of the various reflections may be gathered from the remarks in the previous section. In amplification of these, it is sufficient to point out that in a complex structure each lattice point is associated with an extensive group of atoms; when subgroups are related by translational symmetry elements, exact interleaving at simple fractions of the plane spacing causes certain reflections to have zero intensity; but in each subgroup, each asymmetric unit, there are no symmetry relations between the different atoms, so that interleaving occurs at nonintegral fractions of the plane spacing; and the effect of this is that the waves from different atoms are out of phase by varying amounts, so that the intensity is reduced but not to zero. The phase relations of the various atoms are different for each differently oriented set of reflecting planes, and this is the reason why the different reflections, even those close together in the diffraction pattern, have widely varying intensities. The problem in x-ray analysis is to deduce the atomic positions from the relative intensities of the reflections. First of all, various geometrical and other factors which modify the intensities are allowed for, and the square root of what is left is a wave amplitude known as the *structure factor:* it is this that is directly affected by atomic positions in the way already described. The complete set of structure factors is the raw material for solving the problem of a crystal structure.

The ideal way of doing this would be to put all the structure factors into a set of equations, the solutions of which would be the atomic coordinates. This is equivalent to a physical process of tracing the waves back to the atoms to form an image of them. But to do this it is necessary to know not only the amplitudes but also the phase relations of the waves, for, if the wrong phases were used in the calculations, a totally wrong image would be produced. Unfortunately, there is no experimental way of measuring phases. In certain special circumstances the phase relations may be deduced simply and with certainty: if there is a minority of atoms of relatively high diffracting power (high atomic number) in suitable positions in the structure, they determine the phase relations of the majority of the diffracted waves; it is usually possible to locate the heavy atoms without much difficulty, and it is then a straightforward matter to calculate the phase relations of the waves diffracted by them alone, and, assuming that these phases are imposed on the waves from the structure as a whole, to calculate by Fourier synthesis the image of the complete structure. An extension of this method uses isomorphous crystals containing atoms of different diffracting powers (one of which must be heavy); the differences of intensity of corresponding reflections of the two crystals are used to determine the phases. These methods have been used for a number of important structures, for example, that of cupric tropolone is shown in Figure 5, and it remains true that the best chance of determining the structure of a complex molecule by x-ray analysis lies in preparing and crystallizing heavy atom derivatives (10).

If this cannot be done, or if it is specifically the crystal structure of a substance not containing heavy atoms which is of interest, we must do the best we can with the amplitudes of the waves alone. The oldest procedure is one of trial and error: a set of atomic positions is postulated by utilizing all prior knowledge of interatomic distances and stereochemistry and aiming at satisfying some of the outstandingly large structure factors, and then the rest of the structure factors are calculated and compared with those observed; the magnitudes of the discrepancies between observed and calculated values indicate how far the postulated structure is from the truth, and detailed consideration of them may suggest what changes in postulated atomic positions are necessary to improve the agreement. It is rather like trying to solve a three-dimensional cross-word puzzle in which every clue (every structure factor) involves not merely one corner of the puzzle but the entire structure. Usually, of course, it is best to simplify

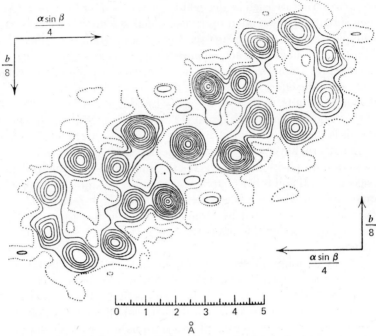

Fig. 5. Cupric tropolone: electron density projection.

matters by working first of all in two dimensions and trying to solve one projection before attempting three-dimensional operations. When a sufficient degree of agreement has been attained to lead to the opinion that the puzzle has been essentially if only approximately solved, the postulated atomic coordinates are used to calculate the phase relations of the majority of the waves; these phases are then used, with the observed structure factors, to calculate by Fourier synthesis an image of the structure.

The contour scale on the central copper atom is reduced by a factor of five.

The image is, of course, a distribution of electron density (since it is the electrons of the atoms which scatter x rays); the centers of the concentrations of electron density are the atomic centers. The new atomic coordinates, which differ a little from those first chosen, are then used to calculate the phase relations of further reflections (and to check the earlier ones), and the process is repeated until all the structure factors are included. A final set of calculations of structure factors from the accepted coordinates should show good agreement with the observed values, and it is the degree of this final agreement that is the criterion by which the proffered solution of the problem is judged. It is customary to express the degree of agreement as the ratio $\Sigma \left| \left| F_o \right| - \left| F_c \right| \right| / \Sigma \left| F_o \right|$, where F_o are the observed and F_c the calculated structure factors. Although this particular expression has no firm theoretical basis and is admittedly open to criticism, it is at any rate a quantitative figure which compares the sets of figures; values between 0.1 and 0.2 are commonly attained; this remaining degree of discrepancy is probably due to nonspherical electron distributions in bonded atoms.

The limitations of the heavy-atom method and the trial method have inspired many attempts, both mathematical and experimental, to develop alternative and preferably more direct methods. A Fourier synthesis of the squares of the structure factors (the Patterson function, in which the phases are lost) gives the distribution of interatomic vectors; it is sometimes possible to deduce actual atomic positions from the vector map, by systematic procedures (see chapter 7 of ref. 9). Optical analog methods, first introduced by Sir W. L. Bragg, have been useful for work on projections. A very small-scale repeating pattern in which atoms are represented by dots is made by photographing one unit of pattern in a multiple microcamera (the "fly's eye"); the optical diffraction pattern of this repeating pattern consists of spots whose relative intensities correspond to those of the x-ray diffraction pattern of the corresponding projection of the crystal structure. The realization that the x-ray diffraction pattern of a crystal represents the diffraction pattern or transform of a molecule (or of the small group of molecules in the unit cell) sampled at the reciprocal lattice points has led to the use of optical transforms, diffraction patterns of sets of holes (representing the atoms of a molecule) in a screen; Lipson and his school have shown that this method facilitates the task of finding the correct orientation of a molecule in the unit cell, and moreover, that for complex molecules the characteristic diffraction features of *parts* of the molecule, such as a ring or a chain, can sometimes be recognized in the complete pattern, so that the orientation of these parts can be deduced (11). The use of digital computers has removed the great burden of calculation and thus enlarged the scope of the standard methods. A useful method of finding out what is wrong with a postulated structure (12), or refining an approximately correct one (13), is the error synthesis or difference synthesis, a Fourier synthesis of the differences between observed and calculated structure factors; it has been used for locating hydrogen atoms, which on account of their small scattering powers are not easy to detect.

All these methods are valuable, but the fundamental difficulty caused by lack of experimental knowledge of the phases of diffracted waves remains. It has been found, however, that for structures containing not too many atoms, the phases can be deduced from the magnitudes of the structure factors. For crystal structures having a center of symmetry, the phases are all either $0°$ or $180°$, and the correct phase for each reflection is deduced by a process which starts with phase relations between the strongest reflections and is then extended to weaker reflections by various mathematical relations which have been discovered. For crystals lacking a center of symmetry, the phases can take any value, and the difficulties are increased; nevertheless, at the time of writing (1969) the structures of noncentrosymmetric crystals composed of molecules of 20 atoms (not counting hydrogens) have been solved by such methods. For centrosymmetric crystals, phase-relation methods can be used for molecules of 50 atoms or more. The power of such methods is still far behind that of the heavy-atom method, which has been used successfully for protein molecules containing thousands of atoms; but the subject is still developing, and with the aid of large rapid computers may be expected to extend its scope still further.

Structures of Noncrystalline Substances. The prospects of determining the arrangements of atoms in noncrystalline substances, gases, liquids, and glasslike solids, are very much more restricted than for crystals. The diffuse diffraction patterns are, it is true, not featureless, for they show definite maxima and minima of intensity, but, except in very simple cases, the ratio of diffraction information to the complexity of the structure is too small to permit detailed interpretation with any great confidence. The problems are simplest for gas molecules containing only a few atoms, since only a few parameters determine the stereochemistry of such molecules; but here the diffraction of electrons (considered below) has been used much more than that of x rays. For solids and liquids, distances between atoms of neighboring molecules as well as those within the molecule are involved; but for simple glasslike solids, such as silica glass, interesting results have been obtained either by postulating particular arrangements of atoms and calculating the diffraction pattern for comparison with that observed, or by converting the observed diffraction pattern into a vector distribution and interpreting this in terms of the spatial arrangement of atoms (14).

Diffraction of Electrons and Neutrons

High-energy radiations other than x rays, having an equivalent wave length on the order of 1 Å (10^{-8} cm), are also diffracted by matter, giving patterns whose details are related to the arrangement of atoms in the irradiated material and can be used like x-ray diffraction patterns to give similar information. A narrow beam of electrons accelerated by an accurately controlled voltage of 50–100 kv, on passing through a thin crystal, gives a pattern of spots (usually recorded photographically) very similar to the x-ray diffraction pattern; the positions of the spots indicate the unit cell dimensions, and their relative intensities are determined by the positions of the atoms in the unit cell. A polycrystalline powder gives a concentric ring pattern, again closely similar to the x-ray powder pattern. Specimens can only be examined in the high vacuum which is essential for such an electron beam; and a solid specimen must be extremely thin (a few hundred angstroms) to transmit a sufficient intensity of diffracted electrons. For solids, the electron diffraction method has been chiefly used for identifying crystalline substances in extremely thin films, and for investigating the orientation of crystals in such specimens; it has not been used much for determining structure. There is, however, another important application of electron diffraction not concerned with solids: this is the determination of the structure of gas molecules. The complexity of the problem, for simple molecules, is much less than for solid or liquid noncrystalline substances. The electron beam passes through a narrow stream of gas molecules entering the high-vacuum apparatus, and a diffraction pattern consisting of diffuse rings with well-defined maxima and minima of intensity is produced. The stereochemistry (interatomic distances and bond angles) of many of the simpler gas molecules has been determined with considerable precision by interpretation of these patterns, usually by trial of postulated configurations (15).

The development of atomic piles has created opportunities for studying the diffraction of neutrons by crystals, and this subject has developed into an important branch of crystallography. The positions of atoms in crystals control the intensities of the diffracted beams as in x-ray diffraction; but the relative intensities of the different diffracted beams of neutrons are often very different from the corresponding x-ray intensities, for the relative diffracting powers of the various atoms do not depend on atomic number as the x-ray diffracting powers do, and in addition some atoms cause phase reversals on scattering. These characteristics are now well established for many atoms. The outstanding contribution of neutron diffraction to crystal structure analysis is the accurate location of hydrogen atoms, a contribution that rests on the fact that the diffracting power of hydrogen atoms is much greater, in relation to that of other common atoms, for neutrons than for x rays. Much larger crystals are needed than for x-ray diffraction: rods 1–2 mm in diameter and 1 cm long are needed. The diffracted neutron beams are measured by counters utilizing either scintillating crystals or boron trifluoride gas under pressure (16).

Chemical Results of Structure Analysis

The general results of structure analysis which are of chemical significance have been described in the article Crystals. It will be sufficient to recall here the information on ionic sizes and the principles which determine the way in which ions are packed in crystals; the contributions to our understanding of the structure of metals and the nature of intermetallic compounds; the entirely new light thrown on the structures of complex minerals, particularly silicate minerals, which has led to the abandonment of earlier "structural formulas" and a virtual rewriting of the subject of mineralogy (17); and, for organic chemistry and molecular compounds generally, the detailed knowledge of the interatomic distances and stereochemistry within the molecule, the contact (van der Waals) distances between atoms in neighboring molecules, and the principles of arrangement of molecules in crystals. Only the last-mentioned aspect, the contribution of structure analysis to organic chemistry and molecular stereochemistry in general, needs some further consideration here.

The lengths of single, double, and triple bonds between carbon atoms show that the bond length decreases with increasing bond order. In aromatic ring molecules the carbon-carbon distances are intermediate between those of single and double bonds. With the increasing accuracy of bond length determination (something like ± 0.01 Å is now obtainable) it has become possible to deal with more subtle questions like the differences between the various bonds in molecules like naphthalene and anthracene; and the results obtained by x-ray analysis are in general agreement with the figures calculated from the wave-mechanical theory of chemical bonding.

The tetrahedral disposition of the four single bonds of a carbon atom, suggested by LeBel and Van't Hoff in 1874, has been quantitatively confirmed; even in compounds like $CHCl_3$ the repulsion between the large chlorine atoms results in opening the tetrahedral angle by only a few degrees (18). In complex asymmetric molecules, where several asymmetric carbon atoms give rise to many possible isomers, x-ray analysis is able to determine which exists in a particular substance. This has been done, for instance, for sucrose and several other sugars, for some of the isomers of hexachloro-cyclohexane (benzene hexachloride, of which the γ isomer is the insecticide Gammexane), for strychnine, and for some of the steroids (19). Still more remarkable, it has been found possible to determine the absolute configuration of the tartaric acid molecule, that is, to discover which of the two mirror-image forms is actually present in ordinary D-tartaric acid; this historic completion of the story begun by Pasteur was achieved in 1949 by Bijvoet, by a special method which makes use of the phase shift which occurs when x rays are scattered by an atom having an absorption edge near the wave length of the incident beam. (Sodium rubidium D-tartrate was irradiated by zirconium $K\alpha$ x rays; see ref. (19).

In chain compounds, such as the natural and synthetic chain polymers, rotation around the single bonds can give rise to a great variety of possible molecular conformations. X-ray analysis is able to decide which is correct in simple cases, in spite of the difficulties of interpretation arising from the fact that only polycrystalline fiber specimens, not single crystals, are available. For paraffin hydrocarbons including polyethylene, and for polyamides (nylons), polyethylene terephthalate (Terylene, Dacron), rubber and gutta-percha and a few others, the general stereochemistry is now known, although atomic positions cannot be determined with the accuracy associated with single-crystal investigations. A structure has been suggested for cellulose which, although less certain than those of the simpler polymers, appears reasonable.

The simpler polypeptides without large side-groups, such as polyglycine, silk fibroin, and fibrous proteins generally, have nearly fully extended chains, but in some of the synthetic polypeptides with large side-groups and in α-keratin, the chain is curled into a particular helical conformation predicted by Pauling. The consideration of molecular transforms and the supporting evidence of polarized infrared spectroscopy on the orientation of C=O and NH groups have played an important part in the discussion of these problems.

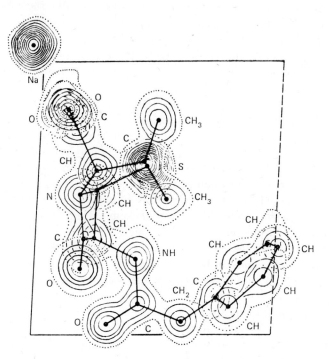

Fig. 6. Sodium benzylpenicillin (22). A composite diagram showing sections through the atoms at different levels, from the three-dimensional electron density distribution. (Reprinted by permission of Princeton University Press from *The Chemistry of Penicillin.*)

The most spectacular and far-reaching achievements of x-ray analysis in the field of molecular stereochemistry are the determination by Watson and Crick (20) of the structure of DNA (deoxyribonucleic acid), the chain polymer which is the chemical basis of genetics, and the determination of the structures of several proteins, following the pioneer work of Perutz and Kendrew on myoglobin and haemoglobin (21). The double helical structure of DNA, deduced from the x-ray diffraction patterns of DNA fibers, has played a key role in molecular biology by demonstrating the stereochemical basis of replication and the transmission of hereditary information; and knowledge of the detailed conformations of the curled-up chains in enzyme proteins, deduced from x-ray diffraction patterns of single crystals, with a thoroughgoing application of the heavy-atom and isomorphous replacement methods, promises to lead to a real understanding of enzyme catalysis which controls the chemistry of life.

For most of the crystal structures that have been solved, the chemical structure of the molecule, the scheme of primary bonds linking the atoms together, has already been known; but in some circumstances crystal structures can be determined in detail on the basis of an incomplete knowledge of the chemical constitution, or even with no assumptions at all about the chemical bonding. When this is possible, the chemical constitution is settled by x-ray crystallography. The prospects are most favorable with heavy-atom derivatives, but success has sometimes been achieved with limited help from heavy atoms. The early example of penicillin (12), in which the ring system in part of the molecule was established by the x-ray work (see Fig. 6), has been followed by successes with more complex molecules; the outstanding example is vitamin B_{12}, a molecule of 93 atoms (not counting hydrogens, see Vol. 21, p. 543) whose structure was solved by Hodgkin and her collaborators though only the structure of degradation fragments was known on chemical evidence (22).

Bibliography

"X-Ray Analysis" in *ECT* 1st ed., Vol. 15, pp. 158–176, by C. W. Bunn, Imperial Chemical Industries Limited.

1. C. W. Bunn, *Chemical Crystallography*, 2nd ed., The Clarendon Press, Oxford, pp. 120–125, 1961.
2. *Ibid.*, pp. 147–184.
3. J. D. H. Donnay et al., *"Crystal Data,"* 2nd ed., American Crystallographic Association, *A.C.A. Monograph No. 5* (1963).
4. Institute of Physics, *X-ray Diffraction by Polycrystalline Materials*, London, 1955.
5. N. F. M. Henry, H. Lipson, and W. A. Wooster, *The Interpretation of X-ray Diffraction Photographs*, The Macmillan Co., London, 1951.
6. T. Ito, *X-ray Studies in Polymorphism*, Maruzen, Tokyo, 1950.
7. M. J. Buerger, *The Precession Method*, John Wiley & Sons, Inc., New York, 1964.
8. *International Tables for X-ray Crystallography*, Vol. 1, Kynoch, Birmingham, England, 1952.
9. H. Lipson and W. Cochran, *The Determination of Crystal Structures*, 3rd ed., G. Bell & Sons Ltd., London, p. 131, 1966.
10. Ref. 1, pp. 374–388.
11. C. A. Taylor and H. Lipson, *Optical Transforms*, G. Bell & Sons Ltd., London, 1964.
12. D. Crowfoot, C. W. Bunn, B. W. Rogers-Low, and A. Turner-Jones, in H. T. Clarke, ed., *The Chemistry of Penicillin*, Princeton University Press, Princeton, pp. 310–366, 1949.
13. Ref. 1, pp. 388–394.
14. Ref. 1, p. 445.
15. Z. G. Pinsker, *Electron Diffraction*, Butterworth's Scientific Publications, London, 1953.
16. G. E. Bacon, *Neutron Diffraction*, Clarendon Press, Oxford, 1962.
17. Sir Lawrence Bragg and G. F. Claringball, *Crystal Structures of Minerals*, G. Bell & Sons Ltd., London, 1965.
18. L. Pauling, *The Nature of the Chemical Bond*, 3rd ed., Cornell University Press, Ithaca, N.Y., 1960.
19. J. M. Robertson, *Organic Molecules and Crystals*, Cornell University Press, Ithaca, N.Y., 1953.
20. J. D. Watson and F. H. C. Crick, *Nature* **171,** 737 (1953).
21. M. F. Perutz, *Proteins and Nucleic Acids*, Elsevier Publishing Company, 1962.
22. D. C. Hodgkin et al., *Proc. Royal Soc. London* **A242,** 228 (1957).

C. W. Bunn
The Royal Institution
London

x-Ray Fluorescence Spectrography

The use of x rays as a physical tool for chemical analysis has greatly expanded since 1945. The powder diffraction method described above is used for the identification of compounds by comparison of the pattern of the unknown specimen with that of known substances, and the x-ray spectrographic method described below directly identifies the constituent elements.

The fundamental difference between the two methods may be illustrated by using a mixture of copper and nickel powders as an example. The x-ray diffractometer pattern of the mixture is a superposition of the patterns of each powder. The patterns are nearly identical because the crystals of each powder have the same atomic crystal structure. Each line originates from one set of atomic planes in the crystals. Since Cu and Ni have different atomic radii, there are small differences in the angular positions of the lines. The relative intensities of lines from the same atomic planes are used to calculate the concentrations when the necessary absorption and other correction factors are applied. If the powders formed a solid solution instead of a mixture, a similar single pattern would appear, and the lines would lie between those of the mixture because the solid solution has the same structure as the end members. Accurate angular measurements of the patterns would show the presence of each phase, or of the solid solution, but would not necessarily identify impurities that might be present. It would be very difficult to determine the identity of several impurities in solid solution only from the net change in the lattice parameter. In only usual cases could the elements be specifically identified without some prior chemical information. The x-ray spectrograph pattern, on the other hand, gives the x-ray spectra for each element present in the specimen. The x-ray spectra of Cu and Ni have different wavelengths which can be readily measured and used for the direct identification of the elements. If impurities are present, each can be identified by its characteristic x-ray spectra and the concentrations determined from the relative intensities. Unlike the diffraction method, however, the spectrographic method gives no information as to the structure or how the elements are combined; hence the mixture and solid solution would give identical spectra. Thus, the two methods supply different basic types of information, and if used together, they provide a complete x-ray method of chemical analysis.

x-Ray fluorescence analysis can be carried out on crystalline or amorphous solids or powders, liquids, coatings, and even gases using suitable containers. The specimens must have a flat surface, are approximately an inch in diameter and all elements from fluorine up may be analyzed quantitatively. The sensitivity and accuracy decrease rapidly for the low atomic number elements (S and below) because of various problems associated with the long wavelength x rays generated by these elements as described below. By using small diaphragms the specimen area analyzed can be reduced to about 1 mm^2 with a considerable loss of intensity. The electron microprobe makes it possible to analyze specimen areas as small as a few square microns by impinging a fine electron beam on the specimen which generates primary characteristic x rays directly; it has been used to analyze elements down to boron and beryllium. In some cases large electron beams are used so that the specimen is essentially the anode of a demountable x-ray tube. The direct electron excitation methods have high sensitivity but accurate quantitative interpretation of the experimental data presents numerous difficulties and many corrections are required. Accuracies of the order of 1% may now be reached using computers to calculate the many corrections. The recent advent of the high resolution solid state detector coupled with a multichannel pulse amplitude analyzer has led to the development of nondispersive analysis with nearly instantaneous readout of the elements present. Fluorescence excitation by radioisotope sources has been used in place of x ray or electron excitation in nondispersive analysis.

x-Ray spectroscopy is now used in a great variety of applications in chemistry, metallurgy, geology, ceramics, biomedicine, etc, both as a laboratory analytical method and for on-line process control. For further details the reader may consult the comprehensive biannual reviews of the literature published in Analytical Chemistry (1). Recent books provide a systematic description of the practical aspects of the method (2–5). Articles on the applications and developments appear frequently in *Advances in X-ray Analysis* (published annually by Plenum Press), *Analytical Chemistry, Applied Spectroscopy, Spectrochimica Acta,* and other journals. Descriptions of the electron microprobe method (which will not be discussed here) have appeared in recent books (6–8), the annual proceedings of the Electron Probe Analysis Society of America and the triennial international conferences on X-Ray Optics and Microanalysis.

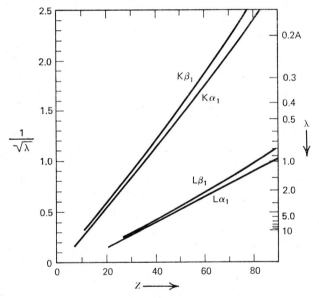

Fig. 1. The characteristic x-ray line wavelengths λ are nearly linearly related to atomic number z (Moseley's law). (Parrish, 2.)

X-Ray Spectra. In 1913, Moseley showed that each element has a characteristic, simple x-ray line spectrum whose wavelengths λ are dependent only upon the atomic number Z of the element. A plot of $(1/\lambda)^{1/2}$ against Z is nearly a straight line for any given spectral line of all the elements as shown in Fig. 1. This fundamental relationship simplified the study of x-ray spectra and provided a sound basis for the Bohr model of the atom. The identification of the new element hafnium, in 1923, was made by measurements of its x-ray spectrum (9). Since x-ray spectra vary in wavelength in a systematic manner from one element to another, the measurement of x-ray wavelengths provides a unique method of identifying the elements. The interpretation and measurement of x-ray spectra, which are nearly identical (except for wavelengths) for

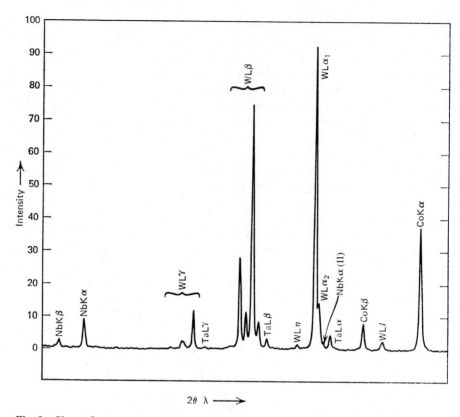

Fig. 2. X-ray fluorescence spectrogram of wolfram carbide alloy showing the K spectra of Nb and Co and the L spectra of W and Ta.   The B could not be analyzed.   Automatic recording using a LiF crystal analyzer and scintillation counter.

all the elements and consist of relatively few lines, is, therefore, much simpler than optical spectra.

The K spectra arise from electron transitions from the L to K shell which give rise to the closely space doublet $K\alpha_1$ and $K\alpha_2$ and M to K transitions which produce $K\beta_1$, $K\beta_2$.   The relative intensities of the lines are $K\alpha_1 > K\alpha_2 > K\beta_1$ and their ratios vary gradually with atomic number of the element.   The L spectra have a dozen or more lines of longer wavelengths which result from transitions from the M and upper shells to the L shell and the three principal groups are $L\alpha$, $L\beta$ and $L\gamma$, each containing several lines.   Typical K and L x-ray spectra are shown in Figure 2.   The higher atomic elements also have M and N spectra.   Condensed tables of x-ray wavelengths are given in several x-ray books, and a complete list has been published by Bearden (10).

The characteristic x-ray lines originate when an electron from one of the outer shells of the atom takes the place of one of the electrons removed from an inner shell. The removal of the inner electron may be caused by a high-energy electron beam as in an x-ray tube, or by irradiation with x rays as in x-ray fluorescence.   The energy required to cause the initial electron vacancy must exceed the binding energy of the electron in its normal shell; this energy increases with atomic number for the same shell. In direct electron excitation the critical voltage V_c (in kilovolts) may be calculated from the relations:

$$V_c = k/\lambda \tag{1}$$

where k is approximately 12.4 and λ is the wavelength (in angstroms) of the absorption edge of the element. This edge has a characteristic wavelength which decreases with increasing atomic number Z and is slightly less than the shortest wavelength line of the series of that element. There is one absorption edge in the K series, three in the L and five in the M series. For example, V_c for the K series increases from 1.6 kv for Al to 115.6 kv for U. The values of V_c corresponding to the L_I absorption edge increase from 1.7 kv for Se to 21.8 kv for U. Once the critical excitation potential is exceeded, all the lines of the series appear, their intensity increases rapidly and the rate of increase is dependent upon a number of factors, such as the voltage applied to the x-ray tube, the method of operating the x-ray tube (half-, full-wave rectification or dc constant potential), selfabsorption in the target, and other factors.

In addition to the line spectrum, a continuous, nearly structureless, broad spectrum of wavelengths is generated by the rapid deceleration of the electrons in the target. The nature of this spectrum is determined primarily by the voltage applied to the x-ray tube. The intensity and proportion of shorter wavelengths generated increase with voltage in a rather complex manner and are roughly linearly proportional to the atomic number of the target element and tube current. The shortest wavelength of the spectrum may be calculated from equation (1). Fluorescence spectra contain only the characteristic lines, but a small percentage of the continuous spectrum from the x-ray tube is usually scattered by the specimen and appears as a low background and as weak lines in the x-ray spectrograph pattern; one usually avoids selecting a target of the same element to be analyzed. In practice, x-ray tubes with W, Mo, Cu, and Cr targets are most frequently used. The window must have a high transmission to efficiently excite the longer wavelengths required for the analysis of the low atomic number elements. Although sealed-off x-ray tubes are preferred because of their simplicity of operation and stability, very thin windows are fragile and demountable (pumped) tubes with easily replaceable windows, and Al and other targets, are often used in this region (11).

The characteristic line spectra are thus produced when the element absorbs an x-ray beam having energies exceeding the inner electron binding energies. This phenomenon, called fluorescence in analogy with the optical case, is generally used in the practical application of the method; hence, the name x-ray fluorescence analysis. In passing from long to short wavelengths (low to high energy) of the x-ray spectrum the absorption increases rapidly and abruptly at the absorption edge of the element and then falls off slowly. The x rays of wavelengths just short of the absorption edge are therefore, most efficient in generating fluorescent x rays. There are many physical factors which determine the intensity of the fluorescent x rays. The photoelectric absorption μ is roughly proportional to λ^3 and Z^4 and the ratio of the μ's very near both sides of the K edge decreases from about 10 for Al to 2 for W. The fluorescent x rays are reduced in intensity as they travel to the surface and the angle-of-view of the specimen, therefore, must be relatively large to minimize errors due to selfabsorption which increases with increasing wavelengths.

Only a fraction of the incident absorbed x-ray photons produce characteristic x-ray fluorescence lines. The fluorescence yield increases with increasing Z and is several times larger for K- than the L-series. The yield is only about 0.006 for OK and increases monotonically to 0.90 for PrK, beyond which the yield increases slowly to 0.96 for UK. The low yield for the low atomic number elements imposes a lower limit in the sensitivity of the x-ray method particular for elements below atomic number 16

(SK 0.05). Although elements as low as fluorine have been analyzed the concentrations must be high and the matrix absorption low. In the low-energy region there is an increasing probability of the atom ejecting one or more electrons from its outer shells (Auger effect) and special spectrographs have been developed for elemental analysis utilizing the energies of the Auger electrons.

Basis of the Method. In the early days of x-ray spectroscopy the specimen was made in the form of a flat target and irradiated by the direct electron beam in a demountable x-ray tube (9). The difficulties of the specimen preparation, the inconvenience of the associated vacuum systems, and the errors arising from heating the specimen by the electron beam possibly with subsequent melting, segregation, or volatilization made the method impractical for a routine analytical procedure and was unsuitable for liquid specimens. The developments of high-power, sealed-off x-ray tubes, the associated commercial x-ray equipment and highly sensitive x-ray detectors made it possible to eliminate the practical difficulties. The specimen is placed outside the x-ray tube, as close to the window as possible, since the primary beam intensity per unit area of specimen irradiated is reduced by the inverse square of the distance between the target and specimen. Although the fluorescent intensity is about one-thousandth of the x rays generated by direct electron excitation, intensities of the order of 10^5–10^6 counts per second are obtainable from large pure specimens with the x-ray tube operated at maximum power, and the background is much lower.

The fluorescent x rays impinge upon a single crystal which acts as an analyzer or monochromator and diffracts ("reflects") the various wavelengths according to the Bragg law:

$$n\lambda = 2d_{hkl} \sin \theta \qquad (2)$$

where n is a small integer giving the order of the reflection, λ the wavelength of the fluorescence x rays, d the lattice spacing of hkl planes of the crystal, and θ the angle between the incident beam and the reflecting planes. The crystal is rotated slowly and reflects each wavelength, one at a time, at an angle 2θ from the incident beam. The reflection phenomena from the crystal are exactly the same as in x-ray diffraction analysis, but, instead of using a single wavelength to measure the many d spacings, there are many wavelengths to be analyzed using the single d spacing (and its higher orders) of the monochromator.

It is apparent that the crystal plays an important role. Not many crystals are suitable as monochromators for this application, since they must be above 1 in. by 3 in. and nearly perfect. If the crystal has a marked mosaic structure, reflections will take place from different portions of the crystal as it rotates. This result in multiple reflections for each wavelength over a wide angular range and cause reduced and variable peak intensities which complicate the interpretation. Naturally, the highest possible reflection intensity is desired, and the crystal itself must not fluoresce in the wavelength region being analyzed, a requirement which restricts the chemical compositions of the available crystals particularly for longer wavelength x rays.

The dispersion of the crystal expressed as $d\theta/d\lambda$ is inversely proportional to $2d$ $\cos \theta$ and, hence, the angular separation of the spectral lines increases with increasing 2θ and decreasing d. In cases of analyses involving a number of closely spaced spectral lines a crystal with a small d spacing should be selected to provide adequate separation, or a higher order reflection from the same crystal (which usually has a much lower reflectivity) may be used if the intensity loss is acceptable. The highest reflection angle that can be reached by the goniometer is approximately 140° (2θ) because of mechanical limitations, and since the crystal cannot reflect wavelengths greater than $\lambda_{max} = 2d$, the d spacing must also be large enough to cover the wavelength region. At the small 2θ angles used to measure short wavelengths the crystal length (and d) must be selected to intercept the entire incident beam to prevent radiation from passing over the crystal and entering the detector. In practice, two or more crystals may be used to cover a wide wavelength region.

Among the crystals frequently used are topaz ($hkl = 303$, $2d = 2.71$ Å), lithium fluoride (200, 4.03), quartz ($10\bar{1}1$, 6.87), ethylenediamine tartrate (020, 8.81), potassium hydrogen phthalate ($10\bar{1}1$, 26.4) and thin Langmuir Blodgett layers such as lead stearate, $2d = 100$ Å. In some cases the crystal surface is lapped lightly to increase the reflectivity without causing much increase in line breadth.

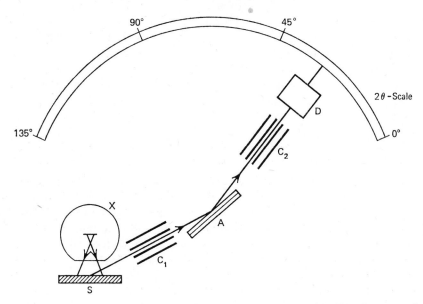

Fig. 3. Schematic representation of flat crystal nonfocusing geometry used for larger specimens in x-ray fluorescence spectrograph.

Soller slits, consisting of thin absorbing equally spaced metal foils are used to limit the angular range of the crystal reflection and to reduce the x-ray background with relatively small loss of peak intensity. The line breadths are equal to the angular aperature of the Soller slits (determined by the length and spacings of the foils) plus the much smaller contribution of the rocking curve of the crystal. The widths of the reflection at one-half peak height are usually in the range of 0.15°– 0.50° (2θ) depending on the collimator used and the reflection angle. The Soller slits are normally placed between the specimen and crystal to reduce the extraneous radiation from the specimen chamber. A second set of slits, usually with a large aperture is used as an antiscatter device to make certain the detector receives only radiation diffracted by the crystal. The resolution can be markedly increased by using a small aperture on the second set although there is a loss of intensity.

Instrumentation. An x-ray spectrograph consists of three major components: an x-ray generator for operating the x-ray tube at 20–60 kv dc constant potential and highly stabilized for constant x-ray output; a goniometer with specimen chamber, collimators, crystal analyzers, detectors and vacuum chamber for the x-ray path; and an electronic circuit panel for operating the detector and measuring the intensities. A computer is frequently used as described below.

The goniometer geometry used for large specimens with surfaces up to 25 mm diameter is the flat crystal nonfocusing arrangement (2) shown in Fig. 3. The x-ray tube X and the specimen S are closely coupled to achieve maximum intensity. The fluorescence x rays are emitted in all directions but only a relatively small solid angle of radiation passing through the first Soller slit collimator C_1 can reach the crystal analyzer A. The radiation reflected by the crystal passes through a second collimator C_2 and into the detector D. The crystal is set at one-half (θ) the angle of the detector (2θ) and is rotated at one-half the angular speed of the detector; this 1:2 relationship is maintained by the goniometer. The crystal reflects each wavelength emitted by the specimen at successive angles according to the Bragg law and the 2θ angles can be read from the goniometer scale. For smaller specimens or for the analysis of small selected areas, diaphragms may be inserted between X and S and between S and A.

If the specimen is small a focusing geometry is advantageous because a larger aperture divergent beam from the specimen can be used to obtain higher intensity than that of the nonfocusing geometry. The collimators are not required and a curved crystal is used as an analyzer to focus the beam on a receiving slit in front of the detector. In the electron microprobe where only a very small area is analyzed at one time, the focusing geometry is essential to obtain adequate intensity.

The instruments described above require a scanning goniometer to analyze a series of elements. x-Ray spectrographs are available in which a number of crystal analyzers and detectors are arranged in a circle and all are pointed toward the center of the specimen. About a dozen small simple nonscanning goniometers may be used, each preset for a different element and all the intensities are read simultaneously. A rotating disc containing a large number of samples and standards may be used to automatically change samples. Simultaneous readout systems, particularly the computer-controlled types, are ideally suited for large scale routine on-line analysis.

A vacuum housing around the x-ray path is required to avoid large intensity losses due to air absorption of the longer wavelengths. The losses increase with increasing wavelength and the vacuum path becomes essential beyond CrK radiation ($\lambda = 2.3$ Å). Alternatively He paths have been used.

In the preliminary study of an unknown, the spectrum may be scanned at a uniform speed (1–2° min^{-1}) to obtain a strip chart recording for qualitative analysis and to determine the best experimental conditions. The goniometer can then be set on each of the peaks and background and intensities measured for a sufficient length of time to obtain good statistical precision.

Computers are widely used in x-ray analysis. Data processing and reduction has been carried out on an off-line computer using punched paper tape, magnetic tape or cards produced at the x-ray unit. Improved methods have been developed in recent years in which a small dedicated computer is interfaced with the spectrograph and used on-line, or a larger computer is time-shared. Progress is being made toward the automation and computer control of whole laboratories to operate the instruments, control experimental conditions and for data processing (12). It is possible to "talk" to the instruments through the computer, to modify the routines as required by the experimenter, and to monitor the analyses on cathode ray screens.

The computer may be used to automatically control the experimental conditions so that they may be optimized for each element. Methods and programs have been developed for the selection of many parameters: 2θ angles, scanning and slewing speeds, samples and standards, crystal analyzers and their reflection order, collimators, detectors, x-ray tube voltages, pulse height analyzer settings, etc. The intensity measurements may be programmed for fixed-time, fixed-count or ratio readings. Data reduction programs have been written for solving the Bragg equation and calculating the elements present, correcting the observed intensities for background, absorption and other effects required to make accurate calculations of the concentrations. The setting up of computer controlled automatic systems requires proper design of the interfacing between the spectrograph, electronic circuits and the computer as well as a considerable amount of programming. Commercial apparatus incorporating many of the required features are now available.

The detectors most frequently used are NaI.Tl scintillation counters, sealed-off proportional counters, thin-window gas-flow proportional counters and recently lithium-drifted germanium or silicon solid state detectors. All of these detectors have the property of producing a pulse whose amplitude is proportional to the incident absorbed x-ray quantum. Electronic pulse-amplitude discrimination techniques are used to limit the range of pulse amplitudes recorded for each element thereby increasing the peak-to-background ratio.

The energy resolution is an important factor in selecting a detector when analyses of closely spaced spectral lines are to be made with the pulse height analyzer. The energy resolution is determined with a monochromatic x-ray beam, a pulse height analyzer and a scaling circuit. The curve relating the pulse amplitudes as a function of the counting rates appears as a Gaussian distribution and the resolution is expressed as the width W of this curve at one-half peak height divided by the average pulse amplitude A (2). The scintillation counter has the poorest resolution, $W/A \approx 0.50$ (300 eV), the proportional counter ≈ 0.18 (1000 eV), Ge(Li) ≈ 0.05 (325 eV), and Si(Li) ≈ 0.02 (175 eV); these are approximate values for 5–10 kV x rays.

The quantum counting efficiency is determined by the transmission of the detector window and the absorption of the x ray in the crystal or gas. Be windows 0.025–0.125 mm thick are commonly used. When the gas-flow proportional counter is used for soft x rays the windows are often 6 μ Mylar coated with a 200 Å Al film (to avoid distortions in the electrical field around the window) or 1 μ polypropylene for softer x rays. Leakage of the counter tube gas (usually argon + 10% methane) through the window is compensated by flowing the gas at about 3 liters per hour through the detector and the pump immediately removes the gas from the spectrometer vacuum chamber. In the short wavelength region up to about 2 Å the window transmission is high and the crystal or gas absorption determines the efficiency. The photomultiplier tube noise limits the efficiency of the scintillation counter in the longer wavelength region. The low absorption of A or Xe gas fillings in the short wavelength region reduces the efficiency of the proportional counter and this built-in discrimination is often useful in certain types of analyses of mixtures of high and low atomic number elements. In practice, the scintillation counter may be used to about 2 Å and the gas-flow proportional counter for longer wavelengths. The two detectors are often mounted in tandem on the spectrometer and since the crystal analyzer is also generally changed when measuring a wide range of wavelengths, the spectrometer is equipped with a lever to allow switching the crystal and detector.

The resolving time of the detectors and their circuits is less than a microsecond which provides a linear response with only a 1% loss at 10^4 Hz. At higher rates a dead-time correction may be applied for precise quantitative data (2). When using a pulse height analyzer at high count rates, the system should also be checked for possible shifts of the pulse amplitude distribution and pile-up of pulses at very high rates which may shift the peak from the channel and cause errors in the recorded intensities.

Many radioisotope sources have become available for use in various types of analyses. Their use in place of the conventional x ray or electron excited source of fluorescence has been tried in a number of applications. A larger variety of α-, β-, and

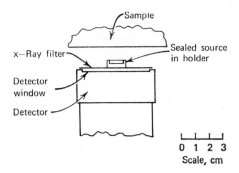

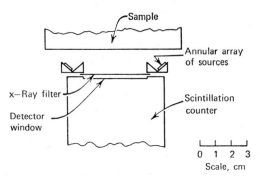

Fig. 4. Source-sample-detector geometries showing a central source (above) and annular source (below) for radioisotope nondispersive analysis (Rhodes, 14).

γ-emitters are available in various forms and their selection is determined by the energies required, convenience in handling, etc. For example, Fe^{55} decays by K-electron capture and generates MnK x rays which are useful for exciting fluorescence in Cr and lighter elements.

Radioisotopes have not been successful in conventional dispersive crystal analyzer apparatus because the usable intensities are far lower than those produced by conventional x-ray tubes or electron beams. In nondispersive analysis the crystal is eliminated and the spectra analyzed with a multichannel pulse height analyzer. When radioisotopes are used in a nondispersive geometry the source, specimen and detector can be closely coupled and the intensities are often ample (13). Special geometries such as those shown in Fig. 4 may be used to further enhance the intensities (14). By selection of the optimum combination of radioisotope source, instrument geometry and detector (scintillation, proportional or solid state) it is possible to tailor the system for many specific types of analyses. The method has the advantage of simple, compact, low-cost instrumentation and the principal limiting factor is the resolution of the detector.

The high resolution Si(Li) detectors are well-suited to this method of analysis if the required cryogenic container does not interfere with the optimum geometry. Even the highest-resolution detectors now available have a lower-resolution capability than a crystal analyzer. However, in working with the higher energy x rays, say above 40 kv, small d spacing crystals are required and since these have low reflectivity and the lines to be resolved have a greater energy separation, the nondispersive method is often used. The use of various selective filters and pulse amplitude discrimination as well as the variation of quantum efficiency of the detector with wavelength, provide additional means of effectively applying the nondispersive technique. The resolution of the recorded data may be improved by deconvolution of overlapping spectral lines with a computer, providing the line shapes and some idea of the relative intensities are known. The method is well adapted to *in-situ* analysis using portable instruments, the composition and thickness of coatings, as well as a wide variety of analyses in which high resolution is not required (1).

There are three major analytical regions based on the instrumentation required for the dispersive crystal fluorescence method. The most convenient region lies between about 0.3–2.3 Å which includes the K spectra of $_{24}$Cr to $_{63}$Eu and the L spectra of $_{62}$Sm to $_{92}$U. These spectra are not highly absorbed in air, are efficiently excited by x-ray tube voltages up to 50–60 kv, a scintillation counter which is simple to use and has nearly 100% efficiency is employed, and many good crystal analyzers are available. The soft x-ray region beyond about 2 Å (for the K spectra of $_9$F–$_{24}$Cr and the L spectra of elements below the rare earths) is more difficult experimentally because a vacuum or helium path is required to eliminate air absorption, a flow counter with thin window is necessary, lower voltage and higher current x-ray tube excitation must be used and the intensities are lower. Commercial apparatus combining both regions are available making it possible to do analyses in a relatively routine manner. The wavelength region below fluorine is not attainable by fluorescence and direct electron excitation has been used for elements down to boron. The K spectra of the elements above the rare earths require voltages above 50 kv to efficiently excite fluorescence and instead the L spectra are used.

Several important related methods have not been discussed. These include the electron microprobe which has become the most important tool for elemental analysis in the one-micron resolution scale, and is essentially the same as the fluorescence method except that the excitation is with a fine electron beam. A number of auxiliary techniques provide information on the composition and the elemental distribution by scanning the electron beam and observing the magnified image on a cathode ray tube screen. The sample current is used to display a two-dimensional distribution of two or three elements in the sample. Other displays include higher energy back-scattered electrons, low-energy electrons emitted by the specimen to obtain the magnified image with a large focal depth as in the scanning electron microscope, and cathodoluminescence in which ultraviolet, visible or infrared radiation emitted by the sample are analyzed. Kossel cameras for microdiffraction are also used for structural studies. Recent instrumentation developments have accelerated Auger and electron spectroscopy applications for the detection of low atomic number elements and for the determination of valence and oxidation states in very thin surface layers.

Correction Factors. Ideally each observed peak intensity should be proportional to the concentration of the element that produced the peak. Rarely is this the case. There are a large number of random and systematic errors which must be corrected to attain accurate quantitative data. The factors include equipment variables such as short- and long-time primary beam stability and drifts in the electronic counting circuits, counting statistics, improper specimen preparation and the many parameters associated with the generation of the fluorescence particularly in the long wavelength region. Present-day equipment minimizes the stability factors and errors are generally traceable to improper use of correction factors and specimen preparation.

x-Ray quanta are emitted randomly and a large number of repeated measurements of a constant source approximates a Gaussian distribution around the true value. Statistical factors must therefore be evaluated to determine the best counting strategy consistent with the available counting time, intensities, peak-to-background ratio and similar factors. It is also evident that longer counting times are required to achieve the same absolute accuracies for the elements present in small concentrations as for the major elements. These factors have been described elsewhere (2,4).

The sample must be homogeneous and the surface representative of the bulk down to the penetration depth to achieve significant results. The primary x-ray beam has a marked intensity distribution and the specimen may be rotated for averaging. Surface roughness and large variations in grain sizes of different components should be avoided; forming a wafer of the powder in a high-pressure press is common practice. Interelement effects may strongly modify the observed relative intensities. For example, when Cu and Co are both present the Cu fluorescence efficiently excites Co fluorescence and Co also strongly absorbs Cu x rays causing the observed intensities to be too low for Cu and too high for Co. A matrix of moderate or heavy elements strongly absorbs light element x rays. A series of carefully prepared standards is therefore essential to correct the observed relative intensities to derive the correct concentrations. Various techniques have been employed including the use of external standards, internal standards with the same or different elements, dilution, borax fusion, and other methods (4).

Applications. Several hundred papers are published each year on the applications of x-ray fluorescence spectrography in research, quality and process control (1). It is nondestructive, precise and rapid once the standards have been prepared (and some experience gained in making the various corrections) and the trend is toward automatic computer control to greatly reduce the cost for analysis, which is already below that of most other methods. The thickness, composition and other properties of thin films and coatings can be determined rapidly and with good precision by x-ray fluorescence, absorption and diffraction methods (15). The method has been used for trace analysis of the sample directly in which case the matrix absorption largely determines the sensitivity, or on chemically or physically extracted portions where the sensitivity may extend to fractional microgram quantities of metals. On-stream analysis of continuously flowing samples of solutions, slurries or powders have been used in process and quality control. Information on the coordination number and valence state may be obtained from changes in the line shapes, reflection angles and relative intensities of the soft x-ray spectra. The large variety of analyses include ores, soils, glasses, catalysts, alloys, clays, dusts, paints, silicates, meteorites, and many others. It is now virtually as widely used as emission spectroscopy and it has become a standard method of chemical analysis.

Bibliography

"X-Ray Fluorescence Spectrography" in *ECT* 1st ed., Vol. 15, pp. 176–185, W. Parrish, Phillips Laboratories.

1. W. J. Campbell and J. D. Brown, *Anal. Chem.* **36**, 312R (1964); *ibid.*, **38**, 416R (1966); *ibid.*, **40**, 346R (1968); *ibid.*, **42**, 248R (1970).
2. W. Parrish, ed., *X-Ray Analysis Papers*, Centrex Publishing Co., Eindhoven, Netherlands, 1965.
3. I. Adler, "X-Ray Emission Spectrography in Geology," Elsevier Publishers, New York, 1966.
4. R. Jenkins and J. L. DeVries, *Practical X-Ray Spectrometry*, Phillips Technical Library, Eindhoven, Netherlands, 1967.
5. R. O. Müller, *Spektrochemische Analysen mit Röntgenfluoreszenz*, R. Oldenburg Verlag, Munich and Vienna, 1967.
6. L. S. Birks, *Electron Probe Microanalysis*, Interscience Publishers, a division of John Wiley & Sons, New York, 1963.
7. T. McKinley, K. F. J. Heinrich, and D. Wittry, eds., *The Electron Microprobe*, John Wiley & Sons, Inc., New York, 1966.
8. K. F. J. Heinrich, ed., "Quantitative Electron Microprobe Analysis," *Nat. Bur. Stand. Special Pub.* **298**, Washington, D.C., 1968.
9. G. von Hevesy, *Chemical Analysis by X-Rays and its Applications*, McGraw-Hill, New York, 1932.
10. J. A. Bearden, *X-Ray Wavelengths*, U.S. Atomic Energy Commission Final Report, Contract AT (30-1)-2543 (1964).
11. B. Henke, *Advances in X-Ray Analysis*, Plenum Press, New York, 1961, Chap. 5, p. 288.
12. H. Cole, *Intern. Bus. Mach. Jour.* **13** (5), 1969.
13. W. J. Campbell, in H. van Olphen and W. Parrish, eds., *X-Ray and Electron Methods of Analysis*, Plenum Press, New York, 1968.
14. J. R. Rhodes, *Analyst*, **91**, 683 (1966).
15. E. P. Bertin, in E. M. Murt and W. G. Guldner, eds., *Physical Measurement and Analysis of Thin Films*, Plenum Press, New York, 1969.

W. Parrish
International Business Machines Corporation

XYLENES AND ETHYLBENZENE

Xylenes and ethylbenzene are eight-carbon-atom aromatic compounds of the benzene family. The term "xylenes" generally applies to a mixture of any two or three of the dimethylbenzene isomers, ie, *ortho-*, *para-*, and *meta*-xylenes. Frequently accepted nomenclature for the xylenes plus ethylbenzene is "mixed xylenes." (The name "xylol" or "xylole," which is to be deplored, is unfortunately still used occasionally in industry.)

A. Cahours is credited with the discovery of xylenes in the middle of the 19th century (1). At about this time, by-product chemicals from coal carbonization were being recovered and this became the principal source of xylenes. Xylenes recovery from by-product coking of coal continued to grow. By the year 1942, the total United States production from this source was 67 million lb (2). At this time, the U.S. Tariff Commission reported for the first time the production of xylenes from petroleum. These xylenes, however, were not materials native to petroleum. They were the products obtained from the newly developed reforming process used in the production of toluene for World War II requirements.

The extensive postwar development of the reforming process by petroleum refineries has resulted in widespread adoption of the industrial process for the manufacture of BTX (benzene-toluene-xylenes) aromatics principally for premium-quality gasoline production. There is an enormous quantity of C_8 aromatics in this product. A conservative estimate of the U.S. production of mixed xylenes from this source is 4.5 billion gallons or more. Of this total amount, only about 10% of the mixed xylenes are recovered by liquid extraction for use as solvents or for chemical uses.

The availability of large volumes of xylenes has had a major impact on the rapid growth of these materials. Since 1950, production has increased at an average of 11% per year. In addition, large amounts of ethylbenzene have been synthesized from benzene and ethylene. The mixed xylenes were originally used for solvents and as components of high-octane quality fuels. The postwar development has resulted in the rapid growth of the use of the individual pure isomers. Each finds its principal outlet to one particular product. Ethylbenzene is primarily employed as a raw material for the manufacture of styrene. Each of the xylene isomers finds its principal outlet as a corresponding dibasic acid. The amount of *meta*-xylene so used, however, is much less than either the ortho or the para isomer.

Properties

Ethylbenzene   *o*–Xylene   *m*–Xylene   *p*–Xylene

Many of the properties of the individual C_8 aromatic isomers are very similar. A consequence of this is that the production of individual components of very high purity becomes difficult. Since there is a great demand especially for pure *p*-xylene, much effort has been directed to accomplish these separations. The study and evaluation of

Table 1.   Physical Properties of C_8 Aromatics

Property	Ethylbenzene	*p*-Xylene	*m*-Xylene	*o*-Xylene
density, g/ml				
at 20°C	0.86702	0.86105	0.86417	0.88020
at 25°C	0.86264	0.85669	0.85990	0.87596
boiling point, 760 mm, °C	136.186	138.351	139.103	144.411
dt/dp, °C/mm Hg	0.04898	0.04917	0.04903	0.04969
freezing point, °C	−94.975	13.263	−47.872	−25.182
refractive index				
n_D^{20}	1.49588	1.49582	1.49722	1.50545
n_D^{25}	1.49320	1.49325	1.49464	1.50295

Table 2.   Critical Properties of C_8 Aromatics

Property	Ethylbenzene	*p*-Xylene	*m*-Xylene	*o*-Xylene
critical temperature, °C	346.4	345.0	346.0	359.0
critical pressure, atm	37	34	35	36
critical density, g/ml	0.29	0.29	0.27	0.28
critical PV/RT	0.26	0.25	0.27	0.26

the physical and chemical processes for this purpose has been in greater depth than for many organic compounds. Therefore, a broader than usual background of information of properties particularly oriented toward isomer separation is given here. Tables 1–6 (3) present physical and thermodynamic values of the individual eight-carbon-atom aromatic isomers.

Of great practical importance, of course, are the distillation characteristics of these isomers. While *ortho*-xylene can be separated readily from *meta*-xylene by distillation, it is only with difficulty that ethylbenzene can be distilled from *para*-xylene in pure form. It is not at all practical to separate *para*-xylene from *meta*-xylene by distillation. Under such circumstances the assumption of "ideal" solutions when making distillation calculations may not be valid. Studies of the nonideality of the mixed xylene isomers have been reported by Redlich, et al. (4). From their data the activity coefficients at infinite dilution can be calculated. These are presented in Table 7, along with the corresponding vapor pressure ratios.

Since deviations from ideality in these systems are symmetrical, the activity coefficient for *para*-xylene at infinite dilution in ethylbenzene is equal to the value of ethylbenzene at infinite dilution in *para*-xylene. Especially in the case of ethylbenzene-*para*-xylene binary, the nonideality can have a significant effect on the relative vola-

Table 3. Heat of Vaporization, Heat of Fusion, and Cryoscopic Constants

Property	Ethylbenzene	*p*-Xylene	*m*-Xylene	*o*-Xylene
heat of vaporization, ΔH_v, kcal/mol				
at 25°C	10.097	10.128	10.195	10.381
at boiling point	8.60	8.62	8.70	8.80
heat of fusion, ΔH, kcal/mol	2.190	4.090	2.765	3.250
cryoscopic constants, mol fraction/°K[a]				
A	0.03471	0.02509	0.02741	0.02659
B	0.0029	0.0028	0.0027	0.0030

[a] For calculating purity, p, in mol %: $\log p = 2.00000 - (A/2.30259)(T_1 - T)[1 + B(T_1 - T)]$, where T_1 is freezing point in °K of the pure component, and T is freezing point in °K of the actual sample.

Table 4. Heat of Combustion of Liquid C_8 Aromatics

C_8 Aromatic	H_c, kcal/mol[a]	Aromatic	H_c, kcal/mol[a]
ethylbenzene	1091.03	*m*-xylene	1087.92
p-xylene	1088.16	*o*-xylene	1088.16

[a] At 25°C to H_2O (1) and CO_2 (g).

Table 5. Heat of Formation, Entropy, and Free Energy of Formation of Liquid Aromatics[a]

C_8 Aromatic	Heat of formation, kcal/mol	Entropy, cal/(day) (mol)	Free energy of formation, kcal/mol
ethylbenzene	−2.977	60.99	28.614
p-xylene	−5.838	59.12	26.310
m-xylene	−6.075	60.27	25.730
o-xylene	−5.841	58.91	26.370

[a] At 25°C.

tility (relative volatility is equal to the product of the activity coefficient ratio multiplied by the vapor pressure ratio). Since the number of theoretical stages is an exponential function of the relative volatility, the small nonideality that exists does have a practical significance.

Table 6. Vapor Pressures of C_8 Aromatics

Pressure, mmHg	Temperature,°C			
	Ethylbenzene	p-Xylene	m-Xylene	o-Xylene
10	25.88	27.32	28.24	32.14
20	38.60	40.15	41.07	45.13
30	46.69	48.31	49.23	53.38
40	52.75	54.42	55.33	59.56
50	57.657	59.363	60.269	64.558
60	61.789	63.535	64.437	68.778
80	68.596	70.383	71.277	75.704
100	74.105	75.931	76.818	81.314
150	84.687	86.583	87.454	92.085
200	92.680	94.626	95.483	100.217
250	99.182	101.167	102.011	106.829
300	104.703	106.719	107.551	112.441
400	113.823	115.887	116.699	121.708
500	121.266	123.366	124.159	129.267
600	127.603	129.732	130.508	135.700
700	133.152	135.304	136.065	141.332
800	138.106	140.278	141.025	146.359
900	142.595	144.787	145.517	150.912
1000	146.71	148.91	149.63	155.08
1200	154.06	156.29	156.98	162.53
1500	163.47	165.73	166.39	172.07

Constants of the Antoine Equation[a]

A	6.95719	6.99052	7.00998	6.99891
B	1424.255	1453.430	1462.266	1474.679
C	213.206	215.307	215.105	213.686

[a] Antoind Equation: $\log P = A - B/(C + t)$.

Table 7. Activity Coefficients of C_8 Aromatic Isomers (4)

Isomer	Activity coefficient at infinite dilution[a]	Vapor pressure ratio[a]
ethylbenzene/p-xylene	1.0072	1.060
ethylbenzene/m-xylene	1.0083	1.081
ethylbenzene/o-xylene	1.0081	1.241
p-xylene/m-xylene	1.0007	1.020
p-xylene/o-xylene	1.0034	1.173
m-xylene/o-xylene	1.0049	1.151

[a] At boiling point at atmospheric pressure.

A comprehensive compilation of vapor–liquid equilibrium data in which one of the components is a C_8 aromatic isomer is found in the text of Hala et al. (5). In addition to this, a great deal of data has been published on azeotropes. Some of these data pertaining to the C_8 aromatics are given in Tables 8–11, inclusive (6).

Table 8. Azeotropes of Ethylbenzene, Boiling Point 136.2°C

Component	Component bp, °C	Azeotrope	
		bp, °C	Ethylbenzene, wt %
water	100.0	33.5[a]	67
formic acid	100.75	≈94.0	32
acetic acid	118.5	114.65	34
propionic acid	140.9	131.1	72
butyric acid	164.0	135.8	96
isobutyric acid	154.6	134.3	
butyl alcohol	117.8	114.8	33
isobutyl alcohol	108.0	107.2	20
2-ethoxyethanol	135.3	127.8	52
isoamyl alcohol	131.9	125.9	51
n-octane	125.75	<125.6	<12

[a] At 60 mm pressure.

Table 9. Azeotropes of p-Xylene, Boiling Point 138.4°C

Component	Component bp, °C	Azeotrope	
		bp, °C	p-Xylene, wt %
formic acid	100.75	≈95	30
acetic acid	118.5	115.25	28
propionic acid	140.9	132.5	66
butyric acid	164.0	137.8	94.5
isobutyric acid	154.6	136.4	87
butyl alcohol	117.8	115.7	32
isobutyl alcohol	108.0	≈107.5	≈17
2-ethoxyethanol	135.3	128.6	50
isoamyl alcohol	131.9	125–126	42
hexyl alcohol	157.8	≈137.7	87

Table 10. Azeotropes of m-Xylene, Boiling Point 139°C

Component	Component bp, °C	Azeotrope	
		bp, °C	m-Xylene, wt %
water	100	92.0	64.2
formic acid	100.75	92.8	28.2
acetic acid	118.5	115.35	27.5
propionic acid	140.9	132.65	64.5
butyric acid	164.0	138.5	94
isobutyric acid	154.6	136.9	85
butyl alcohol	117.8	116.5	28.5
isobutyl alcohol	108.0	107.78	14.5
2-ethoxyethanol	135.3	128.85	49
isoamyl alcohol	131.9	125–126	48
hexyl alcohol	157.8	138.3	85

Table 11.   Azeotropes of *o*-Xylene, Boiling Point 143.6°C

Component	Component bp, °C	Azeotrope	
		bp, °C	*o*-Xylene, wt %
formic acid	100.75	95.5	26
acetic acid	118.5	116.0	24
propionic acid	140.9	135.4	57
butyric acid	164.0	143.0	90
isobutyric acid	154.6	141.0	78
butyl alcohol	117.8	116.8	25
isobutyl alcohol	108.0	nonazeotrope	
2-ethoxyethanol	135.3	130.8	45
isoamyl alcohol	131.9	127	<48
hexyl alcohol	157.8	142.3	≈82

The use of selective azeotropic distillation (see Azeotropy) has been studied for separation of isomers and also for the removal of nonaromatic hydrocarbons from the mixed xylenes.   Lake, et al. (7) and Greene (8) disclose separation of a *meta*-xylene–*para*-xylene mixture from *o*-xylene using ethylene glycol derivatives as the azeotroping entrainer.   Lake also shows that nonaromatic hydrocarbons can first be removed from the para–meta–ortho mixture by azeotropically distilling with the same agent.   Nelson (9) and Berg, et al. (10) studied the effect of numerous azeotropic agents on the separation of ethylbenzene from *p*-xylene.   The best entraining agent, 2-methyl-butanol, reduces the number of theoretical stages required by about 50%.   Berg also studied the separation of *p*-xylene–*m*-xylene and *m*-xylene–*o*-xylene.   Highlights of their data are summarized in Table 12.

Table 12.   Azeotropic Separations (10)

Binary pair	Azeotroping agent	Relative volatility
ethylbenzene–*p*-xylene	none	1.037
	2-methyl-(1?)-butanol	1.079
	4-methyl-2-pentanol	1.074
	n-hexylamine	1.073
p-xylene–*m*-xylene	none	1.020
	2-methyl-(1?)-butanol	1.029
	3-methyl-(1?)-butanol	1.026
m-xylene–*o*-xylene	none	1.105
	formic acid	1.154
	4-methyl-2-pentanol	1.150

Alcohols, it will be noted, are shown to be the best class of compounds for selectively azeotroping the mixed xylenes.

Azeotropes of paraffins with aromatics are of special interest when recovering the C_8 aromatic in high purity from petroleum hydrocarbons.   Literature data on this are scarce, as indicated by the single value listed in Tables 8–11.   As a generalization, one can say that all C_9 paraffins will azeotrope with the mixed xylenes.   Since mixtures of C_9 paraffins and C_8 aromatics are obtained as products from the reforming process, it is impossible to produce high-purity aromatics (98%) by distillation alone.

Extractive distillation has also been studied for removal of nonaromatics from the mixed xylenes, as well as from separation of the individual mixed xylene isomers. Clough, et al. (11) and Black (12) both disclose the use of monohydric phenols to separate nonaromatics from C$_8$ aromatic isomers. Chu, et al. (13) have made extensive studies on the separation of *para*-xylene from *m*-xylene by extractive distillation. A modified Othmer still was employed in these studies to determine relative volatilities. Chu concluded that separation of this pair by extractive distillation was not feasible. Suzuki (14) reports the separation of *p*-xylene from *m*-xylene by extractive distillation. This separation is conducted at temperatures below 110°C in contact with a substantially anhydrous solution of silver salt in phosphoric acid, lower alkanesulfonic acids, or a mixture thereof.

Nixon, et al. (15,16) claimed the use of antimony trichloride as an extractive distillation agent for separating ethylbenzene from xylene. Amir, et al. (17) claim the use of chlorinated aromatics as extractive distillation agents for the recovery of ethylbenzene from xylenes. Some of the compounds tested are the following:

Agent	*Relative Volatility*
p-dichlorobenzene	1.116
2,4,5-trichlorophenol	1.147
2,3,4,5,6-pentachloro-1-butoxybenzene	1.26

The relative volatility data above were obtained at 110°C in a system containing ethylbenzene, *p*-xylene, and solvent in the molar ratio of 1:1:2.

Solubility data of the individual isomers in mixed xylenes are also of great importance, because of the possibility of separation by crystallization. There are three principal equations whereby solubility of the individual isomers may be calculated:

$$-\ln N_1 = \frac{\Delta H_f}{R}\left(\frac{T' - T}{T'T}\right) \tag{1}$$

$$-\ln N_1 = A(T' - T)\,[1 + B(T' - T)\ldots] \tag{2}$$

$$-\ln A_1 = \frac{\Delta H_f}{R}\frac{(T' - T)}{T'T} \frac{C_p}{R}\left(\frac{T' - T}{T}\right) + \frac{\Delta C_p}{R}\ln\frac{T'}{T} \tag{3}$$

where N_1 is the mol fraction component; A_1 is the activity of pure component; ΔH_f is the heat of fusion of pure component; T' is the freezing point of pure component, T is the freezing point of the mixture; R is the gas constant; ΔC_p is the heat capacity of liquid phase minus heat capacity of solid phase; and A and B are cryoscopic constants.

For xylenes, equation 2 appears to give results closer to experimental data than does equation 1. Equation 3, a modified form of equation 2, is suited for use with nonideal as well as ideal systems. For calculations involving nonideal systems, activities must be used instead of mol fractions. In the case of the xylenes, the solutions are essentially ideal. Hence, insofar as solubility calculations are concerned, there is no particular advantage for equation 3 over equation 2. Figure 1 presents a chart showing calculated solubilities based on equation 2. Experimentally determined values for *p*-xylene show good agreement, except for slight deviations in the low-temperature region (18,19). This, however, is the zone of commercial interest.

Mixed xylenes on cooling first form pure solids. Continued cooling then produces eutectic mixtures. As a consequence, the pure xylene isomer can be recovered from a

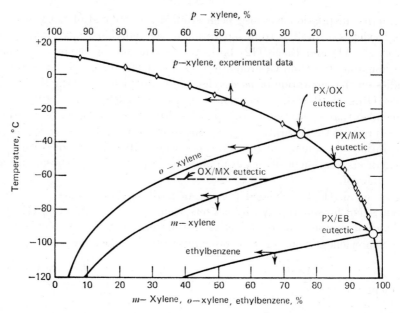

Fig. 1. Calculated C_8 aromatics solubility.

single stage of crystallization and the product will be contaminated only with the associated mother liquor. Because of the eutectic formation, the extent of recovery of a pure isomer is limited. From the solubility chart, the approximate composition of the three binary xylene eutectic mixtures of the ternary mixture can be obtained from properly drawn isotherms. McKay (21) has studied the 3-component system of the three xylenes.

	Xylene Eutectics		*Approximate*
% PX	*% OX*	*% MX*	*Temp, °F*
25	75		−32
14		86	−62
	33	67	−77
9	30	61	−82

Studies of the effect of numerous additives on the eutectic compositions of *m*- and *p*-xylene are disclosed by Szawlowski et al. (20). These studies show that the addition of somewhat less than 3 mol % of so-called "eutectic inhibitors" will allow the recovery of much larger amounts of *p*-xylene.

For example, working with a *p*-xylene–*m*-xylene eutectic with 0.4 wt % methylal, $CH_2(OCH_3)_2$ added, on cooling to −140°F and filtering about 80% of the feed metaxylene was recovered in 95% concentration as a mother liquor stream and the recovered crystals contained 70% of the charged *p*-xylene.

The eutectic inhibitors have been found to be organic compounds which contain at least one atom of oxygen, sulfur, nitrogen, phosphorus, silicon, boron, or halogen and are at least slightly soluble in the xylenes.

The low-temperature characteristics of the quaternary system of the C_8 aromatics have also been studied (22). The system approaches the thermodynamically ideal type. The quaternary eutectic composition was approximately 2% p-xylene, 5% o-xylene, 15% m-xylene, 78% ethylbenzene, mp −101° to −102°C.

Additional crystallization studies have shown that it is possible to achieve selective solid-compound formation with certain aromatic hydrocarbons. Egan and Luthy (23) present data which show the formation of an equimolar compound of p-xylene and carbon tetrachloride, freezing point 24.8°C, heat of fusion of 6.6 kcal/mol, density (solid) 1.37 g/ml at 19°F. No such compounds are formed with m-xylene or o-xylene. The addition of 0.5 volume of carbon tetrachloride to one volume of xylenes reformate feed will increase the p-xylene recovery from a nominal 60% to about 85% of that present in the feed. Bown (24,25) accomplishes essentially the same results by first crystallizing out approximately 60% of the feed p-xylene in the hydrocarbon mixture. The filtrate is then treated with carbon tetrachloride and the overall yield of p-xylene is increased to about 85%, based on the original feed. In addition to carbon tetra-chloride, Bown shows that chloral, CCl_3CHO, can also be used effectively for solid-compound formation of para-methyl aromatics.

The xylene isomers have been shown to form clathrate compounds (26). Schaeffer and co-workers have reported that metal salt complexes of substituted pyridines are a class of clathrate formers that are effective in separating aromatic isomers including xylene isomers. A general formula for the host component of the clathrate is a Werner complex which can be designated by the general formula: XZ_yA_n, where X is a metal atom of atomic number 25–28 (Mn, Fe, Co, N); Z is a basic nitrogen compound, preferably heterocyclic; A is a negative radical; $y = 2 - 6$; and $n = 1 - 3$.

The sharp selectivity of these host compounds is based on the shape and the size of the guest isomer. Schaeffer also discusses ways to make these clathrates and to recover the guest compounds therefrom (27,28). A specific example of such a host compound is tetra-(4-methylpyridine)nickel dithiocyanate. Results of a one-stage separation with this host are shown in Table 13. If a three-stage process is used, it is estimated that a 97% pure p-xylene product would be obtained.

It is reported that hosts involving any of the metallic ions of the generalized formula are selective for p-xylene. However, changes in the other components of a host have a marked effect on its selectivity. This is shown in Table 14.

Table 13. One-Stage Separation with Ni(4-Methylpyridine)₄(SCN)₂ Host (26)

C_8 Aromatic composition, vol %	Component		
	Feed	Clathrate	Xylenes residue
p-xylene	19.3	64.0	2.3
m-xylene	50.6	13.7	64.7
o-xylene	19.1	5.1	24.4
ethylbenzene	11.0	17.2	8.6

Table 14. Host Selectivity for Aromatic Isomers

Host compound	Preferred guest
Ni(4-methylpyridine)₄(SCN)₂	p-xylene
Ni(4-ethylpyridine)₄(CNO)₂	o-xylene
Ni(3-ethyl-4-methylpyridine)₄(SCN)₂	m-xylene
Ni(4-acetylpyridine)₄(SCN)₂	ethylbenzene

De Radzitsky and Hanotier (29) have also reported a number of Werner complexes as hosts for C_8 aromatics. In this instance, substituted primary benzylamines are complexed with nickel thiocyanate. According to these investigators, only amines of the indicated structural formulas will form clathrates with aromatic compounds:

(1)   (2)

In structure (1), R_1 is a primary alkyl group. In (2), R_1 is hydrogen or a primary alkyl group; R_2 is a polar or nonpolar group, but not strongly acid ($-SO_3H$), basic ($-NH_2$), or reactive ($-CHO$). These investigators also report that hosts with particular complexed amines are selective to a particular xylene isomer. See Table 15.

Table 15. Separations with Type I Hosts, Composition, wt %

| | R_1 | | | | | |
| | Methyl | | Butyl | | Hexyl | |
Component	Feed	Guest	Feed	Guest	Feed	Guest
p-xylene	31.8	61.9	34.6	2.6	34.4	8.2
m-xylene	31.1	25.8	31.2	13.6	31.7	79.4
o-xylene	37.1	12.3	34.2	83.8	33.9	12.4
guest in clathrate, wt %	17.8		10.8		8.9	

De Radzitsky and Hanotier conclude that isomer separation can be carried out cyclically, and more effectively than by distillation. They also claim that capacities of the hosts for the isomer are greater than can be achieved in adsorption.

Although there is a good deal of information on the formation of clathrates there is little information on the recovery of the isomers from the clathrate and the recycle of the host compound. Fleck, et al. (30) have reported an apparatus for clathrate separations. Another separation method in which isomer shape and size are of great significance is the use of conditioned membranes. Michaels (31) has reported studies employing a conditioned film of polyethylene. This conditioning is achieved by allowing the polyethylene film to swell in one of the isomers, which results in an increase approximately tenfold in the permeation rate of all the xylenes, and a selectivity for the isomer with which the film was conditioned. The investigators estimate that 10,000 lb/day of an equimolar mixture of o- and p-xylene could be separated into 98%-purity products with a six-stage cascade employing 2000 sq ft of film area per stage.

Adsorption theory and applications in analytical separations are discussed by Mair, et al. (32,33). On the basis of this theory, the separation of benzene and toluene from nonaromatics with silica gel as the absorbent has been studied (34). Xylenes were used as a displacing agent to remove the benzene and toluene from the silica gel. It is also possible to adsorb xylenes from hydrocarbon mixtures and to displace the xylenes from the adsorbent by use of benzene and/or toluene.

Application of the adsorption technique to separation of xylene isomers, in particular the para and meta binary, has been reported by Eberly, et al. (35) and Fleck, et al. (36). Both of these disclose that a 10–13Å molecular sieve is used as the adsorbent in a vapor-phase process. This zeolite has pores of adequate size to admit all components of the xylenes mixture. The separation, therefore, will be related to the heats of adsorption of the key components. As a consequence, such a separation will require a number of stages to give relatively pure products. If the four C_8 isomers are present in the feed, the ethylbenzene will be found with the p-xylene and o-xylene will be associated with the m-xylene. The development of an improved adsorbent that can remove p-xylene in high purity from other C_8 aromatic isomers has been reported (37). This is employed in the Parex process, licensed by Universal Oil Products, for the selective recovery of p-xylene. It is operated as a liquid-phase process (250–350°F) at moderate pressures.

Solubility data for water in the eight-carbon-atom aromatics and vice versa (from smoothed data) are given in Table 16 (38).

Table 16. Solubility Data for Water and C_8 Aromatics, mol fraction (38)

Hydrocarbon	Temperature, °F			
	50	80	100	150
Water in C_8 Aromatics				
xylenes	0.0017	0.0031		
ethylbenzene		0.0026	0.0038	0.0080
C_8 Aromatics in Water				
xylenes	0.000034	0.000033	0.000035	0.000052
ethylbenzene	0.000036	0.000035	0.000037	0.000058

Friedel-Crafts reactions of aromatic systems are important. Fairbrother, et al. (39) have studied the solubility of aluminum chloride in several solvents including the xylenes. See also reference (40).

COMPLEX FORMATION

Aromatic hydrocarbons are capable of forming complexes with a wide variety of electrophilic agents. Because of the importance of these complexes, both in the field and in physical separations and chemical reactions, a discussion of these characteristics follows.

The formation of a 1:1 complex of HBr and ethylbenzene (mp −103°C) (41) and of HBr and m-xylene (mp −77°C) (42) was reported by Maas and co-workers. Studies by Brown et al. (43) of HCl:aromatic complexes show that there is a rapid formation and decomposition of such a complex at −80°C. Klatt (44) has reported a significant solubility of aromatics in liquid HF. This is attributed to some kind of interaction between the hydrogen halide and the aromatic.

C_8 aromatics with hydrogen halide and the aluminum halide (or hydrogen fluoride and boron trifluoride) result in the formation of a very different complex. For example, no significant change in the absorption spectra is noted with the hydrogen chloride complex, but with the HCl:AlCl$_3$ complex, an intense color is noted in the visible region. By contrast with the HCl:aromatic complex, the formation and decomposition of the ternary complex is slow at −80°C. The ternary complex is also reported to have a high electrical conductivity, whereas the HCl binary complex does not. Complexes with HF and BF_3 are known only in one form, but complexes of HBr and $AlBr_3$ have been shown to exist in two forms: m-xylene:HBr:$AlBr_3$, and m-xylene:HBr:Al_2Br_6.

From these and other studies, Nelson et al. (45) state that the properties of HCl and related complexes are consistent with the structure wherein the HCl is closely bound to the pi electron cloud with no definite bond existing between the electrophilic group and any particular carbon atom. The properties of the HCl:$AlCl_3$ complexes are in agreement with a carbonium ion type of structure. In this system the proton has been transferred to the ring: $(ArH)^+ (AlCl_4)^-$.

McCaulay and Lien (46) provide data concerning the relative stability of the aromatic:HF:BF_3 complexes. They conclude that the xylene isomers are more basic (have a tendency to combine with a proton, H^+) than toluene, and m-xylene is the most basic of the three dimethylbenzenes.

Kilpatrick et al. (47) established the relative base strengths of the aromatic hydrocarbons, including the xylenes, in reaction with hydrofluoric acid:

$$Ar + HF \rightarrow ArH^+ F^- \tag{4}$$

$$K = \frac{(ArH^+) (F^-)}{(Ar) (HF)}$$

Conductance measurements on the system lead to determinations of the relative basicities of the xylenes and other methylbenzenes. Ehrenson (49) has made quantum chemical examinations of the equilibrium and compared these with the experimentally derived values from Kilpatrick (47) and also McCaulay (46). He concludes that a combined model including hyperconjugative and inductive effects of the methyl groups is preferred to either single model alone. This confirms the conclusions reported earlier by McCaulay, et al. (46) on the basis of experimental studies. Table 17 presents the

Table 17. Overall Equilibrium Constants, K (eq. 4) (49)

Hydrocarbon	McCaulay and Lien	Kilpatrick and Luborsky	Mackor et al. (48)	HCJ and inductive model
benzene		0.09	2×10^{-4}	1.2×10^{-4}
toluene	0.01	0.63	0.25	0.11–0.13
o-xylene	2	1.1		1.8
m-xylene	20	26	300	90–110
p-xylene	1	1	1	1
pseudocumene (1,2,4-trimethylbenzene)	40	63		110–140
hemimellitene (1,2,3-trimethylbenzene)	40	69		200–310
durene (1,2,4,5-tetramethylbenzene)	120	140		510–810
prehnitene (1,2,3,4-tetramethylbenzene)	170	400		960–1700
mesitylene (1,3,5-trimethylbenzene)	2800	1.3×10^4	2×10^5	2–3.6×10^4
isodurene (1,2,3,5-tetramethylbenzene)	5600	1.6×10^4		0.6–1×10^5
pentamethylbenzene	8700	2.9×10^4		1.1–2.5×10^5
hexamethylbenzene	8.9×10^4	9.7×10^4	1×10^7	0.7–1.5×10^6

data from these several investigators for benzene, toluene and all the (polymethyl) benzenes.

CHEMICAL REACTIONS

Chemical reactions of the C_8 aromatics can be divided into three categories: *Class 1*—reactions involving the number and position of the alkyl groups; *Class 2*—chemical reactions of the alkyl groups; *Class 3*—reactions involving the aromatic ring. The first category is of primary interest in the production of pure individual isomers. The second category dominates the transformation of the pure isomers into products of commercial interest. The third group also produces new products from the parent isomer but these are not of major commercial interest as yet. In any such grouping, of course, there will be overlaps and exceptions will be noted.

Class 1 Reactions. Calculated equilibrium concentrations of the four isomers are shown in Table 18.

Table 18. Calculated Equilibrium Concentration of C_8 Isomers (3)

C_8 Isomer	Temperature, °K		
	300	600	900
ethylbenzene	0.5	5.9	13.8
p-xylene	23.7	22.4	19.2
m-xylene	59.6	50.1	43.9
o-xylene	16.2	21.6	23.1

Experimental studies of the isomerization of xylenes in the liquid phase, using various halide catalysts at 50–120°C, show no ethylbenzene in the products (51–53). The meta isomer was found to be higher than the equilibrium value; this may be influenced by the concentration of the catalyst.

In the presence of large amounts of BF_3, the xylenes form a complex and remain in the acid phase. In the protonated form the xylenes isomerize almost exclusively to the configuration which gives the most basic structure, the meta isomer. The amount of uncomplexed xylenes in the hydrocarbon phase is governed by the thermodynamic equilibrium as shown in Table 18.

Low-temperature (0–30°C) kinetic studies of *o*- and *p*-xylene isomerization in an excess of boron trifluoride show no disproportionation. *p*-Xylene isomerizes about five times as fast as the ortho isomers; the primary product in each case is *m*-xylene (53).

Vapor-phase isomerization of *m*-xylene with different catalysts has been studied at temperatures above 380°C. At 13 atmospheres with Pt on silica-alumina catalyst, an equilibrium mixture of xylene is obtained from *m*-xylene; no ethylbenzene is produced (54). At low pressures (0.1–1.0 atmospheres) with a silica-alumina catalyst, small conversion of *m*-xylene to ethylbenzene is noted (57). At high pressures (40 atmospheres) with tungsten-molybdena on silica-alumina catalyst the presence of ethylbenzene adversely affects the extent of xylene isomerization (55).

The isomerization of ethylbenzene to xylenes does not occur readily. It shows a negative temperature coefficient. This suggests that some intermediate whose concentration is inversely proportional to temperature is involved in this overall reaction. A likely compound would be the hydrogenated aromatic (54).

The mechanism of the xylene isomerization reaction is now considered to be of an intramolecular nature (56,58). Addition of a proton to the aromatic to form a carbonium ion appears to be the first step. On rearrangement and subsequent elimination of the proton the isomerized product results. Additionally the isomerization of ethylbenzene may occur through formation of hydrogenated intermediates which are later dehydrogenated to xylenes.

In studies with a silica-alumina cracking catalyst at 515°C, with a feed of C_8 aromatics, at 90 mmHg there was hardly any disproportionation, only isomerization, but at 760 mmHg there was considerable disproportionation (57).

Disproportionation of ethylbenzene and m-xylene has been studied with an HF-BF_3 catalyst (133,134). The disproportionation is a function of BF_3 concentration up to an equimolar ratio of the BF_3 to aromatic. The migration of ethyl groups is much faster than that of methyl groups (59,60). In the presence of excess HF-BF_3, almost complete disproportionation of ethylbenzene occurs at 0°C. At 66°C, m-xylene is unreactive but at higher temperatures it disproportionates.

Under proper conditions, ethylbenzene will preferentially transalkylate xylenes without disproportionation of the xylenes. Some disproportionation of the ethylbenzene does occur. The rate of disappearance of ethylbenzene by interaction with xylene is greatest with the ortho derivative and least with the para derivative.

Dealkylation processes involving C_8 aromatics may be of two types, ie, those involving dealkylation of C_8 to lower molecular weight compounds and those involving dealkylation of higher molecular weight aromatics to C_8 aromatics. Dealkylation of xylenes in the presence of hydrogen results in many higher-molecular-weight products including dicyclic and polycyclic products.

Silsby, et al. (61) have studied dealkylation of xylenes in the presence of hydrogen. They conclude that: (1) the reaction is principally thermal; (2) pressure has little effect on the degree of conversion; and (3) the xylene reaction products are consistent with the idea of two consecutive reactions:

$$C_6H_4(CH_3)_2 + H_2 \rightarrow C_6H_5CH_3 + CH_4 \tag{5}$$

$$C_6H_5CH_3 + H_2 \rightarrow C_6H_6 + CH_4 \tag{6}$$

Tsuchiya, et al. made extensive studies of the dealkylation rates of the individual C_8 isomers (62,63). These studies encompassed the temperature span of 590–680°C at 10–40 atmospheres. Molar ratios of hydrogen to hydrocarbon of 3:11 were employed at contact times of 10–60 seconds. The xylene isomer rates of dealkylation could be expressed as a reaction of 1.5 order.

Dealkylation of ethylbenzene under similar conditions is much more complex. At least three reactions are encountered:

$$C_6H_5C_2H_5 + H_2 \rightarrow C_6H_6 + C_2H_6 \tag{7}$$

$$C_6H_5C_2H_5 + H_2 \rightarrow C_6H_5CH_3 + CH_4 \tag{8}$$

$$C_6H_5C_2H_5 \rightarrow C_6H_5CH{=}CH_2 + H_2 \tag{9}$$

The production of styrene has been observed only in runs below 20 atm pressure.

The complexity of the reaction makes kinetic analysis difficult. The dealkylation rate of ethylbenzene under these conditions appears to be slightly greater than for the xylenes.

Dealkylation of a highly aromatic extract from a kerosene fraction gives a number of different aromatic products (64). Operating conditions can be adjusted to maximize benzene, toluene or C_8 aromatics. There is a large degradation of feed in such operations since only 50–65% is recovered as liquid product.

Class 2 Reactions. From an industrial viewpoint, the most important reactions of the C_8 aromatic isomers are those involving the oxidation of the methyl groups and the dehydrogenation of the ethyl group.

Ethylbenzene is catalytically dehydrogenated to give styrene and hydrogen. Thermodynamically, high temperatures and low pressures favor the dehydrogenation (Table 19).

Table 19. Calculated Ethylbenzene–Styrene Thermal Equilibrium

Temperature, °F	Conversion to styrene	
	Mol % at 1 atm	Mol % at 0.1 atm
1100	39	75
1300	74	92
1500	90	97

In actual practice, conversion levels of 35–60% are realized with selectivities of approximately 90% or higher, the partial pressure of ethylbenzene being lowered by the addition of large volumes of steam. The use of carbon dioxide in place of steam has been suggested (65). With a proper catalyst the available hydrogen and carbon dioxide are converted via the water gas shift reaction to carbon monoxide and water. As a consequence of the removal of hydrogen by a secondary reaction, the conversion of ethylbenzene to styrene is reported to be substantially increased.

Ethylbenzene can also be reacted by oxidative dehydrogenation to give styrene. Oxygen (66), halogens (especially iodine) (67), and SO_2 (68) have been used in this reaction as hydrogen acceptors. Apparently conversion and selectivity vary quite widely in these processes. Direct oxidation of ethylbenzene has also been developed as a means for the synthesis of styrene (69).

A second oxidation approach for the reaction of ethylbenzene involves the production of the hydroperoxide (70). This oxidation is accomplished by air blowing at low pressures (less than 50 psig) and about 300°F. Low conversions, nominally 10% based on ethylbenzene, are obtained with a selectivity of 80+% to the hydroperoxide. This hydroperoxide is then reacted with propylene to give propylene oxide and 1-phenylethyl alcohol, which is dehydrated to give styrene. See Vol. 19, p. 63.

Oxidations of the xylenes are generally performed on the pure isomers. Mixtures of the xylenes, however, are sometimes oxidized. Liquid-phase oxidations of all isomers have been reported. The desired product in each case is the dicarboxylic acid. In the case of p-xylene, it is sometimes produced in the form of its dimethyl ester. Liquid-phase oxidations are carried out at temperatures of 100–300°C, generally using air as the oxidant. Pressures may vary up to 40 atmospheres and reaction times up to three hours. The reaction is highly exothermic, ie, for p-xylene the heat of reaction is 326 kcal/mole.

Combustion studies of the three xylenes have been carried out to elucidate the differences in these materials when they are employed as motor fuels. The critical compression ratios (71), for example, suggest a difference in oxidation characteristics of *ortho*-xylene as compared with the meta and para isomers.

Compound	Critical Compression Ratio
o-xylene	9.6
m-xylene	13.6
p-xylene	14.2

Combustion experiments show that o-xylene oxidizes much faster than the meta or para isomer and this difference is related to the reactivities of the different radicals involved in the chain branching phenomenon (72,73).

Pyrolysis of xylenes may lead to interesting products. p-Xylene forms p-xylylene, $CH_2C_6H_4CH_2$, when subjected to temperatures above 1000°C. The product is a prototype of a class of hydrocarbons known as Chichibabin hydrocarbons. These compounds may be represented either by a quinonoid structure (A), or by a benzenoid biradical structure (B), as shown in the following equation:

$$ H_3C- \bigcirc -CH_3 \xrightarrow{1000°C} H_2C=\bigcirc=CH_2 \ \ or \ \ \cdot H_2C\bigcirc CH_2\cdot \tag{10} $$

(A)   (B)

Condensation of the gaseous products from this thermal reaction leads to the production of a very stable polymer:

$$ \left[-H_2C-\bigcirc-CH_2- \right]_n $$

Errede and co-workers have conducted extensive research in this field (74–76).

The pyrolysis of o-xylene was expected to yield the corresponding o-xylylene, but none was found. However, the polymer of this product was found from the pyrolysis of 5,6-(diacetoxymethyl) 1,3-cyclohexadiene.

$$ \left[-H_2C-\bigcirc-CH_2- \right]_n $$

Interesting pyrolysis reactions which result in the synthesis of ethylbenzene are those involving alkanes and toluene (77). This reaction is conducted at temperatures in the range of 800–1200°C. With toluene and propane, for example, the combination of benzyl and methyl radicals results in the production of ethylbenzene.

$$ CH_3\cdot + C_6H_5CH_2\cdot \to C_6H_5C_2H_5 \tag{11} $$

The generation of ethyl radicals is in accord with the proposals of Laidler (78). Benzyl radical formation from toluene has been postulated in the mechanism proposed by Steacie (79).

Class 3 Reactions. The classical work of Friedel and Crafts on the alkylation of aromatics with aluminum chloride was reported in the late nineteenth century. Since that time a wide number of catalysts, promoters and alkylating agents have been employed in this reaction. Many of the metal halides have been employed for this reaction, but aluminum chloride is by far the most common catalyst employed. Boron trifluoride with a modified anhydrous alumina can be used. Strong acids such as sulfuric, phosphoric, and hydrogen fluoride are also used. The alkylations of interest insofar as C_8 aromatic chemistry is concerned are those involved in the formation of ethylbenzene. To a more limited extent those concerned with the production of higher molecular weight aromatics by the alkylation of xylenes are also of interest.

$$C_6H_6 \; + \; C_2H_4 \; \xrightarrow{\text{Catalyst}} \; C_6H_5C_2H_5 \tag{12}$$

$$(13)$$

In these cases, especially in the case of benzene alkylation, the alkylating agent is the appropriate olefin. See Alkylation.

Over fifty alkylation reactions of the four C_8 aromatic isomers have been summarized by Price (135). These alkylation reactions include various alkylating agents, catalysts, and operating conditions.

Chloromethylation. A synthesis technique that provides a convenient route to many organic derivatives is that of the chloromethylation reaction. Of the several reagents which can be employed to conduct this reaction, probably the most widely used involves formaldehyde (aqueous or as polymer) and hydrochloric acid (aqueous or gas). The reaction temperature is 80–250°F, with treating times of several minutes to several hours. A mildly acidic catalyst ($ZnCl_2$ or $SnCl_2$) and good agitation are required, particularly when the reaction system uses aqueous formaldehyde. The following equation illustrates the formation of xylyl chlorides from toluene.

$$(14)$$

The major side product encountered (aside from dichloromethylated material) is ditolylmethane (various isomers).

Principally ortho and para products are obtained in the chloromethylation reaction. Studies of operating variables of this system have been reported (80–82). The ratio of the monochloromethyl derivative to the diarylmethane can be altered substantially by the inclusion of water-soluble chlorides in the system.

Formaldehyde Reactions. The reaction of formaldehyde with aromatic compounds to form resins has long been known. It was first reported by von Baeyer in 1872. In general, resins can be prepared by condensing the aromatic hydrocarbons with aldehydes, principally formaldehyde, in the presence of an acid catalyst, generally sulfuric acid. The aromatics employed may be pure compounds or mixtures. *Meta*-xylene, *ortho*-xylene, and *para*-xylene can be reacted separately or in admixture. *Meta*-xylene, however, is the kinetically favored isomer.

The reaction is assumed to proceed by the protonation of the formaldehyde to give the carbonium ion $(CH_2OH)^+$. This intermediate, in turn, adds to the xylene with the regeneration of the proton. The aromatic product from this reaction can further react with itself or with more aromatic or with more formaldehyde. Dehydration may occur, depending on the severity of the reaction conditions. Mechanisms have been proposed by Wegler (83) and Huang (84). The products are mixtures of polymers of hydrocarbons, ethers, acetals, and polyacetals. Polymers are of relatively low molecular weight, probably not more than 1000. A number of complex reactions take place; the formation of a simple ether is shown for illustrative purposes:

Aromatic formaldehyde reactions have also been reviewed by Walker (85).

Reaction conditions vary widely. Molar ratios of aromatic to formaldehyde may range from 2:1 to 1:2 with reaction times up to 5–8 hours. Temperatures are generally mild and in the range of 100–120°C. The concentration of the sulfuric acid is of great importance in these reactions especially since water is a reaction product. The oxygen content of the product ranges up to almost 20%, as reported by Huang (84). It is inversely proportional to the severity of the reaction conditions. Since the reaction of the aromatic with formaldehyde is related to the ease of protonation, the order of increasing reactivity and increasing oxygen content of resin is PX < OX < MX. On the basis of these reactivity differences, Imoto, et al. proposed this as a means of separating the xylene isomers (86).

The *nuclear chlorination* of various methylbenzenes in acetic acid solution has been studied, both with and without a catalyst (87).

De la Mare and Robertson also made a study of the relative rates of nuclear chlorination (88). Condon (89) has proposed a correlation of halogenation rates based on the data of de la Mare, et al. Data are presented in Table 20.

Table 20. Relative Rates of Halogenation of Polymethylbenzenes (Benzene = 1)

	Benzene	Toluene	p-Xylene	o-Xylene	m-Xylene	Mesitylene	Pentamethylbenzene
calculated			2.0×10^3	2.5×10^3	2.4×10^5	1.6×10^8	13×10^8
experimental	1	3.45×10^2	2.2×10^3	4.6×10^3	4.3×10^5	1.8×10^8	7.8×10^8

Sulfonation. Aromatics are readily sulfonated by sulfuric acid, SO_3, or oleum. The reaction is reversible. See Sulfonation.

Extensive studies on the sulfonation of various methylbenzenes have been conducted by Kilpatrick (90,91). Under the experimental conditions employed, the aromatics reacted completely to give the monosulfonic acids. The correlated data show that the reaction rate is first order in regard to the aromatic.

Nitration. The xylenes can be nitrated. See Nitration. Nuclear nitration studies directed specifically to the xylenes have been reported by Kobe, et al. (92–94).

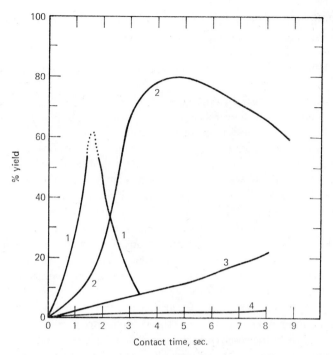

Fig. 2. Ammoxidation of *p*-xylene: 1, *p*-tolunitrile; 2, terephthalonitrile; 3, CO₂; 4, HCN.
Courtesy D. J. Hadley.

Reaction with Ammonia. The production of aromatic nitriles directly from the corresponding alkylbenzenes has been studied (95). It is a vapor phase reaction at high temperatures and requires a contact catalyst. Investigators have termed the reaction *"ammoxidation."*

$$\underset{\substack{\text{NH}_3 + \text{air} \\ 700\text{–}950°\text{F} \\ 5\text{–}30 \text{ psi}}}{\longrightarrow} \qquad + \quad \text{H}_2\text{O} \qquad\qquad (16)$$

p-Xylene gives terephthalonitrile, the expected para product, as the main product. Some *p*-tolunitrile, $CH_3C_6H_4CN$, is formed at first, and there is some formation of CO_2 and HCN. The product distribution is shown in Figure 2. The maximum yield of terephthalonitrile was 89%. *m*-Xylene gives isophthalonitrile, but the optimum yield is under different conditions. With *o*-xylene a mixture of phthalonitrile and phthalimide is obtained. Each xylene appears to react the same whether alone or in admixture with its isomers.

Manufacture

Mixed Xylenes. Mixed xylenes are obtained from petroleum. At present only very small quantities are produced from coal. See Carbonization; Toluene; Tar and pitch.

Mixed xylenes occur naturally in only small quantities in petroleum stocks. Therefore, these materials are produced by the reforming of selected naphtha streams which are rich in naphthenes (alicyclics). In most cases the stream processed contains precursors of benzene, toluene, and xylene commonly referred to as BTX. The reformer effluent consists principally of aromatics and paraffins with only very minor amounts of olefins. If the product is for gasoline use alone, no further processing is needed. But to obtain xylenes for solvent or chemical use, the reformate is purified by liquid-liquid extraction with a selective solvent, giving a BTX fraction of very high aromaticity. The BTX is then fractioned; see Toluene. Reference (96) discusses the chemistry involved in the catalytic reforming process.

The mixed xylenes produced by this combination of reforming, extraction, and distillation may be of very high purity, ie, higher than 99% C_8 aromatic. The impurities are divided between nonaromatics and contiguous aromatics. The bromine number is very low, hence the nonaromatics are largely paraffinic in nature. At times, however, the C_8 aromatics may contain several percent C_{9+} aromatics that are subsequently removed by distillation. If the xylenes are to be used for solvents, no further processing is needed, except perhaps to remove any C_{9+} aromatics that may be present. For chemical feed stocks, however, this mixture must be separated into its individual isomers. It becomes apparent that the sequence of isomer separation can be important in producing these materials.

The overall composition of the C_8 aromatics at this point corresponds approximately to the equilibrium composition at the temperature of reforming. A representative composition is ethylbenzene, p-xylene, m-xylene and o-xylene. Nine-carbon-atom hydrocarbons may be present in amounts of several percent or more, depending on the extent of fractionation of the reformer feed, and of the BTX fraction.

New sources of xylenes, also from petroleum reforming operations, are the processes of Toyo Rayon and Atlantic-Richfield (97–99). These processes involve the disproportionation of toluene or the transalkylation of toluene with trimethylbenzenes. Products are principally benzene and xylenes. If the transalkylation feed stocks are limited to toluene and (polymethyl)benzenes (no ethylbenzene derivatives), then the product xylenes from these operations will contain no ethylbenzene. Hence, these are very good feeds for p-xylene recovery by crystallization and for p-xylene isomerization feed stocks.

The Toyo and Atlantic operations are similar vapor-phase catalytic processes. No information is available on the catalyst compositions used. Atlantic merely states that they employ a nonnoble metal catalyst. Both processes report that either toluene disproportionation or the toluene-trimethylbenzene transalkylation can be employed. Such processes as these require no hydrogen. However, side reactions and catalyst carbonization may dictate the use of hydrogen in the reactor. Both processes also report great flexibility in producing benzene and xylene over a wide range of compositions. Both the disproportionation and alkylation reactions are governed by equilibrium relationships. This requires large volumes of recycle from the product recovery section of the plant to the reactor section.

Atlantic refers to their operation as the "Xylenes Plus" process. One way of integrating this into an aromatics process is shown in Figure 3. Atlantic reports that their process, now (1970) in commercial operation, gives a liquid yield of 95–97 volume %. They report the composition of the product to be 26% p-xylene, 24% o-xylene, and 50% m-xylene.

Ortho-xylene. o-Xylene is the highest boiling of the C_8 homologs. There is enough difference in the boiling points of the meta and ortho isomers to make distillation

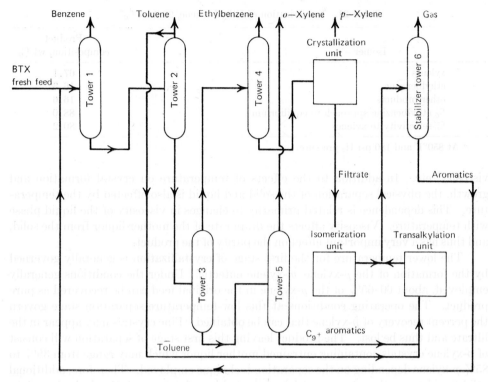

Fig. 3. Aromatics complex employing transalkylation.

economically attractive. The distillation is usually conducted in two stages. In the first step, *m*-xylene, *p*-xylene and lighter material are distilled overhead. In the second step, *o*-xylene is recovered as an overhead product and any C₉ and heavier material is rejected as a bottoms fraction.

As of 1970, the purity of the *o*-xylene product is generally in the range of 95–98%. The C_{9+} material in the product ranges from 0.2–2.0%. Although these heavy materials are much easier to remove than the higher components, their concentration may at times exceed that of the *m*-xylene and lighter materials.

To produce a 95+ % purity material from a typical hydroformate will require a distillation column with nominally 100 theoretical stages to remove the light components. Fewer trays may be employed by use of increased amounts of reflux. To remove the *o*-xylene from the several percent of heavy material, however, will require only about 30 trays. With columns such as these and the appropriate amount of reflux, the recovery of *o*-xylene from the feed should approximate 95%.

***Para*-xylene.** The properties of the C₈ aromatics are such that crystallization is the only technique practiced commercially for the production of *p*-xylene. See Fig. 1. Normal *p*-xylene concentration of feeds to crystallization plants is about 20%, but is frequently lower than this. *p*-Xylene crystal formation begins at about −40°F, and solid-liquid separation is conducted at temperatures of −80°F to −100°F.

Since solid-liquid separations are greatly affected by the size of crystals, process variables governing the nucleation and growth processes are of importance. These include such factors as extent of supercooling, temperature, degree of agitation, solution

Table 21. Isomerization of Ethylbenzene, 99.9 wt %[a]

Isomer	Product composition, wt %
xylenes	67.1
ethylbenzene	16.3
other products	16.6
% ethylbenzine approach to equilibrium	88.0
% selectivity to xylenes	80.2

[a] At 850°F and 180 psi H_2 pressure.

viscosity, etc.   In addition to the effects of temperature on crystal formation and growth, the physical separation of the solid and liquid is also affected by the temperature.   This dependence is related primarily to changes in viscosity of the liquid phase with temperature.   Viscosity affects the drain rate of the mother liquor from the solid, and thus has a very important effect on the purity of the product.

The lower temperature for the first stage of crystallization is generally governed by the formation of the *p*-xylene–*m*-xylene eutectic.   Under the conditions generally employed, about 60–65% of the *p*-xylene in the original feed can be recovered as pure product.   The operating conditions at this low-temperature separation stage govern the percent recovery of *p*-xylene that can be obtained.   Fine crystals may appear in the filtrate and thus be lost.   The product leaving this first stage of separation will consist of *p*-xylene crystals containing entrained mother liquor which may range from 35% to 85% *p*-xylene depending on the separation technique employed.   Therefore, additional solid-liquid separation steps must be conducted in order to reach the desired product purity.   Commercial purities of *p*-xylene are currently above 99%; in most instances product purities about 99.5% are required.

Industrially, a number of techniques have been employed for cooling the xylenes feed down to the first stage crystallization temperature.   Heat exchangers are employed to cool the feed to a temperature somewhat above the crystal point of the liquid.   From there on, the material is refrigerated in one or more stages.   Scraped-surface exchangers in conjunction with the holding tanks or stirred-tank crystallizers are widely employed.   Ethylene is generally used as the refrigerant.   In some cases, however, the ethylene has been employed as a direct autorefrigerating agent.   This eliminates the heat transfer through the walls and the problems arising from crystal deposition on the cold walls of the heat exchanger.

Separation of the solid and liquid phases has been achieved by rotary filters, disc centrifuges, basket centrifuges, screen-bowl centrifuges, solid-bowl centrifuges, combination solid-screen bowl centrifuges, reciprocating-pusher centrifuges, and pulsed columns (see below).   In all instances, two or more stages are required to produce products of the desired purity.

Figure 4 (100) shows a flow diagram of a process employing stirred-tank crystallizers.   A stepwise cooling for the first stage is employed.   Feed containing about 20% *para*-xylene passes through a heat exchanger which is cooled by cold filtrate from the micrometallic filler.   Filtrate from the first- and second-stage centrifuges may be mixed with the chilled solution (still free of crystals) and passed into drum 1, then drum 2 of the first-stage crystallizer.   The crystallizer drums are stirred tanks provided with jackets for refrigerating the system with ethylene.   Holding time in each tank is about three hours, thus providing a total of six hours.   Cold slurry (of C_8 aromatics solution +

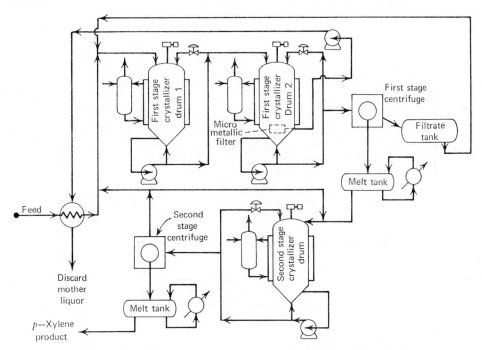

Fig. 4. Crystallization process using stirred tank crystallizers (100).

p-xylene crystals) is withdrawn from the lower side of drum 1. Some is returned to the bottom of drum 1 to fluidize the fines in the slurry, thus ensuring a slurry of large crystals. Another portion of material may be diverted to the top of the drum to ensure proper cooling to the desired temperature. The net flow of chilled feed passes from drum 1 to drum 2. Filtrate is removed from the lower part of drum 2 through a micrometallic filter and leaves the system as product filtrate. The net flow of cold slurry passes from drum 2 of the first-stage crystallizer to the first centrifuge. Filtrate from this centrifuge passes back to drum 1 of the first-stage crystallizer and is mixed with fresh feed, as mentioned earlier. The cake from the first-stage centrifuge is melted and passed into a second-stage crystallizer, along with filtrate from the second-stage crystallizer. After cooling to the desired temperature, the net flow of slurry from the crystallizer goes to the second-stage centrifuge. Filtrate produced therefrom is divided; a portion returns to the second-stage and the balance to the first-stage crystallizer. The volume of filtrate returning to the second-stage crystallizer is adjusted so as to provide a slurry of optimum crystal concentration. The *para*-xylene from the second-stage centrifuge constitutes the high-purity *para*-xylene and, after melting, goes to storage.

The application of a pulsed column in p-xylene production is shown in Figure 5 (21). A sketch of the pulsed column is also shown in this figure.

Feed xylenes from a storage tank are dryer cooled by heat exchange with mother liquor (filtrate from the rotary filter) to just above the crystal point. They are then cooled to filtration temperature (about $-100°F$) in a scraped-surface heat exchanger. Filtration in a rotary filter produces a filtrate of the low p-xylene content and a cake of 65–70% p-xylenes. This goes to a melt tank. Included as part of the charge to this tank is a portion of filtrate from the purification column to be discussed below. This

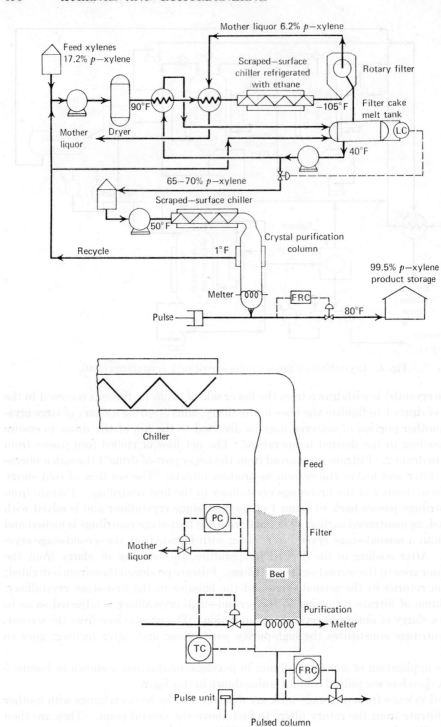

Fig. 5. Fractional crystallization process for *p*-xylene, showing pulsed column.

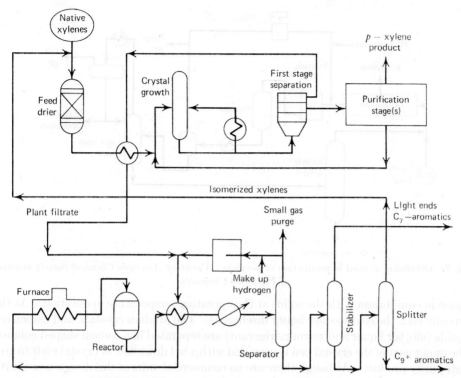

Fig. 6. Isoforming process. Courtesy *Hydrocarbon Processing*.

blend passes through a small storage tank and is then pumped to a second scraped-surface exchanger that is operated at about 0°F. The chilled slurry, containing about 40% crystals, passes into the top of a crystal purification column.

p-Xylene crystals at the bottom of this column are melted and drawn off at a uniform rate by a flow-control device. This constitutes the product from the purification step. A cyclic pulse unit that can displace a controlled volume of material is attached to the column. The liquid displacement of a pulse stroke forces liquid upward through the bed and provides reflux for the descending crystals. A filter area is located in the upper portion of the column. The pulse pressure forces some filtrate out through this filter. Some of this filtrate is returned to the melt tank as mentioned above to control the composition of the material in this melt tank. The balance of the filtrate is recycled to the feed inlet of the plant.

In place of indirect chilling, as employed in these processes, autorefrigeration is also used. In this case, the refrigerant is added directly to the precooled xylenes. Necessary cooling is obtained by controlled autorefrigeration of the material. The Chevron and Maruzen processes are reported to use CO_2 and C_2H_4, respectively, for the autorefrigerant. While the use of such refrigerating techniques eliminates such items as the scraped-surface exchangers, additional facilities to remove the refrigerant from the hydrocarbon products are necessary.

Still another form of cooling has recently been reported (101). This is the countercurrent flow of *p*-xylene in direct contact with an immiscible refrigerant. The warm refrigerant discharge carries along with it the *p*-xylene crystal and associated mother

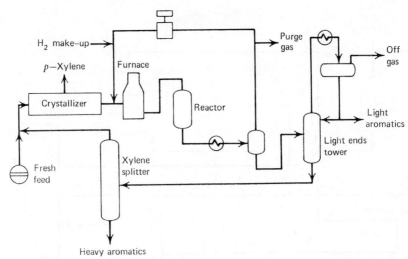

Fig. 7. Octafining as used in production of *p*-xylene. Courtesy *American Chemical Society Division of Petroleum Chemistry.*

liquor in equilibrium with the solids at the operating temperature in question. At the opposite end, the cold mother liquor discharges. The product crystals, and associated liquids (mother liquor and warm refrigerant) are separated in a conical shaped column. The upper part of the crystal bed is provided with a net downflow of crystal melt to give high-purity product. Although there are no commercial units of this design now (1970) in operation, experimental studies reported *p*-xylene purities of 99.8%.

Although all *p*-xylene now produced is obtained by some form of crystallization, it is reported that an adsorption technique (102) is being employed in two commercial plants now under construction. The nature of the adsorbent has not been revealed, but it is stated to be selective for *p*-xylene. The process is reported to be continuous, and to give a high recovery of high-purity *p*-xylene.

To augment the supplies of native *p*-xylene, it has been necessary to develop an isomerization process. The filtrate from the crystallization plant usually serves as the feed stock with this process. Primarily, *m*-xylene and *o*-xylene are isomerized to give a *p*-xylene–*m*-xylene–*o*-xylene mixture essentially at equilibrium. In some instances, *o*-xylene may be a prime product along with the *p*-xylene. In this case, it is recovered by distillation and the *m*-xylene constitutes the principal feed component to the isomerization plant.

Isomerization processes are vapor-phase operations at temperatures of 700–950°F under superatmospheric pressures. They are conducted in the presence of hydrogen. These processes can be classed as (*1*) noble-metal and (*2*) nonnoble-metal catalytic process. There are a number of such processes in commercial operation (103). These may be typified by (*1*) the Isoforming process (104,105), licensed by Esso Research and Engineering Co. and (*2*) the Octafining process (106), licensed by Engelhard Industries Co. Figures 6 (107) and 7 (106) present flow diagrams of these operations.

The Isoforming process is shown as an integral part of a xylenes complex including the Isofining process; Isofining is a crystallization process for recovery of *p*-xylene.

Generally speaking, the noble metal catalysts have good hydrogenation/dehydrogenation potential. As a consequence, the reaction temperature must be maintained at about 800–950°F to avoid the excessive hydrogenation of aromatics. Ethylbenzene isomerization to xylenes has been postulated to be dependent on hydrogenated intermediates (54). Data reported for the Octafining process in Table 21 show a high degree of isomerization of ethylbenzene to xylenes. This may be related to the postulated formation of hydrogenated intermediates for ethylbenzene isomerization.

The nonnoble-metal Isoforming catalyst, by contrast, seems not to bring about ethylbenzene isomerization. It does, however, report a very low make of gas and non-aromatics. This is associated with its low level of hydrogenation activity.

The first Octafining plant went onstream in 1960. Ten additional plants are now reported in operation. Isoforming was commercialized in 1967 and there are three plants in operation as of 1970.

Meta-xylene. *Meta*-xylene has long been the key in the production of xylene isomers. It constitutes about one half of the total mixed-xylenes stream, yet no attractive process for its recovery has yet been developed. Distillation of the filtrate from *p*-xylene crystallization (*m*-xylene:*p*-xylene ratio is about 87:13) can give a product of 85–90% *m*-xylene content. If products containing more than the eutectic concentration of *m*-xylene are obtained by distillation (difficult, but possible), then crystallization can be used to produce high-purity *m*-xylene. This route is not economically attractive, however, for producing *m*-xylene. Other separation techniques that have been tried commercially are primarily based on the greater ease of protonation of the *m*-xylene isomer.

Selective sulfonation of xylenes has been employed. Hetzner (108) prepares a C_8 aromatics feed concentrate by distilling to produce an enriched *m*-xylene/*p*-xylene fraction for a sulfonation unit. This enriched mixture is partially sulfonated at temperatures up to 150°F with 96% sulfuric acid. The material is diluted with water and steam-stripped to liberate the *m*-xylene in high purity. Spence (109) has disclosed an improved process employing a two-stage sulfonation treatment. In the first mixing stage, one volume of feed is mixed with 10 volumes of recycle comprising *m*-xylene, sulfonic acid, and sulfuric acid, of strength adjusted to be below that required for sulfonation. The effluent from this first stage passes to a second stage along with some concentrated sulfuric acid. This mixture is introduced into a reactor-settler system. Hydrocarbon and the recycle stream are withdrawn, and the remaining portion is hydrolyzed to recover pure *m*-xylene. Still another improvement has been proposed (110), in which a paraffinic solvent is added to the system before carrying out the hydrolysis of the *m*-xylenesulfonic acid. On hydrolysis, the *m*-xylene is extracted by the paraffinic wash oil.

Because of the corrosiveness of the reagents, special materials of construction are required for such a plant, and the cost of the *m*-xylene is high. It is understood that one plant using such a process has now discontinued operations.

Selective chlorination as a means of producing *m*-xylene has been patented (111). In this process *m*-xylene is chlorinated selectively in a formic acid solvent. The chlorinated product is recovered by distillation and hydrogenated to convert the derivative back to the original *m*-xylene. As yet, there is no commercial application of this process.

Recovery of *m*-xylene by use of clathrates has been reported on a semicommercial scale (112,113). A feed rich in *m*-xylene (about 80%) is contacted with an aqueous ethanolamine stream containing the clathrate former ($Ni(4\text{-methylpyridine})_4(SCN)_2$) which selectively removes the para isomer. Products recovered from this are a 96.5% *m*-xylene filtrate (about 75% yield based on feed) and a 60% *p*-xylene stream.

A process based on the complex formation of *m*-xylene with hydrofluoric acid and boron trifluoride has been commercialized (Fig. 8) (114,115). In this process, the mixed xylene feed is charged to the lower section of an extraction tower where it is counter-currently treated with a solvent stream of boron trifluoride with an excess of hydro-fluoric acid. *n*-Hexane is used in the extraction as a diluent. The complex, $(MXH)^{+}$-$(BF_4)^{-}$, is formed selectively and remains in the solvent phase. It passes to a distillation tower, the decomposer, where the HF and BF_3 are liberated from the complex and distilled overhead. *Meta*-xylene bottoms may be charged to an isomerizer to increase the production of the *para*- and *ortho*-xylenes. Alternatively, the *meta*-xylene may be charged to a second distillation tower where pure *meta*-xylene is produced (by removal of small amounts of high-boiling material).

The raffinate from the original extractor is stripped of solvent hexane and the hexane is returned to the extraction zone. The C_8 aromatics are distilled to produce high-purity ethylbenzene and *ortho*-xylene. The *para*-xylene from this distillation complex is 95–98% pure. It is purified to 99.5% by a single-stage crystallization. The filtrate is recycled to the distillation zone.

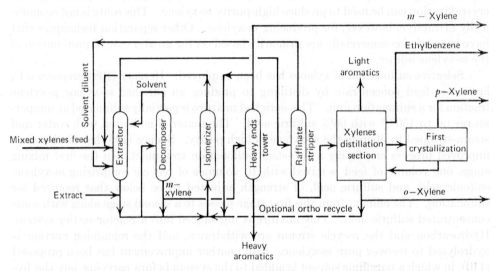

Fig. 8.   Separation of *m*-xylene by complex formation.   Courtesy *Hydrocarbon Processing*.

The xylenes disproportionation and ethylbenzene transalkylation reactions are reportedly held to a very low level. The *m*-xylene product is stated to be 99% pure. The *m*-xylene recovery must also be very high in order to allow the distillation recovery of the other three isomers as shown in the flow diagram.

Japan Gas Chemical Company, the developer of the process, is operating a plant in Japan. The reported capacity is 100,000 metric tons per year of mixed xylenes feed. In addition to making large quantities of *m*-xylene available, this process claims to make the recovery of the other isomers easier and cheaper. It also claims, as noted above, to provide isomerization capacity to convert excess *m*-xylene into *p*-xylene.

Ethylbenzene.   Ethylbenzene is manufactured principally for the production of styrene. The volume produced exceeds the combined volumes of the three xylene isomers. There are two sources of ethylbenzene: recovery from mixed xylenes (somewhat less than 10%) and synthesis from benzene and ethylene. Both are described under Styrene.

Economic Aspects

Prior to World War II, all xylenes were produced from coal-derived sources. The wartime demands for toluene gave impetus to the petroleum industry for the rapid commercialization of the reforming process. Continued postwar expansion of the reforming process to provide high-octane gasoline components has made a reservoir of C_6–C_{10} aromatics potentially available for petrochemicals. At the present time, very large volumes of BTX are produced by petroleum refiners. The xylenes fraction is used as a source of xylene solvents and for the production of individual isomers. Some ethylbenzene is recovered from the C_8 aromatic stream. However, benzene and ethylene are used as the raw materials for the major portion of the ethylbenzene.

The production of xylenes over the past 25 years is shown in Figure 9 (116). The unit reported is the U.S. gallon. One gallon mixed xylenes weighs 7.27 pounds. In this period of time, the xylenes production from coal sources has decreased. This decrease has occurred principally in the past 10 years. By contrast, mixed xylenes from petroleum sources have been increasing rapidly. For the past 20 years, petroleum xylenes have increased at the rate of about 11% per year, equivalent to the output doubling every 6.6 years. The total xylenes increase has been at the rate of about 10% per year. The growth of xylenes sales has been at a somewhat lower rate, indicating a greater captive use of mixed xylenes by producers.

Of the total xylenes produced, about 50,000,000 gal/yr were required by the solvents market in 1965 (117). Present demand is difficult to estimate but there has been

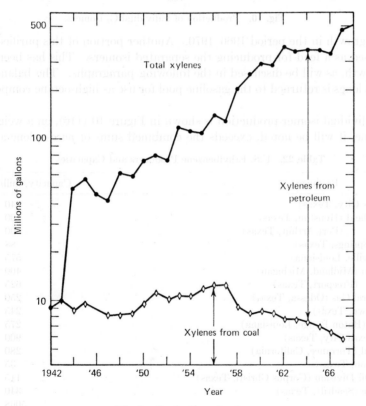

Fig. 9. Production of xylenes.

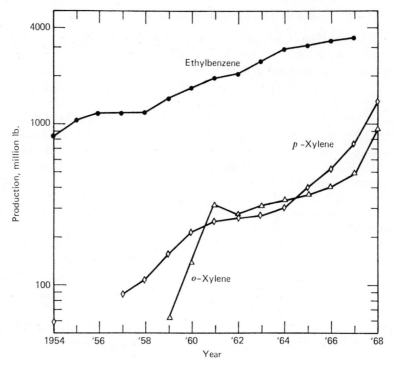

Fig. 10. Production of individual C_8 isomers.

substantial growth in the period 1966–1970. Another portion of this purified xylenes stream is used as a feed for producing the separated isomers. This has been an area of rapid growth, as will be discussed in the following paragraphs. The balance of the extracted xylenes is returned to the gasoline pool for use as high-octane components of motor fuels.

The individual isomer production is shown in Figure 10 (116), on a weight basis. Ethylbenzene, it will be noted, exceeds the combined sums of *para*-xylene and *ortho*-

Table 22. U.S. Ethylbenzene Producers and Capacities

Producer	Capacity, million lb/yr
Amoco (Texas City, Texas)	340
Atlantic-Richfield (Houston, Texas)	100
(Port Arthur, Texas)	500
Cosden (Big Springs, Texas)	88
CosMar (Carville, Louisiana)	575
Dow Chemical (Midland, Michigan)	400
(Freeport, Texas)	625
El Paso Natural Gas (Odessa, Texas)	250
Enjay (Baytown, Texas)	245
Foster Grant (Baton Rouge, Louisiana)	275
Monsanto (Texas City, Texas)	900
Shell Chemical (Torrance, California)	280
Signal (Houston, Texas)	35
Sunray DX Oil Division (Corpus Christi, Texas)	115
Union Carbide (Seadrift, Texas)	340
Total	5068

xylene. Information on the production level of *meta*-xylene is difficult to obtain. This is because most of the *m*-xylene is used captively (for isophthalic acid production). Estimates indicate that this volume now amounts to less than 100,000,000 lb/yr.

Ethylbenzene producers are listed in Table 22, along with plant capacity data. The 1968 production exceeded 4 billion pounds, and the 1973 production has been estimated to exceed 6.8 billion pounds. This will represent a growth approximating 10% per year (118). Essentially all of the ethylbenzene production will be consumed in the manufacture of styrene monomer.

Para-xylene producers are listed in Table 23, along with capacity data. Production in 1968 approximated 1.4 billion pounds, representing a growth of 29% per year for the previous ten years (119). It is projected to grow at the rate of 12% per year and attain a market of 2.4 billion pounds by 1973. Polyester fiber and film represent essentially 100% of the market for the *para*-xylene. Hence, the future *para*-xylene growth will depend on the growth of these markets.

Table 23. U.S. *Para*-xylene Producers and Capacities

Producer	Capacity, million lb/yr
Amoco (Decatur, Alabama)	275
(Texas City, Texas)	275
Atlantic-Richfield (Houston, Texas)	300
Chevron (El Segundo, California)	90
(Pascagoula, Mississippi)	250
(Richmond, California)	110
Enjay (Baytown, Texas)	200
Shell (Houston, Texas)	100
Signal (Houston, Texas)	15
Sunray DX Oil Division (Corpus Christi, Texas)	170
Tenneco (Chalmette, Louisiana)	100
Total	1885

Ortho-xylene, too, has one principal outlet and that is in the production of phthalic anhydride. It differs from *para*-xylene and ethylbenzene, however, in that it is not the sole feed stock for this outlet. Historically, naphthalene has been the raw material for phthalic anhydride manufacture. A shortage of naphthalene about 10 years ago (because of steel strikes) gave impetus to the use of *ortho*-xylene as a feed stock. A list of *ortho*-xylene producers is given in Table 24 along with their capacities. Since *o*-xylene is separated by distillation and distillation capacity can be added quickly or diverted to other uses, these figures can change rapidly. It appears that phthalic producers are favoring *ortho*-xylene as a feed stock. This seems to be true both in the United States and abroad. The U.S. production as shown in Figure 10 has a recent growth approximating that of *para*-xylene.

Meta-xylene, as mentioned earlier, is produced in much smaller amounts than the other three isomers. There are only two or three producers in the United States. Most of the *m*-xylene is consumed captively and is 85–90% or 95+% *m*-xylene.

Basically, raw material prices are set by the value of the xylenes as a motor fuel. This price applies to the xylenes in the hydroformer product containing BTX and nonaromatics. Extraction and distillation gives the pure xylene fraction which can be sold as a solvent or separated for isomer production. Over the past ten years, there has been no change in the basic processes employed in the manufacture of the raw material

Table 24. U.S. *Ortho*-xylene Producers and Capacities

Producer	Capacity, million lb/yr[a]	
Atlantic-Richfield (Houston, Texas)	150	150
Chevron (Richmond, California)	130	175
Cities Service (Lake Charles, Louisiana)	80	120
Coastal States Petrochemical (Corpus Christi, Texas)	30	65
Cosden Oil & Chemical Co. (Big. Springs, Texas)	12	20
Crown Central (Houston, Texas)	20	19
Enjay Chemical (Baytown, Texas)	105	175
Hess Oil (Corpus Christi, Texas)	45	50
Monsanto Chemical (Chocolate Bayou, Texas)	30	30
Pontiac Refining Co. (Corpus Christi, Texas)	12	12
Southwestern Oil (Corpus Christi, Texas)	120	120
Sunray DX Oil Division (Corpus Christi, Texas)	140	160
Tenneco (Chalmette, Louisiana)	155	155
Total	1029	1272

[a] Range of values reported by different sources.

but process improvements have greatly reduced the prices of these products. The following shows data from the Tariff Commission on the average unit price of the merchant sales in 1958 and 1967. The current prices given are approximate values obtained from various sources.

Approximate Prices of C_8 Aromatics, ¢/lb

	1958	1967	1970
use, or component in hydroformer product, as a component of gasoline			2.0
as solvent	3.3	2.5	2.5
o-xylene	6.0 (1960)	3.0	3.0
ethylbenzene	8.0	4.0	4.0
p-xylene	16.0	8.0	7.0
m-xylene (95+%)			14–15

The high price of *meta*-xylene is an obvious deterrent to large-scale use. There is a general consensus that the price of this material must decrease to approximately that of *p*-xylene before large-scale uses of *meta*-xylene will develop.

A forecast of the aromatics demand through 1980 is given in Table 25 (120). Benzene is given here since it is the principal raw material for ethylbenzene and since

Table 25. Estimated Aromatics Demand, Thousands of Metric Tons (2204 lb)

Country	1970		1975		1980	
	Benzene	*o,p*-Xylene	Benzene	*o,p*-Xylene	Benzene	*o,p*-Xylene
U.S.A.	3500	650	4550	1300	5600	2050
W. Europe	2750	740	4200	1140	5600	1725
Japan	850	235	1400	525	2000	900
other[a]	500	65	1000	165	1800	325
Total	7600	1690	11,150	3130	15,000	5000

[a] Free world.

Table 26. ASTM Standard Specifications for Xylene

Property	ASTM Test Method	Nitration, D843-67	Five Degree, D845-67	Ten Degree, D846-67	Industrial, D364-65
specific gravity at 15.56/15.56°C	D-891	0.8650–0.8750	0.860–0.875	0.860–0.875	0.860–0.871
color	D-853	←——————— not darker than 0.0030 g $K_2Cr_2O_7$/liter ———————			not darker than standard
distillation range	D-850	not more than 5°C including 139.3°C	not more than 5°C including 139.3°C	not more than 10°C	
ibp, °C		>137°C	>137°C	>135°C	123°C
% at 130°C, max					5
% at 145°C, min					90
dry point		<143°C	<143°C	<145°C	155
paraffins, max %	D-851	4.0			
acid wash color	D848	←———— not darker than no. 6 color standard ————→			not darker than no. 10
acidity	D-847	←——————————— no free acid ——————————→			
sulfur compounds	D-853	←———————— free of H_2S and SO_2 ————————→			
copper corrosion	D-849	←— Copper strip shall not show iridescence or gray or black discoloration —→			
	D-268				copper strip shall not show greater discoloration than class 2
odor	D-1296				characteristic aromatic odor
water	D-1364				no turbidity at 20°C

almost 40% of all the benzene produced is required for ethylbenzene manufacture. The *ortho-* and *para-*xylenes have been lumped together. Since there is widespread use of isomerization, there is an equivalence of these two isomers. These data show that *ortho-* and *para-*xylene growth combined is predicted to be about 11% per year and that benzene will average about 7% per year growth. These values are in good agreement with the study made by the Pace Company (121).

Specifications and Test Methods

Xylenes are marketed at several quality levels. The American Society for Testing and Materials has four separate specifications representing different quality levels of mixed xylenes. These are summarized in Table 26. This table shows the range of values as well as the test procedures employed. Here, purity is measured indirectly by such tests as gravity and distillation, and impurities by such tests as acidity, corrosion, etc. These specifications apply to mixed xylenes from coal or petroleum sources. Mixed xylenes from petroleum, however, are routinely available at higher quality levels.

Each individual isomer is marketed under a set of specifications stipulating the purity, and impurity levels. These specifications are set by the customer and the manufacturer, and there is some variation throughout the industry. Many of the tests are determined by chromatographic technique.

Table 27. Typical Ethylbenzene Test Data

Test	Range of values	Test method
ethylbenzene, wt %	99.5–99.9	GLC method
nonaromatics, wt %	0.02–0.16	GLC method
benzene, wt %	0.02–0.06	GLC method
toluene, wt %	0.01–0.09	GLC method
hydrocarbons		
heavier than ethylbenzene,[a] wt %	0.01–0.27	GLC method
diethylbenzene, wt %	0.01–0.12	GLC method
isopropylbenzene, wt %	0.01–0.07	GLC method
bromine index, mg/100 g	1–3	ASTM D-1492
color, Pt-Co	<5	ASTM D-1209
distillation, °C		
ibp	135.3–135.5	ASTM D-850
dry point	136.0–137.3	
range	0.7–1.9	
sulfur, ppm (wt)	<1–1.3	

    [a] Including diethylbenzene and cumene.

Table 27 presents typical test data on commercial samples of ethylbenzene from several manufacturers. Note the extremely high purity of the material—about 99.5%. Diethylbenzene and isopropylbenzene contents are specified since these may be dehydrogenated under styrene-producing conditions to form divinylbenzene and alphamethylstyrene, which are undesirable impurities in the styrene monomer. *p*-Xylene quality is shown in Table 28. Purities ranging above 99% are now required, in order to be competitive. *o*-Xylene quality is shown in Table 29. Purity levels are much lower, ranging from 95–98%, which is satisfactory for vapor-phase oxidation to phthalic anhydride.

Table 28. Typical *Para*-xylene Test Data

Test	Range of values	Test method
para-xylene, mol %	99.0–99.7	ASTM D-1016
ethylbenzene, wt %	0.03–0.22	GLC method
meta-xylene, wt %	0.08–0.47	GLC method
ortho-xylene, wt %	0.01–0.15	GLC method
toluene, wt %	0.01–0.02	GLC method
bromine index, mg/100 g	1–3	ASTM D-1492
inorganic chlorides, ppm (wt)	0.2–1.7	
organic chlorides, ppm (wt)	0.1–0.3	
color, Pt-Co	<5	ASTM D-1209
acid wash color[a]	1	ASTM D-848
distillation, °C		ASTM D-850
ibp	137.3–137.9	
dry point	138.1–138.8	
range	0.2–1.2	
paraffin hydrocarbons, vol %	<0.2	ASTM D-851
nvm, g/100 ml	0.001	ASTM D-1353
sulfur, ppm (wt)	<1	

[a] Color in the acid layer when shaken with sulfuric acid.

Table 29. Typical *Ortho*-xylene Test Data

Test	Range of values	Test method
ortho-xylene, wt %	95.0–97.9	GLC method
ethylbenzene, wt %	0.01–0.16	GLC method
para-xylene, wt %	0.17–0.70	GLC method
meta-xylene, wt %	1.04–3.45	GLC method
C_9 aromatics, wt %	0.10–2.30	GLC method
nonaromatics, vol %	0.05–0.59	ASTM D-2360
distillation, °C		
ibp	143.6–143.8	ASTM D-1209
dry point	144.0–146.8	
range	0.4–3.1	
sulfur, ppm (wt)	<1	

Meta-xylene quality can be quite variable. Some manufacturers who employ special processes produce *meta*-xylene up to 98% purity. Sinclair started up a plant to produce material of this quality in 1967 (122). Japan Gas Chemical advertises a 98.5% pure *m*-xylene. Distillation of filtrates from *para*-xylene crystallization produces an 80–90% purity *meta*-xylene. This may be used as an oxidation feed for isophthalic acid production. The product acid may be used as such or it may be purified by removal of terephthalic acid which was coproduced in the oxidation step.

Safe Handling

The safe handling and use of xylene is similar to that of benzene and toluene (see Benzene; Toluene). Xylenes have been classified by the Interstate Commerce Commission as flammable liquids. Thus, they require a red label and must be packaged in authorized containers. Shippers must comply with all ICC regulations regarding handling and labeling. These materials should be kept away from heat and open flame.

Table 30. Toxicity Hazards of the Xylenes and Ethylbenzene

Toxic hazard rating	m-Xylene	o-Xylene	p-Xylene	Ethylbenzene
acute local	slight	slight	slight	moderate
acute systemic	slight	slight	slight	moderate
chronic local	slight	slight	slight	unknown
chronic systemic	slight	slight	slight	unknown
threshold limit for prolonged exposure, ppm in air	200	200	200	200

Containers in which they are stored should be closed.   These materials should be used only with adequate ventilation.   One should avoid breathing the vapor and avoid contact of the material with the skin.

The toxicity hazards of ethylbenzene and the xylenes have been reported in a number of standard references.   A summary of some of this information is presented in Table 30 (123).   The liquid is an irritant to the skin and mucous membrane.   A concentration of 0.1% vapor in air is an eye irritant.   Erythema and inflammation of the skin may result from contact with liquid.   Exposure to vapors causes lacrimation, irritation of the nose, and dizziness.

There is no detectable difference in the toxicity of the three xylene isomers.   However, ethylbenzene is shown to be more hazardous than the dimethylbenzenes.   This is true of both the vapor and the liquid.

It is reported in another reference (124) that several cases of fatal poisonings with xylenes have occurred.   It is also reported that the odor of xylenes is detectable at less than 1 ppm and that the threshold limit is based on sensory data.

Uses

There are two principal uses for xylenes other than as a chemical raw material: (1) as a high-quality octane-blending agent in motor fuels, and (2) as a solvent.

The high quality of xylene as a gasoline component is illustrated by the data of Table 31 (125).   As mentioned earlier, reforming produces a tremendous volume of xylenes for consumption as motor fuel.   If the present concern with pollution leads to the widespread use of leadfree gasoline, the demand for xylene for motor fuel could increase even more rapidly than the demand for gasoline.

Table 31. Blending Octane Numbers of Aromatic Hydrocarbons[a]

Aromatic	Research, ASTM D908-56	Motor, ASTM D357-56
ethylbenzene	124	107
o-xylene	120	102
m-xylene	145	124
p-xylene	146	126

[a] Calculated blending number of hydrocarbon from ASTM rating of 20% hydrocarbon + 80% of a 60:40 mixture of "isooctane" and n-heptane.   Number is extrapolated from rating at 20% concentration to a hypothetical 100% concentration.

Table 32. Some Xylene Properties Pertinent to Use as Solvents

Property	Value
flash point, closed cup, °F	78–85
autoignition temperature, °C	490–550
coefficient of cubic expansion at 20°C	0.001
explosive mixtures with air at 20°C	
upper limit, vol %	5.3
lower limit, vol %	1.0
evaporation rate, ether = 1	13.5
Kauri butanol value	85–90

As a solvent, mixed xylenes are marketed as a component that may be blended with other solvents. See Solvents, also references (126,127). Some properties of xylenes pertinent to their use as solvents are presented in Table 32 (collected from various sources).

Xylene Isomers as Chemical Raw Materials. For each of the four isomers there is one principal derivative which accounts for almost 100% of the production.

Isomer	Product
ethylbenzene	styrene
p-xylene	terephthalic acid and/or dimethyl terephthalate
o-xylene	phthalic anhydride
m-xylene	isophthalic acid

Dealkylation of xylene may be practiced at times to produce high-purity benzene, but toluene is usually more attractive as a feed stock. The theoretical volumetric yield of benzene from xylene and toluene is 72% and 83% respectively; with xylene, therefore, there is increased gas production and increased hydrogen consumption. The heat release per barrel of xylene feed approximates 125,000 Btu as compared to about 75,000 Btu/bbl of toluene feed. This imposes a great added heat load on the reactor (see Toluene).

Small volumes of xylenes are sulfonated to produce surface-active agents. The Tariff Commission reports production of 17,000 lb of ammonium xylenesulfonate and 22,000 lb of sodium xylenesulfonate in 1966.

m-Xylenediamine is currently being manufactured at a 100 metric ton per month capacity (128–130) from m-xylene. This semicommercial plant has been in operation since late 1966. The feed appears to be the meta-para filtrate from the crystallization operation which contains perhaps 85% m-xylene. No significant difference is reported between the properties of the m-xylenediamine and the p-xylenediamine. This diamine is made for fiber applications and is reported to be pure enough for direct polymerization. The manufacturers also report promise in the conversion of the product to diisocyanate, an epoxy curing agent.

Xylene-formaldehyde resins (XF resins) are also produced commercially (131). These materials are made by reaction of xylene, preferably the meta isomer, with formaldehyde as described in an earlier section of this article. Properties of the resin have been reported (132).

Small amounts of the four C_8 aromatic isomers are parent compounds of many derivatives that find end uses in numerous fields, including perfumes, pharmaceuticals, insecticides, inks, adhesives, and dyes (136).

Bibliography

"Xylenes and Ethylbenzene" in *ECT* 1st ed., Vol. 15, pp. 186–194, by Maury Lapeyrouse, Esso Research and Engineering Company.

1. A. Cahours, *Ann. Chemie* **76,** 286 (1850).
2. U.S. Tariff Commission Report, *No. 153* (1941–1943); *No. 154* (1944), Washington, D.C.
3. "Selected Values of Physical and Thermodynamic Properties of Hydrocarbons and Related Compounds," *American Petroleum Institute Research Project 44*, Carnegie Press, 1953.
4. O. Redlich and A. T. Kister, *J. Am. Chem. Soc.* **71,** 505 (1949).
5. E. Hala, J. Pick, V. Fried, and O. Vilim, *Vapor Liquid Equilibrium*, 2nd Eng. ed., Pergamon Press Ltd., London, 1967.
6. "Azeotropic Data," *Advances in Chemistry Series-6*, Am. Chem. Soc., Washington, D.C., 1952.
7. U.S. Pat. 2,456,561 (Dec. 12, 1948), to G. R. Lake and J. M. McDowell (to Union Oil Co.).
8. U.S. Pat. 2,504,830 (April 18, 1950) to R. B. Greene (Allied Chemical and Dye Corp.)
9. Richard L. Nelson, *Am. Chem. Soc. Div. Petrol. Chem., Reprint* **8** (1), 115–123 (1963).
10. Lloyd Berg, Stanford V. Buckland, W. Bruce Robinson, Richard L. Nelson, Thomas K. Wilkinson, and Joseph W. Petrin, *Hydrocarbon Processing* **45**(12), 103–106 (1966).
11. Brit. Pat. 736,856 (Sept. 14, 1955) to Harry Clough and John A. Smith (Imperial Chemical Industries, Ltd.).
12. U.S. Pat. 2,981,662 (April 25, 1961) to Cline Black (Shell Oil Co.).
13. J. C. Chu, O. P. Kharbanda, R. F. Brooks, and S. L. Wang, *Ind. Eng. Chem.* **46,** 754 (1954).
14. U.S. Pat. 3,356,593 (Dec. 5, 1967) to Shigeto Suzuki (Chevron Research Co.).
15. U.S. Pat. 2,532,031 (Nov. 28, 1950) to A. C. Nixon, C. H. Deal, and R. J. Evans (Shell Development Company).
16. U.S. Pat. 2,638,441 (May 12, 1953) to A. C. Nixon and C. H. Deal (Shell Development Co.).
17. U.S. Pat. 3,105,017 (Sept. 24, 1963) to Emanuel M. Amir and William R. Edwards (Esso Research and Engineering Co.).
18. William F. Haddon, Jr. and Julian F. Johnson, *J. Chem. Eng. Data* **9** (1), 158–159 (1964).
19. Roger S. Porter and Julian F. Johnson, *J. Chem. Eng. Data* **12** (3), 392–394 (1967).
20. U.S. Pat. 3,414,630 (Dec. 3, 1968) to Theodore H. Szawlowski and Charanjit Rai (Union Oil Co.).
21. D. L. McKay, G. H. Dale, and D. C. Tabler, *Chem. Eng. Prog.* **62** (11), 104 (1966).
22. V. Kravchenko, *Acta Physicochimica USSR* **20** (4), 567–577 (1945).
23. C. J. Egan and R. V. Luthy, *Ind. Eng. Chem.* **47,** 250 (1955).
24. U.S. Pat. 2,801,272 (July 30, 1957) to D. E. Bown (Esso Research and Engineering Company).
25. U.S. Pat. 2,815,392 (Dec. 3, 1957) to D. E. Bown and W. V. Milligan (Esso Research and Engineering Company).
26. William D. Schaeffer and W. S. Dorsey, *Advances in Petroleum Chemistry and Refining*, Vol. 6, Interscience Publishers, a division of John Wiley & Sons, Inc., New York, 1962, Ch. 3.
27. W. D. Schaeffer, W. S. Dorsey, D. A. Skinner, and C. G. Christian, *J. Am. Chem. Soc.* **79,** 5870–5876 (1957).
28. U.S. Pat. 2,798,103 (July 2, 1957); U.S. Pat. 2,951,104 (Aug. 30, 1960) to W. D. Schaeffer and J. D. Wordie (Union Oil Company).
29. P. De Radzitsky, Jr. and J. Hanotier, *Ind. Eng. Chem., Process Design Dev.* **1** (1), 10–14 (Jan. 1962).
30. U.S. Pat. 2,983,767 (May 9, 1961) to R. N. Fleck and C. G. Wight (Union Oil Company).
31. A. S. Michaels, R. F. Baddour, H. J. Bixler, and C. Y. Choo, *Ind. Eng. Chem., Process Design Dev.* **1** (1), 14–25 (Jan. 1962).
32. Beveridge J. Mair, James W. Westhover, and Frederick D. Rossini, *Ind. Eng. Chem.* **42** (7), 1279–1286 (1950).
33. Beveridge J. Mair, *Ind. Eng. Chem.* **42** (7), 1355–1360 (1950).
34. James I. Harper, John Lee Olsen, and Frank R. Schuman, Jr., *Chem. Eng. Prog.* **48** (6), 276–280 (1952).
35. U.S. Pat. 3,126,425 (March 24, 1964) to P. E. Eberly, Jr. and W. F. Arey, Jr. (Esso Research and Engineering Company).
36. U.S. Pat. 3,114,782 (Dec. 17, 1963) to R. N. Fleck and C. G. Wight (Union Oil Company).
37. Anon., *Chem. Eng. News*, 47, March 10, 1969.

38. *Technical Data Book— Petroleum Refining*, American Petroleum Institute, Div. of Refining, New York, 1966, Ch. 9.

39. Fred Fairbrother, Norman Scott, and Harold Prophet, *J. Chem. Soc. Part 1*, 1164–1167 (1956).

40. Joel H. Hildebrand and Robert L. Scott, *The Solubility of Non-Electrolytes*, Reinhold Publishing Corp., New York, 1950, Ch. 17.

41. O. Maas and J. Russell, *J. Am. Chem. Soc.* **40**, 1561 (1918).

42. O. Maas, E. H. Boomer, and D. M. Morrison, *J. Am. Chem. Soc.* **45**, 1433 (1923).

43. Herbert C. Brown and Howard W. Pearsall, *J. Am. Chem. Soc.* **74**, 191 (1952).

44. W. Klatt, *Z. Anorg. Chem.* **234**, 189 (1937).

45. K. Le Roi Nelson and Herbert C. Brown, *The Chemistry of Petroleum Hydrocarbons*, Reinhold Publishing Corp., New York, 1955, Ch. 56, p. 476.

46. D. A. McCaulay and A. P. Lien, *J. Am. Chem. Soc.* **73**, 2013 (1951).

47. M. Kilpatrick and F. E. Luborsky, *J. Am. Chem. Soc.* **75**, 577 (1953).

48. E. L. Mackor, A. Hofstra, and J. H. Van der Waals, *Trans. Farad. Soc.* **54**, 186 (1958).

49. S. Ehrenson, *J. Am. Chem. Soc.* **84**, 2681 (1962).

50. D. A. McCaulay and A. P. Lien, *J. Am. Chem. Soc.* **73**, 2013 (1951).

51. J. F. Norris and G. T. Vaala, *J. Am. Chem. Soc.* **61**, 2131–2134 (1939).

52. K. S. Pitzer and D. W. Scott, *J. Am. Chem. Soc.* **65**, 803–829 (1943).

53. D. A. McCaulay and A. P. Lien, *J. Am. Chem. Soc.* **74**, 6246–6250 (1952).

54. P. M. Pitts, Jr., J. E. Connor, Jr., and L. N. Leum, *Ind. Eng. Chem.* **47** (4), 770–773 (1955).

55. Karl Becker, Herman Blume, Eberhard Hahner, and Hannelore Grundmann, *Chem. Tech. (Berlin)* **18** (8), 455–459 (1966).

56. G. Baddeley, G. Holt, and D. Voss, *J. Chem. Soc.*, 100–105 (1952).

57. E. R. Boedker and W. E. Erner, *J. Am. Chem. Soc.* **76**, 3591 (1954).

58. P. H. Emmett, Ed., *Catalysis*, Vol. 6, Reinhold Publishing Corp., New York, 1958, Ch. 2, p. 111.

59. A. P. Lien and D. A. McCaulay, *J. Am. Chem. Soc.* **75**, 2407–2410 (1953).

60. D. A. McCaulay and A. P. Lien, *J. Am. Chem. Soc.* **79**, 5953–5955 (1957).

61. R. I. Silsby and E. W. Sawyer, *J. Appl. Chem.* **6**, 347 (August 6, 1956).

62. A. Tsuchiya, A. Hashimoto, H. Tominaga, and S. Masamune, *Bull. Jap. Pet. Inst.* **2**, 85 (1960).

63. A. Tsuchiya, A. Hashimoto, H. Tominaga, and S. Masamune, *Bull. Jap. Pet. Inst.* **1**, 73–77 (1959).

64. S. R. Bethea, R. L. Heinrich, A. M. Souby, and L. T. Yule, *Ind. Eng. Chem.* **50**, 1245 (1958).

65. U.S. Pat. 3,406,219 (Oct. 15, 1968) to Danford H. Olson (Manhattan Oil Co.).

66. Brit. Pat. 986,505 (March 17, 1965) to The Distillers Company.

67. Brit. Pat. 1,004,393 (Sept. 15, 1965) to Shell International Company.

68. U.S. Pat. 3,299,155 (Jan. 17, 1967) to C. R. Adams (Shell Oil Company).

69. Harry F. Keag, Howard S. McCullough, and Howard J. Sanders, *Ind. Eng. Chem.* **45**, 2–14 (1953).

70. Ralph Landau, David Brown, Joseph L. Russell, John Kollar, *Seventh World Petroleum Congress*, Mexico City, 1967, Vol. 5, pp. 67–72.

71. W. G. Lovell, *Ind. Eng. Chem.* **40**, 2388 (1948).

72. J. A. Barnard and B. M. Sankey, *Combustion Flame* **12** (4), 345–352 (Aug. 1968).

73. J. A. Barnard and B. M. Sankey, *Combustion Flame* **12** (4), 354–359 (Aug. 1968).

74. L. Errede and M. Swarc, *Quarterly Reviews*, **12**, 301 (1958).

75. L. Errede, R. Gregorian, and J. Hoyt, *J. Am. Chem. Soc.* **82** (19), 5218 (1960).

76. L. Errede and N. Kroll, *J. Poly. Sci.* **60**, 33 (1962).

77. Erie J. Y. Scott, *Ind. Eng. Chem., Product Res. and Dev.* **6** (1), 72–76 (1967).

78. K. J. Laidler, N. H. Sagert, and B. W. Wojciechowski, *Proc. Roy. Soc.* **270A**, 242, 254 (1962).

79. E. W. R. Steacie, *Atomic and Free Radical Reactions*, 2nd ed., Vol. 1, Reinhold Publishing Corp., New York, 1954, p. 189.

80. U.S. Pat. 2,964,573 (Dec. 13, 1960) to W. G. DePierri and H. W. Earhart (Esso Res. & Eng. Co.).

81. U.S. Pat. 2,973,391 (Feb. 28, 1961) to H. W. Earhart and W. G. DePierri (Esso Res. & Eng. Co.).

82. U.S. Pat. 2,966,523 (Dec. 27, 1960) to W. G. DiPierri and H. W. Earhart (Esso Res. & Eng. Co.).

83. R. Wegler, *Z. Angew. Chem.* **60A**, 88–96 (1948).
84. Ching Yun Huang and Teiichi Tanigaki, *Kobunshi Kagaka* **12**, 335–343 (1955).
85. J. F. Walker, *Formaldehyde*, 2nd ed., Reinhold Publishing Corp., New York, 1953.
86. U.S. Pat. 3,035,023 (May 19, 1962) to Minoru Imoto and Ching Yun Huang (Fine Organics, Inc.).
87. R. M. Keefer and L. J. Andrews, *J. Am. Chem. Soc.* **79**, 4348 (1957).
88. P. B. D. de la Mare and J. Robertson, *J. Chem. Soc.* 279 (1943).
89. F. Condon, *J. Am. Chem. Soc.* **70**, 1963 (1948).
90. M. Kilpatrick, M. W. Meyer, and M. L. Kilpatrick, *J. Phys. Chem.* **64**, 1433 (1960).
91. M. Kilpatrick and M. W. Meyer, *J. Phys. Chem.* **65**, 530 (1961).
92. K. A. Kobe and H. Levin, *Ind. Eng. Chem.* **42**, 352 (1950).
93. K. A. Kobe and W. P. Pritchett, *Ind. Eng. Chem.* **44**, 1398 (1956).
94. K. A. Kobe and H. M. Brennecke, *Ind. Eng. Chem.* **46** (4), 728 (1954).
95. D. J. Hadley, *Chem. Ind.* 238–243 (Feb. 25, 1961).
96. E. L. Pollitzer, J. C. Hayes, and V. Haensel, *Preprints, American Chemical Society Div. Petrol. Chem.* **14** (4), D-8–D-17 (Sept. 1969).
97. Seiya Otani, Shiro Matsuoka, and Maserki Sato, *Japan Chem. Quarterly* **4** (4), 16–18 (Oct. 1968).
98. Anon. *Hydrocarbon Processing* **48** (11), 155 (1969).
99. Joseph A. Verdol, *Oil Gas J.* **67** (23), 63–66 (June 6, 1969).
100. U.S. Pat. 3,177,265 (Apr. 6, 1965) to G. C. Lammers (Standard Oil Company, Indiana).
101. Herbert F. Weigandt and Regis Lafay, *7th World Petroleum Congress*, Mexico City, 1967, Vol. 4, pp. 47–54.
102. Anon, *Hydrocarbon Processing*, **48** (3), 9 (1969).
103. Leo R. Aalund, *Oil Gas J.* 48–52 (Nov. 27, 1967).
104. Leo R. Aalund, *Oil Gas J.* 102–103 (July 15, 1968).
105. James H. Prescott, *Chem. Eng. News* 138–140 (Oct. 7, 1968).
106. H. F. Uhlig and W. C. Pfefferle, *American Chemical Society Div. Pet. Chem.* **14** (4), D-154–D-161 (1969).
107. Anon. *Hydrocarbon Processing* **48**, 250 (1969).
108. U.S. Pat. 2,511,711 (June 13, 1950) to H. P. Hetzner and R. J. Miller (California Research Corp.).
109. U.S. Pat. 2,943,121 (June 28, 1960) to J. A. Spence (California Research Corp.).
110. U.S. Pat. 3,311,670 (March 28, 1967) to W. Smolin (Texaco, Inc.).
111. Austrian Pat. 251,557 (Jan. 10, 1967) to Hermann Zorn.
112. Anon, *Hydrocarbon Processing* **44** (11), 193 (1965).
113. W. D. Schaeffer, H. E. Rea, R. F. Deering, and W. W. Mayes, *6th World Petroleum Congress*, Frankfurt am Main, June 19–26, 1963, Sec. 4, pp. 65–73.
114. Anon., *Hydrocarbon Processing* **48** (11), 254 (1969).
115. Yukio Igaraski, *Japan Chem. Quarterly* **4** (4), 27–30 (Oct. 1968).
116. *Synthetic Organic Chemicals United States Production and Sales: 1945-1967, Inclusive*, U.S. Tariff Commission, Washington, D.C., 1968.
117. *Chemical Economics Handbook*, Stanford Research Institute, Menlo Park, California, 1967, p. 699. 5030Z.
118. *Chemical Profile—Ethylbenzene*, rev. ed. Schnell Publishing Company, 1969.
119. *Chemical Profile—Para xylene*, rev. ed. Schnell Publishing Company, 1969.
120. Anon., *Eur. Chem. News* **14** (349), 4–8 (Oct. 11, 1968).
121. *Free World Aromatics For Chemicals*, Vol. 1 Summary, The Pace Company, Houston, Texas, Feb. 1967.
122. Anon., *Oil Gas J.* 63 (Oct. 2, 1967).
123. N. Irving Sax, *Dangerous Properties of Industrial Materials*, 2nd ed., Reinhold Publishing Corp., New York, 1963.
124. *Handbook of Organic Industrial Solvents*, National Association of Mutual Casualty Companies. Chicago, Ill., 1958.
125. *ASTM Special Technical Publication, No. 225*, American Society for Testing Materials, Philadelphia, Pa., 1958.
126. C. Marsden and S. Mann, *Solvents Guide*, 2nd ed., Interscience Publishers, a division of John Wiley & Sons, Inc., New York, 1962.

127. H. A. Gardner and G. G. Sward, *Paint Testing Manual*, 12th ed., Gardner Laboratories, Inc., Bethesda, Maryland, 1962, Ch. 22.

128. Taijiro Oga, *Hydrocarbon Processing* **45** (11), 174–176 (1966).

129. Anon, *Hydrocarbon Processing* **48** (11), 252 (1969).

130. Taijiro Oga, Masatomo Ito, and Minoru Tashiro, *Am. Chem. Soc. Preprints* **13** (4), A57-A60 (1968).

131. Anon., *Japan Chem. Quarterly* **5** (3), 17 (1969).

132. Harry E. Cier and H. W. Earhart, *Adv. Petrol. Chem. Refining* **8,** 379 (1964).

133. D. A. McCaulay, M. C. Hoff, Norman Stein, A. S. Couper, and A. P. Lien, *J. Am. Chem. Soc.* **79,** 5808–5809 (1957).

134. M. C. Hoff, *J. Am. Chem. Soc.* **80,** 6046–6049 (1958).

135. Charles C. Price, *Organic Reactions*, Vol. 3, John Wiley & Sons, Inc. New York, 1946, Ch. 1.

136. *Chemical Origins and Markets*, Stanford Research Institute, Menlo Park, California, 1967, pp. 18–20.

Harry E. Cier

Esso Research and Engineering Co.

XYLIDINES, $(CH_3)_2C_6H_3NH_2$.   See Aniline, Vol. 2, p. 421.

YEASTS

Although yeast was used for thousands of years for making bread, wine, and beer, its real nature was not known until Leeuwenhoek in 1680, through his newly invented microscope, for the first time observed yeast cells in fermenting liquids. The description and scientific evaluation of yeasts, begun in 1836 (11b) with Cagniard de la Tour, Schwann, and Kützing, was ridiculed by Liebig and others. Following their footsteps, Pasteur, beginning in 1855 (11b), with the aid of more advanced microscopes and chemical and biological knowledge, set the foundation of our present knowledge of yeasts. During the last part of the 19th and the first of the 20th century, Hansen, Guilliermond, Henrici, Jörgensen, Henneberg, Szilágyi, Foth, Kluyver, Westerdijk,

Stelling-Dekker, Lodder, the Lindegrens, Wickerham, and many others defined, described, and classified the yeasts. The ever-increasing literature on the morphology, genetics, biochemistry, classification, industrial manufacture, and utilization of yeasts is in the tens of thousands of publications (1–91).

Characterization and Classification (4,7,11,15,16,20–22,24,25,27,32–34)

According to the most accepted definition, yeasts are true fungi which in standard growth form reproduce hyaline, uninucleate, unicellular individuals that may be pigmented yellow, orange, or red. This definition includes all uninucleate unicellular fungi whether or not they ferment sugar or form sexual spores (see Microorganisms). It excludes fungi which, in their dominant growth form, produce multinucleate cells or multicellular individuals. Under certain conditions some species of *Mucor*, a common mold, and *Amylomyces* (13,17,34) (another mold related to *Mucor* and used for alcohol production in early 1900), produce budding cells that resemble true yeasts in appearance and in the ability to ferment sugars into alcohol; also the fungi *Ashbya gossypii* (Ashby and Nowell) Guilliermond and *Eremothecium ashbyii* Guilliermond resemble yeasts in many respects. None of these is considered a true yeast because they form multinucleate cells or multicellular individuals. Yeasts have also been characterized as unicellular uninucleate fungi with the ability for vegetative reproduction by budding, fission, or both and bearing a yeast-like macroscopic and microscopic appearance. The most recent definition, from the second revised and enlarged edition of Lodder (25), states that "yeasts may be defined as microorganisms in which the unicellular form is conspicuous and which belong to the fungi... ....since the yeasts are not a natural group of organisms, this delimitation is bound to be artificial and arbitrary." In spite of inadequate definitions, yeasts can be distinguished from bacteria and molds by their size, shape, mode of reproduction, and cellular and cultivation characteristics.

The number of yeast strains maintained in industrial and scientific culture collections, frequently identified only by code numbers, are in the thousands. About 1500 strains described in the literature are named. The first edition of Lodder and Kreger-van Rij (25) studied 1307 strains which were classified into 26 genera that included 164 species. The new edition (see below) reports over 4300 strains that were examined and classified into 39 genera incorporating 349 species.

The prototype for the species commonly used in the manufacture of bread, whiskey, and alcohol is *Saccharomyces cerevisiae*.

The various species are identified and classified by their morphological, genetic, and biochemical characteristics. The most up-to-date listing of criteria and of the methods used in classification may be found in the new Lodder (25) in Chapter II (pages 34–113).

Popular classifications have placed yeasts in two or more groups with respect to a specific criterion. "Good" and "bad" signified the yeasts useful or harmful for a certain purpose. A refined version of this grouping is "culture yeasts" (useful) and "wild yeasts" (harmful) depending on their use. A culture yeast useful in a winery, for instance, would be wild yeast in a brewery. The commercial technical man uses these terms in different connotations from those of the scientist, for whom a wild yeast lives freely in nature and a culture yeast is cultivated under controlled conditions in the laboratory.

Depending upon the industry they serve, yeasts are grouped as brewer's, distiller's, rum, wine, champagne, baker's, food, feed yeasts, etc. Sometimes a subdivision is made to conform to the raw material utilized—distiller's yeast may be classified as grain, potato, molasses, sugar beet, fruit, brandy yeast, etc, and grain yeast may be further subclassified as bourbon, corn, rye, milo yeast, etc. Various strains of the same species may even be placed in several groups, and one group may include several species. By the ability to ferment sugars they may be called "strong fermenters," "weak fermenters," and "nonfermenters."

Brewer's yeasts (see also Beer and brewing) are frequently grouped as "Saaz type," for yeasts with low fermenting power, and "Logos type," for those with high fermenting power. Another grouping, principally for brewer's yeasts, is as "top" and "bottom" fermenters (Vol. 3, pp. 303–304, 315–316), depending on the tendency of the yeast to accumulate on the top of the beer or to settle at the bottom during the main fermentation. The tendency of the yeast cells to stay together in flocs or to remain separate classifies them into "flocculating" and "nonflocculating" yeasts (9,46a). Yeasts with a tendency to form a pellicle on the liquid surface are called scum yeasts (Kaumhefe, in German).

For several decades (until ca. 1915) the term "true yeast" denoted the spore-forming yeasts and "not true yeast," the nonspore-forming yeasts. Today "true yeasts" are all species falling within the specification given in the introduction to this section; "not true" yeasts are fungi with morphological or biochemical characteristics similar to true yeasts but falling outside the definition.

The **scientific classification** generally accepted to date had been based on the work of four Dutch scientists, a group of dedicated women who worked at Centraalbureau voor Schimmelcultures in Delft (11,22,24,25). Of the more than 3500 genera of fungi, 26 genera constituted the yeasts and were grouped into 3 families: (1) *Endomycetaceae* including all ascospore-forming yeasts, (2) *Sporobolomycetaceae* including yeasts forming ballistospores, and (3) *Cryptococcaceae* including nonspore-forming yeasts.

The older classifications of yeasts have been reviewed in detail (43–46,149).

The most recent classification, just published (1970), is in *The Yeasts, a Taxonomic Study*, under the editorship of J. Lodder and written by fourteen taxonomists (three from The Netherlands, six from the United States, four from Portugal, and one from South Africa) (25). This new "Lodder" is based on the first edition but incorporates the most recent taxonomic recommendations and describes the history, characteristics, and identification of all presently accepted genera and species. As mentioned above, all yeasts are now placed in 39 genera. These divide into four groups:

I. All ascomycetous (ascospore-forming) yeasts comprising 3 families that include 22 genera: *Citromyces, Coccidiascus, Debaryomyces, Dekkera, Endomycopsis, Hanseniaspora, Hansenula, Kluyveromyces, Lipomyces, Lodderomyces, Metschnikowia, Nadsonia, Nematospora, Pachysolen, Pichia, Saccharomyces, Saccharomycodes, Saccharomycopsis, Schizosaccharomyces, Schwanniomyces, Wickerhamia,* and *Wingea.*

II. Two yeast-like genera belonging to the order *Ustilaginales* (in the *Basidiomycetes*): *Leucosporidium* and *Rhodosporidium.*

III. Three yeast-like genera belonging to the family *Sporobolomycetaceae*: *Bullera, Sporidiobolus,* and *Sporobolomyces.*

IV. Asporogenous yeasts not belonging to the *Sporobolomycetaceae* and comprising 12 genera: *Brettanomyces, Candida, Cryptococcus, Kloeckera, Oosporidium, Pityro-*

sporum, Rhodotorula, Schizoblastosporion, Sterigmatomyces, Torulopsis, Trichosporon, and *Trigonopsis.*

The first edition of *The Yeasts* constituted the internationally accepted classification of yeasts for almost two decades. The new edition, prepared by global cooperation, not only incorporates new recommendations (including those of Gäumann, Wickerham, Phaff, and others) but also helps to clarify several long-disputed questions, such as the *Torulopsis-Candida* problem discussed on page 541.

From the known yeast species, only a few are industrially important: *Saccharomyces cerevisiae, Saccharomyces cerevisiae* var. *ellipsoideus, Saccharomyces carlsbergensis, Saccharomyces fragilis* (reclassified in the new "Lodder" as *Kluyveromyces fragilis*), and the "torula" yeasts (see p. 541).

Interest in pure cultures of yeasts for scientific study and industrial use has increased over the years. The use of pure yeast cultures by industry, and the establishment of pure-culture collections, date back to around the last part of the 19th century. Szilágyi, in 1890 (34), described in detail the industrial utilization of pure cultures. By 1930, in fact, the Centraalbureau voor Schimmelcultures had *ca.* 6000 cultures in its collection (Dr. Johanna Westerdijk, personal communication); by 1939 this number had grown to 6394 (9a). Today, pure cultures are maintained in a number of special laboratories, and several conferences on the problems of culture collections have been held (18).

Morphology

Under specific conditions of cultivation the macroscopic and microscopic appearances of a given yeast strain remain constant for all practical purposes. To a certain degree the identity of a species may be approximated by visual observation. Under the microscope the size and shape of the cells, the manner of vegetative reproduction, spore formation, and spore germination (in the spore-forming yeasts), and the grouping of the cells are considered.

The vegetative cells (grown in malt extract) of the prototype yeast, *Saccharomyces cerevisiae,* are about 4.5 to 10.5 μ in width and 7–21 μ in length. Some strains are much smaller and others much larger. Some strains form filamentous cells as long as 30 μ or more.

Cell structure (6,9,14,32,46a,105,111,115,118,119,142–144,148). A yeast cell under the commonly used bright field illumination exhibits only a cell wall, frequently one or more vacuoles, and possibly a dancing body in a vacuole. The ultrastructure of a yeast cell (based on electron microscopy) as visualized in an ultrathin section exhibits 16 organelles (Fig. 1) (105). They are:

1. The *cell wall,* the outermost layer, gives firmness and shape to the cell, protecting it and holding together the soft inner contents.

2. *Plasmalemma* is the recent name of the cytoplasmic or cell membrane. This semipermeable membrane, with an approximate thickness of 100 Å, surrounds the living cell substance and permits the passage of small molecules by diffusion. Resorption and secretion of larger molecules through the plasmalemma is facilitated by energy-consuming processes.

3. *Ground plasma or matrix.* With the discovery of new organelles and structures, the term "ground plasma" denotes the living plasma remaining, with the exclusion of organelles and structures visible by electron microscope. The ground plasma as part of

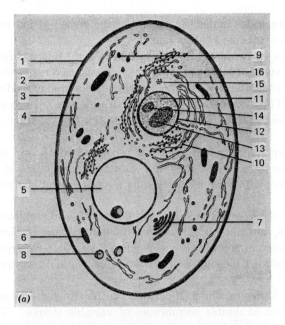

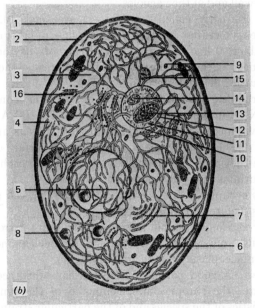

Fig. 1. Concept of organelles of a yeast cell visualized (a) in an ultrathin (0.03 μ) section and (b) in a thicker (0.3 μ) section:

1. Cell wall	9. Lysosome
2. Plasmalemma	10. Ribosome granules
3. Ground plasma or matrix	11. Nucleus
4. Endoplasmic reticulum	12. Nuclear envelop
5. Vacuole	13. Chromosomes
6. Mitochondria	14. Nucleolus
7. Golgi bodies	15. Centrosome
8. Spherosome	16. mRNA and tRNA

the living matter operates under the control of the genes; it has enzymic properties, and a good portion of it consists of proteins.

4. *Endoplasmic reticulum* (E.R.). Invisible by light microscopy, the endoplasmic reticulum is a "system of intercommunicating canals, caverns, vesicles, and cisterns" (14) starting from the outer layer of the nuclear envelop and extending throughout the ground plasma. The strands are 100–200 Å in diameter. In yeast cells the E.R. system increases considerably when the cells switch from aerobic respiration to fermentation.

5. *Vacuoles*, appearing under light microscope as (optically) empty spherical or elliptical bodies, contain metabolites and valuable assimilates as reserve material. The vacuolar reserve substance is enclosed by a semipermeable membrane, the "tonoplast."

6. *Mitochondria* are ovoid or short rods 0.5×2.0 μ in size, consisting of lipoprotein, phospholipids, and ribonucleic acid. Mitochondria, the respiratory organelles of the cell, are necessary in aerobic activities. During aerobic respiration, mitochondria are plentiful in yeast. Under anaerobic conditions, the E.R. system becomes prominent.

7. The *Golgi bodies* or Golgi apparatus, visible by light microscope in stained preparations, consist of double membranes of variable shapes. They are active in formation of the cell wall and of the vacuoles.

8. *Spherosomes* (formerly called microsomes) are small (0.1–1.5 μ) osmophilic spheres and contain enzymes that catalyze the final phase of fat synthesis.

9. *Lysosomes*, about 0.4 μ in size, are small sacs of enzymes enclosed within a lipoprotein membrane. If the protecting membrane breaks, the enzymes hydrolyze the cell substance, causing death of the cell.

10. *Ribosome* granules, around 100–150 Å in diameter, consist of proteins and ribonucleic acid (RNA), in about equal proportions. The RNA contains 1500–4000 subunits. The individual ribosomes produced in the nucleus are either distributed freely in the ground plasma or lined up in groups on the surface of endoplasmic reticulum strands, where they may be temporarily connected to each other by messenger RNA threads, thus forming "polyribosomes" or "polysomes."

11. The *nucleus* occupies about one-fifth of the resting cell and increases in size about three times just before or during budding.

12. The *nuclear envelop*, a double membrane structure surrounding the nucleus, is several hundred Angstroms wide. Pores through the nuclear membrane offer direct contact between the nucleoplasm and the ground plasma.

13. *Chromosomes*. A good portion of the nuclear substance, which stains with basic dyes, constitutes the chromosomes, composed of the genes, which carry the hereditary characteristics of the species and of the particular strain.

14. The *nucleolus*, composed principally of RNA, is produced in cooperation with the deoxyribonucleic acid (DNA) of the chromosomes. It consists of coarse granules of the size of ribosomes, about 150 Å in diameter.

15. The *centrosome* is a dense globular zone located in the ground plasma close to the nucleus, and plays an essential role in nuclear division.

16. *Messenger and transfer ribonucleic acids.* With the cooperation of the DNA of the chromosomes in the nucleus, ribonucleic acid macromolecules are formed which (through the pores of the nuclear envelop) carry the genetic information for protein synthesis into the cytoplasm. These specialized threads of macromolecules, termed

messenger RNA (mRNA), define the sequence of the amino acid units in the synthesis of any specific protein. A soluble type of RNA, smaller than messenger RNA, also migrates from the nucleus into the cytoplasm; because of its ability to transfer amino acids, it is called transfer RNA (tRNA). Each amino acid is carried by a special transfer RNA.

Reproduction

Vegetative reproduction is an asexual form of reproduction in which an existing cell creates a new cell capable of performing all functions of the original (mother) cell. Under favorable conditions the production of a new cell occurs within 50 min, and this may be repeated every 50 min by all the cells in the culture. Thus, a single cell of *Saccharomyces cerevisiae* within 50 min results in two cells; in 100 min, 4 cells will have been produced; in 200 min, 16 cells; and so on. This may continue as long as optimum conditions exist. In some species mother cells may produce simultaneously two or more daughter cells.

Vegetative reproduction in yeasts may follow three distinct patterns—budding, fission, and an intermediate process:

(1) *Budding* begins with a small protuberance on the cell wall, called bud or sprout, which in a steady growth reaches maturity and usually the size and shape of the mother cell. Most yeasts employed for industrial purposes reproduce in this way. Contrary to this monomorphic reproduction characteristic of *S. cerevisiae*, in polymorphic species the mature buds may be smaller, or many times larger, and different in shape from the mother cell (*Nematospora* produce round, oval, elongated, and long pseudo-mycelial cells in the same culture).

(2) During reproduction by *fission* or splitting, the round or elliptical cell elongates, and one or occasionally two cell walls are produced within it, thus forming several new cells from one cell. At the proper stage of growth the newly formed cells separate by splitting off. The prototype of fission yeasts is *Schizosaccharomyces pombe*.

(3) In the *intermediate process*, called "bud-fission" or "budding on broad base" in the new edition of Lodder (25), and mostly typical of species having lemon-shaped cells, reproduction begins with the formation of a bud at the broad terminal of the cell. After the bud reaches a certain maturity, a dividing cell wall is formed between the mother and daughter cell, and the two cells eventually split by fission. In some species two buds are formed simultaneously at both ends of the lemon-shaped cell—this is *bipolar budding*. Species of *Kloeckera*, *Nadsonia*, and *Hanseniaspora* reproduce by the intermediate process.

Some yeasts, species of *Candida*, *Trichosporon*, and *Endomycopsis*, may reproduce by both budding and fission.

Prior to or during reproduction, the cell nucleus duplicates itself, and each cell receives one of the new nuclei. In normal vegetative reproduction, all newly formed cells are genetically identical with the mother cell. Genetic changes (occurring with very low frequency during vegetative reproduction) are the results of spontaneous mutation.

Sexual reproduction in ascospore-forming species is by ascospore formation. The original cell performs as an ascus (spore sac) and may produce up to eight new cells (spores) simultaneously. The genetic changes prior to and during spore formation (known as segregation) and those taking place during fusion of two spores or cells (known as recombination) constitute the most significant biological processes.

Through *segregation* the original characteristics of the mother cell may change significantly, and through *recombination* new strains, hybrids, with highly improved characteristics may be formed. *Hybridization*, which occurs in nature, is also carried out in laboratories for the production of new strains.

Most of the yeasts used industrially reproduce by both vegetative and sexual reproduction. To create new strains in the laboratory, the sexual reproduction is utilized, but for industrial-scale propagation only vegetative reproduction is employed. Optimum conditions for spore formation differ from those for vegetative reproduction. The details of yeast morphology and changes connected with asexual and sexual reproduction of the various yeast species have been reviewed (9,11,13,15,17,20–22,24, 25,32,43,44).

Genetics (23,28a,32,44,46a,111a,112,113)

Yeast genetics is a subject that has been extensively studied from both scientific and practical points of view. The inexpensive and speedy experiments by which scientific evidence or practical results may be produced within days or weeks, and the relative ease of operations, have made yeast one of the classical tools in genetic studies. Stimulated by the classical genetic studies on yeasts made during the period 1930 to 1960, much research was carried out to provide industry with more stable and superior yeast strains. Winge, Laustsen, the Lindegrens, Lederberg, Tatum, Lodder, Kreger-van Rij, Pomper, Burkholder, Mrak, Phaff, Fowell, Ephrussi, Wickerham, Ogur, Mundkur, Roman, Hawthorne, and Mortimer are but a few who have contributed significantly to our present knowledge of yeast genetics.

Only a few basic facts are presented here. All morphological, biological, and biochemical properties characterizing an organism are controlled by genes. Hundreds or possibly thousands of genes lined up in definite order in loose connection, like beads on a string, make up the complex heredity units, the chromosomes. A specific number of chromosomes (haploid, see below) characteristic for each organism constitute the genome, located in the nucleus of the cell. Some characteristics are controlled by more than one gene. An accidental loss of a gene or of a fraction of a chromosome may involve the loss of one or more characteristics. Evidence has been presented to support the belief that some genetic characteristics are controlled by extranuclear, cytoplasmic, factors (12).

When the nucleus divides in vegetative reproduction of the cell, all chromosomes duplicate themselves and one chromosome of each newly formed pair will be part of one of the two new nuclei. This is *mitotic division*, which assures that characteristics for the new cell will be identical with the original cell.

When the nucleus divides in preparation for spore formation, reduction division (*meiosis*) occurs: the chromosomes do not reproduce themselves but separate in two groups, which become part of two nuclei that may undergo one or two additional divisions. The number of chromosomes in the spores and their progeny is one-half that of the chromosomes in the original cell. Cells with a reduced number of chromosomes are in the *haploid* state. Under favorable conditions two haploid cells fuse and their nuclei recombine, thus reestablishing the *diploid* zygote state with two sets of allelic chromosomes in the nucleus. Through segregation (spore formation) and recombination (zygote formation) new strains may be formed.

Ploidy. The minimum number of chromosomes in a species is the haploid number, which for yeast is 14 or more. When two haploid cells recombine, the diploid state is reestablished. The fusion of a diploid and a haploid cell will produce a *triploid* cell, with three sets of chromosomes, and the fusion of two diploid cells will produce *tetraploid* cells with four sets of chromosomes. Cells with a higher number of chromosomes are usually superior (in utilizing diversified sources of nutrients, in higher enzyme activity, ability to survive, etc) to cells with a lower chromosome number within the same species. Artificial diploidization (hybridization), triploidization, and tetraploidization are practiced for developing yeast strains with desirable qualities.

The manner of vegetative reproduction and spore formation, the appearance and number of spores (if existing), the mode of recombination, and ploidy are utilized in characterizing and identifying yeast species. The prototypes of spore-forming yeasts, *Saccharomyces cerevisiae* (and its variant *S. cerevisiae* var. *ellipsoideus*) and *S. carlsbergensis*, species to which most industrially used yeasts belong, reproduce by budding in the diploid state and form one to four ascospores. *Candida utilis* (also known as *Torulopsis utilis* and *Torula utilis*), a well-known fodder yeast (see p. 541), reproduces by budding in the haploid state and forms no spores.

Biochemistry (9,26a,32,45,65,93,126,133,138a,139)

Yeast biochemistry is of major interest to the technical man for two main reasons: (1) the raw materials that can be used and (2) the end products that can be expected. It is intimately involved with the nutrition of yeast and the various ways energy is obtained for the synthetic processes of cell metabolic activities. Yeasts can break down many complex molecules and build up new ones for their needs. A large part of the breakdown process is hydrolytic, or digestive. Another part is oxidative, to produce energy needed for the synthetic reactions in assimilation. The energy-producing oxidative processes in the presence of air may result in completely oxidized products, H_2O and CO_2 from carbohydrates for instance, and in the absence of air, in oxidized products (such as CO_2) in part and reduced products (like alcohols) in part. This last process is referred to as *fermentation* (see Vol. 1, p. 523; Vol. 3, p. 314). All interchanges between the raw material (substrate) and the yeast elicited by the yeast are enzymic, being catalyzed by enzymes produced by the yeast. These enzymic reactions include breaking down of the substrate (for food and energy, using desmolyzing enzymes) and building up of new molecules (for new cells and for the hundreds of enzymes active in the metabolism of every cell). During reproduction, once the yeast cells (the fermenting type) produce enough enzymes needed for alcoholic fermentation, they will continue fermenting far beyond the need for energy, and even long after reproduction has stopped. Up to certain concentrations, fermentation may go on until the sugar is completely used up. The energy liberated and not used for synthesis manifests itself in the form of heat. The readiness of certain yeast strains to ferment sugars far beyond their own need is the reason for their use industrially.

COMPOSITION (9,94)

The composition of a mass of yeast cells, baker's yeast for instance, gives an insight into the assimilative processes:

Water. The water content of a commercial yeast cake or bar varies from 70 to 75%. The external water, occupying the space between the cells, is estimated to be about 25% of the total volume and the remaining portion of water, inside the wall, is a cell constituent that may amount to about 45% of the cell substance.

Minerals. Depending on the composition of the nutrient medium the ash content may vary from 2 to 10% of the dry yeast substance. About 50% of the total ash is P_2O_5. Phosphates, essential in energy transfer, are mainly concentrated in the mitochondria and are also constituents of the ribonucleic acid (RNA) and deoxyribonucleic acid (DNA). The other mineral constituents in the ash (all values approximate) are 30% K_2O, 5% MgO, 5% CaO, and 1% or less each of SiO_2, Na_2O, SO_3, Cl, and FeO; Al, Ba, B, Cr, Co, Cu, Pb, Mn, Mo, Ni, Sn, V, and Zn may be present in trace quantities. The ash constituents are supplied by the raw material, the substrate, which is often supplemented with phosphates as needed.

Proteins. Including some peptides and free amino acids, the protein may amount to 30 to 70% of the solid content. The proteins constitute the main portion of enzymes, and the amino acids are standby reserves (amino acid pool) for quick protein synthesis. Peptides are either hydrolytic products of unneeded proteins or parts of proteins being synthesized for current needs. The bulk of the proteins in yeast is made up of **enzymes** or precursors of enzymes: for instance, glyceraldehyde phosphate dehydrogenase may amount to 20% of the total proteins, carboxylase to 2.5%, and hexokinase to 0.3%. All are essential in the fermentation process. The enzymes of yeast have been reviewed by Woods (56a), Eddy (46b), Boyer et al. (5a), and Dixon and Webb (11a); see also (105a). Specific nomenclature may be found in *Enzyme Nomenclature* (19).

Nucleic Acids. DNA, the major constituent of the chromosomes, is concentrated in the nucleus. RNA is concentrated in the nucleolus and partly is dispersed in the ground plasma in the form of ribosomes, the particles active in the protein synthesis, and in the messenger and the transfer RNA's. The nucleic acid content of yeast cells increases with ploidy.

Carbohydrates. The carbohydrates in yeasts include glycogen, up to 20%, and trehalose, up to 10%, of the dry substance. A starchlike compound, ribose, hexose, and heptose derivatives, glucosamine, mannan, and glucan are usually also present.

Lipids. The lipid content in *Saccharomyces cerevisiae* is about 2–5% of the dry substance. In the fat-producing yeasts, *Torulopsis lipofera* (now *Lipomyces lipofera*), *Endomyces vernalis*, and *Oospora lactis* (*Oidium lactis*, *Endomyces lactis*, *Geotrichum candidum*), lipids may exceed 50% of the dry material. The lipids include triglycerides, glycerophosphatides, cerebrin, sterols, and carotenoids. Most important of the sterols is ergosterol.

Vitamins. In general, yeasts contain generous quantities of most members of the vitamin B group (see Biotin; Inositol; Thiamine; Vitamins). Carotenoid pigments produced by certain yeasts are related to vitamin A, and ergosterol to vitamin D. Yeast contains little or no vitamin C and lacks vitamins B_{12}, E, and K. Most vitamins are constituents of coenzymes (39a): nicotinamide is part of coenzymes I (DPN or NAD) and II (TPN or NADP); pantothenic acid is part of coenzyme A; metabolically active forms of folic acid constitute the citrovorum factor; and pyridoxine (Vitamin B_6) is needed for the coenzyme codecarboxylase. Animal tests indicate that yeasts produce unknown growth factors in addition to the many known vitamins.

Vitamins in yeasts are important for two practical reasons. To a certain extent the fermentative power of a yeast is related to its vitamin content; and the emphasis on the nutritional value of dry yeast and yeast-containing products is based on their contents of vitamins and unidentified growth factors. *Candida guilliermondii*, a true yeast, is useful for the industrial production of riboflavin, and two yeastlike fungi, *Eremothecium ashybii* and *Ashbya gossypii*, have also been used industrially for riboflavin production. The ability of many yeasts to synthesize thiamine (vitamin B_1) from pyrimidine and thiazole has been employed commercially.

NUTRIENT REQUIREMENTS

Vitamins. In spite of the highly complex composition of the cells the nutritional requirements for most yeast are simple. Many species can reproduce in solutions containing sugar, inorganic nitrogen (principally ammonium) compounds, phosphates, trace elements, and traces of the B group vitamins (frequently referred to as "bios" or "nutrilites"). The nutrilites (very small quantities are required) include thiamine, biotin, inositol, nicotinic acid or nicotinamide, pantothenic acid, pyridoxine, and p-aminobenzoic acid. The various species depend on some or all of these substances to a varying degree, sometimes so specifically that the strain can be used as a test organism for the assay of a particular vitamin. Within one species, however, the various strains exhibit different degrees of dependency on one or all of the vitamins.

The ability to utilize precursors to synthesize the needed vitamin also varies according to species or even within the various strains of the same species. Certain strains may adapt themselves to growth on media deficient in the nutrilite components on which they were previously dependent. Often, the vitamins released from a generous inoculum are sufficient to facilitate growth in a nutrilite-deficient medium.

In respect to vitamins, the most important fact is that yeasts produced in industrial processes either principally or as a byproduct (primary or secondary yeast) abound in B vitamins and are thus useful as vitamin sources in human and animal nutrition. The raw materials employed in distilleries, breweries, and wineries contain sufficient amounts of nutrilites. In the manufacture of baker's yeast and feed yeast the raw materials usually contain at least part of the nutrilites needed.

Once he has selected the yeast strain with the desired qualities, the manufacturer will test the relative deficiencies of his strain and of his raw material with regard to the individual vitamins and will supplement the raw materials as needed.

Carbohydrates. The different carbohydrates are utilized by the different species of yeasts in varying degrees as a carbon source in assimilation and a source of energy for synthetic processes. Their ability to ferment or to assimilate the individual sugars is used in species identification. Although other sugars—pentoses and heptoses—may be assimilated, only hexoses and their polymers are of industrial importance, for only hexoses and some of their disaccharides and trisaccharides are fermented by yeasts. *Glucose, fructose,* and *mannose* are assimilated by all yeasts and are fermented by all fermenting yeasts. Yeasts unable to ferment sugars obtain energy through aerobic respiration; the fermenting yeasts may use both the aerobic and the anaerobic processes. *Galactose* is assimilated by the majority of the yeasts but is fermented with ease by a few, weakly by many, and not at all by others. *Sucrose* and *maltose,* the two most important disaccharides from the industrial point of view, are assimilated and fermented by many yeasts, assimilated but not fermented by several, and not utilized at all by others. *Lactose* is not utilized by most yeasts, is assimilated but not fermented by others, and is assimilated and fermented by a few (*Saccharomyces fragilis;* see Vol. 13, p. 572). *Raffinose,* the trisaccharide present in molasses, is not fermented by many yeasts; one-third is fermented by some yeasts and two-thirds or the whole molecule by others. The degree of raffinose fermentation depends on the hydrolytic ability of the yeast and on its ability to ferment galactose.

Higher sugars may be assimilated to varying degrees by some yeasts but are not fermented. Only a few yeasts (eg, *Saccharomyces diastaticus*, ref. 9), can ferment starch, and even then at a very slow rate.

To ferment polymers the yeast must possess the corresponding hydrolyzing enzymes. Yeasts producing maltase can hydrolyze and ferment maltose, those producing amylases can hydrolyze and ferment starch. *Saccharomyces marxianus* (now *Kluyveromyces marxianus*) and *S. fragilis* (*Kluyveromyces fragilis*) produce inulinase and are capable of hydrolyzing and fermenting inulin, a fructose polymer present in Jerusalem artichoke.

Other Carbon Sources. A great variety of organic compounds other than carbohydrates (9) may be used by yeasts in the assimilation process, as glucosides, amino acids, peptides, peptones, organic acids, and ethyl alcohol (131).

Inorganic Substances. Normally the raw materials used in industrial processes contain all inorganic constituents needed by the yeast with the exception of phosphorus and nitrogen, which are supplied, respectively, as phosphoric acid or its salts and as ammonia or ammonium salts. In the production of baker's, food, and feed yeast, and in the production of alcohol from potatoes, tapioca, molasses, sugar juices, low-grade starches, and other incomplete cereal carbohydrate sources, the substrate is supplemented with inorganic salts. Special inorganic compounds are used in the production of glycerol.

Water. Usually water makes up 80–95% of the substrate. As a rule any water suitable for drinking purposes is suitable for yeasts.

Distilled Alcoholic Beverages (13,17,29,32a,34,37,39a,42,55,104,110,153)

The manufacture of distilled alcoholic beverages, whiskeys, vodka, certain gins, beverage alcohol, rum, and brandy produced from various raw materials and having different flavors, aromas, and tastes has one step in common for all—fermentation of the carbohydrates. In this step, yeast plays the major role. See also Alcoholic beverages, distilled.

All mashes for making whiskey, vodka, beverage alcohol, and rum are fermented with specially selected and/or adapted strains of *Saccharomyces cerevisiae*. Fruit juices for brandies are fermented either spontaneously with the yeasts naturally present on the fruits or with specially propagated strains generally selected from *S. cerevisiae* var. *ellipsoideus* (*S. ellipsoideus*). In both cases *S. ellipsoideus* is the main fermenting agent; species of the genera *Kloeckera*, *Hansenula*, and others present on the ripe fruits take secondary part in the fermentation.

For all distilled alcoholic beverages, in common practice, the batches of specially prepared mashes are inoculated (seeded) with an appropriate amount of pre-ferment (1–3% fermenting mash) and are fermented in 1.5 to 4 days or occasionally for a longer time. The success of good fermentation of a properly prepared substrate depends on the quality and the quantity of the seed yeast. It should contain a high concentration of healthy, well-nourished cells in near-pure culture condition of a strain well adapted to fermentation of the specific substrate.

A strain is selected first for its ability to reproduce quickly and to ferment rapidly, leaving no residual sugar or only traces at the end of the fermentation. Second, a relatively high tolerance to sugar and alcohol concentrations is expected; good yeasts produce 10–12% alcohol in the beer with practically no residual sugar. And, finally, the yeast should produce the expected trace products (such as esters, aldehydes, and higher alcohols) in amounts and proportions that will satisfy the aroma requirements. Emphasis is on a relatively low aldehyde content for all types of fermentations, and for high ester and low fusel oil production for most fermentations. Trace products are expressed as grams per 100 l at 100 proof alcohol, g/100 l/100°. The aldehyde content may vary in the distillate from 0.5–5 g/100 l/100°, the ester content from 10–18 g/100 l/100°, the fusel oil content from 120–500 g/100 l/100°. The recent tendency in fermenting bourbon mashes is to utilize yeast strains which produce the least possible amount of fusel oil. Some yeasts, species of genus *Hansenula*, may ferment mashes with extremely high ester content—the ethyl acetate content in the distillate may reach 25,000 g/100 l/100° or more; such distillates may be blended into products with low ester content.

Grain Mashes. *Seed Yeast for Fermenting Grain Mashes and Preparation of the Yeast Mash.* According to general practice in the U.S. about 4 bu of grain out of every 100 bu mashed is used for the preparation of seed yeast. (One distiller's bushel, 1 bu, of any grain is 56 lb.) The final liquid after fermentation is completed is called "beer." The final beer is diluted to 32–40 gal/bu depending upon the final product wanted and the yeast mash is diluted to around 25 gal/bu. The volume of the seed yeast is around 2.0 to 3.3% of the final beer. For inoculation of 100,000 gal beer, 2000–3300 gal seed yeast is prepared. The seed yeast contains around 250–300 million cells/ml, and the beer at the peak of fermentation around 150 million. Every cell supplied with the seed will result in about 20–30 cells in the final beer during main fermentation; this corresponds to the reproduction of four to five generations.

Seed yeast is prepared in 5–6 steps, the last two being made at the plant, and the others in the laboratory:

(1) In the first step, cells from a test tube containing a master culture are transferred with the loop of an inoculating needle into a 100 ml sterile liquid medium; at the same time another master culture is prepared by transferring to a sterile agar slant a loopful of cells from the master culture, or from the 100 ml of liquid culture on the next day.

(2) In the second step, the 24 hr 100 ml liquid culture is transferred to 700 ml agar medium contained in a 2 l Fernbach flask (a wide-mouthed cone-shaped propagating flask).

(3) The cells, developed on the agar surface in 2–4 days, are suspended in 50 ml sterile saline solution and than transferred into 1 l of liquid medium contained in a 2 l Florence flask (a long narrow-necked propagating flask).

(4) The content of the Florence flask after 48 hr propagation time is added to 1–2 gal sterile medium (grain mash).

Steps 1–4 are carried out in the laboratory. Usually the first step is made once or twice a year and steps 2, 3, and 4 are repeated once a week or every 2 or 4 weeks. If step 1 is omitted, a 50 ml portion of the 24–48 hr liquid culture of step 3 is transferred to the 700 ml agar medium contained in the Fernbach flask. Once the culture develops on the agar surface, in 2–3 days, the culture is kept at 4°C until it is needed for step 3. In some practice, step 2 is omitted and the culture is maintained as made in step 3 by transferring a 100 ml portion of it into 1 l sterile medium each time a culture of step 4 is prepared.

For the master culture and for steps 1–3, the media may contain 2.5–5 g glucose, 0.25 g peptone, and 0.25 g yeast extract. The media for the master culture and step 2 also contain 1.5% agar. The medium for step 4 is taken from a standard batch used for the final seed yeast (25 gal/bu). The mash is screened on a coarse (10–12 mesh) screen to remove part of the heavy grain particles and then is diluted, 3 parts mash to 1 part water. The pH is set with lactic acid to 4.2–4.5. The culture in step 4 is allowed to develop for 12–16 hr and is immediately used as starter for step 5.

All media in the flasks (closed with cotton or foam rubber plugs) are sterilized in an autoclave at 120–125°C—the media for steps 1–3 for 30 min and for step 4 for 1 hr.

(5) In step 5 the so-called "dona" tubs should hold about 50 gal (small tub) and 500 gal (large tub) mash, respectively. The compositions of the mashes in steps 5 and 6 are identical; the dona tubs are filled with pasteurized mash (heated to 170°F to destroy the vegetative cells) prepared for step 6, the final seed yeast. The yeast in the dona tubs develops for 8–14 hr. The tubs, made of stainless steel and provided with cooling facilities, are also used as holding tanks for seed yeast.

(6) In standard practice the yeast mash for the dona tubs (step 5) and for the seed yeast (step 6) is made from 1 part rye and 1 part malt. For economic reasons recently, 2 or more parts rye and 1 or less part malt are used. Frequently, in grain alcohol production, the rye is replaced by milo. The ground rye or milo, in a tub (frequently in the yeast tub) provided with agitator, steam inlet, and cooler, is mixed with water and heated with direct steam to 75°C, then cooled to 62°C, when the ground malt is added and mixed. After 30 min agitation, the volume is adjusted to 25 gal/bu, the temperature is set to 54–55°C, and the converted mash is inoculated with 0.25–0.5% ripe lactic acid mash taken from a previous batch. The lactic acid mash contains a rich growth of *Lactobacillus delbrueckii* which, after thorough mixing in the sweet mash at 55°C, will reach the desired growth in 3–6 hr. The lactic acid content may rise to around 0.8% and the pH decrease to 3.5–3.8. The viscosity of the mash is decreased considerably due to the hydrolytic action of the malt and the lactic acid bacteria. The mash during acidification is kept at 54–55°C and is occasionally agitated.

Cooking the rye or milo and conversion of the starch with malt may take place in the yeast tub or in a separate mash tub, from which it is transferred to the yeast tubs. The older yeast tubs are open, made of wood or carbon steel, and provided with heating and cooling facilities and slow agitators. The newer models, made of stainless steel, are closed and provided with fast agitators (lightning mixer type), with automatic and programmed temperature control, temperature and pH recorders, and automatic cleaning and other devices. The tubs and the conducting lines are sterilized by steam.

Bacterial acidification of the yeast mash is interrupted at pH 3.5–4.0 by heating the "sour yeast mash" to 70–80°C. Prior to or during heating, the ring of the dried mash particles adhering to the inner surface of the yeast tubs is washed down to ensure a more thorough pasteurization of the sour mash.

Instead of the widely practiced acidification by *Lactobacillus delbrueckii*, some manufacturers acidify the sweet yeast mash by addition of technical-grade lactic acid, sulfuric acid, or phosphoric acid to a pH of 3.5–4.0.

Just before inoculation, the yeast mash is cooled quickly to about 20°C. If needed, a portion of the pasteurized sour mash is withdrawn to the dona tubs (step 5) and the part remaining in the yeast tub is inoculated with 10–20% ripe yeast produced in the stage-5 large dona tub or in step 6 in a previous batch.

Reproduction (Propagation) in the Dona Tubs. The pasteurized sour mash in the small dona tub is inoculated with about 2–4% pure culture prepared in the laboratory in sterile grain mash (step 4 above) and is allowed to develop for about 14–24 hr. The culture is occasionally agitated by filtered air. The contents of the small dona tub are transferred by air pressure to the large dona tub containing pasteurized sour mash. The inoculum may amount to 5 to 10% of the mash. When the cell count reaches the desired level, the contents of the large dona tub are added to the pasteurized sour yeast mash being held in the yeast tubs. Once the conditions such as pH, temperature, agitation, and time to reach proper cell count are established, only the time is measured. The Balling (see Vol. 3, pp. 317–318; Vol. 6, p. 758) of the yeast mash after the proper cell population is reached is frequently used to monitor the progress of cell reproduction. The initial low temperature, 18–20°C, which discourages development of the bacterial population, is gradually allowed to rise to 28°C, at which point the yeast cells, protected by the CO_2 and alcohol, reproduce more rapidly.

During the period of time until a new laboratory culture is prepared (1–10 weeks), the dona tub is used to store the ripe yeast (taken from the yeast tubs) until used for inoculation of the next yeast tub. The ripe or near-ripe yeast in the dona is kept at a low (4–8°C) temperature.

Propagation in the Yeast Tubs. The yeast tubs, considerably larger than the dona tubs, but essentially serving a similar purpose and having similar construction, are today partially or fully instrumented and automated. Their capacity is selected to produce around 2–3% inoculum (seed yeasts) for the final fermenters to be inoculated. The procedure and the control of reproduction are basically the same as for the dona tubs. The substrate is the same—pasteurized sour yeast mash. When fresh inoculum is prepared, the yeast tubs are inoculated with yeast prepared weekly, semimonthly, or monthly in the dona tub. On other days the inoculum for a new batch is taken from a yeast tub containing ripe yeast or from ripe yeast stored cold in the large dona tub, usually over weekends. The yeast inoculum in the yeast tubs is ripe or ready to use when the cell count reaches its predetermined value, usually 250–300 million/ml. This, once the conditions are standardized, is measured by the time lapse since inoculation— usually 12 hr—or by the Balling, usually 7–10 Balling (7–10°B).

The above example of preparing yeast inoculum in grain distilleries may have a number of variations in the various plants at any of the stages. Yeast inoculum may also be prepared in filtered or settled sterile clear beer with aeration and agitation. Shorter time and smaller yeast tubs are needed. Continuous propagation under pure-culture conditions has been attempted.

Finally, yeast produced from selected and proved strains of distiller's yeast and made by baker's yeast manufacturers are offered and sold in the shape of *compressed yeast*. Universal Foods Corporation and Standard Brands, Inc., produce compressed distiller's yeasts; this eliminates the necessity of preparing yeast in the distillery—for small distilleries and in the future for automated distilleries, such a product could be the ultimate answer.

Cell Reproduction and Fermentation in the Final Grain Mash. The grain, corn or milo or wheat, is ground and cooked with water either at atmospheric pressure or under pressure at 121–150°C. The ratio of water to grain is selected to result in a final mixture of 26–28 gal volume per bushel of grain. The cooked grain mash is cooled to 62°C and a slurry of the calculated amount of 5–20% ground malt in the form of a slurry is added; the mixture is agitated at 62° for 10 or more minutes to liquefy the starchy paste and to convert the bulk of it to maltose by the enzymes contained in the malt. The liquid, which at this stage is called *cooker mash*, is cooled and transferred into the fermenters simultaneously with sufficient amount of thin stillage (screened distillation residue) to produce the required final dilution which, including the seed yeast, is 30–40 gal/bu of grain mashed. If more rye is included in the grain formula than used for the yeast, it is added to the cooker mash after it is cooled to around 67°C to prevent the development of undesired flavor. The inoculation or setting temperature is usually 18–22°C, depending on the climate, to prevent overheating the beer during the main fermentation period. During fermentation it is desirable to keep the temperature below 88°F. The initial pH is 4.9, which gradually drops to 3.9–4.2. Most fermenters are open tubs, but the newer ones are closed, with vents for the escape of CO_2. Mixing the cooker mash with the stillage and yeast is usually done by air. Through surface contact and air agitation in the fermenters, a considerable amount of air is absorbed by the beer to give a good start to the reproducing yeast. The cells begin to multiply quickly, and in about 20–30 hr reach the maximum cell count. As the number of cells increases, fermentation becomes more active. With the increase in alcohol content, reproduction slows down and alcohol production reaches its peak. As the sugar content is reduced, more maltose is formed from the higher polysaccharides by the malt enzymes. Depending on conditions, the fermentation may be completed in 2 to 4 days. Temperature control and agitation are the means of controlling fermentation time. In a well-conducted process using a proper strain and well-prepared seed yeast, 8–9% alcohol is produced in the beer, with as low as 0.1–0.2% fermentable sugar residue. Continuous cooking processes and conversion by acids and reduced amounts of malt have been tested, but in many cases were abandoned. Use of fungal amylases, quite recently of amyloglucosidase (glucamylase), with 1–1.5% malt resulted in higher alcohol yields. Yeast is not able to hydrolyze maltotriose or isomaltose, which are present in malt-converted grain mashes. Amyloglucosidase will hydrolyze these sugars into glucose, which is then fermented by the yeast, resulting in 4–6% more alcohol than obtained for mashes converted by malt alone.

From other starchy raw materials, such as potatoes or byproducts of gluten or starch production, the preparation of the seed yeast and final mash and the conduct of the fermentation of beers follow the fashion described for grain mashes. Usually such mashes must be supplemented with ammonia and phosphate nutrients.

Table 1 lists the numbers of yeast cells produced, from an initial 100 million cells added to 50 ml medium in the laboratory to 57 quadrillion cells (1 quadrillion = 10^{15}) in a 100,000 gal fermenter in about 100 hr.

Fermentation of Molasses (29,37). With the exception of a few tropical species, all yeasts utilized in the fermentation of molasses are strains of *Saccharomyces cerevisiae*. The so-called molasses yeasts are pure cultures, isolated from commercial batches. The differences in the various strains are in the tolerance to higher tempera-

Table 1. Reproduction of Yeast in Distillery Practice (104)

Step No.	Phase	Volume	Yeast population, million cells/ml		Total no. of cells at end of each step	Time, hr
			Initial	Final		
1	laboratory	50 ml	2.0	100	5 billion	16
2	laboratory	1000 ml	5.0	100	100 billion	38
3	laboratory	3 gal	10.0	200	2 trillion	50
4	small dona	30 gal	17.6	300	34 trillion	58
5	large dona	300 gal	30.0	300	34 trillion	66
6	yeast tub	2000 gal	45.0	250	1.9 quadrillion	76
7	fermenter	100,000 gal	5.0	150	57 quadrillion	100

ture and to higher sugar and higher alcohol concentrations; in fermentation speed, flavor production, flocculation, and foaming tendency; in bottom- or top-fermenting tendencies; etc. Each manufacturer has several preferred strains. All are yeasts with high invertase content.

The principles in the preparation of seed yeast are the same as described for grain-fermenting yeasts. The cells from the master culture are, in 2–3 steps, transferred, one or two days apart, successively into larger volumes of sterile synthetic media. For the last 2 gal laboratory stage, the sterile medium is prepared from molasses usually supplemented with 2–4 g ammonium phosphate per 100 g molasses. The medium is diluted to 20% sugar content.

The laboratory culture is propagated in two or three additional plant stages to produce seed yeast for the final batch. The 2 gal laboratory-made pure culture is added in a small dona tub to 100 gal diluted molasses with 12–18% sugar content and supplemented with 1–2% ammonium phosphate; the pH is set to 4.5. In modern plants this step is carried out in closed tanks provided with agitation, aeration, and automatic temperature, pH, and foam control.

Continuous Fermentation (46c,123). The nature of both yeast and molasses makes them applicable for semicontinuous or continuous fermentation. The technology of continuous fermentation of molasses beers was reviewed by de Becze and Rosenblatt (107). For production of rum and beverage alcohol from molasses the main difference is in the dilution, temperature, and amount of stillage used. In Europe, 100 kg of beet molasses with 50% sugar content usually yields up to 32 l of 100% alcohol. Alcohol production from molasses (Vol. 1, p. 630) is reviewed by Hodge and Hildebrandt (37).

Wine (qv) (2,28,49)

Pressed grape juice turns into wine through the reproduction and fermentation activities of yeasts. These activities may be spontaneous, that is, from the yeasts naturally present on the surface of the grape; or they may be carried out by culture yeasts seeded into the grape juice by man.

Spontaneous Fermentation. A great variety of yeasts are present in the soil, especially of orchards and vineyards. The cells in the soil dust are carried by wind and insects to the surface of the ripening grapes. There, in the juice exuded through skin injured by abrasion or insects, the yeast cells begin to multiply rapidly and form colonies. Many of the newly formed cells are washed back by rain to the soil, where they may continue to reproduce in the fruit juice dripping from injured grapes. Many of the spore formers may sporulate and, along with the other yeasts, remain in the soil for another opportunity for colonization.

On the surface of 100 healthy ripe grapes 22 million yeast cells, and on 100 damaged grapes 800 million yeast cells, were found (28) by Pettenkofer. Most frequent are the lemon-shaped cells of species of *Kloeckera* (*K. apiculata*), the oval or long cells of the species of high ester-producing *Hansenula*, the oval or long cells of *Saccharomyces pastorianus* (now *S. bayanus*), and the oval or long cells of the most typical wine yeast, *Saccharomyces cerevisiae* var. *ellipsoideus*. Also present on grapes may be *Saccharomyces oviformis* (also now included in *S. bayanus*), *S. uvarum*, and species of *Pichia*, *Debaryomyces*, *Hanseniaspora*, and *Rhodotorula*. *Saccharomyces belicus*, a spore-forming flor yeast, may form a pellicle on the surface of wines with 12 to 16% alcohol content and is used to impart the flor flavor and bouquet to sherry wines.

Schizosaccharomyces octosporus may occur on grapes, although it was not reported to be present on California grapes. When seeded to must, *S. octosporus* fermented rapidly, producing wine with pleasing aromatic odor (2).

Candida mycoderma, which does not form ascospores, frequently develops a white mat on the surface of weak wine, and contributes to its spoilage. *C. mycoderma* (*Mycoderma vini* in the earlier literature, *Candida vini* in the new Lodder), is highly aerophilic and can utilize alcohol as a source of carbon and energy. In the latter case, it oxidizes alcohol to CO_2 and water.

In addition to yeasts, molds are present on grapes—among them *Dematium pullulans* and *Sachsia suaveolens*, two fungi resembling yeasts, and *Botrytis cinerea*, a gray mold that plays an important role in the fermentation of dessert wines.

In the earliest phase of the spontaneous fermentation of grape juice all these organisms begin to reproduce. As soon as the oxygen is exhausted and CO_2 begins to form, the obligatory aerobic organisms fall out of the race. The fast reproducing *Kloeckera* and many *Hansenula* do not tolerate more than 3–4% alcohol. As the alcohol content increases only the strains of *S. cerevisiae* var. *ellipsoideus* and *S. pastorianus* continue reproducing and fermenting—if sufficient sugar is present, until the alcohol reaches 15–16% by volume. Some specially adapted strains of var. *ellipsoideus* may produce wine with up to 20% alcohol content.

Some strains of *Hansenula anomala* may tolerate alcohol up to 10–12% and may produce 0.25–0.44 g esters per 100 ml, principally ethyl acetate, thus contributing much to the flavor.

The best wine-producing yeasts were isolated from the soils of vineyards where grapes were grown for centuries, in France, Hungary, Germany, Italy, Spain, and other wine-making countries. The strains usually carried the name of the town or district where they were isolated, such as Tokaj (Tokay), Montrachet, Burgundy, Epernay, etc.

At the end of fermentation or toward the last phase of it, the yeast settles with the heterogeneous mixture of undissolved solids known as *lees*. About 80% of the yeast cells in the sediment is var. *ellipsoideus* and the balance, all the other strains. The first phase of spontaneous fermentation, called the main fermentation, may take 4–5 days for musts with low sugar content and 5–9 days for musts with high (22–28%) sugar content. During the main fermentation period the yeast is in suspension in the wine, being lifted up by the CO_2 bubbles.

Single-Strain or Pure-Culture Fermentations. The use of starter yeast containing a pure culture (see Vol. 13, pp. 476–477), or dominant culture, of a specially selected strain of *S. cerevisiae* var. *ellipsoideus* is over 50 years old. The selected strain is propagated in a pasteurized or sterilized must by aeration and then added to the must. The amount of starter yeast culture added to the must may be 0.5–2%. The main purpose is to supply sufficient cells of the selected strain to ensure its dominance over the original mixed yeast population. The selected strain will rapidly populate the must, and the quick development of alcohol will restrain the reproduction of the other yeasts originally in the must.

Saccharomyces cerevisiae var. *ellipsoideus* differs from *S. cerevisiae* mainly on the basis of cell shape (*ellipsoideus* is somewhat more elliptical). Also, the alcohol tolerance of var. *ellipsoideus* in general is higher than that of *S. cerevisiae*.

Some wineries inoculate the must with two distinctly different starters: first, with a pre-ferment starter of *Hansenula* to produce esters at the beginning of the fermentation, then with a pure culture of the selected strain of *S. cerevisiae* var. *ellipsoideus*. They claim that these yeasts, along with the natural inhabitants of the must, ferment wine best, while the school preferring pure-culture fermentation by a selected strain of *S. ellipsoideus* still holds strong in many circles.

When using the same type of must and identical cellar operations, the difference in the yeast strain used as starter will be the factor influencing the quality of the wine. Experience, tradition, and customer acceptance dictate the type of yeast best suited for certain types of wines. On the other hand, it has been proved that the type of yeast is not the *major* determining factor in producing wine with a desired quality of taste and bouquet—the best yeasts cannot produce high-quality wine from mediocre or poor-quality grapes.

Wine yeasts of any specified strain may be produced in the form of bars of compressed yeasts by the same technique used in making baker's yeast. The bars consist of the cells of the specified yeast in nearly pure-culture state. Universal Foods Corporation and Standard Brands Inc. sell compressed yeast made of the strain "Ay," recommended and most frequently used in the U.S. for the production of wines and champagne. The same companies are willing to produce custom-made compressed yeast from any strain of wine yeast.

Different wine yeasts are identified on the basis of the ratio of their secondary fermentation products (esters, aldehydes, higher alcohols) to the amount of alcohol they produce, differences in fermentation speed and efficiencies, sedimentation rate, and quality of the sediment. Most wine yeasts form a powdery sediment; typical champagne yeast forms sticky flocculent masses.

Large industrial wineries in the U.S. generally prefer to start the fermentation with a 1–2% pure culture of active starter yeast. The starter from a small laboratory culture is transferred gradually to increasing volumes of pasteurized or sterilized must in 5–10% proportions at the most actively fermenting stage, which is usually reached when about one-half of the sugar is fermented. In the last stage or the last two stages of the starter, the must is frequently aerated in the beginning to promote rapid cell reproduction.

Continuous or semicontinuous pure-culture seed-yeast propagators are also used. About nine-tenths of the active starter is withdrawn from the pure-culture tank to be used as starter for a commercial batch; it is replaced with freshly sterilized must. When high fermentation activity is reached, again nine-tenths of the culture is withdrawn, and the process is continuously repeated.

Malic Acid Fermentation by Yeast. Malic acid (qv) (2) contributes to the sour taste of many wines made of grapes grown in a cool climate. Early in this century it was recognized that certain lactic acid bacteria may ferment the malic acid in low-alcohol wines into lactic acid (qv) and carbon dioxide. This so-called malo-lactic acid fermentation, occasionally initiated by addition of a selected bacterial culture, reduced the titratable acidity and the harsh sour taste of European wines.

In recent years attention has been directed to the ability of *Schiz. pombe* to ferment malic acid. Recommendations made by several enologists to utilize mixed cultures of *Saccharomyces cerevisiae* var. *ellipsoideus* and *Schizosaccharomyces* species for fermenting wines with high malic acid content undoubtedly will be followed up with experiments and practical applications. It seems more attractive to make wine using only yeasts rather than a mixture of yeast and bacteria.

Summary. Experienced winemakers and scientific researchers, usually independently, have covered a magnitude of topics and collected a treasure of information on yeasts and their function in winemaking. Though the results are gratifying, there is much more work ahead for a coherent treatment of existing scientific evidence and interpretation of practical experiences, and for mapping future research.

The review of winemaking by Joslyn and Turbovsky (49) and the book of Amerine, Berg, and Cruess (2) treat in detail the role of yeast in winemaking.

Beer (qv) (10,32a,39a,45,46a,95,109,113,114,116,117,122,125,128,136,141,147)

In brewing, when converting malt and hops into beer, fermentation by yeast is the fundamental change. The yeast acts on the *wort*, which is the filtered liquid substrate extracted from the enzyme-converted malt and grain. Yeast not only converts the sugar of the wort into alcohol and carbon dioxide but takes a prominent part in the formation of the bouquet that makes beer pleasant and some beers well known and famous all over the world. The aroma and bouquet of beer depend partly on protein assimilation and partly on the formation of esters and higher alcohols by the yeast.

The yeasts utilized in breweries today are carefully selected and adapted strains of two species, *Saccharomyces cerevisiae* and *Saccharomyces carlsbergensis*. In making tropical beers strains of other species are also employed. *S. cerevisiae* sporulates easily and is principally top fermenting, while *S. carlsbergensis* is a bottom-fermenting yeast and seldom sporulates.

The behavior of a brewer's yeast during fermentation is characteristic with regard to the following (10): (1) top or bottom fermentation (Vol. 3, pp. 299, 315–316), (2) strong or weak attenuation, (3) fast or slow fermentation, (4) activity at low or higher temperature, (5) flocculating or dusty consistency.

(1) The top-fermenting yeast, along with the fine particles of hop resins and protein precipitates, is raised to the top of the fermenting wort, where from 12–20 hr after "pitching" (the brewer's term for inoculating or seeding the substrate, wort, with yeast) it forms a dense creamy foam. Part of the yeast is removed from the top at this stage; if not removed, with the advance of fermentation the yeast foam subsides and drops to the bottom. Bottom-fermenting yeasts reproduce and stay at the bottom of the tank during fermentation. There are yeast strains which behave in between the two extremes.

(2) Attenuation, the degree of fermentation of the available sugar, is more the result of the way the fermentation is conducted and the position of the yeast during fermentation than of the ability of the yeast to ferment. Generally all brewer's yeasts are good fermenters. Yeasts remaining in suspension during fermentation fully attenuate the beer. Among bottom-fermenting yeasts the so-called "Saaz" types are weakly attenuating and the "Frohberg" types are strongly attenuating.

(3) The speed of fermentation, too, is the consequence of several factors, principally temperature and the position of the yeast during fermentation. With both types of yeast (top and bottom), fermentation occurs in two distinct steps: In *primary fermentation*, the yeast multiplies and the bulk of the sugar is fermented. In *secondary fermentation*, called lagering, the remaining sugars are fermented, the beer is saturated with carbon dioxide, some precipitates and the yeast settle to the bottom, and the beer is ripened. With bottom-fermenting yeast, the primary fermentation takes about 8 days at 8–12°C. Then the unmatured beer is transferred to closed tanks (made of aluminum, stainless steel, or glass-lined steel) in which, in addition to some physico-

chemical reactions, a secondary fermentation goes on slowly (2 weeks to 3 months) at very low temperature (0°C); the content of higher alcohols and esters increases and of the acids slightly decreases; gradually the beer ripens.

Krausening, a process that originated in Europe, consists of adding to the beer in the lager tank a certain amount of young beer which has just begun to ferment. This second "pitching" of yeast initiates a slow but steady fermentation under pressure at low temperatures. The process imparts a special quality to the beer.

(4) Many yeasts are acclimatized to a certain temperature range: bottom-fermenting yeasts are quite active at 6–8°C, while top-fermenting yeasts ferment at 14–15°C or higher. During fermentation the temperature increases by 4–6°C, but it is not allowed to go above a level predetermined by the brewer for each specific brand of beer.

(5) The formation of clumps or flocs is characteristic for certain strains while a dusty or powdery sediment characterizes others. The composition of the wort and other local factors may influence this phenomenon.

Seed Yeast. The amount of pitching yeast added to the beer to start fermentation is about 0.3–0.75 kg per 100 l. On an average it is about 1 lb/25 gal beer. Seed yeast is prepared in two ways: (a) from pure-culture yeast and (b) from yeast recovered from a previous fermentation:

(a) The master culture is grown on wort-agar in test tubes, from which it is transferred to tubes containing sterile hopped wort. The culture is transferred daily into successively increasing volumes of sterile wort until the culture reaches a 10 l (about 2.5 gal) size. This amount in a 100 gal size pure-culture tank is diluted with sterile wort to 6 million cells/ml. When the cell count reaches 12 million/ml, an equal volume of sterile wort is added, and the process is repeated until the tank is filled to its working capacity. The ripe yeast is then used for inoculating the commercial batch.

(b) Yeast from the middle crop recovered from a beer which has completed the main fermentation is washed, screened, and rewashed until it is free of grain particles, precipitates, and infections. The clarified yeast is cooled and used within 24 hr to pitch a new batch. The excess of recovered yeast is used as secondary yeast for food or feed. It is claimed that the recovered yeast can be used 100 times without difficulty. Generally, after 20 plant generations, or sooner if it seems advisable, the plant yeast is replaced with pure-culture yeast.

To eliminate bacterial contamination the yeast may be washed with inorganic acids at pH 2.8 or with 0.75% ammonium peroxysulfate solution for an empirically determined period of time. The purity and the vigor of reproducing ability of the pitch yeast are tested before use.

Small breweries may purchase purified pitch yeast in compressed form from larger breweries.

Bread (8,26,30,41,50,97,98,124,129,130,134,138,146)

Bread (see also Bakery processes), the prototype and principal food for the bread-eating nations of the world for many thousands of years, is made of at least four ingredients: flour, water, salt, and yeast. Without yeast, the product would be a griddle cake, tortilla, or unleavened bread like matzos. In large mostly undeveloped areas (eg, Africa, the Middle East, and Latin America), yeastless bread products are favored.

Making bread consists of dissolving the salt and yeast in water, adding flour, and thoroughly mixing the four ingredients into a dough (the *straight-dough method*), which is then allowed to rest, usually for two or more hours, so that the yeast can act upon it, shaping the ripe dough into a loaf, and baking the loaf in an oven heated to approximately 400°F until the internal section reaches approximately 212°F. The usual amounts of the four ingredients in simple yeast-raised bread are 100 g flour, 56–60 g water, 2 g salt, and 2.0–3.0 g yeast.

In the *sponge-dough method*, about 70% of the ingredients are added to the water and yeast to form the primary dough, the "sponge," with which, after ripening, the balance of the ingredients are mixed. Commercial bakeries adapted the sponge-dough technique to good advantage. The most recent version of the method uses a pre-ferment built into a continuous process of breadbaking.

In addition to the four basic ingredients—flour, water, salt, and yeast—most breads and all bakery goods such as rolls, buns, coffee cakes, sweet rolls, yeast doughnuts, and similar products contain one or more other ingredients, including yeast food. Amounting to 0.25–0.50% of the flour, yeast food contains 30% calcium sulfate, 9.4% ammonium chloride, 0.3% potassium bromate, 35% sodium chloride, and 25.3% starch. Only the ammonium salt is strictly a food for yeast; the other substances aid in conditioning the dough.

Yeast is a flavoring agent for certain bakery products leavened by chemical compounds (baking powders). The dried yeast powder used is previously preheated to destroy the enzyme activity of the cells.

Action of Yeast in Breadmaking. The yeast converts the heavy mass of dough into a light porous elastic substance which, when baked, is an appetizing and easily digestible nutritious food. The effects of yeast are severalfold—all enzymic in nature: The *invertase* of yeast hydrolyzes the sucrose in the dough rapidly into invert sugar; the *maltase*, acting much more slowly, converts the maltose that was originally present in the flour or added with ingredients or produced by enzymes of the wheat or by enzymes added to the dough; *protoeolytic enzymes* act on the proteins in the flour, principally on wheat gluten. A well-conditioned ripe gluten is stretchable and remains elastic, forming membranes that stretch into numerous little cavities as CO_2 is developed by the yeast. Thus the dough expands, but keeps the gas captured inside the bread.

The main reaction performed by the complex known as yeast *zymase* is the fermentation of the hexose sugars into alcohol and CO_2. Each gram of fermented sugar yields approximately 0.48 g alcohol and 0.48 g carbon dioxide. In addition to the two main products from every 100 g of sugar, about 2–3 g glycerol and more than a score of organic compounds are produced. The total of the trace products may amount to 0.5 g per 100 g fermented sugar. The most important flavor and taste-forming substances are propyl, butyl, amyl, and isoamyl alcohols; ethyl acetate, ethyl formate, ethyl butyrate, diacetyl, acetoin, acetaldehyde, some low-carbon fatty acids, and other compounds.

The fermentation is quite rapid: 2 g yeast in 170 g dough may ferment 0.5 to 1 g sugar per hour if sufficient sugar is present. The CO_2 first saturates the dough by being dissolved; then the excess CO_2 in the gaseous state begins to form gas bubbles in the dough.

The time the yeast acts upon the dough may vary from 2 to 6 hr or even longer depending on the process. In the sponge-dough method used in large commercial bakeries, the sponge containing around 60% of the total flour, the yeast food, the vitamins, enzymes, and the greater part of the water is mixed with the suspended yeast in the mixing bowl. Then the sponge, transferred into troughs, is allowed to ferment at 80°F in a 75–80% humidity room for 3.5 to 5 hr. The volume of the sponge increases and its consistency (viscosity) becomes thin. During this period, the yeast undergoes some reproduction; it ferments a good portion of the available sugar, with formation of flavor components in addition to alcohol and carbon dioxide.

Most fermentation products are volatile. They dissolve in the dough, which, as mentioned above, soon becomes saturated with the CO_2. When a gaseous phase develops in the form of bubbles in the dough, part of the alcohol and other volatiles enter into the gaseous CO_2 phase. When the skin of the sponge breaks, the escaping gas carries the very pleasant scent typical of fermenting dough. Simultaneously with the alcoholic fermentation of the carbohydrates, the yeast and other conditions act upon the gluten to bring it into a ripe stage. When this is reached, the sponge is returned to the bowl and is admixed with the rest of the flour, water, and other ingredients such as nonfat milk, sugar, salt, shortenings, etc. The mixer not only homogenizes the mix but beats out most of the CO_2 captured in the bubbles and saturates the dough with air. The added sugar gives added vigor to the yeast activities during the time available before baking.

After the final mixing, the dough is again transferred to troughs and permitted to rest 15–45 min, still at 80°F. From the troughs the dough goes through the "makeup" process, when it is cut into pieces of exact weight, then fashioned into smooth-surfaced round shapes by appropriate machinery. Following a 15 min additional resting, called overhead proofing, the pieces passing through a molder are placed in the pans. These mechanical manipulations not only shape the loaf and develop the gluten but (by distributing the yeast evenly) enhance fermentation, which progresses continuously. During the next step, the final proofing, the dough rises to about six times its original volume in about 1 hr at 100°F and 80–90% humidity. The final proofing and the next few minutes in the oven give the final size and shape to the loaves. Yeast action at the elevated temperature is at its highest peak, until the enzymes are destroyed and yeast activity terminates (around 130–140°F).

In 1961, for bakery products (primarily bread) in the U.S., the average annual consumption per capita was 1.7 lb yeast and 118 lb flour; in England 2.1 lb yeast and 400 lb flour; in Finland 4 lb yeast and 250 lb flour; and in Austria 2 lb yeast and 190 lb flour (52).

Miscellaneous Applications

The ability of yeast to catalyze reactions other than alcoholic fermentation has resulted in a number of industrial methods for producing glycerol and fat; yeast is also used in sewage disposal and for medicinal purposes.

Glycerol (qv) was produced by yeast in Europe in World War II (37); improved methods were developed in the U.S. as late as 1947 (83,84). The processes, based on acetaldehyde fixation by added chemicals such as sulfites, bisulfites, alkali, lime, and their combination and thus forcing the equilibrium toward glycerol formation, are known as the German process (62,63), ammonium sulfite process (70), sodium carbonate process (67). Because of the low glycerol yields (20–30 g per 100 g sugar) the fermentation glycerol industry today cannot compete with the glycerol obtained in the production of soap and fatty acids or by synthetic processes.

High-fat food supplements (up to 40–60% fat of dry substance) have been obtained by propagating species of *Candida reukaufii* (now *Metschnikowia reukaufii*), *Endomyces vernalis*, *Oospora lactis*, *Rhodotorula glutinis* (*R. gracilis*), and *Torulopsis lipofera* (*Lipomyces lipofera*) under specific conditions. Fat production by yeasts on an industrial scale, practiced during World War I, has been abandoned.

Sewage Disposal. Although the biochemistry of yeast activities in sewage disposal has not yet been studied in detail, yeast is commonly applied in preliminary treatment of household waste waters. Many rural homeowners add a pound of baker's yeast to their cesspools every three to six months, or whenever odorous gases become noticeable. Yeast capable of both anaerobic and aerobic dissimilation will liquefy or oxidize many compounds, and assimilate others, forming new cells. Part of the cells will autolyze and serve as food for other organisms participating in decomposition of the organic matter contained in household liquid waste.

Medicinal Yeast Preparations

Dried Yeast and Yeast Tablets. Live yeast, when consumed, may ferment in the stomach and in the intestines, producing CO_2 and causing gastrointestinal problems. It will also absorb much of the B vitamins present in the intestines, leading to vitamin deficiency of the individual. The yeast must therefore be treated, commonly by heat, to destroy its enzymes and stop its physiological activities.

For nutritional and medical purposes, when in addition to the protein content a high vitamin content is essential, generally primary grown *Saccharomyces cerevisiae* or washed and debittered brewer's yeast, *S. cerevisiae* or *S. carlsbergensis*, is used. The yeast is dried on trays, or belt dryers, or in drum dryers, above 70–80°C to about 90% solid content. The product is ground, tableted, and marketed as dry yeast.

Yeasts readily absorb the water-soluble vitamins, iodine, and iron salts and oxides from solutions added to the yeast cream before drying. The concentration of these substances is predetermined and the products are marketed as vitamin-enriched yeast tablets, ironized yeast (IY) tablets, and iodized yeast tablets.

The United States Pharmacopeia (38) defines "Dried Yeast" ("Saccharomyces Siccum") and "Dried Yeast Tablets" ("Tabellae Saccharomycitis Sicci"); the definition for "Dried Torula Yeast" was dropped with the 16th edition (1960). *The National Formulary* (26b) for 1960 defines "Dried Yeast," "Dried Yeast Tablets," and "Dried Torula Yeast," and drops the torula yeast in a later edition (1965). (The NF also discusses "Brewer's Dried Yeast," "Debittered Brewer's Dried Yeast," and "Primary Dried Yeast" sources.)

Procytoxid. A preparation from yeast, obtained essentially by extraction with 80% alcohol, and found to stimulate the respiration of yeast and tissues, has been called RSF (respiratory stimulating factor) or, more recently, PCO (procytoxid). It was found to offset the depression of respiration caused by germicides and has been used extensively for that purpose in wound-healing preparations to preserve germicidal efficiency without inhibiting tissue repair (145). PCO was believed to exert these effects chiefly by acting at the terminal portion of the respiratory chain to provide an alternate means of electron transport (99,120,121).

More recently, in animal studies, PCO has been found to offset the toxic effects of the antibiotic, kanamycin (137), and the antineoplastic drug, nitrogen mustard (100), without significant impairment of the desired drug activity. With nitrogen mustard, PCO affords protection against damage to bone marrow and maintains the leucocyte count (100). These properties, together with low toxicity and a lowering of the ratio of lymphocytes to neutrophiles in dogs (101), led to the trial of PCO as an adjunct of the immunosuppressive agent, Imuran, in kidney transplants in dogs. Good results were obtained in terms of significantly lengthened survival rates, improved renal function, a lessened histological picture of rejection, and bone marrow protection (132).

PCO is biuret-negative but contains amino acids and, probably, several small peptides. Fractionation studies now under way suggest that the different physiological properties of PCO probably reside in different fractions (102,140). Efforts are being directed to identification of the active substances.

Yeast extracts are used for food and for medicinal and microbiological purposes. Yeast extracts made by different methods give different yields and products that excel in one way or another (53): High yields are obtained by hydrolyzing yeast with HCl under pressure, followed by separation of the dissolved portion by centrifugation, sedimentation or filtration, and concentration of the solution *in vacuo*. The proteins are hydrolyzed and most of the vitamins are lost. The product is neutralized after hydrolysis and is used for *food flavoring*. A method keeping the yeast at 60–70°C for a few hours, extracting it with water, and concentrating the extract *in vacuo* is less destructive to the vitamins but gives a lower yield.

Special types of yeast extracts are the *autolyzates* (108). The yeast, in the form of thick aqueous cream, is kept at 45–60°C for 1–2 days. In autolysis, the yeast's own enzymes cause cell destruction. The part made soluble by enzymic hydrolysis is then extracted with water and is concentrated, usually *in vacuo*. Autolysis also may be effected by chemicals causing initial plasmolysis. Four types of substances are used: neutral salts, principally NaCl; acids—hydrochloric, lactic, or sulfuric; alkali—sodium or potassium hydroxide; and esters—amyl or ethyl acetate. The esters are easily removed by vacuum distillation. Addition of some sugar is also customary in combination with the above methods. Digestion of the yeast cells by *Lactobacillus delbrueckii* at 55°C for 1–2 days produces a special extract. The concentrated extracts are commercialized in the form of a paste or powder, after being dried alone or after mixing with starch.

Yeast *bouillon* is prepared by hydrolyzing yeast with hydrochloric acid at 100°C for 8–10 hr or in an autoclave for a shorter period (1–2 hr). Vegetable and spice extracts are added, the product is filtered, and the filtrate is neutralized, condensed, and shaped into cubes.

Vitamin-enriched yeasts are prepared by adding the vitamins (principally B_1, B_2, and B_6) to the nutrient broth, from which the yeast absorbs the vitamins. Yeast synthesizes vitamin B_1 in high concentrations during reproduction by coupling the two components, thiazole and pyrimidine, when added to the medium. Patents given to Standard Brands, Inc. (85), Anheuser-Busch, Inc. (72) and Hoffmann-La Roche, Inc. (76a) deal with the production of vitamin-enriched yeast. Such yeast may be used in the production of enriched bread or pharmaceuticals.

Other Compounds. Ergosterol (a precursor of Vitamin D_2 and cortisone), glutathione, and nucleic acid are yeast products of therapeutic value which have been produced on an industrial or laboratory scale from yeast. See also Table 2.

Manufacture of Yeast and Yeast Products

A number of manufacturing processes produce marketable products in which yeast is the principal substance or one of the principal substances. Such products are

Table 2. Use of Yeast and Yeast Products

Type	Use	Yeast species
(1) *Primary yeast*		
1. compressed baker's yeast	bread and bakery products	*S. cerevisiae*
2. active dried yeast	"	"
3. compressed wine yeast	wine	*S. cerevisiae* var. *ellipsoideus*
4. compressed distiller's yeast	distd. alc. beverages	*S. cerevisiae*
5. food-grade dried yeast	food ingredient	"
6. food or feed yeast	food or feed ingredient	*Candida utilis*
(2) *Secondary yeasts*		
7. brewery dried yeast	food or feed ingredient, pharmaceuticals	*S. cerevisiae, S. carlsbergensis*
8. distillery dried yeast	food ingredient	*S. cerevisiae*
(3) *Products made of yeast*		
9. pharmaceutical products (ergosterol)	medical	*S. cerevisiae, S. carlsbergensis*
10. yeast extract	food	*S. cerevisiae, S. carlsbergensis*
11. invertase	food	*S. cerevisiae*
12. glucose isomerase	food	*S. cerevisiae*
(4) *Products containing yeast*		
13. distiller's dried grains	feed, microbial fermentations	*S. cerevisiae*
14. brewer's dried grains with yeast	feed	*S. carlsbergensis, S. cerevisiae*
15. whey yeast, with whey proteins	food, feed	*S. fragilis*

classified as (1) primary yeasts, (2) secondary yeasts, (3) products made of either primary or secondary yeast, (4) products containing either primary or secondary yeasts as an essential ingredient (see Table 2). Table 3 lists production data for two important commercial primary yeasts, and Table 4 the principal manufacturers of yeast and yeast products in the U.S.

Table 3. Estimated Yeast Production (1968) in Dry Tons

Country	Baker's yeast[a]	Dried yeast[b]
U.S.	59,000	30,000
Europe	66,000	54,000
U.S.S.R	no data	52,000
Canada	4,000	7,000
Japan	13,000	8,000
South America	8,000	no data
Taiwan	no data	13,000
Cuba	no data	11,000

The consumption of baker's yeast (30% solids) in foods, mainly bread, was somewhat below 2.0 lb yeast per capita in the U.S., and around 2.27 lb in England and Europe. More yeast is consumed in the bread-eating countries; in Finland, for instance, the consumption in 1968 was 4 lb per capita per year.

[a] Including active dried yeast.
[b] All other yeast except baker's yeast.

Table 4. Manufacturers of Yeasts and Yeast Products in the U.S.

Manufacturer	Baker's yeast	Dried yeast	Wine, distiller's etc. yeast	Yeast extracts
Anheuser-Busch, Inc. St. Louis, Missouri	×			
Federal Yeast Co. Baltimore, Maryland	×			
Fleischmann Division, Standard Brands, Inc. New York, New York	×	×	×	×
Red Star Yeast Division Universal Foods Corp. Milwaukee, Wisconsin	×	×	×	×
Yeast Products, Inc. Paterson, New Jersey		×		×
Milbrew, Inc. Milwaukee, Wisconsin		×		×
St. Louis Brewers Yeast Co. St. Louis, Missouri		×		×
Lake States Yeast Division St. Regis Paper Co. Rhinelander, Wisconsin		×		

The manufacturing method used in the production of compressed baker's yeast is the prototype of all methods producing primary yeasts. The differences are in the carbohydrate raw material, in the yeast species, and one or several other factors such as temperature, pH, aeration intensity, etc. Because the amount of baker's yeast produced exceeds the amounts of all other primary yeasts, the production of baker's yeast will be presented in detail.

BAKER'S YEAST (1,3,5,9,31,35,36,39,40,47,48,51–54,96)

"Baker's yeast," "compressed yeast," or "cake yeast," a creamy white claylike mass of yeast cells of selected strains of *Saccharomyces cerevisiae*, is sold in groceries for household baking in $\frac{3}{5}$, $\frac{5}{8}$, 1, and 2 oz packages wrapped in tinfoil or wax paper; and through distributing organizations, yeast is sold to bakeries for the manufacture of bread and bakery products in 1 and 5 lb bar-shaped pieces wrapped in specially treated paper and in 25 and 50 lb bags or cartons. Yeast on special order is also sold commercially to large bakeries in the form of a thick creamy white cell suspension in dilute salt solution. Baker's yeast is stored, shipped, and distributed under refrigeration at 40°F. The wholesale price is around 14¢, and the retail price (in bakeries) 40¢, per lb (in 1970).

The goals in manufacturing baker's yeast are: (1) to produce a concentrated mass of cells of the most suitable strain, free from any contaminating microorganisms, and in a biological state that satisfies the requirements for making bread and other bakery goods; (2) to obtain the highest possible yield from the raw materials in a relatively short period of time at a minimum expense of energy, labor, and human control.

The industrial production of baker's yeast as a commercial commodity, in the form of a pliable substance consisting of the yeast cells substantially free from other materials, is somewhat more than 100 years old. Although the original method gradually underwent a number of changes, each featuring improvements in one or another phase, the basic production steps remained the same: preparation of a suitable nutrient medium, inoculation of the medium with a selected starter and multiplication (propagation or reproduction) of the starter cells, separation of the cells from the fermentation residue (spent nutrients) and washing the cells, and elimination of excess extracellular water. In the U.S. in 1970, the production of baker's yeast is estimated as amounting to 320 million lb.

History. The earliest yeast processes, typified by the Vienna process (Wienerverfahren), developed from grain alcohol production, and manufactured both alcohol and yeast. During the main fermentation, about 20 hr after inoculation, the yeast with the coarse grain particles was raised by CO_2 bubbles to the top of the fermenting liquid; and it was lifted from the top with suitable large spoons and separated from the grain particles through screens. The yield was about 6–8 lb yeast plus 3.75 gal 100° alcohol per bu grain. Fleischmann, in 1870 (69), and Levy in 1883 (75), improved the Old Vienna process.

Four of the basic concepts of modern yeast making were already developed toward the end of the last century: (a) the use of dilute nutrients; (b) continuous addition of the nutrients during propagation (Zulaufverfahren); (c) introduction of a balanced sugar-nitrogen-phosphorus ratio in the nutrients; and (d) aeration of the broth during cell reproduction. The purpose of aeration was to depress the formation of alcohol in favor of cell reproduction. Improvements in programming the nutrient addition, in the method of aeration, and in pH, temperature, and foam control have led to 80–100 lb yields of yeast from 100 lb molasses at 50% sugar basis at the present time.

By 1936 Wagner was able to collect 1086 short abstracts of published technical articles and patents on yeast (39). The number of publications has since steadily increased. Current developments are mainly along the lines of engineering solutions, instrumentation, and automation to improve further the methods for producing primary yeast, and these have almost reached the ultimate level.

The basic steps in each method of making baker's yeast are: (1) preparation of the raw material, (2) preparation of the inoculum (seed yeast), (3) propagation of the yeast cells, (4) separation of the cells from the exhausted broth and from excess water, (5) packaging. Steps 1–4 are discussed below:

Raw Materials. Raw materials must provide carbon, nitrogen, phosphorus, potassium, magnesium, and calcium; traces of sulfur, iron, cobalt, and other minerals; and vitamins and growth factors, in addition to water. Carbon is supplied in the form of glucose, fructose, maltose, and sucrose, although any natural product containing sugar or starch may be used. Today, in the U.S., molasses is employed exclusively. Nitrogen is supplied as ammonium nitrate, ammonium phosphate, or liquid or gaseous ammonia, in addition to the organic nitrogen compounds present in molasses. Phosphorus is supplied in the form of calcium superphosphate, alkali metal phosphate, ammonium phosphate, or phosphoric acid. Today, the raw materials for baker's yeast are, generally, molasses, ammonium hydroxide, phosphoric acid, some trace minerals, and vitamins. Outside the U.S., some grains, malt sprouts, ammonium salts, and phosphates, principally superphosphate, are still used in addition to molasses.

Both sugar beet and sugar cane molasses provide the carbon sources (principally sucrose, invert sugar, and raffinose) for assimilation and energy production, considerable amounts of assimilable nitrogen, mainly in the form of organic nitrogen compounds, and many needed minerals—potassium, magnesium, and trace elements. Cane sugar molasses usually is higher in sugar and vitamin content and lower in nitrogen compounds than beet sugar molasses. Rogers and Mickelson (137a) have reported the vitamin content of the two types of molasses (qv). The two types of molasses are frequently used together in a proportion usually dictated by their composition and price. Molasses are purchased on the basis of their sugar content and the yields are expressed in lb yeast/100 lb molasses at 50% sugar content. Refinery blackstrap molasses is frequently preferred, for it contains less of gummy substances. The total nitrogen content of beet molasses may amount to 2%; about one-half is assimilable.

Most of the nitrogen and phosphorus comes from technical-grade inorganic sources, conveniently from ammonium phosphate prepared at the yeast plant from ammonium hydroxide and phosphoric acid, and added partly in the beginning and partly during the propagation period according to a predetermined program. As needed for pH control, either component is added separately. Nitrogen is required for production of proteins, nucleic acids, peptides, amino acids, and other nitrogen compounds present in the yeast cell. The nitrogen supply is so regulated that under production conditions and at a given yield the crude protein content is at the desired optimum level. Up to a certain limit enzyme activity increases with protein content. On the other hand, high protein content leads to quick spoilage of the yeast. Most manufacturers prefer to make baker's yeasts with 45 to 50% raw protein (N $\times$ 6.25) content on dry basis. It has been contended that organic nitrogen nutrients produce better quality proteins in the yeast than do inorganic salts. Phosphates are essential in coenzymes needed for most enzymic reactions, taking part principally in energy transfers. Phosphorus addition, too, is controlled in a ratio to give a predetermined phosphorus content in the final product. Most manufacturers prefer to keep the P_2O_5 content at 3.00% of the dry yeast substance.

Vitamins, depending on the amount present in the molasses and on the vitamin requirements of the yeast strain used, are added in the form of technical-grade vitamins or precursors. Some yeast strains require no, or only small amounts of, vitamins above that supplied by the raw material. Occasionally corn steep liquor is utilized as a nutrient supplement.

Preparation of the Raw Materials. This step aims to produce an aqueous solution from the carbohydrate raw materials (grains or molasses) as free as possible from any suspended or colloidal substances and microorganisms. Colloidal substances would precipitate during the metabolic processes connected with cell reproduction, and, along with the original suspended material, would be recovered with the yeast as contaminating foreign matter that would, in turn, provide a darker shade to the finished product and also lead to early spoilage.

Starch-containing grain raw materials are first cooked with water to gelatinize the starch, which is then converted to maltose and oligosaccharides by adding a slurry of ground malt at 62°C and maintaining that temperature at a pH of 5.0 for at least 30 min. In general, 10–20 lb malt is used per 100 lb starch. During malt conversion, a considerable amount of the grain proteins are hydrolyzed and dissolved along with most of the inorganic constituents. The converted grain mash is boiled to destroy most of the microorganisms and to precipitate most of the colloidal substances. A pure liquid broth is then obtained by straining the mash through a false-bottom tank in which the coarse grain particles act as a filter medium. The broth thus obtained is subjected, if needed, to centrifugal clarification or to a final filtration (qv) by filter press, continuous drum filters, or clarifying filters.

Preparation of broth from molasses consists of clarification, dilution, and supplementation. Molasses contains small quantities of undissolved material, colloidal substances, and microbial cells which would contaminate and darken the product. The contaminating microorganisms during the propagation period would multiply and cause endless trouble during reproduction and later on in the product. Molasses is clarified by two major methods, chemical and physical clarification, or by combination of the two:

1. *Chemical clarification* consists of forming a precipitate in the partially diluted molasses solution. After adsorbing the undissolved particles, the precipitate settles, and the clear molasses solution is decanted. Among the great variety of precipitating agents used or recommended, the most popular and successful form calcium sulfate, or calcium phosphate, or a mixture of the two. According to the Polish method, 1700 kg molasses (with 0.71% CaO content), 6800 kg water, and 220 kg calcium superphosphate extract (with 3% P_2O_5 content) is heated to 85°C, and the mixture settled for 4 hr. According to the Austrian method, molasses is diluted four times with water, is boiled with H_2SO_4, superphosphate, or phosphoric acid, and is neutralized at 100°C to pH 8 with solutions of NaOH or NH_4OH. Most of the phosphate remains in the solution and is accounted for in the total phosphate used as nutrient. The bulk of the calcium is precipitated in the form of sulfate and the remaining portion as phosphate. After 4–6 hr settling time the supernatant liquid is clear and nearly sterile for practical purposes.

2. *Mechanical clarification* consists of diluting the molasses to a consistency at which the undissolved solids, including microbial cells, will settle, and then separating the nonsolubles by centrifugal force. A final elimination of the trace particles may be done by filtration in clarifying or sterilizing filters. Filtration in filter presses using various precoats is also employed to clarify semidiluted molasses. Today in the U.S. and most other countries, molasses solutions are clarified through specially built molasses centrifugal clarifiers. Most popular types for this purpose are the high-capacity centrifuges made by The DeLaval Separator Co. and Westfalia Separator Inc. and the recently developed desludgers, with up to 260 gal/min capacity.

The individual *inorganic nutrients* (calcium superphosphate, ammonium phosphate, phosphoric acid, ammonia) are dissolved in water to a convenient content of nitrogen and phosphorus and are kept in feeding tanks. When calcium superphosphate is dissolved in hot water under agitation, the undissolved portion, mainly calcium sulfate, first is settled and the supernatant liquid is decanted into the feeding tank, where it is diluted to a specific phosphorus content. The use of phosphoric acid and ammonia as the most convenient inorganic raw materials of today simplifies the handling of nitrogen and phosphorus nutrients; the major portions of the two ingredients are mixed to near neutral point in a feeding tank; small portions are kept separately in diluted form and are supplied to the propagating tanks to maintain the desired pH and the nitrogen-phosphorus ratio. The concentration of the nitrogen and phosphorus is so selected that the feeding ratio is simplified.

Preparation of the Inoculum (Seed Yeast). The seed yeast, produced with special care, differs from the final product in two respects: (1) it is produced under conditions that prevent development of foreign organisms; (2) it is overfed with nutrients—it is higher in proteins and phosphorus compounds than the final product. The seed yeast, especially in the earlier stages, is allowed to reproduce under anaerobic conditions or in the presence of a limited air supply. The yield of seed yeast per unit of molasses is considerably less than the yield of the final baker's yeast.

Seed yeast must be in condition to reproduce for 3–4 generations in the final batch. Three full generations will produce about 8 lb from every pound of seed yeast and four full generations 16 lb from every pound of seed yeast. Depending on the size of the manufacturing unit the seed yeast is produced in 4–7 stages:

Step	Volume	Time, hr	Aeration
1	1 pt agar	48	surface culture
2	1–2 qt liquid	24	none
3	8–10 gal	12–16	little
4	80–100 gal	12	little
5	500–1,000 gal	12	moderate; yeast is separated from nutrient and washed
6	5,000–20,000 gal	12	near full scale, but less than in the final step

In step 7, commercial-size propagators, up to 50,000 gal size, are inoculated with the yeast harvested in the previous stage at a rate of about 1.5–2 g yeast (about 0.6 g solids) per 100 ml. The yeast, while aerated, is allowed to increase 8–10 times its original weight. The cells for the final inoculum are separated from the exhausted nutrient in carefully cleaned and sterilized or disinfected separators and are washed several times with water, followed each time by centrifugal separation.

The first 4–5 steps shown in the table above are carried out in sterile media in sterile tanks using sterile air. The cells reproduce as a pure culture; and the whole culture, cells with the nutrient liquids, is transferred to the next step through sterile pipelines. The last (step 5) or the last two steps usually are carried out, not under pure culture conditions, but in carefully cleaned and steamed propagators to minimize contamination.

Reproduction in the first step, on an agar surface, is aerobic; in the second step it is anaerobic, the energy obtained purely by fermentation. In the consecutive steps, the rate of aeration is gradually increased, but even in the sixth or seventh step the aeration is less than in the commercial batch. The exhausted broth from the fifth and sixth steps usually contains some alcohol; where regulations permit and economy dictates it, the alcohol is recovered by distillation.

The final seed yeast either in compressed form or in the form of a thick slurry (10% solids) is kept at 4–6°C until it is used for seeding the final commercial batch. Although it is best not to keep it too long, seed yeast may be stored for up to 10 days.

The cells of seed yeast are slightly larger than those of the final product, besides having nitrogen and phosphorus contents that are higher, as noted above. In compressed form at room temperature seed yeast spoils (autolyzes) more quickly than the final compressed yeast.

Propagation of the Yeast Cells. The goals are to obtain (1) the highest possible yield in cellular material per unit of raw material, (2) the desired quality of the yeast, (3) the highest possible crop per propagator space per day, and (4) a minimum cost of power needed for aeration. And the following factors are therefore considered very carefully: (1) pH of the medium, (2) concentration of the inoculum, (3) changes in concentration of the yeast cells throughout the period of propagation, (4) initial concentration of nutrient components, (5) concentration of individual nutrient components and their ratio to concentration of the yeast cells at any given time during the propagation period, (6) temperature of the medium and its changes, (7) desired rate of cell reproduction related to initial inoculum at any given time period of reproduction and at the end of reproduction, (8) amount of air blown through the medium, and changes in the rate of supplying air throughout the process, (9) "quality" of air (expressed in the initial size of the bubbles) and amount of air present in the medium at any time, (10) infections, (11) programming the nutrient addition during cell reproduction, (12) cell degeneration, (13) propagation time, and (14) generation time.

pH and temperature can easily be controlled by automatic devices. Most frequently the pH is kept around 4.5–5.0 and the temperature between 28 and 30°C. Variations from these values may be followed, depending on the yeast strain, type of aeration, etc as proved by practical experience.

The concentration of the starting yeast can vary widely. Proponents of the early dilution theory used around 2.5 to 3% seed yeast and let it increase to 5–6%, while doubling the original liquid volume by the stepwise or continuous addition of nutrient liquid. This results in 4–6 cells from every initial cell. Processes employing much higher cell populations have also been reported. In one continuous process, de Becze maintained a cell population between 20 and 24% (20 and 24 g per 100 ml, respectively) for 10 days on a pilot-plant scale. According to a patent (64) obtained by S. C. Darányi, the minimum cell population is 8% during reproduction. Using a higher cell population aims at both a higher yield and a higher production tank capacity. Naturally, a high cell population requires intensive aeration and a large air-to-liquid ratio. However, with fine aeration (very small air bubbles), the air requirement per pound of newly formed yeast is not increased with high cell population.

Generation time, the time required to double the number of cells, is at a minimum during the second to fourth hours. Then the cell population is relatively low per unit of nutrient volume, the nutrient is in excess, and the dissolved oxygen content is high. From then on, the generation time gradually increases until practically no further cell substance is produced. The ideal generation time, 50 min, is never reached in industrial production—doubling the cell number in 2–4 hr average time, depending on the method, is considered good. Most batch systems are designed for 12–14 hr total propagation time; during this time, from every pound of seed yeast, 5–10 lb or slightly more compressed yeast is produced. The limit on time, the longest feasible time, is mainly influenced by the danger of infection, by cell degeneration, by cooling facilities, by the capacity of the aeration system, and other factors.

The addition of nutrients is programmed. At the time of inoculation the liquid may contain 10–25% of the total nutrients with 1% or less initial sugar content. After 2–3 hr, addition of the remaining nutrients, as programmed, begins. Theoretical calculation, following Euler, would suggest a logarithmic increase of nutrients with passage of time, in small batches or continuous flow (see Vol. 13, p. 479). Technically, maintenance of the initial nutritional conditions following a logarithmic program did not prove satisfactory. The limitations of fermenter space and aerating and cooling facilities dictated a program of slower feed addition. Thus distortions and changes from the ideal logarithmic curve were developed, justified by the practical results.

The individual methods for which patents were obtained usually emphasized one or another principle or aspect of successful cell reproduction. They were centered on programming the amounts of nutrients to add for each hour, the ratio of initial cell concentration to the final one, the ratio of cell concentration to amount of air supplied, the changes in the initial liquid volume, the nature of air distribution, and so on. One U.S. patent (65) took advantage of the beneficial effects of initial dilution of the cells and the nutrient concentration; the process allowed a considerable increase in cell population, achieved by adding most of the water (80–90% of the total) initially to the broth in the propagating tank and supplying the nutrients, programmed, in a relatively concentrated form. This led to a rapid rate of reproduction at the start, followed by a most efficient nutrient utilization toward the end. The net result was high yield and excellent quality.

Although the high rate of nutrient utilization under aerobic conditions, with the reduction or near exclusion of alcohol production, was known since the work of Pasteur (11b), it took a long time to understand the *effects of aeration* on cell reproduction in submerged cultures. Aeration of nutrient liquid industrially was started at the end of the last century. During the first decades of the present century the value of aeration in terms of yield came to attention. The high sugar-consuming alcoholic fermentation takes place in the absence of oxygen, while, in the presence of air following the low sugar-consuming respiration process for energy, yeasts use the bulk of the available sugar for cell production with a high overall yield of cell substance per unit of sugar. During the almost 100 year history of aerating wort, the simple method of bubbling compressed air through the open end of a pipe at the bottom of the fermenter underwent a number of considerable improvements and variations. The straight pipe was replaced with an inverted T pipe bubbling air at both open ends. Then the ends of the T pipe were closed and at the bottom of the cross pipe small holes were bored through which the air was far better distributed through the liquid. Later, the air was introduced through the holes of metal rings and spargers, gradually resulting in higher yields in the cell crop and lower power requirement for the air compressor per unit of yeast. There was a limit, however, to the air-holding capacity of the fermenter when it was filled up to $\frac{2}{3}$ to $\frac{3}{4}$ of its total capacity: the air passed through the liquid in large bubbles. In the next step, the cross end of the T pipe was provided with many thin tubes, 2–4 in. apart, covering the entire bottom of the propagating tank; through the tiny holes at the bottom or in two rows on the sides of the tubes the air was more evenly distributed and in much smaller bubbles than before. The yields increased further. The tubes were sterilized by steam and were made removable for better mechanical cleaning.

The high cost of energy for aeration and the recognition of the usefulness of the small-size bubbles led to the concept of fine aeration (Feinbelüftung). Around the 1930s a number of ingenious devices were developed. These introduced air through canvas bags fastened around the aerating tubes or through canvas false tank bottoms through which air entered the broth in finely divided form. Hard rubber heads with tiny holes, stainless-steel tubes with slits on the sides and wound around tightly with fine stainless-steel wire, and porous ceramic bodies through which the air entered the liquid in the form of a fog were tried. The Stich process applied thousands of ceramic candles evenly distributed at the bottom of the propagating tank. The Stich aerating system was popular in Denmark, Holland, and England for several decades (from 1930 on) (87). Some manufacturers still use it.

The difficulties experienced in cleaning and sterilizing these fine-aerating systems led to development of mechanical air distributors or fine-aerating equipment. Early in 1930, simultaneously, Vogelbusch in Austria and de Becze in Hungary developed mechanical fine-aerating equipment adopted in many European countries for the production of baker's yeast, and later on all over the world for many types of aerobic submerged fermentations. The Vogelbusch system, described in Austrian patents (91), consisted of a fast-moving perforated hollow piece, for instance, a propeller blade, rotated by a hollow shaft; compressed air was forced through the perforations of the rotating pieces into the broth; the air entered the liquid as very fine bubbles; the use of baffles in order to prevent rotation of the wort increased the efficiency of the mechanical fine aerators. The de Becze system (66) consisted of establishing a rapid movement of the broth by a fast-moving propeller or other agitating device close to the orifice at which air was introduced into the liquid; one or more such fine air distributors placed in the tank reduced the air requirement per lb of yeast to $\frac{1}{4}$–$\frac{1}{6}$ with a simultaneous 10–20% increase in yield. The topic of aeration in the production of compressed yeast has been reviewed by de Becze and Liebmann (106). They also obtained U.S. patent 2,530,814 (1950) for an apparatus providing fine aeration in liquids for various biochemical processes, including the propagation of microorganisms, without using an air compressor—air enters the liquid by a cavitating effect of the apparatus and is divided into very fine bubbles. Similar aerators were recently applied in sewage disposal processes.

The *concentration of the yeast* at the end of the propagation period depends on three factors: (1) amount of seed yeast, (2) amount of increase in cells (population increase), and (3) increase in initial liquid volume during propagation. All three may vary on a large scale and may be expressed in relation to the liquid volume of the particular phase or to the final liquid volume.

Separation of Cells from the Exhausted Nutrient and from Excess Water. This is done in four steps: (1) preliminary separation, (2) washing, (3) separation from the wash water, and (4) elimination of the excess water. Under primitive conditions as described in the Old Vienna process, the first and third steps consisted of sedimentation and decantations with the adding of water for the washing step. From the creamy yeast sludge the excess water was removed by filter presses. To date in many countries this is the standard method for the final step and hence gives the name "compressed" to yeast.

Separation of cells from the exhausted broth and from wash water has been done, from the early part of this century, by yeast separators (Westfalia, DeLaval). Mashes with high cell concentrations usually are diluted in a continuous manner with water before entering the separator for preliminary washing. The yeast cream coming from the first separator (in a form of about 20% yeast) into a small mixing tank, is continuously diluted (washed) with 4–5 times as much cold water, and is separated into a thick cream in a second separator. This cream, chilled in a heat exchanger, is then separated from the excess water in filter presses at about 40–50 psi. The compressed yeast, a brittle to pliable mass, contains 70% moisture.

FOOD AND FEED YEAST (3,9,22a,32,46a,51,53,56,
92,98a,111b,127,138b,150–155)

Because of their chemical composition, high content of proteins, fat, vitamins, and unidentified growth factors, yeast cells are a highly nutritious natural food and feed ingredient. The ease and speed by which carbohydrates and inorganic salts may be converted in a relatively short time and in a relatively small space into yeast cells provides a convenient method for producing an all-important food or feed supplement at will at any time and any place where manufacturing is possible. No agricultural crop can make a similar claim.

The use of yeast in order to correct nutritional deficiencies was reported as far back as classical times. Delbrück, Hayduck, Wohl, and others at the Institut für Gährungsgewerbe in Berlin, Germany, before and during World War I developed the molasses ammonia process for making compressed yeast mainly for food purposes. Prior to World War II researchers at the same institute and others improved the old method and developed new processes for making food yeast from wood sugars and waste carbohydrates. They introduced *Torulopsis utilis (Candida utilis)*, a non-spore-forming yeast, as the organism for quick and efficient assimilation of cheap materials into high-grade food and feed substance. Contrary to *Saccharomyces*, "torula yeast" can assimilate pentoses. A. C. Thaysen in England isolated two new special strains (144a): (1) *T. utilis* var. *thermophilia*, which tolerates high temperatures and is suitable for propagation in the tropics, and (2) *T. utilis* var. *major*, characterized by large cells (this makes cell separation from the broth easier).

The unfortunate use of the term "torula yeast" (or *Torula* yeast) has caused a good deal of confusion but it is so widely employed commercially that it is unlikely to be abandoned. Even the first edition of *The Yeasts* does not list *Torula* as an accepted genus; *both* editions carry *Torula utilis*, *Torulopsis utilis*, and *Candida utilis* as synonyms. In regard to the *Candida-Torulopsis* problem the new edition (25) comments (pages 894–897):

"The genus *Candida* accommodates a heterogeneous collection of asporogenous yeast species which do not qualify for classification in any of the more homogeneous genera of the imperfect yeasts. The genus *Torulopsis* is the other genus of the imperfect yeasts with a similar heterogeneity." Taxonomic findings made it possible to reclassify a number of their species into other genera, but "at present no information is available upon which a natural classification of the *Candida-Torulopsis* group could be based." Since this group "cannot be resolved at present into natural taxa and since its division into the genera *Candida* and *Torulopsis* is arbitrary and artificial, it appears logical to classify the group as one single huge artificial genus from which natural taxa could be split off in the future whenever sufficient information should become available. This measure, logical though it may be," was deemed "inadvisable since it would make necessary the provisory renaming of a great number of species. This would inevitably lead to confusion and justified irritation among the increasing number of workers in various fields who use or encounter yeasts of this group." *Candida* and *Torulopsis* are therefore maintained as separate genera.

In Sweden intensive studies and developmental work led to commercial production of yeast from sulfite liquor as a substrate (see also Vol. 16, p. 720). In the U.S. efforts had been made to utilize waste carbohydrates from fruits and vegetables (see also Vol. 10, p. 57). Czechoslovakia has produced torula yeast from beet molasses, stillage, waste liquors after alcohol and citric acid production, and sulfite liquors (111c). Recent work in the Philippines (96a) reports on a high-protein yeast, *Rhodotorula pilimanae* Hedrick and Burke, isolated from rotting strawberry fruit, and prepared on a large scale to give a protein with an appetizing flavor that is being tested for various uses as a food supplement.

Depending upon their origin, food and feed yeast form two groups: primary yeasts and secondary yeasts.

Primary food and feed yeast is produced as a principal product with the highest possible yield, usually from *molasses, sulfite liquor, wood hydrolyzate,* or *agricultural wastes.* Specially selected and acclimatized pure cultures of strains of *Torulopsis utilis, Saccharomyces cerevisiae,* and bulk secondary yeast obtained in brewery operations are employed as seed yeasts. Live brewer's yeast after preliminary purification from hops and grain residues when used as an inoculum is propagated for several generations.   In addition to *S. cerevisiae* and *T. utilis*, other species such as *Oospora lactis* (*Oidium lactis, Endomyces lactis, Geotrichum candidum*), *Candida arborea*, and *Saccharomyces fragilis* (*Kluyveromyces fragilis*) have been employed for producing food or feed yeast.

Preparation of Inoculum.   Plant-scale inoculum is prepared in the fashion customary for baker's yeast production.   *Torulopsis utilis* grows rapidly, resists phage infections (see Vol. 13, pp. 460–461), and is easily kept free from serious bacterial infection.   Once a plant-scale inoculum is produced at the plant, or a starter is brought in from another plant, no new starter is needed for a long period of time.   Reproduction of the cells may be maintained in a semicontinuous or a continuous stage.   Once the cell count reaches the desired near-maximum level in the propagation tank, the yeast-containing effluent is withdrawn continuously or in small batches and is replaced by the same volume of fresh nutrients.   The rate of withdrawal is controlled by the cell population in the propagating tanks.   The rate of nutrient addition is based either on a quick sugar determination or a quick yeast content measurement.   The goal is to keep these two factors at a predetermined optimum level.

The basic carbon source (molasses, sulfite liquor, or wood hydrolyzate) is supplemented with nitrogen, phosphorus, and potassium salts and trace metals, according to need.   Molasses needs no potassium and less phosphorus and nitrogen supplement than does sulfite liquor or wood hydrolyzate.   Nitrogen is supplied in the form of ammonia or ammonium salts and phosphorus as phosphoric acid or phosphates in a fashion similar to that for baker's yeast.

Commercial Processes.   The most used and best known are the German Waldhof (61,71,78) and the I. G. Farbenindustrie, the U.S. Sulphite Pulp Manufacturers (SPM) (56), the Carnation-Albers (56,127), and the British Colonial Food Yeast processes.

The *Waldhof* plants use beach-wood sulfite liquor or spruce-wood spent liquor and pentose-containing stillage from producing alcohol from spruce-wood spent liquor. From the spent sulfite liquor, produced at the pulp mill, SO_2 is removed by aeration. The original pH of 1.5–3.0 is raised to 5.0–5.5 by lime.   The liquid in hot storage is clarified for 8 hr.   The clear liquid and the inorganic supplements are continuously fed into the fermenter and aerated by fine-aeration systems.   In the propagating tank the pH is kept at 5.5 as optimum.   The air–liquid emulsion is about 3.3 times the liquid volume, each liter of liquid containing 2.3 l of air during propagation.   The liquid is completely replaced every 2.5 to 4.5 hr.   The foamy yeast, withdrawn from the bottom of the tank, is deaerated in horizontal centrifuges, then separated from the liquid by yeast separators.   From the yeast obtained in the form of a cream, the excess water is removed in drum filters.   The yeast then, after autolysis, is dehydrated by a drum-drying or spray-drying process.   In the Waldhof plants, 8.7 lb ammonia, 1.29 lb phosphorus, and 0.37 lb magnesium were used as supplement per 250 lb reducing sugar, yielding 100 lb dry yeast.   Control of temperature and pH is highly essential.   Cooling coils in the propagators keep the temperature at 32–35°C.   The thermophilic strain of *T. utilis* is propagated at 36–39°C.   The pH is maintained between 4.5 and 6.0.

The I.G. Farbenindustrie process features a number of vertical, water-jacketed columns outside the fermenter.   The columns serve as both aerators and coolers.   The liquid (sulfite-base nutrient containing the reproducing yeast) is continuously lifted from the main tank into the columns by air, which is introduced in finely divided bubbles through a stack of toothed rings at their base. The fully aerated and chilled foamy liquid from the columns is returned into the main fermenter. In other respects the operation is similar to the processing steps described for baker's yeast production.

The British Colonial Food Yeast process (in Jamaica and South Africa) also uses *Torulopsis utilis* (53,144a). The propagation temperature is 36–39°C. The broth is aerated through porous ceramic stoneware which releases the air in very fine bubbles (Stich). Diluted molasses, used as substrate, is clarified after being acidified by sulfuric acid to pH 4.0 and is heated to precipitate the calcium sulfate and the suspended material. The clarified molasses solution supplemented with the inorganic nutrients serves as a medium for propagating the yeast. In a manner similar to baker's yeast production, the yeast is separated, washed, compressed, and dried.

Primary Yeast from Petroleum Products. During the 1960s in the U.S. as well as in Europe, intensive research was carried out on growing yeast on petroleum products supplemented with ammonia, phosphates, and trace nutrients. The goal was to produce proteins (see Proteins in Supplement Volume) for food or feed supplementation. Nickerson and Brown (49a) review this topic especially well.

Chemical & Engineering News (February 2, 1970, pp. 21–22) reports that British Petroleum Co., Ltd., has two plants which will, by the end of 1971, produce a total of 20,000 metric tons of protein annually by fermentation of hydrocarbons by *Candida* yeasts—one at Grangemouth, Scotland, and one at Lavera, France. The former plant uses pure C_{10} to C_{18} alkanes under aseptic conditions; the latter uses a process with gas-oil feedstocks containing about 10% by weight *n*-alkanes and does not exclude atmospheric contaminants. The British Petroleum process has been licensed to Kyowa Hakko Kogyo Co. in Japan. On the basis of the work done by British Petroleum Co., Czechoslovakia (111c) has produced *Candida lipolytica* on paraffins in gas oil. In Russia, a large collection of yeasts is available capable of assimilating hydrocarbons so that it is possible for that country to organize large-scale production of fodder yeast from paraffins (113a). It is interesting to note that the possible use of hydrocarbons for diagnostic purposes in the classification of yeast is mentioned in the new Lodder (25, pages 77–78).

Nutritional Value of Yeast. The nutritional value of yeast as food or feed, beginning in the latter part of the past century, has been studied at an increasing rate; its place among other food and feed products today is accepted (49a). Chemical composition, vitamin content, amount and kind of amino acids in yeast protein (46d), and existence of one or more growth factors not yet identified are well established. Considerable data are published by commercial sources (151–155, for example) as well as by the Brewers Yeast Council, Inc., and others.

Dried yeast products on the market may be grouped as follows:

(1) Primary dried yeast—*Saccharomyces cerevisiae*.
(2) Primary dried torula yeast—*Torulopsis utilis* (*Candida utilis*).
(3) Secondary yeast, brewer's dried yeast—*S. cerevisiae, S. carlsbergensis*.

It is claimed and not yet contested that the nutritional value of yeast protein is equivalent to beef protein.

Primary Yeasts. The composition of commercially available dried primary yeast is specified in the eleventh edition of *The National Formulary* (pages 395–396) as follows:

protein (N × 6.25)	minimum	45%
thiamine hydrochloride (vitamin B$_1$)	minimum	120 μg/g
riboflavin (vitamin B$_2$)	minimum	40 μg/g
nicotinic acid	minimum	300 μg/g
fermenting power	inactive	
fillers	none	
total bacterial count	maximum	7500/g
mold count	maximum	50/g
moisture	maximum	7%
ash	maximum	8%

All commercially marketed primary dried yeast products meet the NF specification with the exception of thiamine and riboflavin contents of basic dried yeast grown on substrate not enriched in vitamins and used primarily for flavoring purposes. The thiamine content of basic primary dried yeast is around 12–15 μg/g and the riboflavin content around 35–45 μg/g. The product was marketed at 0.38–0.40¢/lb in 1968.

A number of primary dried yeast products are marketed with a high vitamin content from propagation in a vitamin-enriched substrate. Sold under code names, their vitamin contents and the proportion of the various vitamins vary on a large scale; in 1968, their prices varied from 50¢ to 95¢/lb, depending on the vitamin content.

A typical analysis of primary dried yeasts marketed for food purposes is as follows:

moisture	5.0%	calcium	0.3%
protein	50.0%	phosphorus	2.4%
fat (ether ext.)	1.2%	potassium	2.6%
total lipids	5.5%	magnesium	0.5%
carbohydrates	31.5%	sodium	0.4%
ash	8.0%		

The energy value of dried yeast is about 3.25 cal/g or 93 cal/oz (155).

The approximate content of amino acids expressed as % of dry protein content is as follows:

alanine	9.0%	lysine	8.2%
arginine	5.0%	methionine	2.5%
aspartic acid	4.0%	phenylalanine	4.5%
cystine	1.6%	proline	2.5%
glutamic acid	13.5%	threonine	5.5%
glycine	0.6%	(and serine)	
histidine	4.0%	tryptophan	1.2%
hydroxyproline	4.5%	tyrosine	5.0%
isoleucine	5.5%	valine	5.5%
leucine	8.0%	others	9.4%

All essential amino acids are present. The vitamin contents of the various products with definite (guaranteed) values are as follows (in μg/g):

thiamine	up to 10,000.0
riboflavin	up to 12,000.0
niacin	up to 30,000.0
pyridoxine	**15.0**
pantothenic acid	**110.0**
biotin	**2.5**
inositol	4,000.0
choline	4,000.0
p-aminobenzoic acid	13.0
folic acid	11.0

A typical analysis of marketed edible-grade dried brewer's yeast is:

protein	50.0%	zinc	38.7 ppm
fat	1.5%	salmonella	**negative**
fiber	1.5%	coliform bacteria	**negative**
ash	7.0%	thiamine	56.6 mg/lb
moisture	6.0%	riboflavin	16.0 mg/lb
nitrogen-free extract	34.0%	niacin	226.5 mg/lb
calcium	0.12%	pantothenic acid	55.2 mg/lb
phosphorus	1.50%	pyridoxine	22.6 mg/lb
potassium	0.86%	choline	2200.0 mg/lb
iron	0.02%	betaine	544.0 mg/lb
copper	35 ppm	biotin	0.5 mg/lb
manganese	5.3 ppm	folic acid	22.2 mg/lb
cobalt	1.5 ppm	inositol	2265.0 mg/lb

Powdered baker's and brewer's dried yeast and yeast extracts in paste or dried powdered form are widely used in a great variety of *food and flavoring products* in addition to bread and bakery products. For such purposes the enzyme activity of the yeast is destroyed; the food value and principally the taste and flavoring effect are utilized. Because yeast extract and autolyzed yeast products are soluble, they are more tasty and enhance the taste of other food ingredients much more than dried yeast. Dried yeast and yeast extracts are added to biscuit mixes, cereal breakfast foods, savory snacks and cocktail savories, potato crisps, sauce mixes, tasty spreads, ready-made dried stuffings, bread crumbs, batter mixes, health foods, vegetarian foods, baby foods, sausages, instant soups in vending machines, canned foods, etc. In such food products 3, 5, or up to 10% of the solid material may be yeast solids. Often 1% yeast autolyzate is added to the water used in restaurant cooking to enhance the flavor of the food. Flavor powders may contain up to 40% yeast product, the balance being made up of spices and other flavoring substances. The use of yeast products in foods is gradually increasing. They are well accepted in Great Britain and Eastern Europe, where they are sold under various trade names (Yestamin, Yestor, Yestabakon, Yesta smoked ham, Yestacheese, Yestamato, etc), with the manufacturers' recommendations and instructions for their use.

Secondary Yeasts. Secondary yeasts are recovered from brewery operations or from molasses alcohol and rum production. Usually after some degree of purification, they are dried to around 90% solids content and marketed as Dried Brewer's or Dried Distiller's yeast.

In the United States the distillation residues from grain alcohol and whiskey production, containing all the yeast cells developed in the process, are concentrated and dried to about 90% solids content and are marketed under the name of "Distiller's Dried Grains with Solubles." Two other products made from grain distiller's residues are "Dried Distiller's Solubles" and "Distiller's Dried Grains"; the former contains the bulk of the soluble portion and fine suspended matter, including the bulk of the yeast. Distiller's dried products (in the U.S. in 1968 they amounted to 450 thousand tons) are highly valued feed ingredients, mainly because of the nutritional value of the yeast they contain (103,135). They were sold at about $70/ton in 1970.

The complete distillation residue in dried form of molasses alcohol and rum plants contains all the yeast formed in the process and is marketed as a feed ingredient under the name of Molasses Distiller's Dried Grains.

Future Outlook

Compressed Yeast. Bread consumption in countries of plenty is gradually declining. With the increase in population, the increase of yeast-to-flour ratio in bread, and the extensive use of food-grade yeast and yeast autolyzate in foods, the production of baker's yeast will probably hold its own within the next 10 years or will slightly increase in the U.S. In many areas of Asia, Africa, and South America, rice, unleavened corn bread, and other starchy food takes the place of yeasted bread; with proper education in those areas about the benefits of yeasted bread and the technology of making baker's yeast, a sharp increase is expected in yeast consumption and possibly in production.

It is obvious that manufacturing methods in the U.S. will eventually adapt mechanical air distributors to aerate wort in closed propagating tanks; the intake air will be sterilized and 80% will be recirculated in the wort, the balance being replaced with fresh sterile air. The yeast concentration in the wort will thus increase and batteries of automated compact units will do the manufacturing with a minimum of human supervision. The needs, possibilities, and technological tendencies indicate that giant bakeries will subject their purchased baker's yeast to a limited reproduction, at their plants, in an edible-grade medium using high yeast concentration and compact propagators provided with fine aerators. The yeast in 2–4 hr time will be increased to twice or four times its original weight and at the same time the fermenting power per unit of yeast will increase by 50–70% over the original seed yeast. The yeast culture thus prepared will serve as a ready inside pre-ferment.

It is likely that distilleries, wineries, and possibly breweries will, as their automation progresses, eliminate their own yeast making and will employ compressed yeast custommade of their own strains. Also, pure cultures of two or more strains mixed in desired proportions will be applied for use as seed yeast. Ester-producing *Hansenula* strains will play more of a part in fermenting alcoholic beverages. Current research indicates that pharmaceutical applications of products made of compressed yeast for both external and internal use will increase.

Food Yeast. In areas with undernourished population, food yeast will play an even more important role than baker's yeast. Food yeast, if available at low cost, could supply part of the proteins and the greatest part of the vitamins needed for maintaining health. Restoration or maintenance of health cannot depend on the yearly 2–3 lb of baker's yeast eaten in bread. A substantial factor in the solution of the nutritional problems of underdeveloped countries would be, per capita, 10 to 30 lb yearly of dried food yeast with 60% protein and a high content of the B vitamins and unknown growth factors. Using their own crops for a carbon source, and imported fertilizer-grade ammonia and phosphate for nutrients, and even foreign-made automatic compact plant equipment, they could produce highly valuable food supplements for their needs; if such plans, already promoted by several international organizations, come to reality, an unprecedented increase may be expected in the world's food-grade yeast production. Using fermenters and processes operating with high concentrations of wort and yeast, effective mechanical air distributors requiring low energy, and automatic feeding and operation control equipment, a compact plant built on several acres may well supplement the nourishment of 10 million people. The conversion of the raw material to a finished high-grade protein can be done in 24 hr.

Bibliography

"Yeasts" in *ECT* 1st ed., Vol. 15, pp. 195–219, by W. J. Nickerson, Institute of Microbiology, Rutgers University, and A. H. Rose, University of Birmingham, England.

Reference Books and Textbooks

1. S. Aiba, A. E. Humphrey, and N. F. Millis, *Biochemical Engineering*, University of Tokyo Press, Tokyo, and Academic Press Inc., New York, 1965.
2. M. A. Amerine and W. V. Cruess, *The Technology of Wine Making*, Avi Publishing Co., Westport, Conn., 1960; M. A. Amerine, H. W. Berg, and W. V. Cruess, *The Technology of Wine Making*, 2nd ed., Avi Publishing Co., 1967.
3. K. Arima, W. J. Nickerson, M. Pyke, H. Schanderl, A. S. Schultz, A. C. Thaysen, and R. S. W. Thorne (contributors), in W. Roman, ed., *Yeasts*, Dr. W. Junk, The Hague, and Academic Press Inc., New York, 1957.
4. E. A. Bessey, *Morphology and Taxonomy of Fungi*, The Blakiston Co., Philadelphia, 1950.
5. N. Blakebrough, ed., *Biochemical and Biological Engineering Science*, 2 Vol., Academic Press, Inc., New York, 1967, 1969.
5a. P. D. Boyer, H. Lardy, and K. Myrbäck, eds., *The Enzymes*, 2nd ed., Vol. 1 (1959)–Vol. 8 (1963).
6. J. Brachet and A. E. Mirsky, eds., *The Cell:* Vol. I, *Methods and Problems of Cell Biology*. Vol. II, *Cells and Their Component Parts*, Academic Press Inc., New York, 1959, 1961.
6a. M. Brook, ed., *Biology and the Manufacturing Industries*, Academic Press Inc., New York, 1967.
7. L. E. Casida, Jr., *Industrial Microbiology*, John Wiley & Sons, Inc., New York, 1968.
8. *Cereal Laboratory Methods*, 7th ed., American Association of Cereal Chemists, St. Paul, Minn., 1965.
9. A. H. Cook, ed., *The Chemistry and Biology of Yeasts*, Academic Press Inc., New York, 1958.
9a. Centraalbureau voor Schimmelcultures (Baarn), *List of Cultures, 1939*.
10. J. de Clerck, *A Textbook of Brewing*, 2 vols., Chapman & Hall Ltd., London, 1957, 1958.
11. H. A. Diddens and J. Lodder, *Die anaskosporogenen Hefen*, 2. Hälfte, North-Holland Publishing Co., Amsterdam, 1942.
11a. M. Dixon and E. C. Webb, *Enzymes*, 2nd ed., Academic Press, New York, 1964.
11b. R. Dubos, *Pasteur and Modern Science*, Doubleday & Co. Inc. (Anchor Book), Garden City, N.Y., 1960. R. Vallery-Radot, *The Life of Pasteur*, Constable & Co., Ltd., London; reprint, Dover Publications, Inc., New York, 1960.
12. B. Ephrussi, *Nucleo-cytoplasmic Relations in Microorganisms*, Clarendon Press, Oxford, New York, 1953.
13. G. Foth, *Handbuch der Spiritusfabrikation*, Verlag Paul Parey, Berlin, 1929.

14. A. Frey-Wyssling and K. Mühlethaler, *Ultrastructural Plant Cytology, with an Introduction to Molecular Biology*, American Elsevier Publishing Co., New York and Amsterdam, 1965.

15. W. Henneberg, *Handbuch der Gärungsbakteriologie*, Verlag Paul Parey, Berlin, 1926.

16. A. Guilliermond, *La Sexualité, le Cycle de Développement, la Phylogénie et la Classification des Levures d'après les Travaux récents*, Masson & Cie., Paris, 1937.

17. J. von Hérics-Tóth and A. von Osztrovszky, *A Szeszgyártás Kézikönyve* (Handbook of Alcohol Production), Dick Manó, Budapest, 1927, 703 pp.

18. H. Iizuka, ed., *Culture Collection of Microorganisms*, University Park Press, Baltimore, Md., 1970 (proceedings of the First International Culture Congress, UNESCO). S. M. Martin, ed., *Culture Collections: Perspectives and Problems*, University of Toronto Press, Toronto, 1963. For a large collection of yeasts, see *The American Type Culture Collection*, 9th ed., Rockville, Md. 20852, 1970, pp. 142–150; see also list of other yeast culture collections, pp. x-xi.

19. International Union of Biochemistry, *Enzyme Nomenclature*, Elsevier Publishing Co., Amsterdam, London, New York, 1965 (1964 recommendations of the IUB on the nomenclature and classification of enzymes and on the units and symbols of enzyme kinetics). The present commission expects to publish a revision in 1971 or later.

20. A. Jörgensen, *Practical Management of Pure Yeast*, 3rd ed., Chas. Griffin & Co., Ltd., London, 1936.

21. A. Jörgensen, *Microorganisms and Fermentation* (rewritten by A. Hansen) Chas. Griffin & Co., Ltd., London, 1948.

22. A. Klöcker, *Die Gärungsorganismen*, 3rd ed., Urban & Schwarzenberg, Berlin and Vienna, 1924.

22a. R. A. Lawrie, ed., *Proteins as Human Food*, Avi Publishing Co., Westport, Conn., 1970.

23. C. C. Lindegren, *The Yeast Cell, Its Genetics and Cytology*, Educational Publishers, Inc., St. Louis, 1949.

24. J. Lodder, *Die anaskosporogenen Hefen*, 1. Hälfte, *Verhandel. Koninkl. Akad. Wetenschap. Afd. Natuurk.*, Sect. II **32,** 1 (1934).

25. J. Lodder and N. J. W. Kreger-van Rij, *The Yeasts, a Taxonomic Study*, North-Holland Publishing Co., Amsterdam, and Interscience Publishers, Inc., New York, 1952 (a condensation in English of refs. 11, 24, and 33) (reprinted in 1967, North-Holland Publishing Co.). J. Lodder, ed., *The Yeasts, a Taxonomic Study*, 2nd ed., North-Holland Publishing Co., Amsterdam and London, 1970.

26. S. A. Matz, ed., *Bakery Technology and Engineering*, Avi Publishing Co., Westport, Conn., 1960.

26a. A. K. Mills, ed., *Aspects of Yeast Metabolism*, Blackwell Scientific Publications, Oxford and Edinburgh, 1968.

26b. *The National Formulary*, American Pharmaceutical Association, Washington, D.C.: 11th ed., 1960; 12th ed., 1965; 13th ed., 1970.

27. M. J. Pelczar, Jr. and R. D. Reid, *Microbiology*, 2nd ed., McGraw-Hill Book Co., New York, 1965.

27a. D. Perlman, ed., *Fermentation Advances*, Academic Press Inc., New York, 1969.

28. S. Pettenkoffer, *Borgazdaság* (Winery), Pátria R. T., Budapest, 1937.

28a. H. J. Phaff, M. W. Miller, and E. M. Mrak, *The Life of Yeasts: Their Nature, Activity, Ecology, and Relation to Mankind*, Harvard University Press, Cambridge, Mass., 1966.

29. S. C. Prescott and C. G. Dunn, *Industrial Microbiology*, McGraw-Hill Book Co., New York, 1949 (3rd ed., 1959).

30. G. Reed, *Enzymes in Food Processing*, Academic Press Inc., New York, 1966.

31. W. Roman, ed., *Yeasts*, Dr. W. Junk, The Hague, and Academic Press Inc., New York, 1957.

32. A. H. Rose and J. S. Harrison, eds., *The Yeasts* (in 3 vol.), Academic Press Inc., New York: Vol. 1, *Biology of Yeasts*, 1969; Vol. 2, *Physiology and Biochemistry of Yeasts*, in press; Vol. 3, *Yeast Technology*, 1970.

32a. F. S. Snell and C. L. Hilton, eds., *Encyclopedia of Industrial Chemical Analysis*, Interscience Publishers, a division of John Wiley & Sons, Inc., New York: Vol. 4, 1967, G. I. de Becze, Alcoholic Beverages, pp. 462–494. Vol. 7, 1968, G. O. Weissler, Brewery Products, pp. 485–495.

33. N. M. Stelling-Dekker, *Die sporogenen Hefen, Verhandel. Koninkl. Akad. Wetenschap. Afd. Natuurk.*, Sect. II **28**, 1 (1931).

34. G. Szilágyi, *Az Érjedés Chémiájának Kézikönyve* (Handbook of Fermentation Chemistry), Franklin Társulat (Franklin Society), Budapest, 1890, 191 pp.

35. *Ullmanns Enzyklopädie der technischen Chemie*, 3rd ed., Urban & Schwarzenberg, Munich and Berlin, 1957: Vol. 8, "Hefe," pp. 449–489.

36. W. W. Umbreit, ed., *Advances in Applied Microbiology*, Academic Press Inc., New York, Vol. 1 (1959)—.

37. L. A. Underkofler and R. J. Hickey, eds., *Industrial Fermentations*, Vol. I, Chemical Publishing Co., New York, 1954.

38. *(The) United States Pharmacopeia*, 13th ed., Mack Printing Co., Easton, Pa., 1947. The most recent edition (18th, 1970) refers back to the 13th ed. for yeast definitions.

39. F. Wagner, *Presshefe und Gärungsalkohole*, published by the author, 1936 (printed by E. Strache, Warnsdorf). The 283 page book is organized into chapters and is provided with an author and subject index. The body of the book consists of abstracts of scientific articles and patents on the two subjects: compressed yeast and fermentation alcohol. It is a valuable reference in establishing prior art in patent proceedings.

39a. A. F. Wagner and K. Folkers, *Vitamins and Coenzymes*, Interscience Publishers, a division of John Wiley & Sons, Inc., New York, 1964.

40. F. G. Walter, *The Manufacture of Compressed Yeast*, 2nd ed., Chapman & Hall Ltd., London, 1953.

41. J. E. Wihlfahrt (in collaboration with R. W. Brooks), *A Treatise on Baking*, The Fleischmann Division, Standard Brands, Inc., New York, 3rd ed., 1935 (reprinted 1948).

42. H. F. Willkie and J. A. Prochaska, *Fundamentals of Distillery Practice*, Joseph E. Seagram & Sons, Inc., Louisville, Ky., 1943.

Reviews

43. G. I. de Becze, "Classification of Yeasts, I. Introduction and Morphology," *Wallerstein Lab. Commun.* **22**, No. 77, 103–123 (June 1959).

44. G. I. de Becze, "Classification of Yeasts, II. Genetics," *Wallerstein Lab. Commun.* **22**, No. 78, 199–225 (Sept. 1959).

45. G. I. de Becze, "Classification of Yeasts, III. Biochemistry," *Wallerstein Lab. Commun.* **23**, No. 81, 99–124 (Aug. 1960).

46. G. I. de Becze, "Classification of Yeasts, IV. Classifications," *Wallerstein Lab. Commun.* **25**, No. 86, 43–64 (April 1962).

46a. H. J. Bunker, "Recent Research on the Yeasts," in D. J. D. Hockenhull, ed., *Progress in Industrial Microbiology*, Vol. 3, Heywood & Co. Ltd., London, and Interscience Publishers, Inc., New York, 1961, pp. 1–41; "Microbial Food," in C. Rainbow and A. H. Rose, eds., *Biochemistry of Industrial Microorganisms*, Academic Press Inc., New York, 1963, pp. 34–67.

46b. A. A. Eddy, "Aspects of the Chemical Composition of Yeast," in A. H. Cook, ed., *The Chemistry and Biology of Yeasts*," Academic Press Inc., New York, 1958, p. 157–249.

46c. P. Gerhardt and M. C. Bartlett, "Continuous Industrial Fermentations," in W. W. Umbreit, ed., *Advances in Applied Microbiology* (ref. 36), Vol. 1, 1959, pp. 215–260. T. Holme, "Biological Aspects of Continuous Cultivation of Microorganisms," *ibid.*, Vol. 4, 1962, pp. 101–116.

46d. H. H. Hall, "Applied Microbiology in Animal Nutrition," *ibid.*, Vol. 4, 1962, pp. 77–99.

47. J. S. Harrison, "Baker's Yeast" in C. Rainbow and A. H. Rose, eds., *Biochemistry of Industrial Microorganisms*, Academic Press Inc., New York, 1963, pp. 9–33.

48. R. Irvin, "Commercial Yeast Manufacture" in L. A. Underkofler and R. Hickey, eds., *Industrial Fermentations*, Vol. I, Chemical Publishing Company, New York, 1954.

49. M. A. Joslyn and M. W. Turbovsky, "Commercial Production of Table and Dessert Wines" in L. A. Underkofler and R. J. Hickey, eds., *Industrial Fermentations*, Vol. I, Chemical Publishing Company, New York, 1954.

49a. W. J. Nickerson and R. G. Brown, "Uses and Products of Yeasts and Yeastlike Fungi," in W. W. Umbreit, ed., *Advances in Applied Microbiology* (ref. 36), Vol. 7, 1965, p. 225 ff.

50. H. K. Parker, "Continuous Bread Making Processes" in S. A. Matz, ed., *Bakery Technology and Engineering*, Avi Publishing Co., Westport, Conn., 1960, pp. 457–478.   H. J. Peppler, "Yeast," *ibid.*, pp. 35–74.

51. H. J. Peppler, "Industrial Production of Single-Cell Protein from Carbohydrates," in R. I. Mateles and S. R. Tannenbaum, eds., *Single-Cell Protein*, M.I.T. Press, Cambridge, Mass., 1968, pp. 229–242.

52. H. J. Peppler, "Yeast Technology" in H. J. Peppler, ed., *Microbial Technology*, Van Nostrand Reinhold Co., New York, 1967, pp. 145–171.

53. M. Pyke, "The Technology of Yeast" in A. H. Cook, ed., *The Chemistry and Biology of Yeasts*, Academic Press Inc., New York, 1958, pp. 535–586.

54. Society of Chemical Industry, *Continuous Culture of Micro-organisms* (Monograph No. 12), London, 1961 (Gordon and Breach Science Publishers, New York): "Manufacture of Bakers' Yeast by Continuous Fermentation," Part I, "Plant and Process," pp. 81–93 (by A. J. C. Olsen), and Part II, "Instrumentation," pp. 94–115 (by H. N. Sher).

55. W. H. Stark, "Alcoholic Fermentation of Grain" in L. A. Underkofler and R. J. Hickey, eds., *Industrial Fermentations*, Vol. I, Chemical Publishing Co., New York, 1954.

56. A. J. Wiley, "Food and Feed Yeast" in L. A. Underkofler and R. J. Hickey, eds., *Industrial Fermentations*, Vol. I, Chemical Publishing Co., New York, 1954.

56a. R. Woods, "Yeast and the Story of Enzymes," *Yeast: Dried Yeasts and Their Derivatives* **2**, No. 4 (Sept. 1953), reprinted in *Wallerstein Lab. Commun.* **17**, No. 56, 47–57 (1954); also a series of reviews in *Borden's Review of Nutrition Research*, 1947–1951.

Patents

57. Aktieselskabet Dansk Gaerings-Industri, "Fremgangsmaade til Fremstilling af Gaer, navnlig Luftgaer," Dan. Pat. 28,507 (Sept. 26, 1921).

58. H. P. Broquist and J. A. Brockman, Jr. (to American Cyanamid Co.), "Production of Lysine," U.S. Pat. 2,965,545 (Dec. 20, 1960).

59. S. Burrows and R. R. Fowell (to Distillers Co., Ltd.), "Improvements in Yeast," Brit. Pat. 868,621 (May 25, 1961).

60. S. L. Chen and E. J. Cooper (to Universal Foods Corp.), "Production of Active Dry Yeast," U.S. Pat. 3,041,249 (June 26, 1962).

61. W. Claus, "Verfahren zur Gewinnung von Hefen und hefeähnlichen Pilzen," German Pats 744,678, 752,725, 759,121 (1943–1944).

62. W. Connstein and K. Lüdecke, "Process for Manufacturing of Propantriol from Sugar," U.S. Pat. 1,368,023 (Feb. 8, 1921).

63. W. Connstein and K. Lüdecke, "Process for Manufacturing of Propantriol from Sugar," U.S. Pat. 1,511,754 (Oct. 14, 1924).

64. S. C. Darányi, "Manufacture of Yeast," U.S. Pat. 2,035,048 (Mar. 24, 1936).

65. G. I. de Becze, "Manufacture of Yeast," U.S. Pat. 2,199,722 (May 7, 1940).

66. G. I. de Becze, Mechanical Air Distributor for Aeration of Yeast Broth, Hungar. Pat. 110,202 (June 14, 1934).

67. J. R. Eoff, Jr. (dedicated to the public), "Process of Manufacturing Glycerol," U.S. Pat. 1,288,398 (Dec. 17, 1918).

68. J. Effront, "Fermentation of Materials Which Have Been Rendered Aseptic," U.S. Pat. 620,022 (Feb. 21, 1899).

69. H. Fleischmann, "Manufacture of Yeast," U.S. Pat. 102,387 (Apr. 26, 1870).

70. E. I. Fulmer, L. A. Underkofler, and R. J. Hickey (to Iowa State College Research Foundation), "Fermentative Glycerol Production," U.S. Pat. 2,416,745 (Mar. 4, 1947).

71. M. Gade and K. L. Schulze, German Pat. 765,434 (1944).

72. H. E. Harrison (to Anheuser-Busch, Inc.), "Culture of Yeast," U.S. Pat. 2,359,521 (Oct. 3, 1944).

73. F. Hayduck (to The Fleischmann Co.): "Process for the Manufacture of Compressed Yeast," U.S. Pats. 1,449,102-4 (Mar. 20, 1923); "Low-Alcohol Yeast Process," U.S. Pats. 1,449,105–10 and 1,449,112 (Mar. 20, 1923); "Process for the Manufacture of Yeast," U.S. Pat. 1,449,-111 (Mar. 20, 1923); "Method for the Precipitation and Preparation of Compressed Yeast,"

U.S. Pat. 1,449,113 (Mar. 20, 1923); "Foam Destroying Device," U.S. Pat. 1,449,114 (Mar. 20, 1923).

74. G. O. W. Heijkenskjöld, "Preparation of Yeast," U.S. Pat. 1,680,043 (Aug. 7, 1928); "Yeast and Its Manufacture," U.S. Pat. 1,703,272 (Feb. 26, 1929); "Method of Manufacturing Yeast," U.S. Pat. 1,881,557 (Oct. 11, 1932).

75. S. Levy, "Preparation of Yeast," U.S. Pat. 255,176 (Mar. 21, 1882), reissue 10,341 (June 12, 1883).

76. Maschinenbau A. G. Golzern-Grimma, "Bewegliche Luftverteilungsvorrichtung für Gärbottiche" (Equipment for Aerating Yeast Mashes), German Pat. 246,709 (May 8, 1912).

76a. M. W. Mead, Jr. and J. Lee (to Hoffmann-La Roche, Inc. and National Grain Yeast Corp.), "Process for the Production of Yeast Having High Vitamin B_1 Potency," U.S. Pat. 2,328,025 (Aug. 31, 1943).

77. M. Moskovits, "Manufacture of Yeast of High Enzymatic Activity," U.S. Pat. 1,962,831 (June 12, 1934).

78. F. Neumann, "Verfahren zur Vorbereitung von Sulfitablangen für die Futter- und Nährhefe-erzeugung," German Pats. 729,842 (Jan. 4, 1943), and 752,723 and 764,401 (1944).

79. Norddeutsche Hefeindustrie A. G., "An Improved Process for Aerating Fermentation Liquids in the Production of Bakers' Yeast," Brit. Pat. 423,331 (Jan. 30, 1935).

80. H. J. Peppler and J. A. Thorn (to Red Star Yeast and Products Co.), "Autolysis Process and Product," U.S. Pat. 2,922,748 (Jan. 26, 1960).

81. J. Rainer, "Verfahren der Presshefe-Fabrikation ohne Alkoholgärung und ohne Erzeugung von Nebenproducten," German Pat. 10,135 (Oct. 12, 1879).

82. S. O. Rosenqvist, "Förfaringssätt vid framställning av jäst," Swedish Pat. 88,558 (Feb. 23, 1937).

83. A. L. Schade (to The Overly Biochemical Research Foundation, Inc.), "Production of Glycerol and Yeast by Fermentation," U.S. Pat. 2,428,766 (Oct. 17, 1947).

84. A. L. Schade and E. Farber (to The Overly Biochemical Research Foundation, Inc.), "Process for the Manufacture of Glycerin," U.S. Pat. 2,414,838 (Jan. 28, 1947).

85. A. S. Schultz, L. Atkin, and C. N. Frey (to Standard Brands, Inc.), "Synthesis of Thiamin and Product," U.S. Pat. 2,262,735 (Nov. 11, 1941).

86. S. Sak, "Method of Controlling Yeast Propagation," U.S. Pat. 1,884,272 (Oct. 25, 1932).

87. E. Stich, "Verfahren zur Feinstbelüftung von Diaphragmen-Gärbottichen," German Pat. 567,518 (Jan. 4, 1933). Norddeutsche Hefeindustrie A. G., addenda to Pat. 567,518: German Pats. 594,192–3 (Mar. 13, 1934); "Gärverfahren durch Feinstbelüftung," German Pat. 594,194 (Mar. 13, 1934); "Vorrichtung zur Feinstbelüftung von Gärbottichen," German Pat. 594,195 (Mar. 13, 1934); "Gärverfahren," German Pat. 594,361 (Mar. 15, 1934). Norddeutsche Hefeindustrie A. G., "Verfahren zur Feinstbelüftung von Diaphragmen-Gärbottichen," German Pat. 594,671 (Mar. 20, 1934), and the following addenda: Wirtschaftliche Vereinigung der deutschen Hefeindustrie, German Pat. 622,962 (Dec. 10, 1935) and "Verfahren zum Belüften von Flüssigkeiten durch Gase," German Pat. 622,963 (Dec. 10, 1935).

88. E. G. Stimpson and H. Young (to National Dairy Products Corp.), "Increasing the Protein Content of Milk Products," U.S. Pat. 2,809,113 (Oct. 8, 1957).

89. R. J. Sumner, W. A. Hardwick, R. D. Seeley, and H. F. Ziegler, Jr. (to Anheuser-Busch, Inc.), "Method of Producing a Bakers' Yeast," U.S. Pat. 3,089,774 (May 14, 1963).

90. F. W. Tanner, Jr. and J. M. Van Lanen (to U.S. Secretary of Agriculture), "Method for the Production of Riboflavin by *Candida flareri*," U.S. Pat. 2,424,003 (July 15, 1947).

91. W. Vogelbusch, Austrian Pats. 122,951, 123,393, 126,573, 127,365, 128,825, 136,969, 142,217 (1933–1935).

Papers

92. Anheuser-Busch Inc., *Yeast: Dried Yeasts and Their Derivatives*, **1**, No. 5 (May 1951), 19 pp.

93. A. Ásvány, Kisérletügyi Közlemények, Sorozaton Kívül (Experimental Publications, Special Issue), Mezögazdasági Kiadó (Agricultural Journal), Budapest, Hungary, 1960.

94. L. Atkin, "Yeast Growth Factors," *Wallerstein Lab. Commun.* **12**, No. 37, 141 ff. (1949); republished in **25**, No. 87, 113–121 (1962).

95. L. Atkin, P. P. Gray, W. Moses, and M. Feinstein, "Growth and Fermentation Factors for Different Brewery Yeasts," *Wallerstein Lab. Commun.* **12**, No. 37, 153 ff. (1949); republished in **25**, No. 87, 122–134 (1962).

96. L. Atkin, M. Feinstein, and P. P. Gray, "Estimation of Yeast Concentration by Means of the Specific Gravity Gradient Tube," *Wallerstein Lab. Commun.* **11**, No. 35, 289 ff. (1948); republished in **25**, No. 87, 103–112 (1962).

96a. L. Baens-Arcega, "Philippine Contribution to the Utilization of Microorganisms for the Production of Foods," *Biotechnol. Bioeng. Symp. No. 1* ("Global Impacts of Applied Microbiology II"), 1969, pp. 53–62.

97. R. A. Bottomley, "Bread for the Millions," *Food Technol. in Australia* **8**, 599–611 (1956).

98. R. A. Bottomley, "Some Observations on the Continuous Process of Breadmaking," *Food Technol.* **15**, No. 10, 423–428 (1961).

98a. H. J. Bunker, "Yeast Production: The Choice of Fat-Producing and Food Yeasts," *J. Appl. Bacteriol.* **18** (No. 1), 180–186 (1955).

99. E. S. Cook and C. W. Kreke, "Studies on the Metabolic and Enzymatic Effects of Tissue Respiration," *Acta Unio Intern. Contra Cancrum* **7**, 545 (1951).

100. E. S. Cook, M. Matsuzaka, L. G. Nutini, and G. Basulto, "Offsetting the Toxicity of Nitrogen Mustard," *Proc. 5th Intern. Congr. Chemotherapy*, Wiener Medizinischen Akademie, Vienna, 1967, p. 593.

101. E. S. Cook and L. G. Nutini, personal communication, 1970.

102. E. S. Cook, K. Tanaka, and L. G. Nutini, personal communication, 1970.

103. J. R. Couch and H. D. Stelzner, "Vitaminlike Unidentified Growth Factors from Corn Distillers Dried Solubles," *Proc. Distillers Feed Research Conf.*, Distillers Feed Research Council, Cincinnati, Ohio, 1961.

104. G. I. de Becze, "Reproduction of Distillers' Yeasts," *Biotechnol. Bioeng.* **6**, 191–221 (1964).

105. G. I. de Becze, "Cell Structure and Cell Reproduction of *Saccharomyces*," 1967 Annual Meeting, American Society of Enologists (Davis, Calif.).

105a. G. I. de Becze, "Food Enzymes," *CRC Critical Reviews in Food Technology* (a new periodical, The Chemical Rubber Co., Cleveland, Ohio), in press.

106. G. I. de Becze and A. J. Liebmann, "Aeration in the Production of Compressed Yeast," *Ind. Eng. Chem.* **36**, 882–890 (1944).

107. G. I. de Becze and M. Rosenblatt, "Continuous Fermentation," *Amer. Brewer* **76**, No. 2, 11–16, 30, 32, 34 (Feb. 1943).

108. E. de la Morena, I. Santos, and S. Grisolia, "Homogeneous Crystalline Phosphoglycerate Phosphomutase of High Activity. A Simple Method for Lysis of Yeast," *Biochim. Biophys. Acta* **151**, 526–528 (1968).

109. B. Drews, G. Bärwald, and H. J. Niefind, "Some Metabolic Products of Beer Yeast and Their Significance in Fermentation Technology," *Tech. Quart. Master Brew. Assoc. Am.* **6**, No. 3, 193–197 (1969).

110. B. Drews, H. Specht, and E. Schwartz, Aromatic Alcohols Produced in the Alcoholic Fermentation of Yeast. Amino Acid Metabolism of Yeasts, *Monatsschr. Brau.* **19**, No. 4, 76–87 (1966) (in German); *Chem. Abstr.* **68**, 86099c (1968).

111. E. A. Duell, S. Inoue, and M. F. Utter, "Isolation and Properties of Intact Mitochondria from Spheroplasts of Yeast," *J. Bacteriol.* **88**, 1762–73 (1964).

111a. W. Duntze, V. MacKay, and T. R. Manney, "*Saccharomyces cerevisiae*: A Diffusible Sex Factor," *Science* **168**, 1472–1473 (1970).

111b. E. East, B. J. Smith, and D. G. Borsden, "Yeast Products and Their Role in Food Development," *Food Manuf.* (*London*) **41**, No. 9, 62, 65, 70 (Sept. 1966).

111c. Z. Fencl, "Production of Microbial Protein from Carbon Sources," *Biotechnol. Bioeng. Symp. No. 1* ("Global Impacts of Applied Microbiology II"), 1969, pp. 63–70.

112. R. R. Fowell, "The Hydridization of Yeasts," *J. Appl. Bacteriol.* **18** (No. 1), 149ff. (1955).

113. R. B. Gilliland, "Yeast Genetics in Industry" in *Recent Studies in Yeast and Their Significance in Industry*, Society of Chemical Industry, London, Monograph No. 3, 1958 (Gordon and Breach Science Publishers, New York), pp. 103–114.

113a. N. B. Gradova, A. P. Kruchkova, G. S. Rodionova, V. V. Mikhaylova, and E. M. Dikanskaya, "Microbiological and Physiological Principles in the Biosynthesis of Protein and Fatty Substances from Petroleum Hydrocarbons," *Biotechnol. Bioeng. Symp. No. 1* ("Global Impacts of Applied Microbiology II"), 1969, pp. 99–104.

114. P. P. Gray, "Some Practical Fermentation Problems in the Light of Recent Yeast Researches," *Wallerstein Lab. Commun.* **14**, No. 46, 185 ff. (1951); republished in **25**, No. 87, 180–189 (1962).

115. D. E. Green and Y. Hatefi, "The Mitochondrion and Biochemical Machines," *Science* **133**, 13–19 (1961).

116. S. R. Green, "A Review of Differential Techniques in Brewing Microbiology," *Wallerstein Lab. Commun.* **18**, No. 63, 239 ff. (1955); republished in **25**, No. 87, 246–255 (1962).

117. S. R. Green and P. J. Sullivan, "Detection of Purity of Brewers Yeast Cultures and of Variations in the Proportions of Bios Types," *Wallerstein Lab. Commun.* **22**, No. 79, 285–295 (Dec. 1959).

118. H. P. Klein, "Nature of Particles Involved in Lipid Synthesis in Yeast," *J. Bacteriol.* **90**, 227–234 (1965).

119. H. P. Klein, "Synthesis of Fatty Acids by Yeast Particles," *J. Bacteriol.* **92**, 130–135 (1966).

120. C. W. Kreke and M. St. A. Suter, "Activity Mechanism of Yeast Extracts in Stimulating Respiration," *J. Biol. Chem.* **160**, 105–110 (1945).

121. C. W. Kreke and M. St. A. Suter, "Action of Yeast Extracts in Stimulating Animal Tissue Respiration," *Studies Inst. Divi Thomae* **4**, 85–95 (1945).

122. H. Kringstad and S. Rasch, "The Influence of the Method of Preparation of Pitching Yeast on Its Production of Diacetyl and Acetoin during Fermentation," *J. Inst. Brewing* **72**, 56–61 (1966); abstr. *Wallerstein Lab. Commun.* **30**, No. 101, 68 (1967).

123. J. M. Lagomasino, "The Melle-Boinot Alcoholic Fermentation Method Using the Yeast Repeatedly," *Intern. Sugar J.* **51**, 338–339 (1949); abstr. *Wallerstein Lab. Commun.* **13**, No. 41, 174–175 (1950).

124. F. K. Lawler and J. V. Ziemba, "How Engineering Revolutionizes Old Art of Breadmaking," *Food Eng.* **28**, 74–79 (Oct. 1956).

125. W. C. Lawrence, "Volatile Compounds Affecting Beer Flavor," *Wallerstein Lab. Commun.* **27**, No. 93/94, 123–152A (1964).

126. S. Y. Lee, Y. Nakao, and R. M. Bock, "The Nucleases of Yeast, II. Purification, Properties, and Specificity of an Endonuclease from Yeast," *Biochim. Biophys. Acta* **151**, 126–136 (1968).

127. F. B. MacKenzie, W. M. Noble, and H. J. Peppler, "Torula Yeast Manufacture on the Pacific Coast," *Chemurgic Digest* **8** (No. 9), 10–13, 15 (1949).

128. V. Mäkinen, "Continuous Fermentation in the Production of Beer," *Mallasjuomat*, No. 10, 352–356 (1966); abstr. *Wallerstein Lab. Commun.* **30**, No. 102/103, 155 (1967).

129. H. G. Milgate, "Automation in the Baking Industry in Australia," *Australasian Baker* **1960** 5–16 (July).

130. B. S. Miller and J. A. Johnson, "Present Knowledge Concerning Baking Processes Employing Pre-Ferments," *Wallerstein Lab. Commun.* **21**, No. 73, 115–132 (1958).

131. J. R. Mor and A. Fiechter, "Continuous Cultivation of *Saccharomyces cerevisiae*, II. Growth on Ethanol under Transient-State Conditions," *Biotechnol. Bioeng.* **10**, 787–800 (1968).

132. G. P. Murphy, *et al.*, "The Effect of PCO (Yeast Respiratory Stimulator) on Treated and Untreated Dog Renal Allografts," *J. Surg. Oncol.* **1**, 247–259 (1969).

133. Y. Nakao, S. Y. Lee, H. O. Halvorson, and R. M. Bock, "The Nucleases of Yeast, I. Properties and Variability of Ribonucleases," *Biochim. Biophys. Acta* **151**, 114 (1968).

134. F. W. Nordsiek, "Bakers' Yeast: Unique Product," *Food Ind.* **23**, No. 2, 101–108 (Feb. 1951).

135. L. C. Norris, "Unidentified Chick Growth Factors in Distillers Dried Solubles," *Proc. Distillers Feed Conf.*, Distillers Feed Research Council, Cincinnati, Ohio, 1954.

136. K. Nosiro and K. Ouchi, "Fermentation Activity of Yeasts and Its Alcohol Tolerance, I. Alcohol Tolerance of the Fermentation Activity of Yeasts," *J. Soc. Brew. Japan* **57**, 824 (1962); abstr. *Wallerstein Lab. Commun.* **28**, No. 95, 87 (1965).

137. L. G. Nutini, R. F. Naegele, and E. S. Cook, "Offsetting the Toxicity of Kanamycin," *Proc. 5th Intern. Congr. Chemotherapy*, Wiener Medizinischen Akademie, Vienna, 1967, p. 225.

137a. D. Rogers and M. N. Mickelson, "Vitamin B Content of Sugar Beets and By-Products," *Ind. Eng. Chem.* **40**, 527–529 (1948).

138. R. J. Robinson, T. H. Lord, J. A. Johnson, and B. S. Miller, "Studies on the Decrease of the Bacterial Population in Preferments," *Cereal Chem.* **35**, 306–317 (1958).

138a. A. H. Rose, "Recent Research in Yeast Biochemistry," *Wallerstein Lab. Commun.* **26**, No. 89, 21–37 (1963).

138b. S. O. Rosenqvist, "Production of Yeast in Sweden for Nutritional Purposes," *Food Ind.* **16**, No. 6, 74–5, 118 (1944).

139. J. F. T. Spencer, "Production of Polyhydric Alcohols by Yeasts," *Progress in Industrial Microbiology*, Vol. 7, The Chemical Rubber Co., Cleveland, 1968.

140. G. S. Sperti, E. S. Cook, L. G. Nutini, and G. P. Murphy, "Use of PCO to Ameliorate Toxic Effects of Nitrogen Mustard and Azathioprine," *Ohio State Med. J.* **66**, 163 (1970).

141. F. B. Strandskov, "Yeast Handling in a Brewery in the United States," *Wallerstein Lab. Commun.* **28**, No. 95, 29–32 (1965).

142. G. Svihla, J. L. Dainko, and F. Schlenk, "Ultraviolet Microscopy of the Vacuole of *Saccharomyces cerevisiae* during Sporulation," *J. Bacteriol.* **88**, 449–456 (1964).

143. H. Tanaka and H. J. Phaff, "Enzymatic Hydrolysis of Yeast Cell Walls, I. Isolation of Wall-Decomposing Organisms and Separation and Purification of Lytic Enzymes," *J. Bacteriol.* **89**, 1570–1580 (1965).

144. P. Tauro and H. O. Halvorson, "Effect of Gene Position on the Timing of Enzyme Synthesis in Synchronous Cultures of Yeast," *J. Bacteriol.* **92**, 652–661 (1966).

144a. A. C. Thaysen and M. Morris, "Preparation of a Giant Strain of *Torulopsis utilis*," *Nature* **152**, 526 (1943). A. C. Thaysen, "Production of Food Yeast," *Food (London)*, **14**, 116–119 (1945). British Colonial Food Yeast, Ltd., England, *Food Yeast*, H.M.S.O., London, 1944.

145. G. W. Thomas, J. C. Fardon, S. L. Baker, and E. S. Cook, "Bacteriological and Toxicity Studies on a Wound-Healing Ointment," *J. Am. Pharm. Assoc., Sci. Ed.* **34**, 143 (1945).

146. J. A. Thorn and J. W. Ross, "Determination of Yeast Growth in Doughs," *Cereal Chem.* **37**, 415–421 (1960).

146a. D. I.-C. Wang, "Developments in Agitation and Aeration of Fermentation Systems," *Progress in Industrial Microbiology*, Vol. 8, The Chemical Rubber Co., Cleveland, 1969.

147. D. M. Watson, "Culture and Maintenance of Lager Yeast," *Brewers' Guardian* **93**, No. 12, 17–24 (1964); abstr. *Wallerstein Lab. Commun.* **28**, No. 97, 254 (1965).

148. R. Weimberg and W. L. Orton, "Elution of Exocellular Enzymes from *Saccharomyces fragilis* and *Saccharomyces cerevisiae*," *J. Bacteriol.* **91**, 1–13 (1966).

149. L. J. Wickerham, "Taxonomy of Yeasts," *U.S. Dept. Agr. Tech. Bull.* No. 1029 (1951).

150. G. Williamson, "Converts Waste Yeast to Profitable Products," *Food Eng.* **28**, No. 2, 64–65 (1956); abstr. *Wallerstein Lab. Commun.* **20**, No. 70, 287 (1957).

Commercial Publications

151. Amber Laboratories, Juneau, Wisconsin, "Amber Fermentation Nutrients," commercial pamphlet, 1969.

152. Cerevisiae Yeast Institute, Chicago, Illinois, "The Food Value and Use of Dried Yeast," commercial pamphlet, 1963, 34 pp.

153. Distilled Spirits Industry, Washington, D.C., "The Distilled Spirits Industry—1968 Statistical Review," a trade report, 1969.

154. The English Grains Co. Ltd., Burton-on-Trent, Staffordshire, England, "How to Use Yeast Products in Food Manufacture and Flavouring," commercial pamphlet, 1968, 30 pp.

155. Universal Foods Corporation, Milwaukee, Wisconsin, "Nutritional Yeasts," commercial pamphlet, 1969.

George I. de Becze
St. Thomas Institute for Advanced Studies

YLANG YLANG. See Oils essential, Vol. 14, p. 214.

YTTERBIUM, Yb, YTTRIUM, Y. See Rare earth elements.

Z

ZECTRAN. See Insecticides, Vol. 11, p. 711.

ZEOLITES. See Ion exchange.

ZEPHIRAN. See Surfactants.

ZETAX. See Rubber chemicals.

ZINC AND ZINC ALLOYS

Zinc, Zn, at wt. 65.38, at. no. 30, is a bluish-white almost silvery metal with a hexagonal crystal structure. It is in Group IIB of the periodic table along with cadmium and mercury and forms divalent compounds only. There are five stable isotopes, 64, 66, 67, 68, and 70 and their abundance has been reported as 48.9%, 27.8%, 4.1%, 18.6%, and 0.6%, respectively, but the composition varies slightly among samples from different areas. There are also ten other radio-active isotopes varying in atomic mass from 64 to 72, all with relatively short half-lives.

Zinc is usually found in nature as the sulfide. Most other zinc minerals have probably been formed as oxidation products of the sulfide. Zinc sulfide is often associated with the sulfides of other elements, especially those of lead, cadmium, iron, and copper. This association is particularly intimate in many cases and, as a consequence, the economic exploitation of some of the more complex ore bodies was delayed until suitable methods for separating the various constituents were developed.

Zinc has been in use as a constituent of brass and bronzes for more than 2000 years. It is probable that in India and China certain workers in metals recognized zinc as a distinct metal as early as 1000 AD but it was not until the time of Paracelsus (1490–1541) that zinc as the metal received any attention in Europe. As a matter of fact its recognition there as a separate metal was rather slow. An ancient method of brass production used in Europe involved heating a chunk of copper with calamine (old-world designation for zinc carbonate) and a reducing agent such as charcoal. The art of smelting zinc ores for their zinc content apparently originated in India, traveled to

China, and was carried from there to Europe by the Portuguese in the early seventeenth century. About a century later, actual commercial production began in England.

Since about 1960, the major application for zinc has become its use in making low-cost, high-quality, high-finish zinc-base diecasting alloys. Previously the most extensive use had been founded upon the ability of zinc to limit and control the corrosion of iron and, to some extent, certain other metals. To accomplish this, various galvanizing and cathodic protection procedures were employed, all based upon the fact that zinc in intimate contact with iron, steel, and certain other metals, corrodes preferentially, thus sacrificially preventing the corrosion of these metals and providing them protection. This now is the second most important use for zinc. Considerably behind in third place is the use of zinc in making brass and bronze products. (See under Uses of zinc metal.)

Properties

Table 1 gives many of the physical properties of zinc. Zinc is a bluish-white metal when freshly fractured, of low to intermediate hardness. The scleroscope hardness of rolled zinc of about 99.94% purity is in the range 13–15; cast zinc of the same impurity content is slightly softer. On the Mohs hardness scale for minerals, zinc rates about 2.5, or between gypsum and calcite, and about the same as copper. The usually encountered impurities, even in small amounts, raise the hardness of zinc at room temperature with the exception of lead.

Zinc is favored as a corrosion retarder for the more vulnerable metals, partly because it corrodes preferentially to them under ordinary conditions. For this reason it is preferred to metals like tin for coating steel, except where corrosion products of the coating metal may be absorbed by foodstuffs. If pinholes appear in the tin coating, the presence of the tin accelerates attack on the steel by the corrosive medium. If holes develop in the zinc coating, the presence of the zinc still protects the exposed steel for several millimeters inside the periphery of the uncoated area.

Zinc is also favored over some other metals as a corrosion retarder because its corrosion products, like those of cadmium and aluminum, are white when visible at all.

The effect on corrosion resistance of removing the last small amounts of impurities is even more pronounced than the effect on hardness, as reported in 1929 by Tainton in describing some of the difficulties encountered in the early development of the electrolytic zinc process:

"When we were making electrolytic zinc about five years ago, we used to make some determinations by dissolving the zinc in sulfuric acid, but after we began to use the lead–silver anodes, the solution in sulfuric acid got slower and slower until finally it took about a month, so we had to give up that method. We then went to solution in hydrochloric acid, but as the zinc got purer, that test began to take from several days to a week to dissolve the sample. Now we use nitric acid entirely for the solution of the zinc."

Although this account typifies the behavior of zinc in acid solutions, no such difference is noted between various grades of zinc in respect to resistance to atmospheric corrosion.

Clearly the physical—as well as the chemical—properties of zinc are greatly influenced by the impurities in the metal. Prior heat and mechanical treatment also affect the properties. Commercial zinc is too brittle at room temperature to be cold

Table 1. Physical Properties of Zinc

Property	Value
melting point,	
°C	419.5
°K	692.7
boiling point, 1 atm	
°C	907.0
°K	1180.0
density, g/cm³	
solid at 25°C	7.133
solid at 419.5°C	6.83
liquid at 419.5°C	6.62
heat capacity, C_p	
solid, 298.16–692.7°K, cal/(mole) (°K)	$5.35 + 2.40 \times 10^{-3} T$
liquid, 692.7–1200°K, cal/(mole) (°K)	7.50
gas (monatomic), cal/mole	4.969
heat of fusion at 419.5°C, cal/mole	1765
heat of vaporization at 907°C, cal/mole	27,430
vapor pressure of liquid, atm	$9.843 - 6755/T - 1.32$
	$\log T - 0.06 \times 10^{-3} T$
thermal conductivity, cal/(sec)(cm²)(°C/cm)	
solid at 18°C	0.27
at 419.5°C	0.23
liquid at 419.5°C	0.145
at 750°C	0.135
coefficient of thermal linear expansion,	
polycrystalline (20–250°C), per °C	$39.7 + 10^{-6}$
electrical resistivity, R, ohm-cm $\times 10^{-6}$	
polycrystalline ($T = 0$–100°C)	$5.46(1 + 0.0042\ T)$
magnetic susceptibility (diamagnetic)	
polycrystalline at 20°C, cgs electromagnetic units	0.139×10^{-6}
surface tension (liquid), dyn/cm	$758 - 0.09(t - 419.5°C)$
crystal structure, close-packed hexagonal, Å	
a	2.664
c	4.9469
c/a	1.856

rolled, but, depending upon the impurity content (particularly iron), it becomes sufficiently ductile between 100 and 150°C to be rolled into sheet or drawn into wire. Pure zinc is ductile even at room temperature. Reports in the early literature that zinc became brittle at above 275°C are thought to be due to eutectics with impurities such as tin. Where ductility is not a prime consideration, zinc is almost always alloyed to increase its strength. Pure zinc does not possess a definite yield point such as do most of the commonly used structural metals. Under a sufficient constant load it creeps even at room temperature. Although it is a rough rule not to subject a zinc alloy of 35,000 psi ultimate strength to more than 10,000 psi stress, it is better to design from actual creep data where possible.

Chemical Properties. Zinc is not significantly attacked by dry air at room temperature but the rate of attack begins to increase rapidly above 225°C. Presence of moisture in the air permits attack to occur at room temperature, the presence of carbon dioxide or sulfur dioxide accelerating it. Zinc burns with a brightly luminous blue-green flame when heated in air, forming the oxide. Zinc dust can spontaneously inflame in moist air. Zinc is oxidized at red heat by steam and carbon dioxide, the latter being reduced to carbon monoxide.

The final product of normal atmospheric corrosion is a hydrated basic carbonate. The ratio of $ZnO:CO_2:H_2O$ in this product has been variously reported as 5:4:8, 5:2:3, 4:1:3, and 4:1:4, strongly suggesting that different external conditions at the time of formation affect the final composition. When zinc is oxidized in moist air, some hydrogen peroxide may be formed. Addition of a reagent preventing the formation of hydrogen peroxide inhibits the reaction $Zn + H_2O_2 \rightarrow Zn(OH)_2$ and hence decreases the corrosion rate by this mechanism. The purest zinc corrodes under water only if air is present, probably because of the action of dissolved carbon dioxide.

Zinc vapor reduces CO_2 to CO; the extent of reduction is dependent upon temperature. Above about 1100°C the equilibrium amount of CO_2 present is negligible in the presence of excess carbon.

Dry fluorine, chlorine, or bromine do not attack zinc at room temperature, but the presence of water vapor causes zinc foil to inflame in these elements. The gaseous hydrohalides, such as HCl, and their aqueous solutions readily attack zinc, forming the zinc halide and hydrogen. Zinc has definite but limited solubility in its molten chloride and bromide. Hydrogen chloride attacks it even in liquid hydrogen cyanide.

Zinc reacts with most mineral acids, the rate of dissolution increasing from sulfuric to hydrochloric to nitric. The purer the zinc, that is, the lower the content of the more inert metals, the more slowly it corrodes. As little as 0.01% of some impurities markedly increases corrosion, as does a water solution of copper sulfate.

Zinc and sulfur are quite reactive under proper conditions ($\Delta H_{25°C} = -48.5$ kcal/mole. A mixture of powdered zinc and sulfur reacts with explosive violence when warmed. When sulfur reacts with larger masses of zinc, however, a protective layer of the sulfide is formed on the metal, considerably reducing the speed of reaction.

Hydrogen sulfide attacks zinc at ordinary temperatures, forming a coating of zinc sulfide which prevents further attack; at higher temperature the reaction is quite vigorous.

Even when zinc is in the vapor state, no reaction with nitrogen can be detected; however, zinc reacts with ammonia at red heat, forming the nitride, Zn_3N_2. Even with zinc powder, the reaction is not complete.

Some typical reactions of zinc and the heat involved are listed below.

$$Zn + \tfrac{1}{2}\,O_2 \rightarrow ZnO \qquad\qquad \Delta H = -83.156 \text{ kcal/(g)(mole)}$$

$$Zn + H_2SO_4 \rightarrow ZnSO_4 + H_2 \qquad\qquad \Delta H = -233.88 \text{ kcal/(g)(mole)}$$

$$Zn + Cl_2 \rightarrow ZnCl_2 \qquad\qquad \Delta H = -99.40 \text{ kcal/(g)(mole)}$$

$$Zn + S \rightarrow ZnS \qquad\qquad \Delta H = -48.5 \text{ kcal/(g)(mole)}$$

$$Zn + 2\,(C_2H_5) \text{—} \rightarrow Zn(C_2H_5)_2 \qquad\qquad \Delta H = -5.0 \text{ kcal/(g)(mole)}$$

Zinc is an active reducing agent for many ions, such as ferric, manganate, and chromate. Another example of its reducing power is seen in the attack on zinc by nitric acid. Instead of only zinc nitrate, by-products, including nitrogen oxides, elemental nitrogen, and even ammonium nitrate, and hydroxylamine, are formed. In hot caustic solutions zinc dissolves to form zincates. See Zinc compounds.

The property that gives zinc its tremendous economic importance is that, with a standard electrode potential of 0.76, it is electropositive to most of the common structural metals, aluminum and magnesium being exceptions. This means that zinc can replace other metals in solution, as follows:

$$Zn + CuSO_4 \rightarrow ZnSO_4 + Cu$$

This overall reaction is the basis of the Daniell cell, at one time an important source of direct current. The reaction

$$Zn + CdSO_4 \rightarrow ZnSO_4 + Cd$$

is the main commercial method for removing and recovering cadmium from impure zinc sulfate solution.

Analysis

Even though zinc has been an article of commerce for many centuries, its analytical chemistry is not as clear-cut, nor as definitive as that of most other industrially useful metals. This is due in considerable degree to the chemical properties of the element, and also to the fact that relatively few of its reactions are even moderately specific and free from interference by the metals commonly associated with it, such as cadmium, copper, lead, iron, etc.

Qualitative. In the usual procedure for qualitative analysis, zinc is precipitated in Group III. The precipitate is treated with 1.5 N HCl; the insoluble material, containing nickel and cobalt, is filtered off and the iron, aluminum, and chromium are precipitated by the addition of NH_4OH or by a slurry of barium carbonate. The residual filtrate is acidified slightly with H_2SO_4, then made alkaline with NaOH (to precipitate the manganese) and filtered; H_2S is added to the final filtrate to precipitate white zinc sulfide. Additional tests can be used to verify the presence of zinc in the white precipitate. Zinc can also be identified by its spark spectrum.

For the detection of zinc in small amounts, the most generally applicable qualitative test involves the use of dithizone in the presence of a complexing agent. It permits the detection of 0.8 ppm in the presence of small amounts of cadmium, copper, lead, and nickel.

Quantitative. The quantitative analytical methods for zinc fall mainly under one of the following classifications: (*1*) wet chemical (titrimetric); (*2*) gravimetric, following separation and precipitation; (*3*) electrochemical, including polarographic; (*4*) spectrographic; and (*5*) photometric, including colorimetric.

The zinc emission spectrum is rather simple but interferences from other elements, such as cadmium, manganese, chromium, etc, can cause complications. The most sensitive lines, 2138.6 and 6362.3 Å, occur in regions of poor photographic properties. The latter line, along with several others, is satisfactory for visual work. With photographic recording, the line ordinarily used is 3345.0 Å.

Zinc also gives a well-developed polarographic half-wave potential in many electrolytes (see Vol. 7, p. 748). It is best determined polarographically in basic, neutral, or very slightly acidic solutions because of interference that can occur due to the fact that zinc's reduction potential is well above that of most of the metal ions. Certain complexing reagents, such as ammonia, hydrazine, pyridine, etc, have proved useful in certain cases, keeping interference with the zinc waves at a minimum. Preliminary chemical separations can also be most helpful.

In wet chemical analysis, there are no known methods which separate zinc easily and effectively from most of the other elements. Hence, many steps are necessary if the material contains other elements. However, when reasonably pure zinc compounds are being analyzed, many of those steps may be eliminated and the procedure simplified considerably.

Most zinc-bearing materials can usually be dissolved in hydrochloric or nitric acid, or a combination of the two. With slags, residues, or complex silicates, it may be necessary to remove the silica by fuming with sulfuric or hydrofluoric acids; in refractory cases, fusion with sodium carbonate is used. The combined solutions are taken to dryness with sulfuric acid.

A common routine method, used where not too many interfering metals are present, consists in dissolving the residue as described above in dilute (1:5) hydrochloric acid, adding ammonium chloride, and precipitating the iron by adding ammonium hydroxide to the hot solution. If considerable iron is present and a high degree of accuracy is desired, ferric hydroxide is filtered off, dissolved in hydrochloric acid, and reprecipitated with ammonium hydroxide (to recover occluded zinc). The combined filtrates from the iron precipitations are made very slightly acid with hydrochloric acid, treated with test lead or aluminum (to plate out the less positive metals), decanted off, and titrated at 70°C with standard potassium ferrocyanide to precipitate zinc ferrocyanide, using uranium nitrate as an indicator. If the filtrates are acidified with sulfuric instead of hydrochloric acid, diphenylamine can be used as an indicator. Any cadmium in the original is titrated with the zinc, so that a correction must be made if the results are to be accurate in the presence of this metal.

For high-precision work, the zinc-bearing solution is adjusted to 0.5 N acidity with sulfuric acid and the heavy metals precipitated with hydrogen sulfide. If necessary, the sulfide precipitate is redissolved and reprecipitated under the same conditions; the filtrates are combined and adjusted to 0.01 N acidity with ammonium hydroxide and sulfuric acid. The zinc is then precipitated with hydrogen sulfide and the resultant zinc sulfide precipitate washed.

The zinc sulfide can be dissolved in acid and titrated as above with potassium ferrocyanide or it can be converted back to the sulfide or to the oxide, sulfate, a suitable phosphate—preferably zinc ammonium phosphate, ($ZnNH_4PO_4$)—or to other compounds and determined gravimetrically. It has been reported that precipitation as either the sulfide or phosphate is as accurate as any gravimetric procedure.

It may be necessary to take additional steps to separate certain interfering elements—depending on the amount present in solution—from the zinc. New methods and specific reagents are suggested in the literature from time to time but the following methods, many of which have been in use for years, have proved to be quite satisfactory.

Nickel. Precipitation of ZnS by passing H_2S through a cyanide solution acidified with acetic acid. Alternatively, the nickel can be precipitated by dimethylglyoxime, the zinc remaining in solution. Another method is the removal of the nickel as salicylaldoximate in the presence of hexamethylenetetramine at pH 5.4.

Alkali and Alkaline Earth Metals. Careful precipitation of ZnS with ammonium sulfide.

Aluminum, Titanium, Zirconium, Niobium. Precipitation of ZnS in ammoniacal tartrate solution with H_2S. Aluminum may also be removed as the benzoate at a pH of 3.5–4.0.

Iron, Vanadium, Tin. Precipitation of these metals from strong (10%) acid solution by cupferron (ammonium salt of nitrosophenylhydroxylamine), leaving zinc in solution.

Copper. Electrolysis of sulfuric or nitric acid solution at 0.5 A and 2 V to remove copper. Copper may also be separated as the sulfide by the acid hydrolysis of thiosulfate and as an amine complexed with ammonium thiocyanate and isoquinoline at pH 4.0.

Tin. Oxidation to SnO_2 by nitric acid and filtering the precipitated SnO_2 from the zinc solution.

Cadmium. Precipitation of CdS in hot 3 N sulfuric acid by H_2S. The zinc remains in solution. Also used is precipitation of the cadmium as the sulfide by the acid hydrolysis of thiocyanate.

Thallium. Reduction of weak acid solution with SO_2 and precipitating the thallous ion with KI as thallous iodide. Zinc remains in solution.

Indium. Precipitation of indium by reaction with a slurry of barium carbonate, leaving zinc in solution.

For determination of small amounts of zinc, a number of methods are in use. As a general rule colorimetric, polarographic, and emission-spectrum methods are most satisfactory. Care must be taken to separate the zinc from interfering elements.

Colorimetric methods are among the most sensitive for trace amounts of zinc, dithizone (diphenylthiocarbazone) being a particularly favorite reagent. The difficulty with this method is that other metals, frequently accompanying the zinc, interfere and there must be appropriate separations. Masking reagents also prove helpful. 8-Hydroxyquinoline is another colorimetric reagent that is sometimes used.

Zincon gives a blue color with small amounts of zinc in aqueous solution at pH 9 but copper and certain other elements interfere and therefore must be removed. However, without interfering elements, a rapid and simple colorimetric determination is possible. Its sensitivity is only about half that of the dithizone procedure.

Turbidimetric methods are no better for zinc than with many other elements. It is difficult to prepare, stabilize, and reproduce the colloidal or near-colloidal suspensions on which this type of determination is based.

The emission-spectrum method is considered to be less sensitive for trace amounts than either the colorimetric or polarographic procedures. The lower limit of sensitivity is said to be about 0.01% zinc although the use of certain techniques in some matrices may permit satisfactory analysis at lower concentrations.

Gravimetric methods have been augmented by a number of organic precipitation reagents and also by certain inorganic complexes. An example of the latter is the use of a thiocyanomercurate complex made with a 10–20% excess of KCNS over the 4:1 theoretical $KCNS/HgCl_2$ molecular ratio, precipitating zinc as the tetrathiocyanomercurate(II), $ZnHg(CNS)_4$, which may be weighed or dissolved in dilute HCl and titrated with KIO_3 solution. However, again small amounts of elements such as cadmium, mercury(I), copper, bismuth, and manganese interfere.

Determination of Impurities in Metallic Zinc. The analysis of metallic zinc for impurities becomes more difficult with increasing purity of the metal. With greater use of purer grade of the metal, precise analytical procedures are necessary not only to guide the production efforts but also to certify as to the purity of the zinc.

The zinc content of slab zinc, high-purity metal, and high-zinc alloys is "determined" by difference, ie the impurities are determined and their sum subtracted from 100% to give the zinc content. Analysis of zinc metal for impurities usually follows the tests recommended by the ASTM. Most of these analyses are conducted spectrographically (ASTM Designation E-26) with a high degree of accuracy. Chemical methods are described in ASTM Designation E-40. In many cases, wet chemical and colorimetric methods are the simplest and they have a high degree of accuracy. Lead is determined either electrolytically (as lead dioxide, PbO_2, deposited on the anode) or gravimetrically as lead sulfate.

Cadmium is separated from the zinc by multiple sulfide precipitations under close pH control. The final cadmium sulfide is either weighed as CdS or dissolved and deposited electrolytically from a suitable electrolyte on a platinum cathode.

Iron may be separated, after solution in hydrochloric and nitric acids, by double precipitation with ammonium hydroxide. The final iron precipitate is dissolved in sulfuric acid, the iron reduced to the ferrous state by the addition of zinc, and the ferrous iron in the solution titrated with potassium permanganate. The iron may also be determined colorimetrically as ferric iron.

Occurrence

Zinc is widely distributed in nature in the form of its various minerals and occurs in small amounts in nearly all igneous rocks, mainly as a substituent for iron. Its abundance is usually estimated to be about 0.013% of the earth's crust. The principal ore mineral of zinc is sphalerite, cubic ZnS, commonly called zinc blende or blende, and by miners "jack." Ideally, it contains 67.1% zinc. A hexagonal form of the sulfide, wurtzite, is much less abundant; however, it is the stable form above 1020°C. Individual specimens of both forms have been known to fluoresce under ultraviolet irradiation, but not all specimens do so.

Sphalerite is resinous in appearance, the color varying from light tan to black in natural specimens (pure artificial ZnS is a white pigment of high hiding power), depending primarily upon the content of iron crystallized substitutionally in the lattice. Above a ratio of Fe:Zn of 1:8 the mineral is usually referred to as marmatite. Above a ratio of 5:6, equivalent to a total iron content of 25%, the sphalerite structure no longer exists.

Next to iron, cadmium is the most common substitutional impurity in the sphalerite lattice; typical cadmium content of zinc concentrates runs about 0.3%. Cadmium is also found associated with zinc as greenockite, CdS (after Greenock, Scotland). Cadmium is about 1/200 as abundant as zinc, the ratio varying somewhat with the source of ore. Germanium and gallium occur in sphalerite deposits formed at relatively low temperatures. Traces of indium and tin occur in sphalerite from high-temperature deposits.

Zinc minerals are commonly associated with lead minerals. The ratio Zn:Pb varies widely, however. Within the state of Missouri the ratio varies from less than 1:7 in the Southeast Missouri lead belt to about 5:1 in the Joplin Tri-state district. It is of interest that the former ores are found in dolomitic limestone, the latter in what was originally limestone but has been altered to a chert usually containing 90% or more SiO_2. Other commonly associated minerals are calcite ($CaCO_3$), dolomite ($CaCO_3 \cdot MgCO_3$), pyrite and marcasite (FeS_2), quartz (SiO_2), chalcopyrite ($CuFeS_2$), and barite ($BaSO_4$), although barite is more commonly associated with lead deposits.

Almost all the oxidized forms of zinc minerals can be considered alterations from the sulfide, including

ZnO	zincite
$ZnSO_4 \cdot 7H_2O$	goslarite
$ZnCO_3$	smithsonite (or calamine in Europe)
$Zn_4Si_2O_7(OH)_2 \cdot H_2O$	hemimorphite (or calamine in America, called electric calamine in Europe)
Zn_2SiO_4	willemite
$(Zn,Mn)O \cdot Fe_2O_3$	franklinite
$2ZnCO_3 \cdot 3Zn(OH)_2$	hydrozincite

Most of these oxidized minerals are minor sources of zinc, although franklinite and zincite are major deposits in the ores of Sussex County, N.J.

Zinc sulfate is more commonly found in limestone, while the carbonate is more commonly found in dolomitic rock. This is perhaps explained on the basis that the sum of the heats of formation of magnesium sulfate and zinc carbonate is greater than that for zinc sulfate and calcium carbonate. Hence, when magnesium is present, as in dolomite, a double decomposition reaction appears to occur between zinc sulfate and magnesium carbonate.

Outline of Zinc Metallurgy. Since practically all zinc ores as mined are too low in zinc content for direct reduction processes, they must first be concentrated. After concentration, the first step in the extractive metallurgy of zinc is virtually always roasting the concentrate. If the roasted concentrate, or calcine, is to be subjected to high-temperature reduction with carbon, it is usually first sintered in order to minimize dusting loss, volatilize impurities, and permit better circulation of reducing gases around the sintered particles. If the calcine is to be reduced electrolytically, it is dissolved directly in the spent electrolyte, consisting primarily of dilute sulfuric acid, just returned from the electrolytic cells.

The choice between pyrometallurgical and electrolytic methods depends on such factors as the purity of raw material, the labor requirements per ton of zinc produced, existing investment in other processes, and the demand for especially pure grades. The method chosen is also influenced by the relative costs of fuel and electric power, but to a lesser extent than for some other metals. For example, aluminum requires 10–12 kWh/lb produced; a typical corresponding figure for electrolytic zinc is 1.5. The direct power requirement for electrothermic zinc smelting (carbon reduction, heat supplied electrically) in vertical shaft furnaces is even less, about 1.3.

Concentration

Crushing of zinc-bearing ore is accomplished by means of standard types of jaw, gyratory, and cone crushers, after which the usual practice is for all of the ore to be ground in ball mills to approx 60% − 325 mesh and separated by flotation. Certain western ores, notably many in British Columbia and Idaho, contain lead and zinc too intimately mixed for satisfactory separation even by flotation; in this case final separation involves sulfuric acid leaching at an electrolytic zinc plant or use of the Imperial furnace. Such ores were the primary reason for the development of electrolytic zinc recovery methods in North America. See Flotation; Gravity concentration; Size reduction; Size separation.

Roasting and Sintering

Naturally oxidized ores of zinc are now so rare that commercial smelters depend exclusively on sulfide concentrates. Therefore roasting in some form is practiced universally on zinc concentrates.

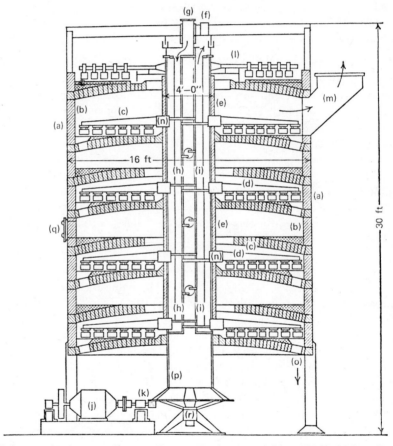

Fig. 1. Wedge roasting furnace. Courtesy Process Equipment Division, The Bethlehem Corporation.

LEGEND:

(a) Furnace shell	(g) Air inlet	(m) Gas outlet
(b) Refractory lining	(h) Supply air duct	(n) Arm holder
(c) Rabble arm	(i) Discharge air duct	(o) Calcine discharge
(d) Rabble blades	(j) Motor	(p) Manhole
(e) Central shaft	(k) Bevel gears	(q) Inspection door (hinged)
(f) Air outlet	(l) Drying hearth	(r) Main bearing

Although calculations indicate that zinc could be obtained directly from the sulfide by a mixture of lime and carbon at very high temperatures, no process involving direct reduction from the sulfide is known to have operated commercially. Hence all sulfide concentrates are roasted (oxidized) to remove sulfide sulfur before smelting. Unlike copper and lead, smelting of zinc cannot be accomplished by reacting the sulfide with an oxide or sulfate. Some of the reactions that occur in roasting are the following:

$$2\ ZnS + 3\ O_2 \rightarrow 2\ ZnO + 2\ SO_2$$
$$ZnO + SO_3 \rightarrow ZnSO_4$$

It is necessary to remove almost all the sulfur originally present as sulfide. In electrolytic zinc plants the formation of some zinc sulfate is desirable, to supply sulfate to compensate for normal operation losses. However, for pyrometallurgy, zinc sulfate must be removed.

Another reaction that occurs is the formation of zinc ferrite (ferrate(III)) with ferric oxide (see Vol. 12, p. 40) of indefinite composition, $ZnO.xFe_2O_3$. In electrolytic plants the formation of ferrite is a disadvantage, for it is almost insoluble in the weak sulfuric acid of the leach liquor, and its zinc content is partly to wholly unrecoverable.

Because the rate of sulfur elimination in a multiple-hearth-type roaster is approximately proportional to the sulfur content of the residue sulfide in the material at the instant considered, the elimination of the last few percent of sulfur requires a disproportionally longer time. Hence a Wedge-type roaster (see Fig. 1) that eliminates all but 6–8% sulfur from up to 350 tons of copper concentrates per day roasts only about 50–60 tons of zinc concentrate per day to about 2% sulfur. This illustrates the difference in the basic objectives of copper and zinc roasting, although not all the variables concerned are considered.

ROASTING

The history of the development of zinc concentrate roasting is closely associated with the development of chemical industries and also with population growth and urban development. In earlier days zinc concentrates were roasted by simply piling on a heap enough wood for ignition and starting a fire. The pile was allowed to stand until it burned itself out and the sulfur dioxide was discharged in the atmosphere. But, as communities became better organized and population more concentrated, legislation was enacted to prevent the wanton release of roasting gases, with the accompanying lung irritation and damage to vegetation. At the same time, communities were growing and the demand for sulfuric acid was increasing.

The first plant built to make sulfuric acid from gases recovered from roasting blende was that of the Rhenania Chemical Works, Stolberg (near Aachen), Germany, in 1855. However, it never recovered more than about 60% of the sulfur content of the blende, and was eventually replaced by the Rhenania furnace, constructed at the same works. This furnace, consisting of a series of long, straight hearths, between which were located firing flues employing coal as the source of heat, permitted much greater conservation of heat and sulfurous gas than did previous models (the outer walls were 20 in. thick).

The Rhenania furnace became the prototype of several more furnaces, each of which represented the trend toward mechanization. In the United States the design of the Hegeler roaster was completed in 1881 by Edwin C. Hegeler of the Matthiessen and Hegeler Zinc Co., LaSalle, Ill. It was the first furnace in the United States to be used for roasting zinc blende to provide SO_2 for sulfuric acid production. It was basically very similar to the Rhenania furnace, except that an attempt had been made to replace hand rabbling by drawing rabbling rakes hitched to long rods through the hearths. It represented a great advance at the time and eventually became the most used single type of roaster for zinc concentrates in the United States. It was not favored in Europe because of lower labor costs there.

Multiple-Hearth Roasters. The Hegeler furnace was eventually replaced by the cylindrical, multiple-hearth roasters, such as the McDougall, Wedge, and Nichols-Herreshoff. Only the Wedge furnace (see Fig. 1) is described here, since the other two differ from it only in minor mechanical details. The furnace consists of a vertical cylinder of boiler plate lined with refractory brick. The hearths are of arched construction. The modern tendency is toward eight, nine, or even more hearths. The hearth floor, walls, and furnace roof are lined with firebrick in order to protect the metal parts, as well as to conserve heat. There is a central vertical shaft of boiler plate (about 4 ft in diam) lined on the outside with brick, to which are attached the rabble arms, two for each hearth. (In the figure, rabble arms are shown only on every other hearth.) The central shaft also contains pipes, which conduct cooling air or water to the arms. To each arm are fixed seven to nine rakes or rabbles. The rotation of this shaft, which is driven by a motor and train of gears beneath the furnace, is on the order of $\frac{1}{3}$–1 rpm. One distinct advantage of this type of furnace lies in the fact that the operators may work in the shaft in order to remove or repair the rabble arms, which, if the temperatures employed are high, tend to warp and corrode.

The ore is fed upon the upper hearth, which serves to dry the crude material. The rabbles are so adjusted that the ore is gradually moved from the outer edge of the upper hearth toward the center and falls through a drop hole upon hearth 1. The rakes there move it across this hearth to a slot near the periphery, through which it drops onto hearth 2, and thus in a zigzag fashion the ore progresses through the furnace until finally it drops into a car or conveyor beneath the lowest hearth. All furnaces have doors on each hearth for visual observation, repairs, and admission of air.

Flash Roasting. It was found, in operating multiple-hearth roasters, that some instantaneous desulfurization took place in suspension during the drop of the concentrates from one hearth to the next. This discovery led to experiments aimed at suspending all the concentrates in a stream of air sufficiently hot to accomplish desulfurization in a comparatively short time.

The development of the Cominco process for the flash roasting of zinc concentrates was carried out by the Consolidated Mining and Smelting Company of Canada Limited at Trail, B.C., Canada. In order to provide a combustion chamber in which the concentrates would be roasted in suspension, the second, third, fourth, and fifth hearths were removed from a standard 25-ft diam, seven-hearth Wedge roaster. The top, or dryer, hearth (not counted as one of the seven), was enclosed, and this hearth, as well as the first roasting hearth, was used to dry the concentrates. Part of the hot gas from the combustion chamber was drawn over these two hearths to provide heat for drying, and the moisture-laden gas was then mixed with combustion air and returned to the main chamber.

In operation, the dried concentrates are discharged from the second drying hearth and fed to a ball mill to disintegrate agglomerations produced by drying and, frequently, to grind finer. The dried, comminuted particles are discharged from the mill and carried by means of an air current or, alternatively, by means of a bucket elevator, to a dry feed bin. Where an air-swept ball mill is used, the air current first carries the particles to an air classifier and then to a cyclone separator. The air, on losing its burden of ore particles, is returned to the mill, and the dried ore particles are discharged from the cyclone to the dried feed bin. Any excess air from the grinding system is bled to the furnace as secondary air.

The dried feed is blown into the roaster through a burner of special design. The feed is forced into a large stream of low-pressure air in the burner by means of small streams of high-pressure air to produce optimum turbulence of the air-concentrate mixture as it enters the furnace.

With converted multiple-hearth roasters, having a long central shaft to drive the rabble arms on the dryer hearths at the top of the furnace, the air-concentrate mixture is directed toward this shaft for extra turbulence. In the newer design, sometimes referred to as the McBean modification, two diametrically opposed burners are used for extra turbulence, and the dryer hearths are at the bottom of the roaster. This modification eliminates the central shaft in the combustion chamber and provides other advantages in design and operation, as all bins and operating controls are at a low level and building requirements are greatly reduced.

The temperature in the combustion chamber is readily maintained at the optimum, which is usually between 1750 and 1850°F, depending on the concentrates being treated.

About 40% of the roasted product settles out on a collecting hearth at the bottom of the combustion chamber. This contains most of the coarser material and is therefore likely to be highest in sulfur. Accordingly it is exposed to further desulfurizing by being rabbled across this hearth and another hearth immediately below before being discharged from the roaster. If desired, all or any proportion of the dust products can be fed to these hearths for further oxidation and to decompose sulfate sulfur or to obtain an overall homogeneous product.

The remaining 60% of the product leaves the furnace with the gas stream which passes first through a waste-heat boiler and then to cyclones and an electrostatic precipitator. About 20% of the suspended dust drops out in the boiler; and the cyclones and precipitator remove about 99.5% of the remainder.

Sulfide sulfur in the final product can be controlled, and consistently obtained, within the range of 0.1–5.0% as required for subsequent processing. Similarly, sulfate sulfur can be controlled from virtually zero to a maximum of about 2.5%.

Roasting rates for each installation can be varied between 50 and 100% of maximum capacity within a few minutes. Most of the older furnaces are treating about 100 tons of concentrates per day, but newer installations have capacities up to 350 tons per day. With oxygen enrichment of the combustion air, an increase of 10 tons per day may be expected for each 100 ft³/min of oxygen added; with sufficient oxygen the capacity can be doubled. The sulfur dioxide concentration in the gas leaving the roaster when "dead" roasting (sulfur taken to a very low level) zinc concentrates is about 9% without oxygen enrichment, and may be increased substantially depending on the amount of oxygen added.

Fluidization Roasting. See Fluidization. Another development of importance has been the adaptation of the technique of fluidization to roasting. Like the suspension-type roaster, fluidization roasting has permitted greatly increased reaction rates, so

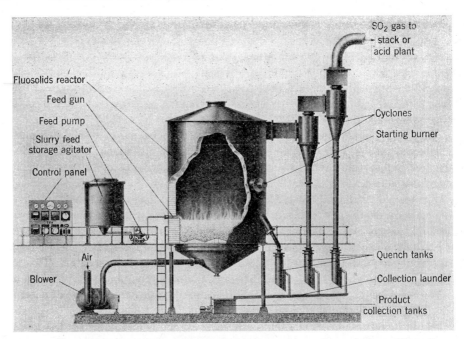

Fig. 2. Flow sheet of the Dorrco FluoSolids roasting system. Courtesy Dorr-Oliver, Inc.

that one fluid reactor may allow the same operating crew to process two to three times the tonnage possible with a hearth roaster of equivalent size. Several such fluidization installations are in operation in North America, roasting zinc concentrates both for pyrometallurgical and electrolytic processes. This type of roaster was originally developed for calcining arsenopyrite gold ores but has been adapted to roasting zinc ores and other metallurgical products requiring similar treatment.

Figure 2 shows a preferred arrangement for a solids fluidization system. Slurry feeding is normally used because of ease in handling, but dry feeding can be employed if desired.

A starting burner is used only to bring the bed up to sulfide ignition temperature (about 1200°F). After this point has been reached, no extraneous fuel is needed, in contrast to multiple-hearth roasters, which ordinarily are supplied with external heat on some lower hearths, and may have to be cooled on higher hearths. Low-pressure air (4–6 psig) is introduced into the wind box under the false bottom of the reactor and passes upward into the bed through a constriction plate. This plate contains holes spaced so as to ensure uniform air distribution across the bed. Club-head tuyeres with horizontal air ports long enough to be nonsifting are in general use. Pressure at the blower is sufficient to overcome the drop across the constriction plate, bed, waste-heat boiler, and cyclones, although it is usually preferred to operate under a pressure just slightly lower than atmospheric through as much of the system as possible. The objectives of this procedure are to minimize migration of SO_3-bearing gas toward the periphery of the roaster, where it could form H_2SO_4 and corrode the steel shell, and to minimize loss of gas and entrained dust out of the system and into the room, especially during cleanout periods; at the same time the negative gage pressure is kept as small as possible to minimize uncontrolled leakage of external air into the system.

Operation of the system is continuous. Feed enters the bed, is fluidized, and is immediately brought to the roasting temperature of about 1600°F. Water injection may be used to control temperature if slurry feeding is not desired. Air or roaster gas moves through the bed at a linear rate of about 1 ft/sec. Attempts to detect a temperature gradient from one zone of the bed to another have been unsuccessful, indicating a very high rate of heat transfer. The sulfide sulfur content of the feed is reduced from about 32% to perhaps 0.3%. The calcine overflows into a pipe placed at "surface level" and passes out of the reactor and through a calcine cooler. Surface level of the expended bed may be about 6 ft above the constriction plate; this much load would collapse to about 4 ft on cessation of the air supply. The fluidized density of the roaster contents is about 90 lb/ft^3, so that at any instant a roaster of 22-ft inside diameter contains about 100 tons. Operation has been found satisfactory when a 20–30% excess of air is supplied over that required for oxidation of both the sulfur and metallic ingredients of the feed.

Dust carryover into the dust-collecting system is claimed to be somewhat less than in flash roasting, but of the same order of magnitude. In some cases it has amounted to 20–40% of the calcine; in others it has risen as high as 77%. The amount varies with the feed rate (and consequently with the air rate), and with the size analysis of the material being roasted. In one operation, roasting 140 tons of dry concentrates per day, 30% of the calcine left the roaster via the overflow pipe, 23% was dropped in the waste-heat boiler, 44% in the cyclones, and 3% reached the hot Cottrell electrostatic precipitator for the flue gases. Flue and Cottrell dusts are not recirculated to the reactor, being sufficiently low in sulfide sulfur. Roasters with pelletized feed yield about 80% to the overflow and 20% carryover as dust.

The SO_2 content of the roaster gas is said to be usually 10–13%, but must be diluted to about 7% before treatment in a contact acid plant. Sulfur dioxide of this strength (13%) is high for a metallurgical sulfur gas producer; 14.6% SO_2 is the theoretical maximum concentration achievable on roasting 100% zinc sulfide concentrates to completion, unless the air supply is enriched with oxygen.

Although the preferred method of regulating temperature within the bed is by water injection, using automatic thermocouple-operated control, water injection and slurry feeding can be dispensed with if it is desirable to minimize the water content of the gas. Relatively low, uniform operating temperatures are said to mean less ferrite formation.

For week-end shutdowns it is claimed to be necessary only to stop feeding and to stop the blower. The bed subsides and becomes a reservoir of heat. Reactors have been down as long as three days and started within $\frac{1}{2}$ hr without addition of fuel.

In spite of the higher tonnage rates of the suspension or fluidization roasters when compared with the multiple-hearth type, the latter holds one advantage: less dust is carried off suspended in the gas stream. When purity of calcine is important, as in the manufacture of higher grades of zinc, the use of the multiple-hearth type permits the more volatile sulfides, such as those of lead and cadmium, to be removed preferentially. Furthermore, if these two metals are to be recovered from the flue dust, the lesser degree of contamination with zinc calcine is an advantage. Flue and Cottrell dusts from multiple-hearth roasters, however, are high enough in sulfide sulfur to require reroasting in a separate smaller roaster before treatment for lead and cadmium recovery.

SINTERING

Roasted zinc concentrates, or calcines, are sintered by several methods, depending upon the smelting method to be used and the purity desired. All known North American and most European methods involve the use of Dwight-Lloyd-type sintering machines (see Vol. 12, p. 217).

For horizontal-retort or rotary-kiln reduction, the sinter need not possess much structural strength. In fact, some producers are content with a soft, friable, poor appearing sinter if, by accepting it, they can increase their cadmium recovery by a point. For vertical furnaces, such as the electrothermic type, however, structural strength becomes important. The individual particle of sinter may be required to support the weight of a column of charge over 38 ft high, despite having become weakened by removal of most of its zinc content. To prevent mechanical collapse of this particle, rather close chemical control of the nonvolatile ingredients of the sinter is necessary, for these eventually become the major constituents as smelting progresses. Sinter for vertical electrothermic furnaces usually has silica added to increase hardness, and is known as hard sinter.

As mentioned earlier, roasting and sintering have been combined on one machine, but sinter produced in this combined process is reported to be sometimes more difficult to reduce in horizontal retorts than sinter produced from preroasted materials, such as calcines from multiple-hearth or fluidized-bed roasters. Current practice for the combined roasting-sintering operation is to return 80% of the product for resintering, so that each new gram of concentrate is assumed to pass over the machine 5–7 times before being discharged as part of a sinter cake with a size larger than $\frac{1}{2}$ in. after crushing and screening.

Where higher grades of zinc are sought, or where recovery of metallic impurities is profitable in itself, or where these impurities must be held to a minimum in certain grades of zinc oxide, the technique of scarifying, or "sinter slicing," may be utilized. It is well known that the sintering process does not proceed simultaneously at all levels in the bed of charge, but begins at the top, where ignition occurs from overhead burners, and proceeds downward as the fuel at each level is consumed. This means that there is repeated volatilization and recondensation of lead and cadmium sulfides and oxides, etc, at progressively lower levels in the bed. A large fraction of these impurities is pulled into the wind boxes directly below the bed of the sinter machine and is caught there or farther on in a cyclone or Cottrell precipitator. However, part of the impurities remains in the sinter cake, concentrated at the lower levels. The lower section of the cake can be sliced away from the upper by several means; it can then be either recycled over the machine or used for purposes not demanding high purity.

It is reported that, in sintering preroasted charges, substitution of fluosolid calcines for those made by older types of roasters has meant handling a more finely divided charge running about 2.8% sulfur, of which about 2.4% is present as sulfate. These calcines, considerably finer than hearth-roaster calcines, are reported to sinter very readily when mixed with 4–5% fine coal. The product averages approximately 0.3% sulfur and 0.05% cadmium.

The trend in machine size appears to run toward eventual abandonment of the once predominant 42-in. wide sintering machines 22 or 33 ft long, and adoption in their place of machines ranging between 42 and 72 in. wide and 55 and 100 ft long. A factor running counter to this trend, however, is the need in many custom smelters for smooth.

ing out their operations. One large machine, if handling calcine from several sources, may require frequent readjustment. With several smaller machines, each handling calcine from only one source and with its own mixing and feeding system, each one may be operated with a minimum of readjustment and probably at full capacity. Preference has also been expressed for a group of machines so sized with respect to the other parts of the smelter that certain sections can be shut down to accommodate the lessened demand for metal, and yet permit the remaining sections to operate at full capacity.

A most important development in sintering machines has been that of a virtually leakless wind-box seal. In some previous machines the volume of air short-circuiting around the bed of charge has been reported to reach as high as 150% of that passing through it. Such leakage lowers the available pressure differences across the bed of charge, especially at the edges of the pallets, wastes fan power, lowers the temperature of recirculated gases if recirculation is used, and seriously lowers the sulfur dioxide concentration in systems designed to recover the SO_2 for acid manufacture. The seal consists essentially of a flat upper bar of steel, situated along the outer edges of each pallet, riding on a flat lower bar of similar shape. Owing to provision for forced, controlled lubrication of the interface between the two bars, the seal is made virtually airtight. Wear on the bars (ordinary hot-rolled steel) is limited to less than $\frac{1}{32}$ in. per year.

Reduction Processes

Reduction and purification of zinc-bearing materials can be accomplished in a single step to a greater extent than for most metals because of zinc's relatively low boiling point, 906°C. Immediate separation from nonvolatile impurities is therefore possible. In fact, if the material is treated pyrometallurgically, there is virtually no alternative, since zinc, even at 857°C (the lowest temperature at which the oxide can be continuously reduced), already has a vapor pressure high enough to cause it to assume vapor form immediately upon reduction. Thermodynamic calculations indicate that the lowest temperature and pressure for the production of liquid zinc are 1020°C and 5 atm, respectively. A 90% yield of the zinc in liquid form would require a temperature above 1300°C and a pressure over 200 atm. Balanced against the easy separation from nonvolatiles, however, are the difficulties in condensing the vapor and the fact that several of the most common impurities, especially cadmium and lead, are also quite volatile.

Processes for the production of zinc oxide also involve a preliminary reduction of the crude oxide calcine to the metal vapor before reoxidation of the purified vapor to zinc oxide.

Thermodynamic calculations and actual tests, especially by the U.S. Bureau of Mines, have shown that it is possible to reduce zinc oxide with hydrocarbons, such as methane, but large-scale processes utilize hydrocarbons only indirectly. The substance almost entirely responsible for direct reduction of zinc in commercial processes is carbon monoxide. The zinc-reduction cycle consists of the two following steps:

$$ZnO + CO \rightleftharpoons Zn \text{ (vapor)} + CO_2 \qquad \text{(actual reduction step)}$$
$$CO_2 + C \rightleftharpoons 2 CO \qquad \text{(regeneration of CO)}$$

However, both of the above reactions are reversible. The second reaction is the slower of the two below about 1100°C and hence controls the rate of reduction in most commercial processes. Above 1100°C the rates of diffusion and heat transfer predominate as rate-controlling factors. Both the above reactions are highly endothermic. The amount of heat theoretically required to produce one pound of metal by this cycle is about 2500 Btu.

Carbon for reduction is normally provided by coal or coke, the choice between the two being determined by the need for structural strength within the smelting column or, if zinc oxide is the product, the danger of possible contamination of product with unburned soot. Carbon in excess of stoichiometric reduction requirements is normally provided in order to furnish extra reaction surface to compensate for the slowness of reduction of carbon dioxide by carbon.

There are five main commercial methods of producing zinc metal.

Metal-Producing Processes

> horizontal (Belgian) retort
> vertical (externally heated) retort
> vertical electrothermic furnace
> blast furnace (Imperial)
> electrolytic

HORIZONTAL RETORT

The horizontal, sometimes called the Belgian, retort process was for many years the most common method for reduction of zinc, although it has been partially replaced by others capable of handling larger volumes of metal per retort, and by the electrolytic process. Closely related to the Belgian retort process are the English, Silesian, Carinthian, and Rhenish processes. All involve the use of more or less tubular refractory receptacles for the charge, the ratio of length to diameter being at least 4 and usually around 7.5. The basic reason for the long, thin shape is the necessity for subjecting the mixture of zinc-bearing material and reduction fuel to prolonged exposure to a highly reducing atmosphere at a high temperature. Heating is accomplished by placing two blocks of retorts back to back and passing gaseous combustion products through the spaces surrounding the retorts. The high ratio of heat-absorbing surface area to volume and the relatively small area of the opening tend to produce the desired objectives, but the capacity per individual retort is quite small, about 40–60 lb of metal per day. A typical retort is about 8 in. in internal diameter and 60 in. long. The retorts are usually cylindrical in cross section, and are composed of a mixture of grog or chamotte (burned, crushed fireclay) and raw (fat) clay in the proportions 50:50 or 60:40, although several variations on this formula exist. Finely divided silica may be added. At Amarillo, Texas, retorts of 65% silicon carbide and 35% clay have been used on a large scale; besides tripling the life of the retorts (clay retorts normally last 20–40 days), the higher heat conductivity of the retorts has improved overall recovery of zinc by 2%. These retorts run noticeably hotter toward the fronts than do clay retorts. Advantage can be taken of the higher conductivity either to improve recovery or by increasing the diameter of the retorts so that a given tonnage of retorts carries a greater daily pay load at the same recovery as with clay retorts.

The optimum composition for retorts is determined partly by the nature of the gang in the roasted concentrate. If the gang is siliceous, the retort should be highly siliceous. If the residue after smelting contains much iron, lime, or other highly basic oxides, the retort material should be higher in alumina and other relatively inert oxides. Fluorspar in the calcine shortens retort life by forming low-melting solutions with the clay. A balance must also be struck in the physical structure of the retort. High grog content produces high mechanical strength but also high porosity, while raising the raw clay content reduces strength but also reduces porosity, resulting in decreased loss of zinc vapor into and through retorts.

Condensers are more or less conical clay receptacles, open at both ends, which are placed over the ends of the retorts to cool and condense the majority of the zinc vapor emerging. They may be made of the same composition as the retorts, or may be cheapened since they are not exposed to such high temperatures as the retorts. Some-times condenser chamotte is made from crushed, used retorts. One company has found that using the same mix for condensers as for retorts more than pays the extra cost in increased condenser life. Belgian condensers, as used in the United States, are usually 18–24 in. long, but may be 36 in. Rhenish and Silesian condensers are usually about 36 in. long. Condensers may be built with depressions in them to aid in holding condensed zinc. Retorts and condensers are made at the smelter by methods similar to those in general use in the ceramics industry. See Ceramics.

The furnaces themselves are double (back-to-back) banks of retort racks arranged approximately 4–7 high and up to 80 retorts over the length. They are now universally fired with natural gas, where coal or producer gas was once used.

One of the principal reasons why the heat economy of the horizontal, batch-type retort does not approach that of the continuous vertical retort is the daily necessity for reducing the firing rate and hence the temperature, toward the end of the run. This is done for two reasons: first, to prevent "butchering" of retorts after there is no longer sufficient charge within to absorb much of the heat in the endothermic reduction reaction, and, second, to reduce the furnace temperature gradually to a point where the "maneuver" of condenser draining and removal, removal of residue from retorts, replacement of broken retorts, and recharging of the retorts can be made more tolerable for the operating crew. During this period relatively little metal is being produced to pay for the fuel consumed.

Before beginning the new distillation cycle, the residue is removed and replaced by a mixture of roasted concentrate or sinter with about 60% of its weight of semianthracite coal or coke. The apparent density of the mixture is about 55 lb/ft^3. Anthracite may be used but ordinarily is not used alone, partly because of its high cost and partly because it is not quite so effective a reducing agent as slightly softer grades of coal. Grades such as semianthracite tend to form a soot during pyrolysis and to leave a solid portion with a rather porous structure. These materials have a much higher ratio of surface area to weight than did the original unpyrolyzed coal. This extra area is advantageous in promoting the re-reduction of CO_2 to CO (this being the rate-controlling reaction in the zinc-reduction cycle), thus speeding the reduction of zinc and minimizing the oxidation of zinc vapor already formed. On the other hand, coal so high in volatile hydrocarbons that they form a large fraction of the retort atmosphere for an appreciable part of the reduction period causes dilution of the zinc vapor and lowers condensation efficiency.

Occasionally about 0.2% NaCl may be added on the contention that it aids the metallic particles in blue powder (see below) to coalesce, by fluxing the oxide shell on them.

In Europe, charging of the retort is done directly through the condensers without removing them, but in America condensers are taken down for emptying and recharging. Charging is now done by machine at all U.S. plants. A charger can fill 210 retorts with 13,000 lb of charge in 20 min, thereby leaving more time available for the firing part of the cycle.

After charging, American practice is to replace the condensers and stuff their mouths with charge or other available high-zinc material, almost completely sealing the outlets. Gas is allowed to escape through one small hole in the stuffing, which is kept open by "spiessing," or frequent rodding. The purpose is to minimize contact between uncondensed zinc vapor and air, increase contact between zinc vapor and liquid, and prolong somewhat the time of residence of the vapor in the condenser.

When the campaign is ready to begin, heavy firing is started, after which condensers are placed over the ends of the retorts. Approx 3 hr is required for expulsion of all water vapor and the bulk of undesirable hydrocarbons from the charge. The temperature inside the retorts is brought up to 1000–1100°C for about 8 hr, after which it may rise slowly to a maximum of 1200°C or more. Temperature of the combustion products outside the retorts reaches a maximum of 1300–1400°C in American practice, and 1400–1500°C in European practice. Temperature drop across the 1-in. thick retort wall is about 100–200°C. The first zinc begins to come off the charge about 6 hr after charging has been completed. The condensers must now perform their function; to do this a condenser handling a retort charge volume of about 1.75 ft³ must have an internal condensing area of about 300 in.², or about 400 in.² if handling 2.0 ft³ of charge. Condensers must be designed to accommodate the maximum rate of vapor production rather than the average, else excessive loss of vapor results. Condenser temperatures range between 415 and 550°C during the campaign, varying of course with the rate of handling of zinc vapor.

Blue powder is an undesirable by-product of all metallic zinc production. The so-called "chemical" blue powder consists of fine globules of zinc metal coated with a film of oxide, which prevents the individual globules from coalescing into the main bath of condensed metal. Most of it is probably formed by a reaction between zinc vapor and carbon dioxide ($Zn + CO_2 \rightarrow ZnO + CO$) at temperatures below about 1100°C. This reaction is exothermic in this range and tends to occur when the zinc vapor is not cooled sufficiently fast from reduction temperature to condensation temperature. As the mixture passes through the range of 900–1000°C, the decomposition of a portion of the carbon monoxide ($2\,CO \rightarrow CO_2 + C$) further adds to the supply of CO_2 available to oxidize the zinc vapor. The portion of chemical blue powder not formed in this way is probably formed by direct oxidation of condensing zinc by entry of external air. Blue powder, apparently formed by freezing of very fine particles (always somewhat oxidized), is referred to as "physical blue powder."

In condensing zinc from horizontal retorts, if the temperature is low enough to prevent the escape of any uncondensed zinc, excessive formation of blue powder takes place. If it is high enough to minimize blue-powder formation, there is appreciable loss of vapor. This fact has led to the use in foreign plants of "prolongs," which are sheet metal tubes attached to the end of the condenser. The condenser is operated at high enough temperature to secure a good recovery of liquid metal; the escaping vapor is recovered as blue powder in the prolong. Prolongs increase the recovery of zinc by 2–3%, but American horizontal retort plants consider the expense of operating with them greater than the value of the extra recovery.

At the end of the campaign, and usually one or more times before, the zinc is removed from the condensers as liquid metal with a hoelike implement. Some blue powder is pulled out with the zinc, but much remains higher up on the condenser walls and is removed separately.

The length of furnace "campaigns" or "maneuvers" varies. (In some plants the term maneuver means the full campaign cycle, while in others this term applies only to the discharging–recharging at the end of each cycle.) The maneuver length (full campaign cycle) is usually either 24 or 48 hr. Three or four "draws" of metal are made each day.

Table 2. Impurity Variation in Metal in a Three-Draw Collecting Program

Draw	Material retorted	Pb, %	Fe, %	Cd, %
1	ore	0.58	0.019	0.460
	blue powder	0.82	0.039	0.330
2	ore	1.05	0.038	0.138
	blue powder	1.48	0.068	0.074
3	ore and blue powder	1.84	0.121	0.030

Table 2 gives the variation in impurity content of metal from various draws, using charges either of roasted ore or of blue powder, the latter representing not only the material just described but also the entire zinc-rich part of the by-products of smelting.

The most volatile impurities, like cadmium, tend to come over first. The first draw is also the purest. Lead and iron are always higher in later draws. The average purity of the metal runs about 98.0–98.5%. Depending upon the metal analysis and the specifications to be met, metal from each draw may be cast separately, or metal from several draws may be combined in a liquating furnace. It is held there just above its melting point, at which temperature impurities are least soluble, until a high-lead layer and a high-iron layer are formed at the bottom. These are removed from below. Cadmium also tends to concentrate in the first blue powder formed; in Europe this is often removed separately and reworked for its cadmium content, this practice raising the purity of the zinc.

Single-pass recovery of zinc from the retorts is only about 65–70%, but, since the material is recycled, overall recovery may run as high as 95%. Maximum attainable recovery rates are not highly influenced by the chemical nature of the gang, that is, whether siliceous, ferrous, or limey, provided the original zinc content is about the same in each concentrate.

In American horizontal-retort practice the entire collection of reworkable residue, including retort residue, condenser and ladle skimmings, floor cleanings, etc, is often referred to as "the blue-powder charge." This collection is often recharged into a separate bank of retorts reserved for it alone. However, since the retort residue may frequently amount to as much as 20–30% of the charge (assaying 13–15% zinc and about 45% carbon), it frequently is not included in the blue-powder charge but treated on magnetic separators. Frequently also the used condensers, which analyze up to 35.5% zinc, are not included in the blue-powder charge but rather are crushed and jigged. Then they are either added to the new charge of calcine or sinter, or incorporated as grog in the new condensers, thus providing vessels already partly saturated with zinc which as a consequence do not absorb much new zinc. It should be remarked that the daily loss of condensers amounts to about 15%.

VERTICAL RETORT, EXTERNALLY HEATED

The obvious disadvantages of batchwise zinc production led to the development of a continuous vertical retort. The calcined and sintered zinc ore to be treated is briquetted. The reduction fuel, coking coal and coke breeze or anthracite, is mixed with the ore and a temporary binder, such as sulfite waste liquor, tar, or pitch, in Chilean mills (edge runner mills). The briquets are then coked in an autogenous coking furnace and this operation also drives off volatiles which serve as fuel. Each unit has a capacity of 100–120 tons of briquets averaging about 40% zinc.

Coked briquets are fed, without intermediate cooling, to the charging extension at the top of each vertical retort, where they are additionally heated by the gas–vapor mixture on its way to the condenser.

The retort itself is rectangular in general cross section, about 1 ft wide and 6 ft to 8 ft long. Walls are of silicon carbide to facilitate heat transfer and minimize penetration by zinc vapor. The heated height of long wall space is up to 35 ft on each side. Joints in the end walls are packed with silicon carbide and graphite to permit differential expansion upon heating. Production rates are reported at an average 40–45 lb of metallic zinc per day for each ft^2 of long wall surface when heated to 1300°C. Retorts are said to have a life of about three years.

Briquets maintain their shape throughout the reduction operation. Residue is removed at the bottom through an automatically controlled roll discharge mechanism into a water seal. During the passage of the briquets through the retort, sufficient displacement gas in the form of air or exhaust combustion products is introduced at the base of the charge, to ensure that no zinc vapor moves concurrently with the charge and eventually condenses on spent residue, rather than move toward the condenser.

ELECTROTHERMIC PROCESSES

A major impediment to the development of a large-scale zinc-smelting retort was the difficulty in heating the charge. Heating by direct contact with gaseous combustion products was practical only for zinc oxide production processes like the Waelz kiln or Wetherill grate methods, but, where any noncondensable gases diluting the zinc vapor had to be removed later in the condensing operation, such practice was high in cost and reduced the condenser capacity and the yield of condensed zinc. External heating methods were for long the only practical ones for metal production, but here heat economy was low because of the temperature drop through the walls and charge. Waste-heat recovery devices were not necessarily profitable. These conditions led to early attempts to develop an electrothermic process.

Vertical Electrothermic Process. Oxide- and metal-producing furnaces are essentially alike internally, differing only in the manner in which zinc vapor is handled after leaving the charge column. Both furnaces are vertical refractory cylinders up to about 8 ft in internal diameter and 37 ft high. Several sets of graphite electrodes protrude into the shaft, each top electrode being linked in a single-phase circuit with a bottom electrode located about 22 ft below and approximately opposite the first, as shown in Figure 3. The furnace is of the charge-resistor type, the coke in the charge carrying the alternating current between each top electrode and the bottom electrode

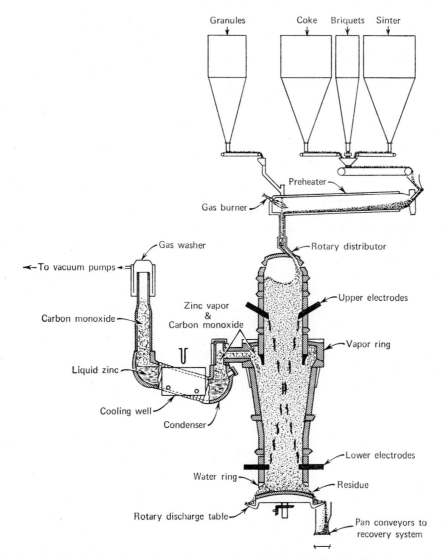

Fig. 3. St. Joseph electrothermic smelting furnace. Courtesy St. Joe Minerals Corporation.

on the opposite side. Contrary to practice in arc-type furnaces, it is not intended to produce temperatures high enough to cause fusion of residue.

Because particles of sinter must maintain their strength and porosity in the tall column even after most of their zinc content is removed, an unusually hard sinter is prepared by addition of silica sand to the sinter mix.

Furnace charge, consisting of sinter several times the stoichiometric requirement of the coke, and granular by-products, such as blue powder, is fed by controlled-rate feeders to a rotary-kiln-type gas-fired preheater, where it is heated close to reduction temperature. It passes then to a rotary feeding top on the furnace; this top is also a closure for the top of the furnace shaft. Sealing of the rotating top is effected by a skirt on the edge of the top, moving in a ring of sand.

Rotation of the top while feeding performs an important function related to the control of the temperature within the charge. The spout of the rotating top is aimed so as to lay the charge down in a ring whose highest point is nearer the periphery of the furnace rather than the axis. This results in a central depression resembling an inverted cone, toward the bottom of which the larger particles tend to roll. Particle size of the feed is controlled so as to provide coke particles larger than sinter, thereby tending to concentrate large coke at the axis of the furnace. In this way the maximum fraction of electric current is directed along the axis, which becomes the region of maximum temperature. This arrangement serves to minimize both the damage to the refractory walls by slagging and the heat loss through them. Temperature distribution within the furnace is about 900°C near the wall, 1200°C in the main body of the charge, and up to 1400°C and higher at the axis.

Current through each of the 12-in.-diam graphite electrodes may range as high as 8000 A. A furnace contains eight individual single-phase circuits, each carrying 800 to 1200 kW. The working range for such a circuit is 260–160 V. The overall power factor of transformers, bus system, and furnaces is 90 to slightly over 95%. Power to furnaces represents about 77% of the total consumed in the plant. Overall plant-power consumption is about 3260 kWh/ton of zinc equivalent produced.

For withdrawal of the mixture of zinc vapor and carbon monoxide (approx 1:1), the generally cylindrical form of the furnace is broken by an enlargement (see Fig. 3) known as a vapor ring, which provides a free space around the periphery of the charge for removal of the gaseous mixture. This leads to a roughly U-shaped condenser (Weaton-Najarian vacuum condenser) constructed of a corrugated steel shell lined with bonded silicon carbide brick with low permeability to zinc. Associated with the condenser is a cooling well through which circulates molten zinc at the rate of around 15

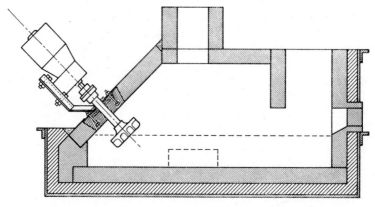

Fig. 4. Splash condenser used with **vertical** retorts and Sterling electric furnaces. Courtesy New Jersey Zinc Co.

ton/min. Water-cooled hairpin coils extract heat from the flowing zinc. This condenser has first been "primed" with about 40 tons of molten zinc. A vacuum of 6–10 in. Hg is applied to the outlet of the condenser, causing the vapor–gas mixture to be drawn through it in large bubbles. Baffles are provided to help break the bubbles and increase the area of vapor–liquid contact. Development of this condenser first made possible the production of over 100 tons of metal per day from a single condensing unit.

Residue is removed from the furnace preferably as discrete solid particles. Besides permitting recovery of residue sufficiently high in zinc and carbon to make retreatment worth while, a minimum of power is consumed in unproductive melting of residue.

Electrothermic Arc Furnace (Sterling Process). This process is named after the Sterling Hill mine, whose ore it was originally developed to smelt. The crude oxidized ore from this mine assays, after calcining to break down carbonates, about 18–20% zinc and an equal amount of iron. Most of the heat was produced by striking an arc between the three 24-in. diameter graphite electrodes and the slag bath. Thus most of the heat was radiated from the arc directly to solid charge floating on the slag bath.

The Sterling furnace is no longer being used, but the condenser which was developed for it has been adapted to the vertical retort with great success.

The condensing unit, which is a splash-type condenser (see Fig. 4) was designed with the following objectives in mind: (a) provision of a large surface area of molten zinc on which vapor can condense directly, thus minimizing formation of blue powder; (b) chilling the mixture of zinc vapor and noncondensable gases as rapidly as possible, which further ensures against excessive formation of blue powder; (c) accomplishment of both the above objectives in a minimum of space.

A graphite impeller, driven by a graphite shaft, scoops molten zinc from the bath, slinging it in a fine sheet, which breaks into droplets, across the path of the incoming gas–vapor mixture. Because the zinc bath is maintained at about 500°C by water-cooled metal coils, both vapor and gas are rapidly chilled below the dewpoint of the vapor, minimizing production of blue powder. Since blue powder is less likely to form when the vapor is condensed directly upon a surface of liquid zinc, rather than on a refractory, the external refractory walls of the condenser are insulated rather than cooled, to discourage condensation there. The intense mechanical agitation also tends to break up any blue powder that does form. It is claimed that 75–80% of the metal value entering the condenser from the Sterling furnace is condensed as liquid metal and that with recirculation of blue-powder skimmings, etc, an overall zinc recovery of 95% is obtained with concentrates of 50% purity or better. On vertical retort gases the condensing efficiency is 95–97%. Because this type of condenser has been found to handle gases leaner in zinc vapor more efficient than older units that depended primarily upon condensation directly on a refractory surface, it has opened the possibility of adding more air, or perhaps oxygen, directly to the charge in the vertical retort. Thermal efficiency is obviously greater when heat is produced within the charge, rather than being produced externally and passed through the retort walls by conduction. The practicality of adding air or oxygen is limited because of the diluting effect of the noncondensable gases produced on the gas–vapor mixture drawn off to the condenser.

The Imperial Smelting Furnace (ISF). This furnace of the Imperial Smelting Corp., Ltd., of Bristol, England, in which zinc and other metals can be produced simultaneously in a blast furnace, is one of the most impressive metallurgical developments of this century. (See Fig. 5.) It had previously been widely believed that zinc

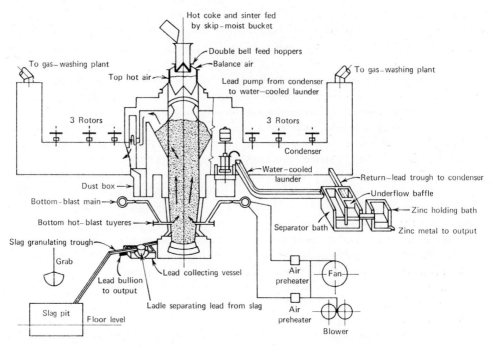

Fig. 5. Flow sheet of the Imperial smelting furnace.

could not be successfully produced in a furnace wherein the combustion of fuel occurred in the same chamber as the reduction of the zinc oxide. Such an arrangement diluted the gases with the nitrogen of the combustion air and the products of combustion so that the zinc vapor reoxidized as the temperature was lowered and could not be condensed as metallic zinc.

The success of the Imperial Furnace depends on maintaining a high temperature of the gases containing the zinc vapor in the top of the furnace and the condenser so that there is no oxidation by reversal of the reaction

$$ZnO + CO \leftrightharpoons Zn + CO_2$$

which goes to the left at lower temperatures. The zinc is recovered from the gases in the condenser, not by cooling but by being absorbed in an intense rain of molten lead produced by mechanical agitators. Subsequently, the zinc is recovered from the lead outside of the condenser either by cooling the alloy and rejecting a zinc-rich layer or by vacuum treatment to vaporize the zinc from the alloy.

Metallic zinc was first produced commercially by this process in 1952 at the Avonmouth Works of the Imperial Smelting Corp., Ltd., and since that time larger furnaces have been developed and installed in almost all of the countries producing zinc with the exception of the United States. Installation of an Imperial Furnace has been proposed for several zinc districts in the United States, but the initial large investment seems to discourage the changeover from presently used processes.

The Imperial Furnace has the obvious advantage of large capacity of the individual unit, the larger, newer furnaces producing as much as 190 tons of slab zinc per 24 hr with a burning rate of 148 tons of carbon per 24 hr. A second advantage is the ability to obtain a simultaneous production of zinc and lead bullion from mixed zinc–lead materials. Plant runs indicate that the process is also capable of handling zinc–lead–copper materials with a Cu:Pb ratio of up to 0.20 with good copper recovery as metal alloyed with the lead. The furnaces follow the standard blast-furnace design with water-cooled tuyeres and a refractory brick-lined shaft. The charge consisting of sintered zinc or zinc–lead concentrates plus coke is preheated and added to the furnace through a double-bell system. The lead oxide is reduced and the molten lead sinks to the bottom of the furnace, resting underneath a slag layer formed by reaction of the gang in the sinter and coke with the fluxes in the charge. Gold, silver, and copper in the charge are almost completely contained in this lead which is tapped periodically. The zinc oxide in the furnace charge is then reduced and forms zinc vapor which moves upward in the furnace and is withdrawn at a temperature of about 1000°C which is above the reoxidation temperature. The zinc vapors plus furnace gases pass through heated passages to the condenser in which vertical motor-driven impellers revolve in a lead pool creating a shower of lead droplets which absorb the zinc. This lead condenser is not to be confused with the lead recovered from the bottom of the blast furnace. The zinc is recovered from the condenser lead either by cooling the alloy or by vacuum treatment. The condenser lead must be circulated at a rate of 420 times the zinc production to maintain thermal balance and absorption capacity.

The ability to treat mixed lead–zinc concentrates has led to the practice of producing bulk lead–zinc concentrates specifically for ISF smelters and thereby simplifying the concentration procedures.

ELECTROLYTIC PROCESSES

The major domestic producers, as reported in the *1969 Yearbook of the American Bureau of Metal Statistics*, are listed in Table 3.

Table 3. Major U.S. Producers of Electrolytic Zinc, Short Tons

Company	Annual capacity[a]	1967	1968	1969
American Zinc Co., East St. Louis, Ill.	84,000	69,075	67,851	73,607
Anaconda Co., Great Falls, Mont.	162,000	78,513	94,610	135,819
Anaconda Co., Anaconda, Mont.	90,000	33,321	48,319	38,216
Asarco,[b] Corpus Christi, Tex.	108,000	101,026	84,707	100,841
Bunker Hill Electrolytic Zinc Plant, Silver King, Ida.	114,300	92,134	102,945	105,700
total	558,300	374,069	398,432	454,183

[a] Estimated.

[b] American Smelting and Refining Co.

The original consideration, prompting much early research on electrolytic zinc, was the economic necessity for finding a process to treat complex lead–silver ores containing zinc, copper, and iron. Concentration yielded a product containing only 30–40% zinc. The high ratio of gang to valuable mineral made retorting difficult, owing to the proportionately lower pay load per retort load, slagging of retorts, etc. Consequently it was necessary for the electrolytic process to develop one or more highly efficient purification stages. As the need for extremely pure solutions increased, treatment of even the "purest" ores became subject to close chemical control. Availability of cheap electric power, although the largest single cost item within the electrolytic process, is not necessarily the most important factor in choosing the electrolytic method.

All electrolytic zinc plants have the following four operations in common: (1) roasting of the concentrate (as described above); (2) leaching of the roasted concentrate or calcine to extract the soluble zinc; (3) purification of the resulting solution; and (4) electrolysis of the solution to obtain metallic zinc.

Leaching. Two general types of leaching processes are used, single and double. *Single leaching* by single batches involves addition of all the calcine of a given bath to sufficient spent electrolyte to give a slight excess, about 0.3–0.5%, of sulfuric acid after all soluble zinc has been dissolved. Milk of lime or finely ground limestone is then added to neutralize the solution and cause precipitation of all the iron, silica, antimony, and arsenic present, and to coagulate the pulp so that all solids settle and the thickener yields a clear overflow. The leaching and neutralization steps must be carried out very carefully, as addition of an excess of calcine results in loss of zinc, while excess lime precipitates zinc from a neutral zinc sulfate solution and causes unnecessary loss of acid as calcium sulfate. The loss of acid might not be economically serious if it could be made up by controlling the degree of sulfation in the roasting step, but the loss of zinc by precipitation and by solution entrainment in the residue could be. Single leaching at Bunker Hill has been described as a continuous process in a series of five stainless tanks heated by steam and with axial flow turbine agitators. About 90% of the calcine and 75% of the electrolyte are added to the first tank, additional acid is added in the second tank, and neutralization is attained in the remaining tanks by careful additions of calcine.

Single leaching is usually justified by savings in plant investment and in operating cost, but these must be balanced against the risk of lower recoveries, higher acid requirements, etc. The Risdon, Tasmania, plant of Electrolytic Zinc Co. of Australasia, has found single leaching preferable owing to the special problem engendered by a high content of soluble silica in its ore.

Double leaching comprises a first or neutral leach, the purpose of which is to extract the easily soluble zinc and precipitate such impurities as silica, iron, alumina, etc, followed by a second, or acid, leach on the residue from the neutral leach. The object of the second leach is to make the maximum recovery of zinc from the residue before discarding. In the first leach it is usual to add sufficient calcine to the spent electrolyte for a large excess of zinc oxide to remain, producing a pH high enough to precipitate the above impurities and promote proper coagulation for fast settling. The residue sent to the acid leaching tanks is consequently a smaller quantity of solids than had to be treated in the first leach. Therefore, when it is treated with enough electrolyte to ensure the solution of all soluble zinc, the tendency for undesirable acid-soluble impurities to dissolve is less, as the smaller quantity of solution involved is nearer to saturation. The final acid strength of the second leach liquor is about 0.3–0.5% H_2SO_4. After removal of residue solids, this liquor is returned to the first leach tanks, either directly or after mixing with spent electrolyte.

Leaching of roasted concentrates has been carried out largely in Pachuca (or Brown agitators) tanks which are vertical cylindrical tanks, usually of wood stave construction, but occasionally of steel, having a conical bottom through which unleached solids can be discharged if desired. A vertical pipe about 6–10 in. in diameter, concentric with the tank, extends from just above solution level to near the bottom. A much smaller pipe supplies air under pressure to the bottom of the pipe. The presence of a number of air bubbles in the central pipe means that the column of fluid in the pipe has a lower average density than any equivalent volume of fluid elsewhere in the tank, and the column is forced upward, to spill back into the tank at the top of the pipe. This is of course the air-lift principle, applied to achieve, instead of true lifting, a violent agitation combined with excellent aeration. Oxygen is required for both cyanidation of precious metals and oxidation of ferrous compounds to ferric in zinc hydrometallurgy. Therefore the Pachuca, with its lack of moving parts, retained its popularity until fairly recently when the better controlled mechanical agitators have tended to take over. Regardless of the type of agitation, some manganese dioxide may be added to the tank to ensure that all iron is in the ferric state.

Whether single or double leaching is employed, the leaching operation can be done batchwise or continuously. If batch leaching is employed approximately twice as many tanks of a given size must be employed, owing to time lost in filling and emptying the tanks. Air at about 90 psig must be available also to clean out accumulating deposits of coarse material at the bottom of the tanks, whereas with continuous leaching it is relatively seldom necessary to use air at pressures higher than the normal operating pressure of 20–35 psig. Continuous leaching requires a larger total volume of air per tank, however, as tanks must be in agitation continually. During actual agitation in either process, consumption ranges from 100 to 150 ft^3 of air per minute. Continuous leaching has the disadvantage of requiring feed of uniform impurity content to make it practical.

The formation of ferric hydroxide during the neutral leach results in a heavy floc, which effectively traps arsenic and antimony (the mechanism of this coprecipitation is not well understood). If the calcine does not contain enough iron to ensure the effective removal of these two impurities, ferrous or ferric sulfate solution is added to the first leach tanks. This may be produced by treatment of scrap iron or other iron-bearing material with hot spent electrolyte. All ferrous iron must be oxidized to ferric before the addition of the calcine.

The primary chemical reaction involved in leaching calcine is:

$$ZnO + H_2SO_4 \rightarrow ZnSO_4 + H_2O$$

Typical of the secondary reactions is the oxidation of ferrous iron to ferric, followed by precipitation of ferric hydroxide as the addition of more calcine raises the pH of the solution:

$$2\,FeSO_4 + 2\,H_2SO_4 + MnO_2 \rightarrow Fe_2(SO_4)_3 + MnSO_4 + 2\,H_2O$$
$$Fe_2(SO_4)_3 + 3\,ZnO + 3\,H_2O \rightarrow 2\,Fe(OH)_3 + 3\,ZnSO_4$$

Tests are made at regular intervals on the neutral leach discharge liquor for iron, arsenic, antimony, and copper. The test for ferrous iron is by titration with potassium permanganate, or it can be roughly determined colorimetrically with potassium thiocyanate after oxidation to the ferric state.

Only about 60% of the acid-soluble zinc is recovered in the first leach because of the large excess of calcine required to guarantee precipitation of silica, iron, etc. Almost half the spent electrolyte used must therefore be added to the second (acid) section of the leaching circuit. If the underflow from the neutral leach thickeners is filtered, some neutral solution can thus be kept from unnecessarily diluting the acid-leach-tank solution. If filtration is not practiced, allowance must be made for pumping and thickening greater volumes of acid leach solution.

Purification.   The neutral leach should have removed silica, iron, alumina, antimony, arsenic, and germanium, although the last three may not have been thoroughly removed if present in appreciable quantity. The major detrimental impurities remaining to be dealt with in most operations are copper, cadmium, and cobalt. The method of treating the solution for these depends partly upon the extent to which they are present. If the ratio of copper to cadmium is not high, both can be precipitated fairly easily with zinc dust only a little in excess of the stoichiometric requirement. However, if the ratio is high, the copper crystals tend to act as cathodes in miniature electrolytic cells and encourage the re-solution of cadmium. To prevent this a large excess of zinc dust must be added. If the cobalt content of the solution is low, it can be precipitated with the zinc dust and later separated from the copper–cadmium cake. If it is higher, it is usually precipitated with 1-nitroso-2-naphthol. Cobalt removal may then be partially paid for by purification and sale of cobalt oxide or metal. As a commercial venture it is not highly profitable in itself because almost 10 lb of 1-nitroso-2-naphthol is consumed for each lb of cobalt recovered; but, in any case, it must be removed from the solution.

Copper, when present in low concentrations, assists in the removal of other impurities. Thus, a solution may have too little copper, and copper sulfate may actually be added to the solution before the zinc dust. This may raise the copper content (intentionally) to a point where much cadmium redissolves, and thus may necessitate a second treatment of the solution with zinc dust to reprecipitate the cadmium.

Silica, if incompletely removed in the neutral leach, is precipitated when zinc dust is added during purification. This immobilizes much of the zinc dust and "blinds" the filter medium. In fact, almost any impurity not completely removed by the thickeners following the leach tanks has a similar effect. The remedy probably lies in more thorough precipitation during the neutral leaching step, or perhaps in increased thickener capacity. Ferric sulfate and free acid causes resolution of cadmium; the presence of these is another indication of incomplete neutralization and precipitation earlier in the process.

Electrolysis.   Early attempts to produce zinc electrolytically failed owing to not recognizing the importance of the purity of the solution. The electrolyzing cell for zinc is unusually sensitive to the presence of elements such as antimony and cobalt lying lower than zinc in the electromotive series. Not only the impurities introduced into the cells by way of the solution must be considered, but also those that may be introduced in the anodes or the tank lining. Solution spray reaching any uncoated copper bus bars overhead can cause the formation of a copper sulfate scale on the bar which must be removed carefully to prevent contamination of the solution. The success of the tank operation is unusually dependent on the thoroughness with which previous processing has been carried through.

Electrolytic plants were formerly divided into two classes: (a) Low-current-density, low-acid circuits, where current density ranges from 20 to 40 A/ft² of cathode area below solution level. The acid strength of the electrolyte should not rise above about 6%.

(b) High-current-density, high-acid circuits, in which current density was normally above 100 A/ft². Acid strength ranged between 22 and 28% H_2SO_4. This is often called the Tainton process after its developer. Modifications of the original high-current-density plants have resulted in intermediate values of current density at about 60–70 A/ft² as being more typical of modern practice.

In both systems care must be taken not to allow the ratio of acid to zinc to become too high or the re-solution rate becomes inordinate. Hence in the second system the zinc content of the electrolyte must be kept higher than in the first. Advantages claimed for the second are as follows: the ability of the strongly acid electrolyte to dissolve any zinc ferrite formed in roasting; solution of more iron, hence better purification; production of more zinc per unit of cathode area; improved filtration; and compactness. Balanced against these are the disadvantages of corrosion of the zinc deposit by strong acid, an equal or greater investment per unit of capacity, and the cost of heating leach solutions. With high-current densities, stripping periods must be more frequent to prevent short-circuiting, because if electrodes are spaced farther apart voltage drops rise proportionately. Stripping periods range from 8 to 72 hr, the shorter intervals being used in the high-current-density plants. The pulling of cathodes by hand from the cells and the manual stripping of the zinc deposit requires a great amount of labor. Extensive efforts have been made to develop mechanical pulling and replacement equipment for the cathodes and to devise either mechanical or hydraulic devices for stripping.

Cell tanks may be of wood, concrete, or pitch concrete. If wood is used, the tanks must be lined with lead, rubber, or an equivalent material capable of resisting attack by a zinc sulfate and sulfuric acid solution, both with and without the presence of direct electric current. The wood must also be heavily treated with a preservative such as creosote to make it acid-resistant. Ordinary concrete is likewise vulnerable to acid-attack. Concrete tanks must be lined with lead, rubber, sulfur sand, etc, or the acid-resistant material may be incorporated directly into the mix itself. In the last case a special patented pitch, known as Prodorite, may be mixed with asphalt, silica rock, and silica sand, the pitch replacing Portland cement as binder. Tanks of pitch concrete can be cast in steel forms to produce a jointless, monolithic structure. More care must be taken in mechanical operations around them, however, because of their greater brittleness and their tendency to hold residual thermal stresses developed during the casting period. Some of the newer acid-resisting plastic materials, for example un-plasticized polyvinyl chloride, should find an excellent field here as coverings over less expensive supporting structures.

Anodes are made of high-purity lead, or of lead alloyed with about 1% silver, the latter being known as the Tainton alloy anode. Pure lead may have the following impurities: 0.0003% Ag, 0.0005% Cu, 0.0007% Sb, 0.0003% Bi, and a trace of As. Anodes are usually cast integrally with a lead-covered bus bar, in such a shape that the top has horizontal extensions for resting on supports over the tank walls, while the submerged portion is essentially rectangular. Anodes may be cast in perforated form either to provide increased surface area for delivery of current to the solution or, where electrodes are spaced so closely that insulating stiffeners must be used, to provide paths for solution flow through the tank. Tainton-type lead–silver anodes are claimed to be less prone to warping and buckling, thus reducing the tendency toward short-circuiting.

It is difficult to obtain zinc in coherent, thin sheets, and so for the cathodes it is necessary to use starters. These starters are made of high-purity aluminum sheet, from about $\frac{3}{32}$ to $\frac{1}{4}$ in. thick. The optimum thickness depends to some extent on the concentration of fluorine (as fluoride) in the circuit, since this is highly corrosive to aluminum. Fluorine may also cause trouble later in the process when the electrodeposited zinc must be stripped from the cathodes, by roughening the cathode so that the zinc adheres more strongly to it, thus making its removal more difficult. Cathode sizes and shapes are similar to those of anodes except that it is customary to provide cathodes slightly wider than the anodes, producing a submerged cathode area of about 9.4 ft², as against 9.2 ft² for the anode (both sides of each included). While this arrangement helps to reduce cathode-current density slightly, its main purpose is to reduce the tendency for current to concentrate along the sharp edges of the cathode and form "trees" (see below).

All anodes and cathodes of a given cell are connected electrically in parallel, the voltage per cell varying from about 3.25 to 4.5 V, respectively, depending on the deposition period, current density, temperature, acidity, and electrode spacing. Cells in a given unit are connected in series, so that the voltage drop from the power source to the last cell may be 500–800 V, depending on the number of cells and the current density employed. Copper requirement for bus bars and tank bars can be estimated by allowing 1 ft² of cross section for each 100 A carried, or 25% more if the current exceeds about 8000 A.

The number of electrodes per cell varies widely. In the Tainton high-acid, high-current-density process of the Bunker Hill Co., twenty-four anodes and twelve cathodes were used. In processes using low- and medium-current densities, one more anode than the number of cathodes is used, so that both end cathodes are surrounded by a source of current on both sides. For example, at Trail, B.C., (Consolidated Mining and Smelting Co.), 23 anodes and 22 cathodes are used; at Corpus Christi, Tex. (American Smelting and Refining Co.), 25 and 24; at Risdon, Tasmania (Electrolytic Zinc Co. of Australasia), 34 and 33 were formerly used, but, by reducing the spacing of electrodes to 3-in. centers, the number has been increased to 45 and 44.

Close spacing of electrodes is desirable to minimize the voltage drop through the solution. The proximity is limited, however, by the tendency of zinc to deposit at localized points on the cathode rather than uniformly over the entire surface. Once such uneven deposition starts it tends to become self-accelerating, since the resistance of the electrolyte is proportional to the distance between electrodes, and the building out of the "tree" reduces this resistance locally, causing the current to concentrate along the path of less resistance and deposit more zinc on the tree, finally resulting in a short circuit. To minimize treeing, a protective colloid such as glue, goulac, gum arabic, agar agar, plus sodium silicate and cresylic acid is added, a typical amount being 0.84 lb/ton of zinc deposited. The colloid, possessing an electric charge, apparently concentrates in areas of high-current density and thereby increases the resistance locally, thus tending to equalize deposition over the entire cathode.

Effect of Impurities. At zinc sulfate concentrations typical of those used in electrodeposition of zinc, the decomposition potential of zinc sulfate is about 2.35 V, whereas that of water is only 1.70 V. From this it appears that zinc could not be deposited from such a solution and that hydrogen could be liberated instead. But hydrogen does not deposit on a zinc surface at the normal decomposition voltage of water; it requires a voltage at least 0.8 V higher, the so-called hydrogen overvoltage on zinc. Because of this, zinc can be plated out provided there are no harmful impurities to ruin this precarious balance which is only slightly in favor of zinc. The elements especially harmful in this manner, because the overvoltage of hydrogen on them is less than on zinc, are germanium, antimony, tellurium, cobalt, and arsenic in decreasing order.

Germanium may exhibit a very pronounced effect on the current efficiency of the solution even when present to the extent of only 1 part in 10 million. When accompanied by other impurities, for example cobalt, it is particularly detrimental. What metal deposited is likely to be black, spongy, and difficult to melt. If present in concentrations much larger than this, it may be commercially impossible to treat such a solution electrolytically, as deposition, if it occurs at all, proceeds only to a point, beyond which re-solution would be more rapid than deposition. For this reason it is usually impractical to treat concentrates from the Tri-state district of Missouri-Kansas-Oklahoma electrolytically; these concentrates are at present America's leading source of germanium.

Antimony is injurious when 1 ppm or more is present. Its effects can be minimized by better solution purification, shortening the deposition period, and lowering the cell temperature. The effects of virtually all impurities can be reduced by the last method, although the problem is not simple when million gallons per day are treated and calcium sulfate scale has to be prevented from impairing the efficiency of the cooling coils. Antimony was actually added to the solution at Risdon at one time to counteract the effects of a high cobalt concentration, but this addition has been abandoned with the adoption of better procedures for the purification of solutions.

Arsenic is, in itself, the least harmful to the current efficiency of the metals mentioned above, but its presence should be viewed as a danger signal that antimony removal has not been thorough. Its presence is manifested by a pronounced corrugation of the deposit and a lack of the usual luster.

Copper is not particularly injurious up to 10 mg/liter. Even a higher concentration can be tolerated if the acidity of the solution is low. Above this figure, however, copper impairs the current efficiency by plating out and forming local miniature cells tending to drive zinc back into solution. The presence of copper is, of course, undesirable from the standpoint of product purity.

Cadmium does not harm current efficiency in concentrations lower than 0.5 g/liter, and is believed by some to raise it slightly. The main objection to its presence is the lowering of the purity of the cathode metal.

Lead is also objectionable primarily because it affects cathode metal purity. If only prime western-grade metal is to be made, this objection becomes void, as most electrolytic zinc must actually be debased with lead metal to make it suitable for galvanizing. However, if the special high-grade (99.99+% Zn) is to be made, cooling the entire electrolyte to as low as 28°C may be necessary to reduce the (already low) solubility of lead sulfate. By doing this, the plant in Risdon, Tasmania, has been able to decrease the lead content of cathode metal from 0.0116% when making high-grade zinc, down to 0.0048% when making the special high-grade required for die castings.

Although most lead is derived from anodes or tank linings, only a very small proportion obtained from this source ends up in the cathode metal. Most of it drops to the bottom of the cell as a sludge along with manganese dioxide, silver, etc.

Patented procedures of adding either $Ba(OH)_2$ or $SrCO_3$ to the electrolyte are effective in reducing the lead content of the cathode metal.

Cobalt is the most serious impurity normally met. While it can be tolerated up to 10 mg/liter (about ten times the threshold tolerance for antimony), it is much more difficult to remove. Its presence also compounds the damage caused by germanium. Some of it is removed in the zinc dust precipitation step if the temperature is high enough, but most is immobilized with 1-nitroso-2-naphthol. Residual amounts of the latter are also detrimental in electrolysis; nevertheless some may actually be added before electrolysis to tie up excess cobalt present. On the other hand, 3–4 mg/liter of cobalt is considered desirable because it tends to reduce the amount of lead in the cathode metal. Its effect depends much on other impurities present and the acid concentration.

Iron can be tolerated up to 2–4 mg/liter. Beyond this it begins to affect current efficiency. Ferrous iron is oxidized to ferric at the anode, then re-reduced to ferrous at the cathode. Only a very minor amount appears in the cathode deposit; high-grade metal from Risdon contains about 0.0009% and the special high-grade less than 0.00001%. High iron content in the solution, like high arsenic, should be viewed as a danger signal indicating incomplete purification.

Nickel is not normally found in zinc-plant solutions in troublesome concentrations. If 0.5 mg/liter or more is present, nickel acts similarly to cobalt, causing re-solution of the zinc deposit and contamination of the cathode metal. If the solution undergoing purification is heated high enough, nickel can be removed with zinc dust. Heating is not ordinarily necessary for this purpose.

Manganese is usually harmless in small amounts; larger amounts produce beads. sprouts, and rough, spongy deposits. Manganese when present as the sulfate, $MnSO_4$, is oxidized electrolytically to permanganate, $MnO_4{}^-$, which reacts slowly with more manganese sulfate to precipitate hydrated manganese dioxide. The precipitate adheres to the anode until it builds up sufficiently to fall to the bottom of the tank. Thick anode coatings are encouraged by low acidity, low temperature, and low current density, the last probably because of its effect on the production of sulfuric acid.

Selenium causes low current efficiency and many holes in the cathode zinc if 1 mg/liter or more is present. Commonly encountered amounts of selenium are removed by regular purification technique.

Tellurium prevents deposition of zinc if 1 mg/liter is present. It is normally removed in the iron-purification step but may also be removed by adding zinc dust to a hot solution.

Thallium deposits with the zinc. It has no appreciable effect on current efficiency but increases the deposition of lead in the zinc in the absence of cobalt.

Tin above 2 mg/liter decreases current efficiency significantly but does not interfere with deposition of zinc up to about 50 mg/liter.

Tungsten is moderately detrimental to current efficiency, 100 mg/liter reducing it by about 2%.

Fluorine is the most troublesome of the strictly nonmetallic impurities because it attacks the aluminum cathodes, shortening their life, and renders stripping of the cathodes difficult. A fluorine concentration of 30–35 mg/liter greatly shortens cathode life by corrosion at the solution line.

Chlorine exceeding about 100 mg/liter increases corrosion of the lead anodes, where it may also be oxidized electrolytically to perchlorate, ClO_4 (but not to chlorate, ClO_3). The age of the anodes affects their oxidation potential and thereby the amount of perchlorate produced. Silver from the feed solution or Tainton-type anodes aids in reducing free chlorine to harmless concentrations.

Nitrates have an effect similar to that of chlorides, increasing corrosion of lead and deposition of it at the cathode.

Sodium, potassium, magnesium, calcium, aluminum, etc, are generally harmless impurities electrolytically, but they may accumulate to a point where they become more or less troublesome mechanically by crystallizing out. If feed solution must be cooled before electrolysis, calcium sulfate may greatly hinder heat transfer through the cooling coils by depositing as a hard scale. A method of electrolytic removal of the scale has been developed.

Experiments have been made on using various kinds of diaphragm materials to prevent migration of impurities to the cathode, but as far as is known, no commercial applications have been made.

Melting and Casting. Stripped cathode zinc is melted and cast into shapes suitable for commercial purposes; at this point alloying elements can also be added to meet special requirements. The older practice used reverberatory furnaces about 30×15 ft in overall dimensions from which the melted zinc was dipped with a 300-lb capacity ladle and cast by hand into molds on a stationary rack. Recent practice utilizes induction heating, possibly combined with a gas-fired furnace, and the molten zinc is cast into 60-lb slabs on an in-line casting machine. Some zinc is cast into 1400- to 2400-lb blocks in stationary molds. Molten zinc exhibits a strong tendency to form dross if not protected by a blanket of flux. Solid zinc at temperatures near its melting point does likewise, the tendency being stronger when the ratio of surface area to volume is high, as in the case of electrolytic cathodes freshly removed from the tanks. Whereas ordinary slab zinc can be melted in a reverberatory furnace with the formation of 1% or less of dross, electrolytic cathodes lose 6–17% under the same circumstances, the actual percentage varying with the presence of glue, cobalt, etc, in the electrolyte.

To achieve an overall melting efficiency of about 96%, dross must be retreated by liquation, centrifugal separation of metal, fluxing, etc, to separate the globules of metal from their oxide shells. Metal recovered in this way may be used to make zinc dust for the purification of the electrolyte, or may be returned to the melting pots.

Zinc Oxide Production Processes

Although there have been improvements over the years in the equipment used in the production of zinc oxide, the fundamental metallurgy remains much the same.

Production of zinc oxide closely resembles that of the metal to the extent that zinc must first be produced in vapor form whether its final destiny is condensation or oxidation. Where the zinc vapor is first condensed to metal before the final vaporization and oxidation, the product is known as indirect, or French, process oxide. Where the zinc vapor is immediately oxidized without ever being condensed, the product is known as direct, or American, process oxide. It is interesting to note that the German term for French process oxide is "zinkweiss" (zinc white), whereas the term for American process oxide is simply "zinkoxyd."

Processes for production of zinc oxide normally are somewhat less difficult to operate than those for metal, in the sense that two of the main difficulties in metal production, namely dilution of the zinc vapor and reoxidation by carbon dioxide, become desirable tendencies when the objective is oxide production. The above problems, however, are replaced by the necessity of producing the oxide in specified particle-size ranges, certain preferred particle shapes, and with an extremely low content of undesirable impurities, notably copper, manganese, and iron, if the oxide is to be incorporated into rubber.

Grate-Type Furnaces. Soon after 1850, a direct process for production of zinc oxide was developed at Newark, N.J., by Samuel Wetherill of the New Jersey Zinc Co. This was the first commercial process to be identified as the "American" process. It consisted of a stationary grate; on it was first spread a bed of coal which, in operation, would be ignited by residual heat from the previous charge. After ignition, a second layer consisting of zinc ore and coal was spread upon the ignited coal, and air blown through the bed from below to provide both heat and carbon monoxide for the reduction of the zinc.

From this point, the process developed in two slightly different directions, although in principle the two were alike.

Eastern Wetherill Furnace. The eastern arrangement is to build a block of four or more furnaces with a common back-wall line and common side walls. All four furnaces are operating at any given time, but each is at a different point in its cycle of charging, burning, and discharging. Vapors from each furnace enter separate combustion chambers, permitting the control of the combustion air supply to match the supply of vapor as it varies over the cycle. Airborne oxide from all chambers discharges to a common duct leading to a bagroom. This method of "blending on the fly" the products from four different stages of the cycle tends to produce a more uniform bagged product.

Western Wetherill Furnace. The essential difference between the eastern and western Wetherill types is that in the western furnaces, a block of which may comprise twelve or more, vapors from all pass into a common overhead flue, which conducts them to the end of the block and into a common combustion chamber. By blending the vapors before oxidation, uniformity in the supply and degree of dilution of the vapors is achieved.

Traveling-Grate Furnace. The hot and laborious job of discharging and recharging the Wetherill grates, plus the necessity for controlling the factors affected by the varying rates of vapor evolution during the cycle, led to the development of a continuous, traveling-grate-type furnace about 1919. The grate itself is a traveling grate stoker first developed for firing steam boilers. It consists of a series of cast-iron grate bars about 12 ft long, attached at each end to supporting chains. These bars support the grate "keys," the actual bearing surface for the charge, through a set of malleable iron "dovetails." The grate traverses the funnel-shaped firebrick furnace chamber, passes over sprockets at the end of the traverse, and is returned to the front for another pass.

Difficulties with accumulation of slag on the grates led early to the development of a briquetted charge. Briquets are made in two compositions, the first consisting of ore and reduction fuel, and the second of fuel only. The preferred type of coal is a fine-sized anthracite which minimizes contamination of the vapor stream with soot, some of which might not be completely burned in the combustion chamber, and prevents caking and slagging.

The briquet press is located directly over the furnace, which permits the briquets to be dried with waste heat from the furnace. The decrease in the amount of handling required also reduces the need for as much binder as would otherwise be necessary. As the traveling grate moves forward into the furnace tunnel, it first receives a bed of fuel briquets about 6–6½ in. deep. This bed is then ignited by contact with previously ignited fuel briquets in a short, hot section and brought to about 1000–1950°C. The grate next travels under an ore hopper, where a bed of ore briquets of controlled depth, usually about 6 in., is deposited on the bed of fuel briquets.

The charge now in place, the grate advances into the reduction zone, where forced upward draft begins to burn the fuel briquets. Reduction of the first-formed carbon dioxide to the monoxide by excess carbon furnishes the reducing atmosphere for liberation of the zinc vapor. The flow of air through the grates can be regulated by means of ten separate wind boxes, permitting a gradual increase in flow as the charge advances. Such division also minimizes difficulties due to short-circuiting of air through accidentally produced open spots in the charge. Once the vapor leaves the furnace tunnel through openings in the roof, the combustion and collection aspects are similar to those of other processes.

The traveling-grate furnace permits close control over operations, enabling the manufacture of various grades of the oxide (particle shape, size, chemical composition) by altering such variables as the composition of the charged ore and the conditions of combustion. In this way, the producer can meet the specifications for the oxide required by such varied consuming industries as paint, rubber, etc.

Rotary (Waelz) Kiln. The Waelz kiln is essentially a large-diameter, long rotary kiln used commonly for pyrometallurgical concentration of residues of a rather mixed character, although it can be used for production of pure end products, of which zinc oxide is one. The kiln has shown satisfactory ability to volatilize zinc, lead, cadmium, arsenic, antimony, tin, and bismuth. Such varied materials as oxidized and sulfide zinc ores, calamine, waste dumps, jig tailings, jig slimes, table concentrate slimes, brass ashes, zinc-bearing iron ores, electrolytic zinc-leach residues, zinc-retort residues, lead-furnace slags, tin ores and slags, and antimonial and arsenical gold ores, have been successfully treated in the Waelz kiln.

The name Waelz is derived from the German word "waelzen," meaning a trundling motion (from which is derived the English "waltz") alluding to the manner of movement through the very slowly rotating kiln, a typical range of rotation being 1–1.5 rpm. A Waelz kiln may be from 35 or 40 ft to more than 100 ft in length and 8 or 10 ft in diameter with a 6-in. lining of super-duty fire brick inside the steel shell. Toward the discharge end 9-in. risers frequently replace every third 6-in. brick so as to provide a "lifting" or "trundling" effect to the solid charge. The kiln is set up with a slight slope, small enough to give sufficient retention time to eliminate zinc from the charge.

A typical charge may consist of 65–75% zinc-containing material and the remainder crushed coke or anthracite (to give about 25% carbon in the charge) plus a small amount of sand (in order to stiffen the bed, if desired). The additional heat necessary for the reduction of the contained zinc is supplied by gaseous fuel. Careful control of draft and of the admittance of air is necessary in order to maintain the correct reducing atmosphere in the kiln. Both cocurrent and countercurrent flow of the gaseous fuel and the zinc-containing charge have been used, countercurrent being more general. Temperature control is one of the most important aspects of kiln operation, because the success of the operation usually depends on the maintenance of intimate contact between the solid, value-bearing part of the charge on the one hand, and the solid fuel in the charge and the reducing atmosphere on the other. Too low a temperature reduces kiln effectiveness and capacity; temperature too high causes coalescence or actual melting of the charge, thereby excluding both the solid fuel and reducing gas from the interior of the molten mass. The latter condition can cause accretion rings to build up, interfering with the time of residence of the charge and the bed depth, which, in turn, affects reaction rates.

Temperature control may be effected by varying one or more of the following factors: (a) fuel-gas consumption, (b) solid-fuel consumption, (c) inflow of external air, (d) rate of kiln rotation, (e) proportion of feed containing endothermically or exothermically decomposable materials or water, and (f) on smaller kilns only, kiln slope. Besides the overall temperature conditions, the zone of maximum temperature can be varied by careful attention to the ratio of several of the above quantities to each other.

The solid residue from the kiln drops into a quench tank and thence to suitable storage. The mixture of zinc vapor and combustion gases passes from the Waelz to a chamber of somewhat larger diameter where entrained dust is dropped because of the decrease in velocity and the change in the direction of the gases. From here the vapor gases go to a combustion chamber where an additional carefully regulated supply of air is admitted and the zinc vapor burned, conditions being carefully controlled so as to yield a maximum of particles of the size and shape desired. The gas-entrained zinc oxide then passes through suitable cooling units, which may also function as a separator of coarser oxide particles, and on to a conventional baghouse.

The recovery of zinc by volatilization in a rotary kiln has its origin in a process patented in 1910 by Edward Dedolph of British Columbia. This patent was taken over by Metallgesellschaft of Frankfurt, Germany. Neither Dedolph nor Metallgesellschaft succeeded in developing the process to a state suitable for large-scale operations before World War I, probably owing mainly to overheating of the charge. After the war, Metallgesellschaft and Krupp Grusonwerk collaborated on development work, bringing the process to commercial workability in 1923.

Vertical Electrothermic Process. The St. Joseph vertical electrothermic furnace (see Fig. 3) may be modified for use as either a metal or an oxide producer; in fact, for the first six years of plant operation, only oxide was produced.

Up to the point of actual removal of the mixture of zinc vapor and carbon monoxide (approx 1:1) from the furnace, the paths of charge materials in both furnaces are similar. But, whereas in the metal furnace a large "vapor ring" or bulge in the furnace barrel is provided about midway between the upper and lower electrodes to collect the vapor for passage to the condenser, the oxide furnace is spotted with openings, or tewels (exit ports), at four levels between electrodes. Each vertically-in-line group of tewels feeds vapor and carbon monoxide into a common duct, which is open at each tewel level to admit air. Zinc vapor and carbon monoxide are immediately oxidized to zinc oxide and carbon dioxide and drawn off through ducts to a cleaning cyclone and thence to bagrooms for collection. Blending and reheating may follow, depending on the product specifications to be met.

Indirect (French) Process. Zinc oxide produced by this process, that is, by remelting slab zinc and oxidizing the vapor, has accounted for about 25% of the total zinc oxide produced in the United States in the 1960s. Of the indirect-process oxide, probably 75% or more is used by the rubber industry, principally where fast-curing zinc is specified, this desirable characteristic being due to the absence of sulfur–oxygen compounds. The remainder goes largely to the paint industry, and a small portion to pharmaceutical manufacturers, where the purity of the oxide is a major consideration.

The equipment used in the manufacture of French process oxide is simpler than that used in the direct American process because in the former the starting material is slab zinc and it is not necessary to go through the reducing stage. In the French process, the purity and grade of the slab zinc is the chief controlling factor in determining the purity of the resulting zinc oxide because the product is not contaminated by the gaseous products of the reduction step which may be present in direct-process oxides.

The simple metallurgy of this process has not altered since the mid-1850s though there have been developments in the matter of equipment. The horizontal retort—the original type of furnace—is still being used to some extent. Other types of furnaces are vertical refining columns, electric-arc vaporizers, and rotary burners. Each of the four types of furnaces has its merit for the operating conditions at the plant involved, and of course the physical properties desired for the zinc oxide must also be taken into consideration. In all cases, the charged zinc has to be melted, vaporized, oxidized, cooled, and collected. The physical characteristics of the zinc oxide bear a distinct relationship particularly to the manner in which the oxidizing, cooling, and collection steps are carried out.

Zinc Oxide Recovery from Lead Blast-Furnace Slag. Lead blast-furnace slag usually contains 10–18% zinc which can be recovered by carbon reduction. If the zinc is to be recovered as metal extra heat must be supplied by means other than oxidation of carbon by air, because of the diluting and oxidizing effect of the combustion products, including that from the additional carbon, on the zinc vapor. To meet this requirement the St. Joseph Lead Co. developed its electrothermic slag-fuming process.

However, if the zinc is to be recovered as oxide, the carbothermic process becomes quite feasible. It should be noted that this oxide fume is impure and for conversion to metal is sent to an electrolytic zinc or zinc retort plant, though it may go to pigment market outlets also. This slag-fuming process, with various modifications, is used by the following plants:

Anaconda Company, East Helena, Mont.;
American Smelting & Refining Co., El Paso, Tex., Selby, Calif., and Chihuahua, Mexico;
Bunker Hill Company, Kellogg, Idaho;
International Smelting & Refining Co., Tooele, Utah;
Hudson Bay Mining & Smelting Co., Flin Flon, Manitoba;
Consolidated Mining & Smelting Co., Ltd., Trail, B.C.

In this process a coal–air mixture is introduced through the tuyeres into a molten slag bath which has been transferred by ladle from the blast furnace to the fuming furnace. Though the coal–air flow to the furnace is continuous, the operation is a batch one in the sense that there is about a two-hour "blowing time" between charging and tapping which is adequate for good zinc elimination from the molten bath. The amount of charge varies with the furnace and the immediate circumstances but can run all the way from thirty to sixty tons. The "pot shell" of slag left in the ladle is usually returned to the blast furnace for remelting. Granulated slag up to about 15% of the total charge may be added if desired and operating conditions permit. The coal, which amounts to about 20% of the slag weight, must be pulverized and suitable auxiliary equipment for handling the coal–air mixture and introducing it into the furnace through the tuyeres must be provided. The amount of air required is about that necessary to burn the coal to carbon monoxide. This represents approximately the lower limit of air, as the charge is in danger of freezing, while more air results in CO being burned to CO_2, thus losing reducing power. The optimum operating temperature is about 2100–2200°F.

The fuming furnace itself is somewhat similar to a reverberatory furnace made of a series of steel water jackets about 8–10 ft wide, varying in length from about 15 to 24 ft and in height from 6 to 10 ft.

Remembering that the air supply is carefully controlled and limited, the reactions in the molten bath may be represented by the following equations:

$$2\,C + O_2 \rightarrow 2\,CO$$

$$ZnO + CO \rightarrow Zn + CO_2$$

The zinc vapor in the mixture with carbon monoxide rising from the slag bath is oxidized to zinc oxide by a tertiary air supply either blown or drawn into the upper part of the furnace chamber. After necessary cooling, which may be accomplished by waste-heat boilers plus whatever auxiliary cooling is necessary, the gas enters automatic baghouse units. These are provided with the usual mechanical discharge and collection equipment. The slag from the fuming furnace is tapped from below the tuyere level and granulated by a jet of water.

The recovery of the zinc as oxide is 90% or better of the zinc in the charge. The recovery of the lead is of the same order. The zinc content of the furnace slag is about 1.0 to 1.5%. The fume has a zinc oxide content of about 70% or better along with 7–10% lead, depending of course upon the character of the furnace feed. If removal of the lead from the fume is desired before retorting or conversion to zinc, this may be accomplished by densifying and fuming off the lead (about 1–2% of coke added) in a kiln treatment at 2200–2300°F.

Economic Aspects

Mine production of recoverable zinc from ores mined in the United States from 1963 to 1969 is listed in Table 4.

Table 4. U.S. Zinc Production, Short Tons

Year	Production	Year	Production
1964	574,858	1967	549,413
1965	611,153	1968	529,446
1966	572,558	1969	544,131

Table 5.   1968 and 1969 Mine Production in Leading Zinc-Mining States, Short Tons

State	1968	1969	State	1968	1969
Tennessee	124,039	127,000	Pennsylvania	30,382	31,510
New York	66,194	58,699	New Jersey	25,668	24,645
Idaho	57,248	58,157	New Mexico	18,686	24,220
Colorado	50,258	54,400	Wisconsin	25,711	23,000
Utah	33,153	33,000	Virginia	19,257	18,859
Missouri	12,301	32,300	Illinois	18,182	14,300

There was also a considerable shift in the 1960s in the standing of the states with respect to the mine production of recoverable zinc. The Tri-state area of Missouri, Kansas, and Oklahoma, which led for so many years (frequently followed by Montana), has been displaced; two states east of the Mississippi River (Tennessee and New York) have now taken top places. However, while Montana does not presently appear on the list of major producing states, Missouri has staged a very decided comeback, currently ranking about fifth or sixth in production due to the increased recovery of zinc in the new lead belt in southeastern Missouri. Whereas the western states formerly accounted for the larger part of the recoverable zinc produced, the eastern states have increased their proportion till it is about half the total.

Table 5 gives the mine production—not to be confused with smelter production—of recoverable zinc in 1968 and 1969 of the twelve leading states. In the latter year, these twelve accounted for almost 92% of the total United States production.

Comparison of the 1969 U.S. production of recoverable zinc with the U.S. consumption of slab zinc, as listed in Table 6, shows a substantial difference, which has to be made up by imports. These imports enter the United States under two chief classifications: (1) as ores and concentrates, and (2) as metal, ie blocks, pigs, or slabs. In each of the years 1968 and 1969 the dutiable zinc of the imported ores and concentrates alone, amounting to 546,382 and 602,120 short tons, respectively, exceeded the U.S. mine production of recoverable zinc. In addition, of course, are the imports of slab zinc which amounted to 329,000 tons in 1969, comprising about 35% of the total zinc in all forms imported in that year. Of this figure Canada supplied somewhat more than 45%, with Japan ranking next. Of the zinc arriving as concentrates, Canada supplied about 60%, Mexico almost 25%, and Peru about 10%.

There is a limited export of zinc metal and zinc-containing material but this is very minor and in 1969 amounted to only slightly more than 11,000 tons. The significance of these figures is that the United States is very much dependent upon imports to handle its substantial industrial requirements for zinc. Table 6 gives salient commercial statistics of zinc.

World mine production in 1968 of zinc (based on content of ore) is reported to have been almost 5,500,000 short tons (see Table 6). Production of the ten most important countries is shown in Table 7, accounting for over 85% of the total world mine production.

The distribution of zinc concentrates in world commerce is shown in Table 8, which gives the smelter production of the ten most important countries in 1968. Four of the countries that appear in Table 7 are replaced by Belgium, France, the United Kingdom, and West Germany, all heavily industrialized countries. The ten countries

Table 6. Commercial Statistics of Zinc, Short Tons

	1967	1968	1969
U.S. production			
domestic ores, recoverable content	549,413	529,446	544,131
slab zinc			
from domestic ores	438,553	499,491	
from foreign ores	500,277	521,400	
from scrap	73,505	79,865	
U.S. imports			
ores and concentrates (zinc content)	534,092	546,382	602,120
slab zinc	222,112	306,540	329,008
U.S. consumption			
slab zinc	1,236,808	1,333,699	1,362,887
all classes	1,591,997	1,728,400	1,729,387
price, prime western,			
East St. Louis, ¢/lb	13.85	13.50	
world production			
mine	5,330,519	5,471,071	
smelter	4,549,667	5,017,196	
price, prime western,			
London, ¢/lb	12.37	11.89	

Table 7. Zinc Production of the Ten Leading Countries, 1969

Country	Short tons	Country	Short tons
Canada	1,273,249	Japan	291,300
U.S.S.R.[a]	595,200	Mexico	264,575
United States (recoverable)	529,446	Poland[a]	174,200
Australia	463,409	Italy	154,102
Peru	340,720	Congo (Kinshasa)	139,473

[a] Estimated.

Table 8. Leading Countries in Smelter Production of Zinc, 1968, Short Tons

Country	Production	Country	Production
United States	1,020,891	Australia	230,216
Japan	667,504	France	228,507
U.S.S.R.[a]	595,200	Poland	223,138
Canada	426,929	United Kingdom	157,491
Belgium	280,315	West Germany	134,481

[a] Estimated.

listed account for 3,964,672 short tons, or almost 80% of the total world smelter production of 5,017,196.

The amount of secondary zinc recovered annually in the United States in the 1960s in all forms is about 35% of the U.S. production of primary slab zinc from both domestic and foreign ores. This is very low when compared with copper and lead which may run as much as 100% for copper and even more at times for lead. This is a good indicator of the extent to which zinc function was a sacrificial corrosion preventive for steel.

Almost one quarter of the zinc scrap processed in the United States is recovered as redistilled or remelted slab zinc. Over 45% is recovered in the form of brass or bronze alloys, with zinc dust, leadfree zinc oxide, zinc-base alloys, zinc chloride, and zinc sulfate following in that order, but all far behind brass.

Table 9 shows the amount of slab zinc used for various industrial purposes in 1968 and 1969.

Table 9. Industrial Uses of Slab Zinc in the United States, Short Tons

Use	1968	1969
zinc-base alloy		
die-casting alloy	551,896	549,895
dies and rod alloy	807	498
slush and sandcasting alloy	10,243	6,654
total	562,946	557,047
galvanizing		
sheet and strip	256,319	254,616
wire and wire rope	36,089	24,233
tubes and pipe	63,621	60,525
fittings for tubes and pipe	13,801	11,489
structural shapes	20,238	16,969
fencing, wire cloth, netting	15,984	15,005
other and unspecified	75,765	66,052
total	481,817	448,889
brass and bronze		
sheet strip and plate	86,185	90,221
rod and wire	49,888	54,707
copper-base ingots	12,153	13,274
other copper-base products	13,680	17,027
total	161,906	175,229
rolled zinc	48,943	46,917
zinc oxide	34,937	41,447
other uses[a]	43,150	42,358
undistributed consumption[b]		51,000
Grand total	1,333,699	1,362,887

[a] Includes zinc used for zinc dust, wet batteries, desilverizing lead, light-metal alloys, and other miscellaneous uses.

[b] Estimated for plants reporting on annual basis only.

In the United States, the term slab zinc applies to various smelter shapes and sizes, which were formerly called "spelter," a term now only seldom used. As stated earlier, zinc in zinc-base alloys has replaced the application for corrosion-protective purposes (galvanizing) as the largest user of slab zinc. In third place is its use in brass and bronze, two alloys which have been known since ancient times.

Grades and Specifications

In the earlier days of the zinc industry, the composition of the ore and the metallurgical procedure used in treating the concentrates had a major influence on the specifications which were established. Nowadays the technical requirements for the zinc metal in its various uses have a very important effect on the specifications. As a matter of fact, the upgrading in the quality of zinc available has made it possible to develop

Table 10. ASTM Impurity Limitations for Slab Zinc[a]

Grade	Max, %			Zinc, min by difference,[b] %
	Pb	Fe	Cd	
special high grade[c]	0.003	0.003	0.003	99.990
high grade	0.07	0.02	0.03	99.90
intermediate	0.20	0.03	0.40	99.5
brass special	0.60	0.03	0.50	99.0
prime western	1.60	0.05	0.50	98.0

[a] When specified for use in the manufacture of rolled zinc or brass, aluminum, if found by the purchaser, shall not exceed 0.005%. Greater amounts may constitute cause for rejection.

[b] Analysis need not regularly be made for copper, tin, and aluminum in any grade. Nevertheless, it is understood that the min % of zinc (by difference) takes into account the copper, tin, and aluminum contents, if any, in addition to the impurities listed in the table.

[c] Analysis need not regularly be made for tin in special high-grade zinc but, if found, shall not exceed 0.001%.

new and important uses, a prime example being the use of zinc castings where the metal must be free of any sensible contamination to prevent damaging intercrystalline corrosion and swelling.

Slab zinc in the United States is produced in five standard grades ranging from about 98.0 to 99.99% zinc. The maximum amount of each of the three chief impurities (Pb, Fe, and Cd) permitted, according to the 1970 ASTM specifications, is given in Table 10. However, the minimum zinc content cannot be lower than the figure given in the final column of the table. This means that all of the impurities cannot be at the maximum for each in any single lot of slab zinc.

The special high-grade zinc meets the stringent requirements for die castings. Production of zinc metal of this purity is a distinct achievement and has contributed in no small degree to the fact that zinc-base alloys now represent the largest single use of slab zinc.

It is an interesting commentary on the ability of the zinc metallurgical industry to produce a high-purity zinc that it is found desirable in some cases to actually add "impurities." As an example, the producer may add lead and even aluminum to the zinc sold for galvanizing purposes. The galvanizer desires some lead in his zinc because it minimizes maintenance problems. The solubility of lead in zinc at galvanizing temperatures (425–460°C) is about 1%. Any excess tends to concentrate at the bottom of the galvanizing kettle, protecting it from attack by the molten zinc and facilitating removal of the zinc–iron alloy dross. Other grades of special zinc may be made in a similar manner; for example, high-purity electrolytic zinc may be "debased" with lead to meet the prime western specifications. When the price of lead is higher than that for zinc, such addition actually increases the cost of the zinc to the producer.

Uses of Zinc Metal

The uses of zinc metal and the trends which have occurred in the 1960s have been referred to from time to time in this article. Mention has been made of the fact that the use of zinc in zinc-base alloys primarily for die-casting has replaced the sacrificial employment of zinc in varied galvanizing operations as the major outlet for zinc. It should be noted that there has been no reduction, but rather an increase in the tonnage of zinc used annually for galvanizing purposes. However, the gain in the employment of zinc for zinc-base alloys has considerably outstripped the increase in the tonnage of zinc used for galvanizing.

The magnitude of the galvanizing industry is shown by the fact that almost 450,000 tons of slab zinc were used for various galvanizing operations in 1969 (481,817 tons in 1968), amounting to about 33% of the total annual slab consumption.

Zinc for galvanizing purposes is now ranking in second place in annual quantity of zinc consumed; a somewhat distant third place, through still very significant in quantity, is occupied, as it has been since the 1940s, by the use of zinc in various brass products, 161,906 and 175,229 tons in 1968 and 1969, respectively.

The fourth area of importance with respect to the amount of slab zinc consumed is its use in rolled zinc products of various types. From the standpoint of physical metallurgy, the broad descriptive all-inclusive term for this area of application is wrought zinc. The total amount of slab zinc consumed for rolled zinc was 48,943 and 46,917 tons in 1968 and 1969, respectively, representing less than 4% of the total annual slab-zinc consumption.

Fifth in yearly quantity of slab zinc consumed is the production of zinc oxide, 41,000 tons in 1969 and about 35,000 tons in 1968. Other miscellaneous and minor uses include the use of zinc in making light-metal alloys, in desilverizing lead, and in wet batteries.

While die-casting (of zinc-base alloys) has soared to first place as a use of slab zinc, the metal can also be processed by any of the usual mechanical operations of rolling, drawing, spinning, extrusion, stamping, etc. However, it must be remembered that a zinc-base alloy which contains a small amount of copper or even fractional percentages of other metals, such as iron, nickel, chromium, titanium, or manganese, cannot be satisfactorily worked. The maximum solid solubilities of these metals in zinc are always small—usually fractional percentages—in the low-temperature range of practical operation; for example, copper dissolved in zinc up to about 2.5% at 400°C and only 1% at 200°C and manganese and aluminum about 0.5 and 1%, respectively, at 200°C.

Zinc-Base Alloys

In the early 1940s the automobile industry recognized the value of zinc-base die-casting alloys. Until then the zinc consumption for this purpose was comparatively small. The rapid increase for this application is shown in Table 11.

A major reason for this tremendous increase was the improvement in overall metallurgical practice in the zinc industry, making increasingly available a very large

Table 11. Consumption of Zinc-Base Die-Casting Alloys, Short Tons

Year	Consumption
1926	13,500
1930	21,500
1935	55,500
1940	58,000
1945	130,836
1968	562,956
1969	557,047

tonnage of high-purity metal meeting extremely stringent specifications. As a matter of fact, of the almost 563,000 tons used for zinc-base alloys in 1968 about 560,000 tons was special high-grade metal. Zinc of such a high degree of purity eliminated the troubles experienced in earlier years with zinc-base castings. The availability of high-purity zinc and the progress made in the die-casting industry itself have made possible products of excellent stability (good resistance to corrosion), easy machinability, attractive appearance with distinct sales appeal, and comparatively low cost. As a result, zinc-base die-castings are now employed extensively in many industries other than automotive, the latter using about 55–60% of the total production in such items as carburetors, speedometer cases, fuel pumps, radiator grills, windshield-wiper motors, door handles, and other interior and exterior hardware. The home-appliance industry (sewing machines, ironers, washing machines, kitchen and food-mixing equipment, radio and television sets, etc) is the next most important consumer, using close to 20% of the production. Some of the other more important uses are in commercial machines and tools of all types, builder's hardware (plumbing and heating), office equipment and business machines, optical and photographic equipment, etc.

The number of zinc-base alloys in general use for die-casting is quite limited, actually only two or three in number. Aluminum is the major alloying constituent, in amounts varying from 3.5 to 4.3%. It imparts increased strength, reducing the grain size and also decreasing the attack of the alloy on iron and steel parts. Aluminum content either above or below the range given affects the properties of the alloy adversely.

Second to aluminum is copper, varying from 0.25% max in one alloy to 0.75–1.25% in another. The effect of copper is to increase tensile strength and hardness. However, it tends to lower the impact strength and dimensional stability on aging. For most commercial uses a copper content in the range of 0.25 to 0.75% is not objectionable but above that the effects, both favorable and unfavorable, imparted by the copper have to be evaluated carefully.

Magnesium is added to zinc alloys in amounts ranging from 0.03 to 0.08% but usually not more than 0.05%, because a larger amount may cause difficulty in casting, mainly hot shortness. This small amount reduces subsurface corrosion and also counteracts the harmful effects of small quantities of impurities such as lead and tin.

There are stringent limits to the amounts of iron, lead, cadmium, and tin that can be present in zinc-base alloys for die-casting. These are 0.100% for iron, 0.007% for lead, 0.005% for cadmium, and 0.005% for tin. Accelerated aging tests indicate that subsurface corrosion does not occur with 0.007% lead if a minimum of 0.03% Mg is present in the alloy. With lead at 0.009%, subsurface corrosion is very likely to occur. No trouble occurs with Cd at 0.005%; indeed, tests indicate a considerable factor of safety. In considerably larger amounts, Cd can be harmful, affecting adversely the castability and the mechanical properties of the casting. Tin, like lead, induces subsurface corrosion; the amount present must not exceed 0.005%.

For melting zinc-base alloys, cast steel, cast iron, or welded steel pots ranging in capacities from about 2000 to 8000 lb are used. The pots are placed in furnaces which are refractory-lined cylindrical steel shells suitably insulated. They are usually gas-fired although oil or electric heat may be employed.

The casting machine is a rather intricate affair of the "hot" chamber or submerged plunger type which injects molten metal into the die at temperatures up to 500°C and at pressure as high as 2500 psi. The operation has become largely automatic because of the development of electronically controlled valves, relays, and limit switches, and the aid of timing devices. The operator only presses buttons to close and open the die and to remove the casting.

The dies themselves with the cores and auxiliary devices and methods of casting may be rather complex, indicating the degree of mechanical development that has been attained. High-quality steel is ordinarily used for the dies, though sometimes low-alloy steels are employed. Since zinc-base alloys have a lower melting point than aluminum, brass, or magnesium, casting problems are somewhat simplified as compared with these of other metals and equipment life, particularly that of the dies, is longer.

Zinc die-castings are comparatively small as may be deduced from their various uses. In fact, the maximum weight of a casting rarely exceeds $\frac{1}{2}$ lb and is usually considerably less. They are used in the "as-cast" state in many applications but since they do tarnish in damp atmospheres, finishes are frequently applied to protect against corrosion from moisture or other atmospheres and also to improve their appearance and thus increase their sales appeal. Various metals, such as copper, chromium, brass, and even gold and silver, may be applied by suitable electroplating operations. Organic finishes are sometimes applied.

Slush- and Sand-Casting Alloys. The amount of slab zinc used for this purpose is small indeed as compared with that required for die-casting; actually it amounted to only 6654 tons in 1969 and 10,243 tons in 1968. Its primary purpose is the production of hollow castings. This is done by freezing a metal shell on the inner surface of a split mold and quickly inverting the mold so that the molten metal flows away from the thin shell-casting left behind.

The principal metallurgical requirements for slush-casting alloys are (a) fairly low freezing temperature, such as a zinc-base alloy provides, and (b) solidification should occur over a range of temperatures rather than at one definite temperature in order to produce the slushy consistency.

The following zinc-base alloys find use in slush casting, using special high-grade zinc in both cases:

Alloy	Zn, %	Al, %	Cu, %
1	94.5	5.5	
2	95.0	4.75	0.25

Slush castings are used primarily in the lighting fixture and novelty field.

Zinc for Galvanizing Purposes. See Galvanizing, Vol. 13, p. 254; Sherardizing, Vol. 13, p. 264; Electroplating, Vol. 8, pp. 46, 66.

Zinc for Brass Products. See Copper Alloys.

Rolled Zinc. The use of slab zinc for rolled products ranks a somewhat distant fourth behind the use for brass products. Production figures are listed in Table 12.

Rolled zinc is produced in all the conventional forms, such as sheet, strip, plate, rod, and wire of various compositions. The trend has been to use special high-grade and high-grade zinc, adding the alloying metals, such as copper, magnesium, manganese, chromium, and titanium in carefully controlled amounts.

Metal of the desired composition is melted in a gas-, oil-, or coal-fired refractory-lined, reverberatory-type melting furnace. Induction furnaces are sometimes used when careful analytical control of the product is necessary. The temperature of the furnace is maintained at about 445–510°C. Molds are heated to a temperature of

Table 12. U.S. Production of Rolled Zinc, Short Tons

Year	Production
1945	97,585
1950	68,444
1955	51,589
1960	38,696
1966	52,612
1968	48,943
1969	46,917

about 80–120°C, depending upon the analysis of the metal and the type of mold. The mold is usually of cast iron of a grade suitable for glass molds. The flat rectangular rolling slabs cast into these molds are about $3/4$ to 4 in. thick, this dimension, as well as the width and length, depending upon the end product desired and the rolling equipment.

The temperature range of 150 to 260°C to which the rolling slabs are heated prior to the rolling operation depends largely upon whether sheet, strip, rod, or wire is the desired product, being the highest in the case of sheet. Composition also affects the choice of temperature. Zinc as cast tends to form low-melting eutectics with such common impurities as cadmium, bismuth, lead, tin, and indium. Because these eutectics are the last to solidify when the slab is cast, they become concentrated at the grain boundaries. On heating the slab to rolling temperature, the grain boundaries become weak points in the coherence of the slab and may permit individual grains to separate from each other under rolling pressures. There may be an optimum concentration of a particular impurity, above and below which the hot shortness of the slab is less. As little as 0.01% of the offending metal has been known to cause serious trouble. Such facts as these emphasize the need for close control of the composition of the rolling zinc slabs and why the trend is toward the use of special high-grade and high-grade zinc as the starting point to which carefully controlled alloying metals can be added.

The precise steps in the rolling operation vary with the end product. In the case of sheet zinc, reductions on the slab roll start at about 10% and are increased to 30% as rolling proceeds, while in the case of strip zinc it starts at from 8 to 30% and is finished at about 15 to 45%. With temperatures of rolling slabs and molds as stated above, the temperature of the rolls is held between 100 and 200°C. The rate of reduction depends on the analysis of the metal, the type of equipment, and the surface and grain desired.

If sheets are being rolled, up to thirty and even more rough-rolled sheets are stacked together and pack-rolled, thus considerably increasing the capacity of the mill. Virtually all the loss in thickness in any given pass is manifested as an increase in the dimension in line with the direction of rolling rather than as an increase in density. (The density of rolled zinc is 7.14, only slightly greater than the 7.133 for pure cast zinc.) Consequently packs are rolled at right angles to the direction of rolling on the slab mill.

Not all sheets within the pads are reduced the same amount in each pass. The amount of restriction on the outer sheets is less than on the inner ones, permitting the outer ones to be reduced further. The additional work done thereby on the outer ones tends also to raise their temperature. Therefore, to equalize the treatment each sheet receives, packs are frequently split, interchanging inner and outer sheets. Roll temperatures are usually low in pack rolling. Consequently, the zinc is cold-worked. If a bright ductile product is required, as is more often the case in strip rolling (not done by the pack method), rolls are held at 120–150°C, and the reduction on the last pass is at the highest limit of from 20 to 40%.

Lubrication of sheet and strip is necessary for all operations. A mixture of paraffin and tallow oil is generally favored, although for special operations vegetable and mineral oils may be used. Requirements for a finish-roll lubricant are more exacting because of the higher temperature employed and the possibility of staining due to pyrolysis of the oils, or of impurities in them. Strip zinc is usually finish-rolled with a vegetable oil such as cotton seed.

Rolled zinc enters into a wide variety of products. Analysis must be controlled so as to give to the zinc the proper working characteristics for the desired products. Among the uses of rolled zinc are photoengraving, lithography (the "offset lithography" process, see Retrography), dry cells (see Batteries), weather stripping and other building and architectural applications, and many other similar uses. It should be noted that in Europe, rolled zinc finds extensive application as a long-lasting roofing material.

Bibliography

"Zinc and Zinc Alloys" in *ECT* 1st ed., Vol. 15, pp. 224–275, by A. Paul Thompson and H. W. Schultz, The Eagle-Picher Co.

General References

Zinc Abstracts, published monthly since 1942 by the Zinc Development Assoc., New York.
C. H. Mathewson, *Zinc: The Science and Technology of the Metal, its Alloys and Compounds*, Reinhold Publishing Corp., New York, 1959.
Mineral Yearbooks and *Mineral Industry Surveys*, published annually by the U. S. Bureau of Mines, Washington, D. C.
Mineral Facts and Problems, published every five years by the U. S. Bureau of Mines, Washington, D. C. (last Vol. 1965).
Annual Review of the Zinc Industry, Zinc Institute, New York.
Metal Statistics, annual publication of Metal Statistics, New York.
Metals Handbook, 8th ed., Vol. 1–5, Am. Soc. of Metals, Cleveland, Ohio, 1961–1969.
Publications of the Intern. Lead Zinc Research Corp., New York.
Annual Book of ASTM Standards, Am. Soc. of Testing and Material, Philadelphia, Pa., 1970.

A. W. Schlechten
Colorado School of Mines
and A. Paul Thompson
Eagle-Picher Industries, Inc.

ZINC COMPOUNDS

Zinc is positive bivalent in its compounds. The colorless ion, Zn^{2+}, is present in solutions. Compounds of zinc are white at room temperature unless the other component, such as the chromate ion, imparts color to the compound. It should be said that in the crystalline form some of the compounds of zinc lack color.

As is evident from the position of zinc in the electrochemical series, it dissolves vigorously in dilute acids, evolving hydrogen (except in the case of nitric acid) and producing zinc salts containing the zinc ion, Zn^{2+}.

$$Zn + 2\,H^+ = Zn^{2+} + H_2$$

With nitric acid, reduction of the nitrogen occurs in whole or in part, and nitrogen in varying states of oxidation results depending upon the concentration of the acid. The formation of nitric oxide may be represented by the equation:

$$3\,Zn + 8\,HNO_3 = 3\,Zn(NO_3)_2 + 2\,NO + 4\,H_2O$$

Zinc is not attacked perceptibly by pure water but it dissolves readily in hot solutions of caustic soda or potash, evolving hydrogen and forming solutions of zincates, the monobasic zincate, $(HZnO_2)^-$, being formed in dilute solutions of the hydroxide and dibasic zincates, $(ZnO_2)^{2-}$, in more concentrated solutions.

Zinc is amphoteric, the hydroxide acting as a weak base as well as a weak acid. Its solution in strong acids and bases may be represented as follows:

$$Zn(OH)_2 + 2\,(OH)^- \rightleftharpoons Zn(OH)_4^{2-}$$
$$Zn(OH)_2 + 2\,H^+ \rightleftharpoons Zn^{2+} + 2\,H_2O$$

Specifically, the white gelatinous zinc hydroxide formed by the addition of sodium or potassium hydroxide to zinc ions in an acid solution is readily soluble in an excess of the reagent with the formation of zincates. In this behavior zinc is similar to aluminum but distinct from iron and manganese. Zinc hydroxide will also dissolve upon acidification.

Zinc forms complex cations, particularly with ammonia. Zinc compounds, the hydroxide, for example, may dissolve in aqueous ammonia forming a complex ion such as $Zn(NH_3)_2^{2+}$.

The most important compound of zinc is the oxide, ZnO. It is a white powder which becomes yellow on heating but regains its original color on cooling. Commercial zinc oxide is prepared almost exclusively by burning a zinc vapor produced by volatilizing zinc metal or by reducing crude zinc oxide to metal, volatilizing, and then burning (see p. 589). Zinc oxide dissolves easily in acids forming zinc ions, and in caustic solutions forming zincates.

Most of the simple salts of zinc are quite soluble in water. However, the oxide, hydroxides, carbonates, sulfides, phosphates, oxalates, and silicates are insoluble or only slightly soluble. All zinc halides except the fluoride are very soluble and are deliquescent. When a solution of zinc chloride is evaporated to dryness, it loses hydrogen chloride and forms a basic chloride of indefinite composition, $ZnCl_2 \cdot xZnO \cdot yH_2O$. As a consequence, nearly all grades of commercial zinc chloride yield turbid aqueous solutions owing to the presence of some basic chloride.

Zinc, as the element, is not inherently very toxic but certain toxic effects of it and its compounds may be noted as a result of any of three causes: (1) inhalation of freshly formed zinc oxide fumes in fairly high concentration; (2) ingestion of zinc compound in sufficient quantity with food or drink; and (3) direct contact with zinc or zinc salts. The fume of freshly formed zinc oxide which upon inhalation can cause a mild disease known as "metal fume fever," "brass founders' ague," "brass chills," "oxide shakes," etc. Continued inhalation of zinc oxide fumes results in fever and nausea along with severe coughing which may become violent enough to cause excessive salivation and vomiting. These effects are not cumulative and removal of the cause of difficulty will result in return to normal physiological reactions. Exposure to zinc chloride fumes can cause damage to the mucous membrane of the nasopharynx and respiratory tract quite possibly due to hydrolysis of the zinc chloride and the resulting hydrochloric acid.

Soluble salts of zinc have a harsh, metallic taste. Small amounts of such salts ingested with food or drink will cause brief illness and in aggravated cases may result in nausea and even purging. As far as can be determined, the continued administration of zinc salts in small doses has no effect in man except those of disordered digestion and constipation.

As for the third possible method, ie, direct contact with zinc or zinc salts, such contact may result in rare and rather limited symptoms of zinc poisoning but in general it can be said that the human skin shows a high degree of tolerance and the difficulty may very likely be due to the anion rather than to the zinc.

It should be noted that it is possible for people to become relatively immune or "habituated" to zinc oxide fumes; also that zinc dust which is not freshly formed is virtually innocuous. Biological studies involving zinc have pretty well established its physiological importance as a trace element in animal nutrition. It has been shown that zinc is a constituent of certain enzyme molecules, one of which, for example, is essential to the elimination of carbon dioxide.

Production and shipment in 1968 of the three most important zinc compounds and pigments, namely zinc oxide, zinc chloride, and zinc sulfate increased by somewhat more than 10% over the previous year, and exceeded the previous record high figures of 1966. Shipments of leaded zinc oxide continued the downward trend of the past 20 years. 1968 production and shipment figures for these four leading zinc compounds and pigments are presented in Table 1 (4). (Figures on lithopone, a coprecipitate of zinc sulfide and barium sulfate, are confidential and not included.)

Table 1. U.S. Production and Shipment of Zinc Compounds, 1968

| | | Shipments | | |
| | | | Value[a] | |
Compound or pigment	Production, short tons	Short tons	Total, $	Average, per ton
zinc oxide[b]	209,963	213,826	58,944,000	$276
leaded zinc oxide[b]	11,125	7,995	2,030,000	254
zinc chloride, 50°Be'[c]	57,914	57,508	w	w
zinc sulfate	57,131	59,647	10,357,000	174

   [a] Value at plant, exclusive of container.

   [b] Zinc oxide containing 5% or more lead is classed as leaded zinc oxide.

   [c] Includes zinc chloride equivalent of zinc ammonium chloride and chromated zinc chloride.

   [w] Withheld to avoid disclosing individual company confidential data.

The major source of zinc and leaded zinc oxide is the direct treatment of zinc ores (both domestic and foreign), with slab zinc and secondary materials being of much less importance. Secondary materials are a major source for zinc chloride whereas both ores and secondary are the major sources for zinc sulfate.

For some of the principal uses of zinc compounds, see such articles as: Colors for ceramics and glass; Luminescent materials; Pigments (inorganic); Reprography; Rubber chemicals; Rubber compounding.

ZINC ACETATE

Zinc acetate, $Zn(C_2H_3O_2)_2$, is white when pure, crystallizing in the monoclinic system; sp gr, 1.84; decomposes at 200°C; solubility in 100 ml of water, 30 g at 20°C, 44.6 g at 100°C; solubility in 100 ml of alcohol, 2.8 g at 25°C, 166 g at 79°C.

Hydrated zinc acetate, $Zn(C_2H_3O_2)_2 \cdot 2H_2O$, is the stable solid phase in aqueous systems below 100°C. It crystallizes in the monoclinic system forming colorless plates or granules; sp gr 1.735; loses all its water of hydration at 100°C; solubility in 100 ml water, 31.1 g of the dihydrate at 20°C; 66.6 g at 100°C. The aqueous solution is neutral or slightly acidic; pH about 5–6.

Zinc acetate is manufactured by heating a slurry of zinc oxide with acetic acid and then filtering. The filtrate is acidified with an excess of acetic acid and evaporated until crystals begin to form. On cooling, crystallization continues; the crystals are separated by centrifuging or filtering. It is used in the preservation and fireproofing of wood, as a mordant in dyeing, and in the manufacture of glazes for painting on porcelain. A 0.1–4% solution of zinc acetate is used as an astringent and mild antiseptic for mucous membranes and infections of the skin. The technical grade of the dihydrate sold in drums in truckload quantities at the works for approximately 32¢/lb in July, 1970, the reagent grade for 53¢/lb.

ZINC BORATE.

Zinc borate, $3ZnO \cdot 2B_2O_3$, forms a white amorphous powder or triclinic crystals depending on the method of preparation, sp gr 3.64 (amorphous), 4.22 (crystalline); mp 980°C. The amorphous form is slightly soluble in water and hydrochloric acid, the crystalline is insoluble in hydrochloric acid. It is used in ceramics, in the fireproofing of fabrics, in fire-retardant paints, and in pharmaceuticals due to its antiseptic and fungistatic properties. The selling price for the technical grade (43% $ZnO \cdot 37\%$ B_2O_3) in 20,000 lb truckload quantities, fob works, was 28¢/lb in July, 1970. The price for the crystallized product, (37% $ZnO \cdot 49\%$ B_2O_3), same basis as the technical grade, was 38¢/lb.

ZINC BROMIDE. See under Zinc halides.

ZINC CARBONATE.

Zinc carbonate, $ZnCO_3$, forms colorless trigonal crystals; indices of refraction 1.818 and 1.618; sp gr 3.98; it loses carbon dioxide at 300°C. It is scarcely soluble in water, one part dissolving in 100,000 parts of water at ordinary temperatures. Zinc carbonate occurs naturally as the mineral smithsonite.

Zinc carbonate is made by treating a zinc sulfate solution with sodium carbonate. It may also be prepared by passing carbon dioxide into a zinc oxide slurry. The slurry is filtered, washed, and dried. It is used as a paint pigment as dried or is first calcined to zinc oxide before using. It is also used in the manufacture of porcelain and pottery. Technical zinc carbonate was priced at approximately 15¢/lb in July, 1970.

ZINC CHLORIDES. See under Zinc halides below.

ZINC CHROMATE. See Chromium compounds; Pigments (inorganic).

ZINC CYANIDE.

Zinc cyanide, $Zn(CN)_2$, is a white powder crystallizing in the orthorhombic system. It has a density of 1.852 and decomposes at 800°C. It is insoluble in water but soluble in alkaline cyanides or hydroxides, also NH_3, insoluble in alcohol. It is used in electroplating. In July, 1970, it sold for 61.2¢/lb in drums at the works.

ZINC, DIETHYL; ZINC, DIMETHYL, etc. See Rubber chemicals.

ZINC 2-ETHYLHEXOATE. See Driers and metallic soaps.

ZINC FLUORIDE. See Fluorine compounds, inorganic, Vol. 9, p. 684.

ZINC FORMATE.

Zinc formate, $Zn(HCOO)_2 \cdot 2H_2O$, forms white monoclinic crystals; indexes of refraction, 1.513, 1.526, and 1.566; sp gr 2.207 at 20°C; it gives off its water of hydration at 140°C. At 20°C, 5.2 g of the dihydrate dissolves in 100 ml water; at 100°C 38 g in 100 ml water. It may be prepared by adding formic acid to a heated and agitated slurry of zinc oxide. The crystals are separated from the solution by evaporation and centrifuging.

ZINC HALIDES.

Zinc fluoride, ZnF_2. See Fluorine compounds, inorganic, Vol. 9, p. 684.

Zinc chloride, $ZnCl_2$, is not easily obtained in an anhydrous form but usually contains approximately 5% water and some basic chloride. It has been called "butter of zinc" because on evaporation it forms a white semisolid mass similar in consistency to butter. It can also be obtained in white, hexagonal crystals with indexes of refraction of 1.681 and 1.713. Its sp gr is 2.91 at 25°C, mp 283°C, bp 732°C. It is highly deliquescent and very soluble in water, 432 g dissolving in 100 ml water at 25°C, 615 g in 100 ml water at 100°C; 100 g is soluble in 100 ml of alcohol at 12.5°C; it is very soluble in ether.

Zinc chloride is prepared by several methods. Roasted zinc sulfide ore, which usually contains lead, is heated with common salt, fuming off the lead and zinc as chlorides which are caught in a suitable bagroom. The fume is treated with sulfuric acid or zinc sulfate to precipitate the lead as lead sulfate. The clear liquor is evaporated to a concentrated solution of zinc chloride. In another method, crude zinc oxide or zinc dross is treated directly with hydrochloric acid with an insufficient amount to dissolve all the metal, and the clear liquor is filtered off. The oxide–dross residue, nearly dry, is then treated with chlorine. The zinc chloride formed is leached with water and added to the first filtrate. The clear liquor is evaporated to a concentrated solution of zinc chloride.   Zinc chloride may be prepared by treating zinc metal with hydrochloric acid and concentrating the zinc chloride solution by evaporation.

Zinc chloride is one of the more industrially important compounds of zinc.   It is used as a deodorant, and in disinfecting and embalming fluids.   It is used alone or with phenol or chromated (see Vol. 10, p. 230) for preserving railway ties and for fireproofing of lumber, with ammonium chloride as a flux for soldering, etching metals, galvanizing; in the manufacture of parchment paper, dyes, activated carbon; and as a mordant in printing and dyeing textiles.   It is also used in mercerizing cotton, sizing and weighting fabrics, vulcanizing rubber, and in chemical synthesis as a dehydrating agent.   It is sold in the following grades: 50% technical solution (50° Bé, 1.53 sp gr), fused, granular, and U.S. Pharmacopeia.   The 50% technical solution sells for 6.4¢/lb in tanks at the works, the fused in drums in car lots at the works for 13.4¢/lb, the granular in drums, car lots at the works for 14.15¢/lb, and the U.S.P. grade, granular in drums at 96¢/lb, all prices as of July, 1970.

Zinc ammonium chloride, $ZnCl_2 \cdot 2NH_4Cl$ (ammonium chlorozincate, $(NH_4)_2$-$ZnCl_4$), forms white thin orthorhombic plates on crystallization; sp gr 1.879.   It decomposes at 150°C, is hygroscopic and very soluble in water with absorption of heat.   It is used as a flux for welding, soldering, and galvanizing.   The July, 1970 price for technical zinc ammonium chloride in bags, carload lots at the works, was 12.65¢/lb.

Zinc bromide, $ZnBr_2$, is, like zinc chloride, a very hygroscopic granular powder with a sharp metallic taste; it crystallizes in the orthorhombic system.   When purified commercially it still contains only 97% $ZnBr_2$, the remainder being chiefly water.   The sp gr is 4.20; mp 394°C, bp 650°C; 447 g dissolves in 100 ml of water at 20°C, 675 g at 100°C.   Zinc bromide is prepared by treating purified zinc with purified bromine in an atmosphere of carbon dioxide and then distilling the zinc bromide.

Zinc iodide, ZnI_2.   It forms colorless hexagonal crystals with a sp gr of 4.7364. It melts at 446°C, and decomposes at 624°C.   It has a solubility of 432 g in 100 ml of water at 18°C.

ZINC HYDROSULFITE.

Zinc hydrosulfite, $ZnS_2O_4 \cdot 2H_2O$, is a white crystalline powder which loses its water of hydration at 100°C, and decomposes at 200°C.   Its price in drums, truckload, works, was 23¢/lb in July, 1970.   See also Vol. 19, p. 420.

ZINC NAPHTHENATE.   See Driers and metallic soaps.

ZINC NITRATE.

Zinc nitrate, $Zn(NO_3)_2$.   The pure salt is unstable and very difficult to obtain.   The system $Zn(NO_3)_2$–H_2O is a complicated one because of the large number of hydrates.

Zinc nitrate trihydrate, $Zn(NO_3)_2 \cdot 3H_2O$, forms colorless needles with a melting point of 45.5°C.   It has a solubility of 327.3 g in 100 ml of water at 40°C.

Zinc nitrate hexahydrate, $Zn(NO_3)_2 \cdot 6H_2O$, forms tetragonal colorless crystals, sp gr 2.065, mp 36.4°C, loses six molecules of water from 105 to 131°C, and is very soluble in water, 184.3 g in 100 ml at 20°C,

The price of the technical grade in less than carload lots was quoted as 21.5¢/lb in July, 1970.

ZINC OXIDE.

Zinc oxide, ZnO, forms white hexagonal crystals, indexes of refraction 2.008 and 2.029, sp gr 5.606, sublimes at 1800°C. Its melting point is given as 1975°C. The pure oxide is yellow when hot. It is insoluble in water but taken up readily by acids and alkalis. It occurs naturally as the mineral zincite, sometimes called red zinc ore, the red color being due to manganese which may amount to as much as 9%. Another important ore is the mineral franklinite, which may be regarded as a mixed oxide of zinc, manganese, and iron, of variable proportions.

On adding alkali to a solution of a zinc salt, a hydrated oxide is precipitated. The precipitate readily dissolves in excess sodium or potassium hydroxide, forming *zincates*. Zinc oxide can also be dissolved in alkaline solutions, and metallic zinc reacts with hot strong caustic solutions, forming zincates and evolving hydrogen. Solid compounds of the hydroxozincate type such as $(Zn(OH)_3)Na.3H_2O$ and $(Zn(OH)_4)Na_2.2H_2O$ can be isolated, and metathesis with barium or strontium hydroxides gives compounds of coordination number 6, as $(Zn(OH)_6)Ba_2$.

Table 2. Distribution of Zinc Oxide and Leaded Zinc Shipments, by Industries (Short Tons)

Industry	1966	1967	1968
zinc oxide			
rubber	104,866	94,388	111,797
paints	27,100	24,547	25,864
ceramics	12,147	9,850	10,226
chemicals	13,678	17,509	22,769
agriculture	1,559	5,048	5,044
photocopying	11,405	14,039	21,564
coated fabrics and textiles	w	w	w
floor covering	w	w	w
other	22,910	16,105	16,562
total	193,665	181,486	213,826
leaded zinc oxide			
paints	10,462	8,644	6,356
rubber, other and unspecified	1,095	1,662	1,639
total	11,557	10,306	7,995

w Withheld to avoid disclosing individual company confidential data, included with "other."

Zinc oxide is the commercially most important compound of zinc. For its manufacture see p. 589. It is used in rubber, paint, ceramics, emollients, and fluorescent pigments, although in the last use zinc sulfide and silicate predominate. In addition, the amount of zinc oxide used in the manufacture of other chemicals, both inorganic and organic, has more than doubled since 1965. An interesting example of its use in the organic field is in the manufacture of zinc-containing organometallic compounds such as accelerators for the curing of rubber. It should be noted also that the amount of zinc oxide consumed in the photocopying industry (see Vol. 17, p. 346) has more than doubled since 1966. Table 2 (4) below gives the distribution of shipments of zinc oxide according to the consuming industries for the latest three years available, thus giving some indication of trends. Leaded zinc oxide is also included in the table, these figures showing the steadily declining use of this formerly important product.

As an illustration of the multiplicity of purposes for which zinc oxide may be incorporated into another base material, the example of rubber, the major use for zinc oxide, may be cited. The following incomplete tabulation lists the functions of zinc oxide in rubber stock, particularly of the types used for tires or electrical insulation: activation of organic accelerator, mechanical reinforcement, high-thermal conductivity, low heat generation (in flexing), high-heat capacity, good adhesive characteristics, improvement of aging (by acting as acid scavenger), high hiding power, increasing of resistance to deterioration by sunlight (due to high opacity to ultraviolet light), and the following electrical characteristics: medium dielectric constant, low power factor, high resistivity, low water absorption. In polysulfide elastomers (see Vol. 7. p. 696; Vol. 16, p. 253) it is used as a vulcanizing agent.

There are many grades of zinc oxide. In American (direct) process oxides, there are rubber grades, paint grades, pharmaceutical grades, etc. Within rubber grades there are fine-particle-size oxides for fast curing and high reinforcement, ranging into coarse oxides with slow-curing properties and giving high resistance to scorching. Most paint grades are of acicular particle shape, rather than spherical, as this is believed to impart superior brushing characteristics to the paint. Pharmaceutical grades are more dependent on chemical analysis than most, except that rubber grades must be extremely low in copper and cadmium to prevent setting off of autocatalytic oxidation within the rubber. Sulfur trioxide may be added to rubber grades to decelerate the rate of cure. Propionic acid may be added for ease of incorporation into rubber.

The French (indirect) process oxides are usually sold in three grades: white seal, characterized by brilliant whiteness, high apparent density, and extreme fineness; green seal, equal in whiteness to white seal but less voluminous; and red seal, neither so fine nor so white as the other two.

Prices of zinc oxide, delivered in carloads in bags, were quoted as follows in July, 1970, at a time when the base price of zinc metal, Prime Western slabs, was $15\frac{1}{2}$¢/lb fob East St. Louis, Ill.:

American process, lead free[a]	$16\frac{1}{2}$¢/lb
French process, regular	17¢/lb
French process, green seal	$18\frac{3}{4}$¢/lb
French process, white seal	19¢/lb
Electrophotographic grade	$19\frac{1}{4}$¢/lb
U.S.P. grade (50 lb boxes)	$20\frac{1}{4}$¢/lb
Leaded 12% (12% basic lead sulfate)	$14\frac{1}{2}$¢/lb
Leaded 18% (18% basic lead sulfate)	$15\frac{1}{4}$¢/lb
Leaded 35% (35% basic lead sulfate)	17¢/lb
Leaded 50% (50% basic lead sulfate)	18¢/lb

[a] The phrase "lead free" usually denotes zinc oxides to which lead compounds have not intentionally been added. The lead oxide content of these zinc oxides is usually under 1%.

ZINC PHOSPHATE.

Zinc orthophosphate, $Zn_3(PO_4)_2$, forms colorless orthorhombic crystals with a sp gr of 3.998 and a mp of 900°C. It is quite insoluble in water. The principal use of zinc phosphate is in providing phosphate coatings of iron, zinc, and aluminum to provide protection against corrosion and to improve paint bonding. Market quotations are not available for zinc phosphates since they are normally prepared as needed by reacting zinc oxide and phosphoric acid to give solutions of the desired compositions.

ZINC PHOSPHIDE.

Zinc phosphide, Zn_3P_2, forms dark gray tetragonal crystals. The compound is poisonous. It has a sp gr of 4.55 and melts somewhere above 420°C. Its boiling point is around 1100°C.

ZINC RESINATE. See Driers and metallic soaps.

ZINC SILICATES.

Zinc metasilicate, $ZnSiO_3$, is said to form colorless rhombic crystals with a sp gr of 3.42 and a melting point of 1437°C. However, its existence is quite doubtful and not substantiated by x-ray diffraction data.

Zinc orthosilicate, Zn_2SiO_4, is the more common form of zinc silicate, and occurs in nature as the mineral willemite. It forms a trigonal crystal, indexes of refraction of 1.694 and 1.723, sp gr 4.103, mp 1509°C, insoluble in water. It has been prepared artificially by heating the proper amounts of zinc oxide and silica at temperatures approaching the melting point of the silicate. The pure zinc silicate may be prepared by starting with a pure precipitated zinc sulfide which is heated in filtered air and converted to the oxide. The zinc oxide and silica in proper amounts are then heated together. Pure zinc silicate is fluoroescent and finds use in this field.

A monohydrate hydroxy silicate mineral of zinc is known as hemimorphite or calamine and may be represented by the formula $Zn_4Si_2O_7(OH)_2 \cdot H_2O$.

Zinc fluorosilicate. $ZnSiF_6 \cdot 6H_2O$ crystallizes in colorless hexagonal prisms with indexes of refraction of 1.3824 and 1.3956. It has a sp gr of 2.104 and decomposes at 100°C. It is very soluble in water. It is usually made by adding zinc oxide or zinc carbonate to fluorosilicic acid (H_2SiF_6) under carefully controlled conditions, keeping the latter in slight excess. It finds use as a wood preservative and in precipitating potash from sugar liquors and other solutions. It is sometimes used to give added strength to plaster. The price quotation in July, 1970 was 17¢/lb in drums, car lots, fob works.

ZINC STEARATE. See Driers and metallic soaps.

ZINC SULFATE.

Zinc sulfate, $ZnSO_4$, forms colorless orthorhombic crystals, sp gr 3.54, indexes of refraction 1.658, 1.669, and 1.670. The mineral zinkosite occurs in nature, crystallizing in the orthorhombic system and having a sp gr of 3.7. Zinc sulfate forms three hydrated compounds, $ZnSO_4 \cdot 7H_2O$, $ZnSO_4 \cdot 6H_2O$, and the monohydrate $ZnSO_4 \cdot H_2O$. A mineral of the composition $ZnSO_4 \cdot 7H_2O$ called goslarite with a sp gr of 2.2 and crystallizing also in the orthorhombic system is known. The monohydrate loses its water of hydration on heating above 238°C. On continued heating, it decomposes into zinc oxide and sulfur trioxide at around 740°C.

The solubility in water is:

Temp, °C	0	10	20	30	40	50	60	80	100
Solubility, g ZnSO₄/ 100 ml H₂O	41.9	47.6	54.2	61.6	70.1	74.0	72.1	65.9	60.6

Zinc sulfate is prepared by leaching roasted zinc ore concentrates with sulfuric acid, filtering out the residue, and treating the clear liquor with zinc dust to remove heavy metals, such as cadmium. After filtering, the clear liquor is evaporated, crystallizing out zinc sulfate crystals which may be separated by filtering or centrifuging. Other sources of zinc such as galvanizers' skimmings are also used as raw materials for zinc sulfate.

One of the major uses of zinc sulfate is in the manufacture of rayon (qv). In the manufacture of viscose rayon, as the filaments are ejected from the spinneret they enter a precipitating bath containing sulfuric acid and sodium sulfate. An addition of glucose is made to prevent crystallization, and an addition of zinc sulfate is made to promote crenellation of the fiber. Another use of zinc sulfate which is gaining in importance is in agriculture. It is added in trace quantities to the soil as a fertilizer and in sprays for the control of certain plant diseases, especially in the citrus industry. See Fertilizers. Other uses include: as a mordant in calico printing, in wood preserving, as a flotation reagent, in clarifying glue, and in electrogalvanizing.

The distribution of zinc sulfate consumption among different industries is listed in Table 3 (4).

Table 3. Distribution of Zinc Sulfate Shipments by Industries[a]

Industry	Dry basis, short tons			
	(Average) 1949–1953	1966	1967	1968
rayon	7,726	16,562	w	w
agriculture	4,732	16,891	14,803	17,631
other	4,846	9,372	24,742	36,470
total	17,304	42,825	39,545	54,101

[a] The weight on the dry basis in the above table varies from about 82 to 90% of the gross weight as shipped.

w Withheld to avoid disclosing individual company confidential data, included with "other".

Zinc sulfate is available in two grades: industrial grade (granular monohydrate, 36% Zn), car lots 10¢/lb, and agricultural grade (powdered), car lots 9.75¢/lb, in July, 1970.

ZINC SULFIDE.

Zinc sulfide, ZnS, crystallizes in two forms, one in the hexagonal system, the mineral being called wurtzite with indexes of refraction 2.356 and 2.378, sp gr 3.98, and the other in the cubic system, called sphalerite or zinc blende, with index of refraction of 2.368 and sp gr of 4.102. At 1020°C, the cubic form transforms to the hexagonal. Wurtzite sublimes at 1182°C. Zinc sulfide is insoluble in water but is taken up by mineral acids with the evolution of hydrogen sulfide gas. When damp, it slowly oxidizes in air to the sulfate.

Zinc sulfide is a white pigment used in the manufacture of paints. Its high index of refraction contributes to its high hiding properties in paint. The chief impurity in the pigment grade is zinc oxide.

Zinc sulfide finds significant usage as a phosphor, the term applied to an important group of products exhibiting luminescent properties. For this use a very high purity material is necessary to which is added an activator such as copper, silver or manganese in very minute amounts (in the range of 0.005%). At times an extension of the properties of the phosphor is obtained by adding cadmium sulfide in amounts of 10% and more. Thus zinc sulfide, suitably prepared, finds its way into the manufacture of x-ray screens, and luminous dials on watches and television screens. It might be mentioned that other compounds such as zinc oxide, zinc silicate, zinc beryllium silicate, zinc aluminate and zinc germanate also find use in this area.

Lithopone is a coprecipitated pigment of zinc sulfide and barium sulfate resulting from the reaction between zinc sulfate and barium sulfide. It is not in as great use as formerly, although it is still an important pigment.

ZINC TALLATE. See Driers and metallic soap.

ZINC 3,5,5-TRIMETHYLHEXOATE. See Driers and metallic soaps.

Bibliography

"Zinc Compounds," in *ECT* 1st ed., Vol. 15, pp. 275–281, A. Paul Thompson and H. Schultz, The Eagle-Picher Co.

1. Gmelins *Handbuch der anorganischen Chemie*, 8th ed. Verlag Chemie, G.m.b.H., Weinheim/ Bergstrasse (1956).
2. H. Remy, translated by J. S. Anderson, and edited by J. Kleinberg, *Treatise on Inorganic Chemistry*, Vol. II, Elsevier Publishing Co., Amsterdam, London, New York, Princeton (1956).
3. C. H. Mathewson, ed., *Zinc*, A.C.S. Monograph, Reinhold Publishing Corporation, New York (1959).
4. 1968 *Minerals Yearbook*, U.S. Bureau of Mines, Washington, D.C.

A. Paul Thompson
Eagle-Picker Industries, Inc.

ZINEB, $Zn(SCSNHCH_2)_2$. See Fungicides, Vol. 10, p 224; Insecticides, Vol. 11, p. 716.

ZIRCONIUM AND ZIRCONIUM COMPOUNDS

ZIRCONIUM METAL AND ITS METAL-LIKE COMPOUNDS

The element zirconium is widely distributed in nature as a component of the lithosphere. It occurs only as the dioxide or as compounds consisting of zirconium dioxide and other oxides. The ores and minerals and their processing are discussed in the section on Other Zirconium Compounds. The chief occurrence of zirconium is in the mineral zircon, $ZrO_2.SiO_2$, which has been known to man from ancient times. It is mentioned in both the Old and New Testaments, and in other ancient writings, sometimes under the name *jargon* or *hyacinth*. Colorless specimens have been called *Matara diamonds*.

It was not until 1789 that it was recognized that zircon contains a distinguishing oxide. In that year, M. H. Klaproth published his analyses of zircons from Ceylon and showed that zircon contained a previously unrecognized oxide which had been mistaken for aluminum oxide. The newly announced oxide did not exhibit the characteristic amphoterism of aluminum oxide. Klaproth named the newly recognized oxide zirkonerde, and he is undisputably the discoverer of the element zirconium.

The element zirconium proved exceedingly difficult to isolate. Indeed, it has provoked advancement of some new thinking on a definition of what constitutes an isolated solid element. Not even a semblance of the liberation and isolation of zirconium from a compound was achieved until 1824, when J. J. Berzelius heated a mixture of potassium metal and potassium fluorozirconate in a closed vessel and obtained a black powder which he regarded as the element zirconium (1). Over eighty years later, Weiss and Neumann showed that the product of Berzelius' procedure consisted only 93.7% of zirconium. They refined the procedure and produced a reaction product assaying 98% zirconium (2).

From a modern point of view, a substance can be identified as zirconium only when it is sufficiently free of impurities to show the essential characteristics of the element. These include an assay of nearly 100% Zr, a steel-gray metallic appearance, and ductility and malleability. A tenth of a percent or so of oxygen, nitrogen, or carbon dissolved in the metal renders it substantially nonductile. As the proportion of these dissolved elements increases, new phases become stable and are likely to form. The new phases belong to the category of compounds, although they may have metallic characteristics.

Zirconium of nearly 100% purity was prepared by Lely and Hamburger in 1914 by the reduction of zirconium tetrachloride with sodium in a bomb (3). Their product could be drawn and pressed, and burnished to a mirror-like surface. Before the end of the last century it was demonstrated that zirconium halides could be reduced to the metal on an electrically heated filament in the presence of hydrogen, but this approach to a practical preparation of zirconium metal was perfected, over a quarter-century later, by van Arkel and de Boer (5). They volatilized zirconium tetraiodide and caused the vapors to impinge on a hot tungsten filament. Elementary zirconium of very high purity was formed. Later production by this method yielded zirconium containing less than 0.01% oxygen.

The perfection of a practical method for the production of zirconium metal in large tonnages at reasonable prices was the accomplishment of William Kroll. He improved upon a procedure that had been disclosed by Troost (6) in 1865, in which zirconium tetrachloride vapor was passed over hot magnesium metal. Troost obtained impure zirconium which was not malleable. Zirconium metal was manufactured by this general procedure by von Zeppelin in 1940. Kroll and his associates altered the procedure so as to conduct zirconium tetrachloride vapor into a reactor containing a pool of liquid magnesium. The reaction

$$ZrCl_4 + 2\,Mg \rightarrow Zr + 2\,MgCl_2$$

yielded a ductile grade of zirconium in sponge form (7). In Kroll's process, the by-product magnesium chloride is distilled out of the sponge. It cannot be completely removed, and tends to form gas and to leave voids during the subsequent melting and casting of the metal. Later studies, using tracer techniques, have shown that when the zirconium tetrachloride is transmitted to a pool of molten magnesium at 800–850°C, liquid magnesium rises by capillary action through the spongy zirconium which forms and maintains a magnesium surface to react with the zirconium tetrachloride which subsequently makes contact (8).

It was not known until after the report of the discovery of hafnium in 1923 that all zirconium found in nature contains some hafnium. Hafnium (qv) is about as similar to zirconium as deuterium is to hydrogen, in chemical properties. In all ordinary processing of zirconium ores, the hafnium remains with the zirconium. One consequence was that the true atomic weight of zirconium was not known and could not have been determined until the presence and certain properties of hafnium had been demonstrated. To this day, for all practical applications of zirconium other than as a fuel element in nuclear reactors the hafnium is not separated from the zirconium. Its presence, commonly at a level of about 2 wt % of the zirconium, has the effect of a heavy isotope in the zirconium. For atomic reactor use, where nuclear as well as chemical properties are important, the presence of hafnium is undesirable due to its relatively high cross section to thermal neutrons.

Zirconium Atom. Zirconium, element no. 40, atomic weight 91.22, is found in nature to be made up of the following isotopes: ^{90}Zr, 51.5%; ^{91}Zr, 11.2%; ^{92}Zr, 17.1%, ^{94}Zr, 17.4%; and ^{96}Zr, 2.8%. Other isotopes have been prepared by nuclear bombardment. The radioisotope ^{95}Zr has been particularly useful in tracer studies. It has been isolated by paper chromatography (17). The low cross section of the zirconium atom toward thermal neutrons, 0.18 barn, has been an important qualification for the use of zirconium in cladding uranium fuel elements. For slow neutrons (0.2–0.78 eV), its cross section is 6.4 barn.

The orbital electron arrangement of the zirconium atom in its ground state is $1s^2 2s^2 2p^6 3s^2 3p^6 3d^{10} 4s^2 4p^6 4d^2 5s^2$. The ionization potentials of the valence electrons ($4d^2 5s^2$) are respectively 6.84, 13.13, 24.00, and 33.8 electron volts. Oxidation potentials of the metal are given by the following relationship:

$$Zr + 2\,H_2O \rightarrow ZrO_2 + 4\,H^+ + 4e \qquad E_{298}{}^0 = 1.43\ V$$

$$Zr + H_2O \rightarrow ZrO^{2+} + 2\,H^+ + 4e \qquad E_{298}{}^0 = 1.53\ V$$

The metallic radius of the zirconium atom at coordination no. 12 is 1.602 Å.

Chemical and Physical Metallurgy of Zirconium

Zirconium is chemically a very reactive element, and its compounds are completely reduced with difficulty. Powerful reducing conditions must be employed to obtain the pure metal, under conditions carefully contrived to preclude possibilities of the retention of oxygen, nitrogen, or other elements in solid solution. Some reducing agents, such as metallic aluminum, are apt to form intermetallic compounds with zirconium. It has proven most satisfactory to reduce halides of zirconium rather than zirconium oxide, conducting the reaction in vacuo or under inert gases, and employing as reducing agent a metal such as sodium or magnesium which does not form intermetallic compounds with zirconium. Acceptable metal produced by the U.S. Bureau of Mines by the Kroll process has contained about 0.12% oxygen and 0.005% nitrogen.

Calcium and calcium hydride react with zirconium dioxide in the temperature range 800–1200°C with formation of a good grade of zirconium metal (18). Low-grade zirconium, contaminated with perhaps 5% carbon, has been made by heating together zirconium dioxide and zirconium carbide. Similar reactions occur between

$$ZrO_2 + ZrC \rightarrow 2\,Zr + CO_2$$

magnesium oxide and zirconium carbide, carbon and zirconium oxide, and calcium carbide and zirconium oxide (80, p. 206). Under some circumstances, the presence of alloying impurities in zirconium is not objectionable and may even be desirable. Then, not only very active metals such as aluminum, but even moderately reactive metals such as zinc may be used to reduce zirconium halides or salt melts containing zirconium halides (19).

Zirconium halides have been reduced to the metal by sodium amalgam (20) and amalgams of other alkali and alkaline earth metals (21). In one procedure, the sodium amalgam was allowed to react with a solution of zirconium tetrachloride in liquid ammonia (22).

Zirconium tetrafluoride is less susceptible to hydrolysis and other contamination with oxygen than the other halides of zirconium, and metal derived from it is less apt to be contaminated with oxygen. It is conveniently reduced to zirconium with calcium (23). Zirconium sulfide has also been reduced with sodium or potassium at 800–1100°C, under inert atmosphere (24). Other reductions of zirconium halides have been accomplished with calcium carbide (25), calcium hydride (26), and calcium metal (23).

The chlorides, bromides, and iodides of zirconium can be decomposed to the metal by pyrolysis (28). But only iodides are decomposable to the metal at sufficiently low temperature—about 1200°C—to be attractive as an industrial process. In the van Arkel-de Boer process, iodine acts as a carrier, forming a volatile iodide from impure, inexpensive metal at a relatively low temperature and decomposing at a surface heated to a much higher temperature with deposition of very pure zirconium. Others have made separate preparations of zirconium tetraiodide and subsequently pyrolyzed it to form zirconium metal (29).

It has proven impossible to prepare zirconium by electrolysis of aqueous solutions of zirconium compounds. But as early as 1865, Troost reported depositing a low-grade zirconium metal by electrolysis of a molten fluorozirconate (6). Aside from the question of purity, it has proven quite difficult to obtain coherent deposits of zirconium. This has been achieved by Mellors and Senderoff by electrolyzing solutions of zirconium tetrafluoride in alkali metal fluoride melts at about 750°C, at a current density of 30 mA/cm² at the cathode. A deposit of metal having the theoretical density was achieved. Chronopotentiometry has indicated that the quadrivalent zirconium was reduced in a single step to zerovalent zirconium metal (30,31). In other electrochemical procedures, zirconium has been deposited from molten potassium hexafluorozirconate at 850°C (32) and from solutions of zirconium tetrachloride in sodium chloride at 600–950°C (33).

It has been demonstrated that when metal or ceramic objects are suspended in slurries of finely divided zirconium metal in a fused salt, deposits of zirconium metal form on their surfaces (34). This phenomenon doubtless plays a role in all electrodepositions in fused salts.

During the melting and casting of zirconium, measures to exclude atmospheric gases must be maintained. The metal can be cold- and hot-worked. Rolling and forging are usually carried out at 850°C (1000°C for zircaloy). The surface is cleaned by sandblasting and/or pickling in nitric-hydrofluoric acid solutions.

Extrusion of zirconium is commonly carried out on billets encased in copper or glass, at 750–1000°C. Drawing is done at room temperature, using metal on which there is a thin casing of zirconia or copper. The metal is machined with carbide-tipped tools at moderate speeds. It can be welded under inert atmosphere, as for example, with the helium arc (15).

Crystallography

Below 863°C, the stable form of zirconium has a hexagonal close-packed lattice, and is designated α-phase. Its parameters at 25°C are $a = 3.23$, $c = 5.15$ Å. Above 863°C, the stable β-phase has a body-centered cubic symmetry, $a = 3.616$ Å at 980°C. The elements oxygen, carbon, and nitrogen are α-stabilizers, having the effect of raising the transition temperature.

Zirconium subjected to 65 kilobars pressure at room temperature forms a new phase designated ω. Release of the pressure causes spontaneous reversion to the α-phase (35,36).

Large crystals of zirconium have been grown by conventional methods for crystal growth of metals, such as the phase-change method (37). Whiskers have been grown on 0.15-mm zirconium wire under high vacuum by cycling through the transformation temperature. Crystal growth by extrusion occurs during the transformation (38).

Properties of Zirconium

Physical properties of zirconium metal are listed in Table 1.

Some properties of liquid zirconium have been studied. It has a surface tension of 1480 dyne/cm at the melting point. Its vapor pressure is given by the equation:

$$\log P_{\mathrm{atm}} = 6.521 - (30,940/T)$$

in the temperature range 2229–2795°K. The density of the liquid is 5.70 g/ml at the melting point.

Graphite dissolves in molten zirconium.

A freshly formed zirconium surface rapidly acquires a protective film when exposed to the atmosphere. Metal from different sources and with different processing history varies considerably in the reactivity of its surface.

Absorption of oxygen into zirconium metal has been observed to occur at 650°C, and of nitrogen at 900°C. Uniform distribution of nitrogen was attained only after annealing at 1200–1300°C (39). Massive zirconium ignites in oxygen at about 1000°C, burning with a brilliant white light. Spontaneous combustion of cleaned zirconium occurs at room temperature in oxygen at 300 psi pressure, or more, and zircaloy-2

Table 1. Physical Properties of Zirconium[a]

Property	Value
boiling point, °C	3650 (estd)
compressibility, $cm^2/kg \times 10^7$	11.77
Debye temperature, °K	250
density, gm/cm^3	6.506
electrical resistance, microohm-cm	41.4
Grueneisen constant	0.82
Hall coefficient, volt-cm/A-Oe $\times 10^{12}$	0.18
hardness, Vickers, kg/mm^2	110
heat of fusion, kcal/g-atom	3.74
heat of sublimation, at 25°C, kcal/g-atom	146
magnetic susceptibility, emu/g-atom $\times 10^6$	119
melting point, °C	1850
Poisson's ratio	0.34
shear modulus, $kg/cm^2 \times 10^{-6}$	0.348
specific heat, cal/(g)(°C)	0.67
strength:	
tensile, psi (0.25 in. rolled bar machined to 0.16 in.)	128,900
ultimate, kg/mm^2	35
yield, kg/mm^2	21
superconductivity transition temperature, °K	0.56
thermal conductivity, cal/(sec)(cm²)(°C/cm)	0.0505
thermal expansion, deg/cm $\times 10^6$	5.78
thermionic work function, eV	4.1
vapor pressure at 1675–1750°C, for equation	
$\log P_{torr} = A/T + B + CT$	
A	31,066
B	10.216
C	-2.415×10^{-4}
Young's modulus, $kg/cm^2 \times 10^{-4}$	0.939

[a] At room temperature unless otherwise noted.

combusts at 400 psi oxygen pressure (40). Zirconium powder ignites at much lower temperatures than the massive metal, and must be handled with extreme caution. The National Fire Protection Association has published a *Guide for Fire and Explosion Protection in Handling Zirconium*.

Heated zirconium is oxidized by carbon dioxide and by sulfur dioxide, and can be regarded as apt to be partly or completely oxidized by any oxygen-containing compound.

Zirconium metal is readily dissolved when heated in ten times its weight of a mixture of three parts of sulfuric acid and two parts of ammonium sulfate (41). The metal is not attacked by aqueous hydrochloric acid in the absence of oxidizing substances. However, it has been observed to dissolve in $6N$ and $10N$ ethanol solution of hydrogen chloride, except for a black residue identified as zirconium hydride. But if the ethanol solution contains as little as 4% water, the metal is unattacked (42).

Hydrofluoric acid is a powerful solvent for zirconium metal. The dissolution reaction is first order with respect to the molecular hydrogen fluoride content of the solution (43,44), and it appears that the reaction is actually with the molecule and not the dissociation products in solution (45). Hydrogen is liberated. When this is not desired, ammonium nitrate is added to the solution to suppress its liberation (46). Liquid bromine trifluoride and bromine pentafluoride have very little effect on zirconium (47).

Both tetragonal and monoclinic zirconium dioxide have been detected in films formed by wet oxidation of zirconium metal, as by boiling in nitric acid. Subsequent heating of the metal in air above 1150°C changes the tetragonal corrosion product to monoclinic (48). Wet corrosion films on zirconium show very little hydration. Hydrogen passes into the metal, but the oxide film is not protonated (49). What appears to be occluded water, phosphate, sulfate, and carboxylate have been found in anodized films on zirconium (50).

Oxide films grown on zirconium at 500°C in steam at atmospheric pressure showed an outer stoichiometric zirconium oxide layer and an oxygen-deficient layer at the oxide–metal interface (51, cf. 52). Oxygen-deficient films range in color through gray to black, and are whitened by inducing their oxidation.

Cubic zirconia is also formed under some conditions. A cubic zirconia layer 300–400 Å thick has been observed to form on zirconium at 270°C. On heating to 650°C, it inverted to tetragonal, and at 750°C to monoclinic zirconia (53). Single crystals of zirconium exposed to deoxygenated water at 360°C at 2700 psi developed black, adherent monoclinic zirconium dioxide scale which became altered to white polycrystalline zirconium dioxide (60).

Zirconium has been observed to be oxidized by molten alkali nitrates and nitrites (54). Zirconium metal heated with strontium titanate was found first to reduce the titania content to Ti_3O_5 and then the strontium to metal which volatilized (55).

Films and scale on zirconium metal can be removed by immersion in molten potassium hydrogen fluoride at 300°C. This is followed by rinsing successively with water, 2–5% hydrofluoric acid, and again water (56).

Zirconium is to be regarded as a metal of outstanding corrosion resistance. A summary of its corrosion behavior is shown in Table 2. It can be generalized that zirconium is not attacked by organic solvents, by aqueous salt solutions other than halides in oxidizing environments, and by dilute acids other than hydrofluoric acid. It is resistant to water at ordinary temperatures, but its resistance to hot water and steam in a nuclear power plant is inadequate, and hence a zircaloy is used. The resistance of zirconium to corrosion by sulfuric acid is little affected by raising the concentration of the acid up to 70%. However, as the concentration increases above 80% the corrosion is accelerated markedly (57).

Boric acid added to high-pressure steam at 500°C inhibited the corrosion of zirconium for hundreds of days. The minimum effective concentration was 0.003 g boric acid/g steam. Boric acid did not inhibit atmospheric oxidation. Anodizing zirconium in a borate solution did not inhibit corrosion in a subsequent exposure to

Table 2. Resistance of Zirconium Metal to Corrosion by Various Agents at Ordinary Temperatures

No Attack

(Hot and cold, unless otherwise stated)

acetic acid	nitric acid, dilute, concentrated, and fuming
alcohol	oxalic acid, aqueous
aluminum chloride, 20–30% aqueous	phenol
alum, aqueous, boiling	phosphoric acid, 10–85% at 25–60°C
ammonium hydroxide	sodium hydroxide, 10–40% aqueous
boric acid, aqueous	sodium hydroxide, fused at 319 and 370°C
bromine water, boiling	sodium hypochlorite, 0.5% aqueous
calcium hypochlorite, aqueous	stannic chloride, 24% aqueous
chromic acid, 1–30% aqueous	sulfuric acid, 1–50% aqueous, cold
formic acid, 90% aqueous	sulfurous acid, 6% aqueous
hydrogen peroxide	tannic acid, aqueous
lactic acid, 10–80%	tartaric acid, aqueous
magnesium chloride, 5–42% aqueous	trichloracetic acid, 10% and 50% aqueous
mercuric chloride, saturated aqueous	zinc chloride, 5 and 20% aqueous
monochloracetic acid	

Some Attack

cupric chloride, 2.5–10% aqueous at 35°C	hydrochloric acid, aqueous aerated
ferric chloride, 1–10% aqueous at 35–60°C	

Severe Attack

aqua regia, cold	hydrochloric acid, aqueous aerated, hot
chlorine saturated with water vapor, cold	hydrofluoric acid
cupric chloride, 2.5–10% aqueous at 60–100°C	phosphoric acid, 75–100%, 100°C to boiling
ferric chloride, >5% aqueous at 100°C and	sulfuric acid, hot concentrated
>15% at 35°C	trichloracetic acid, glacial, boiling

steam. However, saturated ammonium borate containing 1% sodium hydroxide provided inhibition (58).

Aqueous solutions of cupric chloride are typical of chloride salts containing oxidizing components. Progressive pitting of zirconium metal by the action of the solutions occurs without preference for grain boundaries (59).

Zirconium resists attack by molten metals such as lead, lithium, and sodium.

Zirconium Alloys

There is an enormous literature on the systems and the alloys of zirconium with other metallic elements. In general, zirconium forms intermetallic compounds with polyvalent metallic elements, but not with the alkali metals, alkaline earth metals, mercury, or lead. Small amounts of zirconium in aluminum and magnesium, ie, about 0.2–0.3%, refine the grain structures of these metals and increase their strength. The zirconium is usually introduced into the molten metal by placing in contact with the melt alkali fluorozirconates or chlorozirconates. A master alloy of magnesium and zirconium is now sold to the magnesium metal producers for addition to molten magnesium. Zirconium is a powerful deoxidizer, and has been added to iron and steel to serve in this capacity. It stabilizes the carbon, nitrogen, and sulfur in these metals, and reduces the size of austenitic grain in the primary crystallization. Suitable additions of zirconium improve the yield strength, resistance to impact, resistance to underbead cracking (cracking under welds), machinability, and hardenability of steels. They help overcome hot shortness in high-sulfur steels. Zirconium steels have been used as high-strength structural metals, as tool steels, and specifically for axles, crankshafts, rail and rock drills, and pressure-vessel construction. Statistics on the quantity of zirconium added to steel are vague, often being reported along with other additives as "other alloying additions." Some of the best superconducting alloys are made of niobium and zirconium, and contain 20–40% zirconium.

Zirconium alloys are commonly prepared by melting components together, but a variety of less conventional methods have been described. For example, an alloy consisting of the intermetallic compound $ZrAl_3$ has been prepared by reducing zirconium dioxide in molten aluminum sulfide (61).

The reactions of zirconium with metals of the platinum group, platinum, iridium, and osmium, are of an astonishing nature which is shared by its congener, hafnium. When hafnium is heated with platinum, it reacts with explosive violence. The affinity of platinum, iridium, and osmium for zirconium is so great that when the extremely stable and refractory zirconium carbide, ZrC, is heated with these metals, intermetallic compounds of zirconium and the platinum metals form, with liberation of carbon (62). Electrons of the outer shell of the platinum atoms flow into the valence shells of the zirconium atom. In the intermetallic compounds which form, a larger number of electrons become bonding electrons than served as such in the unreacted elements. In $ZrIr_3$, 31 electrons contribute to the bonding, whereas 25 electrons contribute to bonding in the separate metals (62).

The most important alloys of zirconium are zircaloy-2 and zircaloy-4, zirconium containing 1.5% tin, 0.12% iron, 0.10% chromium, and 0.05% nickel in the first, and the same but no nickel in the second. In addition to being more resistant to corrosion than zirconium under conditions of use in an atomic reactor, they are superior in hardness, elongation, yield strength, and ultimate strength.

The aircraft and space industries use magnesium–rare-earth–zirconium alloys for high-temperature conditions, and magnesium–zinc–zirconium alloys for structures that need withstand only ordinary temperatures.

Uses of Zirconium

By far the greatest use of zirconium metal has been in alloy form for the protective cladding of uranium fuel elements used in atomic reactors. The following lists separately the actual uses that have been made of zirconium in industry, and others that have been demonstrated but have not been accepted into industrial practice.

Applied in Industry

(1) Cladding uranium fuel elements (small proportions of alloying elements are used with the zirconium)
(2) Crucibles, particularly for use in analytical laboratories
(3) Getter for gases (see Vol. 8, p. 20)
(4) Explosive primers, based on low combustion temperatures of zirconium powder and its great heat of oxidation (8)
(5) Flashlight powder or foil (9)
(6) Heat exchanger tubes (alloyed with titanium) (10)

(7) Pyrotechnics and fireworks
(8) Refining grain and improving strength of structural aluminum and magnesium
(9) Superconductors (alloyed with niobium)
(10) Welding flux

Demonstrated

(1) Catalyst for chemical syntheses
(2) Coinage metal (11)
(3) Epoxy resin filler, enhancing stability (12)
(4) Propellant, of high-energy type
(5) Prosthetic devices and sutures
(6) Reducing agent, as for europium and samarium oxides (13)
(7) Stylus for marking glass. Zirconium leaves a metallic streak on glass which retains its metallic luster for years (14)

A typical 800-megawatt nuclear power plant uses 32.5 short tons of zircaloy tubing, which must be replaced periodically (22, p. 146). In 1967 about 60 tons of zirconium were used as shredded foil in camera flash bulbs. Three percent of the zirconium metal used that year went into construction in petrochemical plants and in urea, sulfuric acid, and hydrochloric acid processing plants (16, p. 1230). It has been used in the construction of parts of equipment which are particularly sensitive to hydrochloric acid corrosion.

Prices of zirconium metal vary greatly, and the physical condition and purity are reflected in the price. Typical prices per pound in 1970 have been:

Type	Price
zirconium sponge	$5.50–$13.00
zirconium plate	$10.00
zirconium strip, cold rolled	$13.00–$18.00
zirconium powder	$12.00–$13.00

Metal-like Compounds of Zirconium

The nonmetallic first- and second-period elements are small enough to occupy holes in metallic zirconium lattices. In doing so, they can form phases different from those of pure metallic zirconium, yet still metallic in appearance, in having metallic bonding of atoms, and in properties such as electrical conductance which are characteristically metallic.

The borderline between elements of this category and those just outside of it is not sharp. Oxygen dissolves in zirconium metal in considerable quantities without substantial change in the metallic lattice or metallic properties. It forms the metallic phases Zr_3O and Zr_6O. In the presence of certain alloying metals, ternary phases of comparable compositions form, such as $Zr_4Co_2O_x$, $Zr_4Re_2O_x$, Zr_6Fe_3O, $Zr_{4.2}Rh_{1.8}O$, $Zr_{4.2}Ir_{1.8}O$, $Zr_{4.4}Pt_{1.6}O$, Zr_3V_3O, Zr_3Cr_3O, and Zr_3Mn_3O. There are comparable ternary phases in which nitrogen takes the place of oxygen (63). Under some circumstances, certain third-period elements show similar capability of occupying holes or interplanar spaces in zirconium metal lattices. Such is the case with zirconium monochloride and zirconium monosulfide. But the more characteristic interaction of chlorine with zirconium is with formation of zirconium tetrachloride.

Our immediate interest is in compounds of zirconium with elements which form chiefly metal-like compounds with zirconium, that is, in compounds of zirconium with carbon, nitrogen, boron, and hydrogen. Boron and nitrogen form nonmetallic as well as metallic compounds with zirconium.

Zirconium Carbide. The carbon-zirconium system contains but a single phase other than the elements. It contains a maximum of one carbon atom per zirconium atom. Zirconium carbide, ideally ZrC, is a hard, dull metallic gray solid. Its melting point has been variously reported by competent investigators at from about 3200°C to 3900°C. It is, thus, an extremely refractory substance. The ternary tantalum zirconium carbide and tantalum hafnium carbide are the highest melting substances ever reported, their melting points being somewhat over 4200°C. Zirconium carbide is a good electronic conductor of electricity at ordinary temperatures, and it is superconductive at liquid helium temperatures. It is also a good thermionic emitter. Zirconium carbide is not attacked by water or by dilute alkalis or acids other than hydrofluoric acid. It is not attacked by concentrated nitric or hydrochloric acids, but it is dissolved by hot concentrated sulfuric acid and by aqua regia.

The crystal structure of zirconium carbide is of the face-centered cubic sodium chloride type. The carbon atoms are randomly distributed in the octahedral holes of the metallic lattice. For zirconium carbide which is 50 atom % carbon, the value of a_0 is 4.696 Å. The same phase containing only 35 atom % carbon has the parameter 4.677 Å (64). Zirconium carbide that is pure except for 1.5% hafnium exhibited the parameter $a = 4.69764$ Å (65). In general, the phase shows but little difference in properties due to departures from pseudostoichiometry. The lower carbon compositions are powerful getters for nitrogen and oxygen when hot, and their color varies with the content of lattice vacancies and of these elements.

In the industrial preparation of the zirconium carbide phase, it is often found advantageous to use a deficiency of carbon and to permit the retention of some nitrogen and oxygen. The product then usually has a golden color, and is called *zirconium cyanonitride* or *zirconium carbonitride*.

The composition ZrC is to be regarded as pseudostoichiometric rather than stoichiometric in that the composition is unrelated to the valence of the small atom. Boron (trivalent), carbon (tetravalent), and nitrogen (trivalent-tetravalent) all form face-centered cubic 1:1 phases with zirconium.

Pure crystals of zirconium carbide have been prepared by decomposing zirconium halides on hot tungsten filaments in the presence of carbon-containing gases. Zirconium carbide and zirconium cyanonitride are prepared in industry by heating zircon or zirconia with carbon in an electric arc furnace. They are used chiefly in the manufacture of zirconium tetrachloride and zirconium dioxide.

Electrodiffusion studies have shown that the carbon migrates toward the cathode (66).

Zirconium carbide is resistant to molten lithium at 815–1035°C. It decomposes when heated with platinum, forming platinum–zirconium alloy and free carbon. Zirconium carbide has been used as a cladding and structural material in nuclear reactors and as a component of cutting tools (15).

Zirconium Nitride. This, ideally ZrN, is a metallic substance varying in color from yellow to brown, depending on its preparation. It has a density of 6.93 g/cm³, hardness 8–9 (Mohs), and melts at about 3000°C. It is dissolved slowly by cold concentrated hydrochloric and cold dilute sulfuric acid and rapidly by concentrated acids and aqua regia, particularly when hot.

Massive zirconium reacts with nitrogen beginning at about 800°C, but it is extremely difficult, if not impossible, to attain to pseudostoichiometric ZrN by this reaction. Pure ZrN has been obtained by the reaction of zirconium powder with nitrogen at 1800°C, but it is preferable to use zirconium hydride in this preparation (67, p. 629). It has also been obtained by pyrolysis of zirconium halogenide vapors in the presence of nitrogen and hydrogen, and of amminozirconium halogenides.

There are no established uses for zirconium nitride.

Trizirconium tetranitride, Zr_3N_4, is not a metallic phase. See under Other zirconium compounds.

The reaction of zirconium nitride with zirconium dioxide under ammonia gas at 1000–1100°C leads to a number of oxynitride phases (68).

Zirconium Borides. Boron has a slightly larger atomic radius than carbon or nitrogen, and it has an additional difference in its relationship to zirconium in having fewer valence electrons than zirconium while carbon and nitrogen have as many or more, respectively. Boron forms a phase, ZrB, which is similar to the phases ZrC and ZrN, and also some different phases which will be discussed.

Zirconium monoboride, ideally ZrB, is analogous to zirconium carbide and zirconium nitride both in composition and in having the sodium chloride type face-centered cubic crystal lattice. Its parameter a = 4.65 Å; the x-ray density is 6.7 g/cm³. Zirconium monoboride has a silvery-gray metallic luster. It is stable only in the temperature range 950–1250°C (67, pp. 711–717). It has no established uses.

Zirconium borides are prepared by heating powdered boron with powdered zirconium metal or zirconium hydride, or by other procedures which in effect provide the two elements in admixtures suitable for their reaction with one another. Zirconium diboride has been prepared by liberating the component elements in the same molten bath, consisting of fluoroborates and fluorozirconates, electrochemically. In another procedure, zirconium powder was heated with boron carbide and boron oxide:

$$7 \, Zr + 3 \, B_4C + B_2O_3 \rightarrow 7 \, ZrB_2 + 3 \, CO$$

Zirconium diboride, ZrB_2, crystallizes in the hexagonal system and is isomorphous with a large number of other metal diborides. It melts at about 3040°C. It is metallic in appearance and exhibits good metallic conduction of electricity, higher than that of zirconium metal. Its Mohs hardness is about 8. Zirconium diboride powder can be fashioned into useful objects by the techniques of conventional powder metallurgy. Crucibles and thermocouple sheaths have been made available commercially. Tantalum has been coated with zirconium diboride applied in a plasma spray. The uses of zirconium diboride are largely dependent on its unusual resistance to oxidation under conditions of use, for a material composed of such highly oxidizable components.

Zirconium dodecaboride, ZrB_{12}, has face-centered cubic crystalline symmetry, a = 7.408 Å. As a metastable phase it is black and nonmetallic in appearance. Its electrical resistivity is five times that of zirconium diboride. It is stable from 1650°C to its melting point, 2680°C. It decomposes when heated in contact with carbon.

A number of ternary zirconium-containing borides are known: $Zr_3Al_3B_4$, $Zr_2Co_2B_6$, $CrZrB_2$, and $Zr_2Ni_2B_6$.

Zirconium Hydrides. In common with other metals of group IVA, and also metals of group IIIA, zirconium forms metallic hydrides rather than molecular hydrides. This type of compound is sometimes called a *transition metal hydride*. The term *exothermic occluder* has been applied to those metallic elements which absorb extremely high proportions of hydrogen exothermically, with the appearance of one or more new phases.

Hydrogen dissolves in zirconium without formation of a new phase in proportions up to 0.7 atom % in α-zirconium at 300°C and up to 50 atom % in β-zirconium at 850°C (69). The solubility of hydrogen in the metal increases with pressure and decreases with temperature. As the hydrogen in solution increases, phase changes tend to occur. Thus, beginning with pure α-zirconium and forcing hydrogen into solution, the lattice symmetry changes successively from hcp to fcc, then hcp, fcc, and finally fct with $a = 4.964$ Å and $c = 4.440$ Å (70). The presence of elements other than zirconium or hydrogen, and particularly oxygen, affects the morphology of the hydride phases and has been the reason for some inconsistencies in reports in the literature.

The zirconium hydride phases have composition ranges rather than precise compositions. They were first systematized by G. Haegg. The solution of hydrogen in the α-zirconium was called α-phase; the fcc phase of composition approximately $ZrH_{0.25}$ was called β; the hexagonal phase of composition approximating $ZrH_{0.5}$ was called γ; the fcc phase of composition approximately ZrH was called δ; and the fct phase approaching composition ZrH_2 was called ϵ. Later, a previously unrecognized phase designated δ' was observed to form on cooling a specimen that contained 33 atom % hydrogen at 300°C (67, pp. 447–452). Currently, the chief phases of zirconium hydride are considered to be δ of composition $ZrH_{1.4}$ to $Zr_{1.56}$ and ϵ of composition $ZrH_{1.65}$ to $ZrH_{1.965}$. The view has been advanced (69) that the system is essentially built upon a tetragonal lattice. Alterations in the parameters with changes in hydrogen content cause overlap into other symmetries. It is difficult to increase the hydrogen content as the limiting composition ZrH_2 is approached.

The metal lattice expands and the density decreases linearly as the hydrogen content increases (72). $Zr_{1.95}$ has a density of 5.5 g/cm³, as against a density of 6.5 g/cm³ for α-zirconium, approximately a 15% change. The hydrogen embrittles the metal, and it tends to crack as the hydrogen content is increased. The zirconium hydride of commerce is pulverulent, and it has become standard practice to supply it as a fine powder.

Even though there are many zirconium hydride phases, the practice has been in use in industry of calling all of them *zirconium hydride*. It will become apparent from the present discussion that this is a useful and often a necessary convention, particularly when discussing behavior patterns in the hydrogen-zirconium system with changes of temperature and compositions. Only the idealized *zirconium dihydride* will merit especial consideration here, and this name will be employed when the discussion applies uniquely to this phase.

Zirconium dihydride, ideally ZrH_2, is sold in industry as a metallic powder which is stable in air at room temperature. It is not hygroscopic. Its crystal structure is face-centered tetragonal of the fluorite-type, and it is isomorphous with the dihydrides of hafnium and titanium. Zirconium hydride powder differs from zirconium metal powder in presenting little fire or explosion hazard, and it can be handled quite safely. The product of commerce ranges in composition from $ZrH_{1.90}$–$ZrH_{1.95}$. Heating it leads to release of hydrogen gas continuously and reversibly, and hence there is no decomposition temperature but only a variable equilibrium, the constants of which are different for the different hydride phases. No hydrogen is liberated at temperatures lower than 315°C. Quantitative evolution is achieved only at about 900°C under vacuum. The equilibrium composition at 625°C under 400 mm pressure of hydrogen is $ZrH_{1.80}$ (73).

The heat of formation of $ZrH_{1.98}$ is -39.7 kcal/mole; that of $Zr_{1.47}$, -30 kcal/mole; and of $ZrH_{1.23}$, -25.3 kcal/mole (74). Physical and electrical properties such as the Hall constant, electrical resistivity, and magnetic susceptibility vary with the hydrogen content (75). Zirconium hydride of compositions $ZrH_{1.69}$ to $ZrH_{1.96}$ exhibit metallic conductivity from 0 to 1000°C (76). The heat capacity of ZrH_2 at constant pressure is 7.396 cal/deg (77).

Much of the behavior of zirconium hydride is quite similar to that of zirconium metal. It is not attacked by most acids, but is dissolved by hydrofluoric acid and by hot concentrated sulfuric or phosphoric acids. When the hydride is heated in oxygen or nitrogen, it forms zirconium dioxide or nitride.

Industrial production of zirconium hydride is usually by one of these methods: (1) A zirconium compound, usually zirconium dioxide, is reduced with a powerful reducing metal or hydride (eg, calcium hydride) under a hydrogen atmosphere (78); or (2) separately prepared zirconium metal is heated under hydrogen to about 900°C, suitably equilibrated, and then cooled under hydrogen (79). In either case, the reaction products are crushed. In the case of the first process, this is followed by leaching with dilute acid to remove compounds of the element used as reducing agent.

The chief current use of zirconium hydride is by the military in flares, fuses, tracer cartridges, and incendiary compositions. The use of powdered zirconium hydride by the military is due to its greater stability and safety of handling, compared to powdered zirconium metal. While a number of patents have appeared describing this use, it is in the nature of this application that little publicity has been given to what has been found and utilized. The tendency of zirconium hydride to sinter when heated has proven useful in powder metallurgy. Sinters of higher density are obtainable from zirconium hydride powder rather than from zirconium metal powder (80, p. 289). Zirconium hydride has proven to be valuable as a foaming agent in the production of foamy aluminum (27,81). It serves well as a flux or wetting agent in brazing, and forms ceramic–ceramic, ceramic–metal, and metal–metal seals. The zirconium metal liberated by the thermal decomposition of the hydride readily alloys with other metals with which it makes contact (82,83). Zirconium hydride is a suitable feed material for use in forming metallic zirconium surfaces by the flame-spraying technique. The liberated hydrogen shields the hot metal deposit against oxidation.

Zirconium hydride has been used as a getter (see Vol. 8, p. 20) in vacuum tubes. The product of low hafnium content shows promise as a moderator in atomic reactors (73). In this case, the hydrogen is the moderator and zirconium hydride is a convenient compound for introducing the hydrogen into the reactor.

While zirconium hydride is easily handled with safety, certain precautions should be observed. It should not be exposed to flame or to an electric discharge. Should it take fire, the fire should be smothered with sand or other very inert material. Water, carbon dioxide, and soda-and-acid sprays should not be used. Hot zirconium reacts with these substances. The creation of a zirconium-hydride-dust-laden atmosphere should be avoided (84).

Bibliography

"Zirconium and zirconium alloys," in *ECT* 1st ed., Vol. 15, pp. 282–290, by W. B. Blumenthal, National Lead Company.

1. J. J. Berzelius, *Ann. Chim. Phys.* **26**(2), 43 (1824).
2. L. Weiss and E. Neumann, *Z. anorg. allgem. Chem.* **65**, 248 (1910).
3. D. Lely, Jr., and L. Hamburger, *ibid.* **87**, 208 (1914).
4. U.S. Pat. 553,296 (Jan. 21, 1896) to J. W. Aylesworth (Converse D. Marsh Co.).
5. A. E. van Arkel and J. H. De Boer, *Z. anorg. allgem. Chem.* **148**, 345 (1925).
6. L. Troost, *Compt. rend.* **61**, 109 (1865).
7. W. J. Kroll, A. W. Schlechten, and L. A. Yerkes, *Trans. Electrochem. Soc.* **89**, 263 (1946).
8. F. G. Reshetnikov and E. N. Oblomeev, *Sov. J. Atomic Energy* **2**, 561 (1957).
9. U.S. Pat. 3,415,605 (Dec. 10, 1968) to J. C. Van der Tas, W. Werstevelt, and R. M. Kruimink (North American Philips Co., Inc.).
10. Anon., *Chem. Eng. News* **46**(29), 36 (1968).
11. U.S. Pat. 3,373,017 (March 12, 1968) to D. R. Spink and W. W. Stephens (Amax Specialty Metals, Inc.).
12. I. A. Vlasova and Z. N. Samokhvalov, *Zashch, Stoit. Konstr. Korroz., Nauch. Issled. Inst. Betona, Zhelezobetona, Mater, Koord. Soveshch.* **1966**, 236.
13. T. T. Campbell and F. E. Block, *J. Metals* **11**, 744 (1959).
14. U.S. Pat. 2,962,383 (Nov. 29, 1960) to J. Franci and J. S. Logiudice (Owens-Illinois Glass Co.).
15. K. A. Gschneider, Jr., *Technology and Uses of Beryllium, Cesium, Germanium, Hafnium, Niobium (Columbium), Rare Earths (Including Yttrium and Scandium), Titanium, and Zirconium, Identification No. IS-1757*, Ames Laboratory, University of Science and Technology, Iowa State, Jan. 1968, Distributed by Clearinghouse for Federal Scientific and Technical Information, Washington, D.C.
16. J. F. O'Leary, ed., *Minerals Yearbook 1967*, U.S. Govt. Printing Office, Washington, D.C., 1968.
17. C. J. Hardy and D. Scargill, *J. Inorg. Nucl. Chem.* **9**, 322 (1959).
18. U.S. Pat. 2,850,379 (Sept. 2, 1958) to A. S. Hawkes (Ethyl Corp.).
19. I. F. Nichkov, A. V. Volkovich, S. P. Raspopin, and V. E. Alenichev, *Elektrokhimiya* **2**(9), 1101 (1966).
20. Japanese Pat. 3609 (May 12, 1958) to T. Oshiba (Showa Electric Industries Co.).
21. U.S. Pat. 2,813,787 (Nov. 19, 1957) to W. Schmidt (Reynolds Metal Co.).
22. Japanese Pat. 6601 (Aug. 24, 1957) to T. Oshiba (Showa Electric Industries Co.).
23. O. N. Carlson, F. A. Schmidt and H. A. Wilhelm, *J. Electrochem. Soc.* **51**, 6 (1957).
24. U.S. Pat. 2,902,360 (Sept. 1, 1959) to W. Juda (Ionics, Inc.).
25. U.S. Pat. 2,814,561 (Nov. 26, 1957) to H. de Wet Erasmus (Union Carbide Corp.).
26. U.S. Pat. 2,753,255 (July 3, 1956) to P. P. Alexander (Metal Hydrides, Inc.).
27. U.S. Pat. 2,751,289 (June 19, 1956) to J. C. Elliot (Bjorksten Research Laboratories, Inc.).
28. U.S. Pat. 2,820,722 (Jan. 21, 1958) to R. T. Fletcher (inventor).
29. J. T. Martorell and M. L. Rodriguez, *Inst. Hierro Acero* **14**, 232 (1961).
30. G. W. Mellors and S. Senderoff, *J. Electrochem. Soc.* **112**(3), 266 (1965).
31. *Ibid.* **113**(1), 60 (1966).
32. M. J. Barbier and J. Cotteret, *Rev. Intern. Hautes Temp. Réfractaires* **1**(1), 41 (1964).
33. U.S. Pat. 2,985,569 (May 23, 1961) to L. W. Gendvil (National Lead Co.).
34. U.S. Pat. 2,732,321 (Jan. 24, 1956) to C. B. Gill, A. W. Schlechten, and M. E. Straumanis (inventors).
35. B. Tittman, D. Hamilton, and A. Jayaraman, *J. Appl. Phys.* **1**(3), 732 (1964).
36. Yu. F. Bychkov, V. A. Mal'tsev, E. D. Martynov, and V. K. Markob, *Met. i Metalloved. Chistykh Metal., Sb. Nauchn. Rabot.* **1967**(6), 58.

37. M. Billion and J. P. Langeron, *Compt. Rend.* **260**(1), 152 (1965).
38. M. M. Nieto and A. M. Russell, *J. Appl. Phys.* **35**(2), 461 (1964).
39. N. V. Borkhov, *Met. i Metalloved. Chistykh Metal., Sb. Nauchn. Rabot* **1959**(2), 148.
40. F. E. Littman and F. M. Church, *U.S. Atomic Energy Comm. SRIA 29*, U.S. Govt. Printing Office, Washington, D.C., 1960, 55 pp.
41. T. Amako and K. Onoue, *Bunseki Kagaku* **6**, 735 (1957).
42. H. Wertenbach and K. L. Lieser, *Kerntechnik* **8**(7), 307 (1966).
43. R. E. Meyer, *J. Electrochem. Soc.* **112**(7), 648 (1965).
44. T. Sawada and S. Kato, *Bunseki Kagaku* **14**(5), 404 (1965).
45. M. E. Straumanis, W. J. James, and W. C. Custead, *J. Electrochem. Soc.* **107**, 502 (1960).
46. U.S. Pat. 2,992,067 (July 11, 1961) to J. L. Swanson (U.S. Atomic Energy Comm.).
47. L. Stein and R. C. Vogel, *Ind. Eng. Chem.* **48**, 418 (1956).
48. T. Nakayuma and T. Koizumi, *Nippon Kinzoku Gakkaishi* **29**, 472 (1965).
49. J. N. Wanklyn, *Electrochem. Technol.* **4**(3–4), 81 (1966).
50. J. C. Banter, *J. Electrochem. Soc.* **114**(5), 508 (1967).
51. N. J. M. Wilkins, *Corrosion Sci.* **4**(1), 17 (1964).
52. D. Whitman, J. Boghen, and J. Evstyukhin, *Comm. énergie atomique (France), Rappt. CEA No. 771*, 1958, 12 pp.
53. I. I. Korobar, D. V. Ignatov, A. I. Evstyukhin, and V. S. Emel'yanov, *Met. i Metalloved. Chistikh Metal. Sb. Nauchn. Rabot* **1959**, 144.
54. E. I. Gurovich and G. P. Shtokman, *Zhur. Priklad. Khim.* **32**, 2673 (1959).
55. J. Henney and J. W. S. Jones, *UK Atomic Energy Auth., Res. Group, Atomic Energy Res. Estab., Rep. AERE-R-5671*, 1968, 13 pp.
56. U.S. Pat. 2,879,186 (March 24, 1959) to W. Fischer (Aktiengesellschaft für Unternehmungen der Eisen- und Stahlindustrie).
57. K. Okano, Y. Ota, and Y. Kitamura, *Boshoku Gijutsu* **9**, 100 (1960).
58. C. F. Britton, J. V. Arthurs, and J. N. Wanklyn, *J. Nucl. Mater.* **15**(4), 263 (1965).
59. D. C. Stoops, M. D. Carver, and H. Kato, *U.S. Bur. Mines, Rep. Invest. 5945*, U.S. Govt. Printing Office, Washington, D.C., 1962, 12 pp.
60. P. H. G. Draper and J. Harvey, *Acta Met.* **11**(8), 873 (1963).
61. U.S. Pat. 3,386,817 (June 4, 1968) to G. S. Layne and J. O. Huml (Dow Chemical Co.).
62. A. S. Darling, *Platinum Metals Rev.* **11**(4), 138 (1967).
63. H. Holleck and F. Thuemmler, *J. Nucl. Metals* **23**, 84 (1967).
64. S. Nagakura and S. Oketani, *Trans. Iron Steel Inst. Japan* **8**(5), 265 (1968).
65. C. A. Kempter and R. J. Fries, *Anal. Chem.* **32**(4), 570 (1960).
66. Yu. F. Babikova and R. L. Gruzin, *Met. i Metalloved. Chystikh Metal., Sb. Nauchn. Rabot* **1960**(2) 128.
67. J. P. Langeron in P. Pascal, ed., *Nouveau Traité de Chimie Minerale*, Masson et Cie., Paris, 1963, Vol. 9, pp. 711–717.
68. J. C. Gilles, *Bull. Soc. Chim. France* **1962**, 2118.
69. D. A. Vaughan and J. R. Bridge, *J. Metals* 528 (May 1956).
70. G. L. Miller, *Zirconium*, Academic Press, Ltd., London, 1957.
71. E. A. Gulbransen and K. F. Andrew, *J. Electrochem. Soc.* **101**, 474 (1954).
72. J. D. Roach, *Iron Age* **186**(2), 108 (1960).
73. H. M. McCullough and B. Kopelman, *Nucleonics* **14**(11), 146 (Nov. 1956).
74. A. G. Turnbull, *Australian J. Chem.* **17**(10), 1063 (1964).
75. R. A. Andrievsky, E. B. Boiko, and R. B. Ioffe, *Izv. Akad. Nauk SSSR, Neorg. Materialy* **3**(9), 1591 (1967).
76. V. I. Savin, R. A. Andrievsky, E. B. Boiko, and R. A. Lyutikov, *Fiz. Metal. Metalloved.* **24**, 636 (1967).
77. H. F. Flotow and R. Osborne, *U.S. Atomic Energy Comm. TID-15795*, U.S. Govt. Printing Office, Washington, D.C., 1962, 12 pp.
78. U.S. Pat. 2,427,339 (Sept. 16, 1947) to P. P. Alexander (Metal Hydrides, Inc.).
79. D. J. Hurd, *Chemistry of the Hydrides*, John Wiley & Sons, Inc., New York, 1952.
80. B. Lustman and F. Kerze, Jr., eds., *The Metallurgy of Zirconium*, McGraw-Hill, 1955.
81. U.S. Pat. 3,214,265 (Oct. 26, 1965) to W. S. Fiedler (Lor Corp.).

82. C. S. Pearsall, *Materials and Methods* **30,** 61 (July 1949).
83. G. R. Van Houten, *Am. Ceram. Soc. Bull.* **38,** 301 (1959).
84. M. D. Banus, *Chem. Eng. News* **32,** 2424 (June 14, 1954).

WARREN B. BLUMENTHAL and
JOHN D. ROACH (Zirconium Hydrides)
National Lead Company

OTHER ZIRCONIUM COMPOUNDS

In the great majority of its compounds, zirconium exhibits the oxidation number four. It commonly attains to covalencies of five, six, seven, and eight through polymerization or formation of adducts. Monatomic zirconium ions play no perceptible role in the chemistry of this element, but zirconium is commonly the central atom or atoms of complex anions and cations. A useful simplification of the chemistry of zirconium compounds is achieved by treating them in the categories: (*1*) compounds containing complex zirconium cations, (*2*) compounds containing complex zirconium anions, and (*3*) compounds containing zirconium only in unionized groups.

Historically, the hypothetical zirconyl cation, ZrO^{2+}, was believed to be a structural unit in many zirconium compounds, because their analyses indicated the presence of one oxygen atom per zirconium atom, apart from those oxygen atoms that appeared rationally to belong to other groups of atoms, for example, $ZrOCl_2 \cdot 8H_2O$, $ZrO(NO_3)_2 \cdot 2H_2O$, $ZrOF_2$, $ZrO(CNS)_2 \cdot nH_2O$. However, infrared studies of many solid zirconium compounds, the analyses of which suggested the presence of zirconyl ion, indicated none to be present. In a few cases, namely, anhydrous zirconyl chloride (137.10), certain thiocyanates (55.3), and fluorides (55.2a), there has been infrared spectral evidence for the $Zr{=}O$ structure.

Zirconium-containing cations are characteristically oligomers of zirconium dioxide to which protons and neutral ligands such as aquo are chemically bound. In both crystalline zirconyl chloride and its aqueous solutions, we find the structure:

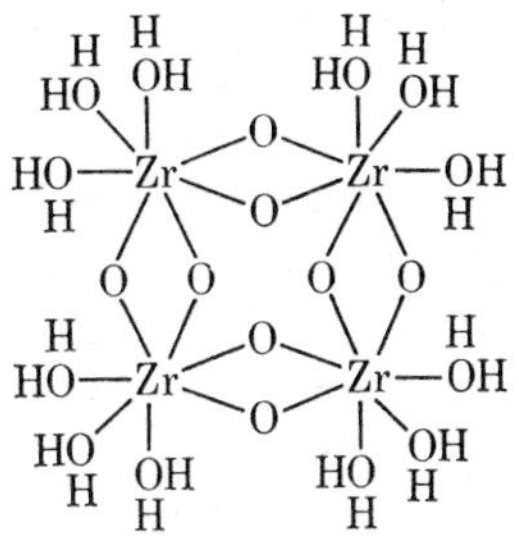

The zirconium atoms lie on a plane, and are bonded, each to the next, by pairs of oxygen atoms. One oxygen atom of each pair lies above and one below the plane of the zirconium atoms (2.0,61.2).

Observations of the pH of zirconyl chloride solutions show them to be nearly those of hydrochloric acid solution of the same molarity (22.4, pp. 127–128). Hence, for each mole of $ZrOCl_2$ in aqueous solution, one mole of hydrochloric acid is formed, and one mole of hydrogen chloride is bound to the zirconium species. X-ray diffraction studies of the aqueous solutions have shown that a chloride ion (similarly, bromide ion) is closely held to each side of the square, doubtless coulombically (61.2). The complex cation is cationic because of the protons that are bound to it, and not because of the presence of a positively charged zirconium atom. Other cationic species appear to be essentially other arrangements of a multiplicity of ZrO_2 units, to which protons are bound.

In compounds such as zirconium alkoxides, zirconium β-diketonates, and massive zirconium dioxide, there are no separations of electrically charged groups and these compounds are regarded as nonionic. The formation of anionic zirconium species is best exemplified by the replacement of electrically neutral ligands by electrically negative ligands. Thus, in acidic media,

$$ZrO_2.xH_2O + 2 SO_4{}^{2-} \rightarrow (ZrO(SO_4)_2.(x-1)H_2O)^{2-} + 2 OH^-$$

Compounds containing anionic zirconium species may be regarded as derivatives of the hypothetical metazirconic acid, H_2ZrO_3, and orthozirconic acid, H_4ZrO_4. Examples are disulfatozirconic acid, $H_2ZrO(SO_4)_2.3H_2O$, diacetatozirconic acid, $HZrOOH(C_2H_3O_2)_2$, and tetramandelatozirconic acid, $H_4Zr(C_6H_5CHOCO_2)_4$. Many of the zirconic acids are polymeric.

Noncomplex compounds of zirconium are rare. Whereas the gaseous tetrahalides consist of simple monomeric molecules, the liquid and solid states often contain ions formed by a kind of disproportionation (22.4, p. 108; 122.5, 144.35):

$$2 ZrCl_4 \rightleftharpoons (ZrCl_3)(ZrCl_5) \rightleftharpoons (ZrCl_2)(ZrCl_6)$$

It is characteristic of zirconium(IV) compounds to be diamagnetic, colorless, and not active as a redox catalyst in the manner of paramagnetic transition elements. Pi-bonded compounds of the transition metal, such as their bis(cyclopentadienyl) compounds are commonly colored, whereas those of zirconium are colorless.

Hafnium is chemically as similar to zirconium as deuterium is to hydrogen, and in any description of areas of chemical uniqueness of zirconium, it is to be understood that hafnium is grouped with zirconium and not with other transition elements. The ratio of the atomic weights of zirconium and hafnium is approximately 1:2, as is that of hydrogen and deuterium.

A peculiarity of zirconium with respect to redox catalysis is that while this element is not active, it commonly promotes the redox catalysis of the paramagnetic-ion-forming elements. Thus, while cobalt and manganese soaps are effective paint driers, and zirconium soaps are not, admixtures of zirconium soaps with those of cobalt or manganese are much more effective than the cobalt or manganese soaps alone (125.12, 125.13). Zirconium is catalytic for many types of reaction other than redox, such as condensation, cracking, esterification, hydrolysis, isomerization, and pyrolysis.

Much of the applied chemistry of zirconium is due to the great stability of oxygen–zirconium bonds and the large number of oxygen atoms that may be simultaneously bonded to zirconium, up to eight. This has had application in the crosslinking of polymers for the enhancement of the toughness and wear-resistance of films (100A.1), the bonding of finishing agents to textiles, the tanning of hides, the hardening of gelatin, and the insolubilization of starch and soluble cellulose derivatives.

Complex compounds of zirconium containing coordinately bound divalent and polyvalent anions, such as sulfate or phosphate, commonly contain zirconium in complex anions. But compounds of zirconium with univalent anions in aqueous solution exhibit zirconium species whose charge is determined by the relative amounts of protons and of anions which are bound to zirconium. Aqueous solutions of zirconyl chloride, zirconyl nitrate, and zirconyl perchlorate at concentrations of about $0.5M$ or less give indications of cationic species of zirconium, only. If the acidity is increased with the parent acid, neutral and anionic species appear. Both are prominent at $1.5N$ acid strength and have been estimated to be present in about equal amounts in $4N$ nitric acid and 8–$9N$ hydrochloric acid. In very dilute and highly acidic solutions, monomeric zirconium species exist, but as the pH is raised hydrolysis and olation give rise to chain and ring structures containing structural units

$$-\text{O}-\overset{|}{\underset{|}{\text{Zr}}}-\text{O}-\overset{|}{\underset{|}{\text{Zr}}}-\quad \text{or}\quad \overset{\text{O}}{\underset{\text{O}}{>}}\text{Zr}\overset{\text{O}}{\underset{\text{O}}{<>}}\text{Zr}<$$

Reagents which react rapidly with monomer species may react slowly or fail to react appreciably with the polymeric species. With further increase of pH, hydrous zirconium dioxide forms as stable micelles having considerable resistance to breakdown by chemical reagents (112.1).

Whereas aqueous solutions of zirconium compounds have been found to be useful in a great variety of industrial applications, the uses of organic solutions have been restricted almost completely to the synthesis of compounds and analytical chemistry. Solvent–solvent extraction has had importance as a technique for separating hafnium from zirconium.

Mineral Sources of Zirconium

Zirconium is widely distributed in nature, and has been estimated to be present in the earth's crust to the extent of about 0.022%, roughly equal to the proportion of carbon. Zirconium does not occur as the element, but as the oxide, baddeleyite, the double oxide, zircon, and other more complex oxides.

The most common ore, *zircon*, has tetragonal symmetry and contains four of the formula units, $ZrO_2 \cdot SiO_2$, per unit cell. Particles of zircon are widely distributed in rocks, particularly in granites and syenites. Commercially important concentrations are found in sands of certain beaches, notably in Kerala (India); Zhdanov (Ukraine); Malaya; near Byron Bay, Australia's east coast; and the coast near Jacksonville, Florida. Massive crystals have been found in Henderson County, N.C., and in Madagascar. Crystals of gem quality occur in many regions, but particularly in Ceylon. A number of minerals are regarded as altered zircon or morphologically closely related to zircon, notably alvite, auerbachite, cyrtolite, malacon, and naegite.

Baddeleyite occurs as monoclinic crystals of composition ZrO_2, principally in the states of Minas Geraes and São Paulo, Brazil. It sometimes occurs as enormous boulders, and at others as alluvial pebbles called *favas*. The ore, caldesite, is made up largely of baddeleyite. It is the chief zirconium ore of Brazil (44.1). An important new source of baddeleyite is the by-product of copper mining at Phalabora, S. Africa (139.1).

The zirconium minerals, other than zircon and baddeleyite, are:

allanite* (97.1)	lavenite
anderbergite	leucosphenite
annerodite	loranskite and wiikite
arfvedsonite or riebeckite	lorenzenite or ramseyite
arrhenite	lovozerite (40.3,82.1)
astrophyllite	mesodialyte* (114.1)
auerlite	monazite
barsonovite* (118.1,40.13)	mosandrite
batisite (40.4)	nohlite and samarskite
calzirtite* (40.7)	oegirite
catapleite (40.7,116.1)	polymignite
cerite	pyrochlor
chalcolamprite	schorlomite
columbite	seidozerite* (75.2,149.1)
dalyite (140.2)	sipylite
elpidite (40.1,75.1)	sogdianite* (40.17)
endeolite	thorianite
erdmannite and michaelsonite	thortveitite and benfanamite
eucolyte and eudialyte	tritonite
euxenite	uhligite
fergusonite	uraninite
guarinite and hiortdahlite	vlasovite* (40.8,81.1)
hainite	wadeite (40.5)
johnstrupite	woehlerite (40.12)
keldyshite* (40.10)	weloganite* (28.1)
kimseyite* (40.9)	xenotime
kobeite (80.1)	zirconolite* (40.7)
	zirkelite (100.1)

The asterisked names are those of minerals relatively recently discovered.   Literature citations are given for those minerals, a large part of the knowledge of which is less than about fifteen years old.   Some of the minerals listed are generally regarded as occurrences of elements other than zirconium, but often contain zirconium.   The minerals ferrihalloyisite, rosenbuschite, and sphene may also contain zirconium.

Zirconium has been found in solution in the waters of many springs, usually in concentrations on the order of tenths of a part per million.   It is a minor constituent of soil, and is found in minute proportions in many living organisms and in coals.

Handling of Ores

Zircon sands are usually concentrated by electrostatic, electromagnetic, and gravitational techniques.   A product 99% in $ZrO_2 \cdot SiO_2$ is obtained.   This is usually either heated with carbon in an arc furnace, with volatilization of silica and recovery of zirconium carbide or cyanonitride, or it is heated with alkalies to form alkali metal silicate plus alkali metal zirconate, or alternatively alkali metal zirconylosilicate, eg, Na_2ZrSiO_5.   The carbide or cyanonitride can be burned in air to make zirconium dioxide, or in chlorine to make zirconium tetrachloride.   The alkali metal silicate can be leached to leave a residue of zirconium oxide. Alkali metal zirconylosilicate can be dissolved in acids to form zirconyl salts or complexes such as sulfatozirconic acids.

World production of zircon in 1965 by free world countries was about 330,000 tons. Most of the metal, ceramics, and chemicals were derived from this source. A small number of companies in eight countries produce nearly all of this zircon. Australia is the chief source, and the United States is the only other sizable source. The United States requirement of zircon has recently been approximately 120,000 tons annually. Well over half of this has been imported. The large number of scientific and technical publications in the Soviet Union in recent years on zirconium compounds suggest that zirconium-containing products have considerable important to it.

Zirconium compounds sold in 1968 for prices ranging from about \$0.85–\$5.00/lb of zirconium dioxide contained. Bulk sale tended to go near the lower end of this price range, and pharmaceutical and specialty products near the higher end. The largest sales volume of all manufactured zirconium compounds was of zirconium dioxide. Considerable volumes of the following compounds were also sold: carbonated hydrous zirconia, ammonium zirconyl carbonate, zirconium 2-ethylcaproate, zirconium tetrachloride, zirconyl chloride, hydroxyzirconyl chloride, zirconium sulfate, sodium zirconium sulfate (sodium disulfatozirconate), and zirconyl acetate.

Analysis

In the absence of metallic elements other than the alkalies, and also in the absence of organic chelate formers, pure hydrous zirconium dioxide can be precipitated quantitatively by the addition of ammonium hydroxide to aqueous solutions of zirconium compounds. After washing, drying, and igniting at about 1000°C, a residue of zirconium dioxide can be weighed.

Phosphoric acid precipitates pure diphosphatozirconic acid quantitatively from 20% sulfuric acid solutions which may contain moderate amounts of other metallic compounds. After washing and drying, the precipitate may be ignited to form zirconium pyrophosphate, ZrP_2O_7, which is weighed. Mandelic acid (phenylglycolic acid, $C_6H_5CHOHCOOH$) precipitates pure zirconium mandelate from hydrochloric acid solutions of zirconium compounds, in the presence of many impurities. Ignition of the precipitate leaves a residue of pure zirconium dioxide.

Zirconium has also been precipitated for quantitative determinations by arsenates, iodates, and fluorides (as barium fluorozirconate). Also, the following organic agents have been used to precipitate zirconium under specified conditions:

α-hydroxybenzylphosphonic acid	nitrophthalic acid
α-hydroxycarboxylic acids	phenylacetic acid
arsinic and arsonic acids	Ponceau Red R
benzenesulfinic acid	purpurogallein
coumarin	sulfonated dyes
cupferron	sulfonic acids
2,5-dihydroxy-*p*-benzoquinone	tannic acid
ethylenediaminetetraacetic acid	tartrazine
Flavazine L	thioglycolic acid
flavianic acid	thiomalic acid
flavones	vanillin
hydroxynaphthoic acids	

Colorimetric and spectrographic determinations have been made of zirconium on reduced phosphomolybdic acid compounds and on those formed with the following dyes or groups of dyes:

Acid Chromium Black	Chromotrope 2C	Nitrosulfophenol S
Alizarine S	Eriochrome Black T	Omega Chrome Black Blue G
Alizarine sulfonates	Eriochrome Cyanine	Orange II
anabasine	Fast Gray RA	phthaleins
anthranilic acid	flavones	Picramine CA
Arsenazo	fluorenones	Picramine R
aurin tricarboxylic acid	gallein	purpurinsulfonate
azo dyes	Gallocyanine MS	Solochrome Violet R
bromanilic acid	Glycerincresol Red	Stilbazo
Catechol Violet	Kaempferol	Xylenol Orange
chlorsulfophenol	Methylthymol Blue	
Chrome Azurol S	Neothorin	

Zirconium can be detected and determined by the techniques of chromatography, fluorescence, flame luminescence (based on effects on alkaline earth spectra), polarography, radiometry (as by precipitation of $^{32}PO_4$), and spectrography. The main lines of the zirconium spectrum are too dense for general use in spectrometry, and it has been found that the lines 2658.7, 2720.6, 2727.0, 3085.3, 3157.8, and 3379.9 Å are useful.

Volumetric determinations of zirconium have been conducted by adding excess of standard ethylenediaminetetraacetic acid and back-titrating with a standard bismuth solution, by forming an oxidizable compound of zirconium such as the tetramandelate and titrating this with potassium permanganate, and by forming the periodate of composition $(ZrO)_3(H_2IO_6)_2$ and determining it iodimetrically. Great care must be taken in volumetric determinations to make sure that the polymeric zirconium species have been broken down to monomeric species, lest spurious values of the equivalent weight of zirconium distort the calculated results of the analysis. Boiling the zirconium solution after making it $5N$ in sulfuric acid has been shown to be satisfactory for depolymerization (11.2).

Biological and Physiological Aspects of Zirconium Compounds

There is no evidence that the element, zirconium, has played a role in the organic life on this planet. An important reason is that zirconium minerals are too insoluble to supply even traces of the element for use in life processes. Nonetheless, minute amounts of zirconium have accumulated in many plants and animals, and radioactive zirconium from fall-out has been added in some cases.

In general, plants have a very high tolerance for zirconium compounds in the environment, and animals have very high tolerance to zirconium compounds ingested orally. In comparison with many other heavy metals, the tolerance to zirconium compounds injected intervenously or intraperitoneally is high (22.4, pp. 39–45). Zirconyl nitrate administered orally to rats was found not to be absorbed into their systems (35.1).

The isotope ^{95}Zr has been found in many plants and animals, including fish (89.1). It has been found to concentrate in algae, but not in sponges growing in the same water

(93.1). Zirconium compounds have been demonstrated to be powerful algaecides (6.1). Zirconium compounds alone have no significant bactericidal effects, but they greatly accentuate the bactericidal effects of compounds of certain metals, such as copper and mercury (113.3; cf. 101.2). Zirconium compounds have shown value as purges for the removal of radioactive heavy metals which have been absorbed into bone or tissue. The incidence of tumor development in animals dosed with ^{90}Sr was found greatly reduced by subsequent administration of a zirconium citrate (52.1).

Stannous fluorozirconate has been shown an effective prophylactic in reducing the incidence of dental caries (20.2), and zirconium silicate has proven of exceptional value as a dental abrasive (125.50). Sodium lactatozirconate (125.51) and hydroxyzirconyl chloride (101.1) are useful as antiperspirants. Some zirconium compounds induce granulomatous growths in the skins of hypersensitive persons (7.1), an effect regarded as allergic rather than toxic.

Zirconium Catalysis and Catalysts

A wide variety of catalytic effects of zirconium and its compounds have been reported. This is largely a consequence of the Lewis acid properties. Zirconium tetrachloride has been demonstrated to catalyze Friedel-Crafts synthesis quite generally, and in some cases with advantageous results. For example, its catalysis of the reaction of 2,5-dimethyl-2,5-hexanediol with benzene gave an 80% yield of 1,1,4,4,5,5,8,8-octamethyl-1,2,3,4,5,6,7,8-octahydroanthracene in comparison with a 20% yield when aluminum chloride was used (50.4). Zirconium is not a redox catalyst after the pattern of the elements which form paramagnetic ions, but it has striking effects in promoting redox catalysis of paramagnetic ions. Superior hydrogenation catalysis has been achieved using a nickel catalyst containing zirconium dioxide, for the conversion of vegetable oil to fat (125.8a). A cobalt–zirconium catalyst has improved the reduction of nitriles to amines (23.18), and cobalt and manganese driers have been improved by the addition of zirconium soaps (125.12, 125.13). Zirconium acts as a primary catalyst in some redox reactions, such as the decomposition of bleach (143.1) and the reduction of disaccharides in acidic and alkaline media (84.4).

There is a large literature on the zirconium catalysis of the following classes of reactions:

alkylation	dehydration	hydrogenation
aromatization	dehydrogenation	hydrolysis
combustion	dephosphorylation	isomerization
condensation	desulfurization	oxidation
cracking	disproportionation	polymerization
curing	esterification	pyrolysis
deamination	Friedel-Crafts	redox
decarboxylation	halogenation	scission
deesterification	hydrocracking	Ziegler-Natta processes

The Zirconium Compounds

ZIRCONIUM ACETATES

Zirconium tetraacetate, $Zr(CH_3CO_2)_4$, has been prepared by boiling a mixture of zirconium tetrachloride and glacial acetic acid (23.9, 66.1), or zirconium tetraethoxide with glacial acetic acid (23.4). Acetic anhydride is evolved by these reactions. The compound is soluble in benzene, and decomposes in water with formation of soluble products. No industrial uses have been established for zirconium tetraacetate.

Zirconium tetra(trifluoroacetate), $Zr(F_3CCO_2)_4$, has been prepared by the reaction of zirconium tetrachloride with trifluoroacetic acid. It is stable up to 280°C (14.1).

Zirconyl acetate or **diacetatozirconic acid**, $H_2ZrO_2(CH_3CO_2)_2$, is prepared by dissolving carbonated hydrous zirconia in its equivalence of acetic acid. It is miscible with water in all proportions. The product of commerce usually is an aqueous solution assaying 22% ZrO_2. When the solution is dried, an amorphous residue is formed. This and other properties indicate the molecule to be a high polymer.

The chief uses of zirconyl acetate have been in rendering textiles water-repellent by aqueous (48.1) or organic solvent (47.4) applications, bonding protective compounds to textiles (113.2), catalyzing the cure of silicone resins applied to textiles, leather, and other materials (77.1, 113.1), hardening gelatin used as a vehicle for photosensitive substances (125.23), and preparing extrudable dopes from which zirconia fibers can be manufactured (23.13a,125.51a).

Zirconyl trichloroacetate or **trichloroacetatozirconic acid**, $H_2ZrO(Cl_3CCO_2)_2$, is but sparingly soluble in water, and is decomposed by hydrolysis. It dissolves promptly in sulfuric acid (41.1).

Dizirconium pentaacetate, $Zr_2O(OH)(CH_3CO_2)_5 \cdot 2H_2O$, is formed when zirconyl chloride is boiled with an excess of nearly 100% acetic acid, followed by evaporation of the solution. It forms an anisotropic solid, $d_{25} = 1.963$, which is unstable in air. It has been observed to begin to decompose at 50°C, first losing one mole of acetic acid and progressively breaking down until a residue of ZrO_2 is formed at 430–450°C (144.13).

ZIRCONIUM ACETYLACETONATES (see also Zirconium ketonates)

Zirconium tetraacetylacetonate, $Zr(C_5H_7O_2)_4$, consists of monoclinic needles which melt at 194–195°C, $d_4^{25} = 1.415$. It is prepared by the reaction of zirconyl chloride with acetylacetone in aqueous solution. It is sparingly soluble in water, cyclohexane, and pinene; moderately soluble in acetone, 1-butanol, chlorobenzene, dimethylformamide, dioxane, nitrobenzene, and ethyl acetate; and highly soluble in benzene.

The structure of zirconium tetraacetylacetonate has been determined by x-ray diffraction and is shown in Fig. 1 (51.1). Uses for zirconium tetraacetylacetonate have been patented; catalysis of the vapor phase polymerization of ethylene (125.24), water-repellent treatment of leather (125.29), and curing of silicone resins (23.5, 23.11).

The zirconium acetylacetonate of commerce is a partially hydrolyzed product, approximating the composition $ZrOH(C_5H_7O_2)_3$.

ZIRCONIUM ALKOXIDES (see also Zirconium polyolates)

The term *zirconium alkoxide* has supplanted the earlier usage of the terms *zirconium ester* and *organic zirconate*. The alkoxides have the generic formula $Zr(OR, OR', OR'', OR''')_4$, and this may be polymeric or solvated with the parent alcohol. Also, there are partially hydrolyzed alkoxides (27.2) and derivatives in which another ligand, such as chloro, takes the place of some of the alkoxide. The polymerization number varies from one to four, depending on steric effects (62.1,62.2,62.3). Examples of partially hydrolyzed alkoxides are the compounds of compositions $(C_4H_9)_6Zr_2O$, $(C_4H_9)_8Zr_3O_2$, $(C_4H_9)_{10}Zr_4O_3$, and $(C_4H_9)_{22}Zr_{10}O_9$ (55.2).

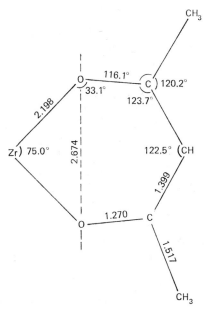

Fig. 1. Crystal Structure of Zirconium(IV) Acetylacetonate. Courtesy American Chemical Society.

Zirconium alkoxides are usually prepared by the reaction of zirconium tetrachloride with anhydrous alcohols in the presence of a base which combines with hydrogen chloride (62.1):

$$ZrCl_4 + 4\ ROH + 4\ NH_3 \rightarrow Zr(OR)_4 + 4\ NH_4Cl$$

Uses of zirconium alkoxides are the subjects of many patents, particularly for the catalysis of the polymerization of ethylene, for paint modifiers, and for use in imparting desired properties to textile and glass surfaces. But little of zirconium alkoxides has been used in industry, and there has been a tendency to keep this secret. A zirconium alkoxide formed from a hydroxyl-substituted branched-chain alcohol is reported to impart unusually high quality to rubber, when used as a lubricant (125.36). See also Alkoxides, metal.

Alkoxides made from aminoalcohols are water-soluble (47.2). An unusual reaction of alkylene oxides with zirconium tetrachloride has been found to provide alkoxides of chlorinated alcohols (125.46a):

$$ZrCl_4 + OC_2H_4 \rightarrow Zr(OC_2H_4Cl)_4$$

Fluoroalkoxyzirconium compounds also are known (32.3). There are also alkoxides in which zirconium and other metals are contained in the same molecule, such as $(C_4H_9)_3SnOZr(OC_3H_7)_3$ (125.56).

Silicon analogs of zirconium alkoxides are well known. $Zr(OSi(CH_3)_3)_4$ and $Zr(OSi(C_2H_5)_3)_4$ have been prepared by the reaction of zirconium alkoxides with trialkylsilylacetates (33.3).

ZIRCONIUM AMMINES, AMIDES, AND IMIDES

When zirconium tetrachloride is treated with ammonia gas or liquid, the ammine of composition $ZrCl_4 \cdot 4NH_3$ forms. On being heated, it decomposes stepwise to zirconium tetraamide, $Zr(NH_2)_4$, zirconium diimide, $Zr(NH)_2$, and zirconium nitride, Zr_3N_4. The last of these is not to be confused with the metallic phase of composition ZrN, which has also been called zirconium nitride. (See Zirconium metal and alloys.)

Amines combine with zirconium tetrachloride to form alkylammino- and arylammino adducts, some of which are readily altered to form zirconium amides and imides. Examples prepared from aniline (60.3), pyridine, bipyridine (68.2), N-benzoylphenylhydroxylamine (133.1), o-, m-, and p-nitroaniline, aminobenzoic acids, acetanilide, benzamide, benzanilide (92.2), imidazoles (145.2), and other amines have been reported. Complexes of pyridine, acetonitrile, bipyridine, and 1,10-phenanthroline have been prepared with zirconium trichloride (31.6,62.9). These have assymetrical structures and probably metal-metal bonding or interaction. Hydrazine complexes, eg, $Zr(OCH(CH_3)_2)_4 \cdot N_2H_4$, are also known (27.1,27.3).

The complex $Zr(bipyridine)_3$, prepared by the reaction of zirconium tetrachloride in carbon tetrachloride followed by reduction with lithium, contains zirconium in a formally zerovalent condition. It is slightly soluble in a number of organic solvents. The zerovalent zirconium can be oxidized to tetravalent by a solution of iodine in pyridine (141.1).

Zirconium tetrachloride tends to form adducts with secondary and tertiary amines, and to undergo aminolysis with primary amines (62.5). A white flocculent solid formed by the reactions of zirconium tetrachloride in liquid ammonia has been found to have the composition $ZrCl_3NH_2 \cdot xNH_3$. Zirconium tetra(dimethylamide), $Zr(NMe_2)_4$, mp 60°C, has been prepared by the reaction of zirconium tetrachloride with lithium dimethylamide. Zirconium tetra(diethylamide), bp 120°C at 1 mmHg pressure, was similarly prepared (90.2). Zirconium tetraanilide was formed by the reaction of zirconium tetra(diethylamide) with aniline (67.10).

Zirconium tetra(dimethylamide) has been observed to be polymeric (62.11a), and this and other observations suggest a parallelism between the polymerization of nitrogen–zirconium compounds and oxygen–zirconium compounds. Zirconium amides form adducts with metal carbonyl, examples of which are the compounds of composition $Zr(N(CH_3)_2)_2 \cdot 2Ni(CO)_4$ (32.2).

Zirconium amides with low-molecular-weight alkyl groups bound to the nitrogen hydrolyze rapidly in water, but others, such as those with tribenzylamine or methyldioctylamine, form sufficiently stable compounds to permit their use in extracting zirconium from aqueous solutions by organic solvents (10.1).

Up to six moles of potassium amide have been found to react with zirconium tetrabromide, and up to 3.5 moles react with zirconium tribromide (67.6), forming amidozirconates of Zr(IV) and Zr(III). The compound $Li_2Zr(dipyridyl)_3(THF)_8$ has been formed in THF (tetrahydrofuran) and recovered as black needles. It decomposes in water (138.3).

Zirconium is precipitated by 8-hydroxyquinoline with formation of a chelate ring in which nitrogen and oxygen are coordinated with zirconium (144.10).

ZIRCONIUM BORATES

The reaction of zirconyl nitrate with sodium tetraborate in water has been observed to precipitate the composition $ZrO_2 \cdot B_2O_3 \cdot 3H_2O$. It lost two moles of water on heating to 200°C and the third mole on heating to 400°C (66.3).

Boratozirconium chloride, $BO_3Zr(OH)Cl$, is formed when a zirconyl chloride solution reacts with boric acid over a wide temperature range. The addition of small proportions of an aqueous solution of this compound to ethanol or methanol causes them to set to transparent jellies. Aquo jellies and other solvent jellies can also be formed with this agent (125.61).

ZIRCONIUM BORIDES (see p. 624)

ZIRCONIUM BOROHYDRIDES

Zirconium borohydride, $Zr(BH_4)_4$, is the most volatile of the known compounds of zirconium. It melts at 29.0°C and boils at 118°C. Its heat of sublimation is 13 kg cal/mole. The vapor pressure at 25°C is 15 mm (60.7).

Zirconium borohydride can be prepared by metathesis of zirconium tetrachloride with aluminum borohydride or lithium borohydride (129.1). It is miscible with cyclohexane, with which it forms a eutectic at 24% zirconium borohydride. The eutectic melts at -41°C (144.49). In the crystalline solid, the zirconium atom of zirconium borohydride is surrounded tetrahedrally by four crystallographically equal boron atoms (31.5). The observed Zr-B distance, 2.35 Å, agrees well with the sum of the covalent radii of zirconium and boron (147.2).

ZIRCONIUM BROMIDES (see Zirconium chlorides, bromides and iodides)

ZIRCONIUM CARBIDE (see p. 624)

ZIRCONIUM CARBONATES

Probably no true carbonate of zirconium has been prepared, in the sense of one containing carbonate anions. The precipitation of zirconium from a zirconyl chloride solution with an alkali carbonate yields a solid of composition $2ZrO_2.CO_2.8H_2O$. A water pulp of this substance, assaying about 50% ZrO_2, is an article of commerce commonly called *carbonated hydrous zirconia*. In 1970 it was sold for $0.85/lb of ZrO_2 contained. A product of composition $ZrO_2.CO_2.2H_2O$ has been prepared by subjecting carbonated hydrous zirconia to the action of carbon dioxide at 30–40 atm pressure.

X-ray diffraction study has shown carbonated hydrous zirconia to contain no carbonate ion, and has supported the assumption that it is hydrous zirconia with adsorbed carbon dioxide (85a.1). Carbonated hydrous zirconia dissolves in many acids with formation of zirconyl salts or complex ions. A number of commercial zirconium compounds are prepared in this way.

Ammonium tricarbonatozirconate or **ammonium zirconyl carbonate,** $(NH_4)_3$-$ZrOH(CO_3)_3.2H_2O$, has been an important article of commerce in the form of an aqueous solution assaying about 10% ZrO_2. Its price in 1970 was approximately $2.30/lb of ZrO_2 contained. It has been used as a component of water-repellent treatments for textiles, in floor polishes (108.1,125.53) and in improving the water-resistance of the adhesive of paper coatings (23.13). It has been shown effective in improving the weather and abrasion resistance of acrylic latex paints (125.14,125.15) and alginate films (125.16), and the adhesion of lithographic films to metal plates (125.20). Also, its use as an agent for deposition of zirconia films by the techniques of solution ceramics has been described (125.14a).

When ammonium carbonate is added to aqueous zirconyl chloride solution, the first precipitate which forms redissolves. The addition of alcohol to the solution causes crystalline *ammonium dicarbonatozirconate*, $(NH_4)_2ZrO(CO_3)_2.3H_2O$, to separate (87.1, 144.20). But when carbonated hydrous zirconia is dissolved in an excess of ammonium carbonate or bicarbonate solution and the solution is evaporated, ammonium tricarbonatozirconate crystallizes. The compositions of the crystals do not necessarily imply the compositions of the ions in the solutions from which they were derived. Dialysis studies of an ammonium carbonatozirconate solution gave evidence of a species $Zr(CO_3)_5{}^{6-}$ (17.1). Crystalline carbonatozirconates of a wide variety of compositions have been isolated (144.34, 144.42). The carbonato ligand on the zirconium atoms appears to be bidentate (144.40). X-ray diffraction analysis of *potassium tricarbonatozirconate*, $K_3ZrOH(CO_3)_3.2H_2O$, has shown the crystal to contain potassium ions and the structural unit $Zr_2(OH)_2(CO_3)_6$, and water (147.1). The carbonato compound $2Sr_2Zr(CO_3)_4.SrCO_3.4H_2O$ occurs in nature as the mineral *weloganite* (28.1).

ZIRCONIUM CARBONYLS

Many efforts to produce carbonyls of zirconium have been made, but without success. Carbonyl adducts of zirconium tetra(dimethylamide) are known (32.2).

ZIRCONIUM CARBOXYLATES · (see also Zirconium acetates and Zirconium formates)

The compositions that have been synthesized from zirconium compounds and higher fatty acids or their alkali salts have been called *zirconium soaps*, but they have never been well defined. Methods of synthesis have been chiefly the mixing together of zirconyl chloride, carboxylic acid, and alkali (125.5, 125.13) or basic zirconyl chloride and alkali soaps (125.17), and the extraction of aqueous solutions of zirconium compounds with immiscible organic liquids in which fatty acid has been dissolved. Uses of zirconium soaps in water-repellents (48.1, 125.2), in paint driers (125.12, 125.13), and for the protection of metal surfaces form corrosion (125.28) have been described. Zirconium soaps sequester vanadium oxide from fuel oils and prevent its volatilization and corrosion of boiler systems (125.46).

Currently it seems most correct to view the zirconium soaps prepared in aqueous media as hydrous zirconia from which the chemisorbed water has been displaced by chemisorbed fatty acid. Measurements of the sizes of the soap particles show them to be of colloidal rather than molecular dimensions (144.28). With the carboxyl group of the fatty acid bonded to the substrate particle of zirconia, the hydrocarbon chain protrudes into the region surrounding the particle and thus imparts a hydrophobic quality to the particle. Such carboxylic adsorbates are formed by monocarboxylic acids (2-ethylhexanoic, oleic, stearic) and dicarboxylic acids (malonic, adipic, pimelic, suberic) (91.1).

A *zirconium propionate* of composition $H_2ZrO_2(C_2H_5CO_2)_2$ has been recovered as a solid on cooling a zirconyl chloride solution to which a large excess of propionic acid had been added (144.39). The *butyrate* of composition $Zr(C_3H_7CO_2)_4$ prepared from the reaction of butyric acid with zirconium tetrachloride, was observed to hydrolyze in air to $H_2ZrO_2(C_4H_9CO_2)_2$ (144.44).

The compositions of empirical formula $Zr(O_2CR)_4$ have been prepared by the reaction of zirconium tetrachloride with anhydrous carboxylic acids (23.9) and by the reactions of zirconium alkoxides (62.4,83.1) or zirconium tetraacetylacetonate (125a.1) with the carboxylic acids. Partially hydrolyzed species such as those of generic formula $(RCO_2)_3ZrOZr(O_2CR)_3$ have also been prepared (33.1). Compositions obtained by the precipitation of zirconium from aqueous solutions by the benzoate ion and some of its substitution products have not been well defined.

ZIRCONIUM-CELLULOSE COMPOUNDS

When a cellulosic material, such as paper or cotton cloth, is immersed into an aqueous solution of a zirconium compound, removed and rinsed, a fraction of a percent of zirconium adheres tenaciously. This effect has been used extensively in applying zirconium-bonded water-repellent finishes. It is surmised that the bonding is due to chelation of the zirconium atoms by pairs of vincinal hydroxyl groups, probably enhanced by partial oxidation of the cellulose surface with formation of carboxyl groups. A study of the ion-exchange properties of oxidized cellulose showed a capacity of 0.29 meq/g of cellulose for zirconium (146.1).

ZIRCONIUM AND CHELATING AGENTS

A great number of chelate compounds of zirconium are known in which the chelation formed five- and six-membered rings. The number of five-membered ring chelates that are known is the greater, and it is suggested that the five-membered rings are the more stable. Oxygen is most commonly the donor atom, and nitrogen is most apt to be the donor atom in the absence of oxygen, or when both oxygen and nitrogen can become members of the ring, as with carboxylated amines (60.15). A notable exception is the chelate with o-phenylene bis(dimethylarsine), in which neither oxygen nor nitrogen is involved (62.6).

ZIRCONIUM CHLORATES, etc., (see Zirconium halogenates)

ZIRCONIUM CHLORIDES, BROMIDES, AND IODIDES

The fluorides are treated separately, p. 645.

Zirconium tetrachloride, $ZrCl_4$, is a colorless solid which sublimes at 331°C and melts at 438°C under 25 atm pressure. It is manufactured industrially by the reaction of chlorine with zirconium carbide or zirconium cyanonitride, or with heated mixtures of zirconia or zircon with carbon. The industrial product is usually a fluffy powder with a light tan tint which is due to the presence of a few hundredths of a percent of ferric chloride. Zirconium tetrachloride is diamagnetic, with a susceptibility of about -0.302×10^{-6} (71.1).

Zirconium tetrachloride vapor is monomolecular, but the liquid and solid state have complex molecular structures. The solid exhibits monoclinic crystalline symmetry in which $ZrCl_6$ octahedra are coupled by edges with formation of zigzag chains (15.2). Infrared and Raman spectra have been interpreted as indicating a dimer as the structural unit (110.1). There is evidence for the presence of the ion $ZrCl_6^{2-}$ in the liquid and solid (122.5, 144.35).

Zirconium tetrachloride reacts violently and irreversibly with water, forming zirconyl chloride:

$$ZrCl_4 + 9\ H_2O \rightarrow ZrOCl_2 \cdot 8\ H_2O + 2\ HCl$$

It dissolves in liquid sulfur dioxide, molten stannous chloride, and molten alkali metal chlorozirconates, from which it may be recovered by heating and volatilization. Zirconium tetrachloride dissolves in many oxygen-containing organic solvents and in liquid ammonia and amines with irreversible formation of solvates or solvolysates.

Zirconium tetrachloride forms stable if readily hydrolyzable adducts with alcohols, amines, ammonia, esters, ethers, ketones, nitriles, nitrosyl chloride, phosphorus oxychloride, phosphorus pentachloride (but not phosphorus trichloride), and a large number of other Lewis bases. Some examples are $ZrCl_4.9(CH_3)_2SO$, $ZrCl_4.2HMPA$ (hexamethylphosphoramide), $ZrCl_4.2CH_3CN$, $ZrCl_4.2C_2H_5CN$, $ZrCl_4.2C_6H_5SeO$, $ZrCl_4.COCl_2$, $ZrCl_4.SOCl_4.SOCl_2$, and $ZrCl_4.2(C_4H_9)_3PO_4$.

Zirconium tetrabromide is chemically very similar to the tetrachloride. **Zirconium tetraiodide** is similar in some respects, but is less stable and is oxidizable by dry air or oxygen at relatively low temperatures, about 200°C. Both the tetraiodide and the tetrabromide have cubic close-packed crystalline symmetry for the halogen atoms, with zirconium atoms occupying octahedral holes (37). The tetrabromide sublimes at 356°C, and has been melted under pressure at 449°C. When the tetraiodide is heated to about 1200°C, it breaks down to lower iodides and finally deposits zirconium metal.

Zirconium(III) chloride or **zirconium trichloride** has been prepared by reduction of the tetrachloride with active metals such as aluminum and zirconium, or by hydrogen under a glow discharge (60.16). Trichloride of 99% purity has been prepared by reduction with zirconium (137.9). It forms hexagonal crystals in which there are infinite chains of $ZrCl_6$ octahedra sharing faces parallel to the c axis. The chains are held together by Van der Waals forces (51.2a). Zirconium trichloride dissolves in water with formation of colored solutions which change in hue and quickly become colorless due to oxidation of Zr(III) to Zr(IV) by the water, with liberation of hydrogen.

Zirconium trichloride forms adducts with monodentate and bidentate ligands, eg, $ZrCl_3.2$ pyridine and $2ZrCl_3.5CH_3CN$ (62.9).

Zirconium triiodide and **zirconium tribromide** are generally similar to the trichloride (60.3). The tribromide forms dark green needles (137.8). The triiodide has been prepared by reducing zirconium tetraiodide with zirconium powder at 510°C, under pressure. It disproportionates at 360–390°C to **zirconium diiodide** and zirconium tetraiodide (68.3).

Zirconium(II) chloride or **zirconium dichloride** is prepared by the disproportionation reaction of zirconium trichloride on heating to 650° (95.1), or by reducing zirconium trichloride with sodium. It dissolves in molten aluminum trichloride, and it forms adducts with 2,2'-bipyridyl and 1,10-phenanthroline (62.9). **Zirconium dibromide** is generally similar to the dichloride.

Zirconium monochloride is not a salt but a metallic phase containing one interstitial chlorine atom per zirconium atom. It is formed by electrolyzing a molten mixture of strontium and sodium chlorides in which zirconium tetrachloride is dissolved (125.32). X-ray diffraction study has shown it to consist of planes of zirconium atoms separated by the interstitial chlorine. The metallic properties of the zirconium are dominant. The composition does not dissolve in water, dilute hydrochloric acid, or common organic solvents (124. 1). Bromine and iodine analogs are also known.

Zirconyl chloride, $ZrOCl_2.8H_2O$, is a colorless crystalline compound, usually forming tetragonal needles. It is an important article of commerce, which sold for about $0.51/lb in 1970. It is prepared by the hydrolysis of zirconium tetrachloride.

Pure grades are prepared by recrystallizing from hydrochloric acid or methanol, or by preparation from pure intermediates. It is stable as the octahydrate at 20°C at 9.75–13.23 mm water vapor pressure. At other pressures, hydrates from about 4–9 moles of water per formula weight have been observed (144.36). The unit of structure in the crystalline octahydrate is a tetramer of composition $4ZrO_2 . 8HCl . 16H_2O$ in which the core is zirconium dioxide tetramer (see p. 629), to which four moles of water are bound at each zirconium atom, and to which an additional four moles of water and two moles of H^+Cl^- are held without covalent bonding to the core (2.0). Although approaching an Archimedean antiprism in form, the coordination polyhedron about the zirconium atom has been shown to have more nearly the symmetry of a dodecahedron with triangular faces. The mean oxygen–zirconium distance in the Zr—O—Zr bridges is 2.14 ± 0.019 Å, and that of $Zr—OH_2$ is 2.72 ± 0.032 Å (27.5). There is no evidence of the zirconyl group, Zr=O (15.1). The core zirconium atoms can be complexed by ligands other than water. Insoluble adducts have been precipitated from solution in formamide, dimethylformamide, acetamide, dimethylacetamide, acetanilide, benzyl-amide, methylacetamide, and acetone (68.5).

X-ray scattering by aqueous zirconyl chloride shows the persistence of the core structure of the crystal in the aqueous solution. Four chloride ions (and similarly bromide in zirconyl bromide solution) are not bound to the zirconium atoms, but are held in definite positions along the sides of the square formed by the sets of four zirconium atoms (61.2). The behavior of zirconium in aqueous zirconyl chloride solution is typical for a complex cation.

Zirconyl chloride has been used to prepare pigments from acid dyes, in the water-repellent treatment of textiles, in tanning, in pharmaceuticals and in the preparation of a large number of other zirconium compounds. Zirconyl bromide and iodide are similar in general chemical behavior and usefulness to the chloride, but their use is often ruled out by their higher cost.

Anhydrous zirconyl chloride, $ZrOCl_2$, cannot be prepared by dehydration of the hydrates. It has been prepared by the reaction of zirconium tetrachloride with antimony trioxide at 180–320°C (54.3):

$$3\ ZrCl_4 + Sb_2O_3 \rightarrow 3\ ZrOCl_2 + 2\ SbCl_3$$

and by the reaction of zirconium tetrachloride with chlorine monoxide in anhydrous carbon tetrachloride at -30°C. It is a white crystalline solid which decomposes at 250°C. Its infrared spectrum suggests the presence of the Zr=O group and a structural unit $ZrOCl_4$ (137.10). It forms adducts with phosphorus oxychloride and with pyridine.

Basic zirconyl chlorides are compounds whose compositions contain less than two chlorine atoms per zirconium atom, examples of which are $Zr_5O_8Cl_4 . 14H_2O$ (23.1), $Zr_2O_3Cl_2 . xH_2O$ (48.2), $ZrOOHCl.xH_2O$ (72.1), and $Zr_2O_3OHCl . xH_2O$ (155.22). The first of these is very soluble in water, and the others are miscible with water in all proportions. A composition $Zr_5(OH)_5O_5 ._5Cl_4 . 17H_2O$ has also been reported (144.43). The compound of composition $Zr_2O_3Cl_2 . xH_2O$ has been found useful in precipitating dyes to form pigments (129.9A), and that of composition $ZrOOHCl . xH_2O$ as the active principle in body deodorants (125.19,125.25). $ZrOOHCl.xH_2O$ precipitates when ether is added to a methanol solution of zirconyl chloride octahydrate. If urea is present in the same solution, a urea adduct of the product is formed (144.23).

When ammonium hydroxide is added under suitable conditions to a solution of zirconyl chloride, a precipitate of composition $ZrO(OH)_{1.75}Cl_{0.25} \cdot xH_2O$ is obtained, which has been regarded as a compound by some investigators (144.2).

Numerous compounds of zirconium containing both chloride and other anions have been prepared in aqueous media. Boiling certain aqueous solutions has given, for example, a precipitate of the composition $Zr_5O_8(OH)_{10}Cl_2(SO_4)_5 \cdot 16H_2O$ (144.43).

Hexachlorozirconates of composition M₂ZrCl₆ are well known and have been widely studied, and some industrial use has been made of them. M^+ may be an univalent ion, including tetraalkylammonium ion. The existence of a chlorozirconate ion has been authenticated by anion-exchange studies in $12M$ hydrochloric acid solution (136.1). Details of the sorption of zirconium from hydrochloric acid solutions on the anion exchangers AV-17 and KU-2 have indicated the optimum HCl concentrations for sorption, and sorption efficiencies as high as 85% (132.2). Raman spectra of hexachlorozirconate anion in aqueous hydrochloric acid have been reported and discussed (62.11).

Cesium hexachlorozirconate has been crystallized from hydrochloric acid solutions and ammonium, potassium, and rubidium hexachlorozirconates from water solutions. The cesium and rubidium compounds are stable at 1000°C. The ammonium compound begins to decompose at about 450°C (144.26).

The hexachlorozirconates are most commonly prepared by heating zirconium tetrachloride with an alkali metal chloride under anhydrous conditions:

$$ZrCl_4 + 2\ NaCl \rightarrow Na_2ZrCl_6$$

Oxochlorozirconates can be prepared by heating hexachlorozirconates with antimony trichloride:

$$3\ M_2ZrCl_6 + Sb_2O_3 \rightarrow 3\ M_2ZrOCl_4 + 2\ SbCl_3$$

where M is Na, K, Rb, or Cs. These compounds have been observed to be stable at 960°C (54.3).

ZIRCONIUM CHROMATE

A compound of composition $Zr_4(OH)_6(CrO_4)_5 \cdot 2H_2O$ has been characterized. It is made up of an infinite chain of zirconium atoms linked by pairs of oxygen atoms. The zirconium coordination figure is an almost regular pentagonal bipyramid (17a.1).

CYCLOPENTADIENYLZIRCONIUM COMPOUNDS

Bis(cyclopentadienyl)zirconium dichloride, $(C_5H_5)_2ZrCl_2$, has monoclinic crystal symmetry and density 1.782 g/cc. (See also Cyclopentadiene.) The cyclopentadienyl radical appears to occupy three coordination sites of the zirconium atom (60.19). The compound melts at 232°C, with volatilization. It is soluble in benzene, chloroform, and ethylene glycol dimethyl ether, sparingly soluble in petroleum ether, and soluble in water, but with slow decomposition (10.2). A large number of derivatives of the compound have been prepared, and some uses have been described, as for example, a catalyst in Ziegler-type condensations of olefins. No industrial use has been publicized.

Bis(cyclopentadienyl)zirconium dichloride is prepared by the reaction of zirconium tetrachloride with cyclopentadienylsodium. Under some conditions, analogs can be prepared in which chloride is replaced by bisulfate (56.1), hydride or borohydride (31.4), cyanate or thiocyanate (19.3), phenylsilyl (31.7), mercaptophenyl or selenophenyl (69.3), and amide (62.10). Cyclopentadienyl can be replaced partly or sometimes entirely by other organic ligands such as allyl (69.2), acetylacetonato (40.14), or pentafluorophenyl (63.1). It appears that the extreme resistance of zirconium to forming sigma bonds with carbon is mitigated by altering the valence electron structure with pi-bonded cyclopentadienyl ligands. The compound of composition $(C_6H_5ZrC_5H_5)_2O \cdot H_2O$ has been recovered from the reaction of bis(cyclopentadienyl)zirconium dichloride with phenyllithium (40.14). *Tetrakis(cyclopentadienyl)zirconium* is also known (56.2), and bis(cyclopentadienyl)zirconium, $(C_5H_5)_2Zr$, has been prepared by reducing the chloride with sodium (60.18a).

ZIRCONIUM FERROCYANIDE

Zirconium ferrocyanide, $ZrFe(CN)_6$, probably hydrated, has been precipitated from $2M$ acid solution. In less strongly acidic solution, only partially hydrolyzed products are obtained. The precipitate is a powerful sorbent for alkali metal ions, the stability for the sorbate increasing with the atomic weight of the metal (103.1).

ZIRCONIUM FLUORIDES

Zirconium tetrafluoride, ZrF_4, is a colorless crystalline solid which sublimes at 901°C. Its specific gravity is 4.4333 at 16°C. Its structure is a three-dimensional array of antiprisms. An α-form is tetragonal and a β-form is monoclinic (67.8, 98.1). The α changes to β at 405°C (36.6). An amorphous form has also been observed and found to change to β at 450–500°C (98.1). Vapor molecules of zirconium tetrafluoride have regular tetrahedral structure, and the Zr-F distance is 1.94 ± 0.02 Å. This is smaller than the calculated distance, and suggests some double bond character (132.3).

Zirconium tetrafluoride has been prepared by a variety of methods (22.4, p. 139), notably by synthesis from the elements (71.2), by reaction of zirconium metal with hydrogen fluoride (67.5), by the reaction of ammonium fluoride with zirconium carbide, nitride, or cyanonitride (36.7), and by pyrolysis of ammonium fluorozirconates (43.1).

Zirconium tetrafluoride dissolves with difficulty in water, probably due to formation of insoluble hydrolyzates which render the surface inactive toward water. It dissolves readily in aqueous hydrofluoric acid and in hot concentrated sulfuric acid. Only minute amounts dissolve in anhydrous hydrogen fluoride. It forms adducts with ammonia and amines, eg, $ZrF_4 \cdot 2bipy$ (bipyridyl) (62.8). A variety of uses for zirconium tetrafluoride have been patented, but it is probable that only its use as a fluorination catalyst has become industrial (125.18).

Zirconium dichloride difluoride, $ZrCl_2F_2$, has been prepared by the reaction of $ZrCl_4 \cdot 2PCl_5$ with arsenic trifluoride. It is a white crystalline substance which hydrolyzes in water (138.2).

Zirconium trifluoride, ZrF_3. Much more difficulty was encountered in preparing zirconium trifluoride than any of the other trihalides. But it was finally prepared by the action of hydrogen fluoride on zirconium hydride. It is said to dissolve in organic acids and hot water without decomposition (137.13). Also, the trivalent zirconium compounds $KZrF_4$ and $NaZrF_4$ have been found in the electrolysis products of alkali fluorozirconates (115.1).

Orthorhombic zirconium difluoride, ZrF_2, has been prepared by the reduction of thin films of zirconium tetrafluoride with atomic hydrogen at about 350°C. It is a black solid which takes fire in air and burns to ZrO_2. It disproportionates at 800°C to zirconium metal and zirconium tetrafluoride (19.2).

It is convenient to regard zirconium tetrafluoride as analogous to an acid anhydride. The addition of hydrogen fluoride forms a complex acid:

$$ZrF_4 + 2\ HF \rightarrow H_2ZrF_6$$

and similarly the oxozirconium fluorides form oxofluorozirconic acids. Many such acids are known: H_2ZrF_6 (67.4); $H_2ZrF_6 \cdot H_2O$ (22.4, p. 137); $H_2ZrF_6 \cdot O \cdot 7H_2O$, hexagonal crystals (67.4); $HZrF_5 \cdot 3H_2O$, orthorhombic (67.4); $HZrF_5 \cdot 1.5H_2O$, face-centered cubic (67.4); H_2ZrOF_4, body-centered cubic (122.2); $H_2ZrO_2F_2 \cdot H_2O$ (22.4, p. 137); HZr_2OF_7 from H_2ZrOF_4 heated above 345°C (122.4); and $H_2Zr_3O_4F_{10}$, orthorhombic (98.1). Also the phases of composition $ZrO_{0.67}F_{2.67}$, cubic; $ZrO_{0.46}F_{3.08}$, rhombic; $ZrO_{0.44}F_{3.12}$, tetragonal; and $ZrO_{0.25}F_{3.50}$, tetragonal, have been observed (98.1).

Fluorozirconates are most commonly prepared by heating zircon or zirconium dioxide with an alkali fluorosilicate:

$$ZrO_2 \cdot SiO_2 + K_2SiF_6 \rightarrow K_2ZrF_6 + 2\ SiO_2$$

or by evaporating solutions in which a fluorozirconic acid and an alkali fluoride have been dissolved. Also, passing fluorine through a molten salt in which zirconium carbide has been suspended yields a fluorozirconate (23.8).

Direct evidence has been adduced from surface tension studies of molten fluorozirconates for the existence of fluorozirconate ions (65.4). Mass spectrography of the vapor over melts of sodium fluoride mixed with zirconium tetrafluoride at 877°K has shown the presence of the ions Na^+, ZrF_3^+, ZrF_2, ZrF^+, and Zr^+ (143.2). Zirconium dioxide has been found to dissolve to a small extent in fluorozirconate melts with formation of new ion species (26.1).

Table 1. The Variety of Fluorozirconates

Anion	Cations forming salts with the anion (moles water per atom Zr)
ZrF_5^-	Na, Na(H_2O), K, K(H_2O), Rb, Cs, Cs(H_2O), Tl, NH_4(H_2O), K + 0.5Ni(4H_2O), Mg(5H_2O), Zn, Mn(5H_2O), Ni(5H_2O), pyridinium, hydrazinium
ZrF_6^{2-}	Li, Li + Na, Na, K, K + Na, K + 0.5Cu(3H_2O), Cs, NH_4, Ba, Sn(II), Ni(12H_2O)
ZrF_7^{3-}	Li, Na, K, Rb, Cs, NH_4, Tl, Cu(8H_2O)
ZrF_8^{4-}	Li, Cu(12H_2O), Zn(12H_2O), Cd(7H_2O), Mn(6H_2O), Ni(12H_2O)
ZrF_{12}^{8-}	6 Li + Be
$Zr_2F_9^-$	Li, Na, Rb
$Zr_2F_{11}^{3-}$	Na
$Zr_2F_{13}^{5-}$	Na, K
$Zr_2F_{14}^{6-}$	Cs(2H_2O), Cu(16H_2O)
$Zr_3F_{14}^{2-}$	Cs
$Zr_3F_{17}^{5-}$	Tl, Tl(H_2O)
$Zr_4F_{19}^{3-}$	Li
$Zr_6F_{31}^{7-}$	Na
$HOZrF_5^{2-}$	NH_4 + H, K + H, Tl + H, Cs + H, 0.5Cd + H(2H_2O)
$ClZrF_6^{3-}$	Na, K
$OZrF_3^-$	K, K(2H_2O)

Potassium fluorozirconate, K_2ZrF_6, is a colorless salt which melts at about 840°C. It dissolves in water to the extent of 1.22 g/100 g of water at 10°C, and 23.53 g/100 g at 100°C, giving a solution with a pH of about 4. The great variety of fluorozirconates that has been reported is indicated by Table 1.

The compound $KZrOF_3 \cdot 2H_2O$ has been made by boiling aqueous potassium hexafluorozirconate with an appropriate addition of potassium hydroxide (125.42, 144.46).

Fluorozirconates have been used as stabilizers in silicone rubber, as a welding flux, in increasing the durability of beryllium lead aluminum fluoride optical glass, improving corrosion resistance of metals, in dental plastics (22.4, p. 144), as wetting agents for metals to be coated with aluminum or its alloys (45.1), in resin-bonded grainding wheels to improve their efficiency (23.12), in bonding light-sensitive resins to metal in the manufacture of lithographic printing plates (125.34), and as a fluorinating agent for carbon tetrachloride (34.1).

The zirconium fluorogermanates $Zr(GeF_6)_2$ and $ZrO(GeF_6)$ have been found to be anticariogenic (125.62).

ZIRCONIUM FORMATES

Diformatozirconic acid, $HZrOOH(HCO_2)_2$, is prepared with difficulty from carbonated hydrous zirconia by dissolving it with formic acid. It is unstable, and has found no use. Evaporation of a solution of zirconyl chloride to which a large excess of formic acid has been added yields crystals of composition $Zr_2OOH(HCO_2)_5 \cdot 2H_2O$. They dissolve in water in all proportions, and are slightly soluble in alcohol, benzene, and chloroform, but insoluble in ether. The compound is stable in air at room temperature (144.29). Infrared studies indicate the formato ligand to be bidentate (144.38).

ZIRCONIUM HALOGENATES

Zirconyl perchlorate, $ZrO(ClO_4)_2 \cdot 8H_2O$ is a well-known compound, generally similar in its chemistry to zirconyl chloride. Zirconyl perchlorate with two moles of water is also known (84.2). Unlike zirconyl chloride, the perchlorate can be dehydrated completely without decomposition, in vacuo over phosphorus pentoxide at 200°C (119.1). Other ligands can take the place of water, as in $ZrO(ClO_4)_2 \cdot 6C_9H_7NO$, which forms in the presence of quinoline oxide (27.4).

Other halogenates are:

$H_2Zr_{10}O_{12}(ClO_4)_{18}$	$H_2Zr_3O_6(IO_4)_2 \cdot 16H_2O$	$KZr_2(IO_3)_9 \cdot 8H_2O$
$H(ZrO)_4(ClO_4)_9$	$H_2Zr_6O_{12}(IO_4)_2 \cdot 19H_2O$	$Na_2Zr(IO_3)_6$
$ZrO(ClO_3)_2 \cdot 6H_2O$		$Zr(IO_3)_4 \cdot 3H_2O$
$HZr_2O_3(ClO_3)_3$		$HZrO(IO_3)_3 \cdot 4H_2O$
		$HZrO_2IO_3 \cdot H_2O$

ZIRCONIUM HYDROXYCARBOXYLATES

Zirconium forms insoluble or sparingly soluble complexes with the aliphatic alpha-hydroxycarboxylic acids, and somewhat analogous compounds with salicylic acid. The ligands are bidentate, and the hydroxyl proton is dissociable.

Trisglycolatozirconic acid, $H_3ZrOH(OCH_2CO_2)_3$, is precipitated slowly when a solution of zirconyl chloride is added to a solution of glycolic acid. The precipitate is dissolved by aqueous alkalies with replacement of one, two, or three protons by alkali ion.

When more than three moles of alkali hydroxide, other than ammonium hydroxide, is added, the structure is irreversibly decomposed, and insoluble hydrolyzates appear. There has been some industrial use of trisglycolatozirconic acid and its sodium salts in body deodorants, and it has been reported useful in curing silicone resin (23.5) and in staining protein matter for electron micrography.

Trislactatozirconic acid, $H_3ZrOH(OCHCH_3CO_2)_3$, is very similar to triglycolato-zirconic acid in method of preparation and in chemical properties.

When sodium or potassium glycolatozirconate or lactatozirconate is calcined, very voluminous ashes of alkali zirconate form.

Tetramandelatozirconic acid, $H_4Zr(OCHC_6H_5CO_2)_4$, is very insoluble in hydrochloric acid solution, and is used in the analytical determination of zirconium. This compound, too, dissolves on the addition of alkalies. Zirconium is also precipitated by the keto acid, phenylpyruvic acid, $C_6H_5CH_2COCO_2H$ (66.2). **Disalicylatozirconic acid** precipitates when salicylate ion is added to aqueous solutions of zirconium chlorides. The precipitate dissolves on addition of alkalies (144.1).

Infrared studies of zirconium tartrate and zirconium citrate have indicated that all carboxyl and hydroxyl groups are involved in the complex formation. Protons of the hydroxy groups ionize. In **zirconium trihydroxyglutarate,** zirconium appears to be bonded to two hydroxyl groups and not to all of the carboxyl groups (144.41). A **potassium citratozirconate** of composition $KZrOC_6H_5O_7 \cdot 2.5H_2O$ has been reported (144.3), and a 1:1 complex of zirconium with trihydroxyglutaric acid (133.2).

When α-hydroxycarboxylatozirconic acids are neutralized, they are apt to be simultaneously hydrolyzed to species containing fewer organic ligands per zirconium atom. Tetramandelatozirconic acid has been observed to hydrolyze to **dimandelato-zirconic acid** at a pH of about 4 (128.4).

Tannic acid, a variable composition of materials which are made up largely of gallic and ellagic acid, (see Leather and leather processing) precipitates zirconium quantitatively (25.1). Since the component acids are hydroxyaryl compounds containing free or esterified carboxyl groups, their behavior relates to that of the simpler hydroxycarboxylic acids. Tannic acid precipitation has afforded separations of zirconium from certain other metals, eg, uranium, vanadium, and thorium (22.4, pp. 373–374).

ZIRCONIUM IODIDES (see Zirconium chlorides, bromides and iodides)

ZIRCONIUM KETONATES

At low temperatures (about $-5°C$), zirconium tetrachloride forms an addition product with acetone. At room temperature, hydrogen chloride is liberated with formation of trichlorozirconium isopropenoxide, $Cl_3ZrOC(CH_3)=CH_2$, (68a.1).

When an aqueous solution of zirconyl nitrate was shaken with mesityl oxide, the zirconium was quantitatively extracted into the organic liquid (10.4). Anthrone forms a 2:1 complex with zirconium tetrachloride (50.1). A large number of 2,4-diketonates have been found to form compounds with zirconium (51.4). See Zirconium acetylacetonates.

ZIRCONIUM MERCAPTIDES (see also Zirconium sulfides)

It was only after many failures to prepare zirconium derivatives of organic sulfur-containing compounds, containing zirconium–sulfur bonds, that such compounds were finally prepared and identified. An adduct with thiophene has the composition $ZrCl_4 \cdot 2C_4H_4S$ (145.1). Tri-n-octylphosphine sulfide, $(C_8H_7)_3PS$, has proven useful as a complexing agent in the solvent–solvent extraction of zirconium from acidic aqueous solutions (11.1).

Zirconium tetrathiophenoxide has been prepared as a blue powder (138.1), and benzene o-dithiol groups have been substituted for the chlorine of bis(cyclopentadienyl)zirconium dichloride (69.3). Zirconium derivatives of N,N-dialkylthiocarbamates have been prepared, having compositions $Zr(S_2CNR_2)_4$, by insertion reactions of carbon bisulfide with zirconium amides. They appear to be examples of zirconium at coordination number 8 with sulfur (31.1, 62.12).

ZIRCONIUM MOLYBDATES AND WOLFRAMATES

Studies of the MoO_3–ZrO_2 system have shown the compound $Zr(MoO_4)_2$ to be the only one that forms (144.47). Single crystals have been prepared and characterized. The compound decomposes above 600°C (137.15). The analogous wolframate, $Zr(WO_4)_2$, forms when the component oxides are heated together at 1000–1400°C (59.2). Monoclinic $Li_2Zr(WO_4)_3$ is also known (81.2).

In aqueous systems, $ZrOWO_4 \cdot 1.5H_2O$ (40.2) and $(NH_4)_4ZrO_4 \cdot 12MoO_3 \cdot 10H_2O$ and its wolfram analog (144.6) have been prepared.

Precipitated zirconium molybdate has been shown to have useful ion exchange properties (59.4, 130.2, 135.1, 148.1).

Phosphomolybdenozirconium complexes form in aqueous solution, and they can be reduced with stannous chloride to form blue solutions, the intensity of which is a measure of the zirconium in solution (150.1).

ZIRCONIUM NITRATES

Anhydrous zirconium nitrate, $Zr(NO_3)_4$, can be prepared by the reaction of zirconium tetrachloride with nitrogen pentoxide at $-200°C$. It can then be warmed to room temperature and sublimed without decomposition (90.2). In fuming nitric acid and aqueous nitric acid solutions, the following compounds are formed: $H_2(Zr(NO_3)_6 \cdot 4H_2O$, $H_2Zr(NO_3)_6 \cdot 3H_3O$, $HZr(NO_3)_5 \cdot 4H_2O$, $Zr(NO_3)_4 \cdot 6H_2O$, $Zr(NO_3)_4 \cdot 5H_2O$, $Zr(NO_3)_4 \cdot 4H_2O$, $Zr(OH)(NO_3)_3 \cdot 3H_2O$, $ZrO(NO_3)_2 \cdot 3.5H_2O$, $ZrO(NO_3)_2 \cdot 2H_2O$, $ZrOOHNO_3 \cdot 2H_2O$, and $ZrOOHNO_3$ (13.2, 22.1, 47.3, 67.9, 144.45).

The compound of composition $((CH_3)_4N)_2Zr(NO_3)_6$ has been formed by the reaction of a chlorozirconate with nitrogen pentoxide in acetonitrile (62.7).

When ether is added to an alcoholic solution of zirconyl nitrate, a product of composition $ZrOOHNO_3 \cdot 2H_2O$ precipitates (144.22). Clear distinctions have not been made between compositions reported $ZrOOHNO_3 \cdot nH_2O$ and others reported $Zr_2O_3(NO_3)_2 \cdot n'H_2O$. Thermal decomposition of $Zr(NO_3)_4 \cdot 5H_2O$ goes through the stages $ZrO(NO_3)_2 \cdot 2H_2O$, $Zr_2O_3(NO_3)_2$, and $Zr_4O_7(NO_3)_2$ (144.45).

In aqueous solutions of zirconyl nitrate which are 1–3 M in concentration, cationic species have been observed to predominate (50.3). Zirconyl nitrate solutions behave chemically very similar to zirconyl chloride solution. Because the nitrates are more expensive, the chlorides are generally preferred for industrial use. The nitrates are specified, however, in some patented processes, as for example, in the hardening of polyvinyl alcohol films (125.27).

ZIRCONIUM NITRIDES (see also p. 624 and Zirconium ammines, amides, and imides)

Zirconium Nitride Halide. On heating an ammonia adduct of zirconium tetra-iodide to 500°C, a compound of composition ZrNI is formed (137.11). It decomposes under an ammonia atmosphere at 750°C with formation of a blue intermediate phase of variable composition, and then brown Zr_3N_4. This is a diamagnetic substance which shows no metallic conductance, and is sharply distinguished from the metallic ZrN. The blue and brown compounds are converted to yellow, metallic ZrN at 1000°C.

Yellow alpha-ZrNBr and yellow–green beta-ZNBr, and yellow–green alpha-ZrNCl and grey–green beta-ZrNCl have been prepared by procedures similar to those for preparing the iodine analogs (137.12).

ZIRCONIUM ORGANIC COMPOUNDS

Zirconium exhibits extreme resistance to forming sigma bonds with carbon, but a considerable number of pi-bonded zirconium–carbon compounds are now known. There is evidence that the formation of the pi-bonded compounds mitigates the resistance to forming sigma bonds between zirconium and carbon. See Cyclopentadienyl-zirconium compounds. Examples of compounds containing pi and sigma bonds between zirconium and carbon are $Zr(CHCH:CH_2)_4$ (31.3), $(C_5H_5)_2Zr(C_3H_5)_2$ (69.1), $(C_2H_5)_2ZrClCH_3$ (31.2), $(C_5H_5)_2Zr(CH_2)_2Zr(C_5H_5)_2$ (69.2).

A dipentylzirconium is said to form at -40°C by a reaction of diazopentane with zirconium hydride (125.26).

ZIRCONIUM OXALATES

Oxalatozirconium compounds with mole ratios of oxalato ligand to zirconium atom ranging from 0.5 to 8 have been reported. Where determinations have been made, the zirconium has been found part of anionic species, and to migrate in aqueous solutions toward the anode (60.13, 94.1). The zirconium is picked up by anion exchangers (117.1).

The structure of tetraoxalatozirconate ion, $Zr(C_2O_4)_4^{4-}$, has been determined by x-ray diffraction study of a single crystal of *tetrasodium tetraoxalatozirconate trihydrate*, $Na_4Zr(C_2O_4)_4 \cdot 3H_2O$. Four oxalato groups are bound to each zirconium atom, as bidentate ligands. The eight oxygen atoms involved are located at the points of a dodecahedron with triangular faces (51.1A).

Dioxalatozirconic acid trihydrate, $H_2ZrO(C_2O_4)_2 \cdot 2H_2O$, precipitates when alcoholic solutions of oxalic acid dihydrate and zirconium tetrachloride are mixed (67.3, 144.11).

In aqueous solutions of high ratios of oxalate ion to zirconium, polyoxalatozirconium compounds form (36.2), of which the following are examples:

$(NH_4)_6Zr(C_2O_4)_5 \cdot 2H_2O$	$Ba_2Zr(C_2O_4)_4 \cdot 4H_2O$
$(NH_4)_3H_3Zr(C_2O_4)_5 \cdot 2H_2O$	$K_2Zr(C_2O_4)_3 \cdot (2.5-4)H_2O$
$K_4Zr(C_2O_4)_4 \cdot (2-3)H_2O$	$(NH_4)_2Zr(C_2O_4)_3 \cdot (1-3)H_2O$
$(NH_4)_2H_2Zr(C_2O_4)_4 \cdot 2H_2O$	$K_2ZrO(C_2O_4)_2 \cdot nH_2O$

From solutions of low oxalate ion to zirconium ratios, polymeric hydrolyzed species are formed: $ZrOC_2O_4 \cdot H_2O$, $ZrOC_2O_4 \cdot 2H_2O$, $Zr_2O_3C_2O_4$ (144.23).

ZIRCONIUM OXIDES

Zirconium dioxide, ZrO$_2$, exists stably at ordinary temperatures as monoclinic crystals, $a = 5.1454$, $b = 5.2075$, $c = 5.3107$ Å, $\beta = 99°14' \pm 0°05'$ (2.3). Reversible transformation between monoclinic and tetragonal symmetry occurs at 1193–1200°C (59.3,124.2). Different investigators report somewhat different transition temperatures (143.3), which reflect small differences in chemical purity, including differences in hafnium content, and also differences in the number and arrangements of lattice defects (131.1).

Tetragonal zirconium dioxide can be formed at 25°C by applying a pressure greater than 37 kilobars to the monoclinic dioxide. A cubic phase forms at about 1900°C. A cubic phase is also frequently found in zirconium dioxide formed by pyrolysis of organic compounds.

Zirconium dioxide melts at about 2750°C. The solid is a fairly good conductor of electricity at a red heat. It is insoluble in hydrochloric, nitric, and dilute sulfuric acid, but dissolves in hot concentrated sulfuric acid and in hydrofluoric acid. It dissolves also in molten cryolite (16.1); but only minute amounts dissolve in alkali fluorides at 1000–1200°C (128.3). It reacts at moderately elevated temperatures with ammonium sulfate to form water-soluble ammonium sulfatozirconate, with alkali fluorosilicates to form water-soluble alkali fluorozirconates, and with alkalies to form zirconates. Some physical absorption of hydrogen has been observed at 50–175°C (127.1). Zirconium dioxide is very resistant to hydrogen reduction, although some investigators have reported reaction as low as 550°C (127.1). Reduction with carbon begins at about 1400°C (53.1). There are patents on wear-resistant products prepared by hydrogen reduction of zirconium dioxide (47.1) and carbon reduction of zirconium dioxide-titanium dioxide mixtures (125.11). Some replacement of oxygen by nitrogen occurs by the reaction of zirconium dioxide with ammonia above 2000°C (24.2). Under some conditions of preparing zirconium dioxide, a monolayer of oxygen is sorbed by the surface and alters the apparent stoichiometry to ZrO$_{2+x}$ (65.3).

During the middle of the nineteenth century, zirconia found use as a glower in gas-light mantles; it was later replaced by thoria. Early in the present century, its usefulness as a ceramic opacifier was demonstrated, and other physical properties were exploited in abrasives, refractories, cements, and fillers. It improved heat resistance when dispersed in certain alloys. Latterly, uses have been developed as components of ceramic stains, components of ceramic elements used in fuel cells and other electrical systems, catalysts, starting material in making zirconium metal and alloys, and internal bonding agent dispersed in rubber and plastics. Zirconia fibers have been prepared (125.37,125.38), titanium dioxide pigment has been surfaced with zirconium dioxide to enhance its stability in organic vehicles (23.16), and the resistance of glass to water, acids, and alkalies has been improved by the incorporation of zirconia (121.1). *Stabilized zirconia*, which contains calcium oxide, rate earth oxide, or occasionally certain other oxides in solid solution, has cubic crystalline symmetry, and is advantageously used in refractories. The crystalline inversion of the unstabilized oxide at about 1200°C leads to failures when ceramic objects are heated or cooled past this temperature.

Next to zircon, zirconium dioxide commands the largest market of all zirconium compounds. The demand has stimulated search for and development of new processes for improving the economy of its production. As it became more available, additional efforts were made to find new applications and to improve the attractiveness of the older applications.

Zirconium dioxide is found in nature as the mineral, baddeleyite, but its separation from contaminants results in no better economy in producing an acceptable grade of zirconium dioxide than the recovery of the compound from mineral zircon. Only the recent production of pure baddeleyite as a by-product of copper mining operations at Phalabora, South Africa, has been an exception (139.1). There are two traditional procedures for deriving technical grade zirconium dioxide from zircon. In the processes of Kinzie and his collaborators, silica was in effect boiled out of the mineral in an arc furnace. They heated mixtures of zircon and carbon in the furnace, evolved silicon monoxide, and recovered pigs of zirconium carbide or cyanonitride. These were burned in air to form zirconium dioxide (125.0a;125.0b;125.0c). The alternative was to heat zircon with alkali and then take advantage of the differences of solubility of the silica and zirconia contents of the reaction product in aqueous alkali and acid. Zirconia was prepared from the acidic solution by one or more processes, most commonly by precipitating a pure basic sulfate, drying and firing it to volatilize water and sulfur trioxide, and recovering a residue of pure zirconium dioxide (125.0;125.0d). In addition to these processes, a large number of alternatives have been published which were operable, but have not yet been made economical. Examples are the combustion and the steam hydrolysis of zirconium tetrachloride. One process has proven particularly useful for manufacturing calcia-stabilized cubic zirconia. Zircon was heated with sufficient calcium oxide (or other specified alkaline earth oxide or mixture of such oxides) to 1250°C or higher. Calcium zirconylosilicate, $CaZrSiO_5$ formed initially, then broke down to zirconium dioxide and calcium orthosilicate. The idealized overall reaction was:

$$2 \; CaO + ZrO_2.SiO_2 \rightleftharpoons ZrO_2 + Ca_2SiO_4.$$

Depending on conditions for carrying out the process, more or less of calcia went into solid solution in the zirconia. Four percent or more was sufficient to produce calcia-stabilized cubic zirconia, admixed with calcium orthosilicate. These solids could be separated by elutriation or sieving (125.8b; 125.11a).

Extremely pure zirconium dioxide has been prepared by a number of processes, and is relatively expensive. It has most commonly been prepared by pyrolysis of zirconium compounds which are readily prepared in very pure state, viz, α-hydroxy-carboxylates (48.3), recrystallized zirconyl chloride, and distilled zirconium alkoxides (59.3a).

The high indexes of refraction of zirconium dioxide (about 2.2) relative to glass (about 1.5–1.6), along with its refractory quality, have rendered it useful as a pigment or opacifier for ceramic glazes and enamels. In some glazes, the zirconium dioxide survives as such as pigmenting agent and in others (acidic glazes and enamels) it is converted during the melting and cooling to zircon. The latter (indexes of refraction 1.9–2.0) then becomes the actual opacifier. Ceramic colorants or *stains* are also prepared from zirconium dioxide. Particularly noteworthy is the beautiful blue stain prepared by heating zirconium dioxide and silica together in the presence of a vandium compound. Zircon crystals form, with vanadia in solid solution. The crystal environment constrains the vanadium to tetravalency, at which oxidation state the ion is blue (120a.1). Zirconium dioxide is also used alone or in conjunction with silica as hosts for other coloring metal oxides. Thus, praseodymium oxide, Pr_6O_{11}, forms orange-yellow stains with zirconia and clear lemon-yellow strains with zirconia plus silica (125.36a).

It is difficult to prepare zirconium dioxide economically in the particle sizes suitable for use as pigment for paint, printing ink, and the like. Therefore, it has not found application in these products even though its higher indexes of refraction than those of zinc oxide and white lead indicate much potential value.

The extremely high melting point of zirconium dioxide attracts attention to its value as a refractory. This qualification tends to be vitiated by the large volume change, about 7.5%, which occurs when the compound passes through its monoclinic-tetragonal inversion point near 1200°C. This gives rise to cracks and failures in ceramic bodies cycled through this temperature. A correction of this deficiency has been achieved by inducing the zirconium dioxide to crystallize in the cubic system, through the influence of certain dissolved oxides. About 5% calcium oxide or 15% yttrium oxide has been used in zirconium dioxide to accomplish this.

Zirconia has been substituted for up to about 20% of the silica in soda–lime glass (101a.1). It increases the acid- and alkali-resistance of the glass. It reduces the tendency of glass to crystallize, and increases adhesion of polar organic polymers to its surface

The name *colloidal zirconia* has been used conventionally and rather loosely to designate permanent fluid suspensions of zirconium dioxide in aqueous media. The name has become a trivial rather than a scientific name for a class of products. The mere addition of alkalies in approximately stoichiometric amounts to solutions of zirconium salts commonly precipitates zirconium dioxide of particle size sufficiently small to be regarded as at the low end of the colloidal range, that is, at the boundary between molecules and colloidal particles. Generally, these fine particles flocculate and form filterable gelatinous precipitates. If a deficiency of alkali is used and the pH is left lower than that of the isoelectric point, flocculation may not occur, and the colloidal suspension will be unfilterable due to clogging the pores of a filter. A deliberate and more useful preparation of a permanent suspension has been achieved by prolonged boiling of a suspension of flocculent hydrous zirconia in very dilute hydrochloric acid for a period of days (125.55a). Monoclinic zirconia particles, about 100 Å in length, form. The suspension does not settle on long standing, and it passes completely through an ordinary filter paper. This is an industrially available product which has been under evaluation for applications to delustering textiles, changing the appearance of electroplated metals, modifying properties of polymers, and medicine.

Studies of systems of zirconium dioxide with other oxides have been extensive, and have embraced oxides of Ag, Al, Am, Ba, Be, Bi, Ca, Cd, Ce, Cr, Cu, Er, Fe, Ga, Gd, Ge, Hf, Ho, La, Mg, Nb, Nd, Pr, Pu, Sc, Sm, Sn, Sr, Ta, Tb, Th Ti, U, W, Y, Yb, and Zn. Noteworthy binary and ternary oxides are:

$Ca_7Al_6ZrO_{18}$	$ScZr_7O_{17}$
$CaZrTi_2O_7$	$Sr_2Al_{12}ZrO_{22}$
$CaZr_3TiO_9$	$Ta_2Zr_6O_{17}$
$CeZr_2O_7$	$ZrNb_{10}O_{27}$
$La_2Zr_2O_7$	$ZrNb_{14}O_{37}$
$Sc_2Zr_2O_7$	

Lower oxides of zirconium are known, but are oddities rarely encountered in the laboratory and never in terrestrial nature. *Zirconium monoxide* was first noted in the sun, existing at about 4900°K. It has been observed in the atmospheres of many stars. It is found in zirconium dioxide vapor formed at about 2500°K, and this has been led into traps at 4–20°K with recovery of solid zirconium monoxide (71.3). Examination of products of reaction of zirconium dioxide with carbon on heating in vacuo has revealed not only the monoxide, but also *zirconium sesquioxide*, Zr_2O_3. The structure of zirconium monoxide is face-centered cubic, with $a = 4.62$ Å (128.1).

The dissolution of oxygen in α-zirconium metal has been shown to give rise to compounds of composition Zr_3O and Zr_6O (54.1). A phase of composition Zr_5O_{24} has also been detected by x-ray diffraction (68.6).

Hydrous zirconium oxide, $ZrO_2 . xH_2O$, is precipitated when aqueous alkalies are mixed with solutions of many zirconium compounds, such as zirconyl chloride or disulfatozirconic acid. If the mixture is somewhat acidic after reaction, small fractions of a mole of anion (chloride, etc) are likely to be bound to the precipitate, while if it is alkaline there will be no appreciable bonding of anions, but there may be sorbed cations.

Hydrous zirconium oxide is commonly called *hydrous zirconia*. It can be formed by dialysis or ion exchange, as well as by adding alkalies to solutions of zirconium compounds (45.2). Its water is not water of hydration, for its vapor pressure and other behavior follow the pattern of adsorbed water and not of chemically bound water (22.4, p. 182).

In acidic suspensions, the sorbed anions tend to impart a negative charge to the particles, while coordinated protons tend to impart a positive charge. Anions other than hydroxyl compete with hydroxyl for sorption sites (60.5). Many of the reactions of hydrous zirconia are resultants of competive sorption reactions. The usefulness of hydrous zirconia as an anion-exchange substrate in acidic environments and as cation exchange substrate in alkaline environments has been demonstrated (33.2,67.2, 125.55). A derivative use is in a salt-rejecting membrane (60.18,106.1). Organic oxygen-containing substances are also strongly sorbed by hydrous zirconia, particularly those containing carboxyl groups and vicinal pairs of hydroxyl groups. The powerful sorption of urushiol, a derivative of 1,2-benzenediol (pyrocatechol) which is the irritant in poison ivy leaves, has led to the use of hydrous zirconia as a prophylactic and treatment for poison ivy dermatitis. This use was invented by Eugene Wainer, a fact obscured by literature which bears the names of formulators of patented compositions (29.1, 125.31).

Hydrous zirconia precipitates as particles with diameters of the order of 25 microns. Modifications which have been designated α, β, and γ have been described, and are distinguished by decreased chemical activity in the stated order (144.32). When a slurry of hydrous zirconia is boiled for a long period, the initial amorphous structure changes to that of monoclinic zirconium dioxide in acidic environment and to cubic zirconium dioxide in alkaline environment. The process is more rapid at the temperatures operable in an autoclave (51.2).

If the precipitated amorphous hydrous zirconia is filtered, washed, dried, and heated slowly, it inverts to tetragonal zirconium dioxide at about 400°C (24.3,101.1). It still retains about one percent of water, by weight. On further heating, it inverts to monoclinic zirconium dioxide, at 600–800°C (40.15). With this inversion, all identity as hydrous zirconia has disappeared, but traces of water persist to about $1000°C$ (40.15).

Zirconium as hydrous zirconia is absolutely insoluble in water. A hypothetical solubility product has been derived, with the value 1.1×10^{-54} (144.8), much less than an atom of zirconium per liter of water. It is soluble in strong mineral acids and insoluble in low concentrations of strong alkalies. It is insoluble in weak acids and weak alkalies. But at high concentrations of strong alkalies, there is some chemical reaction with formation of soluble hydrozirconate ions, $Zr(OH)_5^-$ and $Zr(OH)_6^{2-}$ (41.1). Acids which form insoluble compounds of zirconium, such as phosphoric acid, will convert aqueous suspensions of hydrous zirconia to zirconium compounds of those acids. In the absence of water, glacial phosphoric acid is capable of dissolving appreciable amounts of zirconium phosphate.

Like many zirconium compounds, hydrous zirconia exhibits variations in properties which reflect its history. Depending on conditions of precipitation, when dried it may form a very soft powder or an adamantine glassy residue. Only the most active forms of hydrous zirconia dissolve in ammonium carbonate solution at appreciable rates.

Important observations have been made on the coprecipitation of hydrous zirconia with oxides or hydroxides of certain other metals. Solid solutions are often obtained, which previously had been known to form only at very high temperatures. Particularly noteworthy studies have been reported of coprecipitations of hydrous zirconia with oxides or hydroxides of lead (54.2,144.30), magnesium (36.9, 36.10), silica (54.4), and zinc (144.16).

If hydrogen peroxide is present in aqueous solutions of zirconium compounds to which alkalies are added, *hydrated peroxides* are precipitated (22.4, pp. 198–200). $ZrO_3 \cdot 2H_2O$ has been reported. With addition of excess of strong alkalies, water-soluble peroxozirconates form, such as $Na_4Zr_2O_{11} \cdot 9H_2O$, $K_4Zr_2O_{11} \cdot 9H_2O$, and Na_2ZrO_6. The latter is said to be useful in the manufacture of cracking catalysts (125.54). Peroxides of compositions $ZrO_3 \cdot 2H_2O$, $ZrO_3 \cdot H_2O$, and ZrO_3 are said to form when freshly precipitated hydrous zirconia reacts with hydrogen peroxide (55.1).

Studies of solutions of pH 12–14 have elicited evidence for an ion of composition $OZr(O_2)_2^{2-}$. The compound $Zr(O_2)C_2O_4 \cdot 6H_2O$ has been crystallized from solutions containing zirconyl chloride and oxalic acid, by addition of hydrogen peroxide and aging. Infrared study showed it to contain bidentate peroxo, O_2, ligand (50.5). The compound $Zr_2(O_2)_3SO_4 \cdot 10H_2O$ contains one bidentate peroxo group on each zirconium atom and one bridging peroxo group per two zirconium atoms (G. V. Jere and G. D. Gupta, pending publication).

Zirconium Phenoxides. Reactions of zirconium tetrachloride with phenol and its derivatives yield aryloxides of zirconium which are analogous to the alkoxides. Examples are $Zr(OC_6H_5)_4$, $Zr(OC_6H_5)_4 \cdot C_6H_5OH$, $ClZr(OC_6H_5)_3$, $ClZr(OC_6H_4CH_3)_3$, $Cl_2Zr(OC_6H_5)_2$, and $Cl_2Zr(OC_6H_4CH_3)_2$.

1,2-Benzenediol forms chelates with zirconium, and in alkaline media the compound $K_2Zr(C_6H_4O_2)_3$ has been observed to form (137.5). 1,3-Benzenediol (resorcinol) forms the compound $Cl_2Zr(OC_6H_4)_2$ (39.1). Complexes of highly complex di- and polyhydroxyaryl compounds are known, as of alizarin, 1,2,3-benzenetriol (pyrogallol), and tannic acid. Use has been made in analytical chemistry of the colored alizarin complex $C_6H_4(CO)_2C_6H_2O_2ZrO$ (60.6).

Zirconates

Compounds once called organic zirconates are now regarded as alkoxides, and have been discussed above under that heading. The inorganic zirconates are known only in solid phases, and are crystalline rather than molecular identities. Zirconates of bivalent metals usually have perovskite structures.

Alkali metal and alkaline earth metal zirconates are commonly prepared by sintering zirconium dioxide with a compound of the alkali or alkaline earth metal which provides the oxide on calcination. Alkali metal oxides, particularly sodium oxide, tend to volatilize out of the composition at the sintering temperatures, posing problems in selecting a temperature sufficiently high for reaction but not so high as to cause serious loss of alkali. This and other factors affect the type of zirconate formed during a sintering reaction as illustrated by the following equations (101.4):

$$ZrO_2 + 2\ KNO_3 \xrightarrow{800°C} K_2ZrO_3$$

$$3\ ZrO_2 + K_2CO_3 \xrightarrow{950°C} K_2Zr_3O_7$$

$$2\ ZrO_2 + 2\ KOH \xrightarrow{600°C} K_2Zr_2O_5 + H_2O$$

$$ZrO_2 + 2\ K_2O \xrightarrow{450°C} K_4ZrO_4$$

Zirconates of the follow compositions have been reported:

$BaZrO_3$	$La_2Zr_2O_7$	$Rb_2Zr_2O_5$
$CaZrO_3$	Li_2ZrO_3	$Rb_2Zr_3O_7$
$CeZr_2O_7$	$Li_2Zr_2O_5$	$Sm_2Zr_2O_7$
$Cs_2Zr_3O_7$	Li_4ZrO_4	$SrZrO_3$
Fe_2ZrO_4	Na_2ZrO_3	$Sr_4Zr_3O_{10}$
K_2ZrO_3	$Nd_2Zr_2O_7$	$Sr_3Zr_2O_7$
$K_2Zr_2O_5$	$PbZrO_3$	Sr_2ZrO_4
$K_2Zr_3O_7$	$Pr_2Zr_2O_7$	$Y_2Zr_2O_7$
$K_2Zr_8O_{17}$		$Yb_2Zr_2O_7$
K_4ZrO_4		

There is no magnesium zirconate. It has been calculated that a magnesium zirconate is thermodynamically impossible, and phase studies of the system ZrO_2–MgO support this conclusion.

Zirconates have been prepared by a number of methods other than sintering. Pyrolyses of potassium lactatozirconate or sodium lactatozirconate yield voluminous ashes of potassium or zirconium zirconate, respectively (22.4, p. 333). Crystals of *lead zirconate* 1–2 mm long have been grown in molten lead chloride (105A.1). A melt of iron and zirconia recovered from an arc plasma was found to yield Fe_2ZrO_4 (107.1).

A heated mixture of calcium zirconate and silica reacts with formation of calcium silicate and zirconia, but calcium zirconate reacts with titania to form $CaO.2TiO_2.$-ZrO_2 (86.1). Barium zirconate reacts with aqueous ammonium bicarbonate to form ammonium tricarbonatozirconate (125.60).

Zirconates are used in industry as modifiers for titanates which serve as ceramic components of electrical systems, with high dielectric constants or piezoelectric properties. See Vol. 20, p. 786.

ZIRCONIUM PHOSPHATES

The reaction of zirconium compounds in strongly acidic aqueous solutions with phosphoric acid precipitates amorphous *diphosphatozirconic acid*, $H_2Zr(PO_4)_2 \cdot xH_2O$ (60.1). Definite values for the number of moles of water associated with the precipitate have not been established. If the gelatinous precipitate is boiled for about a day in the presence of strong phosphoric acid solution, crystalline diphosphatozirconic acid is formed (125.59,137.14). The diphosphatozirconic acids are commonly called *zirconium phosphate*. Anhydrous diphosphatozirconic acid, $H_2Zr(PO_4)$, and several hydrates, eg, $H_2Zr(PO_4)_2 \cdot 2H_2O$ are also known (67.11,68.4).

If the aqueous solution from which the precipitation is made is not strongly acidic, only partially hydrolyzed products will be obtained, containing less than two phosphato ligands per zirconium atom, the number decreasing with decreasing acidity. The precipitate also loses phosphato ligands after precipitation if the pH of the environmental liquid is raised, as for example, during washing. The crystalline diphosphatozirconic acid is less readily hydrolyzed than the amorphous (24.4). In strongly acidic solution, the hydrogen of diphosphatozirconic acid can be replaced by zirconium, if it is available:

$$ZrO^{2+} + H_2Zr(PO_4)_2 \rightarrow ZrOZr(PO_4)_2 + 2\,H^+$$

Hydrolysis and ion exchange often affect the apparent stoichiometry of the phosphate precipitate, and for this reason reported compositions of phosphato compounds of zirconium have varied from $Zr:P_2O_5 = 2:1$ to $<1:1$.

Diphosphatozirconic acid has been of considerable interest as an ion exchanger, when amorphous (67.1) and when crystalline (125.59), particularly for Cs, Co, Li, Rb, Sr, and Tl. The crystalline compound rejects cesium ions while accepting other alkali metal ions, as the cesium ion is too large to fit into its cavities. The ion-exchange properties of diphosphatozirconic acid have led to uses in conductive membranes of the type employed in fuel cells (65.1, 99.1).

Calcination of diphosphatozirconic acid at about 1000°C causes loss of all of its hydrogen content as water, and leaves a residue of *zirconium pyrophosphate*, ZrP_2O_7. At still higher temperatures, some phosphorus pentoxide is evolved and a new residue of *zirconyl pyrophosphate*, $(ZrO)_2P_2O_7$, remains (18.1). Both pyrophosphates are dimorphic. A composition $Zr_3(PO_4)_4$ has been reported occasionally without being characterized, and may be regarded as $(ZrO)_2Zr(P_2O_7)_2$. A compound $2ZrO_2 \cdot P_2O_5 \cdot SO_3$ is also known (36.4, 36.5).

When diphosphatozirconic acid is sintered at relatively low temperatures, part of its water remains, and with it, residual ion exchange properties. The sintered product has usefulness as a conductive membrane (5.2).

Ion exchange of diphosphatozirconic acid leads to formation of salts, such as $CaZr(PO_4)_2$. Alkali phosphatozirconates have been prepared not only by such ion exchange, but also by heating mixtures of the component oxides (22.4, p. 302). Other phosphatozirconates have also been prepared, for example $NaZr_2(PO_4)_3$ and its Li, K, Rb, and Cs analogs. Compounds of generic formula $M_2ZrP_2O_9$, where M is an alkaline earth metal, are luminescent. Arsenic analogs are also known (65.2, 124.3, 125.48).

Diphosphatozirconic acids dissolve in alkali metal carbonate solutions, but particularly readily in potassium carbonate solution. Potassium *tetracarbonatodiphosphato-*

zirconate has been crystallized from such a solution (36.1). Other alkali phosphato-zirconates are $Na_4ZrO(PO_3)_6$ and $Na_2(ZrO)_2(PO_6)_6$ (92.1).

In aqueous systems, zirconium can be separated from phosphato ligand by equilibration with ammonium molybdate in the presence of mineral acid, due to separation of phosphomolybdic acid (120.1).

Zirconium compounds with natural and synthetic organic phosphates are known, as with phytin (74.1) and:

$Zr(NO_3)_4 . 2(n\text{-}C_4H_9)_3PO_4$ (96.1) $(NO_3)_2Zr((n\text{-}C_4H_9)_2PO_4)_2$ (122.3)

$ZrCl_4 . 2(n\text{-}C_4H_9)_3PO_4$ $(NO_3)_2ZrCl_2 . 2(n\text{-}C_4H_9)_3PO_4$ (40.11)

$Zr(NO_3)_4 . 4HNO_3 . (n\text{-}C_4H_9)_3PO_4$ (144.7) $(HO)_2Zr(CNS)_2 . 2DIPMP$

                                          (diisopentyl methylphosphonate)

These complexes have been useful in liquid–liquid extraction procedures for separating zirconium from other metals, particularly from hafnium. Zirconyl derivatives of arylphosphonic acids, eg, $ZrO(C_6H_5CH_2PO_3H)_2 . nH_2O$ (132.1) and of phosphinic acids (21.2) are known. Some organic phosphates of zirconium have been demonstrated to improve motor lubricants (125.52).

Both amorphous (67.10a, 109.1) and crystalline (64.1, 67.10a, 141.2) diarsenato-zirconic acids are known. The latter, of composition $H_2Zr(AsO_4)_2 . 2H_2O$, has been prepared similarly to the phosphorus analog, and has similar ion-exchange properties. Distibnatozirconic acid is also known (22.6, p. 112).

Organic derivatives of arsenic acids precipitate zirconium quantitatively from acidic aqueous solutions. The reagent *o*-nitrobenzenearsonic acid precipitates zirconium as a compound believed to have the structure (73.1):

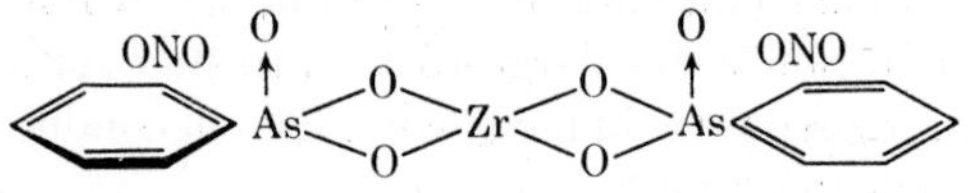

ZIRCONIUM PHOSPHIDES, ARSENIDES, ANTIMONIDES, AND BISMUTHIDES

Zirconium diphosphide, ZrP_2, is a gray, glistening solid which is stable in air and water, resists attack by nitric acid, but reacts with sulfuric acid with liberation of phosphine. **Zirconium monophosphide,** ZrP, forms when the disulfide is heated to about 750°C in vacuo. It is stable at 1000°C, but decomposes at 1100°C. The **subphosphide,** Zr_3P, has been prepared by long heating of zirconium with phosphorus (137.3).

Zirconium diarsenide, $ZrAs_2$, is very similar to the diphosphide, both having orthorhombic morphology and metallic conductance (83.2). **Zirconium subarsenide,** Zr_3As, has been synthesized from the elements (1.1).

Zirconium diantimonide, $ZrSb_2$, is orthorhombic and contains eight formula units in a unit cell. **Zirconium dibismuthide,** $ZrBi_2$, appears similar (83.3).

The phases $ZrFeP$, $ZrFe_4P_2$, and $Zr_2Fe_{12}P_7$ have also been reported.

PHOSPHINE AND ARSINE ADDUCTS WITH ZIRCONIUM COMPOUNDS

The composition $ZrCl_4(diarsine)_2$ (diarsine is *o*-phenylenebis(dimethylarsine)) forms readily from a mixture of its components dissolved in tetrahydrofuran (63.2), and similarly $ZrBr_4 . tas$ (tas is 3-dimethylarsenylpropylarsine).

A solution of trioctylphosphine oxide in cyclohexane extracts zirconium from aqueous zirconyl chloride or nitrate (122.1).

ZIRCONIUM POLYOLATES

It has been observed that when aqueous solutions of zirconyl chlorides containing polyols, such as glycol, glycerol, and pentaerythritol, are made strongly alkaline, there is no precipitation of hydrous zirconia (23.6), although there is precipitation in mildly alkaline solution, ie, below pH about 11–12. This is due to formation of chelates with displacement of protons from hydroxyl groups. Polarimetric observation of such complexes with sugars shows that there are progressive changes in the optical rotation over long periods, and hence progressive changes in the structure of the complexes. Zirconium complexes of zirconium with mannitol in the ratios $Zr:mannitol = 3:7, 4:3, 2:1$, and $5:1$ have been reported (13.2, 36.3).

Zirconium derivatives of glycol have been prepared by methods similar to those used in the preparation of derivatives of monohydric alcohols (47.6). A compound of the shown structure was prepared by the reaction of 2,4-dimethyl-2,4-pentanediol with hydrous zirconia in the presence of butylamine (125.41).

Glyceratozirconium chloride, approximately of composition $HOZr(C_3H_6O_3)Cl$, precipitates when a concentrated zirconyl chloride solution containing glycerol is boiled. The compound is very soluble in water. When a small proportion of the aqueous solution is added to methanol or ethanol, they set to clear, rigid jellies (125.60 a).

ZIRCONIUM SELENATES, SELENITES, TELLURATES, AND TELLURITES

There are no selenato(VI) or tellurato(VI) compounds of zirconium. These would be analogs of sulfato(VI) zirconates.

Diselenato (IV) zirconic acid forms in acidic aqueous solutions containing zirconyl salts and excess of selenious acid. It decomposes at 625°C to ZrO_2 and SeO_2 (144.15). The **polyselenatopolyzirconic acid** of composition $4ZrO_2.3SeO_2.18H_2O$ has long been known for its use in analytical chemistry (221, pp. 85–86).

Precipitates formed on adding sodium benzeneselenate(IV) or naphthaleneselenate to solutions of zirconyl compounds have the composition $ZrO(O_2SeC_6H_5)_2$ and $ZrO(O_2SeC_{10}H_7)_2$, respectively (142.1). A **zirconium benzenetellurate** is also known (30.1).

The compound of composition $ZrTe_3O_8$ has a cubic, distorted fluorite structure (10.3, 59.6). There is also $Zr_4TeO_{10}.8H_2O$ (24.1).

ZIRCONIUM SELENIDES AND TELLURIDES

Zirconium diselenide, $ZrSe_2$, forms as rhombohedral crystals when a mixture of zirconium tetrachloride vapor, hydrogen, and a selenium-containing gas make contact with a heated wire (5, p. 622). More generally, the selenides are prepared by heating mixtures of the elements. The binary selenides Zr_4Se_3, Zr_3Se_4, Zr_3Se_2, and $ZrSe_3$ are known (137.7), and the ternary compound ZrSiSe (84.3), and organic derivative **bis-(cyclopentadienyl) di(phenylselenide)**, $(C_6H_5)_2Zr(SeC_6H_5)_2$ (69.3).

Zirconium ditelluride, ZrTi₂, a black solid, has been prepared by heating the compound $4ZrO_2 \cdot TeO_2 \cdot 8H_2O$ in hydrogen to 500°C (24.1). Tellurides have more generally been synthesized from the elements, and include Zr_4Te_3, Zr_3Te_2, $ZrTe_3$, $ZrTe_2$, and $ZrTe$ (19.1,84.1). Also, zirconium reacts with lead telluride to form a zirconium telluride (4.1).

ZIRCONIUM SILICATE AND ZIRCONYLOSILICATES

The mineral *zircon*, $ZrO_2 \cdot SiO_2$, is the most important occurrence of the element zirconium. The compound can be synthesized by heating mixtures of the component oxides or hydrous oxides (9.1). In the crystal, the zirconium atoms are surrounded by four oxygen atoms at a distance of 2.15 Å and four others at a distance of 2.29 Å. In the silica tetrahedra, the Si-O distance is 1.61 Å (2.2).

When zircon is synthesized in the presence of vanadia under appropriately controlled conditions, vanadium enters into its structure and yields a beautiful blue zircon that is useful as a ceramic stain (125.3A).

Zircon decomposes approximately in the temperature range 1550–1720°C (76.1).

Hydrous silica in aqueous suspension sorbs zirconium-containing ions and hydrous zirconia, and chemical bonding is suggested by the behavior. The sorption complexes have been shown to have ion-exchange properties (42.1). Organic compounds containing Zr—O—Si linkages are also known, such as *tetrakis(trimethylsilyloxo)zirconium*, $Zr(OSi(CH_3)_3)_4$ (38.1).

The reaction of zircon with one mole of alkali metal oxide yields zirconylosilicates, and with two moles, zirconates and alkali metal silicate:

$$ZrO_2 \cdot SiO_2 + Na_2CO_3 \rightarrow Na_2ZrSiO_5 + CO_2$$

$$ZrO_2 \cdot SiO_2 + 2\,Na_2CO_3 \rightarrow Na_2ZrO_3 + Na_2SiO_3 + 2\,CO_2$$

Alkali metal zirconylosilicates can be prepared both by heating dry zircon and alkali, and by heating zircon with aqueous alkalies in an autoclave (144.9). The lithium, sodium, and potassium zirconylosilicates are known. The relatively low temperature reaction:

$$ZrO_2 \cdot SiO_2 + 2\,LiOH \xrightarrow[\;300°C\;]{200-} ZrO_2 + Li_2SiO_3 + H_2O$$

has also been reported (144.25).

Sodium zirconylosilicate is hydrolyzed by boiling water to $NaHZrSiO_5$, without a fundamental change in crystal lattice (144.17, 144.18). Other zirconylosilicates and related compounds of zirconium have compositions $Na_4Zr_2Si_3O_{12}$, $Na_2ZrSi_2O_7$, $Li_8Zr_3Si_5O_{20}$, $Rb_2ZrSi_2O_7$, $CsZrSi_2O_7$, $CaZrSiO_5$, $Ca_3ZrSi_2O_9$, $BaZrSiO_5$, and $Ba_2Zr_2Si_3O_{12}$.

Zircon reacts at a low red heat with fluorosilicates to form alkali metal fluorozirconates and silica. However, its reaction with aluminum fluoride yields zirconium tetrafluoride and aluminum oxide (58.1). Zircon has been found useful as a ceramic opacifier, coating for foundry molds, refractory, electric insulator, gem stone, component of cements and filler for plastics such as polyurethan (23.17) and polytetrafluoroethylene (125.58). The germanium analog of zircon, $ZrO_2 \cdot GeO_2$, is known (36.8).

ZIRCONIUM SILICIDES AND GERMANIDES

Zirconium silicides and **germanides** are prepared by heating mixtures of the elements. **Zirconium monosilicide,** $ZrSi$, has a hexagonal or rhombic crystalline symmetry. The **disilicide,** $ZrSi_2$, forms orthorhombic bipyramidal crystals in which the Zr-Si distance is only 0.56 Å (140.1). It is very resistant to corrosion. Its behavior in boron nitride and in aluminum has been studied (23.7,79.1). Other silicide and germanide phases are:

Zr_6Si_5	$ZrCo_2Si_2$	Zr_5Ge_3	$Zr_4Co_4Ge_7$
Zr_5Si_4	$ZrCuSi$	Zr_3Ge	$ZrCuGe$
Zr_5Si_3	$ZrFe_2Si$	$ZrGe_2$	$ZrPtGe$
Zr_4Si	$Zr_{14}Ni_2Si_9$		$ZrSGe$
Zr_3Si_{17}	$Zr_6Ni_{16}Si_7$		$ZrSeGe$
Zr_3Si	$Zr_4Ni_2Si_5$		$ZrTeGe$
Zr_2Si	$Zr_2Ni_3Si_{18}$		
$ZrSi_2$	$ZrOSi$		
	$ZrPdSi$		
	$ZrPtSi$		
	$ZrSSi$		
	$ZrSeSi$		
	$ZrTeSi$		

SULFATO COMPOUNDS OF ZIRCONIUM

A zirconyl compound in aqueous solution reacts with sulfate ions with formation of sulfatozirconates, eg,

$$ZrOCl_2 + 2\ SO_4{}^{2-} \rightarrow Zr(SO_4)_2{}^{2-} + 2\ Cl^-\ (22.4,\ pp.\ 240\text{--}243;\ 144.4)$$

If fewer or more sulfate ions are available, species may form having fewer or more than two sulfato ligands per zirconium atom, but temperature, concentration, acidity, and other factors play roles in the species that may form. When the sulfato compounds of zirconium crystallize or form amorphous solids, the proximity of the original ions or molecules gives rise to new and additional bond formation. For this reason, it is often useful to regard the solids as reaction products of the species which had been in solution, rather than as the same compounds in solid forms. When the crystals redissolve, the species which go into solution are not necessarily identical either with those from which the solid formed nor with those that are regarded as structural units in the solid. Moreover, they tend to undergo progressive changes after dissolving.

In the solution, ramified series of ionizations, hydrolyses, condensations, olations, and rearrangements occur. A great number of molecular and ion species result (48.2). These are largely arrangements of the fragments:

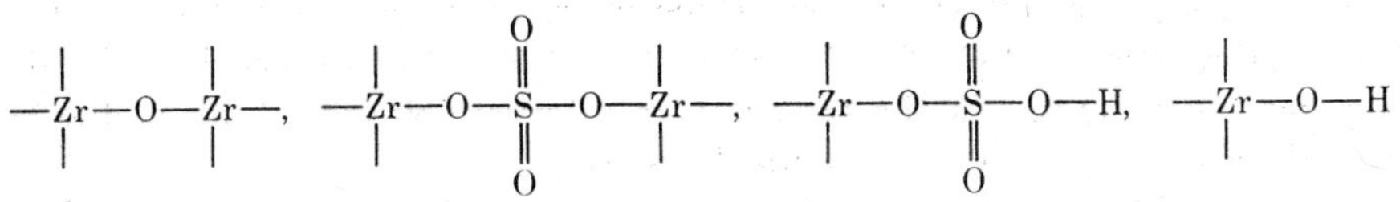

which can be combined in many ways. Hydration occurs to the maximum extent that steric conditions allow. It has already been noted that in zirconyl chloride solutions, the chloride ions are held coulombically rather than by covalency formation, at positions of maximum positive charge on the species. Some sulfatozirconium compounds are found to bond chloride ion from the solution in which they had formed, and it is probable that here, too, the chloride ions are coulombically bound. Species containing the group $ZrOSO_2OH$ are capable of forming salts by replacement of the terminal hydrogen by other cations.

Disulfatozirconic acid, $H_2Zr(SO_4)_2 \cdot 3H_2O$, is very soluble in water. Its solubility decreases when sulfuric acid is added to the solution, and with continued addition of sulfuric acid, the crystalline compound precipitates. Aqueous solutions of disulfatozirconic acid have about the same pH as sulfuric acid solutions of the same molarity. Alcoholic solutions gel on standing. Neutralization of disulfatozirconic acid with sodium bases results in the formation of *sodium disulfatozirconate*, at variable hydration levels. *Potassium sulfatozirconates* are sparingly soluble in water.

It is convenient to group the large number of sulfatozirconic acids into the following categories:

Polysulfatozirconates

The phase diagram of the system ZrO_2–SO_3–H_2O shows no sulfatozirconic acids with more than 3 sulfato ligands per zirconium atom (13.2). However, salts of the compositions $K_9H_3Zr(SO_4)_8 \cdot H_2O$, $K_{7.8}H_{2.2}Zr(SO_4)_7 \cdot H_2O$, $K_{6.5}H_{1.5}Zr(SO_4)_6 \cdot H_2O$, $K_6Zr(SO_4)_5 \cdot 5H_2O$, and $K_4Zr(SO_4)_4 \cdot 3H_2O$ have been prepared (144.37), and also the binuclear $Tl_{14}Zr_2(SO_4)_{11}$ (22.0).

Ammonium and thallium tetrasulfatozirconates, $(NH_4)_4Zr(SO_4)_4 \cdot 5H_2O$ and $Tl_4Zr(SO_4)_4 \cdot 4H_2O$, have been crystallized from aqueous solutions containing their components. Also, $K_4Zr(SO_4)_4 \cdot 3H_2O$ and $Na_4Zr(SO_4)_4 \cdot 4H_2O$ have been prepared by treating hydrous zirconia with the appropriate alkali metal hydrogen sulfate (22.0).

The phase diagram of the system ZrO_2–SO_3–H_2O shows the existence of the trisulfatozirconic acids of compositions $H_3ZrOH(SO_4)_3 \cdot H_2O$, $H_3ZrOH(SO_4)_3$, and $H_2Zr(SO_4)_3$. They form successively as the SO_3 content of the system is increased (13.2). It has been shown that when alkali sulfates are added to aqueous solutions of disulfatozirconic acid, trisulfatozirconate ions form (126.1). The salts $(NH_4)_2Zr(SO_4)_3 \cdot H_2O$ (125.8), $K_2Zr(SO_4)_3 \cdot 3H_2O$ (144.37), and $FeZr(SO_4)_3 \cdot 6H_2O$ (137.2) have been reported.

Disulfatozirconic Acids

The most common of these, the trihydrate of composition $H_2ZrO(SO_4)_2 \cdot 3H_2O$, has already been discussed as to general properties. Its crystals have a layered structure, and the layers appear to be held together by hydrogen bonding. Oxygen of the water of hydration and oxygen of the sulfato ligands form antiprisms around each zirconium atom. The sulfato group, itself, has small but significant departure from tetrahedral symmetry (2.1). The compound loses three moles of water per zirconium atom on heating in the interval 100–160°C. At 400°C, the anhydrous acid, $H_2ZrO(SO_4)_2$, decomposes to $Zr(SO_4)_2$. In the temperature interval 450–800°C, all of the SO_3 is volatilized, leaving a residue of zirconium dioxide (57.1).

Anhydrous disulfatozirconic acid is trimorphic. A dimer is the structural unit in the crystal. Sulfato groups act as linking or bidentate ligands. Three of their four oxygen atoms are in contact with zirconium atoms, and the fourth is probably protonated. Each zirconium atom has 7 fold coordination (31.8). The existence of 1.5, 5, and 7 hydrates has recently been shown (144.27).

The addition of two moles of sulfuric acid to one of sodium zirconylosilicate, Na_2ZrSiO_5, leads to formation of a composition believed to consist of $Na_2ZrO(SO_4)_2$ and $SiO_2.H_2O$ (125.21). It is used in industry as a tanning agent. Sulfatozirconic acids generally tan hides and give high quality, plump white leather. Disulfatozirconic acid also tans gelatin and is useful in the manufacture of gelatin capsules (125.45). It improves the filling power of fowl feathers (125.30).

The anhydrous $Zr(SO_4)_2$, prepared by dehydration of the hydrated disulfatozirconic acids, appears to provide different species immediately upon going into aqueous solution than do the acids.

Sulfatozirconates of Lower Sulfato Contents

Compounds of compositions $nZrO_2.mSO_3.pH_2O$, where $m < 2n$, have been prepared by (1) the reaction of hydrous zirconia with less than two moles of sulfuric acid per mole of ZrO_2, (2) heating, diluting, or partially neutralizing solutions of disulfatozirconic acid, or (3) by achieving indirectly either or a combination of these processes. In the older momenclature, compounds of this category were called *basic zirconium sulfates*. A more appropriate name is *polysulfatopolyzirconic* acids.

Perhaps the most useful compound of this category is that of composition $5ZrO_2.3SO_3.15.5H_2O$, which precipitates when a zirconyl chloride solution containing the stoichiometric proportion of a sulfate salt is heated. The compound often separates in a high state of purity when metal ions other than zirconium complex ions are in the solution, and it has been of particular value in industry in obtaining a zirconium compound free of the iron with which the zirconium had been associated in its ore (125.0). The compound has been shown to be useful in absorbing odoriferous contaminants from air (125.35) and as an additive to glass polishing oxide (20.1).

A compound capable of serving some of the same purposes has the composition $5ZrO_2.2SO_3.14H_2O$. It precipitates when a hot zirconyl chloride solution containing the stoichiometric proportion of sulfate ions is adjusted to pH 1.7 with a base (23.1).

The *monosulfatozirconic* acid of composition $(HZrOOHSO_4.2H_2O)_4$ has been prepared by adding 1.125 moles of sulfuric acid per mole of zirconyl chloride dissolved in methanol. Treatment of this compound with hydrogen chloride gave an adduct with one mole of HCl per zirconium atom (144.33). Infrared studies of some polysulfatopolyzirconic acids indicate them to contain either polydentate sulfato ligands or bridging sulfato ligands, or both, and sometimes pairs of oxygen-bridged zirconium atoms (144.19). The structure of $2ZrO_2.3SO_3.5H_2O$ has been determined by x-ray diffraction. Each zirconium atom is connected by four sulfate groups to adjacent zirconium atoms. Some of the sulfate groups are bonded to three zirconium atoms and others to four. The coordination number of the zirconium atoms is eight, and the coordination figure is dodecahedral (51.3).

Other compounds of this category have the compositions $ZrO_2.SO_3.5H_2O$, $2ZrO_2.3SO_3.4H_2O$, $3ZrO_2.2SO_3.12H_2O$, $4ZrO_2.7SO_3.12H_2O$, $8ZrO_2.5SO_3.2HCl.20H_2O$. Those with terminal $ZrOSO_2OH$ groups form salts, of which the compositions $Na_2Zr_4O_5(SO_4)_4.19H_2O$, $K_2Zr_4O_5(SO_4)_4.16H_2O$, and $Na_4Zr_4O_4(SO_4)_6.12H_2O$ have been reported (144.33).

ZIRCONIUM SULFIDES (see also Zirconium mercaptides)

Zirconium has very little affinity for sulfur, and there is no instance of zirconium–sulfur bonds forming in aqueous systems. When the elements are heated together in the absence of other reactive material, zirconium metal combines with sulfur in a number of proportions, and compounds of the following compositions have been reported: Zr_4S_3, Zr_3S_5, Zr_3S_4, Zr_3S_2, Zr_2S_3, ZrS_3, ZrS_2, and ZrS (137.6). A number of sulfide phases containing zirconium and other metals or metalloids are known: $CuZr_2S_4$, $CuCrZrS_4$, $GeZrS$, and $SiZrS$.

When zirconium dioxide powder is heated to 700°C in hydrogen sulfide, *zirconium oxide sulfide*, $ZrOS$, is formed as a light yellow powder of specific gravity 4.85 (146.0). The same compound has been prepared by the reaction of zirconium oxide with carbon disulfide vapor at 850–1000°C. At 1200–1300°C, zirconium disulfide forms (60.14). Under similar conditions, zirconates are converted to thiozirconates (125.44). The alkaline earth thiozirconates, $BaZrS_3$, $SrZrS_3$, and $CaZrS_3$ have been studied (2.6, 137.5b). The barium compound has a distorted perovskite structure.

ZIRCONIUM THIOCYANATES

The zirconium thiocyanates show considerable similarities in their chemistry to the zirconium chlorides, particularly in aqueous solution, but there are also some distinctive features. Ions with up to eight thiocyanato ligands per zirconium atom have been reported (144.31). Alkali thiocyanatozirconates, and also pyridinium thiocyanatozirconate, have been precipitated from strongly acidic solutions (144.12).

Zirconyl thiocyanate, $ZrO(CNS)_2 \cdot nH_2O$, has been prepared by metathesis of zirconyl sulfate with barium thiocyanate (144.12). Compounds of general formula $ZrO(CNS)_2 \cdot MCNS$ have been identified, where M is NH_4, K, Rb, Cs, and pyridinium. Infrared spectra of solid zirconyl thiocyanates give evidence for the presence of a zirconyl group, $Zr{=}O$. Zirconium is bonded to the thiocyanato ligand through nitrogen (55.3).

ZIRCONIUM VANADATES

The compound of composition $3ZrO_2 \cdot 2V_2O_5 \cdot 7H_2O$ is a yellow solid which is insoluble in water. It has been prepared by the reaction of ammonium vanadate and zirconyl chloride in aqueous solution in the ratios 0.25 to 0.75, in the pH range 2.1–6.9.

Bibliography

"Zirconium Compounds" in *ECT* 1st ed. Vol. 15, pp. 290–312, by Warren B. Blumenthal, National Lead Co.

1. *Acta Chem. Scand.*
    1. **20** (6), 1712–1714 (1966), **66**, 6215b, T. Lundstrom.
2. *Acta Cryst.*
    0. **9**, 955–958 (1956), A. Clearfield and P. A. Vaughan.
    1. **11**, 719–723 (1958), *Chem. Abstr.* **54**, 1967i, J. Singer and D. T. Cromer.
    2. **11**, 896–897 (1958), *Chem. Abstr.* **54**, 3092i, I. R. Krstanovic.
    3. **12**, 951 (1959), *Chem. Abstr.* **55**, 26605a, J. Adam and M. D. Rogers.
    4. **13**, 684 (1960), *Chem. Abstr.* **56**, 84c, H. Templeton.
    5. **14**, 128–132 (1961), *Chem. Abstr.* **55**, 8991f, C. Larson and D. T. Cromer.
    6. **16**, 135–142 (1963), *Chem. Abstr.* **58**, 9695e, A. Clearfield.

3. *Acta Met.*

4. *Advan. Energy Convers.*
    1. **7** (4), 275–287 (1968), *Chem. Abstr.* **68,** 98091g, H. E. Bates, F. Wald, and M. Weinstein.

5. *Am. Chem. Soc. Symposia*
    1. Div. Org. Coatings, Plastic Chem. Preprints **22** (2), 224–235 (1962), *Chem. Abstr.* **61,** 780d, O. Grummitt and A. A. Arters.
    2. Hydrocarbon Fuel Cell Technol. Symp., Atlantic City *1965,* 485–494, *Chem. Abstr.* **65,** 8326c, C. Berger and M. P. Strier.

6. *Am. Dyestuff Reptr.*
    1. **55** (16), 597–600 (1966), *Chem. Abstr.* **65,** 15567e, C. J. Conner, A. S. Cooper, Jr., and W. A. Reeves.

7. *Am. J. Pathol.*
    1. **43** (3), 391–405 (1963), *Chem. Abstr.* **60,** 6017d, W. L. Epstein, J. R. Skahen, and H. Krasnobrod.

8. *Am. J. Sci.*
    1. **239,** 857–898 (1941), *Chem. Abstr.* **36,** 1571³, L. W. Stock.

9. *Am. Mineralogist*
    1. **42,** 759–765 (1957), *Chem. Abstr.* **52,** 2671f, C. Frondel and R. L. Collette.

10. *Anal. Chem.*
    1. **29,** 1660–1662 (1957), *Chem. Abstr.* **52,** 5098h, F. L. Moore.
    2. **30,** 548 (1958), *Chem. Abstr.* **52,** 10683d, H. B. Bradley and L. G. Dowell.
    3. **32** (6), 729 (1960), *Chem. Abstr.* **54,** 11630i, R. P. Agarwala, E. Govindan, and M. C. Naik.
    4. **37** (9), 1158–1159 (1965), *Chem. Abstr.* **63,** 10655f, S. M. Khopkar and S. C. Dhara.

11. *Anal. Chim. Acta*
    1. **33** (3), 237–244 (1965), *Chem. Abstr.* **63,** 9038b, D. E. Elliot and C. V. Banks.
    2. **33** (6), 577–585 (1965), *Chem. Abstr.* **64,** 4255b, E. S. Pilkington and W. Wilson.

12. *Analyst*
    1. **75,** 684–686 (1950), *Chem. Abstr.* **45,** 1902i, A. Purmshottam and Bh. S. V. Raghava Rao.

13. *Ann. Chim.*
    1. **4** (12), 133–195 (1949), *Chem. Abstr.* **43,** 8253g, J. Pierrey.
    2. **16,** 237–325 (1941), *Chem. Abstr.* **36,** 6100², Marie Falinski.

14. *Angew. Chem.*
    1. **65** (9), 376–377 (1964), *Chem. Abstr.* **61,** 1751c, A. Sartori and M. Weidenbruch.

15. *Angew. Chem. Int. Ed. Engl.*
    1. **5** (12), 1041 (1966), *Chem. Abstr.* **66,** 50474r, K. Dehnicke and W. Weidlein.
    2. **8** (2), 146–147 (1969), *Chem. Abstr.* **70,** 81828w, B. Krebs.

16. *Ann. Phys. (Paris)*
    1. **3,** 179–229 (1958), *Chem. Abstr.* **53,** 2880i, P. Mergault.

17. *Ann. Univ. Mariae Curie-Sklodowska Lublin-Polonia, Sect. AA*
    1. **11,** 47–76 (1956), *Chem. Abstr.* **53,** 13743b, B. Frank.

17a. *Arkiv Kemi*
    1. **13,** 59–85 (1958), *Chem. Abstr.* **53,** 6726b, Georg Lundgren.

18. *Atti Accad. Sci. Torino, Classe Sci. Fis. Mat. Nat.*
    1. **94,** 97–106 (1959–1960), *Chem. Abstr.* **60,** 5049h, A. Burdese and M. L. Borlera.

19. *Australian J. Chem.*
    1. **11,** 458–470 (1958), *Chem. Abstr.* **53,** 5936h, J. Bear and F. K. McTaggart.
    2. **17** (7), 727–730 (1964), *Chem. Abstr.* **61,** 9152f, F. K. McTaggart and A. G. Turnbull.
    3. **19** (11), 2069–2072 (1966), *Chem. Abstr.* **66,** 38026p, R. S. P. Coutts and P. C. Wailes.

20. *Belgian Pat.*
    1. 617,559, (Aug. 31, 1962), *Chem. Abstr.* **58,** 6550d, W. J. Baldwin and J. L. Bliton.
    2. 649,780, (Oct. 16, 1964), *Chem. Abstr.* **64,** 7979e, J. C. Muhler.

21. *Ber.*
    1. **B59,** 1890–1893 (1926), *Chem. Abstr.* **21,** 543, G. von Hevesy and M. Logstrup.

2. **95,** 1703–1710 (1962), *Chem. Abstr.* **57,** 8605a, W. Kucher, K. Strolenberg, and H. Buchwald.

22. *Books*

    0. *A Comprehensive Treatise on Inorganic and Theoretical Chemistry,* J. W. Mellor, Longmans, Green and Co., New York, 1930, pp. 98–165.

    1. *Zirconium and its Compounds,* F. P. Venable, Chemical Catalog Co., New York, 1922.

    2. *Inorganic Colloid Chemistry,* Vol. II, H. B. Weiser, John Wiley & Sons Inc., New York, 1935.

    3. *Traité de Chimie Mineral,* Vol. V, P. Pascal, Masson, Paris, 1932.

    4. *The Chemical Behavior of Zirconium,* W. B. Blumenthal, Van Nostrand, 1958.

    5. *Nouveau Traité de Chimie Mineral,* Vol. IX, P. Pascal, Masson, & Cie, Paris 1963.

    6. *Inorganic Ion Exchangers,* C. B. Amphlett, Elsevier Publishing Co., New York, 1964.

23. *British Pat.*

    1. 112,973 (Jan. 29, 1918), R. T. Glazebrook, W. Rosenhain, and E. H. Rodd.

    2. 569,054 (May 2, 1945), *Chem. Abstr.* **41,** 4900f, Titanium Alloy Manufacturing Co.

    4. 784,852 (Oct. 16, 1957), *Chem. Abstr.* **52,** 7344g, Farbwerke Hoechst A. G.

    5. 787,175 (Dec. 4, 1957), *Chem. Abstr.* **52,** 7728d, Carlisle Chemical Works.

    6. 789,566 (Jan. 22, 1958), *Chem. Abstr.* **52,** 13782h, National Lead Co.

    7. 792,733 (April 2, 1958), *Chem. Abstr.* **52,** 17654c, Carborundum Co.

    8. 794,518 (May 7, 1958), *Chem. Abstr.* **52,** 20936f, W. Wainer (to Horizons Titanium).

    9. 800,160 (Aug. 20, 1958), *Chem. Abstr.* **53,** 7016g, Th. Goldschmidt A. G.

    10. 816,679 (July 15, 1959), *Chem. Abstr.* **54,** 896e, S. D. Warren Co.

    11. 881,661 (Nov. 8, 1961), *Chem. Abstr.* **57,** 2462a, Bradford Dyers Association Ltd.

    12. 891,046 (March 7, 1962), *Chem. Abstr.* **56,** 14012h, J. C. Cohen and R. B. Jackson (to Carborundum Co.).

    13. 896,010 (May 9, 1962), S. D. Warren Co.

    13a. 942,103 (Nov. 20, 1963), *Chem. Abstr.* **61,** 12156c, Horizons, Inc.

    14. 956,748 (April 29, 1964), *Chem. Abstr.* **61,** 8528a, Oxford Paper Co.

    15. 969,336 (Sept. 9, 1964), *Chem. Abstr.* **61,** 14900f, Deutsche Advance Produktion. G.m.b.H.

    16. 1,008,652 (Nov. 3, 1965), *Chem. Abstr.* **64,** 2285g, British Titan Products Co., Ltd.

    17. 1,093,173 (Nov. 29, 1967), *Chem. Abstr.* **68,** 30606y, B. G. Hood and R. A. Gardella.

    18. 1,149,251 (April 23, 1969), *Chem. Abstr.* **71,** 21745c, Columbian Carbon Co.

24. *Bull. Soc. Chim. France*

    1. 1946, 176 *Chem. Abstr.* **40,** 6014a, E. Montignie.

    2. 1962, 2113–2117, *Chem. Abstr.* **58,** 9871b, R. Collongues, J. C. Gilles, and Anne Marie Lejus.

    3. 1968 (2), 507–513, *Chem. Abstr.* **69,** 13679c, J. Livage.

    4. 1968 (Spec. No.), 1832–1835, *Chem. Abstr.* **69,** 39000z, V. Vesely, V. Pekarek, and A. Ruvarac.

25. *Bunseki Kagaku*

    1. **7,** 445–449 (1958), *Chem. Abstr.* **54,** 7428a, C. Yoshimura and M. Kiboku.

26. *C. R. Acad. Sci. Paris*

    1. **C246** (9), 768–771 (1967), *Chem. Abstr.* **67,** 15384k, A. Fontana and R. Winand.

27. *Can. J. Chem.*

    1. **40,** 1350–1354 (1962), *Chem. Abstr.* **57,** 10757a, M. S. Bains and D. C. Bradley.

    2. **39,** 1434–1443 (1961), *Chem. Abstr.* **55,** 26522c, D. C. Bradley and D. G. Carter.

    3. **44** (4), 534–538 (1966), *Chem. Abstr.* **64,** 15358f, M. S. Bains.

    4. **44** (8), 972–975 (1966), *Chem. Abstr.* **64,** 18946h, V. Krishnan and C. G. Patel.

    5. **46** (22), 3491–3497 (1968), *Chem. Abstr.* **70,** 7280v, T. W. C. Mak.

28. *Can. Mineralogist.*

    1. **9** (4), 468–477 (1968), *Chem. Abstr.* **69,** 108561x, Ann P. Sabina, J. L. Jambor, and A. G. Plant.

29. *Canadian Pat.*

    1. 503,469 (June 1, 1954), W. B. Blumenthal.

30. *Chem. Anal. (Warsaw)*

    1. **2**, 222–227 (1957), *Chem. Abstr.* **52**, 6065a, I. P. Alimarin and V. S. Sotnikov.

31. *Chem. Commun.*

    1. 1965 (13), 289, *Chem. Abstr.* **63**, 9427b, D. C. Bradley and M. H. Gitlitz.

    2. 1965 (22), 567, *Chem. Abstr.* **64**, 6686c, J. R. Surtees.

    3. 1966 (10), 302–303, *Chem. Abstr.* **65**, 3706a, J. K. Becconsall and S. O'Brien.

    4. 1966 (23), 849–850, *Chem. Abstr.* **66**, 38023k, B. D. James, R. K. Nanda, and M. G. H. Wallbridge.

    5. 1967 (8), 403, *Chem. Abstr.* **67**, 37122n, P. H. Bird and M. R. Churchill.

    6. 1967 (13), 646–647, *Chem. Abstr.* **67**, 87347y, G. W. A. Fowles, B. J. Russ, and G. R. Willey.

    7. 1967 (20), 1035, *Chem. Abstr.* **68**, 13057c, D. J. Cardin, S. A. Keppie, B. M. Kingston, and M. F. Lappert.

    8. 1969 (5), 230–232, *Chem. Abstr.* **70**, 100646d, Isabel J. Bear and W. G. Mumme.

32. *Chem. Ind. (London)*

    1. 1964 (17), 713–714, *Chem. Abstr.* **61**, 1340g, T. N. Waters.

    2. 1965 (41), 1730, *Chem. Abstr.* **63**, 16202c, D. C. Bradley, J. Charalambous, and S. Jain.

    3. 1968 (39), 1314, *Chem. Abstr.* **69**, 102583j, P. N. Kapoor, R. N. Kapoor, and R. C. Mehrotra.

33. *Chem. & Ind. (London)*

    1. 1958, 68, *Chem. Abstr.*, **52**, 5858h, R. N. Kapoor and R. C. Mehrotra.

    2. 1958, 1200–1202, *Chem. Abstr.*, **53**, 11940g, C. B. Amphlett and J. Kennedy.

    3. 1958, 1231–1233, *Chem. Abstr.* **53**, 4996d, D. C. Bradley and I. M. Thomas.

34. *Chemiker Ztg.*

    1. **92** (5), 137–142 (1968), *Chem. Abstr.* **69**, 2435u, B. Cornils, M. Rasch, and G. Schiemann.

35. *Comm. Energie At. (France) Rappt. CEA*

    1. 767, 5 pp. (1958), *Chem. Abstr.* **53**, 9460a, M. T. Guilloux and G. Michon.

36. *Compt. Rend.*

    1. **200**, 1668–1669 (1933), *Chem. Abstr.* **29**, 4279[8], A. Karl.

    2. **203**, 87–90 (1936), *Chem. Abstr.* **30**, 6269[4], J. Boulanger.

    3. **214**, 27–29 (1942), *Chem. Abstr.* **37**, 2291[2], A. Tchakirian.

    4. **224**, 654–665 (1947), *Chem. Abstr.* **41**, 5004f, R. Stumper and P. Mettelock.

    5. **224**, 122–124 (1947), *Chem. Abstr.* **41**, 4061f, R. Stumper and F. Classen.

    6. **246**, 2266–2268 (1958), *Chem. Abstr.* **52**, 15182i, A. Chrétien and B. Gaudreau.

    7. **251**, 875–877 (1960), *Chem. Abstr.* **55**, 196g, J. Miemiec.

    8. **251**, 1016–1018 (1960), *Chem. Abstr.* **55**, 2328h, J. Lefevre and R. Collongues.

    9. **257** (20), 2993–2994 (1963), *Chem. Abstr.* **60**, 3698g, G. Montel.

    10 **259** (6), 1337–1339 (1964), *Chem. Abstr.* **62**, 1121e, J. Livage, J. Cabane, and C. Mazières.

37. *Congr. intern. chim. pure et appl. 16*ᵉ, Paris, *1957*, Mem. sect. chim. minerale, (Pub. 1958) 43–46, *Chem. Abstr.* **54**, 12862e, W. Klemm, E. Holze, and W. Basualdo.

38. *Coord. Chem. Rev.*

    1. **2** (3), 299–318 (1967), *Chem. Abstr.* **68**, 87770b, D. C. Bradley.

39. *Current Sci. (India)*

    1. **29**, 222–223 (1960), *Chem. Abstr.* **55**, 1261c, S. S. Sandhu, J. S. Sandhu, and G. S. Sandhu.

40. *Dokl. Akad. Nauk SSSR*

    1. **114**, 1101–1103 (1957), *Chem. Abstr.* **52**, 178a, I. P. Tikhonenkov, E. I. Semenov, and M. E. Kazakova.

    2. **125**, 120–123 (1959), *Chem. Abstr.* **53**, 19656i, V. I. Spitsyn, L. N. Komissarova, and Z. A. Vladimirova.

    3. **131**, 176–179 (1960), *Chem. Abstr.* **54**, 13997g, V. V. Ilyukhin and N. V. Belov.

    4. **133**, 657–660 (1960), *Chem. Abstr.* **54**, 24150c, S. M. Kravchenko, E. V. Vlasova, and N. G. Pinevich.

    5. **134**, 920–923 (1960), *Chem. Abstr.* **55**, 8188c, I. P. Tikhonenkov, M. V. Kucharchik, and Yu. A. Pyatenko.

6. **136,** 350–353 (1961), *Chem. Abstr.* **56,** 6871g, L. N. Komissarova, L. I. Yuranova, and V. E. Plyushchev.

7. **137,** 681–684 (1961), *Chem. Abstr.* **55,** 26871g, T. B. Zdorik, G. A. Sidorenko, and A. V. Bykova.

8. **137,** 944–946 (1961), *Chem. Abstr.* **55,** 23197h, R. P. Tikhonenkova and M. E. Kazakova.

9. **141,** 1454–1456 (1961), *Chem. Abstr.* **57,** 3094b, L. S. Borodin and A. V. Bykova.

10. **142,** 916–918 (1962), *Chem. Abstr.* **56,** 15176e, V. I. Gerasimovskii.

11. **143,** 1413–1416 (1962), *Chem. Abstr.* **57,** 5359e, A. M. Reznik, A. M. Kozen, S. S. Korovin, and I. A. Apraksin.

12. **146,** 897–900 (1962), *Chem. Abstr.* **58,** 3976f, R. P. Shibaeva and N. V. Belov.

13. **153** (5), 1164–1167 (1963), *Chem. Abstr.* **60,** 9026d, M. D. Dorfman, V. V. Ilyukhin, and T. A. Burnova.

14. **156** (6), 1375–1378 (1964), *Chem. Abstr.* **61,** 7034b, E. M. Brainina, G. G. Dvoryantseva, and R. Kh. Freidlina.

15. **160** (5), 1065–1068 (1965), *Chem. Abstr.* **63,** 3702a, A. G. Boganov, V. S. Rudenko, and L. P. Makarov.

16. **165** (1), 117–120 (1965), *Chem. Abstr.* **64,** 2811d, Yu. A. Zolotov, V. G. Lambrev, M. K. Chmutova, and N. T. Sizonenko.

17. **182** (5), 1176–1177 (1968), *Chem. Abstr.* **70,** 21730y, V. D. Dusmatov, A. F. Efimov, Z. T. Kataeva, L. A. Koroshilova, and K. P. Yanulov.

41. *Dopovidi Akad. Nauk Ukr. RSR*

1. 1960, 1090–1094, *Chem. Abstr.* **55,** 17334e, I. A. Sheka and Ts. V. Pevzner.

2. 1963 (9), 1201–1205, *Chem. Abstr.* **60,** 5049f, I. A. Sheka, S. A. Kacherova, and L. O. Malinko.

42. *East German Pat.*

1. 40,952 (Sept. 15, 1965), *Chem. Abstr.* **64,** 9287a, D. Naumann.

43. *Energie Nucl.*

1. **1,** 155–160 (1957), *Chem. Abstr.* **55,** 1961, C. Decroly, D. Tytgat, and J. Gerard.

44. *Eng. Mineracao e Met.*

1. **27,** 265–269 (1958), *Chem. Abstr.* **52,** 16992c, G. E. Tolbert.

45. *French Pat.*

1. 1,059,648 (Mar. 26, 1954), *Chem. Abstr.* **52,** 15010d, J. Day and M. Calis (to Companie Général de Telegraphie sans Fil).

2. 1,447,640 (July 29, 1966, *Chem. Abstr.* **66,** 49589a, J. G. Smith and F. T. Fitch

46. *Gazz. Chim. Ital.*

1. **72,** 77–83 (1942), *Chem. Abstr.* **37,** 573^2, G. Peyronel.

47. *German Pat.*

1. 925,276 (Mar. 17, 1955), *Chem. Abstr.* **52,** 1577b, W. Dawihl (to W. Dawihl).

2. 941,430 (April 12, 1956), *Chem. Abstr.* **53,** 4134h, D. F. Herman and H. H. Beachman (to National Lead Co.).

3. 1,035,630 (April 7, 1958), *Chem. Abstr.* **54,** 21680h, W. Brugger (to Th. Goldschmidt Akt.-Ges.).

4. 1,048,866 (Jan. 22, 1959), *Chem. Abstr.* **54,** 25873d, A. Metzger (to Elektrochemische Fabrik Kempten Gmbh).

5. 1,116,335 (Nov. 2, 1961), *Chem. Abstr.* **56,** 13806i, C. S. Oliver and A. J. Haltner (to General Electric Co.)

6. 1,124,933 (Mar. 8, 1962), *Chem. Abstr.* **57,** 7109d, M. Reuter and L. Orthner (to Farbewerke Hoechst Akt.-Ges.).

48. *Ind. Eng. Chem.*

1. **42,** 640–642 (1950), *Chem. Abstr.* **44,** 5599e, W. B. Blumenthal.

2. **46,** 528–539 (1954), *Chem. Abstr.* **48,** 6893c, W. B. Blumenthal.

3. **55** (4), 50-57 (1963), *Chem. Abstr.* **59,** 2601a, W. B. Blumenthal.

49. *Ind. Eng. Chem., Anal. Edition*

1. **9,** 371–373 (1937), *Chem. Abstr.* **31,** 7356^1, W. C. Schumb and E. J. Nolan.

50. *Indian J. Chem.*

1. **3** (6), 277–278 (1965), *Chem. Abstr.* **63,** 11453f, R. C. Paul, R. Parkash, and S. S. Sandhu.

2. **3** (10), 448–451 (1965), *Chem. Abstr.* **64**, 5813h, K. S. Venkateswarlu, V. Subramanyan, M. R. Dhaneswar, R. Shanker, M. Lal, and J. Shankar.

3. **4** (12), 535–536 (1966), *Chem. Abstr.* **67**, 6198a, N. Souka and A. Alain.

4. **5** (3), 126 (1967), *Chem. Abstr.* **67**, 53289r, S. Geetha and P. T. Joseph.

5. **6** (1), 54–55 (1968), *Chem. Abstr.* **69**, 7940q, G. D. Gupta and G. V. Jere.

51. *Inorg. Chem.*

1. **2**, 243–249 (1963), *Chem. Abstr.* **58**, 9702c, J. V. Silverton and J. L. Hoard.

1a. **2**, 250–256 (1963), *Chem. Abstr.* **58**, 9702d, G. L. Glen, J. V. Silverton, and J. L. Hoard.

2. **3** (5), 146–148 (1964), *Chem. Abstr.* **60**, 7519c, A. Clearfield.

2a. **5** (2), 281–283 (1966), *Chem. Abstr.* **64**, 7456b, J. A. Watts.

3. **5** (2), 284–289 (1966), *Chem. Abstr.* **64**, 9022h, D. B. McWhan and G. Lundgren.

4. **7** (3), 502–508 (1968), *Chem. Abstr.* **68**, 74753y, T. J. Pannavaia and R. S. Fay.

52. *Intern. J. Radiation Biol.*

1. **8** (5), 427–437 (1964), *Chem. Abstr.* **62**, 15046e, G. E. Zander-Principati, and J. F. Kuzma.

53. *Issled. po Zharoproch. Splavum, Akad. Nauk SSSR, Inst. Met.*

1. **8**, 224–259 (1962), *Chem. Abstr.* **57**, 16264e, V. P. Grechin, K. K. Chuprin, A. V. Frolov, and A. P. Sonyushkina.

54. *Izv. Akad. Nauk SSSR, Neorg. Materialy*

1. **1** (10), 1834–1837 (1965), *Chem. Abstr.* **64**, 7471h, V. V. Glazova and I. I. Kornilov.

2. **1** (4), 591–596 (1965), *Chem. Abstr.* **63**, 3864e, T. F. Limar, V. I. Andreeva, and K. A. Uvarova.

3. **1** (4), 600–603 (1965), *Chem. Abstr.* **63**, 6588a, A. I. Morozov and I. S. Morozov.

4. **4** (9), 1607–1609 (1968), *Chem. Abstr.* **70**, 43444k, Yu. M. Polezhaev.

55. *Izv. Akad. Nauk SSSR, Otd. Khim. Nauk*

1. 1961, 958–964, *Chem. Abstr.* **58**, 2127d, C. Z. Makarov and L. V. Ladeinov.

2. 1961, 1595–1599, *Chem. Abstr.* **56**, 3499a, E. M. Brainina and R. Kh. Freidlina.

2a. 1962, 393–401, *Chem. Abstr.* **57**, 8082f, Yu. Ya. Kharitonov and Yu. A. Buslaev.

3. 1962, 402–407, *Chem. Abstr.* **57**, 8163e, Yu. Ya. Kharitonov and I. A. Rozanov.

56. *Izv. Akad. Nauk SSSR, Ser. Khim.*

1. 1964 (8), 1417–1421, *Chem. Abstr.* **65**, 3905b, R. Kh. Freidlina, E. M. Brainina, M. Kh. Ninacheva, and A. N. Nesmeyanov.

2. 1965 (10), 1877–1879, *Chem. Abstr.* **64**, 3594b, E. M. Brainina, M. Kh. Minacheva, and R. Kh. Freidlina.

57. *Izv. Vysshikh. Ucheb. Zavedanii Khim. i Khim. Tekhnol.*

1. 1958 (1), 37–42, *Chem. Abstr.* **52**, 16946i, L. M. Komissarova, V. E. Plyushchev, and L. I. Yuranova.

58. *Izv. Vysshikh. Ucheb. Zavedanii, Tsvetn, Met.*

1. **11** (5), 45–48 (1968), *Chem. Abstr.* **70**, 39337k, A. I. Lainer, G. B. Borisov, and L. V. Mirkin.

59. *J. Am. Ceram. Soc.*

1. **37**, 277–280 (1954), *Chem. Abstr.* **48**, 8505f, D. E. Harrison, H. H. McKinstry, and F. A. Hummel.

2. **42**, 570 (1959), *Chem. Abstr.* **54**, 3029i, J. Graham, A. D. Wadsley, J. H. Weymouth, and L. S. Williams.

2a. **49** (5), 286–287 (1966), *Chem. Abstr.* **65**, 5180c, K. S. Mazdiyasni, C. T. Lynch, and J. S. Smith.

3. **44**, 147–148 (1961), *Chem. Abstr.* **55**, 11791f, C. T. Lynch, F. W. Vahldiek, and L. B. Robinson.

4. **51** (4), 227–228 (1968), *Chem. Abstr.* **69**, 13126v, C. A. Martinek and F. A. Hummel.

5. **51** (10), 582–584 (1968), *Chem. Abstr.* **69**, 99033x, G. L. Kulcinski.

6. **51** (12), 674–677 (1968), *Chem. Abstr.* **70**, 22612y, C. A. Sorrel.

60. *J. Am. Chem. Soc.*

1. **47**, 2540–2544 (1925), *Chem. Abstr.* **19**, 3404, G. von Hevesy and K. Kimura.

2. **53**, 921–923 (1931), *Chem. Abstr.* **25**, 2071, S. G. Simpson and W. C. Schumb.

3. **53**, 1276–1278 (1931), *Chem. Abstr.* **25**, 2930, H. S. Gable.

4. **53**, 2148–2153 (1931), *Chem. Abstr.* **25**, 4193, R. C. Young.

5. **57**, 2131–2135 (1935), A. W. Thomas and H. S. Owens.

6. **69**, 1130–1134 (1947), *Chem. Abstr.* **41**, 4650d, H. A. Liebhafsky and E. H. Winslow.

7. **71**, 2488–2492 (1949), *Chem. Abstr.* **43**, 8937f, H. R. Hoekstra and J. J. Katz.

8. **71**, 3179–3182 (1949), *Chem. Abstr.* **43**, 8959d, E. H. Huffman and L. J. Beaufait.

9. **71**, 3182–3192 (1949), *Chem. Abstr.* **43**, 8929d, R. E. Connick and W. H. McVey.

10. **72**, 1226–1230 (1950), *Chem. Abstr.* **45**, 1073b, W. J. Roberts and A. R. Day.

11. **72**, 3610–3614 (1950), *Chem. Abstr.* **44**, 10460g, B. G. Schultz and E. M. Larsen.

12. **73**, 1958–1959 (1951), *Chem. Abstr.* **46**, 847a, S. R. Patel.

13. **74**, 6154–6155 (1952), *Chem. Abstr.* **47**, 12103h, T. R. Sato, H. Diamond, W. P. Norris, and H. H. Strain.

13a. **75**, 1560–1562 (1953), *Chem. Abstr.* **47**, 7941b, E. M. Larsen and G. Terry.

14. **80**, 6511–6513 (1958), *Chem. Abstr.* **53**, 5937c, A. Clearfield.

15. **82**, 358–364 (1960), *Chem. Abstr.* **54**, 9590e, B. I. Intorre and A. E. Martell.

16 **82**, 2113–2115 (1960), *Chem. Abstr.* **54**, 18146a, I. E. Newnham and J. A. Watts.

17. **83**, 4293–4295 (1961), *Chem. Abstr.* **56**, 9541f, J. L. Hoard, G. L. Glen, and J. V. Silverton.

18. **88**, 5744–5746 (1966), *Chem. Abstr.* **66**, 32260p, A. E. Marcinkowsky, K. A. Kraus, H. O. Phillips, J. S. Johnson, Jr., and A. J. Shor.

18a. **88**, 5926–5927 (1966), G. W. Watt and F. O. Drummond, Jr.

19. **91** (11), 2890–2894 (1969), J. Stezowski and H. A. Eick.

61. *J. Chem. Phys.*

1. **20**, 1050–1051 (1952), *Chem. Abstr.* **46**, 10770i, B. Post and F. W. Glaser.

2. **33** (1), 194–199 (1960), *Chem. Abstr.* **55**, 53b, G. M. Muha and P. A. Vaughan.

62. *J. Chem. Soc.*

1. 1951, 280–285, *Chem. Abstr.* **45**, 7513f, D. C. Bradley and W. Wardlaw.

2. 1952, 5020–5023, *Chem. Abstr.* **47**, 6296a, D. C. Bradley, R. C. Mehrotra, and W. Wardlaw.

3. 1953, 2025–2030, *Chem. Abstr.* **47**, 12082b, D. C. Bradley, R. C. Mehrotra, J. D. Swanwick, and W. Wardlaw.

4. 1959, 422–426, *Chem. Abstr.* **53**, 9043a, R. N. Kapoor and R. C. Mehrotra.

5. 1960, 1498–1502, *Chem. Abstr.* **54**, 20852a, J. E. Drake and G. W. A. Fowles.

6. 1962, 2460–2465, *Chem. Abstr.* **57**, 4299a, J. H. Clark, J. Lewis, and R. S. Nyholm.

7. 1964 Suppl. No. 1,5523–5525, *Chem. Abstr.* **63**, 235a, K. W. Bagnall, D. Brown, and J. G. H. Du Preez.

8. A1967 (2), 258–261, *Chem. Abstr.* **66**, 70575c, R. I. H. Clark and W. Errington.

9. A1968 (6), 1435–1438, *Chem. Abstr.* **69**, 15496q, G. W. A. Fowles and G. R. Willey.

10. A1968 (8), 1940–1945, *Chem. Abstr.* **69**, 67510j, G. Chandra and M. F. Lappert.

11. A1968 (10), 2560–2564, *Chem. Abstr.* **69**, 11573h, J. E. Davies and D. A. Lung.

11a. A1969 (1), 980–984, *Chem. Abstr.* **70**, 119723y, D. C. Bradley and M. H. Gitlitz.

12. A1969 (7), 1152–1156, D. C. Bradley and M. H. Gitlitz.

63. *J. Chem. Soc. A, Inorg. Phys. Theoret.*

1. 1966 (7), 838–841, *Chem. Abstr.* **65**, 13755h, M. A. Chaudhari and F. G. A. Sone.

2. 1966 (8), 989–900, *Chem. Abstr.* **65**, 11743e, R. J. H. Clark, W. Errington, J. Lewis, and R. S. Nyholm.

64. *J. Chromatog.*

1. **30** (2), 584–592 (1967), *Chem. Abstr.* **68**, 43508k, E. Torracca, U. Constantine, and M. A. Massucci.

65. *J. Electrochem. Soc.*

1. **109**, 746–749 (1962), *Chem. Abstr.* **57**, 13524f, R. P. Hamlen.

2. **110**, 23–28 (1963), *Chem. Abstr.* **58**, 3997f, D. E. Harrison, T. Melamed, and E. C. Subbarao.

3. **111** (9), 1020–1027 (1964), *Chem. Abstr.* **61**, 7736f, T. Smith.

4. **111** (12), 1355–1357 (1964), *Chem. Abstr.* **62**, 2293c, G. W. Mellors and S. Sendoroff.

66. *J. Indian Chem. Soc.*

1. **35**, 157–160 (1958), *Chem. Abstr.* **53**, 11079i, R. N. Kapoor, K. C. Pande, and R. C. Mehrotra.

2. **40** (6), 491–492 (1963), *Chem. Abstr.* **59**, 8341f, M. Katyal and R. P. Singhd.

3. **42** (5), 301–306 (1965), *Chem. Abstr.* **63**, 14080b, M. H. Kundkar, A. S. Khan, and P. K. Das.

67. *J. Inorg. Nucl. Chem.*

1. **6,** 220–235 (1958), *Chem. Abstr.* **52**, 14280i, C. B. Amphlett, L. A. McDonald, and M. J. Redman.

2. **6,** 236–245 (1958), *Chem. Abstr.* **52**, 14281b, C. B. Amphlett, L. A. McDonald, and M. J. Redman.

3. **11,** 169–170 (1959), *Chem. Abstr.* **54**, 2070f, A. Clearfield.

4. **15,** 320–328 (1960), *Chem. Abstr.* **55**, 5077c, T. N. Waters.

5. **18,** 148–153 (1961), *Chem. Abstr.* **55**, 23146e, E. L. Muetterties, and J. E. Castle.

6. **25,** 67–74 (1963), *Chem. Abstr.* **58**, 9856g, M. Allbutt and G. W. A. Fowles.

7. **25,** 237–240 (1963), *Chem. Abstr.* **58**, 10962b, A. Clearfield and E. J. Malkiewich.

8. **26,** (11), 2038–2039 (1964), *Chem. Abstr.* **62**, 3485e, V. Amirthaligam and K. V. Muralid-havem.

9. **28** (10), 2404–2408 (1966), *Chem. Abstr.* **66**, 16139a, C. J. Hardy, B. O. Field, and D. Scargill.

10. **28** (10), 2448–2449 (1966), *Chem. Abstr.* **66**, 16138z, R. K. Bartlett.

10a. **30** (1), 277–285 (1968), *Chem. Abstr.* **68**, 62995r, A. Clearfield, G. D. Smith, and B. Hammond.

11. **30** (8), 2249–2258 (1968), *Chem. Abstr.* **69**, 90780f, A. Clearfield, R. H. Blessing, and J. A. Stynes.

68. *J. Less-Common Metals*

1. **3,** 149–154 (1961), *Chem. Abstr.* **55**, 18413b, J. E. Drake and G. W. A. Fowles.

2. **5** (6), 510–512 (1963), *Chem. Abstr.* **60**, 1324e, G. W. A. Fowles and R. A. Walton.

3. **9** (1), 60–63 (1965), *Chem. Abstr.* **64**, 2997d, F. R. Sale and R. A. J. Snelton.

4. **10** (2), 130–132 (1966), *Chem. Abstr.* **64**, 9021g, L. Sedlakova and V. Pekarek.

5. **16** (3), 288–289 (1968), *Chem. Abstr.* **70**, 16730u, R. C. Paul, S. L. Chadha, and S. K. Vasisht.

6. **17** (4), 429–436 (1969), *Chem. Abstr.* **70**, 91505c, S. Steeb and A. Reikert.

68a. *J. Org. Chem.*

1. **24,** 1371–1372 (1959), *Chem. Abstr.* **54**, 24350d, P. T. Joseph and W. B. Blumenthal.

69. *J. Organometal. Chem.*

1. **6** (4), 373–382 (1966), *Chem. Abstr.* **65**, 18459h, H. Sinn and E. Kolk.

2. **14** (1), 149–156 (1968), *Chem. Abstr.* **69**, 87148a, H. A. Martin, P. J. LeMaire, and F. Jellinek.

3. **14** (12), 1353–1358 (1968), *Chem. Abstr.* **70**, 4260c, H. Koepf.

70. *J. Pharmacol. Exptl. Therap.*

1. **94,** 1–6 (1948), *Chem. Abstr.* **43**, 312g, L. T. McClinton and J. J. Schubert.

71. *J. Phys. Chem.*

1. **62,** 1422–1426 (1958), *Chem. Abstr.* **53**, 5788f, W. R. de Monsabert and E. A. Boudreaux.

2. **65,** 1168–1172 (1961), *Chem. Abstr.* **55**, 23023c, E. Greenberg, J. L. Settle, H. M. Feder, and W. N. Hubbard.

3. **69** (10), 3488–3500 (1965), *Chem. Abstr.* **63**, 15734d, W. Weltner, Jr., and D. McCleod, Jr.

72. *J. Prakt Chem.*

1. **11** (2), 219–222 (1875), H. Endemann.

73. *J. Proc. Inst. Chemists (India)*

1. **36** (3), 144–148 (1964), *Chem. Abstr.* **61**, 12629e, G. K. Sharma.

74. *Khim. Redkikh Elementov, Akad. Nauk SSSR, Inst. Obshch. i Neorg. Khim.*

1. 1957 (3), 114–118, *Chem. Abstr.* **52**, 2659b, I. P. Alimarin and L. Z. Kozel.

75. *Kristallografiya*

1. **9** (6), 828–834 (1964), *Chem. Abstr.* **62**, 9887a, N. N. Neronova and N. V. Belov.

2. **10** (5), 591–596 (1965), *Chem. Abstr.* **63**, 17696c, Z. Skrzat and V. I. Simonov.

76. *La Ceramica (Milan)*

1. **12** (8), 45–48 (1957), *Chem. Abstr.* **51**, 18526g, A. Cocco and Nora Schromek.

77. *Melliand. Textilber.*
　　1. **41**, 1125–1129 (1960), *Chem. Abstr.* **55**, 1002g, O. Glenz.
78. *Met. i Metalloved. Chistykh Metal., Sb. Nauch. Rabot.*
　　1. 1960 (2), 128–133, *Chem. Abstr.* **55**, 16332i, Yu. F. Babikova and P. L. Gruzin.
79. *Metall.*
　　1. **12**, 6–12 (1958), *Chem. Abstr.* **52**, 5254g, H. Huschka and H. Nowatny.
80. *Mineral J. (Tokyo)*
　　1. **3** (3), 139–147, *Chem. Abstr.* **59**, 9677f, K. Masutomi, K. Nagashima, and A. Kato.
81. *Mineral Mag.*
　　1. **36** (278), 233–241 (1967), *Chem. Abstr.* **67**, 75274j, S. G. Fleet and J. R. Cann.
　　2. **36** (279), 436–437 (1967), *Chem. Abstr.* **69**, 55130x, L. Chang.
82. *Mineral Pegmatitov. Gidroterm. Shchelochnykh Massivov, Akad. Nauk SSSR, Inst. Mineral., Geokhim. Krystallokhim. Redk. Elem.*
　　1. 1967, 3–13, *Chem. Abstr.* **68**, 88938z, E. I. Seminov.
83. *Nature*
　　1. **172**, 74 (1953), *Chem. Abstr.* **47**, 10873f, R. C. Mehrotra.
　　2. **204** (4960), 775 (1964), *Chem. Abstr.* **62**, 4719e, F. Hulliger.
　　3. **204** (4962), 991 (1964), *Chem. Abstr.* **62**, 4719h, F. Hulliger.
84. *Naturwissenschaften*
　　1. **44**, 534 (1957), *Chem. Abstr.* **52**, 5109d, H. Hahn and P. Ness.
　　2. **48**, 693 (1961), *Chem. Abstr.* **56**, 9685d, P. R. Murthy and C. C. Patel.
　　3. **49**, 103 (1962), *Chem. Abstr.* **57**, 1836a, F. Jellinek and H. Hahn.
　　4. **49**, 420 (1962), *Chem. Abstr.* **58**, 572g, E. Bamann, H. Trapmann, R. Riehl, and S. Sethi.
　　5. **52** (12), 344 (1965), *Chem. Abstr.* **63**, 12433c, H. Onken.
85. *Nauchn. Tr. Mosk. Tekhnol. Inst. Legkoi Prom.*
　　1. 1956 (7), 3–11, *Chem. Abstr.* **54**, 11528c, G. A. Arbusov, P. I. Starol'skii, and Z. I. Kuznetsova.
85a. *Nippon Kagaku Zasshi*
　　1. **86** (10), 1038–1039 (1965), *Chem. Abstr.* **64**, 13743c, S. Tagagi.
86. *Ogneupory*
　　1. **26**, 581–586 (1961), *Chem. Abstr.* **56**, 9727c, E. K. Keler and A. B. Andreeva.
87. *Osaka Furitsu Kogyo-Shoreikan Hokoku*
　　1. **19**, 16–70 (1958), *Chem. Abstr.* **52**, 19664h, Y. Ogawa.
88. *Physica*
　　1. **4**, 286–301 (1924), *Chem. Abstr.* **19**, 1359[1]), A. E. van Arkel.
89. *Pochvovedenie*
　　1. 1958 (3), 1–15, *Chem. Abstr.* **52**, 17587e, V. M. Klechkovskii and I. V. Gulyakin.
90. *Proc. Chem. Soc.*
　　1. 1959, 225–226, *Chem. Abstr.* **54**, 4228a, D. C. Bradley and I. M. Thomas.
　　2. 1962 (Feb.), 76–77, *Chem. Abstr.* **61**, 1494a, B. O. Field and C. J. Hardy.
91. *Proc. N. Dakota Acad. Sci.*
　　1. **15**, 93–99 (1961), *Chem. Abstr.* **56**, 13771f, G. P. Dinga and J. E. Maurer.
92. *Proc. Natl. Inst. Sci., India*
　　1. **16**, 59–65 (1950), *Chem. Abstr.* **44**, 8278f, R. C. Mehrotra and N. R. Dhar.
　　2. **A34** (3), 181–184 (1965), *Chem. Abstr.* **63**, 1465c, S. Prasad and K. S. Devi.
93. *Radioecol. Concent. Processes, Proc. Int. Symp., Stockholm*
　　1. 1966, 735–752 (Pub. 1967), *Chem. Abstr.* **68**, 17963t, F. G. Lowman, R. A. Stevenson, R. McC. Escalera, and S. L. Ufret.
94. *Radiokhimiya*
　　1. **1**, 400–402 (1959), *Chem. Abstr.* **54**, 8217a, V. P. Shredov and N. A. Pavlova.
95. *Razdelanie Blizkikh po Svoistvam Redkikh Metal*
　　1. 1962, 51–62, *Chem. Abstr.* **58**, 2111a, V. S. Emel'yanov, A. I. Evstyukhin, I. P. Barinov, and A. M. Samonov.
96. *Rec. Trav. Chim.*
　　1. **75**, 730–736 (1956), *Chem. Abstr.* **50**, 16510b, T. V. Healy and H. A. McKay.

97. *Rep. Suid-Afrika, Dept. Mynwese, Geol. Opname Bull.*
    1. **37,** 1–65 (1961), *Chem. Abstr.* **57,** 1881a, P. J. Hugo.
98. *Rev. Chim. Minerale*
    1. **2** (1), 1–52 (1965), *Chem. Abstr.* **63,** 17245d, B. Gaudreau.
99. *Rev. Energ. Primaire*
    1. **2** (2), 5–7 (1966), *Chem. Abstr.* **68,** 92314k, C. Berger.
100. *Rentgenogr. Mineral'n. Syr'ya, Vses. Nauchn-Issled. Inst. Mineral'n Syr'ya, Akad. Nauk SSSR.*
    1. **4,** 25–38 (1964), *Chem. Abstr.* **63,** 12439a, Z. V. Pudvokina and Yu. A. Pyatenko.
100a. *Rubber World*
    1. 56–59 (January 1967), W. B. Blumenthal.
101. *S. African Pat.*
    1. **66** 04,591 (Jan. 24, 1968), *Chem. Abstr.* **70,** 50429n, (Armour Pharmaceutical Co.).
    2. **66** 04,902 (Jan. 18, 1967), *Chem. Abstr.* **70,** 59152z, J. J. Parran.
101a. *Sb. Nauchn. Rabot, Beloruss. Politekh. Inst. im. I. V. Stalina, Khim.-Tekhnol. Fak.*
    1. 1956 (55), 46–53, *Chem. Abstr.* **53,** 7539i, M. A. Bezborodov and A. I. Zelenskii.
102. *Sb. Nauchn. Tr. Permsk. Politek. Inst.*
    1. 1961 (10), 155–159, *Chem. Abstr.* **58,** 9852c, V. V. Pechkovskii, L. P. Kostin, and A. G. Zvezdin.
103. *Sb. Ref. Celostatni Radiochem. Konf., 3, Lidice, Czech.*
    1. 1964, 4–5, *Chem. Abstr.* **64,** 15023h, V. Kourim.
104. *Sci. Ceram.*
    1. **4,** 381–388 (1967), *Chem. Abstr.* **70,** 99274x, Michel Devalette, Claude Fouassier, G. LeFlem, M. Tournoux, and P. Hagenmuller.
105. *Science (Japan)*
    1. **20,** 184 (1950), *Chem. Abstr.* **45,** 10467i, Y. Yasue.
105a. *Segnetoelektriki, Rostovsk-na-Donu Gos. Univ., Nauchn.-Issled. Fiz.-Mat. Inst., Sb. Statei.*
    1. 1961, 3–6, *Chem. Abstr.* **58,** 6261f, M. L. Sholokhovitch, O. P. Kramarov, and V. I. Varicheva.
106. *Separation Science*
    1. **2** (5), 617–623 (1967), *Chem. Abstr.* **68,** 97104, J. R. Kuppers, A. E. Marcinkowsky, K. A. Kraus, and J. S. Johnson, Jr.
107. *Silicates Ind.*
    1. **34** (1), 17–20 (1969), *Chem. Abstr.* **70,** 99272v, D. Becherescu and Fr. Winter.
108. *Soap Chem. Specialities*
    1. **43** (9), 73, etc (1967), *Chem. Abstr.* **67,** 101097a, B. G. Gower, D. L. Marion, R. F. Poss, L. R. Hanson, and R. C. Strand.
109. *Soosazhdenie i Adsorbsiya Radioaktivn. Elementov., Akad. Nauk SSSR, Otd. Obshch. i. Tekh. Khim.*
    1. 1965, 140–144, *Chem. Abstr.* **63,** 9418h, Yu. D. Sinochkin and D. A. Perumov.
110. *Spectrochim. Acta*
    1. **24A** (3), 253–258 (1968), *Chem. Abstr.* **68,** 100140t, J. Weidlein, U. Mueller, and K. Dehnicke.
111. *Surgery*
    1. **26,** 682–684 (1949), *Chem. Abstr.* **47,** 10741b, S. W. Hunter, M. Miree, and H. Bloch.
112. *Talanta*
    1. **11** (8), 1197–1202 (1964), *Chem. Abstr.* **61,** 6394b, R. Pribil and V. Vesely.
113. *Textile Res. J.*
    1. **30,** 171–178 (1960), *Chem. Abstr.* **54,** 9304c, C. J. Conner, W. A. Reeves, and L. H. Chance.
    2. **33** (8), 600–608 (1963), *Chem. Abstr.* **53,** 14159a, E. J. Gonzales, C. M. Welch, and J. D. Guthrie.
    3. **36** (4), 359–367 (1966), *Chem. Abstr.* **65,** 4019e, C. J. Conner, G. S. Danna, A. S. Cooper, Jr., and W. A. Reeves.
114. *Tr. Buryatsk. Kompleksn. Nauch-Issled. Inst. Akad. Nauk SSSR, Sibirsk. Otd.*
    1. 1964 (15), 120–124, *Chem. Abstr.* **63,** 16042g, G. V. Andreev and E. M. Denezhkina.

115. *Tr. Inst. Elektrokhim. Ural'sk. Filial Akad. Nauk SSSR*
    1. 1961 (2), 71–78, *Chem. Abstr.* **59**, 2396e, L. E. Ivanovskii and O. S. Petenev.
116. *Tr. Inst. Mineralog Geokhim. i Kristallokhim. Redkikh Elementov, Akad. Nauk SSSR*
    1. 1962 (9), 88–93, *Chem. Abstr.* **58**, 8781d, E. I. Semenov and I. P. Tikhonenkov.
117. *Tr. Komissii Anal. Khim. Akad. Nauk SSSR, Inst. Geokhim. i. Anal. Khim.*
    1. **9**, 144–147 (1968), *Chem. Abstr.* **53**, 3842e, V. P. Shvedov and N. A. Pavlova.
118. *Tr. Mineralog. Muzeya, Akad. Nauk SSSR*
    1. **16**, 219–224 (1965), *Chem. Abstr.* **63**, 4023e, M. D. Dorfman, V. V. Ilyukhin, and T. A. Burovo.
119. *Tr. Tomskogo. Gos. Univ., Ser. Khim.*
    1. **185**, 161–167 (1965), *Chem. Abstr.* **66**, 43273w, M. M. Bel'kova, L. A. Alekseeno, V. V. Serebrennikov, and Yu. V. Indukaev.
120. *Tr. Ural. Politekh. Inst. Sb.*
    1. 1956 (57), 45–49, *Chem. Abstr.* **53**, 5018a, R. F. Khovyakova.
120a. *Trans. Brit. Ceram. Soc.*
    1. **58**, 532–564 (1959), *Chem. Abstr.* **54**, 9238i, F. T. Booth and G. N. Peel.
121. *Trans. Indian Ceram. Soc.*
    1. **27** (3), 111–112 (1968), *Chem. Abstr.* **70**, 108712t, N. K. Mitra, P. K. Ganguly, and R. K. Datta.
122. *U.S. Atomic Energy Comm.*
    1. *ORNL 2498*, 37 pp. (1958), *Chem. Abstr.* **52**, 15200e, J. C. White and W. J. Ross.
    2. *IDO 14455*, 12 pp. (1958), *Chem. Abstr.* **53**, 8762a, R. L. Wells.
    3. *IDO 14543*, 18 pp. (1961), *Chem. Abstr.* **55**, 23148g, A. J. Moffat and R. D. Thompson.
    4. *TID 19478*, 15 pp. (1962), *Chem. Abstr.* **61**, 13009d, J. Rynasiewicz.
    5. *ORNL P-707*, 50 pp. (1964), *Chem. Abstr.* **63**, 10747b, A. J. Shor.
123. *U.S. Bur. Standards Circ.*
    1. **500** (1952), *Chem. Abstr.* **46**, 5417f, F. D. Rassini, D. D. Wagman, W. H. Evans, S. Levine, and I. Jaffe.
124. *U.S. Dept. Comm., Off. Tech. Serv. PB. Repts.*
    1. 148,863, 7 pp. (1961), *Chem. Abstr.* **57**, 3061c, F. L. Scott, H. Q. Smith, and I. Mockrin.
    2. 171,948, 27 pp. (1960), *Chem. Abstr.* **57**, 13427a, F. W. Vahldick, C. T. Lynch, and L. R. Robinson.
    3. AD 406,473, 11 pp. (1962), *Chem. Abstr.* **60**, 6310e, M. T. Melamed and D. E. Harrison.
125. *U.S. Pat.*
    0. 1,316,107 (Sept. 16, 1919), E. J. Pugh (to Pennsalt Co.).
    0a. 2,072,889 (Mar. 9, 1937), *Chem. Abstr.* **31**, 3222^1, C. J. Kinzie and D. S. Hake.
    0b. 2,168,603 (Aug. 8, 1939), *Chem. Abstr.* **33**, 9159^6, C. J. Kinzie and D. S. Hake.
    0c. 2,270,527 (Jan. 20, 1942), *Chem. Abstr.* **36**, 3108^9, C. J. Kinzie, R. P. Easton, and V. V. Effimov.
    0d. 2,294,431 (Sept. 1, 1942), *Chem. Abstr.* **37**, 861^8, E. Wainer.
    1. 2,401,859 (June 11, 1946), *Chem. Abstr.* **40**, 4876, L. A. Clarke (to The Texas Co.).
    2. 2,402,857 (June 25, 1946), *Chem. Abstr.* **41**, 602h, H. L. van Mater.
    3. 2,412,762 (Dec. 17, 1946), *Chem. Abstr.* **41**, 1694c, A. R. Workman (to Titanium Alloy Mfg. Co.).
    3a. 2,441,447 (May 11, 1948), C. A. Seabright (to Harshaw Chemical Co.).
    4. 2,469,663 (May 10, 1949), *Chem. Abstr.* **44**, 2567h, F. Moser (to Standard Ultramarine Co.).
    5. 2,482,816 (Sept. 2, 1949), *Chem. Abstr.* **44**, 2253c, H. L. van Mater (to National Lead Co.).
    6. 2,507,128 (May 9, 1950), *Chem. Abstr.* **44**, 6585f, E. Wainer (to National Lead Co.).
    7. 2,514,115 (July 4, 1950), *Chem. Abstr.* **44**, 9127h, A. H. Angerman (to U.S. Atomic Energy Commission).
    8. 2,525,474 (Oct. 10, 1950), *Chem. Abstr.* **45**, 1312d, W. B. Blumenthal (to National Lead Co.).
    8a. 2,564,331 (Aug. 14, 1951), H. K. Hawley (to Procter & Gamble Co.).
    9. 2,600,360 (June 10, 1952), *Chem. Abstr.* **46**, 8369g, A. V. Gross (to A. V. Gross).

9a. 2,626,255 (Jan. 20, 1953), W. B. Blumenthal (to National Lead Co.).

10. 2,635,037 (April 14, 1953), *Chem. Abstr.* **47,** 7743e, H. A. Wilhelm and R. A. Walsh (to U.S. Atomic Energy Commission).

11. 2,653,107 (Sept. 22, 1953), *Chem. Abstr.* **48,** 972a, W. B. Blumenthal (to National Lead Co.).

12. 2,739,902 (Mar. 27, 1956), *Chem. Abstr.* **51,** 6184g, G. P. Mack and E. Parker (to Carlisle Chemical Works).

13. 2,739,905 (Mar. 27, 1956), *Chem. Abstr.* **51,** 6184i, G. P. Mack and E. Parker (to Carlisle Chemical Works).

14. 2,758,102 (April 7, 1956), *Chem. Abstr.* **51,** 740g, O. J. Grummitt and A. A. Arters (to Sherwin-Williams Co.).

14a. 2,763,569 (Sept. 18, 1956), *Chem. Abstr.* **51,** 1570d, S. W. Bradstreet and J. S. Griffith (to Armour Research Foundation).

15. 2,773,850 (Dec. 11, 1956), *Chem. Abstr.* **51,** 4732a, V. M. Willis (to Sherwin-Williams Co.).

16. 2,780,555 (Feb. 5, 1957), *Chem. Abstr.* **51,** 11732c, E. P. Budewitz (to Sherwin-Williams Co.).

17 2,802,847 (August 13, 1957), *Chem. Abstr.* **52,** 5008g, W. B. Blumenthal (to National Lead Co.).

18. 2,805,121 (Sept. 3, 1957), *Chem. Abstr.* **52,** 3205h, C. Woolf (to Allied Chemical and Dye Corp.).

19. 2,814,584 (Nov. 26, 1957), *Chem. Abstr.* **52,** 5759e, E. W. Daley (to Procter and Gamble).

20. 2,814,988 (Dec. 3, 1957), *Chem. Abstr.* **52,** 4072b, W. Bradstreet and J. S. Griffith (to Armour Research Foundation).

21. 2,826,477 (Mar. 11, 1958), *Chem. Abstr.* **52,** 12439g, W. J. Rau and I. C. Somerville (to Rohm and Haas).

22. 2,837,400 (June 3, 1958), *Chem. Abstr.* **53,** 1656f, W. B. Blumenthal (to National Lead Co.).

23. 2,842,451 (July 8, 1958), *Chem. Abstr.* **52,** 21164d, O. J. Grummitt and A. A. Arters (to Sherwin-Williams Co.).

24. 2,846,426 (Aug. 5, 1958), *Chem. Abstr.* **52,** 21243f, W. E. Larson and R. E. Edmonson (to Dow Chemical Co.).

25. 2,854,382 (Sept. 30, 1958, *Chem. Abstr.* **53,** 2544i, M. Grad (to Procter and Gamble Co.).

26. 2,864,842 (Dec. 16, 1958), *Chem. Abstr.* **53,** 7014d, H. A. Walter (to Monsanto Co.).

27. 2,882,161 (April 14, 1959), *Chem. Abstr.* **53,** 12077c, J. R. Dann, J. W. Gates, Jr., D. A. Smith, and C. C. Unruh (to Eastman Kodak Co.).

28. 2,883,289 (April 21, 1959), *Chem. Abstr.* **53,** 11186h, M. J. Furey and E. O. Forster (to Esso Research & Engineering Co.).

29. 2,884,393 (April 28, 1959), *Chem. Abstr.* **53,** 16572b, J. W. Gilkey (to Dow Corning Corp.).

30. 2,962,061 (Feb. 23, 1960), *Chem. Abstr.* **54,** 15764i, V. Z. Pasternak and R. M. Lollar (to U.S. Dept. of the Army).

31. 2,930,735 (Mar. 29, 1960), *Chem. Abstr.* **54,** 15854h, B. Vogel (to B. Vogel).

32. 2,941,931 (June 21, 1960), *Chem. Abstr.* **54,** 19234c, R. S. Dean (to Chicago Development Corp.).

33. 2,942,975 (June 28, 1960), *Chem. Abstr.* **54,** 18144b, J. D. Eerde (to Keuffel & Esser Co.).

34. 2,946,683 (July 26, 1960), *Chem. Abstr.* **54,** 20598e, I. Mellan and R. Gumbinner (to Polychrome Corp.).

35. 2,956,856 (Oct. 18, 1960), *Chem. Abstr.* **55,** 4840h, S. F. Urban (to National Lead Co.).

36. 2,985,607 (May 23, 1961), *Chem. Abstr.* **55,** 24073g, J. O. Koehler and H. Lamprey (to Union Carbide Corp.).

36a. 2,992,123 (July 11, 1961), *Chem. Abstr.* **55,** 24046d, C. A. Seabright.

37. 3,065,091 (Nov. 20, 1962), *Chem. Abstr.* **58,** 1973c, R. G. Russell, W. L. Morgan, and L. F. Schleffer (to Owens-Corning Fiberglass Corp.).

38. 3,082,051 (Mar. 19, 1963), *Chem. Abstr.* **58,** 12279c, E. Wainer and R. M. Beasley (to Horizons Inc.).

39. 3,087,949 (April 30, 1963), *Chem. Abstr.* **60,** 2769c, J. Rinse (to J. Rinse).

40. 3,096,153 (July 2, 1963), *Chem. Abstr.* **59,** 6628g, R. C. Horrigan (to National Lead Co.).

41. 3,098,861 (July 23, 1963), *Chem. Abstr.* **60,** 1590b, R. W. Russell (to Stauffer Chemical Co.).

42. 3,114,600 (Dec. 17, 1963), *Chem. Abstr.* **60,** 6521q, M. A. Hobin and R. A. Foos (to National Distillers and Chemical Corp.).

43. 3,133,873 (Aug. 19, 1964), W. L. Miller and S. Tudor (to W. L. Miller and S. Tudor).

44. 3,148,998 (Sept. 15, 1964), *Chem. Abstr.* **61,** 12979d, A. Clearfield (to National Lead Co.).

45. 3,201,353 (Aug. 7, 1965), *Chem. Abstr.* **63,** 13578d, L. Corben (to American Agricultural Chemical Co.).

46. 3,205,053 (Sept. 7, 1965), *Chem. Abstr.* **63,** 12953f, A. T. McCord (to Carborundum Co.).

46a. 3,203,812 (Aug. 3, 1965), H. G. Emblem and A. K. Harrison (to Unilever Ltd.).

47. 3,208,910 (Sept. 28, 1965), *Chem. Abstr.* **63,** 16125b, H. E. Cassidy (to H. E. Casaidy).

48. 3,210,289 (Oct. 5, 1965), *Chem. Abstr.* **63,** 17296b, E. C. Subbarao and D. E. Harrison (to Westinghouse Elec. Corp.).

49. 3,242,028 (Mar. 22, 1966), *Chem. Abstr.* **64,** 19999d, R. T. Hart (to Oxford Paper Co.).

50. 3,257,282 (June 21, 1966), *Chem. Abstr.* **65,** 5310a, J. C. Muhler (to Indiana University Foundation).

51. 3,259,545 (July 5, 1966), *Chem. Abstr.* **65,** 8664e, W. K. Teller (to Wallace & Tiernan Co.).

51a. 3,270,109 (Aug. 30, 1966), *Chem. Abstr.* **65,** 16656d, R. Kelsey (to Horizons, Inc.).

52. 3,297,573 (Jan. 10, 1967), *Chem. Abstr.* **66,** 48100x, G. H. Hain and A. J. Revukas (to Cities Service Oil Co.).

53. 3,320,196 (May 16, 1967), *Chem. Abstr.* **67,** 22976h, J. R. Rogers (to S. C. Johnson & Co.).

54. 3,329,480 (July 4, 1967), *Chem. Abstr.* **67,** 55730b, D. A. Young (to Union Oil Co. of California).

55. 3,332,737 (July 15, 1967), *Chem. Abstr.* **67,** 85337, K. A. Kraus (to K. A. Kraus).

56. 3,361,775 (Jan. 2, 1968), *Chem. Abstr.* **68,** 50618g, A. J. Gibbons, Jr., and R. E. DeMarco (to SCM Corp.).

57. 3,395,976 (Aug. 6, 1968), *Chem. Abstr.* **70,** 5581v, O. Glemser and A. von Baeckmann (to O. Glemser and A. von Baeckmann).

58. 3,409,584 (Nov. 5, 1968), *Chem. Abstr.* **70,** 20769f, F. X. Buschman, J. E. Dillon, and J. E. Sloat (to Markel, L. Frank & Sons).

59. 3,416,884 (Dec. 17, 1968), *Chem. Abstr.* **70,** 49121z, J. A. Stynes and A. Clearfield (to National Lead Co.).

60. 3,418,073 (Dec. 24, 1968), W. B. Blumenthal (to National Lead Co.).

60a. 3,418,251 (Dec. 24, 1968), J. A. Stynes (to National Lead Co.).

61. 3,423,193 (Jan. 21, 1969), *Chem. Abstr.* **70,** 49122a, J. A. Stynes (to National Lead Co.).

62. 3,441,371 (April 29, 1969), J. C. Muhler, (to Indiana University Foundation).

125a. *U.S.S.R. Pat.*

1. 133,873 (Dec. 10, 1960), *Chem. Abstr.* **55,** 14312h, R. Kh. Freidlina, E. M. Brainina, and A. N. Nesmeyanov.

126. *Uchenye Zap. Belorussk. Gos. Univ. Ser. Khim.*

1. 1958 (42), 271–279 *Chem. Abstr.* **53,** 21338b, N. F. Ermolenko and G. I. Vasil'eva.

127. *Uchenye Zap., Tomsk. Gos. Univ.*

1. 1959, 37–41, *Chem. Abstr.* **56,** 8039d, L. G. Maidanovskaya, E. F. Filatova, and M. V. Kalinova.

128. *Ukr. Khim. Zh.*

1. **23,** 287–296 (1957), *Chem. Abstr.* **52,** 1826f, G. V. Samsonov.

2. **27,** 290–295 (1961), *Chem. Abstr.* **56,** 3105g, A. K. Babco and N. V. Ul'ko.

3. **28,** 565–570 (1962), *Chem. Abstr.* **58,** 3960c, Yu. K. Delimarskii and G. G. Buderskaya.

4. **34** (12), 1215–1221 (1968), *Chem. Abstr.* **70,** 118691z, I. V. Pyatnitskii and T. I. Kravchenko.

129. *Univ. Microfilms (Order No.)*

1. 68–6014, 138 pp. (1968), *Chem. Abstr.* **69,** 49017z, R. E. Walsh, Jr. Oregon State Univ.

130. *University Theses*
    1. "Extraction of Zirconium and Hafnium with Various Diketones," Univ. of Wisconsin Press 1952, E. M. Larsen and G. Terry.
    2. "Sub-solidus Phase Equilibria and Thermal Expansion Properties of the Compounds in the System ZrO_2-WO_3-P_2O_5," Pennsylvania State Univ. 1968, C. A. Martinek.

131. *Vacuum Microbalance Tech.*
    1. **4,** 141–157 (1964), *Chem. Abstr.* **63,** 17124b, W. C. Tripp, R. W. Vest, and N. M. Tallan.

132. *Vestn. Mosk. Univ. Ser. II, Khim.*
    1. **16** (4), 49–54 (1961), *Chem. Abstr.* **56,** 10908a, I. P. Alimarin and V. I. Fadeeva.
    2. **23** (2), 131–134 (1968), *Chem. Abstr.* **69,** 70097k, T. A. Belyarskaya and G. D. Dobryshina.
    3. **23** (1), 113–114 (1968), *Chem. Abstr.* **69,** 71358h, V. P. Spiridinov.

133. *Visn. Kiivs'k Univ., Ser. Astron., Fiz., ta Khim.*
    1. **5** (1), 118–119 (1962), *Chem. Abstr.* **60,** 7458f, F. G. Zharovskii, E. A. Shpak, and E. V. Piskunova.
    2. **5** (1), 132–135 (1962), *Chem. Abstr.* **60,** 12879g, V. V. Grigor'eva and M. M. Kalibabchuk.

134. *Vyskotemperaturnye Metallokeram. Materialy Akad. Nauk Ukr. SSR, Inst. Metallokeram. i Spets. Splavov.*
    1. 1962, 87–95, *Chem. Abstr.* **58,** 6300d, L. D. Dudkin.

135. *Wiss. Tech. Tagung Deut. Atomforums, 2nd Munich*
    1. 1963, 252–258 (Pub. 1964), *Chem. Abstr.* **65,** 14474c, H. J. Riedel.

136. *Z. Anal. Chem.*
    1. **156,** 258–265 (1957), *Chem. Abstr.* **51,** 16202d, G. B. Larrabee and R. P. Graham.

137. *Z. Anorg. Allgem. Chem.*
    1. **128,** 96–116 (1923), *Chem. Abstr.* **17,** 2842, O. Ruff and R. Wallstein.
    2. **209,** 335–336 (1932), *Chem. Abstr.* **27,** 1290, H. Trapp.
    3. **239,** 216–224 (1938), E. F. Strotzer, W. Biltz, and K. Meisel.
    4. **252,** 323 (1944), *Chem. Abstr.* **40,** 4311⁵, H. Funk and E. Rogler.
    5. **293,** 92–99 (1957), *Chem. Abstr.* **52,** 9993a, R. N. Kapoor and R. C. Mehrotra.
    5b. **288,** 269–278 (1956), *Chem. Abstr.* **51,** 7210c, H. Hahn and Ursula Mutschke.
    6. **292,** 82–96 (1957), *Chem. Abstr.* **52,** 5921d, H. Hahn, B. Harder, Ursula Mutschke, and P. Ness.
    7. **302,** 37–49 (1959), *Chem. Abstr.* **54,** 8404f, H. Hahn and P. Ness.
    8. **316,** 15–24 (1962), *Chem. Abstr.* **57,** 10760e, H. L. Schlaefer and H. Skoludek.
    9. **327,** (3–4), 253–259 (1964), *Chem. Abstr.* **61,** 1487d, H. L. Schlaefer and H. W. Willie.
    10. **331,** (3–4), 121–128 (1964), *Chem. Abstr.* **61,** 11593f, K. Dehnicke and K. U. Meyer.
    11. **332,** (1–2), 1–4 (1964), *Chem. Abstr.* **62,** 6116h, R. Juza, A. Rabenau, and I. Mitschke.
    12. **332** (3–4), 159–172 (1964), *Chem. Abstr.* **62,** 6117a, R. Juza and J. Heners.
    13. **333** (4–6), 209–214 (1964), *Chem. Abstr.* **62,** 8639h, P. Ehrlich, F. Ploeger, and E. Koch.
    14. **346** (1–2), 92–112 (1966), *Chem. Abstr.* **65,** 14807c, A. Winkler and E. Thilo.
    15. **347** (5–6), 289–293 (1966), *Chem. Abstr.* **66,** 32680a, W. Kleber and D. Doerschel.

138. *Z. Chem.*
    1. **6** (6), 227 (1966), *Chem. Abstr.* **65,** 14819e, H. Funk and M. Hesselbarth.
    2. **6** (9), 347–348 (1966), *Chem. Abstr.* **66,** 16078e, L. Kolditz and P. Degenkolb.
    3. **6** (10), 382–383 (1966), *Chem. Abstr.* **66,** 34423t, S. Herzog and H. Zuehlke.
    4. **7** (9), 352–353 (1967), *Chem. Abstr.* **67,** 104732q, P. Muehl.

139. *Z. Erzbergbau Metallhuetenw.*
    1. **19** (5), 232–235 (1966), *Chem. Abstr.* **65,** 5233b, H. Behmenberg, H. Borchert, and A. Anger.

140. *Z. Krist.*
    1. **67,** 295–328 (1928), *Chem. Abstr.* **22,** 2861, H. Seyfarth.
    2. **121** (5), 349–368 (1965), *Chem. Abstr.* **63,** 5054b, S. G. Fleet.

141. *Z. Naturforsch.*
    1. **B15,** 466 (1960), *Chem. Abstr.* **55,** 2338g, S. Herzog and H. Zuehlke.
    2. **B22** (11), 1100–1102 (1967), *Chem. Abstr.* **68,** 63000z, E. Michel and A. Weiss.

142. *Zh. Analit. Khim.*
    1. **13,** 332–336 (1958), *Chem. Abstr.* **53,** 130d, I. P. Alimarin and V. S. Sotnikov.

143. *Zh. Fiz. Khim.*

1. **28**, 1999–2005 (1954), *Chem. Abstr.* **50**, 4609d, A. Yu. Prokopchik and I. V. Yanitskii.
2. **38** (1), 146–150 (1964), *Chem. Abstr.* **60**, 10039g, L. N. Sidorov, P. A. Akishin, V. I. Belousov, and V. B. Shol'ts.
3. **41** (11), 2958–2959 (1967), *Chem. Abstr.* **68**, 63392d, Yu. M. Polezhaev.

144. *Zh. Neorgan. Khim.*

1. **1** (1). 69–75 (1956), *Chem. Abstr.* **50**, 13642f, O. E. Svyagintsev, B. N. Sudarikov, and Z. S. Barsukova.
2. **3**, 1273–1280 (1958), *Chem. Abstr.* **53**, 18712b, I. V. Tananaev and M. Ya. Bokmel'der.
3. **4**, 367–371 (1959), *Chem. Abstr.* **53**, 16790g, L. N. Sheronov and B. V. Ptitsyn.
4. **6**, 1319–1325 (1961), *Chem. Abstr.* **56**, 1123g, B. I. Nabivanets.
5. **5**, 1413–1417 (1960), *Chem. Abstr.* **55**, 2332h, L. N. Komissarova, Yu. P. Simanov, and Z. A. Vladimirova.
6. **6**, 330–333 (1961), *Chem. Abstr.* **56**, 3099e, Z. F. Shakhova, E. N. Semenovskaya, and E. N. Timofeeva.
7. **6**, 489–492 (1961), *Chem. Abstr.* **56**, 5455c, Z. N. Tsevkova, A. S. Solovkin, N. S. Povitskii, and I. P. Davydov.
8. **6**, 534–538 (1961), *Chem. Abstr.* **56**, 5453b, P. N. Kovalenko and K. N. Bagdasarov.
9. **6**, 1332–1337 (1961), *Chem. Abstr.* **56**, 1121i, V. G. Chukhlantsev and A. K. Shtol'ts.
10. **6**, 1338–1341 (1961), *Chem. Abstr.* **56**, 6887h, A. A. Vinogradov, and V. S. Shpinel.
11. **7**, 1552–1558 (1962), *Chem. Abstr.* **57**, 9443c, L. M. Zaitsev and G. S. Buchkarev.
12. **7**, 1854–1859 (1962), *Chem. Abstr.* **57**, 16116h, I. V. Tananaev and I. A. Rozanov.
13. **8** (1), 56–62 (1963), *Chem. Abstr.* **58**, 10958b, L. M. Komissarova, M. V. Savel'eva, and V. E. Plyushchev.
14. **8** (11), 2483–2489 (1963), *Chem. Abstr.* **60**, 3535g, T. F. Limar and K. P. Shatskaya.
15. **8** (12), 2821–2822 (1963), *Chem. Abstr.* **60**, 4862c, V. Tananaev and T. N. Kuz'mina.
16. **9** (2), 451–455 (1964), *Chem. Abstr.* **60**, 15202d, V. I. Plotnikov.
17. **9** (5), 1123–1128 (1964), *Chem. Abstr.* **61**, 2708f, Yu. M. Polezhaev and V. G. Chukhlantsev.
18. **9** (6), 1358–1362 (1964), *Chem. Abstr.* **61**, 5188c, V. G. Chukhlantsev and Yu. M. Polezhaev.
19. **9** (7), 1617–1623 (1964), *Chem. Abstr.* **61**, 11598c, Yu. Ya. Kharitonov, L. M. Zaitsev, G. S. Bochkarev, and O. N. Evstaf'eva.
20. **10** (1), 115–120 (1965), *Chem. Abstr.* **62**, 7354c, T. F. Limar and K. P. Shatskaya.
22. **10** (3), 643–646 (1965), *Chem. Abstr.* **62**, 14155h, V. E. Plyushchev, L. I. Yuranova, and L. N. Komissarova.
23. **10** (5), 1088–1096 (1965), *Chem. Abstr.* **63**, 3868a, L. M. Zaitsev, G. S. Bochkarev, and V. N. Kozhenkova.
24. **10** (7), 1581–1584 (1965), *Chem. Abstr.* **63**, 12644f, L. M. Zaitsev.
25. **10** (7), 1585–1587 (1965), *Chem. Abstr.* **63**, 12670a, V. G. Chukhlantsev and Yu. M. Polezhaev.
26. **10** (10), 254–261 (1965), *Chem. Abstr.* **63**, 17455c, G. M. Toptygina and L. B. Barskaya.
27. **10** (10), 2215–2219 (1965), *Chem. Abstr.* **63**, 17454c, I. G. Atanov and L. M. Zaitsev.
28. **10** (12), 2764–2773 (1965), *Chem. Abstr.* **64**, 7423e, M. Kyrs, P. Selucky, and P. Pishtek.
29. **11** (2), 266–271 (1966), *Chem. Abstr.* **64**, 15350g, L. N. Komissarova, S. V. Krivenko, Z. N. Prozorovskaya, and V. E. Plyushchev.
30. **11** (3), 464–467 (1966), *Chem. Abstr.* **64**, 15344g, I. N. Belyaev and S. A. Artamonova.
31. **11** (4), 770–774 (1966), *Chem. Abstr.* **65**, 1463a, A. M. Golub and V. N. Sergun'kin.
32. **11** (7), 1684–1692 (1966), *Chem. Abstr.* **65**, 13180g, L. M. Zaitsev.
33. **11** (8), 1798–1806 (1966), *Chem. Abstr.* **65**, 16740c, G. Bochkarev, I. G. Atanov, and L. M. Zaitsev.
34. **11** (8), 1863–1879 (1966), *Chem. Abstr.* **65**, 19659c, L. A. Pospelova and L. M. Zaitsev.
35. **11** (10), 2185–2188 (1966), *Chem. Abstr.* **66**, 14570k, N. D. Denisova, E. K. Safronov and O. N. Bystrova.
36. **12** (2), 302–306 (1967), *Chem. Abstr.* **66**, 108483m, Ya. G. Goroshchenko and T. P. Spasibenko.

37. **12** (2), 363–376 (1967), *Chem. Abstr.* **66,** 101156y, I. G. Atanov and L. M. Zaitsev.

38. **12** (6), 1492–1499 (1967), *Chem. Abstr.* **67,** 96375h, V. T. Spitsyn, L. N. Komissarova, Z. N. Prozorovskaya, and V. F. Chuvaev.

39. **12** (10), 2553–2558 (1967), *Chem. Abstr.* **68,** 35411r, Z. N. Prozorovskaya, L. N. Komissarova, and V. I. Spitsyn.

40. **12** (10), 2635–2644 (1967), *Chem. Abstr.* **68,** 44405u, Yu. Ya. Kharitonov, L. Pospelova, and L. M. Zaitsev.

41. **12** (10), 2725–2728 (1967), *Chem. Abstr.* **68,** 44411t, A. N. Ermakov, I. N. Marov, and L. P. Kazanskii.

42. **13** (1), 102–105 (1968), *Chem. Abstr.* **68,** 99498g, Yu. E. Gorbunova, V. G. Kuznetsov, and E. S. Kovaleva.

43. **13** (2), 400–405 (1968), *Chem. Abstr.* **68,** 92545m, I. G. Atanov, L. M. Zaitsev, and T. N. Shubina.

44. **13** (3), 706–711 (1968), *Chem. Abstr.* **68,** 118941q, Z. N. Prozorovskaya, L. N. Komissarova, and V. I. Spitsyn.

45. **13** (4), 956–964 (1968), *Chem. Abstr.* **68,** 118896d, V. E. Plyushchev, L. I. Yuranova, L. N. Komissarova, and V. K. Trunov.

46. **13** (11), 2974–2979 (1968), *Chem. Abstr.* **70,** 33870g, I. A. Sheka, A. A. Lastochkina, and A. A. Malinko.

47. **13** (12), 3170–3171 (1968), *Chem. Abstr.* **70,** 53483m, G. P. Novoselov and O. A. Ustinov.

48. **14** (2), 412–418 (1969), *Chem. Abstr.* **70,** 83776b, L. A. Malinko, K. F. Karlysheva, and I. A. Sheka.

49. **14** (2), 593–594 (1969), *Chem. Abstr.* **70,** 81464z, V. V. Volkov, K. G. Myakishev and G. I. Bagryantsev.

145. *Zh. Obshch. Khim.*

    1. **31,** 2451–2456 (1961), *Chem. Abstr.* **56,** 5623a, O. A. Osipov and Yu. B. Kletenik.

    2. **37** (2), 312–317 (1967), *Chem. Abstr.* **67,** 28852m, V. T. Panyushkin, A. D. Garnovskii, O. A. Osipov, A. L. Sinyavin, and A. F. Pozharskii.

146. *Zh. Prikl. Khim.*

    0. **40** (1), 7–11 (1967), *Chem. Abstr.* **66,** 101157z, G. N. Dubrovskaya and G. V. Samsonov.

    1. **40** (6), 1279–1283 (1967), *Chem. Abstr.* **68,** 16415d, V. B. Aleskovskii, T. I. Kalinina, and N. A. Tyutina.

147. *Zh. Strukt. Khim.*

    1. **9** (5), 918–919 (1968), *Chem. Abstr.* **70,** 32381m, Yu. E. Gorbunova, G. Kuznetsov, and S. E. Kovaleva.

    2. **10** (1), 133–135 (1969), *Chem. Abstr.* **70,** 100656g, V. P. Spiridonov and G. I. Mamaeva.

148. *Zh. Vses. Khim. Obshchestva. im. D. I. Mendeleeva*

    1. **8** (2), 227–229 (1963), *Chem. Abstr.* **59,** 4753b, A. K. Lavrukina, V. V. Malyshev, and S. S. Rodin.

149. *Zap. Vses. Mineralog. Obschestva.*

    1. **87,** 590–597 (1958), *Chem. Abstr.* **53,** 997c, E. I. Semenov, M. E. Kazakova, and V. I. Simonov.

150. *Zavodsk. Lab.*

    1. **26,** 927–929 (1960), *Chem. Abstr.* **54,** 24126f, R. M. Veitsman.

WARREN B. BLUMENTHAL
National Lead Company

ZMBT. See Rubber chemicals.

ZONE REFINING

Zone refining is a modification of the classical purification technique of fractional crystallization. To effect purification, fractional crystallization utilizes the difference in solubility of a solute in the solid and liquid phases of the solvent. The difference between zone refining and fractional crystallization lies in the mode of melting. In zone refining, only a part of the material is melted at any one time; a short length of molten zone is passed through a relatively long solid charge to bring about purification. The classic application of the technique to germanium by Pfann (1) was made as recently as 1952. Since that time numerous theoretical and experimental zone-refining studies have been reported. The technique has been applied to other semiconductors, to metals, alloys, intermetallic compounds, and to organic and inorganic chemical compounds.

Although purity is usually thought of in terms of foreign atoms, it is now known that line defects (dislocations), point defects (vacancies, interstitials), and interior surface defects (grain or twin boundaries) affect the physical and mechanical properties of materials. This is particularly true in high-purity materials. The structural integrity of a material is directly related to the mode of crystal growth. In practice, many of the methods of crystal growing do in fact constitute zone-refining conditions; under these circumstances, even though emphasis is placed on zone refining, the aspect of crystalline perfection is inevitably taken into account.

In general, zone refining is a costly operation in terms of both time and money. It is therefore appropriate to question the need for high-purity materials. The need for high purity in semiconductor materials for the transistor industry is immediately apparent; only advanced techniques, such as zone refining, can bring about the required degree of purification and perfection. In atomic-reactor technology, the lifespan of a number of metals is critically dependent on the impurity content and the nature of the impurity. In attempting to understand the role of alloying additions on mechanical behavior, and thereby provide a scientific basis for alloy development, it is essential that both the solvent metal and solute additions each be of high purity. From a scientific standpoint, the quest for ultrahigh purity is readily justified on the basis that only in this way can the intrinsic properties of the material be determined. The need for high-purity materials is particularly crucial in studies of electronic structure involving phenomena such as cyclotron resonance and magnetoresistance. In the alkali halides ionic conductivity, fluorescence, thermal conductivity, and radiation damage are sensitive functions of the purity content.

A number of excellent reviews are available covering the principles, theory, and techniques of zone refining (2–7) and also the intimately associated aspects of crystal growth (8,9).

Principles and Theory

The Normal Mode of Freezing. In the normal mode of freezing the alloy (binary) is allowed to freeze from one end (see Fig. 1a). Solute redistribution takes place because of a difference in the concentration of solute in the liquid phase from that in the solid phase at equilibrium. This difference is controlled by the partition coefficient k, defined as the ratio of the solute concentration in the frozen solid, C_s, to that in the bulk liquid, C_L. Three values of k are possible: (1) the equilibrium value $k_0 = C_s/C_L$, taken from the equilibrium diagram; (2) unity; and (3) the effective value k which lies between k_0 and unity. Solutes having $k_0 < 1$ lower the melting point, and the solute is concentrated in the last regions to freeze (Fig. 2a). For $k_0 > 1$, the solute raises the melting point, and concentration of solute occurs in the first regions to freeze (Fig. 2b).

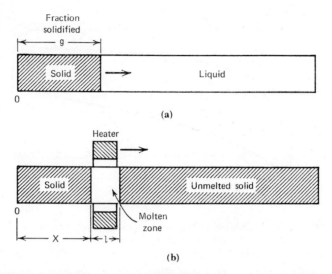

Fig. 1. (a) Solidification by normal freezing; (b) melting and freezing conditions in zone refining.

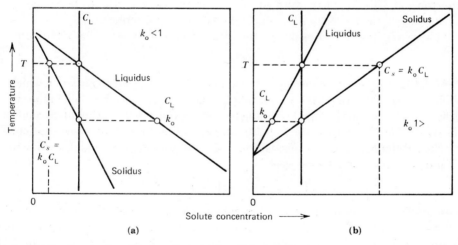

Fig. 2. Idealized parts of equilibrium diagrams in which the solute (a) lowers and (b) raises the melting point.

Under normal rates of freezing equilibrium conditions are not attained, and the effective partition coefficient k is operative; for $k < 1$ a solute-enriched layer exists in the liquid at the liquid–solid interface, whereas for $k > 1$ there is a corresponding layer depleted in solute. The relation between k and k_0, as postulated by Burton et al. (10), is given by

$$k = k_0/[k_0 + (1 - k_0)e^{-f\delta/D}] \tag{1}$$

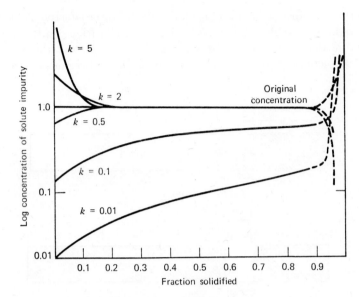

Fig. 3. Concentration and distribution of impurity after one zone pass for various values of the distribution coefficient k.

where f is the freezing velocity, D is the solute diffusion coefficient in the liquid, and δ is the thickness of the enriched or depleted solute layer at the solid–liquid interface. 8ohnston and Tiller (11) have verified this relationship in the case of Pb–Sn alloys subjected to magnetic stirring in the plane of the solid-liquid interface.

The segregation occurring during the normal freezing process (3) is of the form

$$C_s = kC_0(1 - g)^{k-1} \tag{2}$$

in which C_0 is the initial solute concentration in the melt and g is the fraction which has solidified (Fig. 1a). In deriving equation 2 it is assumed that diffusion in the solid is negligible, diffusion in the liquid is complete, and k is constant. The first two conditions must be fulfilled in order to use k_0 for k, corresponding to maximum segregation.

Zone Refining. The melting and freezing conditions in zone refining are illustrated in Figure 1b. For a single zone pass,

$$C_s/C_0 = 1 - (1 - k)\, e^{-kx/l} \tag{3}$$

where x is the distance from the starting end and l is the zone length. The assumptions are those of normal freezing and a constant value of l. Curves showing the solute distribution in the solid, following one zone pass, are shown in Figure 3. Curves for various values of the distribution coefficient k are shown in Figure 4.

Considering the case in which $k < 1$, the passage of a second molten zone in the same direction as the first pass leads to a further decrease in solute concentration at the starting end of the material. Solute enrichment progresses backward one zone length for each succeeding pass. Multiple passes lower the concentration in the initial region, raise the concentration in the end region, and decrease the length of the intermediate region (see Fig. 4). The three regions merge into a smooth curve after a sufficient number of zone passes. Analytical solutions for describing the impurity distribu-

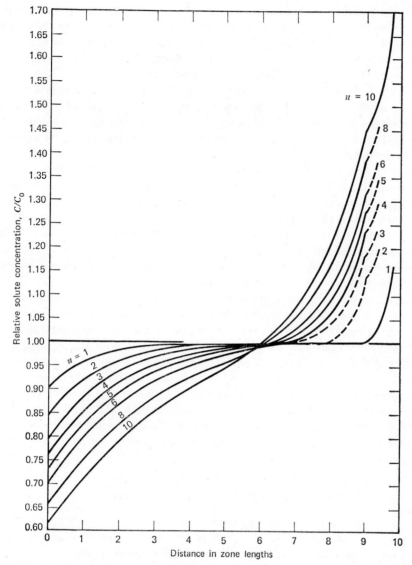

Fig. 4. Relative solute-concentration distance profiles for various numbers of passes.

tion after n zone passes have been given by Lord (12), Reiss (13), Reiss and Helfand (14), and Gold and Johnson (15). In the limit ($n = \infty$), the forward solute flux is equal to the backward solute flux at all points in the material, and

$$C_s = A e^{Bx} \tag{4}$$

where $k = Bl/(e^{Bl} - 1)$ and $A = C_0 BL/(e^{BL} - 1)$; L is the ingot length and l the zone length. In a typical example, $L = 10$, $l = 1$, $k = 0.1$, and $C_s(n = \infty)$ at $x = 0$ is $\sim 10^{-14} C_0$.

In many instances, the solute impurity is volatile, so that there is an exchange of solute between the liquid and gaseous phases. This being the case, the solute concentration in the liquid (C_L) is a function of g (eq. 2). Modified forms of equation 2 have been solved using iterative methods to determine concentration profiles after one and two passes of unidirectional zone melting in a semi-infinite bar of uniform initial concentration (16–18). Taking account of evaporation of the solute, good agreement exists between theory and experiment for the distribution of P, As, and Sb in Ge (16) and Mn, C, S, and O in Fe (17). Boomgaard (19) has also treated the case of zone melting with a volatile solute present. The theory indicates that it is possible to distribute a volatile impurity homogeneously in an ingot by zone melting under a constant vapor pressure of the solute; experimental results on the distribution of Se in Ge are in agreement with the theory.

Zone leveling is a modification of the zone refining technique in which a number of zone passes are made in alternate directions, making sure that the end zone containing the high impurity content is excluded. As the term implies, zone leveling brings about a homogenization of the remaining impurities in a zone-refined bar. The degree of homogenization achievable can be calculated from the known values of k, L, and l. Zone refining coupled with zone leveling can be used to bring about wide variations of solute concentration in a bar of solvent. Solute elements can be introduced deliberately or dispersed to give specific impurity gradients.

Although simple solid solubility systems have been considered here, the principles of zone refining apply equally well to eutectic or peritectic reactions in which compounds exist in equilibrium with other compounds, solid solutions, or elements. Details of these systems have been discussed by Pfann (3) and by Goodman (20).

Field-Aided Zone Refining. If an electric field is applied to a conducting liquid solution, changes in concentration usually occur because of differences in ionic mobility. Applications of this effect in conjunction with the processes of freezing and zone refining have been considered by Angus et al. (21), Pfann and Wagner (22); Hay and Scala (23,24), and Verhoeven (25,26). In the case of a direct current passing through a solid–liquid interface, the relationship (10) between k and k_0 (eq. 1) is modified to the form

$$k = \frac{1 + f'/f}{1 + [1/k_0(1 + f'/f) - 1] \exp[(-f\delta/D)(1 + f'/f)]} \tag{5}$$

where f' is the terminal velocity of solute away from the interface produced by the field of the imposed direct current. If B is the solute mobility, e is the electronic charge, E the applied electric field, and Z^* the effective valence, then

$$f = BeEZ^* \tag{6}$$

Examination of equation 5 shows that the value of k is largely determined by two competing solute fluxes. The one flux arises from the rejection of solute at the advancing solid-liquid interface and is proportional to f, $(1 - k_0)$, and concentration. The other flux arises from a difference in mobilities between solute and solvent ions in the electric field at the diffusion layer; this flux is proportional to f' and concentration. The flux due to freezing is either away from the solid for $k_0 < 1$, or toward it for $k_0 > 1$. However, the field flux can be set in either direction. If f' is positive (solute flux toward interface), the field increases k, whereas for f' negative k is decreased.

The calculations of Pfann and Wagner (22) make it clear that k can be varied over a wide range, including values outside the usual range between k_0 and unity. Control of the distribution coefficient by the application of an applied field is extremely useful. This allows for an overall increase in the efficiency of the process, and since a normal k_0 value of unity can be changed to $k_0 \gtrless 1$, previously inseparable solutes may be transferred along the charge. Also, if some of the solutes present raise the melting point while others lower the melting point, the individual k values can be adjusted in order to concentrate all solutes at the same end of the ingot. Experimental results on field-aided zone refining of tungsten (23,24) and field-freezing of Bi–Sn (27) confirm the value of electrotransport as a means of enhancing purification. The general usefulness of the technique does depend on the degree to which convection within the liquid can be controlled (22,26).

Zone-Refining Techniques

The Zone-Refining Parameters. In designing zone-refining apparatus, the objective is to optimize the controlling parameters in order that the required amount of material of specified purity be obtained with a minimum of time and expense. The parameters that have to be taken into account are the following: zone length l, interzone spacing i, number of zone passes n, travel rate f, and diffusion-layer thickness δ. A small value of n reduces the time, but at the expense of the usable ingot length. A small zone length l leads to better separation, particularly at large n values. Reducing l and i lowers the time per pass. Increasing f decreases the time per pass, but there is a concomitant increase in k so that the efficiency per pass is reduced.

The assumption that solid-state diffusion is negligible in comparison to liquid diffusion usually holds true. Within the liquid zone, however, a concentration gradient is common; rejected solute atoms entering the molten zone at the interface must be continuously transported into the liquid by diffusion, convection, or by a stirring action. This means that the rate of solidification has to be much more rapid than the rate of solid diffusion, but not so fast as to prevent reasonable diffusion of the solute impurity into the molten zone. Enhancement of liquid diffusion (eg stirring, agitation) allows for a significant increase in the efficiency of the process. In practice, it is found that zoning speeds can be increased by at least one order of magnitude, and still give the same degree of purification, provided efficient stirring takes place. It is necessary that the interface be reasonably smooth and approximately perpendicular to the direction of zone refining. This places an upper limit on the speed of zoning, since at too high a zoning speed irregular crystallization can occur, and this is usually accompanied by the entrapping of pockets of solute-rich material. The more viscous liquid metals are particularly sensitive to this phenomenon.

The choice of zone length l is governed largely by practical considerations, and by the physical properties of the material. A narrow molten zone with a sharp demarcation between the liquid and solid phases is most desirable. Given a reasonable focusing of the heat input, such zones can be obtained in materials with high melting points and low thermal conductivities. The length of solid material i between molten zones is governed entirely by practical considerations; this parameter does not enter into any of the zone-refining calculations. An interzone spacing on the order of one zone length is common.

Optimization of Zone-Refining Efficiency. Pfann (3) has outlined a procedure for selecting the value of the zone travel rate f and the number of passes n which minimize

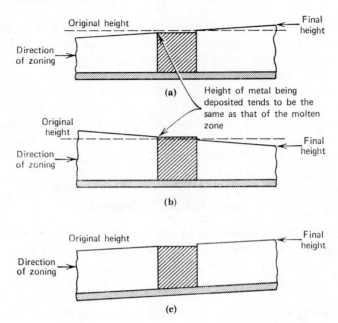

Fig. 5. Scheme of the method used to counteract mass transport during zone refining. (**a**) Material which contracts upon melting. (**b**) Material which expands upon melting. (**c**) Material which expands upon heating.

the time for a given degree of purification at any given point in the material. The procedure involves the determination of the optimum partition function k_0. Harrison and Tiller (28) have extended Pfann's original analysis and describe the efficiency of the process in terms of the normalized growth velocity $f\delta/D$, where δ is the thickness of the diffusion-boundary layer at the solid–liquid interface and D is the diffusion coefficient of the solute in the melt (9). They find that k_{opt} is not sensitive to the relative purification desired, or to the position in the material at which this purification is specified. As a result of this insensitivity, a normalized growth rate $f\delta/D \approx 1$ gives the optimum zone-refining efficiency, irrespective of the value of k_0. In the case of a lead-rich lead–tin alloy, subjected to magnetic stirring in the plane of the solid–liquid interface, the calculation gives $f_{opt} \approx 12$ in./hr (28). This is a somewhat idealized situation because of the efficient stirring and the associated reduction in δ; in general, zoning speeds for metals are slower, and fall in the range 0.25 to $\sim$6 in./hr.

Davies (29) has shown that for a monotonic distribution of solute impurity, the optimum conditions for solvent and solute purification are identical. If the solvent is initially distributed uniformly, and the partition coefficient k is not known, the optimum ratios of zone length l to ingot length L are 1 and 0.3, respectively, for the first and second zone passes; thereafter l/L should be 0.1.

Transport of Matter. Since most materials undergo a change of density on melting, the traverse of an ingot in an open horizontal boat with a molten zone leads to tapering of the ingot (see Fig. 5). This phenomenon is called matter transport. The mechanism of matter transport is very similar to that of solute transport, and the mathematical relationships describing the phenomenon (3) are similar to those derived for distribution of the solute. The severity of tapering increases with increasing number of zone passes. If the material expands on melting, matter transport is in a direction opposite to that of zone movement (Fig. 5a); contraction on melting leads to matter transport in the direction of zone travel (Fig. 5b). Matter transport can be eliminated by tilting of the ingot, relative to the horizontal, to the appropriate angle θ_c, given by

$$\theta_c = \tan^{-1} \frac{2\,h_0\,(1-\alpha)}{l} \tag{7}$$

where h_0 is the height of the solid, l the length of the molten zone, and α is the ratio of the density of the solid to that of the liquid. The tilting procedure is illustrated schematically in Figure 5c.

If zone refining is carried out in a vertical container with a closed end, precautions must be taken to prevent the transport of matter from cracking the container. Four conditions exist, depending on whether the material expands or contracts, and whether the zone is moved up or down. If a material contracts on melting, the passage of a zone down the charge cracks the container; this happens because the first solid to freeze at the top of the container freezes at the level of the liquid, thereby tending to compress the charge into a smaller volume. It follows that if the material expands on melting, then the procedure is to pass the zone down the charge.

Secondary Effects. Although the removal of an impurity from a material has been considered in terms of the partition coefficient k, other means of purification exist, and are in fact frequently operative during zone refining. These include volatilization, sublimation, and outgassing. In vertical zone refining (floating zone or in a containing vessel), it is possible to derive purification from the flotation or segregation of inclusions or self-slagging impurities. Vertical zone refining is also conducive to the transport of impurity atoms away from the solidifying face by convection. For the relative importance of these secondary means of purification, see below under Application.

Zone-Refining Techniques Involving Material Containment

The majority of the low-melting metals and alloys, and many organic and inorganic compounds, are amenable to zone refining in a containing vessel. Equipment design depends for the most part on the nature of the material to be purified and the operator's ingenuity in experimental innovation. The various aspects of the technique to be considered are the following: (1) the container material, (2) the shape of the container, (3) the travel mechanism, (4) heating and cooling of the material, (5) the method of stirring, and (6) containment of the atmosphere. Each of these has been discussed in some detail by Pfann (3) and by Parr (5). The necessary criteria for materials selection and equipment design are reviewed here, keeping in mind that we are primarily concerned with zone refining as a purification technique in materials research as opposed to large-scale production considerations.

Conventional Techniques. *Container Material.* The important physical and chemical characteristics of the container material are:

1. Inertness in relation to the material being purified. This includes not only the basic container material but also any impurities (solids or gases) likely to be present either as entrapped elements or in solid solution. Porous materials are undesirable.

2. The condition that the molten material not wet the container. Wetting leads to sticking of the bar in the container or to actual fracture of the container. A con-

tainer material with a coefficient of thermal expansion greater than that of the charge alleviates the fracture problem; however, this is not always possible, in which case the thermal coefficients of expansion of the container and charge have to be matched as closely as possible.

3. The condition that the thermal conductance of the container material be comparable to or less than that of the charge, in order to minimize heat-transfer problems.

The materials most commonly used for containers in zone refining are plastics, metals, glass, silica, mullite, alumina, zirconia, beryllia, silicon nitride, and graphite. In general, the choice becomes more limited as the temperature of the molten zone increases. Extremes in the useful working-temperature range of container material are provided by plastics and graphite; the upper limit on melting point for metals and alloys is $\sim 1500°C$, since above this temperature reaction occurs between the metal and the container.

Shape of the Container. The simplest and most frequently used vertical container has either a circular or square cross section. In horizontal zone refining, a circular or rectangular cross section is widely used. The circular form exposes a minimum of surface to the container and surrounding atmosphere, thereby minimizing contamination. At the same time, the small surface-to-volume ratio makes it difficult to obtain short, sharp molten zones with a small interzone spacing.

In materials research, the requirement is usually for a simple straight cylinder or bar of purified material. Where larger quantities of purified material are required, space limitations may be overcome by the use of circular, spiral, or helical charges, with the appropriate heater arrangement.

Thin-walled containers offer definite advantages. Heat conductance is minimized, so that a sharper molten zone is possible. Also, a thin-walled container can better withstand the thermal stress gradients associated with the zone-refining process.

Travel Mechanisms. The major requirement of the travel mechanism is a smooth constant motion of the charge material relative to the heat source(s). Movement of the heat source (which is possible with a small furnace) is conveniently arranged in the laboratory; in this way, vibration of the melt is avoided. Movement of the specimen is more convenient when large furnaces or induction heating is used. There are three main methods for passing n molten zones along the charge material, irrespective of whether the charge or heater moves: (*1*) n separate passes with one heater, (*2*) one pass through multiple heaters, and (*3*) multiple heaters with a reciprocal stroke. The first method is time-consuming, and the second approach calls for large space requirements; in the third method, a small number of heaters travel a set distance and then reciprocate rapidly in such a way that the molten zone is transferred from one heater to the next. Notwithstanding the disadvantages, the first two methods find wide application in the research laboratory.

The actual travel mechanisms in research equipment are usually very simple. Motion is obtained through a straight lead screw, cord and drum arrangement, direct motor drive, or cam drive. The power requirements are small, even for relatively large charges.

Heating and Cooling. In designing a heat source, the objective is to arrive at a stable molten zone having planar interfaces. This requires sufficient heat flux to provide the heat of fusion for the charge, and a degree of focusing that establishes relatively short molten zones with a sharp planar temperature gradient at both the melting and freezing interfaces. A high temperature gradient stabilizes the molten zone, since the influence of variations in longitudinal cooling due to the variation in ingot length outside the heater is reduced. Also, a high temperature gradient reduces the effect of temperature fluctuations on the actual zone length.

The two most common modes of heat input are resistance heating and induction heating. Disadvantages of resistance heating are: (*1*) the air (or gas) gap and the container wall provide a large thermal resistance; (*2*) the container is hotter than the charge material, so that the possibility of contamination from the chamber is increased. Induction heating is ideally suited to metals and semiconductors, since the heat is generated entirely within the charge itself, thereby minimizing contamination from the container. The eddy currents produced by induction heating give rise to some degree of stirring. However, the expense, space requirements, and critical nature of the timing must be considered. Typical frequencies are in the range 3 kc to 6 Mc, with power outputs in the range 5–50 kW.

Other methods of producing a molten zone in a material in a containing vessel include: gas heating, electrical discharge (30,31), radiation heating (32–34), and joule heating (ohmic heating) (3). Electron-bombardment heating is discussed below under Floating-zone refining.

In the case of a material which melts at or below room temperature, zone refining is carried out in a refrigerated atmosphere, or, alternatively, a solid zone can be moved along the bar or column of liquid material (35).

Stirring. The increase in efficiency in zone refining with superimposed stirring comes about primarily through a reduction in the thickness of the diffusion layer δ at the freezing interface, and from an improvement in heat transfer to the liquid. Successful methods of stirring include stirring as a result of inductance heating, mechanical stirring, agitation or vibration, magnetic stirring (10,36), and pumping of the liquid in the zone (37). Experimental difficulties do exist, particularly in small-scale laboratory zone refiners.

Atmosphere. In many cases, the material being purified reacts with the surrounding air; reaction need not necessarily be restricted to the material in the molten zone. Under these circumstances, zone refining must be carried out under vacuum, or in an inert or reducing atmosphere. The simplest approach consists of enclosing the material and container in a suitable ampoule and sealing off the assembly either under vacuum or with the required pressure of inert gas. Zone refining under a dynamic vacuum, or in a constant stream of inert or reducing gas, requires a more elaborate system of sealing, cooling, pumping, and gas purification (5).

In the zone purification of metals, the use of a hydrogen atmosphere is limited to those metals which do not form hydrides. If the impurity elements react with hydrogen, then this constitutes a beneficial secondary purification process; examples are the reduction of metallic oxides, or the formation of volatile impurity metal hydrides. Pure nitrogen can be used as the inert atmosphere provided compounds do not form, or the gas is not dissolved in the metal lattice. Metals amenable to a nitrogen atmosphere (Ni, Pb, Zn, Co, Cu, Ag) have a close-packed lattice structure.

Zone Refining in a Water-Cooled Container. High-melting and/or reactive metals, compounds, and semiconductors can be zone refined in a containing crucible if the latter is water-cooled. A number of ingenious cold crucible designs have been reported (77,78). This technique may be considered as an alternative to floating-zone refining, with the advantage that it is much simpler to carry out in practice. In general, the zone-purified material is in polycrystalline form.

Floating-Zone Refining

Many materials can be zone refined successfully when supported in crucibles, boats, or tubes of refractory material. However, serious problems of contact contamination arise in the case of reactive metals (eg titanium, uranium), high-melting metals (melting point $\gtrsim 1500\,^{\circ}$C), and some elemental or compound semiconductors. The advent of floating-zone refining, in which a molten zone is held in place by its own surface tension between two vertical colinear solid rods, constituted a major advance in high-purity materials research and technology. Originally developed by Kech and Golay (38), Emeis (39), and Theuerer (40), for the preparation of high-purity silicon, the technique has since been modified for use with a wide spectrum of materials, particularly the refractory metals and alloys. The starting material for floating-zone refining can be in the form of powder compacts, rods, or tubes, or a rod around which is wound a helix of wire of the required alloying addition.

The forms of heating adopted include induction heating, electron bombardment, radiation heating (39), joule heating (5), dielectric heating (5), solar energy (5), and arc image (41,42). However, induction heating or electron-bombardment heating are by far the most common. Accordingly, it is convenient to consider the experimental details of the floating-zone technique in association with these two forms of heating. Before making this subdivision, however, zone stability will be discussed.

A number of factors must be considered in preventing collapse of the molten zone. These include the diameter of the specimen in relation to the zone length, which, in

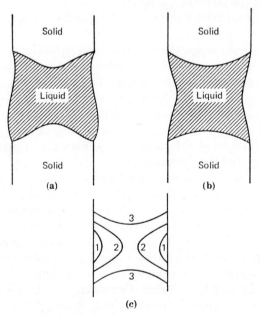

Fig. 6. Contour of the molten zone, with zone moving (**a**) down and (**b**) up, relative to the specimen; and (**c**) successive liquid-solid interface contours.

turn, is a function of the degree of focusing of the heat source, the surface tension, and density of the liquid. Heywang (43) and Heywang and Ziegler (44) have analyzed the conditions of stability of a vertical floating zone in a cylindrical rod assuming that (1) the solid–liquid interfaces are flat and horizontal, (2) the liquid causes complete wetting of the solid faces, (3) there is no change in density on melting, and (4) there is no motion of the zone. For a zone traveling under such steady-state conditions that the rate of solidification is equal to the rate of melting, conditions (3) and (4) compensate each other. The maximum length of zone that can be supported by surface tension increases linearly with rod radius for small radii, and then approaches a limiting value at large radii. The maximum zone length is determined by the surface tension and density of the liquid, through the parameter $(\gamma/p)^{1/2}$, where γ is the surface tension and ρ is the density. The Heywang relation predicts maximum stable zone lengths of $\sim$4.5 and $\sim$4.3 mm respectively for 3 mm diam rods of tungsten and tantalum. The constant height value for silicon is $\sim$15 mm. Kech, Green, and Polk (45) describe the shape of the zone as a function of length and rod radius. For a moving zone, the vertical tangent to the liquid surface at the solidifying interface must be perpendicular to the radius vector. If the initial volume of liquid is too small, the solidifying rod first decreases in radius and then increases until the steady state is reached. At the other extreme, too much liquid in the zone leads to bulging out and spillage.

Experience has shown that the theory can only be used as a guide to zone stability. In practice, the solid–liquid interfaces are not flat (Fig. 6), and the zone shape is a function of direction of zone travel, extent of stirring in the molten zone, and thermal conductivity along the rod. In general, it is found that for small diameter rods ($\lesssim$5 mm), floating-zone refining can be carried out with a zone length approximately equal to the rod diameter.

Benson (46,46a) has described methods that avoid the problem of melting through a cross section without exceeding the maximum zone height. The solution lies in the use of cross sections that are small in thickness and large in width, such as plates, discs, or pipes. Wide, stable molten zones have been produced in these forms which are as high as the zones in large diameter rods. This approach to the production of high-purity material in strip form is most attractive in the area of semiconductor-device fabrication. Pfann and Hagelbarger (47) and Oliver and Schaler (48) have also discussed methods of increasing the stability of the molten zone.

Since variations in the temperature of the molten zone do occur (due to changes in heat input or to outgassing), it is advantageous to be able to adjust the length of the molten zone mechanically. This is usually accomplished by means of an up-or-down movement of the lower solid rod. Uniformity of heating, stirring of the molten zone, and a more uniform cross section of zoned rod are enhanced by providing rotation of the lower solid rod during zone refining.

Induction Heating. Induction heating for floating-zone refining offers many advantages. It is possible to work under high vacuum, in a reducing or oxidizing atmosphere, or at a positive pressure of inert gas. The latter condition is required if the material has a high vapor pressure at the melting point. A wide range of materials has been purified, including silicon, tungsten, beryllium, and iron, in diameters up to $\sim$1 in. and the technique is easily adapted to the seeding of single crystals of specific orientation. The utility of induction heating in zone refining has been increased by the use of low-frequency electromagnetic fields in conjunction with induction coil design.

Table 1. Summary of Zone Refining as Applied to Low-Melting Metals

Metal	Experimental details	Remarks	Reference
Al	graphite container, vacuum $\sim 10^{-6}$ mm Hg, 18 passes at 6 in./hr, induction heating	cellular structure indicative of impurities present in last quarter of bar, but absent in first half of bar, recrystallizes at room temperature	81
	tilted graphite boat, argon atmosphere, 4.9 in./hr, 20 passes, induction heating	$R_{4.2°K}/R_{278°K}$ increases from 0.00015 along bar to 0.002 in last zones to freeze	52
	graphite boat, argon atmosphere, 30 passes ~ 4 in./hr	purity increases from 99.992% to 99.9995% along bar; recrystallizes at room temperature	82
	alumina boat, ~ 0.2 in./hr, 10 passes, resistance heating	0.005% Zn and 0.007% Cu added. Up to 78% Zn and 97% Cu removed by zone refining	83
	alumina boat, 0.2 in./hr, 3 passes	$R_{298°K}/R_{20°K}$ increases from 1000 to 1400; purity $> 99.998\%$; Cu, Na, Fe, Si, and rare earths segregate to same end	84
	electron-beam floating zone; zone length 0.5 cm, rod diam 0.95 cm, power input 84 W		85
Ag	graphite container, helium atmosphere, 18 passes at 6 in./hr, induction heating		81
	graphite boat, argon atmosphere, induction heating, 10 passes at ~ 1 in./hr	$R_{298°K}/R_{4.2°K}$ increases from 360 to 770; possible removal of volatile impurities or impurity oxides	86
Au	graphite container, vacuum $\sim 10^{-6}$ mm Hg, 24 passes at 6 in./hr, induction heating		81
	graphite boat, argon atmosphere, 12 passes at 3.6 in./hr, induction heating	some zone refining, but also pickup of impurities. $R_{4.2°K}/R_{273°K}$ increases from 0.03 along the bar to 0.05 in last zones to freeze	52
Cd	tantalum sheet boats, resistance heating ~ 1 in./hr, boats tilted 1–1.5°, argon atmosphere, 15 and 30 passes	Te, Bi, Pb, Cu, Sb, Zn, segregate in direction of zoning $(k < 1)$, $R_{4.2°K}/R_{273°K}$ increases from 5×10^{-5} along 60% of the rod to $\sim 10^{-3}$ in the last zones to freeze	87
	purified hydrogen atmosphere	purity > 99.9999; $R_{273°K}/R_{4.2°K}$ in range 27,000–35,700	88
Zn	6 passes at 2 in./hr, zone refining in air	Pd and Cd reduced to ~ 1 ppm from initial concentrations of ~ 12 ppm Pb and ~ 2 ppm Cd, respectively; Tl and Bi effectively removed	3
	purified hydrogen atmosphere	$R_{273°K}/R_{4.2°K}$ in range 30,000–50,000	88
	pointed boats of graphite or fused quartz		89
Bi	45 passes at 1.8 in./hr in pyrex tube at vacuum. 2.5×10^{-2} mm Hg, induction heating, reciprocating zone refiner with 4 heating coils	effective segregation of Ag, Cu, Pb, Sn, Ni, Mg, Ca $(k < 1)$, and Fe $(k > 1)$. Total metallic impurity content < 10 ppm	90
	glass tube, vacuum, resistance heating, 3 heaters, 20 zone passes, $l/L = 0.1$		91
	horizontal circular tubular container, ~ 0.5 in./hr, multiple heaters	effective segregation of Zn, Cu, Ag $(k < 1)$, and Sb $(k > 1)$	92

Table 1. (*continued*)

Metal	Experimental details	Remarks	Reference
Sn	pointed boats of graphite or fused quartz		89
	annular crucible of alumina, or quartz tubes, in air, 0.8 in./hr, 50 and 60 passes	effective segregation of Sb, Ca, Mn, ($k > 1$), and Pb, Cu, Bi, Cd, Fe, Ag, In, Zn, Au, Ni, Al, Mg, Si ($k < 1$); Si, Al, Fe, most effectively removed; purity > 99.9999, with $R_{4.2°K}/R_{273°K} \approx 1.6 \times 10^{-5}$	93
	graphite boat, argon atmosphere, 0.2 in./hr, 9 passes, vibration	purification factor ~2000 for Ag at an initial concentration of 0.003%	83
	pyrex container, vacuum, 40 passes	99.9% tin converted to 99.999% purity; segregation of Pb, Cu, and Fe	94
	graphite boat	purity ~99.999%	95
	pyrex, quartz, or graphite boats, argon atmosphere, single and multiple heater units, 2 in./hr, up to 500 passes	graphite boat container material, k (theoretical) not in good agreement with k (experimental), most impurities below limit of detection; exceptions are Al, Cd, Fe, Mg, Si, Sb	96
Sb		all impurities with $k < 0.2$ removed	97
	7 passes, nitrogen atmosphere	Ni, As, Pb, Ag, Cu reduced by factor of 10. As ($k \simeq 1$) not removed.	94
	graphite boat, 25 passes		98
Pb	96% silica glass tube coated with carbon film, induction heating, argon atmosphere	estimated impurity content < 10^{-6} at. % for $k_0 = 0.001$; 10^{-5} at. % for $k_0 = 0.01$; 10^{-4} at. % for $k_0 = 0.1$; 10^{-3} at. % for $k_0 = 0.5$	99
Pb–10 wt % Sn	graphite boat, resistance heater, magnetic stirring, gas atmosphere, 1 pass	verification of theoretical relationship between k and k_0	9, 10
Ga	sealed ferrosilicate glass tube, resistance heating, water-cooling coils, reciprocating unit	Pb is the only impurity found	100
	zone refining of $GaCl_3$ (20 passes) and conversion to Ga	Cu, Fe, Ca, Mg, Si, Al, Ag, reduced from ~10–70 ppm to levels < 1 ppm	101
K	zone refining, followed by Bridgman crystal growth	$R_{273°K}/R_{4.2°K} = 6800$	102
Li	molybdenum plate boat, helium atmosphere, resistance heating, 1 in./hr, 12 or 20 passes	$R_{4.2°K}/R_{273°K}$ increases fom 2×10^{-3} along the bar to 3.9×10^{-2} in the last zones to freeze; starting purity 99.3%, zoned purity > 99.95%; effective segregation of Na, Mn, Ca, Fe, Cu, ($k < 1$)	103
Mg	graphite boat, resistance heating, SO_2 atmosphere	Ni, Cu, Ca, Pb, Sn, Si, Zn, reduced below the detectable limits for spectrographic analysis	104
Cu	graphite crucible, induction heating, nitrogen atmosphere, 7 passes	zone refining effective, but continuous pickup of impurities. $R_{298°K}/R_{4.2°K}$ ~1000 for zoned copper	86
	graphite crucible, pure dry hydrogen, induction heating, $l/L \simeq 0.1$, 2–5 in./hr	effective segregation of Fe, Co, Ni, ($k > 1$), and Sb, Cr, Mn, Si, Ag, Sn ($k < 1$)	105
	tungsten strip heaters, graphite boat, hydrogen atmosphere, ~4 in./hr, 10 passes	segregation of Ag ($k < 1$), and Fe, Ni, ($k > 1$)	106
	electron-beam floating zone, vacuum ~3×10^{-6} mm Hg, 3 mm diam rod	$R_{273°K}/R_{20°K}$ ~ 1700	61, 70, 107

These fields interact with the induced eddy currents in the liquid and give rise to a beneficial levitating force (49). Several zone-refining units incorporating induction heating have been described in the literature (5,38,40,50–60).

Electron-Beam Heating. Calverley et al. (61) first used electron-bombardment heating for the floating-zone refining of metals. Since that time, various modifications and improvements have been reported (30,62–75). Lawley (67) and Schadler (68) reviewed all aspects of the electron-beam floating-zone technique. The method can

Table 2. Application of Zone-Refining Techniques to High-Melting Metals

Metal	Experimental details	Remarks	Reference
Be	beryllium oxide boats, argon atmosphere, induction heating, 1.5 in./hr and 3.0 in./hr, 12 passes	evaporation of Mg, Ca, Zn; Al < 10 ppm, pickup of Fe	108
	floating zone, argon atmosphere, induction heating, 1 in. and 1.25 in. diam rod, 0.5 in./hr, 2 passes	effective removal of BeO, total interstitials (C, N, O, F) $\simeq$ 16 ppm (atomic); total metallics $\simeq$ 17 ppm (atomic)	109
Ni	water-cooled copper hearth, induction heating	no contamination from copper crucible	77
	floating zone, induction heating, 2.25 in./hr, 3/16 in. diam rod, vacuum $\sim 10^{-5}$ mm Hg, 4 passes	purification by volatilization, $R_{4.2°K}/R_{273°K} \simeq 0.0014$	52
	floating zone, induction heating	~ 20 ppm carbon, $R_{4.2°K}/R_{273°K} \simeq 0.0005$	51,110
	electron-beam floating zone, vacuum $\sim 10^{-5}$ mm Hg		61,69,70,85
	electron-beam floating zone; up to 10 passes at 0.001 or 0.0005 in./sec; vacuum 4×10^{-7} to 2×10^{-5} mm Hg	Fe + C < 5 ppm	111
	oxidize specimen before zone refining	$R_{273°K}/R_{4.2°K} = 4000$	
Ti	water-cooled copper hearth, induction heating	no contamination from copper crucible	77,79
	floating zone, induction heating, helium atmosphere, 0.25 in./hr, 2 passes	distribution of Au and Zn determined	51
	floating zone, induction heating, vacuum $\sim 10^{-5}$ mm Hg, 6 passes, 1.25 in./hr	purification mainly due to volatilization	52
	floating zone, argon atmosphere, ~ 2 in./hr, 1 cm diam bar, $l/L \simeq 1$, induction heating		112
Fe	water-cooled hearth, induction heating, 4–8 in./hr, 6 and 12 passes, gas atmosphere	effective segregation of phosphorus (P $\approx$ 0.016 ppm), requires ~ 2 times more zone passes than floating zone for same degree of purification	77,79

Table 2.   (*continued*)

Metal	Experimental details	Remarks	Reference
	ceramic boat, induction heating, argon atmosphere, up to $\sim$ 12 in./hr	distribution of Cr, Si, P, Mn studied in iron and steel	113
	floating zone, induction heating, gas atmosphere	quantitative determination of segregation of several impurities with $k < 1$ and $k > 1$	17, 18, 51, 55
	levitation zone melting, wet hydrogen atmosphere	studies of immobilization of C in Fe, zone leveling of B in Fe	48, 56, 114
	electron-beam floating zone, 0.64 cm diam bar, 0.2 in. zone length, 138 W for zoning		85
Zr	water-cooled hearth, induction heating, gas atmosphere		77
	floating zone, induction heating, helium atmosphere, 1.4 in./hr, 1 pass	k for Au in Zr believed to be $\sim$ 1	51
	displacement of β phase along bar of α phase induction heating	pronounced segregation of O and N	115
			116–118
Y	floating zone, induction heating, vacuum 5×10^{-6} mm Hg, 2.25 in./hr, 9 passes	zone refining has little effect on gaseous impurities, metallic impurities reduced from 4300 to 1800 ppm in 6 passes	59
Pd, Pt, Rh, Ir, Ru, Os	electron-beam floating zone, and induction-heating floating zone	purification by evaporation and segregation, resistance ratio measurements	75, 85, 119–124
Cr, Mn, Co	vertical water-cooled copper crucible, induction heating	sound, high-purity ingots	77, 79
V, Nb, Mo, Ta Re, W, and binary alloys	electron-beam floating zone, vacuum in range 10^{-5} to 10^{-9} mm Hg, zoning speeds up to $\sim$60 in./hr, rod diam up to $\sim$0.5 in.; number of zone passes usually $\lesssim$12	effective purification by outgassing, volatilization, some purification by zoning; maximum $R_{273°K}/R_{4.2°K}$ values are: 70,000 (W), 14,000 (Mo), 9500 (Ta), 4000 (Re). H_2, N_2, O_2, usually $\lesssim$1 ppm by weight, purity $\geq$99.995	23, 24, 30, 52 61, 69, 71–75 85, 125, 126–138
	water-cooled copper hearth, induction heating, vacuum or gas atmosphere		77

only be used in a vacuum (better than $\sim 10^{-4}$ mm Hg), but control of the length of the molten zone is much easier than with induction heating, since a high concentration of heat input is possible.

Application

In order to give a comprehensive review of the application of zone refining, and at the same time to allow for easy reference to a particular material, most of the data are presented in tabular form. As a further convenience, the following grouping of materials is made: low-melting metals, high-melting metals, semiconductors and group III–V compounds, and alkali metal halides.

Low-Melting Metals. Most of the low-melting metals are amenable to zone refining in a containing vessel. In this classification, copper (mp 1083°C) is considered to represent the upper limit of the low-melting metals; zone refining of copper has been carried out in horizontal boats, and by the floating-zone technique. Experimental details and pertinent purification data are summarized in Table 1.

High-Melting Metals. The refractory metals (mp > 1775°C) were first purified by electron-beam floating-zone refining (61); numerous examples of the technique have been described in the literature. These have been reviewed up to 1961–1962 by Schadler (68). It has been established that in electron-beam floating-zone refining, purification occurs primarily by volatilization of gaseous elements and compounds, and by vacuum distillation of low-melting elements. The degree to which impurity segrega-

Table 3. Application of Zone-Refining Techniques to Semiconductors and III–V Compounds

Material	Experimental details	Remarks	Reference
Ge	inclined carbon or carbon-coated boats, dry nitrogen, induction or resistance heaters, zoning speeds ~0.1–0.2 in./hr	impurity concentration of electrically detectable elements (Cu, Fe, Ni)) $<10^{-10}$, others ~10^{-7} except O_2 and H_2 (~100 ppm)	2, 5, 139
	~3 in./hr, atmosphere of selenium at pressures up to 40 mm Hg	study of distribution of Se in Ge as function of Se vapor pressure	19
	floating zone, induction heating, 10^{-5} mm Hg, 1–10 in./hr	$k_0 = 0.09$ for As in Ge, vapor pressures and evaporation rates of P, As, Sb, in Ge	16
	electron beam, horizontal molten zone (disc) (~2 cm diam) in Ge sheet		50
Si	fused clear thin-walled quartz boats, argon atmosphere, induction heating, 5 heaters, 5.5 in./hr	donors either volatilize or are segregated (not B), increased lifetime due to substantial reduction of lifetime suppressors	140, 141
	water-cooled copper hearth, induction heating		38
	floating zone, induction heating, argon, vacuum, or wet hydrogen for removal of B ($k \approx 1$)	B, P, 10^{-13} to 10^{-4}, O ~1, Cu ~10^{-2}, Zn ~10^{-3}, Sb ~2×10^{-3}, As ~10^{-4}, Fe 0.1 (all in ppm)	5
	6 in./hr, induction heating, floating zone, 5 passes, H_2 + water vapor	removal of B	44
Si	floating zone; argon, helium, vacuum, or H_2 and water vapor. 67 passes, 1.3 cm diam rod, 3.6 in./hr	$<10^{-9}$ concentration of electrically active impurities, resistivity ~16,000	50
	strip-floating zone in slab 1 in. $\times$ 4 in. $\times$ $^3/_{16}$ in., $^1/_4$ in. high zone, induction heating		46a
	carbon arc image, floating zone, $\lesssim$$^7/_8$ in. diam rods, ~2 in./hr, argon atmosphere	purity and structure comparable to that in Si produced by other methods	41

Table 3 (*continued*)

Material	Experimental details	Remarks	References
Te	fused quartz boats, hydrogen atmosphere, resistance heaters, $\sim$2 in./hr	purity $>$ 99.999%, Mg and Si not removed	142
	several hundred zones	extrinsic-hole density $\lesssim 10^{14}/cm^3$	143
	sealed silica tubes		144
HgTe	96% silica glass boat, induction heating, 15 atm Hg pressure, $\sim$0.5 in./hr		145
HgTe, CdTe, and HgTe–CdTe	2 in./hr, ambient temperature within 50°C of melting point of compound due to high vapor pressure		146
Bi_2Te_3	96% silica glass boat, induction heating $\sim$0.7 in./hr, 1 pass, high ambient temperature $\sim$875°C	constant resistivity along zoned bar	147
$CdSnAs_2$	stoichiometric proportions of compounds melted and zone purified		148
As_2Te_3	vertical sealed 96% silica glass tube resistance rod heaters	large homogeneous crystals	147
$CoSi_2$	water-cooled copper hearth, induction heating, 10 in./hr, 14 passes	no contamination from the hearth, reduction in conc. of Ni, Fe, Mn, Al, Ag, Ca, Ga, Mg, Na	80
InSb	quartz boat or crucible, hydrogen atmosphere or vacuum, require critical temperature gradients	$k \approx 1$, several impurities, oxide not volatile at melting point	149–152
GaAs	floating zone, induction heating, completely sealed system, 0.9 atm pressure of As. $\sim$3 in./hr, 6 mm diam crystals	dislocation density in zone-refined crystals $\sim 5 \times 10^3$ cm^{-2}	53,153–155

tion takes place is dependent on the zoning speed; in many cases, speeds employed ($>$ 12 in./hr) have been far too high to allow for impurity segregation. It must be remembered, however, that the zoning speed is largely controlled by the vapor pressure of the material at the melting point, and a compromise is usually necessary. Floating-zone refining of refractory metals at low zoning speeds is possible in a controlled atmosphere with induction heating (76) and it seems probable that this technique will find increasing use. The water-cooled hearth technique (30,77–80) is particularly useful and effective for the refractory metals.

At the lower end of the melting-point scale, beryllium can be zone-refined in a container or, more frequently, by the floating-zone technique in a controlled atmosphere. The reactive metals iron, titanium, zirconium, and nickel are zone-refined either by the floating-zone technique or in a water-cooled hearth. Applications are summarized in Table 2. It should be pointed out that in many cases high-purity binary alloys can be produced by all of these approaches.

Semiconductors and III-V Compounds. Germanium can be zone-refined in graphite containers or by floating-zone techniques. Silicon is more difficult to zone-refine than germanium since the element reacts with carbon, alumina, zirconia, and the refractory metals. High-purity fused silica has been used as a container material, but

Table 4. Application of Zone-Refining Techniques to Alkali Metal Halides

Halide	Experimental details	Remarks	Reference
KCl	dilute halogen atmosphere	Na < 1 mg/g, Br < 5 mg/g	156
NaCl, KCl	quartz container, respective halogen atmospheres, 3.7 mm/hr	determined distribution coefficients of monovalent and divalent impurities	157
KCl, NaCl, KBr, KI, AgCl, AgBr, CsI LiF	graphite or quartz boats, 1–3 heaters, ~0.2–4 in./hr, $l/L \simeq 0.2$, up to 70 passes, pure halogen gas, or acid hydrogen halide atmosphere	most work on KCl, NaCl, KBr; data for $HCl:OH^- < 2 \times 10^{-9}$, $O_2^- < 2 \times 10^{-9}$, heavy metal ions 10^{-9}, heterovalent ions $< 10 \times 10^{-9}$	158–166 $\lesssim 2 \times$
	floating zone, induction heating, $^3/_8$ in. to $^1/_4$ in. diam rods, $\lesssim 15$ in./hr, atmosphere of O_2, N_2, H_2, or HCl	all impurities except alkaline earth metals purged from melt; ultimate purity not yet determined	54

there is a slow reaction between the silicon and fused silica, resulting in a pickup of boron and oxygen. Best results have been obtained with floating-zone techniques.

InSb, GaSb, and AlSb are compounds with low dissociation pressures at their melting points; these compounds do not present undue difficulties in zone refining. GaAs, InAs, InP, and GaP have a high dissociation pressure at their melting points, and the volatile component is released. In order to maintain stoichiometry, zone refining must be carried out under a pressure of the volatile component. Because of the reactivity of Al, As, and P, special crucible materials are required, unless, of course, floating-zone methods are used. Data on the zone refining of semiconductors and III–V compounds are presented in Table 3.

Alkali Metal Halides. Problems arise in the zone refining of alkali metal halides due to (*1*) adherence of the melt to the container, (*2*) the large thermal expansion of salt, and (*3*) the high strength of the salt under compression. A solution to the adherence problem lies in the removal of any alkali metal oxides or hydroxides. The floating-zone technique has also been applied to the alkali metal halides. A brief summary is given in Table 4.

The examples in Tables 1–4 illustrate a growing volume of literature and summarize the state of the art for those categories of materials. Each monthly issue of major abstracting services lists new apparatus and applications of zone refining to new materials. The list extends to a multitude of organic and inorganic compounds, as well as elementary biological systems. For low-melting compounds in these categories the ease of zone refining has produced a large list of materials which precludes tabulation.

Examination of Zone-Refined Material

The detailed characterization of zone-purified material in terms of the level and distribution of impurities constitutes a real problem. Conventional methods of analysis are usually not sufficiently sensitive to measure residual impurity levels. The magnitude of the problem may be appreciated when it is realized that in the case of zone-refined semiconductor materials concentrations in the range of 10^{-9} are quite common.

Notwithstanding this problem, a number of suitable techniques are now in everyday use; these may be classified as direct or indirect. Direct methods involving the chemical and physicochemical determination of trace elements include: radioactivation analysis (neutron, photon, or charged-particle irradiation), emission spectrography, mass spectrography, spectrofluorimetry, and polarography. Indirect methods of analysis rely on the measurement of the physical, mechanical, or magnetic properties of the material. Such methods are more rapid and less expensive than the direct methods, but generally give an indication of the total rather than individual impurity levels. Low-temperature resistance measurements provide a rapid, sensitive, and nondestructive means for studying impurity distributions in zone-refined ingots (85).

In mechanical-property evaluations, the parameters which give a measure of the material purity are yield strength, tensile strength, ductility, and hardness. Apart from a determination of mechanical properties, macroscopic and microscopic examination is beneficial. The surface appearance after etching indicates whether the material is in single or polycrystalline form. Optical microscopy of polished and etched cross sections reveals the presence and condition of internal boundaries, flaws, inclusions, and precipitates. The electron microscope is a particularly powerful tool in the examination of internal structure. Transmission-electron microscopy and electron microscopy of high-resolution replicas allow for a detailed characterization of the dislocation substructure in the material; this, in turn, is a sensitive function of the level and distribution of impurities, inclusions, or other internal defects.

Bibliography

1. W. G. Pfann, *Trans. AIME* **194,** 861 (1952).
2. W. G. Pfann, *Met. Rev.* **2,** 29 (1957).
3. W. G. Pfann, *Zone Melting,* 2nd ed., John Wiley & Sons, Inc., New York, 1966.
4. E. F. G. Herington, *Endeavour* **19,** 191 (1960).
5. N. L. Parr, *Zone Refining and Allied Techniques,* George Newnes, London, 1960.
6. J. H. Wernick, in *Ultra-High Purity Metals,* American Society of Metals, Cleveland, O., 1962, p. 55.
7. H. Schildknecht, *Zone Melting,* Academic Press Inc., New York, 1966.
8. J. J. Gilman, ed., *The Art and Science of Growing Crystals,* John Wiley & Sons, Inc., New York, 1963.
9. E. A. D. White, *Brit. J. Appl. Physics* **16,** 1415 (1965).
10. J. A. Burton, R. C. Prim, and W. P. Slichter, *J. Chem. Phys.* **21,** 1987 (1953).
11. W. C. Johnston and W. A. Tiller, *Trans AIME* **221,** 331 (1961).
12. N. W. Lord, *Trans. AIME* **197,** 1531 (1953).
13. H. Reiss, *Trans. AIME* **200,** 1053 (1954).
14. H. Reiss and E. Helfand, *Trans. AIME* **32,** 228 (1961).
15. L. Gold and V. Johnson, *J. Franklin Inst.* **270,** 367 (1960).
16. J. R. Gould, *Trans. AIME* **221,** 1154 (1961).
17. T. Ooka, H. Mimura, S. Yano, and S. Soeda, *Proc. Japan. Acad.* **39,** 294 (1963).
18. E. J. Koepel and B. Park, "A New Method of Using the Zone-Refining Technique," *Am. Soc. Metals Tech. Rept.* **1964,** 17.
19. J. van den Boomgaard, *Phillips Res. Rept.* **10,** 319 (1955).
20. C. H. L. Goodman, *Research (London)* **7,** 168 (1954).
21. J. Angus, D. V. Ragone, and E. E. Hücke, in C. R. St. Pierre, ed., *Physical Chemistry of Process Metallurgy,* Part II, Interscience Publishers, New York, 1961, p. 833.
22. W. G. Pfann and R. S. Wagner, *Trans. AIME* **224,** 1139 (1962).
23. D. R. Hay and E. Scala, in R. Bakish, ed., *1st Intern. Conf. Electron Ion Beam Sci. Technol.,* John Wiley & Sons, Inc., New York, 1965, p. 550.
24. D. R. Hay and E. Scala, *Trans. AIME* **233,** 1153 (1965).
25. J. D. Verhoeven, *Trans. AIME* **233,** 1156 (1965).

26. J. D. Verhoeven, *J. Metals* **18,** 26 (1966).
27. R. S. Wagner, C. E. Miller, and H. Brown, *Trans AIME* **236,** 554 (1966).
28. J. D. Harrison and W. A. Tiller, *Trans. AIME* **221,** 649 (1961).
29. L. W. Davies, *Trans. AIME* **215,** 672 (1959).
30. G. A. Geach and F. O. Jones, *J. Less-Common Metals* **1,** 56 (1959).
31. R. D. Burch and C. T. Young, *U.S. At. Energy Comm. Rept. NAA-SR-1735,* April 1957.
32. R. Handley and E. F. G. Herington, *Chem. Ind. (London)* **1956,** 304.
33. *Ibid.,* **1957,** 1184.
34. L. R. Weisberg and G. R. Gunther-Mohr, *Rev. Sci. Instr.* **26,** 896 (1956).
35. H. Rock, *Naturwissenchaften* **43,** 81 (1956).
36. W. G. Pfann and D. Dorsi, *Rev. Sci. Instr.* **28,** 720 (1957).
37. B. N. Aleksandrov, B. I. Verkin, and B. G. Lazar'ev, *Fiz. Metal i Metalloved.* **2,** 93 (1956); **2,** 105 (1956).
38. P. H. Kech and M. J. E. Golay, *Phys. Rev.* **89,** 1297 (1953).
39. R. Emeis, *Z. Naturforsch.* **9a,** 67 (1954).
40. H. C. Theuerer, *Trans. AIME* **206,** 1316 (1956).
41. R. P. Poplawsky and J. E. Thomas, Jr., *Rev. Sci. Instr.* **31,** 1303 (1960).
42. R. E. DeLa Rue and F. A. Halden, *Rev. Sci. Instr.* **31,** 35 (1960).
43. W. Heywang, *Z. Naturforsch.* **11a,** 238 (1956).
44. W. Heywang and G. Ziegler, *Z. Naturforsch.* **9a,** 561 (1954).
45. P. H. Kech, M. Green, and M. L. Polk, *J. Appl. Phys.* **24,** 1479 (1953).
46. H. C. Gatos, ed., *Properties of Elemental and Compound Semiconductors,* Interscience Publishers, New York, 1960.
46a. K. E. Benson, in reference 46, p. 17.
47. W. G. Pfann and D. W. Hagelbarger, *J. Appl. Phys.* **27,** 12 (1956).
48. B. F. Oliver and A. J. Shaler, *Trans. AIME* **218,** 194 (1960).
49. E. C. Okress, D. M. Wroughton, G. Comenetz, P. H. Brace, and J. C. R. Kelly, *Appl. Phys.* **23,** 545 (1952); *J. Electrochem. Soc.* **99,** 205 (1952).
50. E. Buehler, *Rev. Sci. Instr.* **28,** 453 (1957).
51. R. L. Smith and J. L. Rutherford, *J. Metals* **9,** 478 (1957).
52. J. H. Wernick, D. Dorsi, and J. J. Byrnes, *J. Electrochem. Soc.* **106,** 245 (1959).
53. F. A. Cunnell and R. Wickham, *J. Sci. Instr.* **37,** 410 (1960).
54. R. W. Warren, *Rev. Sci. Instr.* **33,** 1378 (1962).
55. W. M. Williams, G. B. Craig, and W. C. Winegard, *Can. Mining Met. Bull.* **55,** 35 (1962).
56. B. F. Oliver, *Trans. AIME* **227,** 960 (1963).
57. E. Buehler, *Trans. AIME* **212,** 694 (1958).
58. S. J. Silverman, *J. Electrochem. Soc.* **108,** 585 (1961).
59. W. C. Necker, in R. F. Bunshah, ed., *Transactions of the Vacuum Metallurgy Conference, 1960,* Interscience Publishers, New York, 1961, p. 289.
60. J. L. Rutherford, R. L. Smith, M. Herman, and G. E. Spangler, *New Physical and Chemical Properties of Metals of Very High Purity,* Gordon and Breach, New York, 1965, p. 345.
61. A. Calverley, M. Davis, and R. F. Lever, *J. Sci. Instr.* **34,** 142 (1957).
62. E. F. Birbeck and A. Calverley, *J. Sci. Instr.* **36,** 460 (1959).
63. R. G. Carlson, *J. Electrochem. Soc.* **106,** 49 (1959).
64. A. Lawley, *Electronics* **32,** 39 (1959).
65. H. W. Schadler, *Trans. AIME* **218,** 649 (1960).
66. M. Cole, C. Fisher, and I. Bucklow, *Brit. J. Appl. Phys.* **12,** 577 (1961).
67. A. Lawley, in R. Bakish, ed., *Introduction to Electron Beam Technology,* John Wiley & Sons, Inc., 1962, p. 184.
68. H. W. Schadler, in reference 8, p. 343.
69. H. G. Sell and W. H. Grimes, *Rev. Sci. Instr.* **35,** 64 (1964).
70. R. Brownsword and J. P. G. Farr, *J. Sci. Instr.* **41,** 350 (1964).
71. *Abstracts, 2nd Intern. Conf. Electron Ion Beam Sci. Technol., New York 1966.*
72. L. C. Skinner and R. M. Rose, in reference 71.
73. R. E. Reed, in reference 71.
74. E. B. Bas and H. Stevens, in reference 71.
75. H. Prekel and A. Lawley, in reference 71.

76. G. J. London, "The Flow and Fracture of Zone Refined Tungsten," *Franklin Inst. Final Rept. FB 1973 (Watertown Arsenal, DA-36-034-ORD-3747 RD)*, April 1964.
77. A. Berghezan and E. Bull-Simonson, *Trans. AIME* **221**, 1029 (1961).
78. E. Bull-Simonson, *J. Iron Steel Inst.* **200**, 193 (1962).
79. V. G. Epifanov and A. G. Lesnik, *Akad. Nauk Ukr. SSR* **20**, 185 (1964).
80. R. M. Ware, *J. Sci. Instr.* **38**, 166 (1961).
81. R. Schaefer, Y. Nakada, and B. Ramaswami, *Trans. AIME* **230**, 605 (1964).
82. A. W. Demmler, *Trans. AIME* **206**, 958 (1956).
83. P. Albert, F. Montariol, R. Reich, and G. Chaudron, in J. E. Johnson, ed., *Radio Isotope Conference*, Vol. 2, Academic Press, Inc., New York, 1954, p. 75.
84. F. Montariol, R. Reich, P. Albert, and G. Chaudron, *Compt. Rend.* **238**, 815 (1954).
85. D. K. Donald, *Rev. Sci. Instr.* **32**, 811 (1961).
86. J. E. Kunzler and J. H. Wernick, *Trans. AIME* **212**, 856 (1958).
87. B. N. Aleksandrov and B. I. Verkin, *Phys. Metals Metallog. USSR (Engl. Transl.)* **9** (38), 1 (1960).
88. J. H. Wernick and E. E. Thomas, *Trans. AIME* **218**, 763 (1960).
89. P. J. Schlicta, *J. Sci. Instr.* **39**, 392 (1962).
90. J. H. Wernick, K. E. Benson, and D. Dorsi, *Trans. AIME* **209**, 996 (1957).
91. W. W. Mullins, *Acta Met.* **4**, 421 (1956).
92. N. P. Sajin and P. Y. Dulkina, *Intern. Conf. Peaceful Uses Atomic Energy, United Nations, New York, 1956*, p. 265.
93. B. N. Aleksandrov, *Phys. Metals Metallog. USSR (Engl. Transl.)* **9** (46), 1 (1960).
94. M. Tannenbaum, A. J. Gross, and W. G. Pfann, *Trans. AIME* **200**, 762 (1954).
95. D. Walton, W. A. Tiller, J. W. Rutter, and W. C. Winegard, *Trans. AIME* **203**, 1023 (1955).
96. A. F. Armington and G. H. Moates, in M. S. Brooks and J. K. Kennedy, eds., *Ultrapurification of Semiconductor Materials*, The Macmillan Co., New York, 1962.
97. H. A. Schell, *Z. Metallk.* **46**, 58 (1955).
98. J. H. Wernick, J. N. Hobstetter, L. C. Lovell, and D. Dorsi, *J. Appl. Phys.* **29**, 1013 (1958).
99. W. A. Tiller and J. W. Rutter, *Can J. Phys.* **34**, 96 (1956).
100. D. P. Detweiler and W. M. Fox, *Trans. AIME* **203**, 205 (1955).
101. J. L. Richards, *Nature* **177**, 182 (1956).
102. H. J. Foster and P. H. E. Meijer, *J. Res. Natl. Bur. Std.* **71A**, 127–132 (1967).
103. I. G. D'yakov and I. R. Khvedchuk, *Phys. Metals Metallog. USSR (Engl. Transl.)* **17**, 139 (1964).
104. A. S. Yue and J. B. Clark, *Trans. AIME* **212**, 881 (1958).
105. E. D. Tolmie and D. A. Robins, *J. Inst. Metals.* **85**, 171 (1956–1957).
106. E. D. Tolmie, *J. Sci. Instr.* **37**, 175 (1960).
107. J. LeHeircy, *Compt. Rend.* **251**, 1509 (1960).
108. W. R. Mitchell, J. A. Mullendore, and S. R. Maloof, *Trans. AIME* **221**, 824 (1961).
109. G. J. London and M. Herman, *Conference Internationale sur la Metallurgie du Beryllium, Grenoble, France, May 1965.*
110. V. J. Albano and R. R. Soden, *J. Electrochem. Soc.* **113**, 511 (1966).
111. R. R. Soden and V. J. Albano, *J. Electrochem. Soc.* **115** (7), 766–769 (1968).
112. F. J. Darnell, *Trans. AIME* **212**, 356 (1958).
113. W. A. Fisher and R. Uberoi, *Arch. Eisenhuettenw.* **33**, 661 (1962).
114. B. F. Oliver and F. Garofalo, *Trans. AIME* **233**, 1318 (1965).
115. P. Aillard and J. P. Langeron, *J. Nucl. Mater.* **25**, 69–74 (1968).
116. G. D. Kneip, Jr., and J. O. Betterton, Jr., *J. Electrochem. Soc.* **103**, 684–689 (1956).
117. J. P. Langeron, *Mem. Sci. Rev. Met.* **61**, 637–644 (1964).
118. D. Mills and G. B. Craig, *Trans. AIME* **236**, 1228–1229 (1966).
119. D. W. Rhys, *J. Less-Common Metals* **1**, 269 (1959).
120. D. W. Rhys, *Symposium on Electron Bombardment Melting, S.E.R.L. Baldock, England, 1959*, p. 22.
121. W. P. Allred, R. C. Himes, and H. L. Goering, in J. Hetherington, ed., *Proc. 1st Symp. Electron Beam Technology, Boston, 1959*, p. 38.
122. E. Buehler and E. Berry, *Trans. AIME* **224**, 874 (1962).
123. H Hieber, B. L. Mordike, and P. Haasen, *Platinum Metals Rev.* **8**, 102 (1964).

124. J. T. Schriempf, *J. Less-Common Metals* **9**, 35 (1965).
125. J. A. Belk, *J. Less-Common Metals* **1**, 50 (1959).
126. D. P. Ferriss, R. M. Rose, and J. Wulff, *Trans, AIME* **224**, 975 (1962).
127. T. E. Mitchell and W. A. Spitzig, *Acta Met.* **13**, 1169 (1965).
128. L. I. Van Torne and G. Thomas, *Acta Met.* **14**, 621 (1966).
129. T. E. Mitchell, R. A. Foxall, and P. B. Hirsch, *Phil. Mag.* **8**, 1895 (1963).
130. E. Votava, *Physica Status Solidi* **5**, 421 (1964).
131. E. Votava, *J. Less-Common Metals* **9**, 409 (1965).
132. F. H. Cocks, R. M. Rose, and J. Wulff, *J. Less-Common Metals* **10**, 157 (1966).
133. J. L. Orehotsky and R. Steinitz, *Trans. AIME* **224**, 556 (1962).
134. R. G. Garlick and H. B. Probst, *Trans. AIME* **230**, 1120 (1964).
135. P. Beardmore and D. Hull, *J. Less-Common Metals* **9**, 168 (1965).
136. I. Drangel and G. Murray, *AIME Annual Meeting, New York, Febr, 1964.*
137. I. Drangel, P. McMahon, and S. Weinig, *Sixth Alloyed Flectron Beam Symposium, Cambridge, Mass., April 1964.*
138. G. T. Murray, S. Weinig, and I. Drangel, *Trans. Vacuum Met. Conf., 1964,* Am. Vacuum Soc., Boston., 1965.
139. W. G. Pfann and K. M. Olsen, *Phys. Rev.* **89**, 322 (1953).
140. E. A. Toft and F. H. Horn, *J. Electrochem. Soc.* **105**, 2 (1958).
141. D. H. Hartmann and P. L. Ostapkovich, *Metal Progr.* **70**, 100 (1956).
142. N. F. Shvartsenau, *Soviet Phys.-Solid State* (*Engl. Transl.*) **2**, 797 (1960).
143. J. S. Blakemore, K. C. Nomura, and J. W. Schultz, *J. Appl. Phys.* **31**, 2226 (1960).
144. G. Fisher, G. J. White, and S. B. Woods, *Phys. Rev.* **106**, 480 (1957).
145. T. C. Harmon, M. J. Logan, and H. L. Goering, *J. Phys. Chem. Solids* **7**, 228 (1958).
146. W. D. Lawson, S. Neilsen, E. H. Putley, and A. S. Young, *J. Phys. Chem. Solids* **9**, 325 (1959).
147. T. C. Harmon, B. Paris, S. E. Miller, and H. L. Goering, *J. Phys. Chem. Solids* **2**, 181 (1957).
148. A. S. Borschchevskii, N. A. Goryunova, G. A. Sikharulidze, V. M. Tuchkevich, and Yu. V. Shmartsev, *Dokl. Acad. Nauk SSSR* **171**, 830–832 (1966).
149. M. Tannenbaum and J. P. Maita, *Phys. Rev.* **91**, 1009 (1953).
150. T. C. Harmon, *J. Electrochem. Soc.* **103**, 128 (1956).
151. K. F. Hulme, *J. Electron. Control* **6**, 397 (1959).
152. K. F. Hulme and J. B. Mullin, *Solid-State Electron.* **5**, 211 (1962).
153. L. R. Weisberg, F. D. Rosi, and P. G. Herkart, in reference 46, p. 25.
154. J. M. Whelan and G. H. Wheatly, *J. Phys. Chem. Solids* **6**, 169 (1958).
155. J. L. Richards, *J. Appl. Phys.* **31**, 600 (1960).
156. S. Susman, *J. Chem. Phys.* **47** (1), 83–85 (1967).
157. M. Ikeya, N. Itoh, and T. Suita, *Japan. J. Appl. Phys.* **7**, 837–845 (1968).
158. H. Gründig, *Z. Physik* **157**, 232 (1960).
159. H. Kanzaki and K. Kido, *J. Phys. Soc. Japan* **15**, 529 (1960).
160. H. Kanzaki, K. Kido, and T. Ninomiya, *J. Appl. Phys.* **33**, 482 (1962).
161. M. Inoue and H. Mizuno, *J. Phys. Soc. Japan* **16**, 128 (1961).
162. S. Anderson, J. S. Wiley, and J. Hendricks, *J. Chem. Phys.* **32**, 949 (1960).
163. T. J. Neubert and S. Susman, *Rev. Sci. Instr.* **35**, 724 (1964).
164. R. W. Warren, *Rev. Sci. Instr.* **36**, 731 (1965).
165. R. W. Dreyfus, in reference 46, p. 410.
166. H. Gründig and E. Wassermann, *Z. Physik* **176**, 293 (1963).

ALAN LAWLEY
AND D. ROBERT HAY
Drexel University